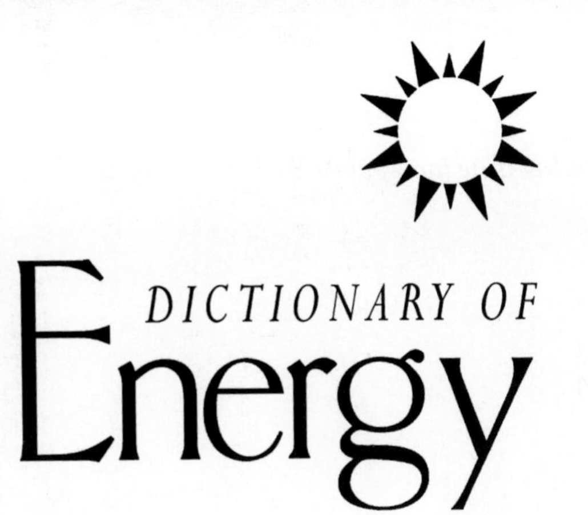
DICTIONARY OF
Energy

# DICTIONARY OF Energy

Editors-in-Chief

**Cutler J. Cleveland**
*Boston University, Boston, Massachusetts, United States of America*

**Christopher Morris**
*Lexicographer, Morris Books, Escondido, California, United States of America*

ELSEVIER

AMSTERDAM • BOSTON • HEIDELBERG • LONDON • NEW YORK • OXFORD
PARIS • SAN DIEGO • SAN FRANCISCO • SINGAPORE • SYDNEY • TOKYO

Elsevier
The Boulevard, Langford Lane, Kidlington, Oxford OX5 1GB, UK
Radarweg 29, PO Box 211, 1000 AE Amsterdam, The Netherlands



First edition 2006
Reprinted 2006



**British Library Cataloguing in Publication Data**
A catalogue record for this book is available from the British Library

**Library of Congress Cataloging-in-Publication Data**
A catalog record for this book is available from the Library of Congress

ISBN–13: 978-0-080-44578-0
ISBN–10: 0-080-44578-0

For information on all Elsevier publications
visit our website at books.elsevier.com

Printed and bound in *Italy*

06 07 08 09 10 10 9 8 7 6 5 4 3 2



**Illustration Credits**

**air pollution:** Courtesy of Litton Industries; **animal power:** Morris Books; aqueduct: HBJ Photo/Craig Collins; **Arctic National Wildlife Refuge:** U.S. Fish & Wildlife Service; assembly line: Courtesy of Ford Motor Company; astrolabe: HBJ Photo/Metropolitan Museum of Art; atomic bomb: U.S. Army; **autotroph:** Illustration by Sharron O'Neill; **Babbage:** HBJ Photo; **Bell:** Courtesy of American Telephone & Telegraph; **Bernard:** National Library of Medicine; **biocontrol:** U.S. Dept. of Agriculture/Scott Bauer; **biofuel:** Courtesy of Georgia Pacific; **biplane:** British Information Services; **block cutting:** U.S. Forest Service; **von Braun:** Courtesy of IBM Corporation; **caravel:** The Mariners Museum; **clipper ship:** HBJ Photo/Granger Collection; **coal face:** University of California, Riverside; **cold-rolling:** HBJ Photo; **company town:** U.S. Environmental Protection Agency; **controlled burning:** U.S. Fish & Wildlife Service; **converter:** Pennsylvania State Museum/Bethlehem Steel Company; **Copernicus:** Courtesy of The American Museum of Natural History; **cultivator:** Courtesy of John Deere Company; **Curie:** Courtesy of American Institute of Physics, Emilio Sergre Visual Archives; **dam:** HBJ Photo; **de Broglie:** American Institute of Physics; **deforestation:** American Forest Product Industries; **derrick:** Morris Books; **desertification:** U.S. Soil Conservation Service; **dike:** Image After; **draft animal:** HBJ Photo;

(Credits continued on p.503)

**Subject Areas of the Dictionary**

Biographies
Biological Energetics
Biomass
Chemistry
Climate Change
Coal
Communication
Consumption and Efficiency
Conversion
Earth Science
Ecology
Economics and Business
Electricity
Environment
Geothermal
Global Issues
Health and Safety
History
HVAC (heating/ventilation/air conditioning)
Hydrogen
Hydropower
Lighting
Materials
Measurement
Mining
Nuclear
Oil and Gas
Organizations
Photovoltaic
Physics
Policy
Refrigeration
Renewable/Alternative Forms
Social Issues
Solar
Storage
Sustainable Development
Thermodynamics
Transportation
Wind

## Editors-in-Chief

**Cutler J. Cleveland** is the Director of the Center for Energy and Environmental Studies at Boston University, where he also holds the position of Professor in the Department of Geography and Environment. Dr. Cleveland is Editor-in-Chief of the *Encyclopedia of Energy* (Elsevier, 2004), winner of an American Library Association award, and Editor-in-Chief of the journal *Ecological Economics*. Dr. Cleveland is a member of the American Statistical Association's Committee on Energy Statistics, an advisory group to the Department of Energy, and a participant in the Stanford Energy Modeling Forum. He has been a consultant to numerous private and public organizations, including the Asian Development Bank, Charles River Associates, the Technical Research Centre of Finland, the U.S. Department of Energy, and the U.S. Environmental Protection Agency. The National Science Foundation, the National Aeronautics and Space Administration and the MacArthur Foundation have supported his research. Dr. Cleveland's research focuses on the ecological-economic analysis of how energy and materials are used to meet human needs. His research employs the use of econometric models of oil supply, natural resource scarcity, and the relation between the use of energy and natural resources and economic systems. Dr. Cleveland publishes in journals, such as *Nature, Science, Ecological Modeling, Energy, The Energy Journal, The Annual Review of Energy, Resource and Energy Economics, the American Association of Petroleum Geologists Bulletin, the Canadian Journal of Forest Research,* and *Ecological Economics.* He has won publication awards from the International Association of Energy Economics and the National Wildlife Federation.

**Christopher Morris** is owner of Morris Books and a professional lexicographer who has edited more than 20 different dictionaries on a wide variety of subjects. He is Editor-in-Chief of the award-winning *Academic Press Dictionary of Science and Technology* (Academic Press, 1992), which provides the largest vocabulary of science yet compiled and features special essays by 120 eminent scientists, including nine Nobel laureates. He served as Chief Editor of the Macmillan school dictionary series, which includes several of the largest-selling educational dictionaries in U.S. history. He has also been an author of school and college textbooks and has compiled many different scientific glossaries, for fields, such as ecology, endocrinology, microbiology, oncology, reproductive biology, and toxicology. He and Cutler J. Cleveland previously collaborated on the *Encyclopedia of Energy* (Elsevier, 2004), winner of an American Library Association award, for which Dr. Cleveland was Editor-in-Chief and he served as chief development editor.

## Special Essays on Important Energy Terms

**acid rain (p. 4)**
Eugene Likens
Institute of Ecosystem Studies, USA

**agriculture (p. 9)**
David Pimentel
Cornell University, USA

**alternative fuel (p. 14)**
Danilo J. Santini
Argonne National Laboratory, USA

**battery (p. 34)**
Elton J. Cairns
University of California, Berkeley, USA

**biomass (p. 42)**
Donald L. Klass
Biomass Energy Research Association, USA

**blackout (p. 45)**
Stephen Connors
Massachusetts Institute of Technology, USA

**boiler (p. 48)**
W. Randall Rawson
American Boiler Manufacturers Association, USA

**bomb calorimeter (p. 49)**
Lisardo Núñez Regueira
Universidade de Santiago de Compostela, Spain

**boom and bust (p. 50)**
Roger M. Olien
University of Texas of the Permian Basin, USA

**Calvin cycle (p. 61)**
Joyce Diwan
Rensselaer Polytechnic Institute, USA

**cap and trade (p. 62)**
Carolyn Fischer
Resources for the Future, USA

**carbon (p. 64)**
Cutler Cleveland
Boston University, USA

**carbon sequestration (p. 67)**
Howard Herzog
Massachusetts Institute of Technology, USA

**climate change (p. 79)**
L. Danny Harvey
University of Toronto, Canada

**coal (p. 82)**
Mildred B. Perry
National Energy Technology Laboratory, USA

**combustion (p. 88)**
Chenn Q. Zhou
Purdue University, Calumet, USA

**DDT (p. 108)**
Cutler Cleveland
Boston University, USA

**distributed energy (p. 121)**
Neil D. Strachan
Pew Center on Global Climate Change, USA

**ecological energetics (p. 129)**
Charles A. S. Hall
State University of New York, Syracuse, USA

**ecological footprint (p. 130)**
Mathis Wackernagel
Global Footprint Network, USA

**elasticity (p. 134)**
Tom Tietenberg
Colby College, USA

**emergy (p. 139)**
Mark T. Brown
University of Florida, USA
Sergio Ulgiati
University of Siena, Italy

**endotherm (p. 142)**
Thomas H. Kunz
Boston University, USA

**energy (concept of) (p. 143)**
Joseph Priest
Miami University of Ohio, USA

**energy conservation (p. 143)**
William Chandler
Pacific Northwest National Laboratory, USA

**energy-GDP ratio (p. 145)**
David Stern
Rensselaer Polytechnic Institute, USA

**energy transitions (p. 146)**
Arnulf Grubler
International Institute for Applied Systems Analysis, Austria

**entropy (p. 148)**
Kozo Mayumi
University of Tokushima, Japan

# Preface to the Dictionary

While energy has always been a driving force in the evolution of human culture, its importance has reached new heights in the first decade of the 21st century. Due to its overarching macroeconomic importance, energy is now a precious commodity in global financial markets. Energy issues pervade global geopolitics, and will continue to do so in light of the increasing concentration of oil supplies in the Middle East coupled with rising global energy demand. Energy is central to global environmental change as emissions from energy use contribute significantly to the human component of climate change. Finally, and most importantly, access to modern energy services is a fundamental prerequisite to the alleviating the poverty from the lives of the three billion people living below subsistence level.

In education, energy is the common link between the living and non-living realms of the universe, and thus it is an integrator across all fields in science, technology, engineering, and mathematics education and research. Virtually every discipline investigates some aspect of energy, including history, anthropology, public policy, international relations, human and ecosystem health, economics, technology, physics, geology, ecology, business management, environmental science, and engineering.

The concept of energy is also applicable at all levels of formal education and lifelong learning, including many topics and tools that motivate successful learning. Kindergarteners get some of their first exposure to science by learning about heat from the sun. Middle school students learn ecology by examining the flow of energy through a food chain. College students may use a physics textbook that is organized around the concept of energy, learn about international relations by studying the geopolitics of oil, or learn about the effect of taxation on human welfare by studying a tax on carbon dioxide emissions. Consumers are increasingly confronted with decisions about the cost and availability of different forms of household energy, while diverse groups in civil society organize around political, social, and environmental ramifications of the energy system.

Despite its overarching importance, there is no coherent field of energy studies; the scientific, engineering, economic, and sociological communities share no common energy journals, conferences, or identity. There is a wealth of information about energy, but it is spread across many books, journals, websites, disciplines, ideologies, and user communities. Information about energy tends to target either a particular form of energy or a specific audience. There is no central repository of information that meets diverse user communities, and no primary machinery of communication among those communities. These characteristics of energy information define the formidable barrier to successful teaching and learning related to energy, a barrier we seek to significantly erode, if not eliminate, with the *Dictionary of Energy*.

A commonly agreed upon set of terms and definitions is essential to build communication among disparate groups, and to improving the general public's understanding of energy issues. This is especially true as new words are generated, and old ones are discarded, and as technologies, institutions, and behaviors change. An authoritative dictionary is important for an area in which identical words mean very different things. For example, "efficiency" and "elasticity" mean different things to an economist and an engineer.

The distinguishing features of the *Dictionary* are its integration of the social, natural, and engineering sciences and its breadth of coverage. It covers all academic disciplines and all the multifaceted aspects of the concept of energy. It uses an integrated approach that emphasizes not only the importance of the concept in individual disciplines, such as physics and sociology, but also how energy is used to bridge seemingly disparate fields, such as ecology and economics. The *Dictionary* covers all the environmental, engineering, and physical science topics found in existing dictionaries. It also covers entirely new areas, such as the economic and sociological aspects of energy use, energy flows in the biological realm, methods of energy modeling and accounting, energy and materials, energy and sustainable development, energy policy, net energy analysis, and energy in world history, among others.

An interdisciplinary dictionary can never be all things to all people. Specialists may be disappointed not to find important but quite narrow terms from their field. But a number of subject-specific dictionaries exist to meet this need. This dictionary is intended to aid students, researchers and the general public in their search for words, ideas, and information about energy from outside their own areas of expertise. There is also the challenge of deciding just how much technical information a definition should include to be accurate, while at the same time being clear to non-specialists. Striking a balance between these sometimes competing ends was the most difficult challenge we faced. These compromises will without doubt not please everyone. However, the breadth and depth of this dictionary far exceed anything attempted to date on the subject of energy. We are convinced that the *Dictionary* will provide a comprehensive, organized body of knowledge for what is certain to continue as a major area of scientific study in the 21st century.

My co-editor Chris Morris added what can only be described as an uncanny combination of vision, enthusiasm, and professionalism for the project. His editorial experience with more than 20 different dictionaries on a wide variety of subjects insured the lexicographic excellence of this work. I learned an immense amount from our collaboration.

Henri van Dorssen at Elsevier was an important champion of the project from the outset, and provided outstanding editorial guidance throughout. Margo Leach and her team at Elsevier put together a highly effective and creative marketing plan. Emily Griset, Brad Strode, Noam Reuveni, Richard Weiss, Tybe Goldberg, Keith Williges, Robyn Kenney, and Alejandra Roman provided invaluable data gathering and analysis.

The Editorial Board was extremely helpful in defining subject areas and providing subject-specific guidance on many terms. The Board is an outstanding collection of scholars from the natural, social, and engineering sciences that are recognized leaders in their fields of research. To all members of our Board, I extend my thanks and congratulations.

Cutler Cleveland
*Boston University*
*Boston, Massachusetts, USA*

# Introduction to the Use of the Dictionary

## Entry Selection

The *Dictionary of Energy* contains about 8100 main entries and about 10,000 entries in all. It is our belief that this entry selection provides a fully comprehensive description of the vocabulary currently used in the field of energy.

In order for a word to qualify for inclusion in the Dictionary, it has to be either a term that is specific to one or more of the subject areas of energy covered in the book (see discussion of Subject Areas following), or it has to be a general term that has a specialized use in energy (e.g., the use of the terms *sweet* and *sour* to describe oil and gas components). It also has to be sufficiently qualified in terms of two criteria, frequency and range. Frequency is the sheer number of times that the term occurs in the literature, and range is the variety of reputable sources in which it appears.

## Entry Sources

A dictionary of this size draws its entry list from a vast number of sources. Important sources for this book include other reference works in the field; in particular we made an exhaustive study of the vocabulary employed by the hundreds of authors who appear in the six-volume *Encyclopedia of Energy* (2004). We also examined existing energy dictionaries and glossaries, books on energy, scientific journals dealing with energy topics, and many energy-related Websites, both scholarly and popular.

## Subject Areas

In the planning stage of the Dictionary project, we identified 40 different subject areas that collectively encompass the vocabulary of the field of energy. (See p. v of this section for a complete list of these subject areas.)

Subject areas include familiar industrial forms of energy (e.g., Coal, Electricity, Nuclear, Oil and Gas) and also alternative forms (e.g., Geothermal, Hydrogen, Solar, Wind). Disciplines of general science that have special relevance to energy are also included (e.g., Chemistry, Earth Science, Ecology, Physics). Other subject areas deal with energy-related processes (e.g., Conversion, Lighting, Refrigeration, Transportation). In addition, the Dictionary includes subject areas that are outside the conventional realm of energy but that are impacted by human energy use (e.g., Climate Change, Global Issues, Health and Safety, Sustainable Development).

## Alphabetical Order

Entries in the Dictionary are arranged in an alphabetical order according to standard lexicographic convention, which dictates that only the letter-by-letter spelling of a given entry is considered in alphabetization. The sequence is not affected by the presence or absence of a word space or punctuation. For example:

**acid**
**acid-based**
**acid deposition**
**acidolysis**
**acid rain**

When two entries have the same spelling, the more common of the two will appear first.

**bit** *Oil & Gas.* a boring device used in drilling oil and gas wells . . .

**bit** *Communications.* the basic item of information in a digital computer . . .

## Entry Format

The typical entry in the Dictionary consists of three elements: the headword, the subject area, and the definition. Many entries also have a fourth element, the word origin. For example:

**Reitwagen** *History.* a two-wheeled vehicle built (1885) by automotive pioneer Gottlieb Daimler, employing what is considered to be the first successful version of the modern gas engine. [A German term for "riding carriage".]

## Headwords

The headword, also known as the main entry or entry word, is the particular term being defined, such as *langley* in the following example. The headword appears in **boldface type** and is written exactly as it would appear in print.

**langley** *Solar.* a measure of solar energy flux per area . . .

Thus, the headword is capitalized if it is written that way in context (as is the case with the previous example *Reitwagen*) but terms are not artificially capitalized as a style convention.

## Area Labels

The area label appears after the headword in *italic type,* as an indication of the subject area in which the given term is primarily used. This does not mean it is used exclusively in this area; many terms have significant levels of usage in two or more areas. For example the term *pollution credit* has the area label Policy, but it also has relevance in Economics and Environment.

### Definitions

Definitions in the Dictionary follow the principle that a term should not be defined using terminology that is more difficult to understand than the term itself, so that the reader will not have to look up other terms in order to fully comprehend the original definition. Even beyond this, we have tried to make sure that all definitions provide scientific accuracy within a context that is concise and understandable.

Many entries in the Dictionary have more than one definition. In that case, the definitions are sequenced on the basis of which is the more common, more basic, or more general meaning.

> **particle** *Physics.* **1.** any finite object that may be considered to have mass and an observable position in space. **2.** specifically, a minute subdivision of matter . . . .

### Word Origins

Selected entries in the Dictionary include the word origin (etymology) of the given term. The word origin indicates how this term came into existence in the English language. The word origin appears at the end of the entry and is set off by brackets. For example:

> **Boyden turbine** *Hydropower.* a type of water turbine . . . . [Developed by U.S. mechanical engineer Uriah *Boyden,* 1804–1879.]

### Special Essays

A distinctive feature of the Dictionary is the inclusion of special essays written by eminent scientists. There are more than 100 of these essays providing detailed descriptions of key energy terms, such as *alternative fuel, biomass, emergy, horsepower,* and *resource curse.*

We refer to these as "window" essays, first because their appearance on the page gives the impression of a window (they are set off from the rest of the text by a color tint), and second because they are windows in the metaphorical sense, in that they provide greater insight into the meaning of important words. (See p. vii for a table of contents for the window essays.)

### Cross References

Many entries in the Dictionary have cross references to other entries elsewhere in the book. A cross reference is indicated by being printed in SMALL CAPITAL LETTERS. The most common type of cross reference is an alternate form of a preferred or more common term. For example:

> **flaming spring** see BURNING SPRING.
> **footprint** see ECOLOGICAL FOOTPRINT.

Other cross references appear at the end of an entry to indicate that additional relevant information can be found under another entry.

> **extensive** *Physics.* describing a property of a system, such as its mass or total volume, that is a function of its extent . . . . Compare INTENSIVE.

A third type of cross reference appears within a definition to indicate that an important or possibly unfamiliar term is explained under its own alphabetical entry.

> **igneous** . . . one of the three principal classifications of rocks, along with METAMORPHIC and SEDIMENTARY.

### Acknowledgments

I wish to thank my co-editor, Cutler Cleveland, for the knowledge, dedication, and enthusiasm he has brought to this endeavor. At all times the experience of collaborating with him has been an enjoyable and rewarding one for me.

In the course of the project Cutler and I received strong support from the publishing staff at Elsevier, especially Margo Leach and her predecessor Clare Marl for marketing strategy and Debbie Clark for production services. Above all, the contributions of sponsoring editor Henri van Dorssen have been crucial to the project from its inception all the way to publication.

I also would like to thank Andrew Morris for help with research and for general editorial assistance, Nancy Tobin for suggestions on image sources, and Peter Jovanovich for initiating my involvement with scientific dictionaries.

Christopher Morris
*Morris Books*
*Escondido, California, USA*

# Common Abbreviations in Energy Usage

|  |  |
|---|---|
| A | ampere |
| Å | angstrom |
| A-h | ampere-hour |
| amu | atomic mass unit |
| API | American Petroleum Institute |
| atm | (standard) atmosphere |
| AU | astronomical unit |
| bbl | barrel (42 U.S. gallons) |
| bq | bequerel |
| Btu (BTU) | British thermal unit |
| C | Celsius (centigrade) |
| C | coulomb |
| cal | calorie |
| cd | candela |
| CDD | cooling degree-day |
| Ci | curie |
| cm | centimeter |
| CFS | cubic feet per second |
| d | day |
| EMU | electromagnetic unit |
| ESU | electrostatic unit |
| eV | electron volt |
| F | Fahrenheit |
| F | farad |
| ft | foot (feet) |
| $ft^2$ | square foot |
| $ft^3$ | cubic foot |
| g | gram |
| gal | gallon |
| gr | grain |
| Gy | gray |
| h | hour |
| H | henry |
| ha | hectare |
| HDD | heating degree-day |
| hp | horsepower |
| Hz | hertz |
| in | inch |
| J | joule |
| k | kilo (thousand) |
| K | Kelvin |
| kcal | kilocalorie |
| kg | kilogram |
| kJ | kilojoule |
| km | kilometer |
| kPa | kilopascal |
| ktoe | thousand tons of oil equivalent |
| kVa | kilovolt-ampere |
| kW | kilowatt |
| kWh | kilowatt-hour |
| kWp | kilowatt peak |
| L | liter |
| lb | pound |
| lx | lux |
| lm | lumen |
| m | meter |
| $m^2$ | square meter |
| $m^3$ | cubic meter |
| M | mega (million) |
| mcf | thousand cubic feet |
| MeV | million electron volts |
| mg | milligram |
| mi | mile |
| min | minute |
| MJ | megajoule |
| mm | millimeter |
| MPa | megapascal |
| mpg (MPG) | miles per gallon |
| mT | metric ton |
| MW | megawatt |
| MWe | megawatt electric |
| MWh | megawatt-hour |
| Mx | maxwell |
| N | newton |
| nm | nanometer |
| Nm | newton-meter |
| Ω | ohm |
| Oe | oersted |
| oz | ounce |
| Pa | pascal |
| psi | pounds per square inch |
| Q | quintillion Btu ($10^{18}$ Btu) |
| quad | quadrillion Btu ($10^{15}$ Btu) |
| R | Rankine |
| R | roentgen |
| rad | radiation (absorbed dose) |
| rem | roentgen equivalent man |
| s | second |
| S | siemens |
| Sv | sievert |
| t (T) | ton (tonne) |
| T | tesla |
| toe | tons of oil equivalent |
| V | volt |
| W | watt |
| yd | yard |

**A** ampere.

**ab-** *Electricity.* a prefix indicating electromagnetic units of the centimeter-gram-second system; e.g., abohm, abvolt.

**abandoned** *Mining.* describing an excavation that is deserted and in which no further mining is intended. Thus, **abandonment, abandoned working** or **mine.**

**abatement** *Environment.* a reduction in the amount of pollution generated, or in the effects of pollution.

**abatement cost** *Economics.* the cost of reducing emissions from some established reference level.

**Abdel Aziz** 1906–1975, King of Saudi Arabia from 1964 to 1975. He ruled in a period when the Arab oil exporting nations gained increasing control over the world oil market, and was seen as a major force behind the oil embargo against the U.S. in 1973, which helped produce the first oil price shock.

**aberration** *Physics.* an apparent displacement in the position of a celestial object, due to the time it takes for light from the object to reach an earthbound observer and to the orbital motion of the earth during that time.

**abiogenic** *Earth Science.* not resulting from the activities of living organisms.

**abiogenic theory** *Oil & Gas.* the theory that hydrocarbon deposits have a (mainly) non-biological origin. According to this, such materials became trapped far below the earth's crust when the basic structure of the planet evolved, and have subsequently migrated into reservoirs and to the surface through openings in the earth's crust. Contrasted with the more generally accepted BIOGENIC THEORY that hydrocarbon deposits derive from the remains of living organisms.

**abiotic** *Environment.* relating to or caused by nonliving environmental factors, such as light, temperature, water, soil, pH, salinity, and atmospheric gases.

**abrasion (abrasive) drilling** *Oil & Gas.* a rotary oil-drilling method in which abrasive material under pressure rotates while being pressed against the rock.

**absolute advantage** *Economics.* the ability of a country, individual, company, or region to produce more of a good or service with the same amount of resources, or the same amount of a good or service with fewer resources, than the cost at which any other comparable entity produces that good or service. See also COMPARATIVE ADVANTAGE.

**absolute entropy** *Thermodynamics.* the entropy of a substance existing at a temperature of absolute zero (0 degree Kelvin); any pure substance can be assumed to have an entropy of zero at this point.

**absolute expansion** *Thermodynamics.* the true expansion of a liquid with a change in temperature, allowing for the expansion of the container holding the liquid in calculating this measurement.

**absolute humidity** *Earth Science.* a statement of humidity that describes the mass of water vapor present in relation to the unit volume of space that it occupies; usually expressed in grams per cubic meter.

**absolute permeability** *Oil & Gas.* a measurement of the ability of a fluid, such as oil, gas, or water, to flow through a rock formation when the formation is at complete saturation.

**absolute pressure** *Measurement.* the total pressure of a gas system measured with respect to a pure vacuum (zero pressure).

**absolute temperature scale** *Measurement.* a temperature scale whose zero point corresponds to absolute zero (see next), such as the Kelvin scale or Rankine scale. Thus, **absolute temperature.**

**absolute zero** *Measurement.* the zero point on a temperature scale of ideal gases, denoted by 0°K on the Kelvin scale (or 0°R on the Rankine scale), –273°C on the Celsius scale, or –459°F on the Fahrenheit scale. At a condition of absolute zero, all molecular motion (in the classical sense of motion) is assumed to stop.

**absorb** *Chemistry.* to take up or receive matter; undergo a process of absorption. Thus, **absorbent.**

**absorbance** *Physics.* the ability of a medium to absorb radiation, which depends on temperature and wavelength; expressed as the negative common logarithm of the transmittance (ratio of transmitted energy to incident energy).

**absorbed dose** *Nuclear.* the amount of energy from ionizing radiation that is absorbed by a unit mass of any material; it is expressed in grays or rads. One gray is defined as one joule per kilogram; one rad is defined as 100 ergs per gram.

**absorbed glass mat battery** see AGM BATTERY.

**absorber** *Solar.* **1.** a material or device that absorbs solar radiation. **2.** *Photovoltaic.* the material that readily absorbs photons to generate charge carriers (free electrons or holes).

**absorber plate** *Solar.* a metal sheet in a flat plate solar collector whose primary function is to maximize the transfer of solar radiation reaching it through the glazing to the heat transfer fluid.

**absorptance** *Solar.* the ratio between the radiation absorbed by a surface (absorber) and the total amount of solar radiation striking the surface.

**absorptiometer** *Measurement.* a device used to measure the ability of a sample substance to absorb a given material.

**absorption** *Chemistry.* **1.** the process by which a liquid or gas is drawn into the permeable pores of a solid material. **2.** the action of energy or matter penetrating or being assimilated into a body of matter with no reflection or emission.

**absorption chiller** *Refrigeration.* a device that transfers thermal energy from a heat source to a heat sink through an absorbent fluid and a refrigerant. Most commercial absorption chillers use lithium bromide (a salt) and water as the fluid pair, with lithium bromide being the absorbent and water the refrigerant.

**absorption coefficient** *Measurement.* a measure of the amount of incident energy that is absorbed per unit distance or unit mass of a substance.

**absorption cycle** *Refrigeration.* a process within a refrigeration system during which the primary fluid (the refrigerant) and the secondary fluid (the absorbent) mix after the refrigerant leaves the evaporator. Thus, **absorption refrigeration, absorption cooling.**

**absorption spectroscopy** *Chemistry.* a technique for determining the concentration and structure of a substance, by measuring the amount of electromagnetic radiation that the sample absorbs at various wavelengths.

**absorptive glass mat battery** see AGM BATTERY.

**absorptivity** *Thermodynamics.* the ratio of the energy absorbed by a body to the total energy incident upon the same body.

**abutment** *Mining.* the part of a mine structure that directly receives thrust or pressure, as from an arch or vault.

**AC** alternating current.

**acceleration** *Physics.* the rate of change in velocity with respect to time.

**accelerator** *Transportation.* **1.** a pedal or lever designed to control the speed of an engine by actuating the carburetor throttle valve or fuel-injection control. **2.** *Nuclear.* short for PARTICLE ACCELERATOR; i.e., a device used to produce high-energy, high-speed beams of charged particles.

**accelerometer** *Measurement.* an instrument used to measure and record acceleration in a given direction, as in an aircraft or a spacecraft.

**accent light(ing)** *Lighting.* light that illuminates specific features of a room or building, as for enhanced light, aesthetic purposes, safety, and so on.

**acceptor** *Chemistry.* **1.** a molecule, or part of a molecule's structure, that accepts an electron pair from a donor. **2.** *Photovoltaics.* a dopant material, such as boron, having fewer outer shell electrons than required in an otherwise balanced crystal structure, providing a hole that can accept a free electron; this is an important attribute of photovoltaic materials.

**accessed** *Coal.* describing a coal resource that has been prepared for mining.

**accessible** *Coal.* describing a coal resource for which there are no land-use restrictions and for which it can be assumed that ownership or leaseholds will be obtainable for mining. Thus, **accessibility, accessible resources, accessible (reserve) base.**

**access to safe water** *Global Issues.* **1.** the fact of having a reasonable means of obtaining safe water, either treated water (e.g., a municipal water supply) or clean untreated water from springs, wells, and so on. **2.** a measure of the percentage of a given population having such access.

**accident** see NUCLEAR ACCIDENT.

**acclimation temperature** *Biological Energetics.* the ambient temperature that an organism has become accustomed to by a period of constant exposure to this condition over time.

**accumulated dose** *Health & Safety.* a total dose resulting from repeated exposure to a toxic substance or radiation.

**accumulation** *Oil & Gas.* one or more reservoirs of petroleum that have distinct trap, charge, and reservoir characteristics; this may encompass several fields or be equivalent to a single field.

**accumulator** *Refrigeration.* a device that prevents liquid from entering the suction of the compressor in a refrigeration cycle.

**accumulator battery** *Storage.* a term for any battery that stores (accumulates) energy in chemical form and discharges it as electrical energy.

**AC/DC** alternating current/direct current.

**ACEEE** American Council for an Energy-Efficient Economy (est. 1980), a nonprofit organization influential in advancing energy efficiency as a means of promoting both economic prosperity and environmental protection.

**acetylene** *Chemistry.* $C_2H_2$, an odorless, colorless gas obtained from limestone and coal; its combustion in pure oxygen produces the highest achievable flame temperature, over 3300°C, releasing 11,800 J/g. A widely used welding fuel that also has been used in illumination and in the synthesis of industrial compounds.

**Achnacarry agreement** *History.* a cartel agreement reached by major international oil companies in 1928 at Achnacarry Castle in Scotland, to control the downstream marketing of oil and divide it according to market shares existing at that time.

**acid** *Chemistry.* a general term for compounds that when dissolved in water increase the concentration of hydrogen ions (H+). Acids are proton donors and have a pH less than 7. Thus, **acidic, acidification, acidizing.**

**acid-based** *Storage.* describing batteries that use an acid such as sulfuric acid as the major component of the electrolyte; e.g., automobile batteries.

**acid deposition** *Environment.* the environmental deposit of acidic or acidifying components from the atmosphere by rain or other precipitation (cloud droplets, fog, snow, or hail); precipitation that has an excessive concentration of sulfuric or nitric acids, as a result of chemical pollution of the atmosphere from such sources as automobile exhausts and the industrial burning of coal or oil. Generally defined as any precipitation with a pH of less than 5.6. Also known as ACID RAIN or ACID PRECIPITATION.

**acid fracturing** *Oil & Gas.* a method to stimulate oil production in which an acid-in-oil emulsion is used to fracture and dissolve carbonate reservoirs; called **acid-frac** for short.

**acid gas** *Chemistry.* a term for any of the acidic constituents of natural gas, such as hydrogen sulfide or carbon dioxide.

**acid hydrolysis** see ACIDOLYSIS.

**acid mine drainage (AMD)** *Mining.* drainage with a pH of 2.0 to 4.5 from mines and mine wastes, especially coal mines, resulting from the oxidation of sulfides exposed during mining, which produces sulfuric acid and sulfate salts. The acid dissolves minerals in the rocks, causing the water **(acid mine water)** to have elevated concentrations of sulfate and dissolved iron. The low pH water can harm aquatic ecosystems and degrade aquifers.

**acidolysis** *Chemistry.* acid hydrolysis; the decomposition of a molecule with the addition of the elements of an acid to the molecule.

**acid precipitation** another term for ACID DEPOSITION.

**acid rain** *Environment.* the popular term for ACID DEPOSITION; i.e., rain or other precipitation that has an excessive concentration of sulfuric or nitric acids. See below.

**acid rain** Acid rain is a popular term for the atmospheric deposition of acidified rain, snow, sleet, hail, acidifying gases and particles, as well as acidified fog and cloud water. The increased acidity of these depositions, primarily from the strong acids, sulfuric and nitric, is generated as a by-product from the combustion of fossil fuels. The heating of homes, using electricity, and driving vehicles, all rely primarily on fossil fuel energy. When fossil fuels are combusted, acid-forming nitrogen and sulfur oxides are released to the atmosphere. These compounds are transformed in the atmosphere, often traveling thousands of kilometers from their original source, and then fall out on land and water surfaces as acid rain. As a result, pollutants from power plants in New Jersey or Michigan can impact pristine forests or lakes in undeveloped parts of New Hampshire or Maine. Acid rain was discovered in 1963 in North America at the Hubbard Brook Experimental Forest in the White Mountains of New Hampshire, in rain that was some 100 times more acidic than unpolluted rain. Innovations for reducing fossil fuel emissions, such as scrubbers in tall smoke stacks on power plants and factories, catalytic converters on automobiles, and use of low-sulfur coal, have been employed to reduce emissions of $SO_2$ and $NO_x$. As a result of increasing global economies, fossil fuel combustion is increasing around the world, with concomitant spread of acid rain.

**Gene E. Likens**
Institute of Ecosystem Studies
Millbrook, New York

**acid rock drainage (ARD)** *Materials.* a similar process to ACID MINE DRAINAGE, in which the condition results from pyrite exposure that is not mining-related.

**acid snow** see ACID DEPOSITION.

**acid treatment** *Oil & Gas.* a method to stimulate production from a well in which an acid is injected into a carbonate formation and then dissolves the reservoir rock. Also known as **acidizing** or **acid job.**

**acid wash** *Oil & Gas.* an operation and maintenance procedure to improve production from an oil or gas well, in which acid is pumped into the well tubing to remove scalings and other contaminants from the well.

**acoustic** *Physics.* **1.** having to do with sound or the study of sound. **2.** *Materials.* describing a material intended to absorb sound. Thus, **acoustic ceiling, acoustic insulation,** and so on. Also, **acoustical.**

**acoustical engineering** *Consumption & Efficiency.* the study of the practical applications of sound and the control of sound and vibration, as in structural design.

**acoustical logging** *Oil & Gas.* a determination of the physical properties or dimensions of a borehole by acoustical means.

**acoustics** *Physics.* the scientific study of sound, including its production, propagation, and effects.

**acre-foot** *Hydropower.* the volume of water covering one acre to a depth of one foot; used to describe a quantity of storage in a reservoir, lake, and so on.

**actinide** *Chemistry.* a classification for any of the heavy radioactive metallic elements having atomic numbers in the range from 90 (thorium) to 103 (lawrencium); actinium (89) is sometimes also included. Thus, **actinide series.**

**actinium** *Chemistry.* a radioactive chemical element having the symbol Ac, the atomic number 89, an atomic weight (in its most stable isotope) of 227, and a half-life of 21.7 years, with a melting point of 1050°C and a boiling point of about 320°C; a rare silvery-white metal found in compound form in uranium ores or obtained from radium by neutron bombardment.

**actinometer** *Solar.* an instrument used to measure direct radiation from the sun.

**activated alumina** *Materials.* a highly porous type of aluminum oxide that readily absorbs moisture and odors, used in the petroleum industry, refrigeration, and water purification.

**activated carbon** *Biomass.* a highly porous form of carbon, typically from wood, lignite, coal, or coconut shell; widely used as a filtration medium; e.g., to remove taste and odor from water by adsorbing organic compounds. Also, **activated charcoal.**

**activated shelf life** *Storage.* the length of time (at a specified temperature) that a charged battery can be stored before its capacity falls to an unusable level.

**activated sludge** *Environment.* the semiliquid, microbe-rich sediment that is added to secondary-stage sewage material in the activated-sludge process (see next).

**activated-sludge process** *Environment.* a widely used process for sewage treatment that raises the level of biological activity by increasing the contact between the wastewater and the actively growing microorganisms. Thus, **activated-sludge effluent.**

**activation** *Nuclear.* the creation of a radioactive element from a stable one by the absorption of neutrons or protons, occurring when a particle interacts with an atomic nucleus, shifting the nucleus into an unstable state and causing it to become radioactive.

**activation analysis** *Nuclear.* a method for identifying and measuring chemical elements in a sample of material that is made radioactive. Newly formed radioactive atoms in the sample then give off characteristic radiations indicating what kinds of atoms are present, and how many.

**activation energy** *Chemistry.* the energy needed to initiate a chemical reaction.

**activation enthalpy** *Thermodynamics.* the internal energy contribution to the free energy barrier that an atom in a metastable position must surmount in order to participate in a thermally activated process.

**activation entropy** *Thermodynamics.* the entropic contribution to the free energy barrier that an atom in a metastable position must surmount in order to participate in a thermally activated process.

**active material** *Storage.* describing a material that reacts chemically to produce electric energy when a battery discharges, and which is restored to its original state during the charge.

**active power** *Electricity.* the real power supplied by a generator set to the electrical load; it does the work of heating, turning motor shafts, and so on; measured in watts.

**active solar cooling** *Solar.* the converse of active solar heating; the use of the sun's radiant energy to power a cooling appliance.

**active solar heat(ing)** *Solar.* a solar energy system that uses mechanical devices and an external energy source in addition to solar energy, to collect, store, and distribute thermal (heat) energy. Thus, **active solar energy, active solar system,** and so on. Compare PASSIVE SOLAR ENERGY.

**activity** short for RADIOACTIVITY.

**actual emissions intensity** *Climate Change.* a ratio of the amount of emitted greenhouse gas over the associated production of common units of economic output.

**actuator** *Conversion.* any device that is moved a predetermined distance to operate or control another mechanical device; e.g., a controlled motor that converts voltage or current into a mechanical output.

**acute exposure** *Health & Safety.* a short interval of usually heavy exposure to radiation or a toxic substance. Exposure received within a short period of time.

**acute radiation syndrome** *Health & Safety.* a syndrome resulting from a whole-body dose of ionizing radiation in excess of 1 gray, and characterized by multiple symptoms such as diarrhea, vomiting, fever, and bleeding. Given a sufficiently large dose, death may result within hours or weeks of exposure. Also, **acute radiation exposure.**

**acute respiratory infection (ARI)** *Health & Safety.* the rapid onset of an infection of any part of the respiratory tract, usually classified as either an **upper respiratory infection (URI),** which involves the ears, nose, throat, or sinuses, or a **lower respiratory infection (LRI),** involving the trachea, bronchial tubes, and the lungs themselves.

**acute toxicity** *Health & Safety.* a toxic reaction that occurs over a relatively short period of time, directly following exposure to a single, typically large dose of the toxic substance. Similarly, **acute health effect.**

**Adair, Paul (Red)** 1915–2004, U.S. engineer known for his ability to control spectacular oil well blowouts and fires, noted especially for his efforts during the Persian Gulf War, when he and his teams extinguished oil well fires ignited by Saddam Hussein's troops.

**Adams, William Grylls** 1836–1915, English scientist who observed that a solid material (selenium) produced electricity when exposed to light. This became known as the PHOTOELECTRIC EFFECT, a major discovery in the field of electricity.

**adaptation** *Ecology.* **1.** a particular developmental, behavioral, anatomical, or physiological modification in a population of organisms, based on genetic changes and occurring as a result of natural selection. **2.** the general capacity of a species to undergo evolutionary change and thus enhance its ability to survive. **3.** *Climate Change.* the fact or policy of making adjustments in practices, structures, or systems in response to projected or actual changes in climate, such as the protection of coastal areas from sea-level rise.

**additionality** *Consumption & Efficiency.* an energy efficiency improvement above that required by building codes and efficiency standards.

**additive** *Materials.* **1.** any substance that is added to another substance, usually in a small quantity, in order to produce a desired effect in the primary substance. **2.** a chemical compound added to gasoline to improve performance; e.g., to reduce engine knock or emissions.

**Adelman, Morris** born 1917, U.S. economist noted for his founding work in the economics of oil supply and the world oil market. Central to his work is the premise that oil is a commodity like any other; i.e., that it should not have a special status as a depleting resource or an increasing-cost good.

**Adelman's rule** *Economics.* a practical approach to valuing oil reserves based on current prices and extraction costs; the principle that the in situ value ($V$) of an oil reserve ($R$) can be reasonably approximated by the formula $V = 1/2\ (p - c)R$, where $p$ is the price of oil and $c$ is the unit cost of extraction.

**adenosine triphosphate (ATP)** *Biological Energetics.* an energy-bearing molecule formed during light reactions through the phosphorylation of adenosine diphosphate. The energy in ATP is the primary energy source for most biological reactions.

**Ader, Clement** 1841–1926, French engineer noted for his pioneering work in aviation. In 1886 be built the Éole, a bat-like machine run by a lightweight steam engine that drove a four-blade propeller. In 1890 a second version of the Éole was built, which managed to take off into the sky and fly a distance of more than 40 yards.

**adhesion** *Materials.* **1.** a static attractive force at the contacting surface between two bodies of different substances in contact with each other. **2.** the sticking together of structural parts by means of cement or glue.

**adiabatic** *Thermodynamics.* **1.** without loss or gain of heat. **2.** specifically, describing a process (e.g., the expansion of a gas) in which there is no transfer of heat into or out of the system in question. Thus, **adiabatic process, adiabatic temperature change,** and so on. Compare DIABATIC.

**adiabatic envelope** *Thermodynamics.* a surface surrounding a thermodynamic system across which there is no heat transfer; disturbances to the envelope can only be made by long-range forces or by motion of part of the envelope.

**adiabatic flame temperature** *Thermodynamics.* the temperature of the products in a combustion process that takes place with no heat transfer and no energy exchange; this is the maximum possible temperature for these products.

**adiabatic (lapse) rate** *Earth Science.* the rate at which temperature decreases as a mass of air rises, or increases as the air falls.

**adiant exposure** *Solar.* radiant energy incident on a unit surface over some specified time period; the units are joules per square meter.

**adit** *Mining.* a level, or nearly level, access passage from the surface of a mine, used to excavate or drain the main tunnel.

**admittance** *Electricity.* the measure of how readily an alternating current flows through a circuit; the reciprocal of impedance.

**adsorber** *Environment.* **1.** a material or device that adsorbs; i.e., that will take up and hold a substance on its surface. **2.** specifically, an emissions control device that removes volatile organic compounds from a gas stream by means of attachment onto a solid matrix such as activated carbon.

**adsorption** *Chemistry.* the adhesion of the molecules of gases, dissolved substances, or liquids to the surface of solids or liquids with which they are in contact; distinguished from ABSORPTION, a process in which one substance actually penetrates into the inner structure of the other. Thus, **adsorb, adsorbent.**

**ADT** average daily traffic.

**ad valorem** *Economics.* per unit of value (i.e., divided by the price); many states and federal governments tax energy extraction in this manner. Thus, **ad valorem tax.**

**advanced gas-cooled reactor (AGR)** *Nuclear.* a type of graphite-moderated power reactor that uses helium as a coolant and graphite as a moderator, operating at substantial temperatures.

**Advanced Research Projects Agency** see ARPA.

**advanced vehicle technology (AVT)** *Renewable/Alternative.* the engineering and design processes that (will) lead to vehicles with high energy efficiencies and low emissions, such as direct-injection, hybrid electric, fuel cell, and battery-powered electric vehicle systems, as well as improved materials and vehicle design.

**advance mining** *Mining.* a system of mining in which the service roadways move forward along with the working face, as opposed to being formed beforehand (RETREAT MINING).

**adverse hydro** *Hydropower.* describing conditions that are unfavorable to the generation of hydropower; e.g., low rainfall or snowfall, lack of runoff from higher elevations.

**AEC** Atomic Energy Commission.

**AEEI** *Economics.* autonomous energy efficiency index; the rate at which the economy becomes less energy intensive, with the price of energy remaining constant relative to the prices of other inputs. This indicates productivity improvements in energy use that are not caused by increases in energy prices.

**aeolian** *Wind.* having to do with the wind; produced or affected by the action of winds.

**aeolipile** *History.* an ancient device operating on steam power, developed about 2000 years ago, reportedly by the Greek scholar HERO of Alexandria; considered the earliest working steam engine and a predecessor to modern jet engines. It is described as a sealed, water-filled boiler mounted over a source of heat; as the water turned to steam it rose into a hollow sphere above. The steam would then escape from two curved outlet tubes on the sphere, causing it to rotate. [From a Greek term for "wind ball".]

**aerate** *Physics.* to expose a substance to air or another gas, e.g., the passing of air through a liquid substance, or the process of air entering the soil. Thus, **aeration.**

**aeration cell** *Electricity.* a device that generates electromotive force across electrodes that are made of the same material but located in different concentrations of dissolved air.

**aerator** *Consumption & Efficiency.* **1.** a device used to aerate a substance or medium; i.e., expose it to air or another gas. **2.** specifically, a device installed in a faucet or showerhead to add air to the water flow, thus maintaining an effective spray while reducing overall water consumption.

**aerobe** *Biological Energetics.* an organism, especially a bacterium, that requires atmospheric oxygen to live. Contrasted with an ANAEROBE, which does not require oxygen.

**aerobic** *Biological Energetics.* requiring or occurring in the presence of oxygen ($O_2$). Thus, **aerobic bacteria, aerobic metabolism.**

**aerobic respiration** *Biological Energetics.* respiration in which molecular oxygen is consumed through its use as a terminal electron acceptor, and which produces carbon dioxide and water.

**aeroderivative** *Conversion.* an aviation propulsion gas turbine (jet engine) used in a non-aviation application (e.g., an electric power plant) to provide shaft power.

**aerodynamic drag** *Physics.* the opposing force encountered by a body moving relative to a fluid; e.g., an aircraft in flight displacing the air in its path.

**aerodynamics** *Physics.* **1.** the scientific study of gases in motion and the forces that affect this motion. **2.** specifically, the study of the effects of air in motion on an object; either objects moving through air, such as aircraft or automobiles, or stationary objects affected by moving air, such as bridges or tall buildings.

**aerogenerator** *Wind.* a device that generates electricity from the kinetic energy produced by wind.

**aeromechanics** *Physics.* the scientific study of air and other gases in motion or in equilibrium, including the two distinct branches of aerodynamics and aerostatics.

**aerometer** *Measurement.* an instrument used to measure the weight or density of air and other gases.

**aeronautics** *Transportation.* the science and technology of flight, especially the design, construction, and operation of aircraft.

**aerosol** *Environment.* **1.** the suspension in a gaseous medium of very fine particles of a solid or of droplets of a liquid. Fog, smoke, and volcanic dust are examples of aerosols in the atmosphere. **2.** the particles themselves.

**aerosol propellant** *Environment.* a liquefied gas that is used as the driving force to expel a liquid from a container, such as an aerosol spray from a can; e.g., hair spray, deodorants, spray paints, or various household cleaning products.

**aerospace** *Transportation.* a collective term for activities involving flight either in the atmosphere or in space.

**aerostat** *Transportation.* a lighter-than-air vehicle, such as a balloon or dirigible.

**aerostatics** *Physics.* **1.** the scientific study of air and other gases in equilibrium, and of bodies suspended or moving within such gases. **2.** the study of lighter-than-air aircraft.

**aerothermodynamics** *Thermodynamics.* **1.** a branch of thermodynamics that studies the effects of heating and the dynamics of gases. **2.** the analysis of aerodynamic phenomena at high gas speeds, incorporating the essential thermodynamic properties of the gas.

**aesthetic impact** *Social Issues.* the effect that a change in land use has or would have on the visual appeal or other aesthetic qualities of the given setting; e.g., the aesthetic impact of a wind farm placed on a hillside or offshore location.

**aestivation** another spelling of ESTIVATION.

**AFC** alkaline fuel cell.

**afforestation** *Environment.* a direct, human-induced conversion to forest of land that had not been forested for a significant preceding period of time, through planting, seeding, or the human-induced promotion of natural seed sources.

**afterburner** *Transportation.* **1.** an auxiliary combustion chamber placed behind a jet engine turbine to gain extra thrust by injecting additional fuel into the turbine's hot exhaust gases. **2.** *Environment.* a catalytic or thermal combustion device used to control air contaminant emissions.

**aftercooling** *Nuclear.* the cooling of a nuclear reactor after it has been shut down.

**afterdamp** *Mining.* the residue mixture of gases in a mine following a mine fire or an explosion of firedamp.

**afterheat** *Nuclear.* heat that is liberated by the decay of radioactive materials in a reactor that has been shut down.

**aftermarket conversion** *Renewable/Alternative.* the alteration of a standard factory-produced vehicle having a conventional fuel system, by adding equipment that enables the vehicle to operate on alternative fuel. Thus, **aftermarket converted vehicle.**

**aftertreatment** *Transportation.* a method of controlling emissions from internal combustion engines by applying pollution control technologies to the engine exhaust stage (as opposed to treatments applied during the intake or combustion stages).

**AFUE** annual fuel utilization efficiency.

**Agenda 21** *Sustainable Development.* **1.** a comprehensive plan of action to be taken globally in the interest of sustainable development, agreed upon at the Earth Summit in Rio de Janeiro, Brazil in 1992. Major issues of environment and development were examined, including resource management, poverty, population, and human health. **2.** any of various local programs modeled after this initiative. [So called because it is described as a plan for the 21st century.]

**Agent Orange** *Health & Safety.* a toxic chemical agent used as a herbicide; investigated for possible health effect in humans. See TCDD.

**agglomerate** *Materials.* **1.** a mass or cluster of fine particles, gathered together in a body of larger size than the original particles, but not in a coherent pattern. **2.** of materials, to become grouped together in this manner. Thus, **agglomerated, agglomeration.**

**agglomerating character** *Coal.* a classification for coal based on its caking properties; coals are considered **agglomerating** if the coke button remaining from a test for volatile matter will support a weight of 500 grams without pulverizing, or if the button swells or has a porous cell structure.

**aggregate intensity** *Economics.* the ratio of total energy use to total output measured at a comprehensive level, as for an entire industry or a national economy; e.g., the ratio of energy use to gross domestic product for a nation.

**AGM battery** *Storage.* absorbed glass mat battery; a type of sealed lead-acid battery in which the electrolyte is absorbed in a matrix of glass fibers, which holds the electrolyte next to the plate, and immobilizes it to prevent spills. AGM batteries tend to have good power characteristics and low internal resistance.

**agrarianization** *Ecology.* the establishment of agriculture; the formation and spread of a system structured around the domestication of plants and animals.

**agribusiness** *Economics.* **1.** those aspects of agriculture involving issues of finance, sales, marketing, and the like. **2.** larger corporate entities of agriculture as opposed to smaller family-run operations; e.g., Archer Daniels Midland corporation.

**agricultural energetics** *Biological Energetics.* **1.** the various forms of energy involved in the process of agriculture, either as inputs (e.g., human labor, animal power, electricity) or as useful output (e.g., food, manure). **2.** specifically, the relationship between energy in the form of food produced and the energy input required to achieve this production.

**agricultural residue** *Biomass.* a fuel source composed of plant parts, primarily stalks and leaves, that are not harvested for use as food or fiber; e.g., corn stalks and husks, wheat straw, or rice straw.

**Agricultural Revolution** *History.* a term for the period in history beginning shortly before the Industrial Revolution, when significant improvements in agricultural production were achieved through such means as land reform, crop rotation, livestock improvements, and technological innovation (e.g., improved plows).

**agriculture** *Consumption & Efficiency.* the process, business, or science of producing food, feed, fiber and other desired products by the cultivation of certain plants and the raising of domesticated animals (livestock). See below.

**agriculture** Humans began to cultivate their food crops about 10,000 years ago. Prior to that time, hunter-gatherers secured their food as they traveled in the nearby environment. When they observed some of the grains left behind at their campsites sprouting and growing to harvest, they began to cultivate these grains. From these humble beginnings agriculture began. Slash and burn, an early type of crop culture remains today a truly sustainable agriculture, one that is independent of fossil energy. In such a system, about 10 hectares of productive land is held in fallow for each planted hectare. With this rotation system, a hectare is planted once in 20 years, allowing the soil to reaccumulate vital plant nutrients. Although the practice requires large acreages and large labor inputs, the crop yields are adequate. For example, corn with ample rainfall can yield about 2000 kg/ha. Over time human labor in agriculture has decreased because of the use of animals and finally with machinery powered by fossil fuel. Currently plentiful and economical fossil energy supports an era of machinery and agricultural chemicals. Now about 1000 liters of oil equivalents are used to produce a hectare of corn with a yield of 9000 kg/ha. One-third of this energy is used to replace labor, one-third for fertilizers, and one-third for others. Worldwide more than 99.7% of human food (calories) comes from the land. Serious environmental impacts, such as soil erosion, water runoff and pesticide pollution, result from fossil-fuel intensive agriculture. A critical need exists to assess fossil energy limits, the sustainability of agriculture, and the food needs of a rapidly growing world population.

**David Pimentel**
Cornell University

**agroecology** *Ecology.* the study of the relationship between an agricultural system and its surrounding environment.

**agroecosystem** *Ecology.* the biotic and abiotic components of an agricultural system, including not only the livestock and cultivated crops but also, for example, the water supply, other plant and animal species, soil characteristics, climate, and human inputs.

**agroforestry** *Ecology.* the practice of growing a combination of forest growth products and agricultural crops on the same area of land.

**agronomy** *Consumption & Efficiency.* the scientific study of agricultural crops and soils.

**AIM** Action Impact Matrix.

**AIP** American Institute of Physics.

**air** *Chemistry.* the invisible, odorless, and tasteless mixture of gases forming the earth's atmosphere. At normal sea-level pressure, dry air consists of (percentage by volume) nitrogen 78%, oxygen 20.95%, argon 0.93%, carbon dioxide 0.033% (currently; thought to be increasing), neon 0.0018%, helium 0.0005%, methane 0.0002%, krypton 0.0001%, and smaller amounts of nitrous oxide, hydrogen, xenon, and ozone.

**AIR** air-injection reactor.

**air basin** *Environment.* a defined land area that has generally the same quality and properties of air, because of similar meteorologic and geographic conditions throughout the area; a geographic unit used in the regulation of air pollution and air quality.

**air battery** *Storage.* a type of battery having an electric storage cell **(air cell)** in which one electrode is activated chemically by the oxygen in the ambient air.

**air blast(ing)** *Mining.* **1.** a method of blasting by the use of a jet of compressed air at very high pressure. **2.** a disturbance in underground workings accompanied by a strong rush of air.

**airborne fraction** *Climate Change.* the portion of carbon dioxide that remains in the atmosphere after its release from energy consumption and land-use activities, as opposed to the portion absorbed by plants and oceans.

**air brake** *Transportation.* a mechanism operated by compressed air acting on a piston, used to stop or slow a moving element, as in a motor vehicle.

**air-breathing** *Transportation.* describing an engine or vehicle that operates on the basis of utilizing air for combustion.

**air change** *HVAC.* the amount of air required to completely replace the air in a room or building (as opposed to simply recirculating it).

**air change efficiency** *HVAC.* a measure of how quickly the air in a given space (a room or building) is completely replaced.

**air change rate** *HVAC.* the replacement of a quantity of air in a given space within a certain period of time, typically expressed in units per hour; e.g., if a building has one air change per hour, this is equivalent to all of the air in the building being replaced in that time.

**air conditioner** *HVAC.* **1.** any device that modifies or controls one or more aspects of air, such as its temperature, relative humidity, purity, or motion. **2.** such a device used in a building, room, vehicle, or other enclosed area to maintain the air therein at a comfortably cool and dry level.

**air conditioning** *HVAC.* the simultaneous control of the temperature, relative humidity, purity, and flow of the air in a given enclosed space, as required either for the thermal comfort of people who live or work in the respective space, or for specified technological prescriptions. This can involve either cooling or heating of air, but in popular use is assumed to refer only to cooling.

**air-conditioning ton** see REFRIGERATION TON.

**air-cooled** *Transportation.* describing an engine that is cooled by a stream of air rather than by water or another liquid coolant.

**aircraft** *Transportation.* any weight-bearing vehicle designed for navigation in the air, supported by the action of air upon its surfaces or by the vehicle's own buoyancy.

**air curtain** *Consumption & Efficiency.* a continuous broad stream of high-velocity, temperature-controlled air that is circulated across an opening, in order to reduce airflow in or out of the space, minimize heat loss or gain, contain a fluid (e.g., an oil spill), and so on.

**airdox** *Coal.* a system for breaking down coal through the use of compressed-air blasting.

**air drilling** *Mining.* a form of drilling in which compressed air or gas is the circulation medium; used in coal mining, and also in petroleum extraction instead of mud drilling because of its greater speed.

**air exchange** see AIR CHANGE.

**air exfiltration** see EXFILTRATION.

**air film** *Consumption & Efficiency.* a layer of still air adjacent to a surface, providing some thermal resistance.

**air filter** *Consumption & Efficiency.* a device attached to an air intake mechanism to remove solid impurities from an airstream; may be used with ventilating mechanisms or to prevent pollutants from entering an instrument or engine.

**airflow** *Measurement.* a rate of movement for air, computed by volume or mass for a certain time unit.

**airfoil** *Transportation.* a body, part, or surface designed to provide a useful reaction on itself, such as lift or thrust, during motion through the air.

**air-fuel ratio** *Consumption & Efficiency.* the relative amounts of fuel and air in a combustion chamber; a method of expressing the composition of a mixture of fuel and air by the measurement of either weight or volume. Thus, **air-fuel mixture.**

**air handler** *HVAC.* the interior of an air-conditioning system that contains the blower, cooling (evaporator) coil, and heater. Also, **air-handling unit.**

**air-injection reactor (AIR)** *Transportation.* a system installed in an automotive engine to mix fresh air with exhaust gases in the exhaust manifold, causing reaction with any escaped and unburned or partially burned fuel from the cylinders.

**air lock** or **airlock** *HVAC.* a compartment that serves to control air exchange into or out of a conditioned space.

**air mass** *Solar.* a measure of how far light travels through the earth's atmosphere. One air mass, or AM1, is the thickness of the earth's atmosphere. Air mass zero (AM0) describes solar irradiance in space, i.e., radiation not attenuated at all by the atmosphere. The power density of AM1 light is about 100 $mW/cm^2$; the power density of AM0 light is about 136 $mW/cm^2$.

**air pollution** *Environment.* a general term for the discharge or release of harmful or unwanted substances to the atmosphere; the extraction, processing, and use of energy, particularly fossil fuels, is a major source of such releases. Thus, **air pollutant.**

***air pollution*** *View of midtown Manhattan, New York City, late 1960s; an era of more relaxed emission standards.*

**air pressure** *Physics.* the force per unit area that air exerts on any surface in contact with it, due to the motion of air molecules.

**air quality** *Environment.* a measurement of the relative presence (or absence) of pollutants in the air; the properties and degree of purity of air to which humans and other organisms are exposed.

**air quality index (AQI)** *Environment.* a numerical index used to describe the extent to which harmful pollutants (e.g., carbon monoxide, sulfur dioxide) are present in the atmosphere at any given time. The higher the index value, the higher the level of pollutants and thus the greater the likelihood of health effects.

**airshed** another term for AIR BASIN.

**airship** *Transportation.* a term for a dirigible or other such lighter-than-air craft.

**air-source heat pump** *HVAC.* a type of heat pump that transfers heat from outdoor air to indoor air during the heating season, and works in reverse during the cooling season.

**airspeed** *Transportation.* the velocity of an aircraft or other airborne body relative to the velocity of the surrounding air; it differs from ground speed to the extent that the air is also in motion.

**air-standard cycle** *Thermodynamics.* a cycle involving a working fluid that is considered to be an ideal gas with properties of dry air; an ideal thermodynamic cycle used to assess the performance of actual devices; e.g., internal-combustion engines.

**air-standard refrigeration cycle** *Refrigeration.* an air-standard cycle (see previous) used to assess the performance of actual refrigeration systems or substances.

**airtight** *Consumption & Efficiency.* a relative term indicating the extent to which a system (e.g., a building envelope) is resistant to the passage of air. Thus, **airtightness.**

**air-to-air heat exchanger** *HVAC.* a device with separate air chambers that transfers heat between the conditioned air being exhausted from a building and the outside air being supplied to it.

**air vapor barrier** *HVAC.* a layer of material impervious to moisture, applied to the surfaces enclosing a space to limit moisture migration.

**AIT** Asian Institute of Technology.

**ALARA** *Nuclear.* as low as reasonably achievable; a requirement in U.S. federal law that facilities possessing radioactive material licenses must keep all doses, releases, contamination, and other risks to the lowest level that is reasonably possible.

**Alaska National Wildlife Refuge** *Environment.* another name for the ARCTIC NATIONAL WILDLIFE REFUGE.

**albedo** *Earth Science.* the fraction of incident light that is reflected in all directions from an uneven surface, especially the surface of the earth.

**alchemy** *History.* **1.** a form of inquiry that existed from about 500 BC to about 1600 AD, practiced especially in Europe in the Middle Ages and primarily concerned with attempts to transform base metals into gold. Not a productive effort in itself, but regarded as a forerunner to modern chemistry because it involved the experimental analysis of various materials. Thus, **alchemist. 2.** a derogatory term for an investigative activity in the modern world that is regarded as based on spurious principles.

**alcohol** *Chemistry.* **1.** any of a general class of organic compounds formed by the attachment of one or more hydroxyl (–OH) groups to carbon atoms in place of hydrogen atoms; e.g., methanol, ethanol, propanol, butanol. **2.** specifically, the transparent, colorless liquid that is the intoxicant in beverages such as wine, beer, or whiskey, known technically as ETHANOL.

**alder** *Biomass.* any of various trees and shrubs of the genus *Aldus.* Alder trees grow rapidly and cover a site quickly and thus are a useful source of fuelwood; they also have the ability to increase the nitrogen content of the soil by means of nitrogen fixation.

**Alfvén, Hannes** 1908–1995, Swedish physicist, known for his work in magnetohydrodynamics, and its applications in plasma physics.

**Alfvén wave** *Physics.* a transverse hydromagnetic wave that propagates along magnetic field lines and is generated by the low-frequency oscillation of ions in a fluid medium, typically a plasma.

**algae** *Ecology. singular,* **alga.** any of a large group of mostly aquatic organisms that contain chlorophyll and other pigments and can carry on photosynthesis, but lack true roots, stems, or leaves; they range from microscopic single cells to large multicellular structures, including nearly all seaweeds. Research is currently being conducted into the use of algae as biomass fuel.

**algal bloom** *Ecology.* **1.** an unusual concentration of algae in or on a body of water, especially as a result of pollution from the runoff of fertilizers, industrial wastes, and so on. **2.** *Climate Change.* the proposal that by sprinkling a relatively small amount of iron into certain areas of the ocean, large algal blooms can be created that would increase the uptake of carbon from the atmosphere and thus mitigate the effects of climate change.

**algal coal** another name for BOGHEAD COAL.

**aliphatic** *Chemistry.* **1.** describing a major class of compounds in which carbon and hydrogen molecules are arranged in a straight or branched chain. **2.** a compound of this type.

**alkali** *Chemistry.* **1.** any substance having basic, as opposed to acidic, properties and that reacts with and neutralizes an acid; a term often used interchangeably with *base.* **2.** *Earth Science.* a bitter-tasting salt consisting of sodium or potassium carbonate, found in soils in arid or semiarid regions and generally unproductive for agriculture.

**Alkali Act** *History.* a measure enacted in Great Britain in 1863, calling for a reduction in harmful emissions from alkali manufacturers; considered to be the first formal environmental protection law. The Act established an official regulator, the **Alkali Inspectorate,** to enforce this legislation.

**alkali metal** *Chemistry.* any of the elements found in group 1A of the periodic table, which form highly alkaline solutions in water and burn vigorously in air; i.e., lithium, sodium, potassium, rubidium, cesium, and francium. They have a valence of one and are softer and less dense than other metals.

**alkaline** *Chemistry.* having the properties of an alkali; having a pH greater than 7.0. Thus, **alkalinity.**

**alkaline battery** *Storage.* a type of battery having a primary cell (**alkaline cell**) in which the electrolyte consists of an alkaline solution, usually potassium hydroxide.

**alkaline earth (metal)** *Chemistry.* any of the metallic elements found in group 2 of the periodic table; i.e., beryllium, magnesium, calcium, strontium, barium, and radium. They are highly reactive and thus are not found free in nature.

**alkaline fuel cell (AFC)** *Hydrogen.* a type of hydrogen/oxygen fuel cell using an electrolyte that is an aqueous solution of potassium hydroxide retained in a porous stabilized matrix. AFCs operate at relatively low temperatures and are among the most inexpensive and efficient fuel cells. They have been used for spacecraft but have not had wide application elsewhere because they are highly sensitive to carbon dioxide and thus must operate in a closed environment.

**alkane** *Chemistry.* any of various aliphatic hydrocarbons that have the general formula $C_nH_{2n+2}$. The first (lightest molecular weight) four are gases; higher members are liquids, and those above $C_{16}H_{34}$ are waxy solids. Thus, **alkane series.**

**alkylate** *Oil & Gas.* the product of an alkylation reaction, especially a high-octane product from alkylation units that is blended with motor and aviation gasoline to improve the antiknock value of the fuel.

**alkylation** *Oil & Gas.* a refining process for chemically combining isobutane with olefin hydrocarbons (e.g., propylene, butylene) through the control of temperature and pressure in the presence of an acid catalyst, usually sulfuric acid or hydrofluoric acid.

**allobar** *Nuclear.* an isotope with a different atomic weight than the naturally occurring form of the same element.

**allochromy** *Physics.* the radiation emitted from a substance at a particular wavelength, resulting from the absorption of incident radiation of a different wavelength.

**allochthonous** *Coal.* describing a type of coal formed from accumulated plant material that was transported from its original place of growth and deposited elsewhere.

**allowance** *Economics.* a unit of trade in emissions trading systems that grants the holder the right to emit a specific quantity of pollution once (e.g., one ton).

**alloy** *Materials.* any of various materials having metallic properties and composed of two or more closely mixed chemical elements, of which at least one is a metal; e.g., brass is an alloy of copper and zinc. Alloys are produced to obtain some desirable quality such as greater hardness, strength, lightness, or durability.

**alpha particle** *Nuclear.* a particle emitted from the nucleus of an atom, containing two protons and two neutrons, identical to the nucleus (without the electrons) of a helium atom. Thus, **alpha decay, alpha emission, alpha radiation.**

**ALRI** acute lower respiratory infection.

**Altair** *Communication.* the brand name for the first personal computer to have a significant public impact, introduced in 1975 with 256 bytes of memory. It was sold by MITS, Inc. of Albuquerque, New Mexico.

**Altamont Pass** *Wind.* the site of one of the world's first large wind farms, an area of low hills east of Oakland separating the relatively cool San Francisco Bay area from the hotter San Joaquin (Central) Valley.

**alternating current (AC)** *Electricity.* an electric current that reverses its direction of flow at regular intervals. AC is easier to transmit over long distances than DIRECT CURRENT (DC), and it is the form of electricity used today in most homes and businesses.

**alternative** *Renewable/Alternative.* having to do with or employing alternative energy or an alternative fuel source (see following).

**alternative energy** *Renewable/Alternative.* any energy system other than the traditional fossil, nuclear, and hydropower energy sources that have been the basis of the growth of industrial society over the past two centuries; e.g., solar, wind, or hydrogen energy.

**alternative fuel** *Renewable/Alternative.* a non-petroleum energy source used to power transportation vehicles, especially road vehicles, such as ethanol or hydrogen. Thus, **alternative fuel vehicle.** See below.

**alternator** *Electricity.* a machine or device that generates alternating current.

**altimetry** *Earth Science.* the measurement of altitude; i.e., the elevation of the land, sea, or ice surface, or of an aircraft moving above the surface. An **altimeter** is a device used for this purpose.

**alumina** *Materials.* the oxide of aluminum, which occurs in nature as corundum and which is used in its synthetic form for the production of aluminum metal.

**aluminum** *Chemistry.* a soft, lightweight, silver-white metallic chemical element, the third most common element, having the symbol Al, the atomic number 13, an atomic weight of 26.9815, a melting point of 650°C, and a boiling point of 2450°C. It is highly ductile, malleable, conductive, and resistant to corrosion and wear, and it is widely used in alloys.

**AM** amplitude modulation.

**ambient** *Environment.* of or relating to a condition of the environment surrounding a body or object (such as an organism or a building), especially a condition that affects this body or object but is not significantly affected by it. Thus, **ambient air, ambient temperature,** and so on.

**ambient light(ing)** *Lighting.* surrounding light; light that provides general background illumination for daily activities. Contrasted with TASK LIGHT.

**ambient standard** *Policy.* a law or regulation that sets the minimum desired level of air or water quality, or the maximum level of a pollutant. Usually expressed as a concentration; e.g., the national ambient air quality standard for sulfur oxides in the U.S. is an annual arithmetic mean concentration of 0.03 ppm.

**ambipolar diffusion** *Physics.* the diffusion of charged particles in a plasma, resulting from the combined influences of a density gradient and the internal electric field set up by the

**alternative fuel** Any fuel that can technically be used to reduce the amount of dominant fuel required. Common usage: a fuel that can be used to reduce the amount of petroleum-derived fuels used in transportation. Thus, alternative fuel in common usage means alternative transportation fuel (ATF). In the United States the Energy Policy Act of 1992 (EPAct) legally defined alternative fuels for purposes of the Act. EPAct defined alternative fuels to include: natural gas and liquid fuels derived (domestically) from natural gas; blends of 85% or more (by volume) of alcohol with gasoline; liquefied petroleum gas; coal-derived liquid fuels; hydrogen; electricity; biodiesel. Under EPAct, alternative fuels may be thought of as a limited set of fuels that may technically be used "instead of" petroleum derived fuels, rather than broadly "to reduce the amount of" such fuels. Replacement fuels, also defined in the U.S. EPAct, are fuels that can reduce the amount of dominant fuel required by being blended into petroleum-derived fuels. The most important of such fuels include: U.S. gasohol (10% by volume ethanol, 90% gasoline), and methyl tertiary butyl ether (MTBE). Brazil blends ethanol into gasoline at about 23% volume. Worldwide, an important future transportation fuel that is not an alternative fuel according to EPAct is Fischer-Tropsch diesel fuel derived from natural gas outside the U.S. Alternative fuels as legally specified in the U.S. EPAct are only a subset of ATFs as defined here. Examples given here include the ATFs that most nations, vehicle producers and energy companies consider most feasible to reduce use of petroleum.

**Danilo J. Santini**
Argonne National Laboratory

separation of the electrons and the positive ions. Such electric fields are self-generated by the plasma and act to preserve charge neutrality.

**AMD** acid mine drainage.

**American boring (system)** *History.* a system of percussive boring in which a string of boring tools is attached to a rope suspended from a derrick, thus allowing the tools to be raised clear of the hole in order to facilitate cleaning the hole; widely used in the 19th-century U.S. oil industry.

**American Institute of Physics** (est. 1931), a nonprofit membership corporation whose purpose is to promote the advancement and diffusion of the knowledge of physics and its application to human welfare.

**American Petroleum Institute (API)** (est. 1919), the primary trade association of the U.S. oil and gas industry. It seeks to influence public policy in support of the U.S. oil and natural gas industry. See also API GRAVITY.

**American Physical Society** (est. 1899), an academic and professional society dedicated to the advancement of physics education and research.

**American (farm) windmill** *Wind.* a multiblade mechanical wind pump, initially made of wood and later of metal, developed during the late 19th century in the midwestern U.S. and subsequently used worldwide. Used principally to pump water from underground reservoirs. Also known as a **Chicago windmill.**

**American wire gauge** *Electricity.* a standard system for designating the size of electrical wire; the higher the number, the smaller the wire. Most electrical house wiring is #12 or #14 gauge.

**americium** *Chemistry.* a human-made radioactive element created by the neutron bombardment of plutonium; it is a crystalline silver-white transuranic element of the actinide series, having the symbol Am, atomic number 95, an atomic weight of 243 in its most stable isotope, and a half-life of 475 years. Its longest-lived isotopes, **Am 241** and **Am 243,** are alpha-ray emitters used as radiation sources in research.

**ammeter** *Measurement.* an instrument used to measure and indicate the rate of flow of an electric current, usually expressed in amperes.

**ammonia** *Chemistry.* $NH_3$, a colorless gas that has a strong, highly irritating odor and an alkaline reaction in water; lighter than air and extremely soluble in water. Formed in nature as an end product of animal metabolism and also commercially produced for a variety of industrial uses; e.g., as a fertilizer, cleaning agent, and refrigerant.

**Amoco-Cadiz** *Environment.* a supertanker that ran aground just off the coast of Brittany, France in March of 1978, spilling an estimated 69 million gallons of crude oil; this was the second largest tanker spill in history, and among the most environmentally damaging at the time.

**amorphous** *Materials.* not having a definite shape or structure; describing substances that are solids but not crystals.

**amorphous silicon** *Photovoltaic.* an alloy of silica and hydrogen with a disordered, noncrystalline internal atomic arrangement; it can be deposited in thin-film layers by a number of deposition methods to produce photovoltaic cells on glass, metal, or plastic substrates.

**amp** short for AMPERE.

**amperage** *Electricity.* a measure of the strength of an electrical current, expressed in amperes.

**Ampère, André Marie** 1775–1836, French physicist and mathematician, a key figure in the development of the fields of electricity and magnetism. Ampère produced a definition of the unit of measurement of current flow, now known as the *ampere.*

**ampere (A)** *Electricity.* the basic unit of electric current in the meter-kilogram-second system; equivalent to one coulomb per second. One ampere is the current that, if held constant in two parallel conductors of infinite length at a distance one meter apart in a vacuum, will produce a force of $2 \times 10^{-7}$ newton per meter of length.

**ampere-hour** *Electricity.* the quantity of electricity that passes through a circuit in one hour when the rate of flow is one ampere; equal to 3600 coulombs.

**Ampère's rule** *Electricity.* the statement that the direction of the magnetic field lines surrounding a conductor is counterclockwise when viewing the conductor with the current flow coming toward the observer, and clockwise if the current is moving away from the observer.

**Ampère's theorem** *Electricity.* a quantitative relationship for the strength of a magnetic field in relation to an electric current.

**amplifier** *Electricity.* a device that magnifies the strength of a relatively weak input signal without altering the characteristics (such as waveform) of that signal, as in a radio or television set or as part of a stereophonic sound system.

**amplitude modulation (AM)** *Communication.* the deliberate processing of a carrier signal so that its amplitude varies in accordance with the level of the modulating signal; used in ordinary radio broadcasting **(AM radio)** and in transmitting the video portion of a television signal.

**AMU** or **amu** **1.** atomic mass unit. **2.** air mileage unit.

**anabatic** *Earth Science.* of air, moving upward. An **anabatic wind** is a local wind blowing up a hill or mountain due to local surface heating; this is caused by the difference in density between the warm ground air and the cooler air in the free atmosphere.

**anadromous** *Ecology.* describing fish born in fresh water that descend to live in the ocean before ascending to the freshwater source again to spawn and reproduce; e.g., salmon of the U.S. Pacific Northwest.

**anaerobe** *Biological Energetics.* an organism, especially a bacterium, that does not require atmospheric oxygen to live. Contrasted with an AEROBE that does require oxygen.

**anaerobic** *Biological Energetics.* occurring or existing in the absence of oxygen. Thus, **anaerobic bacteria, anaerobic metabolism, anaerobic respiration.**

**anaerobic digester** *Biomass.* a device that optimizes the anaerobic digestion (see next) of biomass, such as animal manure or wastes from food processing; the biogas that is a byproduct of the digestion process (mostly methane) can be used as a commercial fuel source.

**anaerobic digestion** *Biological Energetics.* the process by which complex plant and animal compounds are broken down into simpler compounds in the absence of oxygen, producing a variety of gaseous and soluble products. Also, **anaerobic decomposition.**

**anaerobic lagoon** *Biomass.* a holding pond for livestock manure that is designed to stabilize the manure anaerobically; the effluent from the lagoon is a source of nitrogen for plants and it also may be used to capture biogas, with the use of an impermeable floating cover.

**anahydrous** another spelling of ANHYDROUS.

**analytical engine** see CALCULATING ENGINE.

**anemoclinometer** *Earth Science.* an instrument used to measure the direction of wind in relation to the horizontal plane.

**anemometer** *Earth Science.* a device for measuring or indicating the speed of the wind; a typical version consists of three or four hemispherical cups attached to the ends of arms that radiate from a vertical spindle, so as to be rotated by an air current.

**anerobic** another spelling of ANAEROBIC.

**aneutronic** *Nuclear.* not producing fusion neutrons.

**angle of draw** *Coal.* a measurement for coal mine subsidence; the angle between a vertical line drawn upward to the surface from the edge of the underground opening and a line drawn from the edge of the opening to a point of zero surface subsidence.

**angle of incidence** *Physics.* the angle between a ray (e.g., incoming solar radiation) that strikes a surface and the perpendicular to that surface at the point of incidence.

**angle of reflection** *Physics.* the angle formed by the normal line to a surface and the direction of propagation of waves that are reflected from that surface.

**angle of refraction** *Physics.* the angle between the direction of propagation of a refracted ray and the perpendicular to the interface at the point of refraction.

**Anglo-Persian Oil Company (APOC)** the predecessor organization to BRITISH PETROLEUM, founded in 1909 to develop the oil resources of Persia (Iran).

**Angström, Anders** 1814–1874, Swedish physicist, noted for his study of light, especially spectrum analysis. He discovered that hydrogen is present in the sun's atmosphere, and he was the first to map the solar spectrum. The *angstrom,* a unit of length used to measure light waves, is named for him.

**angstrom** *Physics.* a unit of length used for atomic dimensions and light wavelengths, abbreviated Å; equal to $10^{-8}$ cm (one-hundredth of a millionth of a centimeter). An angstrom is approximately the size of an atom. Also, **Angström unit.**

**ANGTA** *Oil & Gas.* Alaska Natural Gas Transportation Act; a 1976 measure calling for a pipeline to transport natural gas from deposits in Alaska through Canada to the lower 48 United States.

**angular momentum** *Physics.* **1.** for a single particle moving about an axis, the moment of its linear momentum; i.e., the vector product of the particle's position and its linear momentum at the moment it passes a given point. **2.** for a system of particles, the vector sum of the individual angular momentum vectors of all particles in the system.

**anhydrous** *Chemistry.* containing no water or very little water.

**animal power** *Biological Energetics.* the purposeful use by humans of other animal species in physical activities that are beneficial to humans; e.g., using horses or oxen to pull a plow or wagon.

***animal power*** *A fundamental energy source throughout the world for most of human history, animal power is still employed in many contemporary societies, as in this use in Burma (Myanmar).*

**animal waste** *Biomass.* a term for waste products of domestic animals; mostly manure (dung), but may also include carcasses and industrial processing wastes (e.g., whey from milk separation).

**animate energy** *Biological Energetics.* humans and draft animals as energy sources, especially as contrasted with the inanimate energy provided by machines.

**anion** *Chemistry.* a negatively charged ion.

**Annapolis Royal** *Renewable/Alternative.* a power station on the Bay of Fundy coastline of Nova Scotia, the first large-scale tidal power plant established in North America (1984). The Bay of Fundy is known for the largest tidal changes in the world.

**annealing** *Materials.* the sustained heating of a material, such as metal or glass, at a known high temperature, followed by a gradual cooling, in order to reduce hardness or brittleness, eliminate various stresses and weaknesses, or produce other desired qualities.

**Annex I** *Policy.* a term for the 36 industrialized countries and economies in transition listed in Annex 1 of the United Nations Framework Convention on Climate Change. Their responsibilities include a non-binding commitment to reduce greenhouse gas emissions. The industrialized countries listed in **Annex II** have a special obligation to help developing countries reduce their greenhouse gas emissions.

**annihilation** *Nuclear.* a nuclear event in which a particle and its corresponding antiparticle collide, converting into annihilation radiation (see next).

**annihilation radiation** *Nuclear.* the energy produced from the joining of a positron and an electron, in which the positive and negative charges neutralize each other and become electromagnetic radiation.

**annual dose** *Nuclear.* the amount of radiation absorbed in a year, due to external exposure plus the committed dose from intakes of radionuclides.

**annual fuel utilization efficiency (AFUE)** *HVAC.* a description of heating efficiency, measured as the ratio of the amount of heat transferred to a given conditioned space to the fuel energy supplied over one year to accomplish this.

**annual solar savings** *Solar.* a statement of the reduction in energy use by a solar building compared to a non-solar building.

**annulus** *Oil & Gas.* the space in a well between the outer wall of the drill string and the wall of the well bore; it provides a passageway for fluids and drilled rock cuttings to return to the surface.

**anode** *Electricity.* **1.** the electrode that is positive with respect to the cathode in a electrochemical cell. It is the electrode at which oxidation occurs, toward which anions generally migrate as they carry current, and from which electrons leave the system. **2.** the negative electrode of a battery or storage cell that is delivering current.

**anthracene** *Chemistry.* a colorless crystalline solid that melts at 217°C and boils at 340°C, insoluble in water and soluble in alcohol and ether; used to produce the red dye alizarin and in the manufacture of wood preservatives, insecticides, and coating materials.

**anthracite** *Coal.* a hard, slow-burning coal of the highest rank, used primarily for residential and commercial space heating; black in color with a bright, often submetallic luster, having a high percentage of fixed carbon (over 86%) and a low percentage of volatile matter (under 14%), a low moisture and ash content, and a hardness of 2 to 2.5 on the Mohs scale. Anthracite has the highest heating value of any type of coal, typically containing 22 to 28 Btu per ton.

**anthropogenic** *Ecology.* describing conditions or phenomena in nature that occur mainly or entirely because of human influences; e.g., acid rain.

**anthropogenic emissions** *Environment.* emissions that are caused, directly or indirectly, by human activities; the emission of sulfur dioxide from the use of fossil fuels is a direct cause, and the emission of nitrogen oxides from farmland as a function of fertilizer application is an indirect cause.

**anthropogenic heat** *Environment.* heat generated by humans or by human activity, such as the heating (and cooling) of buildings, the operation of machinery, appliances, and transportation vehicles, and various industrial and manufacturing processes.

**anthroposphere** *Environment.* the part of the biosphere that has been affected by human activity.

**anticlinal trap** *Oil & Gas.* a formation at the top of an anticline (see next) where crude oil or natural gas accumulates.

**anticline** *Earth Science.* an upward fold of stratified rock in which the sides slope down and away from the crest; the oldest rocks are in the center, and the youngest rocks are on the outside. Oil may occur in the crest of anticlines.

**antigravity** *Physics.* a theoretical phenomenon in which gravitational force causes two bodies to repel each other, rather than to attract.

**antiknock** *Transportation.* any of various compounds that are added to gasoline to reduce engine knocking. Tetraethyl lead was the most widely used antiknock compound, but because of its contribution to air pollution it has been replaced by a nonmetallic compound in lead-free gasolines.

**antimatter** *Physics.* any matter that is entirely composed of antiparticles, such as positrons, antiprotons, and antineutrons.

**antimony** *Chemistry.* a silver-white, opaque metallic element having the atomic number 51, an atomic weight of 121.75, a specific gravity of 6.7, and a hardness of 3 to 3.5 on the Mohs scale; melts at 630.5°C and boils at 1750°C. It has low thermal conductivity and is used for electric cable covers and storage batteries and in semiconductors.

**antineutrino** *Physics.* the antiparticle of the neutrino; having zero mass, spin 1/2, and positive helicity. An antineutrino may be emitted from a nucleon when it undergoes beta-decay.

**antineutron** *Physics.* an uncharged particle of mass equal to that of the neutron but with a magnetic moment in the opposite direction, relative to its spin.

**antiparticle** *Physics.* an elementary particle that is identical to another elementary particle in mass and spin but opposite in electric and magnetic properties and that, when brought together with its counterpart, produces mutual annihilation.

**antiproton** *Physics.* an antiparticle of the proton having the equivalent mass of a proton, a unit negative charge, and a spin of 1/2;

mutually annihilating with a proton, it yields mesons.

**antistatic** *Materials.* describing an agent or material that serves to attract moisture from the air to a surface, thus improving the surface conductivity and reducing the likelihood of a spark or discharge.

**ANWR** *Environment.* (pronounced "an-wahr") short for ARCTIC NATIONAL WILDLIFE REFUGE.

**AOC** approximate original contour.

**AOSIS** Alliance of Small Island States (est. 1991); a coalition of small island and low-lying coastal countries that share similar development challenges and concerns about the environment, especially their vulnerability to the adverse effects of global climate change.

**aperture** *Solar.* **1.** the opening in an exterior wall surface through which unconcentrated solar radiation enters a collector. **2.** in general, any device that controls the amount of light admitted in a scanner, photometer, telescope, camera, or the like.

**aperture area** *Solar.* **1.** the maximum projected area of a solar collector through which unconcentrated solar radiant energy may be admitted to an absorber. **2.** such an area projected normal to the sun's rays and corrected for any shading.

**aphelion** *Earth Science.* the point in the orbit of the earth that lies farthest from the sun.

**API** American Petroleum Institute.

**API gravity** *Oil & Gas.* the specific gravity, or weight per unit of volume, of petroleum products or other liquids, as determined by the recommended procedure of the American Petroleum Institute.

**apparent consumption** *Consumption & Efficiency.* a measure for the consumption of a product or material (e.g., natural gas, foodstuffs), defined as domestic production of the item, plus imports, minus exports, plus any net change in inventories. Called "apparent" because consumption itself is not directly measured but instead is derived from the net of its individual components that are measured.

**apparent day** *Earth Science.* the interval between two successive passages of the sun over a given meridian; referred to as "apparent" because the length of this is not exactly equal to 24 hours as measured by a clock, due to slight variations in the relative position of the earth and sun.

**apparent (solar) time** see TRUE SOLAR TIME.

**appliance** *Consumption & Efficiency.* **1.** in general, any tool or machine that is used to carry out a specific task or produce a desired result. **2.** specifically, an electrical device that is used for some household purpose, such as a washing machine, refrigerator, or toaster.

**appliance label** see ENERGY LABEL.

**appliance saturation** *Consumption & Efficiency.* a measure of the extent to which the households in a given area own or have direct access to a certain appliance (e.g., a clothes dryer or room air conditioner); can be employed as a predictor of energy consumption.

**approximate original contour (AOC)** *Mining.* a regulatory requirement that the final topographic configuration of reclaimed surface-mined land be approximately the same as the pre-mining slopes and the general aspect of the pre-mine topography.

**APS** American Physical Society.

**AQI** air quality index.

**aquaculture** *Ecology.* the purposeful farming of aquatic organisms, including fish, mollusks, crustaceans, and aquatic plants; this includes management of the rearing process to enhance production, such as regular stocking, feeding, and protection from predators, as well as fertilization to enhance growth processes.

**aquatic ecology** *Ecology.* the study of the relationship between aquatic organisms and their environment.

**aqueduct** *Consumption & Efficiency.* a conduit used for carrying water over long distances; a technology first developed in the Middle East about 700 BC and widely employed in the Roman Empire.

**aquifer** *Earth Science.* **1.** a water-bearing geological formation; a body of underground water. **2.** specifically, a large, permeable body of rock or other geologic structure that contains and conducts economically significant quantities of groundwater to supply wells and springs.

**aquifer storage** *Storage.* a form of thermal energy storage that uses a natural underground layer for the temporary storage of

*aqueduct The Los Angeles Aqueduct brings water to the city from distant sources and has been a major factor in the growth of Los Angeles into a vast, sprawling metropolis with huge energy demands.*

heat or cold. The transfer of thermal energy is realized by extracting groundwater from the layer and by reinjecting it at a modified temperature level at a separate location nearby. Typically used to store winter cold for the cooling of large office buildings and industrial processes.

**Arab Light** *Oil & Gas.* a major classification of crude oil, representing the largest crude stream produced by Saudi Arabia.

**Aramco** *Oil & Gas.* Arabian-American Oil Company; the historic Saudi-U.S. venture to develop the oil resources of Saudi Arabia; established in 1933 and gradually nationalized by Saudi Arabia from the 1970s onward until finally taken over fully in 1988 as the Saudi Arabian Oil Company (Saudi Aramco).

**arbitrage** *Economics.* the simultaneous purchase of one commodity against the sale of another, in order to profit from fluctuations in the usual price relationships; a combination of transactions designed to profit from an existing discrepancy among prices, exchange rates, or interest rates in different markets; frequently used in electricity, oil, and natural gas markets.

**arbitrageur** *Economics.* a person or group engaging in arbitrage (see previous).

**arc** *Electricity.* a luminous, sustained discharge of electricity across an insulating medium. Thus, **arc discharge.**

**arch dam** *Hydropower.* a masonry or concrete dam that utilizes an arched shape in the horizontal plane to transfer the forces of the retained water to the sides of a gorge or canyon. Typically employed in narrow, rocky sites; e.g., the large U.S. Hoover Dam in the Black Canyon of the Colorado River. A **multiple-arch dam** employs a series of structures of this shape.

**Archimedean** *History.* based on the principles or discoveries of Archimedes (see next).

**Archimedes** c. 287–212 BC, Greek scholar known as one of the greatest mathematicians and scientists of antiquity. He discovered basic laws of hydrostatics and the principle of the lever and the pulley. He also is known as the founder of integral calculus and is credited with designing many war machines used in the defense of Syracuse.

**Archimedes' principle** *History.* the statement that a body immersed in a fluid is buoyed up by a force equal to the weight of the fluid it disperses; this explains the buoyancy of ships.

**Archimedes screw** *History.* a helix-shaped screw in an inclined cylinder, a device that can be used to raise water from a lower level to a higher one by turning a handle; employed in ancient times to lift water from rivers for irrigation.

**arcing** *Electricity.* a condition in which electricity flows through the air from one pole of an electric circuit to another, or jumps from its source to ground without flowing through a desired circuit. **Arc resistance** is the ability of a material, usually a dielectric, to resist this.

**arc lamp** *Lighting.* an early, brilliant form of illumination produced by passing an electric arc between two carbon rods; later generally replaced by filament lamps except in certain specialized uses requiring very bright light. Thus, **arc light(ing).**

**Arctic haze** *Climate Change.* a smog-like layer of haze observed in the Arctic atmosphere at certain times of the year, of unknown origin but thought to be caused by the transport of industrial pollutants from lower latitudes.

**Arctic National Wildlife Refuge (ANWR)** *Environment.* a vast area of land (more than 19 million acres) in northeastern Alaska that is designated by the U.S. as a wildlife sanctuary; ANWR is currently a focus of debate between those who argue that the entire area should be maintained as wilderness and those who argue that a portion of it (the Coastal Plain) should be opened to oil and gas exploration and production.

***Arctic National Wildlife Refuge*** *The Coastal Plain of ANWR borders on the Beaufort Sea.*

**are** *Measurement.* a unit of measure for land, equal to 100 square meters or 119.6 square yards.

**area heating** *HVAC.* the process of heating only a certain room or space by means of a local source such as a wood-burning fireplace or stove, or a self-contained electric heater. Thus, **area heater.** Contrasted with CENTRAL HEATING.

**area mining** *Mining.* a form of surface mining used to mine coal on flat to moderately rolling terrain; the overburden is excavated down to a coal seam, and then the mining area is enlarged horizontally to expose and remove the coal. Also, **area stripping.**

**area-wide source** *Environment.* **1.** pollution involving many different individual sources that are dispersed over a wide area, as opposed to a specific waste pipe, factory smokestack, and so on; usually refers to informal household emissions, as from wood fireplaces and kilns, burning of yard waste, agricultural runoff, and so on. **2.** another term for NON-POINT SOURCE.

**Argand lamp** *Lighting.* a historic advance in lighting, patented in 1784 in England; this lamp was the first basic change in lamps in thousands of years. Its design featured the incorporation of a hollow cylinder within a circular fuel-soaked wick, which allowed air to flow both inside and outside the flame at the upper edge of the wick. The addition of a cylindrical glass chimney created greater draft, promoting steadiness and greater brilliance in the flame. [Developed by Swiss scientist Aimé *Argand,* 1755–1803.]

**argon** *Chemistry.* a nonmetallic chemical element, one of the noble gases, having the symbol Ar, the atomic number 18, and an atomic weight of 39.948; freezes at –189.2°C and boils at –185.7°C; a colorless, odorless, inert gas that makes up 0.93% of the atmosphere and that is not known to form any chemical compounds. It is used to fill lightbulbs, in welding, and in lasers.

**argon-39** *Earth Science.* an isotope of argon that is useful as a tracer in ocean studies; a radioactive inert gas with a half-life of 269 years, produced in the atmosphere by cosmic ray interaction with **argon-40.**

**ARI** acute respiratory infection.

**aridity** *Environment.* the fact of being arid; a lack of moisture; i.e., the presence of desert conditions.

**Aristotle** 384–322 BC, Greek philosopher who extended many areas of knowledge and provided much of the foundation for modern principles of scientific inquiry. He provided the first technical definition of ENERGEIA, the source of the modern word *energy.*

**arithmetic growth** *Measurement.* (pronounced "air-ith-MET-ik") an increase in a quantity at a constant rate per unit time over a specified time period.

**Arkwright, Richard** 1732–1792, English inventor, a manufacturing pioneer and central figure of the Industrial Revolution in England and Scotland. Working with clockmaker John Kay, Arkwright developed a mechanical machine for spinning cotton, a process that was traditionally done in small homes and farms.

**armature** *Electricity.* a movable part in an electromagnetic mechanical device, such as the winding in which the electromotive force is induced in an electric motor.

**Armstrong, Edwin** 1890–1954, U.S. engineer and inventor of basic electronic circuits underlying modern radio, radar, and television.

**Armstrong system** *Electricity.* a frequency modulation (FM) system in which a low-frequency carrier is modulated at a low level, and the signal is passed through several amplifying stages to reach the desired high level and high carrier frequency.

**aromatic** *Chemistry.* **1.** any of a major class of unsaturated cyclic hydrocarbons characterized by the presence of one or more rings; so called because they have a strong odor; e.g., benzene. **2.** relating to or being a compound of this type. Thus, **aromatic (hydro)carbon.**

**ARPA** *Communication.* Advanced Research Projects Agency; an agency formed within the U.S. Defense Department in 1958 to promote the development of new technologies with military application (in the aftermath of the USSR's launching of the first artificial satellite); ARPA became a catalyst for the development of the modern computer, the Internet, and other information technology.

**Arps (decline) curve** *Oil & Gas.* a predictable pattern describing the rate at which oil production from a single well will decline over time; i.e., production will increase for a short period, then peak and follow a long, slow decline. [Described by U.S. geologist J. J. *Arps.*]

**Arps-Roberts model** *Oil & Gas.* a model of oil discovery whose premise is that for each additional wildcat well drilled, the probability of finding a field of a certain size class is proportional to the number of remaining undiscovered fields of that class, and to the size of fields in that class. The model is based on the fundamental principle that since large fields tend to have the largest areas, they are more likely to be found earlier in the exploration process. [Named for J. J. *Arps* and T. G. *Roberts.*]

**array** see PHOTOVOLTAIC ARRAY.

**arrester** *Electricity.* a device used to protect an installation from lightning current.

**Arrhenius, Svante August** 1859–1927, Swedish chemist noted for his electrolytic theory of dissociation and his early description of the greenhouse effect in climate change. In chemistry.

**Arrhenius acid** *Chemistry.* a definition of an acid as any substance that increases the concentration of hydrogen ($H^+$ ions when added to a water solution. Similarly, an **Arrhenius base** is any substance that increases the concentration of hydroxide ($OH^-$ ions when added to a water solution.

**Arrhenius equation** *Chemistry.* an equation that describes the rate of a chemical reaction at a certain temperature, in which the rate is exponentially related to the temperature; used for many chemical transformations and processes,

**Arrhenius law** *Chemistry.* the principle that the rate of chemical reaction is approximately doubled for every 10°C increase in temperature.

**Arrhenius theory** *Electricity.* the theory that electrolytes are separated, or dissociated, into electrically charged particles (ions) even when there is no current flowing through the solution.

**artificial intelligence (AI)** *Communication.* a field of study concerned with the development and use of computer systems that have some resemblance to human intelligence, including such operations as natural-language recognition and use, problem-solving, selection from alternatives, pattern recognition, generalization based on experience, and analysis of novel situations.

**artificial light** *Lighting.* a general term for any form of illumination other than that produced by natural sources such as sunlight.

**artificial neural network** *Communication.* an array of artificial neurons interacting with one another in a concerted manner to achieve "intelligent" results through many parallel computations, without employing rules or other logical structures.

**artificial photosynthesis** *Biological Energetics.* a human-made system that mimics or enhances the energy conversion processes occurring in natural photosynthesis; a current research effort with the goal of producing hydrogen or other alternative fuels and reducing the amount of carbon dioxide in the atmosphere.

**artificial radioactivity** see INDUCED RADIOACTIVITY.

**artificial refrigeration** *Refrigeration.* a general term for any form of cooling by means other than naturally lower temperatures in the environment; e.g., ice.

**artificial satellite** see SATELLITE.

**artificial sky** *Lighting.* an enclosure that simulates the luminance distribution of a real sky for the purpose of testing physical models under daylight conditions.

**asbestos** *Materials.* a group of fibrous metamorphic minerals used in buildings for their flame-retardant and insulating properties, tensile strength, flexibility, and resistance to chemicals; asbestos is now known to be carcinogenic and its use as a building material is now banned in many countries.

**ash** *Materials.* the solid matter left as a residue after an organic substance has been burned.

**ash-free basis** *Coal.* a standard for evaluating coal quality, assuming that all ash has been removed from it; a hypothetical standard because the ash is actually not separated until the incineration of the coal.

**ash fusibility** *Coal.* a measure of the fusion of coal ash prepared and tested under standard conditions.

**Asian Institute of Technology** (est. 1961), an international graduate institution of higher learning whose mission is to develop qualified and committed professionals who will play a role in the sustainable development of the region and its integration into the global economy.

**Asian Tiger** *Economics.* one of the four Asian economies that were the first to show rapid economic development in this region after the success of Japan; i.e., Hong Kong, South Korea, Singapore, and Taiwan.

**askarel** *Materials.* any of a group of fire-resistant, synthetic, aromatic hydrocarbons used as insulating liquids for electrical applications.

**as-mined coal** see RUN-OF-MINE COAL.

**asphalt** *Materials.* **1.** a dark, tarry, bituminous material found naturally or distilled from petroleum. **2.** a mixture of this material with sand, gravel, or similar additives, widely used in paving.

**asphaltene** *Oil & Gas.* any solid, dark brown bitumen that is either dissolved or dispersed in crude oil. Asphaltenes are large complex molecules, often with molecular weights in the many thousands, composed of carbon, hydrogen, oxygen, and other heteroatoms. They are responsible for most of the color of crude and heavy oils.

**asphaltine** *Oil & Gas.* a black to dark brown bitumen that melts above 110°C.

**asphaltite** *Materials.* a naturally occurring black to dark brown solid hydrocarbon that is highly soluble in carbon disulfide and less fusible than native asphalt (fusing above 110°C).

**assaying** *Mining.* the process of analyzing and testing a mineral substance to determine its commercial value.

**assembly line** *History.* a historic advance in industrial production, in which the production stages are arranged in sequential order, with unfinished items moving along the line through each production stage; first employed in the early 20th century by auto-

***assembly line*** *Workers on a radiator soldering line at a Ford assembly plant, 1923.*

motive pioneers such as Ransom Olds and Henry Ford.

**assigned amount** *Policy.* a term for the national emission allowance for carbon dioxide assigned under the Kyoto Protocol that an industrialized country (Annex 1 country) is permitted to emit over a certain commitment period. Thus, **assigned amount unit (AAU).**

**assimilation efficiency** *Biological Energetics.* **1.** in animals, the percentage of the energy content of ingested food that is absorbed by the digestive system. **2.** in plants, the percentage of solar visible light that is fixed by photosynthesis.

**assimilative** *Environment.* describing the ability of the atmosphere, a water body or the land to receive a pollutant without significantly degrading the structure or function of that system.

**associated natural gas** *Oil & Gas.* the volume of natural gas that occurs in crude oil reservoirs either as free gas or as gas in solution with crude oil. Also, **associated-dissolved natural gas.**

**assured refill (curve)** *Hydropower.* a rule implemented to deal with variations in streamflow, indicating the minimum elevations that must be maintained in a given reservoir to ensure refill even if the third lowest historical flow sequence should occur.

**ASTAE** Asia Alternative Energy Program (est. 1992); the World Bank initiative that seeks to mainstream renewable energy and energy efficiency in the Bank's power sector lending operations in Asia.

**asthenosphere** *Earth Science.* the zone or layer of the earth's upper mantle that lies below the lithosphere, which is capable of plastic deformation, and in which magmas may be generated and the velocity of seismic waves reduced.

**ASTM** American Society for Testing and Materials (est. 1898), now ASTM International; a voluntary standards development group that provides a source for technical standards for materials, products, systems, and services.

**astral lamp** *Lighting.* a type of Argand lamp having a flattened annular oil reservoir that causes the lamp to cast a shadow across a table when placed centrally.

**astrolabe** *History.* an instrument for measuring the altitude of the sun and stars, widely used by astronomers until replaced by the sextant. Invented by the ancient Greeks and improved by the Arabs, it was an important navigation aid for explorers such as Columbus.

***astrolabe*** *An astrolabe of Arab design, reportedly used on the Red Sea about the year 1300.*

**astronomical unit (AU)** *Measurement.* the mean distance of the earth from the sun, which is 149,597,870 kilometers, or about 93 million miles.

**asymmetric** *Economics.* describing a situation in which an energy price increase produces a different response than an energy price decrease of the same magnitude; e.g., when a 1% increase in the price of motor gasoline causes a 0.5% decrease in the demand for gasoline, while a 1% decrease in price produces a greater than 0.5% increase in demand. Thus, **asymmetry, asymmetric price response.**

**asynchronous** *Physics.* **1.** not at the same rate or time. **2.** *Electricity.* describing a machine operating at a speed that is not the same as the existing power supply, either above that speed (super-synchronous), such as an **asynchronous generator** producing electric current (e.g., in a wind turbine), or below it (sub-synchronous), such as an **asynchronous motor** powered by current.

**Atanasoff, John Vincent** 1903–1995, U.S. scientist credited with the design of the first electronic-digital computer (see next), working with Clifford Berry at Iowa State University. The ENIAC machine of Mauchly and Eckert had been given this distinction for many years, until a court ruled their patent invalid and credited Atanasoff.

**Atanasoff-Berry computer (ABC)** *Communication.* a pioneering machine identified as the world's first true electronic-digital computer (1939); it represented several innovations in computing, including a binary system of arithmetic, parallel processing, regenerative memory, and a separation of memory and computing functions.

**ATgas** *Coal.* a high-Btu synthetic gas produced by dissolving coal in a bath of molten iron

**Atlantic heat conveyor** *Earth Science.* a heat transfer from south to north by near-surface waters moving across the equator in the Atlantic. An inhibition in the strength of this warming current. A disruption of this current caused by global warming could affect the relatively mild climate of northern Europe.

**atmometer** *Earth Science.* an instrument measuring the rate at which water evaporates into the atmosphere.

**atmosphere** *Earth Science.* **1.** the envelope of gases surrounding the earth and held to it by the force of gravity. It consists of four distinct layers, whose boundaries are not precise: the troposphere (extending from sea level to about 5–10 miles above the earth), the stratosphere (up to about 30 miles), the mesosphere (up to about 60 miles), and the thermosphere (up to about 300 miles or more). The upper region of the troposphere is often regarded as a separate region, the exosphere. **2.** the pressure of the earth's atmosphere at sea level; see ATMOSPHERIC PRESSURE.

**atmospheric cooling** see NATURAL-DRAFT COOLING.

**atmospheric distillation** *Oil & Gas.* a refining process in which crude oil components are separated at atmospheric pressure by heating to temperatures of about 600–750°F and subsequent condensing of the fractions by cooling.

**atmospheric (steam) engine** *History.* a term for early steam engine designs (e.g., Papin, Savery, Newcomen), so called because the pressure of the steam was the same as or near the pressure of the surrounding atmosphere.

**atmospheric fixation** see NITROGEN FIXATION.

**atmospheric lifetime** *Environment.* **1.** the length of time that a given chemical substance remains in the atmosphere in its original form. **2.** specifically, the approximate length of time required for an atmospheric pollutant to fall to a specified fraction of its original level, as a result of either being converted to another (nontoxic) chemical compound or being removed from the atmosphere by means of a sink.

**atmospheric pressure** *Earth Science.* a unit of pressure that is taken to be the standard pressure of the earth's atmosphere at sea level; equal to the pressure of a column of mercury 760 mm high and expressed as 101.325 kilopascals ($1.01325 \times 10^5$ newton per square meter), or about 14.7 pounds per square inch.

**atmospheric turbidity** *Solar.* haziness in the atmosphere due to aerosols such as dust. If turbidity is zero, the sky has no dust, as measured by a sun photometer.

**atom** *Physics.* the basic component of all matter; the smallest particle of an element that can still retain all the chemical properties of that element. An atom has a dense central core (the nucleus) consisting of positively charged particles (protons) and uncharged particles (neutrons). Negatively charged particles (electrons) are scattered in a relatively large space around this nucleus and move about it in orbital patterns at extremely high speeds. An atom contains the same number of protons as electrons and thus is electrically neutral and stable under most conditions. See also ATOMIC THEORY.

**atomic** *Physics.* of or relating to atoms or the energy that exists in atoms. Thus, **atomic energy, atomic fission, atomic mass,** and so on.

**Atomic Age** see NUCLEAR AGE.

**atomic bomb** *Nuclear.* a very powerful bomb that derives its explosive power from the sudden release of huge amounts of atomic energy.

***atomic bomb*** *A "cloud chamber" effect is observed after the underwater explosion of an atomic bomb at Bikini atoll in the South Pacific, October 1952.*

**atomic charge** *Physics.* the electric charge of an ion, calculated by the number of electrons that the atom gains or loses during ionization, times the charge of an individual electron.

**atomic clock** *Measurement.* an extremely precise electronic clock that uses the oscillations of individual atoms or molecules, especially the cesium-133 atom, to regulate its movement.

**atomic cloud** see RADIOACTIVE CLOUD.

**Atomic Energy Act** *Policy.* a U.S. law passed in 1954 to govern the use of nuclear materials. It required that civilian uses of nuclear materials and facilities be licensed, and it empowered the Atomic Energy Commission and its successor, the Nuclear Regulatory Commission, to establish and enforce health and safety standards related to nuclear materials.

**Atomic Energy Commission (AEC)** (est. 1947; abolished 1975), an independent civilian agency of the U.S. government with statutory responsibility for atomic energy matters, unique among federal agencies in having responsibilities to both promote and regulate a technology. Succeeded by the Department of Energy.

**atomic heat** *Physics.* the amount of heat required to raise the temperature of a gram-atomic weight of an element by one degree.

**atomic mass unit (AMU)** *Physics.* a unit of mass used to describe the masses of atoms and molecules, employing a scale in which the most abundant isotope of carbon has a mass of 12; i.e., one AMU is 1/12 of the mass of this carbon atom.

**atomic number** *Physics.* the number of protons in the nucleus of an atom; this number uniquely characterizes a nuclear species and determines its place in the periodic table.

**atomic temperature** *Measurement.* a temperature corresponding to the mean kinetic energy of the neutral atoms in a plasma (or what would be the temperature of a gas such as air in the absence of ions or electrons).

**atomic theory** *Physics.* the theory that all matter is composed of minute, distinct particles called atoms, and that these are the smallest particles of matter which cannot be further divided; first developed by the Greek philosopher Democritus in about 400 BC and formally proposed by the English chemist John Dalton in 1808; later modified as it became established that atoms actually are divisible and that they consist of even smaller particles.

**atomic weight** *Measurement.* the mass of one atom of an element, based on a scale in which the isotope carbon-12 weighs 12.00 atomic mass units.

**atomism** *Physics.* an earlier name for the atomic theory (see above); i.e., the idea that all natural bodies are composed of minute indivisible units called atoms.

**atomization** *Materials.* **1.** the process of reducing a liquid or meltable solid to fine particles or spray by forced passage through a nozzle or jet. **2.** the reduction of liquid fuel to a fine spray or mist that readily ignites in an automotive engine. Thus, **atomize, atomizer.**

**Atoms for Peace** *Policy.* a speech made by U.S. President Dwight D. Eisenhower in 1953 to the UN General Assembly in which he called on the U.S. and the USSR to make contributions from their stockpiles of nuclear materials to an international agency that would use these materials for peaceful purposes.

**atom-smasher** an earlier term for a PARTICLE ACCELERATOR.

**ATP** adenosine triphosphate.

**attemperation** *Thermodynamics.* the control of any excessive superheat in a steam boiler, either by admixture between the superheated steam and cooling steam, or by forcing cooling steam across superheated steam tubes, thus regulating the final steam temperature. Thus, **attemporator.**

**attenuation** *Physics.* **1.** a reduction in amplitude, density, or energy as the result of such effects as friction, absorption, or scattering. **2.** *Solar.* a loss of solar irradiance as it passes through the atmosphere to the surface of the earth. **3.** *Health and Safety.* a reduction in the toxicity of a pollutant substance in the environment, especially as a result of natural processes.

**atto-** *Measurement.* a prefix meaning one quintillionth ($10^{-18}$).

**attribution** *Climate Change.* the process of establishing the most likely causes for a detected change in climate with some defined level of confidence; a concept employed in understanding the human versus natural causes of climate change. Compare DETECTION.

**Atwater system** *Biological Energetics.* a description of the available caloric value of different types of foods; this provides a useful estimate of the energy content of the daily diet. The Atwater system uses the average values of 4 Kcal/g for protein, 4 Kcal/g for carbohydrate, 9 Kcal/g for fat, and 7 Kcal/g for alcohol. [Developed by Wilbur *Atwater* and colleagues.] Also, **Atwater general factors.**

**A-type** *Lighting.* a term for the traditional standard type of incandescent lightbulb that is the most widely used artificial lighting source.

**AU** astronomical unit.

**auction** *Policy.* a method for issuing emission permits to firms in a domestic emissions trading regime, based on a willingness of the firm to pay for the permits.

**audio** *Communication.* the sound portion of a television signal, as opposed to the visual portion (video).

**Audion** *History.* the name for Lee de Forest's first vacuum tube, a three-element tube (generally known as a triode today) that was a key factor in making practical radio broadcasts a reality in the 1920s.

**auger** *Mining.* a rotary drilling device used in earth boring.

**Auger effect** *Nuclear.* a process in which an atom ionizes without emitting radiation; this occurs when an electron in the outer regions fills a vacancy in an inner orbit, and at the same time another outer electron is ejected by the atom. [Described by Pierre *Auger,* 1899–1993, French physicist.]

**auger machine** *Mining.* a large-diameter horizontal drill used in surface mining, able to remove coal at a rapid rate.

**auger mine** *Mining.* a relatively inexpensive form of mining in which coal is recovered through the use of a large-diameter drill driven into the coal bed, typically employed after contour surface mining when the overburden is too costly to excavate.

**aurora** *Earth Science.* a form of sporadic radiant emission occurring in the upper atmosphere over the middle and high latitudes and seen most often in the Arctic and Antarctic regions; thought to be caused by charged particles from the sun that collide with and excite atoms in the upper atmosphere, which then emit light as they return to their ground state.

**autoignition** *Transportation.* in an internal combustion engine, the spontaneous ignition of fuel when introduced into the combustion chamber, either due to the heat of compression or to glowing carbon in the chamber.

**automation** *History.* the transition to an industrial process in which machines operate automatically with relatively little human involvement.

**automobile** *Transportation.* any self-guided, motorized passenger vehicle used for land transport, usually with four wheels and an internal combustion engine; a category typically including passenger cars, sport-utility vehicles, minivans, and light trucks.

**autopoiesis** *Thermodynamics.* the ability of a system to regenerate itself by self-maintenance of its own elements and of the same network of processes which produced them.

**autothermal reforming** *Oil & Gas.* an energy-efficient reforming process that uses heat generated from combining partial oxidation and catalytic steam reformation in a single step; this can significantly reduce air emissions.

**autotroph** *Biological Energetics.* literally, self-feeder; an organism that obtains its nutrients directly from inorganic sources in the environment, rather than by consuming nutrient material from other organisms; e.g., green plants. Thus, **autotrophy.** Contrasted with HETEROTROPH.

***autotroph*** *Green plants such as the oak tree are autotrophs (self-feeders); other organisms that are heterotrophs (other-feeders) feed on plant parts such as leaves and fruit.*

**autotrophic** *Biological Energetics.* describing organisms that obtain their food by synthesis from the environment; e.g., green plants, certain bacteria.

**autotrophic respiration** *Biological Energetics.* carbon loss to the atmosphere as a result of internal plant metabolism; typically amounts to half of the carbon fixed by plants through photosynthesis.

**available energy** *Thermodynamics.* the potential to do work; the maximum amount of energy able to be utilized for work.

**available heat** *Thermodynamics.* the maximum amount of heat energy that can be obtained in the combustion of a given fuel under ideal conditions.

**available potential energy** *Physics.* the difference in potential energy between the actual physical state and a reference state representing the minimum potential energy that can be reached through reversible adiabatic processes.

**available resources** *Coal.* coal resources having no land-use, environmental, or technological restrictions.

**average daily traffic (ADT)** *Transportation.* a measure of the daily volume of traffic passing a given point on a highway, usually during a typical weekday period.

**average rated life** *Consumption & Efficiency.* the number of hours at which half of a large group of product samples (such as lamps) fail under standard test conditions.

**averted dose** *Nuclear.* the amount of radiation absorption prevented by the application of a countermeasure or set of countermeasures.

**avgas** short for AVIATION GASOLINE.

**AVHRR** *Measurement.* advanced very high resolution radiometer; the main sensor on U.S. polar orbiting satellites, used for a wide range of large-scale analyses of planetary phenomena, such as deforestation and climate change.

**avian mortality** *Wind.* a term for the death of birds from being struck or otherwise injured by a wind turbine.

**aviation** *Transportation.* the science, technology, and business of designing, producing, and operating heavier-than-air aircraft.

**aviation gasoline** *Transportation.* gasoline fuel for use in reciprocating piston engine aircraft (as distinguished from jet fuel). Also, **aviation spirit.**

**Avogadro, Amadeo** 1776–1856, Italian scientist noted for fundamental contributions to atomic theory; he made a distinction between atoms and molecules, which today seems clear, but which was largely ignored at the time because John Dalton rejected it.

**Avogadro's law** *Chemistry.* the statement that equal volumes of all gases at the same temperature and pressure contain the same number of molecules regardless of their chemical nature and physical properties. Also, **Avogadro's hypothesis** or **principle.**

**Avogadro's number** *Chemistry.* the number of molecules in a gram mole of any chemical substance ($6.0221367 \times 10^{23}$); a fundamental constant.

**avoidance cost** *Economics.* an actual or imputed costs for preventing environmental deterioration by alternative production and consumption processes, or by the reduction of or abstention from certain economic activities.

**avoided cost** *Economics.* the cost that an electrical utility would incur if it were not for the existence of an independent generator or an alternative energy service option; often used as the rate utilities offer to independent suppliers for their power.

**avoided emissions** *Environment.* emissions that are not produced through the use of lower-emitting or non-emitting technologies instead of the emitting technologies previously or usually employed; e.g., electricity produced by wind farms rather than by a coal-burning plant.

**avoirdupois** *Measurement.* describing a system of weights traditionally used in Great Britain, the U.S., and other English-speaking countries, based on a pound of 16 ounces.

**AVT** advanced vehicle technology.

**A&WMA** Air & Waste Management Association (est. 1907), a nonprofit professional organization that seeks to provide a neutral forum for technology exchange, professional development, networking opportunities, public education, and outreach to environmental professionals on an international basis.

**axial** *Physics.* relating to or operating along an axis.

**axial flow** *Conversion.* any flow parallel to an axis, especially a flow in turbomachinery in which the flowing fluid always moves parallel to the length of the rotating shaft, as in an axial-flow turbine. Thus, **axial-flow compressor, engine,** or **pump.**

**axial-flow turbine** *Conversion.* a common type of turbine with axial flow (see previous) through the runner blades axially to the turbine shaft.

**azeotropic mixture** *Refrigeration.* a liquid mixture whose boiling point is constant, so that the vapor produced in distillation or partial evaporation has the same composition as the liquid phase; used in refrigeration. Thus, **azeotrope.**

**azimuth angle** *Earth Science.* the angle between the horizontal direction of a body (such as the sun) and a reference direction (usually north).

**Baba Gurgur** *Oil & Gas.* an oil field near Kirkuk, Iraq, about 140 miles north of Baghdad, the original site of the Iraqi oil industry.

**Babbage, Charles** 1791–1871, British scientist regarded as the originator of the concepts behind the modern computer, through his development of the CALCULATING ENGINE.

*Charles Babbage*

**backcast stripping** *Mining.* a stripping method that uses one dragline to strip and cast the overburden, while a second line recasts a portion of the overburden.

**backdraft(ing)** *HVAC.* the flow of air down a flue or chimney and into a house, caused by lowered indoor air pressure; can occur from the use of a fan or if the house is relatively airtight.

**backfill** *Mining.* **1.** mine waste or rock used for support in an excavation after coal removal. **2.** any previous cut material used to fill in an excavation. **3.** to fill in an excavation in this manner.

**background count** *Nuclear.* a level of radioactivity recorded by a radiation detection system that is due to some source other than the specific source being measured.

**background irradiance** *Solar.* the irradiance at the entrance aperture of an infrared sensing system that is not radiated directly from the object being investigated.

**background radiation** *Nuclear.* the radiation in the natural environment, including cosmic rays and radiation from naturally radioactive elements, both outside and inside the bodies of humans and animals. The usually quoted average individual exposure from background radiation is 360 millirem per year.

**background station** *Environment.* a facility used to monitor concentration levels of air pollutants, situated in an area that is distant from major sources of pollution, so that it will not be affected by day-to-day fluctuations in pollutant level and thus can measure air quality on a long-term, regional or global basis.

**back pressure** *Physics.* **1.** air pressure in a pipe exceeding atmospheric pressure. **2.** reverse pressure opposing forward flow, as in a refrigerating system. **3.** *Mining.* resistance encountered by the drill stem in a rock drill that causes the bit to be fed faster than it can cut.

**back-pressure turbine** *Conversion.* a type of turbine used in combined heat and power operations, in which the turbine generator is employed in a steam distribution pipeline in place of a pressure reduction valve; this extracts more energy from low-pressure steam.

**backscatter(ing)** *Physics.* **1.** the scattering of radiation, particles, or light waves at an angle greater than 90°. **2.** any radio waves or light waves produced by propagating in a direction approximately reverse to that of the incident wave.

**backstop (technology)** *Renewable/Alternative.* a substitute energy supply assumed to be the functional equivalent of current sources at a higher but constant marginal cost; its development and commercialization would be triggered by the depletion and cost increase of current energy sources.

**back surface reflector (BSR)** *Solar.* a mirror formed at the back surface of solar cells to increase the reflection of unabsorbed sunlight from the back surface.

**backwardation** *Economics.* an energy market condition in which future prices are lower than the current spot prices, or in which the near futures price exceeds the distant futures price. Compare CONTANGO.

**Bacon, Francis Thomas** 1904–1992, British engineer who developed the first practical hydrogen-oxygen fuel cells.

**BACT** best available control technology.

**bacteria** *Ecology. singular,* **bacterium.** a major group of living organisms that are microscopic and mostly unicellular, with a relatively simple cell structure typically contained within a cell wall and lacking a cell nucleus; bacteria reproduce by fission, and occur in spherical, rodlike, spiral, or curving shapes. They are found in virtually all environments and are the most abundant organisms. Some types are important agents in the cycles of nitrogen, carbon, and other matter, while others cause diseases in humans and animals.

**bacterial photosynthesis** *Biological Energetics.* the process of photosynthesis carried out by bacteria rather than by plants; e.g., *Rhodospirillum rubrum.*

**BAF** *Health & Safety.* bioaccumulation factor, a measure of the tendency of an environmental toxin to accumulate in living organisms.

**baffle** *Consumption & Efficiency.* a barrier or partition used to obstruct light, sound, or air or another gas; e.g., an object to restrict or modify the flow of flue gases.

**baffle chamber** *Consumption & Efficiency.* a chamber in an incinerator used to settle fly ash and coarse particulate matter by changing the direction and reducing the velocity of the combustion gases.

**bagasse** *Biomass.* the residue remaining after juice is extracted from sugarcane in a milling process; used as a fuel.

**bag filter** *Environment.* a cloth bag filter used for the recovery of impurities that are suspended in a gas, especially waste combustion gases. A **baghouse** is a chamber or room employing such filters.

**Baghdad battery** *History.* a name for a device speculated to be the first known battery, discovered in Khujut Rabu just outside Baghdad and estimated to be about 2000 years old. It is a clay jar with a stopper made of asphalt, through which extends an iron rod surrounded by a copper cylinder. When filled with an electrolytic solution, the jar has an electrical potential of about 1.1 Volts. It is thought to have been used for electroplating.

**Baird, John Logie** 1888–1946, British television innovator, responsible for the first publicly demonstrated television broadcast in 1926. Among his pioneering ideas were early versions of color television, the video disc, large screen television, televised sports, and pay television by closed circuit.

**Baku** *Oil & Gas.* a city on the western shore of the Caspian Sea, the capital of modern Azerbaijan; a historic site of oil production, possibly dating back to the time of Alexander the Great. Reportedly the world's first commercially successful oil well was drilled here in 1846.

**balance point** *HVAC.* the lowest outdoor temperature at which the refrigeration cycle of a heat pump will supply heating requirements without the aid of a supplementary heat source.

**ballast** *Lighting.* a device that provides starting voltage and also limits current during normal operation in electrical discharge lamps such as florescent lamps.

**ballast efficacy factor (BEF)** *Lighting.* the ratio of the ballast factor to the active power (in watts), usually expressed as a percentage; used as a relative measurement of the system efficacy of the fluorescent lamp/ballast combination. Also, **ballast efficiency factor.**

**ballast factor** *Lighting.* the ratio of lumen output on a particular lamp ballast as compared to that lamp's rated lumen output on a reference ballast under test conditions.

**ballast water** *Oil & Gas.* water added to the empty tanks of an oil tanker after it has discharged its cargo, to stabilize the vessel; this typically contains some residual oil and thus can be a source of pollution if improperly released.

**ballistics** *Physics.* the scientific study of the motion and behavior of projectiles, missiles, rockets, bullets, and similar objects, and of the accompanying phenomena.

**ballistic trajectory** *Physics.* the path of an unpowered object that moves only under the influence of gravity and atmospheric friction, with its surface providing no significant lift to alter its course. Thus, **ballistic vehicle.**

**balneology** *Geothermal.* the use of hot spring mineral water for healing or therapeutic purposes; considered to be the oldest historic use of natural geothermal waters.

**bamboo** *Biomass.* a group of large, fast-growing woody grasses widely used for fiber and food, especially in Asia; may also have potential as a bioenergy source.

**banded coal** *Coal.* a heterogeneous coal composed of thin bands of highly lustrous coalified wood within layers of striated coal; a common variety of bituminous and subbituminous coal.

**band gap** *Electricity.* in a semiconductor, the minimum energy separation between the highest occupied state and the lowest empty state; this determines the temperature dependence of its electrical conductivity.

**band gap energy** *Electricity.* the amount of energy (in electron volts) required to free an outer shell electron from its orbit about the nucleus to a free state, and thus promote it from the valence to the conduction level.

**bandpass** *Electricity.* a range that indicates the difference between the limiting frequencies at which a desired fraction of maximum output is obtained; usually expressed in hertz.

**bandwidth** *Communication.* the range or difference between the limits of a continuous frequency or wavelength band; the greater the bandwidth, the more data that can be transmitted in a given amount of time.

**banking** *Policy.* the saving (as if in a bank) of emissions credits for future consumption or trading, as by applying reductions earned from one period to another, or applying reductions earned before the first compliance period.

**barge** *Transportation.* a typically flat-bottomed vessel, either towed or self-propelled, used to carry products in navigable waterways; oil barges used in inland and coastal waters generally hold about 20,000 to 30,000 barrels, while ocean-going barges are much larger.

**barley coal** *Coal.* a description for stream-sized anthracite coal sized on a round punched plate with approximately quarter-inch holes.

**barney** *Mining.* a small car or truck attached to a rope or cable, used to push cars up a slope or inclined plane.

**barometer** *Measurement.* an instrument for measuring atmospheric pressure; used in determining height above sea level and predicting changes in the weather.

**barothermograph** *Measurement.* an automatic instrument used to record temperature and pressure.

**barotropic** *Earth Science.* **1.** relating to or characterized by an atmospheric condition in which surfaces of equal pressure coincide with surfaces of equal density. **2.** *Physics.* having a density that is a function solely of pressure.

**barrage** *Hydropower.* a low dam or barrier that increases the depth of a river or water course, or diverts it for irrigation or navigation.

**barrel** *Oil & Gas.* the international standard of measure for crude oil and oil products, equivalent to 42 U.S. gallons or 0.15899 cubic meters; used since first defined in American oil fields in about 1870. [Based on a historic standard size for actual wooden oil barrels, similar to a wine barrel.]

**barrel of oil equivalent (BOE)** *Measurement.* a unit of measure for energy content that equates one barrel of crude oil to 6000 cubic feet of natural gas or 1.5 barrels of natural gas liquids.

**barrels per stream day** *Oil & Gas.* the maximum number of barrels of input that a refinery can process during a 24-hour period.

**barren** *Mining.* describing rock, vein, or strata material that has no minerals of sufficient value to be commercially viable.

**barricading** *Mining.* the enclosure of part of a mine to lessen or confine the effects of a mine fire or explosion.

**barrier** *Economics.* a factor that explains why energy efficiency measures are not undertaken in the industrial sector despite the fact that they appear economically desirable according to company criteria; e.g., a factor such as lower interest rates for energy producers compared to consumers, incomplete markets for energy efficiency, and so on.

**barrier energy** *Photovoltaic.* the energy given up by an electron in penetrating a photovoltaic cell barrier; a measure of the electrostatic potential of the barrier.

**Bartholomew, George** U.S. biologist, the developer of important concepts in biological energetics. His work on energy use in animals, especially in mammals and insects, are cornerstones of the field of physiological ecology.

**baryon** *Physics.* any of the class of the heaviest subatomic particles composed of protons, neutrons, and other particles whose eventual decay products include protons.

**basal energy expenditure (BEE)** *Biological Energetics.* the amount of energy required by an organism to maintain the body's normal metabolic activity, e.g., respiration, body temperature, and so on.

**basal metabolic rate (BMR)** *Biological Energetics.* in higher vertebrates (mammals and birds), the lowest rate of heat production by an adult individual, measured when body temperature is normal and energy costs for activity, temperature regulation, digestion, and other expenditures are low or zero. Thus, **basal metabolism.**

**base** *Chemistry.* a general term for compounds that when dissolved in water increase the concentration of hydroxide ions (OH). Bases are proton acceptors and have a pH greater than 7.

**baseboard heat(ing)** *HVAC.* a type of heating system in which electric resistance coils, or finned tubes carrying steam or hot water, are mounted behind shallow panels along baseboards (boards forming the foot of an interior wall).

**base gas** *Oil & Gas.* the minimum quantity of natural gas that must be retained in storage to ensure production and deliverability at a sufficient rate and pressure throughout the withdrawal season.

**baseline** *Measurement.* a value that represents the background level of a measurable quantity, used as a basis for comparison with subsequent values, in order to assess the effect of changed conditions; e.g., the impact of an energy efficiency measure. Thus, **baseline forecast.**

**baseline emissions** *Environment.* the average amount of emissions that would occur over a given period (e.g., 1 year) if no changes in policy take place that would affect emission level either negatively or positively, such as the building of a new factory or the implementation of an emission-reduction program.

**baseload** *Electricity.* the constant load that a power station or system must continuously provide in order to cover minimum or normal needs, as distinguished from the peak load generated during intermittent periods of heavy usage.

**baseload plant** *Electricity.* a plant that is normally operated to take all or part of the minimum load of a system, and which therefore produces electricity at an essentially constant rate and runs continuously.

**basic oxygen process** *Materials.* a steelmaking process in which hot metal and steel scrap are charged into a **basic oxygen furnace (BOF);** high-purity oxygen is then blown into the metal bath, combining with carbon and other elements to reduce the impurities in the molten charge and convert it into steel.

**basic sediment** *Oil & Gas.* a term for soil impurities and salt water that are mixed with crude oil; they typically settle and collect in the bottom of a petroleum storage tank and are removed in the oil separation process.

**basin** see SEDIMENTARY BASIN.

**basin-centered accumulation** *Oil & Gas.* a regionally extensive and typically thick zone or unit of hydrocarbon saturated low-permeability rock, in the deep, central part of a sedimentary basin.

**basket** *Oil & Gas.* **1.** a group of oil or gas types considered collectively; see OPEC BASKET. **2.** *Environment.* a term for a group of greenhouse gases regulated under the Kyoto

Protocol: carbon dioxide ($CO_2$), hydrofluorocarbons (HFCs), methane ($CH_4$), nitrous oxide ($N_2O$), perfluorocarbons (PFCs), and sulfur hexafluoride ($SF_6$).

**basket pricing** *Economics.* the fact of setting the price of an energy commodity based on an aggregation of several individual prices. See also OPEC BASKET PRICE.

**batch heating** *Solar.* a passive solar hot water system consisting of one or more storage tanks placed in an insulated box that has a glazed side facing the sun. Thus, **batch heater.**

**batching** *Oil & Gas.* **1.** in a pipeline operation, the separate pumping of different grades of oil or gasoline that are next to one another and of different densities, to prevent mixing of deliveries. **2.** in a refining operation, the mixing of two grades of petroleum to improve the distillation of one.

**bathythermograph** *Earth Science.* an instrument that records ocean water temperature in relation to depth.

**battery** *Storage.* **1.** a direct-current voltage source consisting of two or more electrochemical cells connected in series or parallel to convert chemical energy into electrical energy. **2.** a single electrochemical cell in such a system. See below.

**battery acid** another term for SULFURIC ACID.

**battery bank** *Storage.* a group of batteries wired together to store the energy produced for a solar energy system.

**battery capacity** see CAPACITY.

**battery cell** see CELL.

**battery electric vehicle (BEV)** *Renewable/Alternative.* an electric-drive vehicle that derives the power for its drive motor(s) from a battery pack.

**battery electrode** see ELECTRODE.

**battery life(time)** *Storage.* the period during which a battery is capable of operating at or above a specified capacity or efficiency performance; this may be measured in cycles or years.

**battery rating** *Storage.* a system to measure the rate of energy discharge from a battery. Automotive and marine batteries are rated in cold cranking amps, while batteries used in photovoltaic systems are rated in ampere-hours.

**baud** *Communication.* a unit measuring the number of signal elements (bits of information) that can be transmitted per second in a given transmission channel. Thus, **baud rate.**

**bauxite** *Materials.* the principal ore used in producing aluminum; a residual sedimentary rock formed by the leaching of groundwater

**battery** Alessandro Volta's invention of the Volta Pile in 1800 represents the beginning of batteries. The Pile consisted of a pile of series-connected cells made up of alternating layers of zinc, electrolyte soaked into cardboard or leather, and silver. Following Volta's report of his invention, many investigators constructed electrochemical cells for producing and storing electrical energy. For the first time, relatively large currents at high voltages were available for significant periods of time. Batteries, strictly speaking, are composed of more than one electrochemical cell. The electrochemical cell is the basic unit from which batteries are built. A cell contains a negative electrode, a positive electrode, an electrolyte held between the electrodes, and a container or housing. A battery is an assemblage of cells connected electrically in series and/or parallel to provide the desired voltage and current for a given application. In contrast to the alternating current available in our homes from the electric utility company, batteries deliver a direct current that always flows in one direction. There are a few different types of batteries: Primary batteries can be discharged only once and then are discarded; they cannot be recharged. Secondary batteries are rechargeable. Forcing current through the cells in the reverse direction can reverse the electrochemical reactions that occur during discharge. Both primary and secondary batteries can be categorized based on the type of electrolyte they use: aqueous, organic solvent, polymer, ceramic, molten salt, and so on. Colloquially, the term *battery* is often used in place of the more proper term *cell*. The common 1.5 volt flashlight battery is actually a single cell.

**Elton J. Cairns**
University of California, Berkeley

and consisting of one or more aluminum hydroxides, along with variable amounts of clay and some iron and titanium oxides. [From *Les Baux,* in southern France, where it was discovered.]

**Bayer process** *Materials.* the first stage in the extraction of aluminum from bauxite, consisting of digesting the ore in a sodium hydroxide solution. [Developed by Austrian chemist Karl Joseph *Bayer,* 1847–1904.]

**Bay of Fundy** see FUNDY, BAY OF.

**BCS model** *Communication.* a description of why superconductors behave as they do; i.e., the electrons in a superconductor condense into a quantum ground state and travel together collectively and coherently. [From the last names of the three U.S. physicists who described this, John Bardeen, Leon Cooper, and J. Robert Schrieffer.]

**beam** *Physics.* **1.** any concentrated, unidirectional stream of particles or radiation, as of light or sound. **2.** an electromagnetic signal transmitted in a certain direction, as by a beacon or radio tower.

**beam angle** *Lighting.* the angle between the two directions for which the intensity (candlepower) is 50% of the maximum intensity, as measured in a plane through the nominal beam centerline.

**beam-plasma reaction** *Physics.* a fusion reaction that occurs in neutral-beam heated plasmas from the collision of a fast beam ion with a thermal plasma ion.

**beam radiation** another term for DIRECT RADIATION.

**beam spread** *Lighting.* the angle between the two directions in a plane in which the candlepower is equal to a stated percentage (usually 10%) of the maximum candlepower in the beam.

**Beart drill** *History.* the first rotary oil drill (1844); a device that employed many of the basic principles of modern drilling. It provided a more effective method of drilling than prior techniques, increasing the speed, depth, and efficiency of drilling for oil and gas. [Named for English engineer Robert *Beart.*]

**Beaufort, Francis** 1774–1857, British oceanographer noted for his pioneering work in nautical surveying and charting and for his development of the Beaufort scale.

**Beaufort scale** *Wind.* a system for estimating and reporting wind speeds that equates wind speed, a descriptive term, and visible effects upon land objects or the sea surface, based on the **Beaufort force** or **number.** A 0 on the Beaufort scale describes a wind speed of less than 1 knot and calm seas; a 12 describes hurricane conditions. The Beaufort scale originally was based on the effect of various wind speeds on the amount of canvas that a full-rigged 19th-century frigate could carry. Also, **Beaufort wind scale.**

**Beaumont, Huntingdon** c. 1560–1624, British mining entrepreneur who built the first documented railway in 1603–04, a two-mile wooden wagonway from his mining concessions near Strelley (northwest of Nottingham) down to Wollaton.

**Becquerel, Antoine César** 1788–1878, French physicist, noted for his pioneering work in electrochemistry.

**Becquerel, Antoine Henri** 1852–1908, French physicist, noted for his discovery of the phenomenon of natural radioactivity. The becquerel unit is named for him.

**becquerel (Bq)** *Nuclear.* a unit of radionuclide activity, equivalent to 1 disintegration per second; this term is now usually used in place of the term *curie.*

**Becquerel cell** *Storage.* a photoelectric cell consisting of two identical, unequally illuminated electrodes that are placed in an electrolyte, causing a voltage to flow between them.

**bed** see COAL BED.

**BEE** basal energy expenditure.

**behavioral change** *Social Issues.* a change in the activities of a person or organization that affects the level of energy conservation, either positively (e.g., turning off lights when not in use) or negatively (e.g., using an electric dryer to replace a clothesline).

**behavioral thermoregulation** *Biological Energetics.* in an endothermic organism, the use of behavior in the regulation of body temperature, such as avoiding or seeking sources of heat (as opposed to the usual regulation by means of internal physiological mechanisms).

**Bell, Alexander Graham** 1847–1922, Scottish inventor known as the first person to patent the telephone and also the founder of the first company to bring telephone services successfully to the marketplace. Bell's telephone grew out of improvements to his **harmonic telegraph** which could send more than one message at a time over a single telegraph wire.

*Alexander Graham Bell*

**bellows** *Consumption & Efficiency.* a flexible, accordion-sided device designed for pumping air, consisting of a chamber that expands to draw in air through a valve and contracts to expel it through a tube.

**bell pit** or **bellpit** *Coal.* a type of excavation used in early coal mining, a shallow underground mine so called because of the shape of the excavation; i.e., a narrow vertical shaft below which a small chamber was opened out in a bell shape.

**bench** *Coal.* **1.** a term for a horizontal layer of coal that is separated from other layers in the same seam by sections of non-coal rock, or simply separated in the process of cutting the coal, and that is mined separately from the other layers. Thus, **bench coal. 2.** to cut coal into layers in this manner.

**benchmark crude** see MARKER CRUDE.

**beneficiate** *Materials.* to remove impurities from, or otherwise refine, a material such as coal or an ore prior to its processing. Thus, **beneficiation.**

**benefit-cost ratio** *Economics.* the ratio of the economic benefit of an energy-related project (e.g., the construction of a new power plant) to the project cost, computed in present value terms.

**benefits charge** *Electricity.* a per unit tax on sales of electricity, with the revenue generated being used for investments in energy efficiency measures or renewable energy projects.

**benthic** *Earth Science.* relating to or involving the benthos (see next).

**benthos** *Earth Science.* the biogeographic region at the bottom of a body of water, especially the ocean, and the complex of living organisms and their specific assemblages that inhabit the bottom substrates.

**bentonite** *Materials.* a soft plastic clay formed by the chemical alteration of volcanic ash, which expands on the absorption of water; used chiefly as a suspending agent and to thicken oil-well drilling muds. [Named for Fort *Benton,* Montana.]

**Benz, Karl** 1844–1929, pioneering German automobile engineer who registered a patent for his three-wheeled vehicle on January 29, 1886, a date now widely recognized as the "birth certificate" of the modern automobile. The vehicle was powered by a single-cylinder water-cooled engine that was fueled by the vapor of ligroin (benzine). See also DAIMLER-BENZ.

**benzene** *Chemistry.* a colorless, volatile, flammable, and toxic liquid aromatic hydrocarbon that freezes at 5.5°C and boils at 80°C. It is a component of products derived from coal and petroleum and is found in gasoline and other fuels. It is used in the manufacture of plastics, detergents, pesticides, and other chemicals. Benzene has been shown to be a carcinogen (cancer-causing agent).

**benzin** or **benzine** *Materials.* a clear, colorless, highly flammable, volatile liquid mixture of hydrocarbons (mostly of the methane series) that is obtained from the distillation of petroleum and used as a solvent for many organic compounds.

**Bergius process** *Coal.* a method of producing an oil similar to crude petroleum by the hydrogenation and liquefaction of coal or cellulosic materials, carried out at high hydrogen pressures in the presence of a catalyst.

[Developed by German chemist Friedrich *Bergius*, 1884–1949.]

**berm** *Earth Science.* **1.** a temporary bench, shelf, ledge, or narrow terrace on the shore of a beach above the high tide line, formed by the action of waves. **2.** a stabilizing earthwork, such as the shoulder of a road, or a mound of earth placed against a building wall for stabilization, wind protection, or insulation.

**Bernard, Claude** 1813–1878, French physiologist who made fundamental early contributions to the mechanisms of human energetics, such as the principle of HOMEOSTASIS.

*Claude Bernard*

**Berners-Lee, Tim(othy)** born 1955, British scientist often credited as the foremost individual in the creation of the Internet's World Wide Web, through his development of innovative methods to identify, code, and display documents.

**Bernoulli, Daniel** 1700–1782, Dutch-born mathematician working in Switzerland, known for his theoretical contributions to the field of aerodynamics.

**Bernoulli's principle** *Physics.* a statement of the conservation of energy per unit mass for an ideal fluid in steady motion, in which decreased pressures are associated with greater velocities, and, conversely increased pressures are associated with lower velocities; i.e., the sum of pressure and potential and kinetic energies per unit volume is constant at any point. Also, **Bernoulli's equation.**

**Berry, Clifford E.** 1918–1963, U.S. computer scientist and co-creator of the first electronic-digital computer. See ASTANOFF.

**Bertalanffy, Ludwig von** 1901–1972, Austrian biologist who first proposed the basic tenets of general systems theory, and was among the first to apply the system methodology to the social sciences.

**Berthelot, Pierre** 1827–1907, French chemist known for his early work in thermochemistry and for the invention of the BOMB CALORIMETER.

**Berthelot principle** *Chemistry.* the principle that of all possible chemical processes that can proceed without the aid of external energy, the one that will occur is the one accompanied by the greatest evolution of heat; this applies at low temperatures only and does not account for endothermic reactions.

**Berthollet, Claude** 1748–1822, French chemist who helped design the first system for naming chemical compounds, which serves as the basis of the modern system of chemical nomenclature.

**beryllium** *Chemistry.* a chemical element having the symbol Be, the atomic number 4, an atomic weight of 9.01218, and a melting point of about 1280°C, the lightest structural metal known. Used in the manufacture of ceramics, as a moderator in nuclear reactions, and as an alloy in various metals.

**Bessemer, Henry** 1813–1898, noted British inventor and engineer who developed the first process for mass-producing steel inexpensively, the BESSEMER PROCESS. Modern steel is still made using technology derived from this process.

**Bessemer process** *Materials.* a steel refining method in which air is blown through the molten charge to oxidize impurities such as carbon, silicon, and manganese; an improved process operating up to ten times faster than previous methods and using no fuel once the charge has been melted, thus dramatically reducing the energy and monetary cost of steel production. Thus, **Bessemer steel, Bessemer ore, Bessemer (pig) iron,** and so on.

**best available control technology (BACT)** *Policy.* a regulation for the control of air pollution from a stationary source, such as a power plant; the permissible level is derived by considering the maximum degree of reduction of each pollutant that could be achieved by applying the best available technology.

**beta particle** *Nuclear.* a small electrically charged particle thrown off by many radioactive materials. It is identical to the electron and possesses the smallest negative electric charge found in nature. Beta particles emerge from radioactive material at high speeds, sometimes close to the speed of light.

**beta radiation** *Nuclear.* a type of radioactive decay in which the parent nucleus emits either an electron or a positron, thus raising or lowering the atomic number by 1 while leaving the atomic mass unchanged. Also, **beta decay, beta emission.**

**beta ray** *Nuclear.* a stream of beta particles.

**betatron** *Nuclear.* an electron accelerator characterized by a circular geometry, with acceleration achieved by magnetic flux increase.

**Bethe, Hans** born 1906, German-American physicist noted for his discovery of the reactions that supply the energy in stars (the carbon-nitrogen cycle for massive stars and proton-proton reaction that powers fainter stars such as the sun).

**Betz formula** *Wind.* a principle describing the maximum theoretical efficiency of extracting energy from the wind, equal to 16/27 (about 59%) of the kinetic energy in the wind. Actual wind turbine rotors convert much less than this **Betz limit**, because of energy losses in transmissions, generators, and power conditioning. Other power losses are caused by changes in temperature and in wind speed and direction. [Developed by German physicist Albert *Betz,* 1885–1968.]

**BEV** battery electric vehicle.

**bicycle** *Transportation.* a two-wheeled, human-powered vehicle, with one wheel in front of the other, usually driven by a rider turning footpedals attached to the rear wheel by a chain.

**bi-fuel vehicle** *Transportation.* a vehicle capable of operating either on a conventional fuel, an alternative fuel, or both, either by simultaneously using two separate fuel systems for the conventional and alternative fuels, or by using a mixture of gasoline or diesel fuel and an alternative fuel in the same fuel tank (the latter is usually called a flexible-fuel vehicle).

**bifurcation** *Thermodynamics.* a change in the stability or type of solutions of a dissipative dynamical system as a parameter of the system is varied.

**Big Bang** *Earth Science.* a term for the theory that the present universe originated about 15 billion years ago when a gigantic explosion (the "Big Bang") released huge amounts of energy.

**Bigas process** *Coal.* a two-stage process utilizing a superpressure oxygen-blown slagging gasifier to produce a raw gas from coal that can be converted to high-Btu pipeline-quality gas.

**Big Inch** see INCH LINES.

**Big Muskie** *Coal.* a gigantic machine once used in surface coal mining, described as the world's largest earth-moving machine at over 200 feet in height, weighing some 13,500 tons, with a bucket weighing 550 tons when fully loaded. A type of walking dragline, i.e., a machine designed to expose deeply buried coal seams that smaller machines could not reach.

**bilge** *Oil & Gas.* a space or well inside a double-bottomed hull into which seepage **(bilge water)** drains to be pumped out; in an oil tanker this water tends to be mixed with oil and other contaminants to form a complex wastewater that can be harmful to the marine environment.

**bimetal** *Materials.* a material composed of two metals of different coefficients of thermal expansion welded together so that the piece will bend in one direction when heated, and in the other when cooled; can be used to open or close electrical circuits, as in thermostats. Thus, **bimetallic, bimetallic thermometer.**

**binary** *Communication.* describing a system or property (e.g., a computer program) having two (and only two) possible conditions or choices.

**binary cycle** *Conversion.* a combination of two turbine cycles employing two different working fluids **(binary fluids)** for power production; the second turbine cycle operates by employing waste heat from the first cycle, thus providing higher overall efficiency. Thus, **binary (fluid) system.**

**binary cycle plant** *Geothermal.* a system employing two fluids in separate adjoining pipes; the heat from a geothermal liquid is used to vaporize a secondary working fluid with a lower boiling point; this vapor then drives a turbine to generate electricity. Used in geothermal areas with moderate-temperature water (typically below 400°F). Also, **binary power plant.**

**binary-vapor cycle** *Thermodynamics.* a power cycle that is a combination of two cycles, one in a higher temperature region and the other in a lower temperature region.

**binding energy** *Physics.* **1.** the minimum amount of energy required to extract an individual particle from a system of particles. **2.** the minimum amount of energy required to disassemble a system of particles and separate them at infinite distances so that there is no interaction among them.

**binding target** *Policy.* an agreed or mandated emission standard that is to be met at a specific point of time or within a given period.

**bioaccumulation** *Health & Safety.* **1.** any increase in the concentration of a chemical in a biological organism over time, compared to the chemical's concentration in the environment. **2.** an increase in concentration of a pollutant from the environment to the first organism in a food chain.

**bioamplification** another term for BIOMAGNIFICATION.

**bioassay** *Health & Safety.* **1.** a determination of the concentration of a substance, especially a toxin, by its effect on the growth of a test organism under controlled conditions. **2.** the use of a living organism or cell culture to test for the presence of a toxic substance.

**biobased** or **bio-based** *Biomass.* having a biological source; describing materials, especially plastics, that are fully or partially produced from biomass-derived feedstocks.

**biocapacity** *Ecology.* the potential productivity of all the biologically productive space within a specified country, region, or territory.

**biochemical conversion** see BIOCONVERSION.

**biochemical oxygen demand (BOD)** *Environment.* the amount of dissolved oxygen (DO) consumed by microorganisms in the biochemical oxidation of organic and inorganic matter; a commonly used measurement to determine water quality, since the amount of oxygen used is an indicator of the level of organic waste that is present.

**biochemistry** *Chemistry.* the scientific study of the chemical composition and reactions of biological systems; the chemistry of living organisms.

**biocide** *Ecology.* a substance that is capable of killing living organisms, such as an insecticide or herbicide.

**bioclimatology** *Ecology.* the scientific study of the relationship between living organisms and climate.

**bioconcentration** another term for BIOACCUMULATION.

**biocontrol** *Ecology.* the deliberate use by humans of one species of organism to eliminate or control another, as in the control of undesirable plants or insects by the use of natural parasites, diseases, or predators, rather than by herbicides and pesticides.

***biocontrol*** *Biocontrols involve the use of natural predators rather than industrial pesticides, as in this view of beetles specially selected to control aphids that infest wheat and other grain crops.*

**bioconversion** *Conversion.* **1.** the microbial conversion of a chemical to a compound of economic importance. **2.** any of various processes that use plants or microorganisms to change one form of energy into another; e.g., the fermentation of carbohydrates into alcohol; the digestion of organic wastes or sewage by microorganisms to produce methane.

**biocrude** *Biomass.* a crude oil similar to petroleum that can be produced from biomass under high pressure and temperature; it can then be treated with hydrogen to upgrade it to a transportation fuel for use in place of conventional diesel fuels.

**biodegradable** *Materials.* describing a material that can be decomposed by natural biological processes, as by the action of microorganisms, plants, or animals. Thus, **biodegradation.**

**biodiesel** *Biomass.* any liquid biofuel suitable as a diesel fuel substitute or diesel fuel additive or extender; typically made from oils such as soybeans, rapeseed, or sunflowers, from animal tallow, or from agricultural byproducts such as rice hulls.

**biodiversity** *Ecology.* **1.** the existence of a wide variety of different species within a given area or at a given point in time. **2.** the relative number of different species in a given area at a certain time.

**bioelectricity** *Biological Energetics.* **1.** the presence of electric current within muscular and neural tissue. **2.** the production of electric power directly from biomass.

**bioenergetics** *Biological Energetics.* the study of the flow and transformation of energy within and between organisms and their environment.

**bioenergy** short for BIOMASS ENERGY.

**bioethanol** *Biomass.* a liquid fuel consisting of ethanol produced from biomass, capable of being used for the same purposes as light oil.

**biofuel** *Biomass.* any solid, gaseous, or liquid fuel obtained from biomass; this may be in its natural form (e.g., wood, peat) or a commercially produced form (e.g., ethanol from sugarcane residue, diesel fuel from waste vegetable oils).

***biofuel*** *A biomass fuel machine known as "Jaws" harvests forest waste materials to be processed and used as fuel.*

**biogas** *Biomass.* a gaseous fuel of medium energy content, composed of methane and carbon dioxide; produced from the anaerobic decomposition of organic material in landfills.

**biogenic** *Biological Energetics.* resulting from the activities of living organisms.

**biogenic hydrocarbon** *Environment.* naturally occurring hydrocarbon compounds, including VOCs (volatile organic compounds) that are emitted from trees and vegetation. High VOC-emitting tree species such as eucalyptus can contribute to smog formation, and thus the relative emission rates of various species can be a consideration in large-scale tree plantings.

**biogenic theory** *Oil & Gas.* the theory that fossil fuels represent the altered remains of ancient plant and animal life deposited in sedimentary rocks, and thus have a biological origin. Generally accepted in preference to the ABIOGENIC THEORY that hydrocarbon deposits became part of the earth as it formed.

**biogeochemical** *Earth Science.* of or relating to biogeochemistry or the biogeochemical cycle.

**biogeochemical cycle** *Earth Science.* the exchange of elements (e.g., oxygen, carbon, nitrogen, and so on) in the environment between storage pools such as the atmosphere, biota, oceans, soils, the earth's crust, and human society.

**biogeochemistry** *Earth Science.* the branch of science that studies the biological, chemical and geological aspects of environmental processes.

**biological amplification** see BIOMAGNIFICATION.

**biological concentration** see BIOCONCENTRATION.

**biological control** see BIOCONTROL.

**biological dose** *Nuclear.* the amount of radiation absorbed in biological material.

**biological dosimetry** *Nuclear.* an area of radiation dosimetry that uses the biological damage produced by radiation to estimate radiation dose.

**biological energy** see BIOENERGY.

**biological energy conversion** see BIOCONVERSION.

**biological fixation** see NITROGEN FIXATION.

**biological half-life** *Health & Safety.* the time required for the quantity of a material in a specified tissue, organ, or region of the body (especially a toxin) to reduce in amount by half as a result of biological processes.

**biological hazard potential (BHP)** *Health & Safety.* a total measure of the danger to living organisms presented by a certain quantity of radioactive materials, accounting for the variation in biological effects on different individuals within the given population.

**biologically effective dose (BED)** *Health & Safety.* **1.** the amount of a substance that is sufficient to bring about some significant physiological change in the affected organism. **2.** specifically, the level of exposure to a toxic substance that is required to produce a harmful effect.

**biologically productive** *Ecology.* **1.** able to sustain organic life. **2.** specifically, a term for areas of land and water capable of supporting photosynthesis at sufficient rates to provide economically useful concentrations of biomass. Marginal and unproductive regions are generally excluded, such as deserts, tundra, and the deep oceans.

**biological magnification** see BIOMAGNIFICATION.

**biological oxidation** *Biomass.* the decomposition of complex organic materials (e.g., in water) by microorganisms.

**biological oxygen demand (BOD)** *Biological Energetics.* a measurement of the amount of oxygen required by aerobic organisms to carry out oxidative metabolism in a given volume of water containing organic material (e.g., waste matter in a water supply).

**biological shield** *Nuclear.* a mass of absorbing material (e.g., concrete or lead) placed around a reactor or radioactive material to reduce the radiation to a level safe for humans.

**biomagnification** *Ecology.* the increase in concentration of a pollutant from one trophic level in a food chain to another; this usually occurs when the pollutant is metabolized and excreted much more slowly than the nutrients that are passed from one trophic level to the next. Such pollutants are long-lived, mobile, soluble in fats, and biologically active. Biomagnification is a particular threat for species living at the top of food chains.

**biomarker** *Ecology.* a physiological or pharmacological indicator of a particular biological property, especially one that is used to predict a disease or toxic event.

**biomass** **1.** a collective term for all organic substances of relatively recent (nongeological) origin that can be used for energy production, including industrial, commercial, and agricultural wood and plant residues; municipal organic waste; animal manure; and crops directly produced for energy purposes. Biomass can be solid (e.g., wood, straw), liquid (biofuels), or gaseous (biogases). **2.** a quantitative estimate of the entire assemblage of living organisms, both animal and vegetable, of a given habitat, measured in terms of mass, volume, or energy in calories.

**biomass combustion** *Biomass.* a technology that extracts heat energy from biomass so that it can then be used for a variety of heat and power applications.

**biomass energy** *Biomass.* a general term for renewable energy produced from biomass, such as wood and wood wastes, agricultural crops and wastes, or municipal and industrial wastes. See next page.

**biomass fuel** see BIOFUEL.

**biomass energy** Biomass energy was utilized in 1860 to meet over 70% of the world's total energy needs, mainly via the conventional combustion of wood fuel for heating and cooking. Biomass energy is the chemical energy content of non-fossil, energy-containing forms of carbon such as land-based and water-based vegetation, and waste materials such as municipal solid wastes, biosolids, forestry and agricultural residues, and some industrial wastes. In 2000, the percentage contribution of biomass energy to the world's energy demand had decreased to about 10% of the total. In terms of millions of barrels of oil equivalent consumption per day, biomass energy usage had increased from about 5 out of a total consumption of 7 in 1860, to about 20 out of a total consumption of 200 in 2000. The scope of biomass energy applications also increased manyfold during this period and included a wide variety of fuels, organic chemicals, and products. Various projections of the practical global energy potential of biomass energy using advanced combustion, gasification, and liquefaction processes for integrated biomass production-biorefinery systems that supply heat, steam, electricity, fuels, chemicals, and bioproducts on a sustainable basis range up to 100 million boe/day. Commercial systems of this type will be essential in the 21st century if the global community decides that carbon-based fuels and commodity organic chemicals, as well as many specialty chemicals, must be manufactured from renewable biomass resources to maintain the living standards of a modern energy economy, improve environmental quality, and counteract the inevitable shortages and supply disruptions of natural gas and petroleum crude oils expected to start in the first and second quarters of this century. Without large-scale waste and virgin biomass conversion to multiple products in biorefineries, biomass energy utilization will be limited to niche markets, and coal will be the primary source of carbon-based energy, fuels, and commodity chemicals.

**Donald L. Klass**
Biomass Energy Research Association, USA

**biomass (integrated) gasification** *Biomass.* a composite system used to convert biomass feedstock into gas fuel for an electricity-generating unit that consists of one or more gas turbines, with a portion of the required energy input provided by exhaust heat of the turbine to increase efficiency.

**biomass oil** *Biomass.* a biomass energy feedstock in the form of lipids from animal fats, fish and poultry oils, plant oils, or recycled cooking greases.

**biomass resource** *Biomass.* any form of organic material that can be used as an energy source, such as forest, mill, and agricultural residues, urban wood wastes, and dedicated energy crops. A **biomass resource assessment** estimates the quantities of such resources available by location and price level.

**biomass supply curve** *Biomass.* an estimate of the quantity of biomass resources that could be made available, as a function of their price.

**biome** *Ecology.* **1.** a complex biotic community existing in a given region, produced by the interaction of climatic factors, living organisms, and substrate. **2.** specifically, a community that has developed to climax vegetation, such as tundra, coniferous forest, or grassland.

**bionavigation** *Biological Energetics.* the ability of certain animals to travel to a precise distant location (e.g., a breeding or wintering site) without any evident use of landmarks.

**biophotolysis** *Hydrogen.* the action of light on a biological system (e.g., certain algae and bacteria) that results in the dissociation of water to produce hydrogen (known as **biohydrogen**).

**biophysical economics** *Economics.* a school of economic thought involving analysts from diverse fields who use basic ecological and thermodynamic principles to analyze the economic process.

**biopollutant** *Health & Safety.* a collective term for polluting agents that are living organisms, or the products of organisms, such as airborne microorganisms, inflammatory agents (endotoxins), and indoor allergens such as house dust mites.

**biopolymer** *Chemistry.* a polymer found in nature; examples include starch, proteins and peptides, and DNA and RNA.

**bioproductive** *Ecology.* **1.** able to produce and sustain living organisms. **2.** specifically, describing land area that is capable of providing natural substances that support human activities; e.g., land used for growing food crops.

**biorefinery** *Biomass.* a factory employing various processing steps for the production of chemical and fuel products from biomass, including pretreatment, separation, and catalytic and biochemical transformations.

**bioremediation** *Environment.* **1.** the natural capacity of living organisms to degrade or transform hazardous organic contaminants. **2.** the purposeful addition of organic materials to a contaminated environment to cause an acceleration of such natural processes.

**biosolid** *Biomass.* a nutrient-rich organic byproduct of municipal wastewater treatment; when treated and processed, biosolids can be safely recycled; e.g., for fertilizers.

**biosphere** *Ecology.* those regions of the earth and its atmosphere that are capable of supporting life; the earth's living system as a whole.

**biota** *Ecology.* **1.** the totality of all forms and species of living organisms within a certain area or habitat at a given time. **2.** plant and animal life in general.

**biotechnology** *Consumption & Efficiency.* **1.** the practical application of biological knowledge and techniques for industrial purposes; e.g., fermentation. **2.** specifically, the use of contemporary biological techniques to produce new substances or perform new functions; e.g., recombinant DNA technology.

**biotic** *Ecology.* having to do with life or living organisms.

**Biot number** *Measurement.* a dimensionless constant that is equivalent to the ratio of the heat transfer coefficient for convection at the surface of a solid to the specific conductance of the solid. [Named for Jean Baptiste *Biot,* 1774–1862, French physicist.]

**biotoxin** *Health & Safety.* a toxic substance that is a product of the cells or secretions of a living organism. **Biotoxicology** is the scientific study of such substances.

**Biot–Savart law** *Electricity.* the statement that the intensity of the magnetic field set up by a current flowing through a wire varies inversely with the square distance from the wire. [Named for French scientists Jean-Baptiste *Biot* and Félix *Savart.*]

**biplane** *Transportation.* an early type of aircraft having two sets of wings rather than one (a monoplane).

***biplane*** *British aviators John Alcock and Arthur Brown take off in their biplane for the first flight to cross the Atlantic Ocean (Newfoundland to Ireland, June 1919). Their flight preceded the more famous Lindbergh crossing by 8 years.*

**bipolar plate** *Electricity.* a dense electronic (but not ionic) conductor that electrically connects the anode of one cell to the cathode of another. It also distributes fuel or air to the electrodes.

**bipropellant** *Transportation.* a propellant system consisting of two different liquids, a fuel and an oxidizer, that are kept separate before combustion.

**Birkeland–Eyde process** *Materials.* an electric-arc process for nitrogen fixation, one of the first processes used in the large-scale manufacture of nitrogen fertilizer from atmospheric nitrogen. [Developed by Norwegian scientists Kristian *Birkeland,* 1867–1917, and Samuel *Eyde,* 1866–1940.]

**bit** *Oil & Gas.* a boring device used in drilling oil and gas wells, consisting of a cutting element and also a circulating element that permits the passage of drilling fluid and utilizes the force of the fluid stream to improve drilling efficiency.

**bit** *Communication.* the basic item of information in a digital computer system, having a value of either 1 or 0. [An acronym for binary digit.]

**bitumen** *Oil & Gas.* a naturally occurring, viscous mixture of hydrocarbons heavier than pentane that may contain sulfur compounds; in its natural state it is not recoverable at a commercial rate through a well. At room temperature, it looks and flows like molasses. Bitumen comprises up to 20 per cent of oil sands, where it can be mined and converted to commercial liquid fuels and products. A substance known since ancient times and sometimes called *asphalt, tar,* or *pitch.*

**bitumenize** *Oil & Gas.* **1.** to change a substance into bitumen. **2.** to mix with bitumen. Thus, **bitumenization.**

**bituminous** *Oil & Gas.* **1.** characteristic of or containing bitumen. **2.** see BITUMINOUS COAL.

**bituminous coal** *Coal.* a dark brown to black banded coal of intermediate rank, ranking above lignite (and sub-bituminous) and below anthracite; typically having a heating value between 19 and 30 Btu per ton on a dry, mineral-matter-free basis and containing 15–20% volatile matter. A general-purpose fuel used to generate steam electric power and for power and heat applications in manufacturing, as well as to produce coke for use in steel-making.

**BKB** *Coal.* Braunkohlenbriketts, a composition fuel manufactured by crushing, drying, and molding lignite (brown coal) into an even-shaped briquette, under high pressure without the addition of binders.

**Black, Joseph** 1728–1799, French-born Scottish chemist noted for his fundamental work on latent and specific heats and for his discovery of carbon dioxide. He also appears to have directly influenced Watt's work on steam engines.

**black body** or **blackbody** *Physics.* a theoretical object that is simultaneously a perfect absorber and emitter of radiant energy (i.e., it absorbs all the radiation striking it and reflects no radiation), and whose energy distribution is dependent only on its temperature. Most stars behave approximately like black bodies. [So called because such a body would reflect no radiation and thus appear to be black.] Thus, **black-body energy, radiation, temperature,** and so on.

**black-bulb thermometer** *Measurement.* a type of thermometer in which the sensing element, when covered with a shade, approximates a black body (see previous).

**black carbon** *Environment.* a term for soot and other such minute carbon particles in the atmosphere, resulting from incomplete combustion of various fuels; e.g., diesel exhaust. Such airborne particles can absorb sunlight and thus **black carbon aerosols** are a topic of interest to climate scientists.

**black coal** *Coal.* a term for anthracite or bituminous coal, as opposed to lignite (brown coal).

**blackdamp** *Mining.* a nonexplosive mixture of carbon dioxide and other gases, especially nitrogen, which is deficient in oxygen and therefore unable to sustain life. It is heavier than air and thus may be found along the floor of a mine, especially after a fire or explosion. [So called because its presence reduces light in a mine; i.e., makes it black.]

**black hole** *Physics.* an object in space so dense that no light or radiation can escape from it. Black holes, which cannot be observed directly, are believed to be formed when a star collapses inward upon itself.

**black liquor** *Biomass.* a type of wood-derived fuel formed as a byproduct during the pulping of wood in the papermaking industry; it consists mainly of the lignin that remains after the separation of cellulose to form paper fibers. This is concentrated and burned, and the spent chemicals are collected and reused.

**black locust** *Biomass.* a fast-growing hardwood tree, *Robinia pseudoacacia,* that has significant potential as a fuel source.

**black lung (disease)** *Health & Safety.* the common name for coal workers' pneumoconiosis, a chronic respiratory disease caused by the inhalation of coal dust.

**Blackman, F. F.** 1866–1947, British plant physiologist who discovered that photosynthesis is a two-step process, only one of which uses light directly. See DARK REACTIONS.

**Black Mesa** *Coal.* a pipeline transporting coal slurry from a mine on Navajo land in northern Arizona to a generating station in Nevada; currently the only coal slurry pipeline operating in the U.S.

**blackout** *Electricity.* **1.** a sudden and unexpected failure of electrical power, extending over a widespread area and resulting in the loss of lighting and other electrical services. **2.** a purposeful reduction of electric lighting; e.g., in a city during wartime as a precaution against a bombing attack. See below.

**black start** *Electricity.* a power source's ability to power up from a cold shutdown condition to fully operational status through a dedicated auxiliary source that is independent of external power systems.

**blackout** An interruption of electric service, often being larger in duration and geographic scope than other power outages. Interruptions of electric service occur for many reasons. Most frequent are local power outages on the distribution (low and medium voltage) network, resulting from equipment failures, or storm-related events causing tree limbs to interact with overhead wires. Less common are interruptions on the high voltage system, which transport bulk power over long distances from large central station generators to population centers. When one of these power lines, or large generators, unexpectedly fails, the power network automatically attempts to redistribute the flow of power between supply and demand. If the redirected flows overload the remaining power lines, then a cascading failure can occur. This happens when the breakers at numerous transformer stations "trip" to prevent the damage of equipment. Enough of these can result in a large regional blackout that can affect millions of people for a period of hours to days. Notable North American blackouts include the August 2003 Northeast blackout that began in northern Ohio, and affected service from Michigan to New York City, including Ontario. Others include the Northeast Great Blackout of November 1965, which initiated the formation of the North American Electric Reliability Council (NERC), the New York City Blackout of July 1977, and numerous Blackouts in the 1990s on the West Coast when power flows from Pacific Northwest to California were interrupted.

**Stephen Connors**
Massachusetts Institute of Technology

**blade** *Wind.* **1.** the flat arm of a fan, turbine, or propeller. **2.** a propeller arm or rotary wing, especially the part that cleaves the air and has an efficient airfoil shape.

**blade load(ing)** *Wind.* the force sustained by a rotor blade, as of a wind turbine, helicopter, and so on.

**blade pitch** see PITCH (def. 4).

**blanket** *Nuclear.* a term for the part of a fusion reactor where fusion energy is converted to thermal energy, where the fusion fuel tritium is generated, and which provides shielding and neutron moderation.

**blanket fuel** *Nuclear.* a term for nuclear reactor fuel containing the fertile isotopes that are bred into fissionable isotopes.

**blast furnace** *Conversion.* a vertical shaft furnace in which the combustion of a solid fuel, such as coke, is intensified by an air blast to smelt iron ore in a continuous operation.

**blasting cap** *Mining.* a small sensitive charge placed in a larger explosive charge to detonate the larger charge.

**bleeding** *Mining.* the purposeful and controlled draining of air and other gases away from the active workings of a mine and into a ventilation system.

**blended cement** *Materials.* a powder obtained by mixing Portland cement with an admixture (usually fly ash) for the purpose of obtaining the normal properties of the Portland cement.

**blending** *Oil & Gas.* the technique or process of combining two or more petroleum liquids to produce a product with specific characteristics. See also GASOLINE BLENDING.

**blending components** *Oil & Gas.* hydrocarbons that are mechanically mixed in a refinery to produce a fuel such as motor gasoline or jet fuel with specific characteristics. In the case of motor gasoline, these components include napthas and oxygenates, among others. Two basic types are **gasoline blending components** and **aviation gasoline blending components.**

**blendstock** *Oil & Gas.* **1.** a complex mixture of relatively volatile hydrocarbons that are combined to produce various types of motor gasoline. See also BLENDING COMPONENTS. **2.** *Nuclear.* uranium that is mixed with more highly enriched uranium in a downblending operation.

**blind coal** *Coal.* a term for any type of coal that burns without a noticeable flame.

**BLM** Bureau of Land Management.

**block cut(ting)** *Mining.* **1.** a form of surface mining in which an initial (block-shaped) cut of overburden is temporarily removed, and subsequent overburden cuts are each then placed in the previous open cut. This is repeated until mining is completed and the final cut is then backfilled with the overburden from the initial cut. **2.** *Environment.* a forestry procedure in which only specific sections of a forest are harvested at a given time and other sections are left with their natural structure; a technique that favors maintenance of wildlife habitat and more favorable regeneration of the forest landscape.

***block cutting*** *Block cutting technique in an area of old-growth Douglas fir and Western hemlock, Olympic National Forest, state of Washington.*

**block rates** *Electricity.* an electric rate schedule having different unit costs for incremental blocks (quantities) of demand. Typically, unit costs progressively decline as consumption increases.

**blondel** *Lighting.* a unit of luminance equal to pi lumens per square meter per steradian (1 pi cd/$m^2$). [Named for French physicist A. E. *Blondel,* 1863–1938.]

**bloomery** *History.* an earlier name for a furnace in which ore is smelted to produce iron products; so called because the mass of iron that is the intermediate stage in this process was known as a **bloom.**

**blow-by** or **blowby** *Transportation.* **1.** in an internal-combustion engine, a leakage of the air-fuel mixture or combustion gases past a piston and into the crankcase during periods of maximum pressure in a particular cylinder; caused by an excessive gap or seized rings. **2.** a device fitted to the crankcase and designed to route such leaking substances back to the cylinders for combustion.

**blowdown** *Consumption & Efficiency.* **1.** the use of pressure in order to remove solids or liquids from a process vessel, such as the removal of sludge or concentrated feedwater from the boiler of a steam-generating plant **2.** the difference between the opening and closing pressures of a safety valve.

**blower** *Consumption & Efficiency.* **1.** a fan that is enclosed so that the inlet gas is compressed to a higher pressure at discharge. **2.** *HVAC.* the device in an air conditioner that distributes filtered air from the return duct over the cooling coil/heat exchanger.

**blower door** *HVAC.* a device that fits into a doorway of a building, containing a powerful fan that can extract or supply a measured rate of air flow; used for testing air leakage.

**blown-in** *HVAC.* describing an insulation product composed of loose fibers or fiber pellets that are sprayed into a building cavity or attic by means of pneumatic equipment.

**blowout** *Oil & Gas.* a sudden, unplanned, and dangerous eruption of gas or oil from a well being drilled; a wild well.

**blowout preventer** *Oil & Gas.* an arrangement of heavy-duty valves secured to the top of a well casing to control pressure; this prevents loss of pressure during drilling operations in the space between the drill pipe and casing or in the open borehole.

**Blucher** *History.* the first practical steam locomotive, built by George Stephenson in England in 1814; a vast improvement over horse-drawn wagons. Also spelled **Blutcher.**

**blue-green algae** *Ecology.* another term for the CYANOBACTERIA (see), based on the fact that they resemble algae in having aquatic habitats and being capable of photosynthesis.

**blueshift** *Physics.* **1.** the phenomenon that the wavelength of light emitted by an object moving closer to the observer is shifted toward shorter wavelengths (i.e., the bluer end of the visible spectrum), due to a Doppler shift. **2.** specifically, this phenomenon as observed with respect to celestial bodies.

**bluewater gas** *Materials.* a gas obtained in a cyclic process in which steam is passed over red-hot coke, a characteristic of the combustion of carbon dioxide; so called because it burns with a blue flame.

**BMI** body mass index.

**BMR** basal metabolic rate.

**board foot** *Measurement.* a unit of measure for logs and lumber, equivalent to a piece of wood that is 1 inch thick, 12 inches wide, and 12 inches long.

**Bobrka** *History.* a town in the Gorlice region of Poland, in which an oil extracting business was established in 1854; often described as the beginning of the modern oil industry, since it predated the DRAKE field by about five years.

**BOD** **1.** biochemical oxygen demand. **2.** biological oxygen demand.

**BOD5** *Environment.* biochemical oxygen demand/five; the amount of dissolved oxygen consumed over a period of five days by biological processes breaking down organic matter. A test used to predict the effect of wastewater on receiving streams and to determine their capacity for assimilating the organic matter.

**body mass index (BMI)** *Biological Energetics.* a measurement of the amount of fat and muscle in a given human body, determined by weight (kg) divided by height (m) squared; used to determine whether a person may be overweight or underweight by comparing this index against a set of reference values.

**body-on-frame** *Transportation.* the traditional form of automobile construction technology in which a separate body is mounted on a rigid frame that supports the drivetrain. Trucks and other heavy-duty vehicles still employ this method, but it has been generally replaced by UNIBODY construction for passenger vehicles.

**BOE** barrel of oil equivalent.

**bog** see PEAT BOG.

**boghead coal** *Coal.* a variety of nonbanded, translucent bituminous or sub-bituminous, derived from organic residues of algae, containing a high percentage of volatile matter, and providing high yields of tar and oil upon distillation. Resembles cannel coal in appearance and behavior during combustion. Thus, **boghead cannel, boghead shale.**

**Bohr, Niels** 1885–1962, Danish physicist who identified the fundamental structure of atoms and laid the groundwork for the field of quantum mechanics, which underlies modern physics. He escaped Denmark during the Nazi occupation and eventually worked on the Manhattan Project in the U.S. He later became a leading advocate for the peaceful use of atomic energy.

**Bohr magneton** *Physics.* an electromagnetic constant defined as being equal to the moment of one electron spinning about its own center.

**Bohr theory** *Physics.* an early theory in quantum mechanics that postulated the structure of an atom as consisting of a positively charged nucleus around which revolve one or more electrons in discrete circular orbits of constant energy; an increase or decrease in an electron's energy must be accompanied by a transition to another orbit. Similarly, **Bohr atom.**

**boiler** *HVAC.* a closed, pressurized vessel in which water or other liquid is heated, either to be utilized in its liquid state or to generate steam energy; e.g., a device that utilizes the heat from a furnace to convert water into high-pressure steam. See next page.

**boiler feedwater** see FEEDWATER.

**boiler horsepower** *Measurement.* a measure of the evaporation of water into dry steam, expressed as the rate of evaporation per hour of 34.5 pounds of water at a temperature of 212°F; a traditional method of rating a steam boiler's output.

**boiler slag** see SLAG.

**boiling point** *Chemistry.* the temperature at which a liquid boils; the temperature at which the liquid's vapor pressure equals the atmospheric pressure of its environment. The normal boiling point of pure water at sea level is 100°C or 212°F.

**boiler** A closed vessel in which water is heated, steam is generated or superheated, or any combination thereof, under pressure or vacuum by the application of heat resulting from the combustion of fuel or the recovery and conversion of normally unused energy. There are two basic types of boilers—watertube, a boiler in which the tubes contain water and steam, the heat being applied to the outside surface; and firetube, a boiler with straight tubes, which are surrounded by water and steam and through which the products of combustion pass. Boilers date back at least to steam production undertaken by Greek scientist/mathematician, Hero, in 200 BC. Out of the late 18th century grew a manufacturing base in the United States that became the envy of the world—steam boilers were the spark to a successful Industrial Revolution. In 1888, to combat early poor construction practices that led to unsafe operational conditions, the founders of the American Boiler Manufacturers Association initiated and promoted construction-related and safety-related boiler codes and standards. These, along with many innovative technological advances, have stimulated the use of safe, clean, and efficient boilers and boiler-related systems. The boiler industry provides the products that drive the engine of U.S. industrialization, yet also heats and cools the hospitals, schools, churches, offices, gathering places, and homes of America. It is no exaggeration to say that few technologies devised by man have produced so much to advance mankind as has the safe and dependable generation of steam made possible by the boiler.

**W. Randall Rawson**
American Boiler Manufacturers Association

**boiling water reactor (BWR)** *Nuclear.* a nuclear power reactor in which water, used as both coolant and moderator, is allowed to boil in the reactor core. The resulting steam can be used directly to drive a turbine and electrical generator, thus producing electricity. After being condensed, the water returns to the reactor vessel as feedwater.

**boil-off** *Consumption & Efficiency.* **1.** the process by which a liquid is allowed to reach its boiling point and is then vaporized, as when liquid oxygen is exposed to ambient (room) temperature. **2.** the amount of gas that vaporizes in such a process. Thus, **boil-off loss.**

**Boisguillebert, Pierre** 1646–1714, French economist who was a precursor of the Physiocrats and an advocate of economic and fiscal reforms for France during the reign of Louis XIV. See also PHYSIOCRACY.

**bolometer** *Electricity.* an instrument that measures minute amounts of radiant energy by detecting changes of electrical resistance on a thin, heat-sensitive metal conductor; can be used to measure the earth's radiation budget.

**Boltzmann, Ludwig** 1844–1906, Austrian physicist known as the founder of the field of statistical mechanics. He provided the framework for relating the microscopic properties of individual atoms and molecules to the macroscopic or bulk properties of ordinary materials.

**Boltzmann's constant** *Measurement.* a value (k) relating the average energy of a molecule to its absolute temperature; $k = 1.3803 \times 10^{-23}$ Joules per molecules degree Kelvin.

**Boltzmann's entropy hypothesis** *Thermodynamics.* a hypothesis stating that the entropy of a system of particles is a linear function of the logarithm of the statistical probability of the distribution; expressed as $S = k \log W$, where S is entropy, k is Boltzmann's constant, and W is probability.

**Boltzmann's H-theorem** *Thermodynamics.* a description of the evolution of molecules in a gas in an initial state, stating that if no external forces are present, then the gas should tend to reach equilibrium over a long period of time; this can be interpreted to indicate that the entropy of a system will always tend to increase.

**bombard** *Nuclear.* to subject a target to the impact of an intense stream of high-energy particles, such as electrons, nucleons, alpha particles, or other atomic nuclei, as for the study of the resultant reactions or for the conversion of one substance to another. Thus, **bombardment.**

**bomb calorimeter** *Measurement.* an apparatus that can measure heats of combustion, used in various applications such as calculating the calorific value of foods and fuels. See next page.

**bomb calorimeter** An apparatus primarily used for measuring heats of combustion. The reaction takes place in a closed space known as the calorimeter proper, in controlled thermal contact with its surroundings, the jacket, at constant temperature. This set, together with devices for temperature measurement, heating, cooling, and stirring comprise the calorimeter. The calorimeter proper is usually a metal can with a tightly fitting lid containing water, stirred continually, in which the bomb itself is situated. It consists of a sealed heavy-walled container in which the reactants are allowed to react, under constant volume conditions, following the ignition of the combustible matter in an oxygen atmosphere. Gases at high pressures are frequently used, hence the name. In 1878, Paul Vieille (1854–1934) developed the first bomb calorimeter which was used for measuring heats of explosion at the French service of explosives in Paris. However, this bomb was attributed by many authors to M. Berthelot (1827–1907). For many years, the use of the bomb (static) was limited to studies on C, H; C,H,O, and C, H, O, N compounds and could not be used to study those containing sulfur or halogen atoms. It was not until the use of the moving bomb technique, in 1933, that these substances could be studied. The method was improved from 1948 and onward in the universities of Lund (Sweden) and Bartlesville (U.S.). The use of oxidants other than oxygen was introduced in 1961.The use of bomb calorimetry has recently been extended to industries relating to foodstuffs, animal feed, cement, and combustible waste. Bomb calorimeter data are increasingly applied to environmental studies concerned with prevention of forest fires and fire fighting through the design of energy and risk index maps.

**Lisardo Núñez Regueira**
Universidade de Santiago de Compostela, Spain

**Bonbright, James** 1891–1985, U.S. economist and political scientist, a pioneer in the field of public utility regulation, advocating private over public ownership, limited public power as a means of assessing performance of investor-owned utilities, and state regulation of utilities based on original cost of property.

**bond** *Chemistry.* the attractive force that links atoms together in a molecule or an ionic structure, so that they act as a unit.

**bond enthalpy** *Chemistry.* the enthalpy change per mole for a particular substance when a bond is broken in the gas phase.

**bonding energy** *Chemistry.* the force that maintains a chemical bond, expressed as the average amount of energy per mole required to break bonds of the same type per molecule.

**bone coal** *Coal.* **1.** a hard, compact coal with a high ash content. **2.** a thin sedimentary layer found in coal seams.

**bone dry** *Biomass.* having zero per cent moisture content. Wood heated in an oven at a constant temperature of 212°F or above until its weight stabilizes is considered bone dry or oven dry. Thus, **bone dry ton.**

**bone dry unit** *Biomass.* a quantity of wood residue (chips) that weighs 2400 pounds at zero percent moisture content.

**bone seeker** *Health & Safety.* a term for a radioactive substance that tends to accumulate in the bones when it is introduced into the body; e.g., strontium-90, which behaves chemically like calcium.

**Bonneville Dam** *Hydropower.* a large dam in the gorge of the Columbia River about 40 miles east of Portland, Oregon, and Vancouver, Washington, a supplier of electrical power to the Pacific Northwest since its construction in 1938.

**Bonneville Power Administration** (est. 1937), a U.S. federal agency that sells approximately 50% of the electricity consumed in the Pacific Northwest, and also owns about 75% of that region's transmission line.

**Bonny Light** *Oil & Gas.* a crude oil type of Nigeria, one of a group (basket) of crudes used by OPEC in its pricing system. See OPEC BASKET.

**bonus bid** *Policy.* a method used by the U.S. government to lease the right to explore and develop offshore oil; a company makes a cash bid for the right to explore a tract of land, and if oil or gas is found, the company agrees to pay a royalty on the value of all production.

**boom** *Economics.* **1.** a period of extremely rapid economic growth. **2.** *Oil & Gas.* a floating device for corralling, or sometimes absorbing, oil on water.

**boom and bust** *Economics.* a pattern of performance over time in an economy or industry that alternates between extremes of rapid growth and extremes of slow growth or decline, as opposed to sustained steady growth. Thus, **boom and bust cycle** or **sequence.** See below.

**boom town** *Social Issues.* a town that develops, or rapidly expands, as a result of a population influx in response to the discovery of a valuable resource (e.g., oil, gold) in the area.

**boom and bust** The boom-and-bust sequence has recurred in coal, timber, oil and other industries. Ordinarily, the term "boom" refers to population growth much greater than the rate of natural increase would cause in a locality or region. In some instances, new settlements have been created during booms while, in others, existing towns and villages expanded. During the early years of the oil industry, Oil City and Bradford, Pennsylvania grew as oil expanded their economic bases beyond lumber and agribusiness. Pithole, a new town, also in northeastern Pennsylvania, boomed and disappeared in a few years. Since then, entrepreneurs have continued to create new boomtowns when the location of activity was distant from existing centers. Like the earlier towns, they boomed and busted as a result of economics and geology. Thus, high prices of oil and gas prompted development, while lower prices led to diminished activity and employment. Geology has always determined the intensity and extent of a boom because the size and accessibility of oil and gas reservoirs determines the extent of development in the area, much as the size and extent of gold and silver veins determined the duration of a mining boom. Developments in related sciences and technologies have also been important because they have facilitated the location and exploitation of fields. Geophysics, for example, lowered risks in exploration, prompting exploration where it might not have occurred otherwise. Reservoir engineering has extended the lives of oil fields, sustaining employment and population in producing localities and regions.

**Roger M. Olien**
University of Texas of the Permian Basin

**bootstrapping** *Consumption & Efficiency.* the use of materials or energy generated by the activity of a system to continue the operation of the system. [From the traditional expression of a person "pulling himself up by his bootstraps".]

**boreal** *Environment.* relating to or located in the geographical zone between the temperate and the subarctic zones. Thus, **boreal forest.**

**borehole** *Mining.* a hole made in the ground by a drill, auger, or other device to explore strata in search of minerals, or for other purposed such as water supply, blasting, or relief of underground pressures.

**borehole storage** *Storage.* a type of thermal energy storage in which vertical heat exchangers are inserted into the ground and bring about the transfer of thermal energy toward and from the ground.

**Borelli, Giovanni** 1608–1679, Italian physiologist and physicist who was the first to explain muscular movement and other body functions according to the laws of statics and dynamics.

**boring** *Mining.* the extraction of minerals in the form of gas or liquid from the earth, through the use of suction pumps and boreholes.

**boron** *Chemistry.* a highly reactive nonmetallic chemical element, having the symbol B, the atomic number 5, an atomic weight of 10.811, a specific gravity of 2.34 (amorphous) or 2.46 (crystalline), a valence of 3, and a melting point of 2300°C. Used in the manufacture of enamels, as a shield for nuclear reactions, in advanced aerospace structures, and as the dopant in photovoltaic cell material.

**Bosch, Carl** 1874–1940, German chemist and engineer known as the co-discoverer of the HABER–BOSCH PROCESS, the basis of modern fertilizer production.

**Bosch, Robert** 1861–1942, German automotive engineer who developed what is considered to be the first successful fuel injection system (about 1912) and also was a pioneer in the development of the automotive spark plug and magneto.

**Bose, Satyendra Nath** 1894–1974, Indian physicist noted for his collaboration with Albert Einstein in developing a theory regarding the behavior of fundamental particles (see next).

**Bose–Einstein statistics** *Physics.* the statistical mechanics law obeyed by a system of particles whose wavefunction is unchanged when two identical particles are interchanged. It is assumed that all identical particles be regarded as absolutely indistinguishable, and that any number of identical particles can have the same set of quantum numbers in a given system. Particles obeying Bose–Einstein statistics are known as *bosons.*

**boson** see BOSE–EINSTEIN STATISTICS.

**bottled gas** another term for LIQUEFIED PETROLEUM GAS.

**bottleneck** *Consumption & Efficiency.* **1.** any point in a system that acts as an obstruction to the flow of materials, such as an area of reduced diameter in a drill pipe. **2.** *Electricity.* a transmission line through which all electricity must pass to get to its intended buyers; also known as a **bottleneck facility. 3.** *Transportation.* a section of a road that experiences traffic congestion, due to a lower traffic capacity than preceding and following sections; may be permanent, due to road design or traffic patterns, or temporary, due to road work or an accident.

**bottom ash** *Coal.* a residue of noncombustible ash that is left after coal or other solid fuel has been burned.

**bottomhole** or **bottom hole** *Oil & Gas.* the section of an oil well with the greatest depth. Thus, **bottom-hole pump, bottom-hole separator.**

**bottom-hole pressure** *Oil & Gas.* **1.** the recorded gas-drive pressure at the base of an oil or gas well, expressed in pounds per square inch, and used to determine oil-reservoir performance and downhole equipment productivity. **2.** the well pressure determined at a point opposite the producing formation.

**bottoming cycle** *Conversion.* a cogeneration system in which steam is used first for process heat and then for electric power production. Compare TOPPING CYCLE.

**bottom-up** *Economics.* describing a modeling or analytical approach that arrives at economic conclusions from an analysis of the effect that changes in specific parameters have on narrowly defined parts of the total system. Compare TOP-DOWN.

**Boulder Dam** an alternate name for the HOOVER DAM.

**Boulding, Kenneth** 1910–1993, U.S. economist noted for his emphasis on the social, moral, and ecological implications of economic growth. He popularized the terms SPACESHIP EARTH and COWBOY ECONOMY (see) to emphasize the limits imposed on the economy by energy, material, and environmental constraints.

**boundary** *Physics.* the set of points at which the inputs to and outputs from a physical system take place; the area separating a system from its surroundings.

**boundary layer** *Physics.* a layer in a fluid medium (e.g., air) whose momentum, mass, or heat characteristics are affected or dictated by the fluxes of respective quantities at the bounding surface. Thus, **boundary-layer flow.**

**Bourdon gauge** *Measurement.* an instrument used to measure the pressure of gases or liquids, consisting of a curved hollow metal tube closed at one end that tends to expand and straighten under pressure, moving a needle on an attached indicator. [Named for French inventor Eugene *Bourdon,* 1808–1884.] Also, **Bourdon tube.**

**Bowen ratio** *Earth Science.* the ratio of heat loss through air circulation to heat loss from evaporation; this ratio is high when the evaporation rate is low (in conditions of low water supply); it varies from about 0.10 for oceans to about 2.00 for dry regions such as desert interiors; the global average is 0.29, meaning that latent heat is more significant than sensible heat on a global scale. [Formulated by U.S. astrophysicist Ira *Bowen,* 1898–1973.]

**Boyden turbine** *Hydropower.* a type of water turbine first built in 1844, employing an outlet diffuser to recover part of the kinetic energy exiting the device, thus increasing efficiency. This was an improvement on the Fourneyron design and the first turbine to be manufactured in quantity in the U.S., but later it was generally superseded by the Francis turbine. [Developed by U.S. mechanical engineer Uriah *Boyden,* 1804–1879.]

**Boyle, Robert** 1627–1691, Anglo–Irish scientist often described as the founder of modern chemistry because of the many milestones associated with his work; e.g., establishing the concept of a chemical element, performing controlled experiments, publishing his findings with elaborate detail, and assembling what today would be called a research group.

**Boyle's law** *Chemistry.* the statement that the product of pressure and volume is exactly a constant for an ideal gas.

**BP** British Petroleum.

**BPA** Bonneville Power Administration.

**Bradley, James** 1693–1762, English astronomer who discovered the ABERRATION of light (see) in 1729. He reasoned that this is due to the finite speed of light relative to the velocity of an observer on earth.

**bradymetabolic** *Biological Energetics.* describing an animal whose metabolism slows to a low level of activity when resting. Compare TACHYMETABOLIC.

**brake** *Consumption & Efficiency.* any device designed to slow or stop the motion of a vehicle or machine by the use of friction in a controlled manner.

**brake horsepower** *Transportation.* an engine's horsepower measured without the loss in power caused by the gearbox, generator, differential, water pump, and other auxiliaries; determined by a brake attached to the drive shaft and recorded on a dynamometer.

**brattice** *Mining.* **1.** a board or other partition used to divert air into a particular working place or section of a mine. **2.** an airtight partition in a mine shaft, designed to separate intake from return air.

**Braun, Karl Ferdinand** 1850–1918, German physicist who shared the Nobel Prize with Guglielmo Marconi for the development of wireless telegraphy.

**Braun, Wernher Von** 1912–1977, German engineer who was an important rocket developer from the 1930s to the 1970s. During World War II he led the research team at Peenemünde, Germany that developed the V1 and V2 rockets for use as weapons. At the war's end Braun and his team surrendered to the U.S. Army, and they were taken to the

*Wernher Von Braun*

U.S. to establish a guided missile program. The launch vehicle used for U.S. lunar missions was the Saturn rocket developed by this group.

**Braun tube** *Communication.* a cathode-ray oscilloscope developed by Karl Ferdinand Braun in the late 1890s; regarded as the precursor of the modern cathode ray tube used in televisions because it was the first to contain all the elements of the modern device.

**Braunkohlenbriketts** see BKB.

**Brayton cycle** *Thermodynamics.* an ideal gas cycle used as a standard for the actual performance of a simple gas turbine, consisting of four processes: a reversible adiabatic (no heat transfer) compression at constant entropy; a heat transfer at constant pressure up to the maximum temperature; an adiabatic expansion at constant entropy back to the original pressure; and a heat transfer at constant pressure back to the original volume and entropy. [Proposed by U.S. engineer George *Brayton,* 1830–1892.]

**breadbox system** *Solar.* another name for a BATCH HEATING system, from the appearance of the insulated box used in this.

**breaker** see CIRCUIT BREAKER.

**breaker boy** *History.* a young coal mine worker employed in former times to separate coal from unwanted stone and slate; this was a grueling task requiring a boy to spend the day sitting suspended above a conveyer belt as the coal rushed by.

**breakeven** *Nuclear.* a quantitative measure that fusion scientists use to measure progress towards achieving fusion at a practical level; the plasma density multiplied by the energy confinement time, given a certain temperature. When this number is large enough, the fusion reactions release the same amount of energy that was used to start the reactions.

**breast** *Hydropower.* a term for the point on a water wheel midway between the top and the bottom.

**breast wheel** *Hydropower.* a type of water wheel into which water enters at or near the mid-point (breast) of the wheel and is kept in buckets until being discharged at or near the lowest point on the wheel; the weight of the water in the buckets turns the wheel to provide power. A more efficient design than an undershot wheel and operable in sites not suited for an overshot wheel. Also, **breast shot wheel.**

**breed** *Nuclear.* **1.** to produce fissile material from fertile material, usually in excess of the amount consumed. **2.** specifically, to produce plutonium-249 and uranium-233 from, uranium-238 and thorium-232, respectively. Thus, **breeding, breeder material.**

**breeder reactor** *Nuclear.* a type of fission reactor that uses neutrons produced in the fission process to make (breed) new fuel, usually at a faster rate than it consumes fuel.

**breeze** *Materials.* a byproduct of coke manufacture or the burning of charcoal; can be used as a residual fuel.

**bremsstrahlung** *Physics.* a stream of electromagnetic radiation produced by the rapid deceleration of a high-speed charged particle (usually an electron or beta particle) in the electrical field of another high-speed charged particle (usually a nucleus). [A German term meaning "braking radiation".]

**Brent crude** *Oil & Gas.* a combination of crude oil from 15 different oil fields in the Brent and Ninian systems located in the North Sea. Brent is a marker crude oil and is traded on the NYMEX or on the London International Petroleum Exchange (IPE). Also, **Brent blend.**

**bright coal** *Coal.* a type of jet-black, pitch-like, banded coal that contains more than 5% vitreous constituents and less than 20% opaque matter.

**bright spot** *Oil & Gas.* a term for a seismic amplitude anomaly or high amplitude that may indicate the presence of hydrocarbons; so called because it appears as a white dot on a seismographic recording strip.

**brimstone** *History.* a former name for SULFUR.

**brine** *Earth Science.* a general term for sea water, or any water with a high content of dissolved salts.

**briquetting** *Coal.* the fact of making coal more suitable for burning by processing it into small brick-like pieces of regularized shape and size **(briquettes)**; this provides greater thermal stability and cleaner combustion.

**British Petroleum (BP)** (est. 1954), the world's second largest integrated oil company, behind Exxon Mobil. In 1901 Englishman William Knox D'Arcy obtained a concession from the Shah of Persia (Iran) to explore for and exploit the oil resources of the country. Discoveries there led to the creation of the Anglo–Persian Oil Company (1909). The name was changed to BP with nationalization of the international oil concessions in Iran in 1954.

**British thermal unit** *Measurement.* a widely used unit of measure, generically defined as the average amount of energy required to produce a change in temperature of 1°F in one pound of pure liquid water; often specified as occurring at standard atmospheric pressure and a specified temperature increase (e.g., 39°F–40°F, or 59°F–60°F); equivalent to about 1055 joules or 252 calories. Variously abbreviated as Btu, BTU, or btu.

**broadband irradiance** *Solar.* the solar radiation arriving at the earth that is limited to the electromagnetic spectrum in the range of 300 nm–3000 nm wavelength.

**bronze** *Materials.* any of various brown metallic alloys composed mostly of copper, with up to 11% tin and sometimes small amounts of zinc or other metals.

**Bronze Age** *History.* the period in human history between the Stone and Iron Ages, characterized by the manufacture and use of bronze tools and weapons. This smelting required large volumes of wood.

**brow** *Mining.* a low place in the roof of a mine, offering insufficient headroom.

**brown coal** *Coal.* another name for LIGNITE, a low-rank coal that is intermediate between peat and bituminous coals; so called because of its brown or brownish-black color.

**brownfield** *Environment.* an area of previously developed property whose expansion, redevelopment, or reuse may be complicated by the presence of a hazardous substance, pollutant, or contaminant; e.g., an abandoned chemical plant, former gasoline station, unused oil storage facility, or historic coal mine site.

**Brownian motion** *Physics.* the random movement of microscopic particles suspended in a fluid, caused by the interaction of these particles with the molecules of the fluid; described by Einstein and others as a confirmation of the existence of atoms. [Identified by Scottish botanist Thomas *Brown,* 1773–1858.]

**brown kelp** see KELP.

**brownout** *Electricity.* **1.** a deliberate reduction of line voltage by a power system in order to lessen load demands, often with the intent of preventing a blackout. **2.** any unintended reduction or curtailment of electric power short of a full blackout, such as may occur during a storm.

**BRT** bus rapid transit.

**Brundtland Commission** a name for the World Commission on Environment and Development chaired by Norwegian Prime Minister Gro Harlem Brundtland. The Commission's report *Our Common Future* (1987) popularized the concept of sustainable development.

**Brunel, I. K.** 1809–1859, English engineer noted for his planning and construction of bridges, tunnels, railways and steamships; e.g., the expansion of the Great Western Railway, and a series of ocean-going ships that were firsts of their kind, such as the GREAT WESTERN.

**brush** *Electricity.* a conductor that maintains contact between stationary and moving parts of a generator or motor.

**Brush, Charles F.** 1849–1929, a founder of the U.S. electrical industry, known for milestones in municipal and residential lighting. His Brush Electric Company merged with Thomas Edison's company under the name General Electric Company, now one of the world's largest energy corporations. He developed the BRUSH ARC LAMP and BRUSH DYNAMO.

**Brush arc lamp** *Lighting.* a historic innovation in lighting; the first practical illuminating arc light, first used for street illumination in 1879 in Cleveland, Ohio and soon thereafter in general use.

**brush discharge** *Electricity.* a luminous discharge of electricity into the air from a conductor when its potential exceeds a predetermined value but remains too low for spark formation.

**Brush dynamo** *History.* a highly efficient direct-current electric dynamo, the first electrical generator capable of delivering power for large-scale use.

**BTEX** *Environment.* a collective term for benzene, toluene, ethylbenzene, and xylene, a group of volatile organic compounds (VOCs) found in petroleum hydrocarbons, such as gasoline, and also in other common environmental contaminants.

**Btu** British thermal unit. Also often written as **BTU** or **btu.**

**Btu per cubic foot** *Measurement.* a common measure of the heat content of natural gas, equal to the heat that would be produced by the combustion of a gas occupying a volume of one cubic foot under standard conditions of temperature, pressure, gravitation, and so on.

**Btu rating** *HVAC.* a measure of the cooling capacity of an air conditioner measured by the amount of heat it can remove per hour; the higher the number, the greater the cooling capacity.

**Btu tax** *Economics.* an energy tax levied at a rate based on the Btu energy content of the given fuel.

**BTX extraction** *Oil & Gas.* the process of removing benzene, toluene and xylene from reformate or pyrolysis gasoline.

**bubble** *Economics.* **1.** a short-term period of great economic success or expansion followed by a decline or collapse; e.g., the U.S. stock market bubble of the late 1990s fueled by "dot.com" offerings. [So called because such an economic boom is bound to come to an end because of fundamental problems, like a bubble bursting.] **2.** *Environment.* an option in the Kyoto Protocol that allows a group of countries to meet their targets jointly by aggregating their total emissions. The member states of the European Union currently utilize this option.

**bubble tower** *Oil & Gas.* a distillation tower in which the rising vapors pass through layers of condensate, bubbling under caps on a series of plates.

**bubbling fluidized bed** see FLUIDIZED BED COMBUSTION.

**bucket** *Hydropower.* one of a set of blades, scoops, or containers employed in a water wheel or water turbine to capture the force of flowing water so that it can be employed to generate power.

**bucket thermometer** *Measurement.* a water-temperature thermometer whose bulb is surrounded by an insulated container, used to measure ocean temperatures.

**bucket-wheel excavator** *Mining.* a power-operated shovel consisting of a series of buckets attached to a wheel, which allows for continuous digging; the buckets scoop up material, then empty it onto a conveyor.

**buckwheat coal** see COAL SIZING.

**bucky ball** or **buckyball** *Materials.* a large molecule of 60 carbon atoms resembling a soccer ball; an extremely durable material that is the third known form of pure carbon (in addition to graphite and diamonds). [So called in tribute to Buckminster "Bucky" FULLER.]

**budget** **1.** see CARBON BUDGET. **2.** see ENERGY BUDGET.

**buffalo** *Biological Energetics.* a general name for several species of oxen in the family Bovidae, including the water buffalo *(Bubalus bubalis, B. carabenesis)* which has been used as a draft animal since ancient times.

**buffer factor** *Climate Change.* the ratio of the instantaneous fractional change in the partial pressure of $CO_2$ exerted by seawater to the fractional change in total $CO_2$ dissolved in the ocean waters. Used to study the relative distribution of $CO_2$ between the atmosphere and the ocean, and to measure the amount of $CO_2$ that can be dissolved in the mixed surface layer. Also called Revelle factor because of its description by the noted U.S. oceanographer Roger REVELLE.

**Buffon, Comte de** 1707–1788, French scientist and author whose views influenced future naturalists, such as Jean-Baptiste, Lamarck and Charles Darwin. In his monumental *Historie Naturelle,* a 44-volume encyclopedia, Buffon described everything then known about the natural world.

**bug dust** *Coal.* a term for fine particles of coal or other material left over from a boring or cutting operation; e.g., cutting the coal face by drill or machine.

**building-associated symptoms** another term for SICK BUILDING SYNDROME.

**building code** see CODE.

**building envelope** *HVAC.* a collective term for all the components of a building that enclose its conditioned space and separate conditioned spaces from unconditioned spaces (e.g., an unheated garage) or from outside air.

**building-integrated** *Photovoltaic.* describing the design and integration of photovoltaics into a building envelope, typically in place of conventional building materials.

**building-related illness (BRI)** *Health & Safety.* a specific disease that has identifiable symptoms and whose cause can be directly attributed to toxic agents within a given building; may be an infectious disease such as Legionnaires' disease, an allergic condition such as asthma, or a toxic condition such as carbon monoxide or asbestos poisoning. See also SICK BUILDING SYNDROME.

**bulb turbine** *Conversion.* a type of water turbine in which the entire generator is mounted inside the water passageway as an integral unit with the turbine.

**bulk boat** *History.* a wooden barge used for bulk transportation of oil in the early period of the U.S. oil industry.

**bulk power** *Electricity.* the actual power and the infrastructure (generating plants, transmission lines, interconnections with neighboring systems, and associated equipment) producing power that is made available for sale on the wholesale power market or directly to retail customers. Similarly, **bulk system, bulk market,** and so on.

**bulk terminal** *Oil & Gas.* a facility that receives petroleum products by tanker, barge, or pipeline, and stores them for eventual shipment to refineries or other marketing outlets. Also, **bulk station.**

**bullet train** *Transportation.* **1.** a name for Japan's Shinkansen train, the world's first high-speed railway system (1964). **2.** any such high-speed train.

**bundling** *Economics.* **1.** a process or policy in which multiple products or services (e.g., gas, electricity) are combined into a single package that is sold at a set package price. **2.** a process in which all electricity generation, transmission, and distribution services provided by one entity for a single charge. Thus, **bundled rate, charge, service,** and so on.

**bunker** *Storage.* a tank or other container used to store fuel for later use in a furnace or engine, especially such a storage unit on a ship. Thus, **bunkering.** [Originally a term for a receptacle used to hold coal on a steam-fired ship.]

**bunker coal** *Coal.* coal consumed by ocean steamers, tugboats, ferries, or other steam watercraft; so called because stored in bunkers before use.

**bunker fuel** *Oil & Gas.* a term for marine fuel, especially for international shipping.

**Bunsen burner** *Consumption & Efficiency.* a gas burner having an adjustable air inlet that allows the heat of the flame to be modified; widely used in laboratories. [Developed by German chemist Robert Wilhelm *Bunsen,* 1811–1899.]

**burden** *Environment.* a term for the total mass of a gaseous substance of concern in the atmosphere.

**Bureau of Land Management** an agency within the U.S. Dept. of the Interior that administers more than 260 million surface acres of federal and Indian lands, located primarily in 12 Western states.

**burnable poison** *Nuclear.* a nuclide of large neutron absorption cross section, such as boron, that is incorporated into a nuclear reactor's fuel, to compensate for loss of reactivity as fuel is consumed and fission-product poisons accumulate.

**burner** *Consumption & Efficiency.* any device used for the final conveyance of a combustible gas, or a mixture of gas and air, to the combustion zone.

**burner reactor** *Nuclear.* a nuclear reactor with little or no fertile material; thus there is no conversion of fertile material into fissile material.

**burnertip** *Consumption & Efficiency.* a term for the ultimate point at which natural gas is used by a consumer; e.g., a furnace, cooking device, or engine. Thus, **burnertip price.**

**burning mirror** *History.* a legendary ancient weapon using solar energy, described as a giant parabolic mirror of polished metal combined with a large series of smaller mirrors to generate a powerful beam of heat; supposedly invented by Archimedes and used to defend the harbor of Syracuse against a Roman siege by setting fire to the Roman ships from shore.

**burning speed** *Chemistry.* the speed at which an area of burning gas travels through a combustible gas mixture toward the unburnt gas. Also, **flame velocity.** Distinct from *flame speed,* which is the sum of the burning speed and displacement velocity of the unburned gas mixture.

**burning spring** *History.* a historic term for a natural gas vent in the earth, which could be ignited by a flame or by lightning.

**burnout** *Nuclear.* in a water-cooled reactor, a term for a rupture in fuel cladding, with a release of fission products into the coolant, caused by localized heat buildup.

**burnup** *Nuclear.* a measure of the consumption of nuclear fuel in a reactor, usually expressed as the ratio of the fissile material consumed to that originally present.

**Burqan (Al Burqan)** *Oil & Gas.* an important oil field in southern Kuwait, the site of one of the first major discoveries in this region (1938); hundreds of wells here were destroyed or set afire by Iraqi forces during the Gulf War in 1991.

**burst** see SOLAR BURST.

**Burton, William** 1865–1954, U.S. chemist and oil industry executive who developed a thermal cracking process that doubled the yield of gasoline from crude petroleum. This evolved into the first commercially successful process for cracking crude oil into gasoline and other products.

**bus** *Transportation.* **1.** short for *omnibus;* a large, elongated motor vehicle that is fitted with seats and used as a public conveyance. **2.** *Electricity.* a noninsulated conductor used to carry a large current or to make a common connection between several circuits. **3.** *Communications.* a high-speed structure in a computer system that is shared by the processor, memory, and peripherals, transferring data between them.

**busbar** *Electricity.* the power conduit of an electric power plant; the starting point of the electric transmission system.

**busbar cost** *Electricity.* the cost of producing 1 kilowatt per hour of electricity and delivering it to, but not through, the transmission system.

**Bush, Vannevar** 1890–1974, U.S. engineer whose work in the 1940s laid the groundwork for the Internet. He described a theoretical machine called a **memex,** which would enhance human memory by allowing the user to store and retrieve documents connected by associations (a form of linking similar to what is known today as hypertext).

**business as usual (BAU)** *Economics.* **1.** a prediction or projection of future conditions on the assumption that there will be no significant changes in current, normal operating conditions. **2.** specifically, an estimate of a company's emissions under normal operating circumstances. Depending on the scope of the BAU scenario, this may incorporate some emission reduction regulatory controls, including carbon taxes.

**bus rapid transit (BRT)** *Transportation.* a program that seeks to improve bus service by reducing travel time and providing enhanced rider convenience.

**butadiene** *Chemistry.* a colorless gas that is a commercially important compound used in making nylon, latex paints, and synthetic rubbers **(buna rubber).**

**butane** *Oil & Gas.* a colorless, highly flammable gas; it occurs in natural gas and is produced by cracking petroleum; widely used as a fuel for lighters and other household products, as a component of LPG (liquefied petroleum gas), as a gasoline additive, and for other industrial purposes.

**butene** another name for BUTYLENE.

**butterfly effect** *Earth Science.* an expression developed by U.S. meteorologist Edward Lorenz to describe the concept of uncertainty and chaos in weather forecasting. He said that something as small as a butterfly flapping its wings in China could change the weather in the U.S. a few days later, because the butterfly would move a little bit of air that moved more air, and so on until the moving air reached the other side of the world. This effect, known technically as the "sensitive dependence on initial conditions", is the essence of chaos.

**button cell** *Storage.* a miniature battery having a circular cross-section in which the overall height is less than the diameter.

**buttress dam** *Hydropower.* a dam consisting of a watertight upstream face supported at intervals on the downstream side by a series of flat or curved supports (buttresses) that resist the forces of the reservoir water. Typically employed in a valley to hold back a wide river or lake.

**butylene** *Chemistry.* the alkene hydrocarbon group $C_4H_8$, including three known isomeric forms; all are flammable and easily liquefied gases. Used mainly in making synthetic rubbers.

**buyback tariff** *Economics.* a price paid to the owner of a distributed energy (DE) resource for electricity that is sold back to the electricity grid.

**buythrough** *Economics.* an agreement between a utility and customer to import power when the customer's service would otherwise be interrupted.

**BWR** boiling water reactor.

**bypass diode** *Photovoltaic.* a diode connected across one or more solar cells in a photovoltaic module in such a way that the diode will conduct if the cells become reverse biased.

**bypass system** *Hydropower.* a structure in a dam that provides a route for fish (e.g., migrating salmon) to move safely through or around the dam without going into the turbine units. A **collection and bypass system** collects and holds the fish approaching the dam for later transportation.

**byproduct** or **by-product** *Materials.* a secondary or additional product resulting from an industrial process, especially a useful material that can be employed in a subsequent process. Thus, **byproduct fuel.**

**byte** *Communication.* the smallest group of bits (discrete items of data) representing a transmission character. usually eight bits.

**C** carbon; Celsius (centigrade); coulomb.

**C3 plant** *Biological Energetics.* a plant that produces a three-carbon compound during photosynthesis; a classification that encompasses the greater percentage of plant species on earth, including trees and most important food crops, such as rice, wheat, barley, soybeans, potatoes, beans, and other vegetables.

**C4 plant** *Biological Energetics.* a plant that produces a four-carbon compound during photosynthesis; a classification that includes a small percentage of plant species, mainly tropical and subtropical grasses such as the agricultural crops corn (maize), sugarcane, millet, and sorghum. Because of the different photosynthetic processes of C3 and C4 plants, global warming and increased carbon dioxide levels can affect their relative abundance and distribution on earth, though precisely in what manner is not certain.

**cable car** *Transportation.* **1.** an engine-driven continuous track used to carry coupled bulk-transport cars; historically one of the first steps to replace horses with mechanical power in urban transit systems, but now limited to a few locations (e.g., San Francisco, Wellington, NZ). **2.** a similar conveyance suspended from an overhead cable, as for transportation across a canyon or up a mountainside. Thus, **cable railway** or **tramway.**

**cable tool** *Mining.* a bottom-hole tool in which the drilling bit is connected by cable with the machine on the surface, allowing a percussive action to drill the boreholes. Thus, **cable-tool drilling.**

**cable tool drilling** *Oil & Gas.* an older type of oil well drilling in which the hole is drilled by dropping a sharply pointed bit on the bottom of the hole. The bit is attached to a cable and the cable is picked up and dropped repeatedly to deepen or drill the hole.

**cable yarding** *Biomass.* the process of transporting cut logs to a landing or yarding area by means of an overhead cable and winch system.

**CADDET** Centre for the Analysis and Dissemination of Demonstrated Energy Technologies (est. 1988); an international information network that informs decision-makers about renewable energy and energy-saving technologies that have been successful in other countries. Part of the International Energy Agency.

**Cadiz** see AMOCO–CADIZ.

**cadmium** *Chemistry.* a rare element having the symbol Cd, the atomic number 48, an atomic weight of 112.4, a melting point of 320.9°C, and a boiling point of 767°C. It is a white, ductile metal obtained from zinc ores, and is used as an anticorrosive and in making alloys. Compounds of cadmium are widely used as components of nickel cadmium batteries and as solar energy materials; e.g., cadmium sulfide (CdS), cadmium telluride (CdTe).

**cadmium hydroxide** *Storage.* an active material used at the negative electrode of a nickel-cadmium battery.

**CAES** compressed air energy storage.

**CAFE** *Transportation.* Corporate Average Fuel Economy; a U.S. standard officially defined as the sales-weighted average in miles per gallon achieved by a manufacturer's fleet of passenger cars or light trucks having a gross vehicle weight rating of 8500 lbs. or less, manufactured for sale in a given model year. Enacted in 1975 as a part of the Energy Policy Conservation Act.

**cage** *Mining.* an enclosed or semi-enclosed structure in a vertical mine shaft, similar to an elevator car, that is used for transporting personnel and materials.

**caking** *Coal.* the fact of coal becoming fused together in a coherent, solidified mass when heated. Thus, **caking character, caking test.**

**caking coal** *Coal.* a type of coal that softens and then cakes upon heating, producing a hard, gray cellular mass of coke.

**calcination** *Materials.* the heating of a solid to a high temperature, below its melting point, to create a condition of thermal decomposition or a phase transition other than melting or fusing. Also, **calcining.**

**calculating engine** *Communication.* a name for either of two machines designed in the 1820s by English scientist Charles Babbage, the table-making **difference engine** or the more ambitious **analytical engine,** a flexible, punch-card controlled general purpose calculator, embodying many features found in the modern computer. The machines were never actually built due to a lack of funds.

**caldera** *Earth Science.* a large, more or less circular volcanic crater, usually resulting from the collapse of underground lava reservoirs, having a diameter that is many times greater than that of the vent.

**Calder Hall** *Nuclear.* the world's first commercial nuclear power station (opened in 1956), a complex of four reactors in West Cumbria on the coast of northwest England. British Nuclear Fuel closed this facility in 2003.

**Callendar, G. S.** 1897–1964, British engineer who was the first to empirically connect rising carbon dioxide concentrations in the atmosphere and the increase in the earth's temperature.

**caloric** *Thermodynamics.* **1.** having to do with heat transfer, or the measurement of this. **2.** a theoretical substance once thought to be the source of the phenomenon of heat. See CALORIC THEORY. **3.** *Biological Energetics.* having to do with calories or the calorie content of a given food.

**caloric theory (of heat)** *History.* a historic concept of heat as a substance that flows into a body as it is heated and flows out as it is cooled; from about 1800 onward this was refuted by Count Rumford, Joule, Mayer, Clausius, and others, and then replaced with the concept of heat as the kinetic energy of molecules.

**caloric value** *Biological Energetics.* the sum of the calories provided by the energy-containing nutrients in a given food: i.e., protein, carbohydrate, fat, and alcohol. Because carbohydrates contain some fiber that is not digested and utilized by the body, the fiber component is usually subtracted before calculating the caloric value.

**calorie** *Measurement.* a unit of energy, defined as the amount of heat transfer required to raise the temperature of one gram of pure water by one degree Celsius (from 14.5°C–15.5°C) at standard atmospheric pressure (sea level); equivalent to 4.184 J. Used as a description of the energy content of a given food. Also called a **small calorie** in contrast with a KILOCALORIE or large calorie (1000 small calories).

**calorific value** see CALORIC VALUE.

**calorimeter** *Measurement.* an instrument that is used to measure the amount of heat generated during a physical process such as burning, change of state, or friction. See BOMB CALORIMETER.

**calorimetry** *Measurement.* the measurement of heat transfer.

**Calvin cycle** *Biological Energetics.* the complete route that carbon travels through a plant during photosynthesis. [Named for U.S. biochemist Melvin *Calvin,* 1911–1977.] Also called the **Calvin-Benson cycle.** See next page.

**CAM (plant)** *Biological Energetics.* crassulacean acid metabolism; a descriptive term for plants that close their stomata during the day to reduce water loss and open them at night for carbon dioxide uptake.

**camel** *Biological Energetics.* an animal used historically for transportation, especially in desert regions of Africa, the Middle East, and central Asia; either the one-hump Bactrian camel *(Camelus dromedarius)* or the two-hump dromedary *(Camelus dromedarius).*

**CAN** Climate Action Network (est. 1979); a global network of over 280 nongovernmental organizations (NGOs) working to promote government and individual action to limit human-induced climate change to ecologically sustainable levels.

**canal** *Transportation.* **1.** an artificial waterway that is dug to connect two adjacent bodies of water to allow for the passage of shipping between them. **2.** a similar waterway dug to conduct water across an extent of land for irrigation or drainage.

**canary** *Coal.* **1.** a small songbird of the finch family, having bright yellow plumage and often kept as a pet. For centuries caged canaries were brought into coal mines to signal

**Calvin cycle** The metabolic pathway by which carbon dioxide ($CO_2$) is incorporated into carbohydrate. Nobel Laureate Melvin Calvin had a major role in elucidating this cyclic series of enzyme-catalyzed reactions. The enzyme RuBisCO (ribulose bisphosphate carboxylase/oxygenase) catalyzes the initial reaction of $CO_2$ with a five-carbon compound ribulose-1, 5-bisphosphate (RuBP). The resulting 6-carbon intermediate splits into two copies of a 3-carbon compound that is converted in subsequent steps to the carbohydrate glyceraldehyde-3-phosphate. Some of this product exits the pathway to be used for synthesis of more complex carbohydrates or other carbon compounds. The rest is converted back to RuBP (the substrate for the initial $CO_2$ fixation reaction), completing the cycle. Most carbon compounds in the biosphere are derived from the carbohydrate product of the Calvin Cycle. The abbreviated structure of a typical carbohydrate is $(H–C–OH)_n$. Due to unequal sharing of electrons in a C–O bond, the carbon atom in $CO_2$ is electron deficient relative to a carbon atom in a carbohydrate, that bonds with only one oxygen atom. Carbon in $CO_2$ is thus said to be more oxidized, while carbon in a carbohydrate is more reduced. The Calvin Cycle does not directly utilize light energy, but is part of the process of photosynthesis. Some Calvin cycle reactions require ATP (adenosine triphosphate), a compound that functions in energy transfer, and NADPH (reduced nicotinamide adenine dinucleotide phosphate), a source of hydrogen atoms for reduction reactions. ATP and NADPH are formed during light-energized reactions of photosynthesis.

**Joyce Diwan**
Rensselaer Polytechnic Institute

the presence of dangerous gases, the idea being that a canary would succumb quickly to the gas and thus warn the miners to get out. **2. canary in a coal mine.** based on this practice, a metaphor for an early warning of oncoming danger or disaster.

**cancer** *Health & Safety.* a collective term for a wide variety of diseases that are generally characterized by improperly regulated cell growth and differentiation, resulting from specific alterations in the function of one or more genes. Various forms of cancer have been linked to environmental pollutants.

**candela** *Lighting.* the SI unit of luminous intensity, symbol cd; it is equal to the luminous intensity in a given direction of a source that emits monochromatic radiation of frequency $540 \times 10^{12}$ Hz and that has a radiant intensity in the direction of (1/683) W per steradian. [From the Greek word for *candle.*]

**candle** *Lighting.* **1.** a cylinder of wax or other solid fuel, with a central wick that burns to produce light. **2.** an older name for the CANDELA or CANDLEPOWER units of measure.

**candlepower** *Lighting.* a unit formerly used for measuring the light-radiating capacity of a lamp or other light source. One candlepower represents the radiating capacity of a light with the intensity of one **international candle,** or about 0.981 candela as now defined.

**candlepower distribution** *Lighting.* a curve that represents the variation in luminous intensity in a plane through the light center of a lamp or luminaire; each lamp or lamp/luminaire combination has a unique set of candlepower distributions that indicate how light will be spread.

**CANDU** *Nuclear.* Canadian Deuterium Uranium (Reactor); a pressurized heavy-water, natural-uranium power reactor designed in the 1960s by a partnership between Atomic Energy of Canada Limited and the Hydro-Electric Power Commission of Ontario, as well as several private industry participants.

**cannel coal** *Coal.* **1.** a variety of nonbanded, fine-grained, highly volatile coal derived mainly from spores and other plant remains, characterized by a dull to greasy luster and conchoidal fracture. **2.** a former term for any coal that burns with a steady luminous flame.

**Canola oil** *Biomass.* another name for RAPESEED OIL, especially as used in commercial food products. [Short for "Canadian oil".]

**canopy (cover)** *Ecology.* **1.** the leafy cover or top layer formed by the uppermost branches of trees in a forest. **2.** another term for LEAF AREA INDEX.

**cap** *Economics.* **1.** a supply contract between a buyer and a seller, according to which the buyer is assured that they will not have to pay more than a given maximum price. **2.** the maximum amount of a pollutant that can be emitted under a specific regulatory regime. **3.** SEE CAP-AND-TRADE. **4.** see EMISSIONS CAP.

**capacitance** *Electricity.* the ability of conductors separated by dielectric material to store energy in the form of electrically separated charges; a value described as the ratio of a quantity of electricity to a potential difference.

**capacitative** *Electricity.* having to do with or employing capacitance or a capacitor.

**capacitor** *Electricity.* an electric circuit element used to store an electric charge temporarily, consisting in general of two metallic plates separated and insulated from each other by a dielectric (e.g., air or mica).

**capacity** *Electricity.* **1.** the electrical charge effectively stored in a primary or secondary battery and available for transfer during discharge. Usually expressed in ampere-hours (Ah) or milliampere-hours (mAh). **2.** *Transportation.* the maximum number of vehicles per hour that a highway can carry under normal driving conditions; exceeding this level will tend to produce congestion and delays rather than greater traffic volume.

**capacity factor** *Conversion.* the ratio of the actual energy output of an energy converter (power plant or wind turbine) to the energy that could have been generated at continuous full-power operation over a specified period of time.

**cap and trade** *Policy.* an emission trading program in which the government creates a fixed number of allowances and requires regulated sources to surrender them to cover actual emissions. See next column.

**capillarity** *Physics.* the general behavior of fluids acting with surface tension on interfaces or boundaries, by which the fluid is either elevated or depressed.

**capillary** *Physics.* a thin hollow tube through which a liquid can rise by means of capillary action (see next).

**capillary action** *Physics.* the attraction of the surface of a liquid to the surface of a solid, which either elevates or depresses the liquid depending upon molecular surface forces; e.g., crude oil clings to the surface of each pore in a rock formation, making it difficult to recover the oil.

**cap and trade** A policy to limit overall emissions from a group of sources in a cost-effective manner by using market forces. A fixed number of allowances, equal to the desired cap on total emissions, are distributed or sold to participating firms. Each firm must retire allowances corresponding to its emissions for that time period. Firms that find abatement more difficult can purchase excess allowances from others that reduce emissions more cheaply. Such trades benefit both firms while reducing overall compliance costs. When supply meets demand for allowances, in theory, the market price reflects the marginal cost of abatement among the firms, the cap is met, and total costs are minimized. The idea of using a system of tradable property rights to manage emissions was first articulated by economist Tom Crocker and political scientist J. H. Dales in 1966 and 1968, respectively. In practice, the 1990 U.S. Clean Air Act Amendments initiated a cap and trade program for sulfur dioxide emissions to combat acid rain. Since then, emissions trading has been used in California's RECLAIM program and for nitrogen oxide emissions in the Northeastern states. Recently, the European Union began a cap-and-trade program for carbon dioxide emissions. Several countries, most notably New Zealand, use a similar program of individual transferable quotas to manage allowable catch of fish stocks. Cap and trade is most effective when abatement costs are disparate, individual emissions are easily measured, and their environmental impacts are not significantly affected by the location of the source.

**Carolyn Fischer**
Resources for the Future, USA

**capital** *Economics.* **1.** the buildings, equipment, tools, and other manufactured items used in producing goods and services. Thus, **capital asset, capital stock,** and so on. **2.** an accumulation or supply of financial assets.

**capitalism** *Economics.* an economic system in which capital assets are chiefly held by private individuals and companies rather than by public or quasi-public entities, and in which prices and the production and distribution of goods are chiefly determined by conditions in the marketplace rather by government policy.

**capping** *Oil & Gas.* **1.** the process of sealing or closing a borehole, such as a spouting gas or oil well, to prevent the escape of fluids. **2.** the device used to accomplish this closure. **3.** *Mining.* the overburden situated atop a valuable seam or bed of mineral.

**cap rock** *Mining.* **1.** consolidated barren rock material overlying a mineral or ore deposit, which must be removed before mining; e.g., a layer of hard rock overlying a coal bed. **2.** *Oil and Gas.* an impervious layer of rock overlying an oil or natural gas deposit.

**captive** *Electricity.* **1.** a term for a customer who does not have a reasonable alternative to buying power from the local utility. **2.** a battery that has an immobilized electrolyte (that is, gelled or absorbed in a material). **3.** *Mining.* a mine that produces coal for direct use by the same company that owns the mine, rather than its being marketed to others. Thus, **captive coal.**

**capture** see CARBON DIOXIDE CAPTURE AND STORAGE.

**caravel** *History.* a historic type of small sailing vessel, usually with lateen sails on two or three masts; used in the Middle Ages and later and noted especially for its use by explorers of Portugal and Spain; e.g., the exploration of the west coast of Africa and the voyages of Columbus.

***caravel*** *Portuguese caravels entering Lisbon harbor, 15th century (detail).*

**carbide** *Materials.* any of various compounds made up of carbon and another element (other than hydrogen); typically a metal such as iron, calcium, tungsten, silicon, or boron; usually produced by heating the reacting substances in an electric furnace.

**carbide lamp** *Mining.* a miner's lamp burning acetylene that forms as a result of charged calcium carbide and water; widely used historically but now largely superseded by electric lamps.

**carbide nuclear fuel** *Nuclear.* a ceramic formulation of uranium in the form of uranium or plutonium monocarbide or dicarbide that has desirable thermal conductivity characteristics; used in advanced reactors that operate at very high temperatures.

**carbohydrate** *Biological Energetics.* a collective term for chemicals composed of carbon, hydrogen, and oxygen and typically having a hydrogen-to-oxygen ratio of 2:1; e.g., sugars such as glucose or xylose, starches, and cellulose or hemicelluloses.

**carbon** *Chemistry.* a very common nonmetallic element having the symbol C, the atomic number 6, an atomic weight of 12.01115, and a melting point about 360°C; the active element of photosynthesis and the key structural component of all living matter. See next page.

**carbon-12** *Chemistry.* an isotope of carbon that has a stable nucleus; it is the naturally occurring, dominant form of carbon and thus makes up most of the carbon found in nature. Carbon-12 is the basis of the atomic weights of other elements ($^{12}C$ = atomic weight 12).

**carbon-14** *Chemistry.* a naturally occurring, radioactive isotope of carbon; commonly used in demonstrating the metabolic path of carbon in photosynthesis and in dating ancient materials (see next).

**carbon-14 dating** *Measurement.* a widely used method for estimating the age of an ancient carbonaceous specimen by measuring the radioactivity of its carbon-14 content; this will indicate how long ago a once-living artifact ceased to be in equilibrium with the atmosphere. Carbon-14 is continuously produced by cosmic-ray bombardment and decays with a half-life typically described as 5568 years; dating is accomplished by

**carbon** Carbon is the sixth most abundant element in the universe and is unique due to its dominant role in the chemistry of life and in the human economy. It is nonmetallic element having the symbol C, the atomic number 6, an atomic weight of 12.01115, and a melting point about 360°C. There are four known allotropes of carbon: amorphous, graphite, diamond, and fullerene. A new fifth allotrope of carbon was recently produced, a spongy solid called a magnetic carbon "nanofoam" that is extremely lightweight and attracted to magnets. Due to carbon's unusual chemical property of being able to bond with itself and a wide variety of other elements, it forms nearly 10 million known compounds. Carbon is present as carbon dioxide in the atmosphere and dissolved in all natural waters. It is a component of rocks as carbonates of calcium (limestone), magnesium, and iron. The fossil fuels (coal, crude oil, natural gas, oils sands, and shale oils) are chiefly hydrocarbons. The isotope carbon-12 is used as the basis for atomic weights. Carbon-14, an isotope with a half-life of 5730 years, is used to date such materials as wood and archeological specimens. Organic chemistry, a major subfield of chemistry, is the study of carbon and its compounds. Because carbon dioxide is a principal greenhouse gas, the global carbon cycle has become a focus of scientific inquiry, and the management of carbon dioxide emissions from the combustion of fossil fuels is a central technological, economic, and political concern.

**Cutler Cleveland**
Boston University

comparing the carbon-14 activity per unit mass of the specimen with that of a contemporary sample.

**carbonaceous** *Chemistry.* containing or composed of carbon.

**carbonaceous coal** *Coal.* coal derived from the accumulation of undecayed plant matter.

**carbon arc** *Electricity.* an electrical discharge between two carbon electrodes; used in welding and in high-intensity lamps. Thus, **carbon (arc) lamp.**

**carbonate** *Chemistry.* **1.** a compound formed by the reaction of carbonic acid with either a metal or an organic compound. See CARBONIC ACID. **2.** relating to or containing such a compound.

**carbonate pump** *Earth Science.* a process in which various organisms in the ocean (phytoplankton and zooplankton species) form calcium carbonate ($CaCO_3$) skeletal coverings. When these organisms die, some fraction of this $CaCO_3$ is eventually remineralized back to calcium and carbonate ions in the deeper parts of the water column and in sediments. This carbon cycle or pump leads to a reduction in surface ocean levels of dissolved inorganic carbon (DIC) and alkalinity; an increase in the strength of the pump will serve to increase levels of $CO_2$ in surface water.

**carbon black** *Materials.* a black colloidal substance consisting wholly or mainly of amorphous carbon, produced commercially by thermal or oxidative decomposition of hydrocarbons. Used as a reinforcing agent in rubber products, as a black pigment in inks, paints and plastics, in the manufacture of dry-cell batteries and electrical conductors, and in high-temperature insulating material.

**carbon budget** *Climate Change.* the balance of the exchanges of carbon between carbon reservoirs (atmosphere, oceans, biota, fossil fuel deposits, soils, and society) in the carbon cycle. It is defined in terms of a "balance sheet" of gains and losses of carbon in the atmosphere, where fossil fuel combustion is a gain of carbon, while forest regrowth is a loss. The net of gains and losses produces an increase or decrease in the level of atmospheric carbon dioxide concentrations. Thus, **carbon debt** or **deficit, carbon purchase, carbon surplus,** and so on.

**carbon capture** see CARBON DIOXIDE CAPTURE AND STORAGE.

**carbon compensation depth** *Earth Science.* the level in the ocean below which the solution rate of calcium carbonate exceeds its deposition rate; i.e., the preservation of calcium carbonate shells is negligible.

**carbon conversion factor** *Measurement.* the amount of carbon per unit mass of a fuel (e.g., kg/ton).

**carbon cycle** *Environment.* **1.** the movement of carbon among its reservoirs by various chemical, physical, geological, and biological processes. The major reservoirs are the atmosphere, terrestrial biosphere (usually including freshwater systems), oceans, and sediments (including fossil fuels). **2.** *Nuclear.* the process in which a carbon-12 atom engenders a succession of thermonuclear reactions which result in the release of massive amounts of energy, believed to be the source of energy in the sun and other stars.

**carbon dioxide** *Chemistry.* $CO_2$, a colorless, odorless, noncombustible gas that is uniformly distributed over the earth's surface at a concentration of about 0.033% or 330 ppm of air. In the atmosphere it results from the combustion of organic matter, including fossil fuels such as oil, natural gas, and coal. It is also produced by various microorganisms from fermentation and cellular respiration. $CO_2$ is used in carbonated beverages, as a pressurized gas, in fire extinguishers, and as a solvent. Plants uptake $CO_2$ in photosynthesis. $CO_2$ is a greenhouse gas, and its concentration is rising as a result of the combustion of fossil fuels and land use changes; thus it has become the focus of public concern in recent years.

**carbon dioxide capture and storage** *Environment.* a method to reduce atmospheric concentrations of carbon dioxide, in response to climate change concerns, in which $CO_2$ is collected prior to release to the atmosphere and placed in long-term storage. Capture is most amenable to large stationary sources of $CO_2$, such as power plants. Potential storage reservoirs include depleted oil and gas fields, deep saline formations, unmineable coal seams, and the deep ocean. This is one of a suite of technologies known as CARBON SEQUESTRATION.

**carbon dioxide challenge** *Global Issues.* a term for the global phenomenon of rising levels of atmospheric carbon dioxide and the resulting affects of this increase on world climate.

**carbon dioxide emissions** see EMISSIONS.

**carbon dioxide equivalent** *Measurement.* an expression of the concentration of an effective greenhouse gas in relation to the amount of carbon dioxide that would be required to achieve the same effect; e.g., the global warming potential for carbon dioxide is 1 and for methane is 21; thus 1 ton of atmospheric methane is equivalent to 21 tons of carbon dioxide.

**carbon dioxide fertilization** *Climate Change.* an enhancement of plant growth by $CO_2$ enrichment; this could occur in natural or agricultural systems because of an increase in the atmospheric concentration of $CO_2$.

**carbon dioxide flooding** *Oil & Gas.* a method of enhanced oil production in which carbon dioxide is injected into a producing formation to stimulate production.

**carbon dioxide flux** see CARBON FLUX.

**carbon dioxide intensity** see CARBON INTENSITY.

**carbon dioxide mitigation** *Environment.* any of various strategies or technologies intended to limit human-related emissions of carbon dioxide.

**carbon dioxide sequestration** see CARBON SEQUESTRATION.

**carbon dioxide storage** see CARBON DIOXIDE CAPTURE AND STORAGE.

**carbon flux** *Environment.* the rate of transfer of carbon dioxide between carbon pools; positive values are flows to the atmosphere, and negative values represent uptake from the atmosphere; described in units of measurement of mass per unit area and time.

**carbonic acid** *Chemistry.* a weak acid formed by combining carbon dioxide and water; inorganic carbonates (salts of carbonic acid) are formed from it by reaction with metals or metal oxides, and organic carbonates (esters of carbonic acid) are formed by reaction with organic compounds.

**Carboniferous** *Earth Science.* **1.** a geologic division of the Upper Paleozoic era from about 345 to 280 million years ago, during which coal was formed. **2.** referring to the rocks formed during that time. **3. carboniferous.** containing or producing carbon.

**carbon intensity** *Measurement.* **1.** the amount of carbon dioxide produced or emitted per unit of energy produced or consumed. **2.** the amount of carbon dioxide produced or emitted per unit of economic output; e.g., kilograms of carbon released per dollar of gross domestic product. Thus, **carbon-intensive.**

**carbon isotope ratio** see CARBON RATIO (def. 3).

**carbonization** *Chemistry.* **1.** the conversion of a carbon-containing substance into carbon or carbon residue, either by natural processes or by heating or partial burning to remove other components, as in the destructive distillation of coal. **2.** the buildup of carbon or carbon residue as a result of natural processes; e.g., the slow decomposition of organic matter.

**carbonize** *Chemistry.* to undergo or cause to undergo a process of carbonization (see above).

**carbonized coal** *Coal.* coal decomposed into solid coke and gaseous products by heating it in an oven in a limited air supply or in the absence of air.

**carbon knock** see KNOCK.

**carbon monoxide** *Chemistry.* CO, a colorless, almost odorless, highly poisonous and flammable gas, widely produced in industrial applications when a carbon-containing fuel is burned with an insufficient amount of oxygen; it is also an exhaust product of motor vehicle engines. CO is a primary pollutant that is more severe in colder temperatures and at higher altitudes. If inhaled in sufficient quantity, it can cause asphyxiation by combining with blood hemoglobin to prevent the flow of oxygen. Thus, **carbon monoxide poisoning.**

**carbon nanofiber** *Materials.* any of various carbon-based materials that have structures made of carbon fibers with diameters of about 10–100 nm; possible future uses of these materials include low-pressure hydrogen storage.

**carbon nanotube** *Materials.* a cylinder of carbon atoms arranged in a hexagonal pattern, with a diameter measured in nanometers.

**carbon-neutral** *Climate Change.* describing an activity or process that does not generate a net increase in the level of carbon in a given system, especially the amount of carbon dioxide in the atmosphere. Thus, **carbon neutrality.**

**carbon-nitrogen-oxygen (CNO) cycle** *Nuclear.* one of two fusion reactions by which stars convert hydrogen to helium, the dominant source of energy in heavier stars. The carbon, oxygen, and nitrogen nuclei serve as catalysts in a cycle whose net result is to fuse four protons into an alpha particle plus two electrons and two neutrinos, releasing energy in the form of gamma rays. See also PROTON-PROTON CHAIN REACTION.

**carbon pile** *Electricity.* a variable electrical resistor that is composed of a number of carbon disks stacked between a fixed metal plate and a movable one. The resistance changes when the carbon is compressed.

**carbon pool** *Climate Change.* the sum of all reservoirs containing carbon as a principal element in the geochemical cycle (the atmosphere, oceans, soils, plant and animal life, and fossil fuels).

**carbon ratio** *Chemistry.* **1.** the percentage of fixed carbon in coal. **2.** the ratio of fixed carbon in coal to fixed carbon plus volatile hydrocarbons. **3.** the ratio of carbon-12 to either of the other, less common, carbon isotopes, carbon-13 or carbon-14.

**carbon sequestration** *Environment.* a suite of technologies that can sequester (store) carbon dioxide in reservoirs other than the atmosphere. See next page.

**carbon sink** *Environment.* **1.** any site or reservoir that absorbs or takes up released carbon from another part of the carbon cycle; e.g., forests and other woody vegetation. **2.** more generally, any process, activity, or mechanism that removes carbon dioxide from the atmosphere.

**carbon source** *Environment.* any process, activity, or mechanism that releases carbon dioxide to the atmosphere.

**carbon stock** *Earth Science.* the absolute quantity of carbon held within a given stock at a specified time. The units are mass (e.g., t C).

**carbon storage** see CARBON DIOXIDE CAPTURE AND STORAGE.

**carbon tax** *Policy.* a tax on fossil fuels based on the individual carbon content of each fuel. Under a carbon tax, coal is taxed the highest rate per MBtu, followed by petroleum and then natural gas.

**carbon-thread lamp** *Lighting.* an early type of electric lamp patented by Thomas Edison in the U.S. and Joseph Swan in England, employing a carbonized cellulose thread as the incandescing filament.

**carbureted water gas (CWG)** *Materials.* a manufactured gas produced by a process in

**carbon sequestration** One of the three major pathways for reducing atmospheric concentrations of $CO_2$ in response to climate change concerns. The other two pathways are (1) lowering energy intensities through improved efficiency and conservation, and (2) switching to low or no carbon fuels, such as renewables or nuclear. Carbon sequestration refers to the suite of technologies that "sequesters" or stores the $CO_2$ in reservoirs other than the atmosphere, such as trees, soils, the oceans, and underground geologic formations. One form of carbon sequestration involves removing the $CO_2$ from the atmosphere, primarily through biological means. This is also referred to as enhancing natural sinks. Examples include reforestation, afforestation, no-till agriculture, and iron fertilization of the oceans. Another form of carbon sequestration involves capturing the $CO_2$ before it is emitted into the atmosphere and storing it in a secure reservoir. This is also referred to as carbon dioxide capture and storage. Capture is most amenable to large stationary sources of $CO_2$ like power plants. Potential storage reservoirs include depleted oil and gas fields, deep saline formations, unmineable coal seams, and the deep ocean. Carbon sequestration has the advantage of being compatible with today's energy infrastructure, which is highly dependent on fossil fuels. These technologies are viewed as a bridge to a more climate friendly future energy system.

**Howard Herzog**
Massachusetts Institute of Technology

which steam is injected through an incandescent bed of coke or anthracite coal to yield water gas, which is then directed into a carburetor where it is enriched with light hydrocarbons. CWG historically provided the majority of manufactured gas used for heating and lighting in the U.S., offering a cheaper and more efficient alternative to the previously existing coal gas process.

**carburetion** *Chemistry.* **1.** the process of enriching a combustible gas by adding volatile carbon compounds. **2.** the process of mixing air and fuel in a carburetor.

**carburetor** *Transportation.* a device that is used to provide and regulate the mixture of air and fuel that is burned inside the cylinders of an internal-combustion engine.

**carcinogen** *Health & Safety.* an agent that will induce cancer; i.e., any substance (e.g., asbestos), exposure (e.g., radiation or tobacco), imbalance (e.g., hormones), deficiency (e.g., inadequate consumption of fruits and vegetables), or other factor that has been shown to increase the probability of cancer in epidemiological studies. Thus, **carcinogenic.**

**carcinogenesis** *Health & Safety.* the process of forming cancerous cells; i.e., development of the capacity for unregulated proliferation and metastasis that is characteristic of cancers.

**Carnot, Nicolas Léonard (Sadi)** 1796–1832, French physicist who developed a fundamental theory of heat engines; he laid out principles that would influence the most fundamental laws of science and engineering; i.e., the law of conservation of energy and the first and second laws of thermodynamics. See also following entries.

**Carnot cycle** *Thermodynamics.* an ideal cycle of maximum efficiency that is used as a standard for actual heat engine cycles, consisting of four successive reversible processes: a constant-temperature expansion and heat transfer to the system from a high-temperature reservoir; an expansion with no heat transfer; a constant-temperature compression and heat transfer from the system to a low-temperature reservoir; and a compression with no heat transfer that restores the system to its original state. Thus, **Carnot (cycle) engine.**

**Carnot efficiency** *Thermodynamics.* the maximum efficiency with which thermodynamic work can be produced from thermal energy flowing across a temperature difference; equal to 1 Tc/Th, where Tc is the lowest temperature and Th is the highest temperature.

**Carnot principle** *Thermodynamics.* **1.** the efficiency of an irreversible heat engine is always less than the efficiency of a reversible one operating between the same two reservoirs. **2.** the efficiencies of all reversible heat engines operating between the same two reservoirs are the same.

**Carnot theorem** *Thermodynamics.* the statement that no engine operating between two heat reservoirs can be more efficient than a Carnot cycle engine operating between the same reservoirs.

**Carrier, Willis Haviland** 1876–1950, U.S. engineer considered the founder of the modern air conditioning industry; among his many innovations were the first air conditioning system for large commercial spaces (e.g., movie theaters) and the first residential air conditioning system. In 1915 he and others formed the Carrier Engineering Corporation, which remains the largest air conditioning firm in the world.

**carrying capacity** *Sustainable Development.* the maximum number of individuals of a particular population that a given area can maintain indefinitely.

**carry market** another term for CONTANGO.

**Carson, Rachel** 1907–1964, U.S. writer and ecologist, author of *Silent Spring,* the first widely read work to describe the effect on health and the environment of human-produced toxic chemicals, such as the insecticide DDT. This book is often given credit for launching the modern environmental movement at the popular level.

**cartel** *Economics.* a group of companies within the same industry who act in concert to reduce production and increase prices, in order to maximize the wealth of the group; e.g., the oil cartel known as OPEC (Organization of Petroleum Exporting Countries).

**Carter, Jimmy** (James Earl Carter, Jr.) born 1924, president of the U.S. (1977–1981), known as the first president to directly confront the issue of U.S. energy consumption and reliance on imported oil. His administration saw the creation of the Department of Energy and the National Energy Act, but soaring oil prices contributed to his defeat by Ronald Reagan in 1980.

**Cartesian** *History.* having to do with or based on the philosophy of René DESCARTES (see).

**cascade aeration** *Biomass.* the aeration of an effluent stream through the action of falling water.

**cascade system** *Refrigeration.* a refrigeration system comprised of at least two compressors, each circulating a refrigerant within an isolated loop. A heat exchanger called a **cascade condenser** links adjacent circuits. Cascade systems produce useful cooling at quite low temperatures while rejecting heat just above ambient temperature. Also, **cascade cycle.**

**cascading** *Consumption & Efficiency.* the multiple use of an energy source by various applications, with each successive one using an increasingly lower temperature in order to maximize the efficiency of the system.

**cascading failure** *Consumption & Efficiency.* a disruption at one point in an energy network that causes a disruption at other points, leading to a major system breakdown.

**casing** *Oil & Gas.* a heavy steel pipe or tubing that is screwed or welded together, then lowered and secured into a borehole by cementing; used to stop liquids, gas, or rocks from entering the hole and to prevent the loss of circulation liquid.

**casinghead** *Oil & Gas.* the top of a casing where the control valves and flow pipes are attached; it protrudes above the surface of the well, thus permitting the pumping operation to occur along with the separation of gas and oil.

**casinghead gas** *Oil & Gas.* natural gas that bubbles out of crude oil at the surface due to a decrease in pressure at the surface compared to the reservoir.

**casinghead gasoline** *Oil & Gas.* liquid hydrocarbons that are removed from casinghead gas through refrigeration, absorption, or compression.

**cask** *Nuclear.* a heavily shielded container, as of lead, used to store or ship radioactive materials.

**Caspian Pipeline Consortium (CPC) pipeline** *Oil & Gas.* a 1700-kilometer pipeline originating in Tennis, Kazakhstan, and terminating near the port of Novorossiysk, Russia, on the Black Sea. The CPC pipeline is the first to connect the oil rich-region of central Asia with significant export terminals.

**cassava** *Biomass.* a shrubby tropical plant of the genus *Manihot,* native to South America and widely grown there and in other regions (e.g., Africa) as a food crop for its large, tuberous, starchy roots. Now also grown for its potential value as a fuel crop. Also known as manioc or yuca.

**Castillo de Bellver** *Environment.* an oil tanker that caught fire and split into two off the coast of South Africa in 1983, causing an oil spill estimated at 78.5 million gallons, one of the largest oil spills in history and the largest tanker-related spill to that date.

**cast iron** *Materials.* any of many iron-based cast alloys containing more than 2% carbon; i.e., more carbon than steels; known for hardness and heaviness; widely used to make engine blocks, machine parts, stoves, cookware, and various other products.

**Catalan forge** *History.* a precursor to the modern blast furnace, able to produce significantly higher temperatures than existing furnaces and thus allow larger amounts of ore to be smelted at one time. Thought to have been introduced in Andorra in the 700s AD and widely used thereafter in Spain; considered the first major advance in iron smelting after classical times. [So called because developed by the *Catalan*-speaking (Catalonian) people of this area.]

**catalysis** *Chemistry.* a change (typically an acceleration) in the rate of a chemical reaction, brought about by the presence of a catalyst (see below).

**catalyst** *Chemistry.* **1.** a substance that noticeably affects the rate of a chemical reaction without itself being consumed or essentially altered by the reaction. **Positive catalysts** (the great majority of catalysts) accelerate reactions; **negative catalysts** retard them. **2.** in popular use, any effect that puts in motion a significant change or series of events; e.g., the U.S. energy crisis of the 1970s was the *catalyst* for efforts to increase fuel efficiency.

**catalyst-coated membrane** *Renewable/Alternative.* a membrane in a protein exchange membrane (PEM) fuel cell system having surfaces that are coated with a catalyst layer to form the reaction zone of the electrode.

**catalyst load(ing)** *Renewable/Alternative.* the amount of catalyst incorporated in a fuel cell per unit area.

**catalytic** *Chemistry.* taking place by means of a catalyst; involving catalysis.

**catalytic converter** *Transportation.* an antipollution device in an automotive exhaust system that uses a catalyst to chemically convert pollutants in the exhaust gases, such as carbon monoxide and unburned hydrocarbons, into harmless compounds.

**catalytic cracking** *Oil & Gas.* a process of breaking down the heavy hydrocarbons of petroleum, using heat and silica or alumina as a catalyst; used to convert heavy oils into lighter and more valuable products.

**catalytic hydrocracking** *Oil & Gas.* a refining process that uses hydrogen and catalysts with relatively low temperatures and high pressures for converting middle boiling or residual material to high-octane gasoline, reformer charge stock, jet fuel, and/or high grade fuel oil.

**catalytic hydrotreating** *Oil & Gas.* a refining process for treating petroleum fractions from atmospheric or vacuum distillation units (e.g., naphthas, middle distillates, reformer feeds, residual fuel oil, and heavy gas oil) and other petroleum fractions in the presence of catalysts and substantial quantities of hydrogen.

**catalytic oxidation** *Environment.* the process of oxidizing unburned hydrocarbons and carbon monoxide by means of a catalytic reaction, in order to reduce pollution.

**catalytic reforming** *Conversion.* the process of improving the octane number of straight-run gasoline by increasing the proportion of aromatic and branched chain alkanes.

**catastrophism** *Earth Science.* an older theory to account for the present geological features of the earth as being the result of a limited number of extreme events, such as the Great Flood described in the Bible. Compare UNIFORMITARIANISM.

**catchment** *Hydropower.* the act or fact of capturing or collecting water, or the reservoir or container that does this. Thus, **catchment area** or **basin.**

**cat cracking** short for CATALYTIC CRACKING.

**catfeed** *Oil & Gas.* catalytic feed; a term for feedstock supplied to a catalytic cracking process.

**cathode** *Electricity.* the electrode in an electrochemical cell that injects electrons into the electrolyte; the electrode at which reduction occurs, usually negative with respect to the anode.

**cathode-ray tube (CRT)** *Communication.* a tube whose electron beam can be focused to present alphanumeric or graphical data on an electroluminescent screen; widely used in television receivers, computer monitors, radar screens, and the like.

**cathodic protection** *Storage.* the protection of a metallic material from corrosion, by coupling the material with a small electrical current that opposes the flow of electrons from a corrosive medium (e.g., the ground).

**cation** *Chemistry.* a positively charged ion; i.e., a species that migrates to the cathode during a process of electrolysis.

**caulking** *Materials.* the process of applying a puttylike material in order to seal seams, cracks, or similar openings of a window frame, doorway, and so on, in order to reduce heat loss or gain, Thus, **caulking, caulk, and seal.**

**caustic wash** *Oil & Gas.* a process in which distillate is treated with a caustic solution (e.g., sodium hydroxide) to remove impurities or contaminants.

**Caux, Saloman de** (also spelled **Caus**) 1576–1626, French engineer and physicist, the first to describe some the basic principles of the steam engine, and also developer of one of the first solar energy devices (see next).

**Caux fountain** *History.* an early solar-powered device consisting of a set of glass lenses, a supporting frame, and an airtight metal vessel containing water and air. This would produce a small water fountain when the air heated up during operation.

**cave** *Nuclear.* a term for a heavily-shielded room for storing or handling highly-radioactive materials

**Cavendish, Henry** 1731–1810, English chemist and physicist who identified key chemical substances such as hydrogen, carbon dioxide, nitrogen, and nitric acid, and described the composition of air and water.

**Cavendish balance** *Measurement.* a sensitive torsion balance used by 17th century English scientist Henry Cavendish to measure the value of the gravitational constant *G*. This allowed him to calculate the mass of the earth.

**cavitation** *Chemistry.* **1.** the formation of gas bubbles in a liquid, due to pressure variations, heating, or vibration. **2.** *Hydropower.* specifically, the formation of gas pockets or bubbles on the blade of an impeller or the gate of a valve; collapse of these pockets or bubbles drives water with such force that it can cause pitting of the surface.

**Cayley, George** 1773–1857, British engineer who was among the first scientific investigators to understand the underlying principles and forces of flight, such as lift, drag, wing and body shape, propulsion, and vehicle weight. In 1849 he built the first successful human-carrying glider.

**CBCP** center beam candlepower.

**CBM** coal bed methane.

**CCA** cold cranking amps.

**CCE** cost of conserved energy.

**CCT** **1.** clean coal technology. **2.** correlated color temperature; see COLOR TEMPERATURE.

**cd** candela.

**CDE** carbon dioxide equivalent.

**CDIAC** Carbon Dioxide Information Analysis Center (est. 1982); the primary global-change data and information analysis center of the U.S. Department of Energy.

**CDM** clean development mechanism.

**CDP method** *Oil & Gas.* common-depth-point method; a signal-to-noise-enhancing technique in subsurface exploration that has greatly improved the ability of geologists to interpret information gained from seismic surveys, and thus improve the efficiency of the oil exploration process.

**cell** *Biological Energetics.* **1.** the fundamental microscopic unit of which all living things except viruses are composed, consisting of a nucleus and cytoplasm and bounded by a membrane; the minimal structural unit of life that is capable of functioning independently. **2.** *Electricity.* a primary galvanic unit that converts chemical energy directly into electric energy. It typically consists of two electrodes of dissimilar material isolated from one another electronically in a common conductive electrolyte. See also BATTERY. **3.** a unit designed to convert radiant energy into electrical energy; e.g., a photovoltaic (PV) cell.

**cell barrier** *Photovoltaic.* a thin region of static electric charge along the interface **(cell junction)** of the positive and negative layers in a photovoltaic cell, which inhibits the movement of electrons from one layer to the other, creating a current and thus a voltage across the cell.

**cell proliferation** *Health & Safety.* the fact of cells increasing in number by the process of division; a carefully regulated process in a healthy organism, but various factors (such as environmental pollution) can cause a mutation of the process to allow cells to multiply at an uncontrolled rate (i.e., become cancerous).

**cell reversal** *Storage.* an excessive discharge of a battery that causes the cells with least capacity to be partly recharged in the reverse direction; this tends to result in cell damage.

**cellulase** *Biological Energetics.* a combination of enzymes that catalyze the reaction of water with cellulose to release shorter chains and ultimately soluble glucose sugar.

**cellulose** *Materials.* a polysaccharide that is the major complex carbohydrate material in plants, especially their cell walls.

**cellulose ether** *Oil & Gas.* a water-soluble polymer in powder form that forms a stable viscous colloid after its dissolution in water; used as a drilling fluid.

**cellulose insulation** *HVAC.* a type of insulation composed of waste paper (e.g., used newspapers and boxes) shredded into small particles with chemicals added to provide resistance to fire and insects.

**cellulosic** *Chemistry.* **1.** having cellulose as a major component. **2.** a compound derived from cellulose.

**Celsius scale** *Measurement.* a standard temperature scale based on 0 degrees for the ice (freezing) point of water and 100 degrees for the steam (boiling) point. [Developed in 1742 by Swedish astronomer Anders *Celsius,* 1701–1744.]

**Centennial Bulb** *History.* a specific hand-blown light bulb with a carbon filament, first lit at a firehouse in Livermore, California; so called because as of this writing, it has been burning continuously for more than 100 years.

**center beam candlepower** *Lighting.* the intensity of light produced at the center of a reflector lamp, expressed in candelas.

**center of mass** *Physics.* the point in a body that moves as though the body's entire mass were concentrated there, and as though all forces were applied there. Also, **center of inertia.**

**centi-** *Measurement.* a prefix meaning one hundredth, as in *centigrade.*

**centigrade scale** *Measurement.* another name for the CELSIUS SCALE, based on the fact that it uses 100 units for the range from freezing to boiling for water.

**centimeter** *Measurement.* a basic unit of length in the metric system, equal to one-hundredth of a meter, or 0.39 inch.

**centimeter-gram-second** see CGS.

**central heating** *HVAC.* a system of heating a building in which the heat is generated by a central source (such as an oil or coal burning furnace) and transmitted as heated air, water, or steam throughout the building by a series of pipes or vents and radiators. Contrasted with AREA HEATING.

**central receiver system** *Solar.* a type of solar thermal energy system in which a field of heliostats (solar-focusing mirrors) transfers radiation to a heat-transfer fluid such as molten salt that flows through a receiver atop a tower. The salt's heat energy is then used to make steam to generate electricity in a conventional steam generator, located at the foot of the tower.

**centrifugal** *Physics.* acting or moving in a direction away from the axis of rotation. **Centrifugal fans, pumps,** and **compressors** (and so on) use this action to create pressure differences that result in the flow of fluids.

**centrifugal blade pitch(ing)** *Wind.* an overspeed protection system in a wind turbine, employing a spring and weight mechanism attached to propellers to change the angle, or pitch of the blade in high-velocity winds.

**centrifugal chiller** *HVAC.* a machine that produces cold water by using centrifugal action in its compressor to raise the pressure level of the refrigerant gas; used in large commercial buildings to supply chilled water to HVAC systems.

**centrifugal compressor** *Refrigeration.* a machine that uses centrifugal force to compress and discharge large volumes of gas, as in a refrigeration cycle.

**centrifugal force** *Physics.* **1.** a postulated inertial force in a rotating coordinate system, directed outward from the axis of rotation and, for a given magnitude of velocity, becoming weaker with increased distance (inversely proportional to the distance from the axis of rotation). **2.** in popular use, the force that seems to pull an object outward as it moves in a circle, thus acting as an equal opposing effect to the actual centripetal force.

**centrifugal separation** *Chemistry.* the use of centrifugal force to separate two fluids with different densities, or to separate two materials of different phases, such as a solid from a liquid.

**centrifuge** *Materials.* **1.** a rotating device that uses centrifugal force to separate substances of different densities, to remove moisture, or to simulate gravitational effects. **2.** to separate substances by means of such a device. Thus, **centrifugation.**

**centripetal** *Physics.* acting or moving in a direction toward the axis or center of rotation.

**centripetal force** *Physics.* the force that is required to keep an object moving around a circular path; it is directed toward the center of the circle. In the absence of this effect, the object would move in a straight line tangential to the circle.

**ceramic fuel** *Renewable/Alternative.* a technology to convert hydrocarbon fuels into electrical energy by means of an electrochemical reaction in a solid oxide fuel cell, rather than by conventional combustion.

**CERCLA** *Policy.* Comprehensive Environmental Response, Compensation, and Liability Act, commonly known as **Superfund;** a U.S. law passed in 1980 that established liability for those responsible for illegal waste dumping, as well as a trust fund to clean up sites when the responsible parties cannot be found or determined. Most of the funds have come from a tax on the chemical and petroleum industries.

**CERN** European Organization for Nuclear Research (est. 1954), the world's largest center for particle physics research, founded as a joint venture of various European countries. It is located near Geneva on the Swiss border with France. [An abbreviation for Centre Européen pour la Recherche Nucleaire.]

**certified emission reduction (CER)** *Policy.* a unit of greenhouse gas reduction that has been generated and certified under the provisions of Article 12 of the Kyoto Protocol, the clean development mechanism (CDM).

**CES** constant elasticity of substitution.

**cesium** *Chemistry.* an alkali metal element having the symbol Cs, the atomic number 55, an atomic weight of 132.905, a melting point of 28°C, and a boiling point of 705°C; a soft solid that becomes liquid at about room temperature. The most reactive of all elements, it decomposes water to produce hydrogen that ignites spontaneously. Used in photoelectric cells and as a radiotherapy agent.

**cetane** *Chemistry.* a colorless liquid hydrocarbon of the alkane series, $C_{16}H_{34}$, used in cetane number determinations and as a solvent.

**cetane number (CN)** *Oil & Gas.* the performance rating of a diesel fuel, corresponding to the percentage of cetane in a cetane-methylnaphthalene mixture with the same ignition performance. A higher cetane number indicates greater fuel efficiency.

**ceteris paribus** *Economics.* a Latin phrase meaning "all else being equal" or "holding other things constant"; used to analyze a complex future event by isolating one or more variables (e.g., price or supply), with the assumption that the other variables that are not considered will not affect the outcome.

**CFC** chlorofluorocarbon.

**CFD** computational fluid dynamics.

**CFL** compact fluorescent lamp.

**cf/s** or **cfs** cubic feet per second.

**CGE** computable general equilibrium.

**CGS system** *Measurement.* centimeter-gram-second system; a system of measurement based on fundamental units of a centimeter for length, a gram for mass, and a second for time. Introduced by the British Association for the Advancement of Science in 1874, it was the system commonly used in science for many years, until eventually replaced with the MKS system and then the International System of Units (SI).

**Chadwick, James** 1891–1974, English physicist who discovered the neutron (1932).

**chained dollars** *Economics.* a measure used to express real prices, i.e., prices that have been adjusted to remove the effect of changes in the purchasing power of the dollar; usually a reflection of buying power relative to a reference year.

**chain pillar** *Mining.* one of a series of supporting pillars of coal left in place in a mine to provide roof support for the gangway and airways.

**chain reaction** *Physics.* **1.** a self-sustaining sequence in which neutrons are produced in a fission reaction, during which a large amount of energy and several neutrons are emitted, leaving behind fission fragments; these neutrons induce subsequent fission reactions in neighboring nuclei, which in turn emit more neutrons and energy, causing the next generation or chain of fissions. **2.** in popular use, any of a succession of changes in a system, each of which induces a subsequent similar change.

**Chalk River** *Nuclear.* a nuclear facility on the banks of the Ottawa River in Ontario, Canada, the site of a nuclear accident in 1952 when human error caused an unintended reaction. The resulting explosion released radioactive water and set off warnings of lethal radiation levels; although no casualties occurred this was an early indication of potential dangers of nuclear plants.

**Challenger** *History.* **1.** a British ship that undertook an exploratory voyage (1872–1876) that laid the scientific foundation for the modern science of oceanography. **2.** a U.S. space shuttle vehicle that exploded in 1986, killing the entire crew.

**Champion, Albert** 1878–1927, French-born U.S. pioneer in spark plug design. His AC Spark Plug Co. (AC=Albert Champion) became a renowned and highly successful manufacturer of spark plugs for automobiles and aircraft, as well as other products vital to industry. AC later became a division of General Motors.

**channeling** *Physics.* the flow of particles or material into a medium that contains voids or regions of lower density, so as to equalize density throughout the medium.

**chaos** *Thermodynamics.* **1.** the property of a real-world, dynamical system that is random, irregular, aperiodoc, and unpredictable. **2.** an older word for a gas, from the sense that it is uncontrolled.

**chaos theory** *Thermodynamics.* the study of complex, nonlinear dynamic systems. Also, **chaotic dynamics.**

**char** *Materials.* the noncombusted remains of organic matter that has burned incompletely.

**characteristic hazardous waste** *Health & Safety.* a classification of a product as having one or more of the following properties that render it dangerous to life and health; that is, ignitable, corrosive, reactive, or toxic.

**charbon de terra** *History.* a French term meaning "carbon of the earth;" an early word for coal.

**charcoal** *Materials.* a wood product that is made by heating wood in the absence of sufficient air for full combustion to occur; this heating releases the wood's volatile compounds, leaving behind a lightweight and cleaner burning fuel that is 70–90% carbon.

**charge** *Physics.* **1.** a property of the elementary particles of matter giving rise to electric and magnetic interactions; electrons have negative charge and protons have positive charge. **2.** *Electricity.* the electrical energy stored in an insulated object, such as a battery or capacitor. **3.** the conversion of electric energy into chemical energy within a cell or battery. **4.** to direct electrical energy into a cell or battery for such a conversion. A device that does this is a **charger. 5.** *Oil & Gas.* the occurrence of conditions of hydrocarbon generation and migration adequate to cause an accumulation of minimum appropriate size.

**charge capacity** *Oil & Gas.* the input (feed) capacity of an oil refinery.

**charge conservation** *Physics.* the observation that electric charge is conserved in any process of transformation of one group of particles into another.

**charge cycle** *Storage.* the use of all of a battery's power and a recharging to its full capacity; this is not necessarily a single complete cycle; i.e., two instances of using half a battery's power and then recharging it fully would also constitute a charge cycle. Also, **charge-discharge cycle.**

**charge-discharge** *Nuclear.* **1.** a process in which the fuel in a nuclear reactor core can be removed and replaced without shutdown of the facility. **2.** see CHARGE CYCLE.

**charge factor** *Storage.* the time in hours during which a battery can be charged at a constant current without damage to the battery; usually expressed in relation to the total battery capacity.

**charge fluid** *Storage.* the heat transfer fluid used to remove heat from a thermal storage device or generator during the charge or build period, or used to add heat to a heat storage.

**charge rate** *Storage.* the amount of current applied to a cell or battery to restore its available capacity.

**Charles** *Oil & Gas.* a vessel sailing from Antwerp, Belgium that was the forerunner of the modern oil tanker, the first ship to carry oil from the U.S. to Europe in iron tanks (1869).

**Charles's Law** *Physics.* the observation that the volume of a dry ideal gas at constant pressure and quantity is proportional to the absolute temperature of the gas; i.e., as the temperature of the gas increases. so does its volume. [Formulated by French mathematician and physicist Jacques *Charles,* 1746–1823.]

**Charlotte Dundas** *History.* the first vessel built for the specific purpose of being propelled by a steam engine, launched in Scotland by William Symington (1801).

**chassis** *Transportation.* **1.** the frame upon which an automobile body is mounted. **2.** any major part or framework of an assembly to which other parts are attached.

**check dam** *Hydropower.* a small dam constructed across a gully or other drainageway to reduce concentrated flows in the channel and thus minimize erosion, promote deposition of sediment, and protect vegetation in the early stages of growth. Commonly used in the bottom of channels that will be stabilized at a later date.

**Chelyabinsk** *Nuclear.* a nuclear facility in the former Soviet Union that was the site of a notable accident in 1957; the cooling system of a radioactive waste containment unit malfunctioned and exploded, releasing about two million curies and exposing a quarter of a million people to radiation.

**chemical element** see ELEMENT.

**chemical energy** *Chemistry.* energy produced or absorbed in the process of a chemical reaction.

**chemical equilibrium** *Chemistry.* a condition in which a chemical reaction and its opposing reaction occur at the same rate, producing constant concentrations of the reacting substances and products.

**chemical exergy** *Thermodynamics.* the maximum work obtainable from a substance when it is brought from the environmental state to the dead state by means of processes involving interaction only with the environment.

**chemical oxygen demand (COD)** *Environment.* the quantity of oxygen required for the complete oxidation of organic chemical compounds in water; used as a measure of the level of organic pollutants in natural and waste waters.

**chemical pulp** *Materials.* pulp obtained by the digestion of wood with solutions of various chemicals.

**chemical reaction** *Chemistry.* a chemical change in which the atoms or molecules of two or more substances are rearranged to form one or more additional substances, often having different properties.

**chemical recombination** see RECOMBINATION.

**chemical vapor deposition (CVD)** *Photovoltaic.* a method of depositing thin semiconductor films to make photovoltaic devices; a substrate is exposed to one or more vaporized compounds that contain desirable constituents, and a chemical reaction is initiated that deposits the desired material on the substrate.

**chemical weathering** *Earth Science.* the breakdown and alteration of rocks at or near the earth's surface as a result of chemical processes, including solution, hydrolysis, ion exchange, oxidation, and biochemical reactions.

**chemiluminescence** *Chemistry.* any process in which a chemical reaction produces visible light. Thus, **chemiluminescent.**

**chemiosmotic theory** *Biological Energetics.* the principle that the stepwise transfer of electrons through electron carriers to oxygen results in the release of energy; this energy is used to pump protons across the inner mitochondrial membrane into the space between the inner and outer mitochondrial membranes.

**chemisorption** *Chemistry.* a process in which the atoms or molecules of a gas or liquid are held to the surface of a solid material (adsorbed).

**chemoautotroph** *Biological Energetics.* an organism that uses carbon dioxide as a source of carbon, but rather than using light energy obtains its energy from the oxidation of inorganic compounds. Thus, **chemoautotrophic, chemoautotrophy.**

**chemosynthesis** *Biological Energetics.* **1.** a process in which carbohydrates are manufactured from carbon dioxide and water, using chemical nutrients as the energy source, rather than the sunlight used for energy in photosynthesis. **2.** any process that produces organic compounds by means of the energy contained in inorganic molecules.

**chemurgy** *Chemistry.* the use of agriculturally derived substances to produce new, nonfood industrial products; e.g., the use of soybean oil for paints and varnishes. Thus, **chemurgic, chemurgist.**

**Chernobyl** *Health & Safety.* a city in Ukraine (former Soviet Union), the site of the world's most severe nuclear accident to date. Human error caused a core meltdown at the Chernobyl nuclear power plant; 31 people died in the accident or soon after, and another 134 were treated for acute radiation syndrome. Total related deaths probably number in the thousands, and the long-term impacts **(Chernobyl effects)** are still being assessed.

**chestnut coal** see COAL SIZING.

**Chevron** *Oil & Gas.* a corporate name for Standard Oil of California, one of the historic leaders of the U.S. oil industry (founded in 1879 as Pacific Coast Oil Co.); now operating as ChevronTexaco through its merger with another historic firm, Texaco (founded in 1901, merged in 2001).

**Chile(an) nitrate** *Materials.* a commercial name for the sodium nitrate mineral natratine that occurs naturally in caliche in northern Chile, widely used as a source of nitrogen for explosives. Also, **Chile niter.**

**chilled water system** *HVAC.* a type of air conditioning system in which the refrigerant is not contained in the unit itself, but rather in a remotely located chiller, from which cooled water is piped to the air conditioner.

**chiller** *HVAC.* **1.** a device that cools water for eventual use in cooling air. **2.** any similar device used to cool a substance for some industrial purpose.

**chimney** *HVAC.* a vertical passageway in a building that draws up combusted particles from a furnace, stove, or fireplace.

**Chinese tallow** *Biomass.* an Asian seed-oil tree, *Sapium sebiferum,* that grows and spreads rapidly over large areas, tending to take over regions in which it is introduced by out-competing native plants. It is the subject of various studies for use as a biomass energy crop.

**chipper** *Biomass.* a machine that produces wood chips (see below) of a small uniform size by a cutting action.

**chips** *Biomass.* small fragments of wood chopped or broken by mechanical equipment, used as a raw material for pulping and fiberboard or as biomass fuel.

**chlorofluorocarbon (CFC)** *Environment.* one of a family of inert, nontoxic, and easily liquefied chemicals used in refrigeration, air conditioning, packaging, insulation, or as solvents and aerosol propellants. Due to their role in depleting the ozone layer, their use has gradually been phased out according to international agreements made in the Montreal Protocol in 1987.

**chlorophyll** *Biological Energetics.* a magnesium chlorin pigment that is found in all higher plants and that gives plants their green color by absorbing red and blue-violet light and reflecting green light. The solar radiation energy absorbed by chlorophyll transforms carbon dioxide and water into carbohydrates and oxygen.

**Chlorophyta** see GREEN ALGAE.

**chloroplast** *Biological Energetics.* a large cellular organelle that is bounded by two membranes; it contains a complex system of membranes, and it functions as the site of photosynthesis in green algae and green plant cells.

**choke** *Transportation.* **1.** a valve in a carburetor intake to reduce the air supply and thus give a rich mixture for starting purposes while the engine is still cold. Also, **choke valve. 2.** any restriction or narrowing in a system to reduce fluid or energy flow.

**chokedamp** *Mining.* another name for BLACKDAMP, because of the tendency of this gas to cause choking and suffocation.

**choke price** *Economics.* the maximum price that anyone would be willing to pay for a certain good or service (e.g., oil); i.e., the price level at which demand for this item would disappear and the market would move on to substitutes.

**chord** *Wind.* **1.** in a wind turbine blade, a straight line joining the mean thickness line between the blade's leading edge and the trailing edge. **2.** the length of such a line. Also, **chord line.**

**CHP** combined heat and power.

**chromosphere** *Solar.* the lower level of the solar atmosphere between the photosphere and the corona.

**chronic exposure** *Health & Safety.* a long-term period of usually limited exposure to radiation or a toxic substance.

**chronic obstructive pulmonary disease (COPD)** *Health & Safety.* a collective term for various chronic lung conditions that obstruct the airways; the most common form is a combination of chronic bronchitis and emphysema. COPD is a widespread condition of global distribution; smoking is a major cause and other causes include occupational exposures and indoor air pollution (e.g., from biomass energy).

**chronic toxicity** *Health & Safety.* a toxic effect that becomes evident after repeated or continued exposure over a long period of time (e.g., months or years). Similarly, **chronic health effect.**

**chulha** *Biomass.* a traditional cooking device that employs biomass as a fuel, widely used in rural areas of Asia.

**Churchill, Winston** 1874–1965, British political leader, noted in the field of energy for his early recognition of the superiority of oil over coal as a military fuel. As First Lord of the Admiralty, he hastened the conversion of British battleships from coal to oil prior to World War I. The Royal Navy eventually overcame the largely coal-based German fleet.

**CIE** Commission Internationale de l'Eclairage, the French title for the International Commission on Illumination (est. 1914), a nonprofit organization that promotes international cooperation and exchange of information among its member countries on all matters relating to the science and art of lighting.

**CIF** see COST, INSURANCE, FREIGHT.

**cinder coal** *Coal.* **1.** coal that has been naturally cindered by the heat from adjoining rock that has melted and then solidified. **2.** a similar product resulting from a coal mine fire or explosion.

**CIP** critical infrastructure protection.

**circuit** *Electricity.* **1.** an interconnection of electrical elements forming one or more complete paths for the flow of current. **2.** a network of media, usually conductors and semiconductors, providing a closed path for electromagnetic phenomena to occur.

**circuit breaker** *Electricity.* a current-sensitive safety switch that will automatically open a circuit when the current exceeds a certain predetermined value; unlike a fuse, a circuit breaker can be reset after being activated.

**circuitry** *Electricity.* the entire combination of circuits in a particular electrical system or device.

**circular collider** *Nuclear.* any of a category of particle accelerators in which counter-rotating beams are brought into collision at experimental areas, where they generate intense bursts of energy; used to study the fundamental properties of subatomic matter.

**circulating fluidized bed (CFB)** see FLUIDIZED BED COMBUSTION.

**circulation cell** *Earth Science.* large areas of air movement created by the rotation of the earth and the transfer of heat from the equator toward the poles.

**circumscription theory** *Social Issues.* a theory advanced by U.S. anthropologist Robert Carneiro to explain the origin of states; i.e, the principle that threats to the survival of a society (stemming from population pressure on resources and also from competition from other societies and environmental constraints) will lead to the increasing complexity of the sociopolitical organization.

**circumsolar energy** *Solar.* radiation scattered by the atmosphere in such a way that it appears to originate from an area of the sky immediately adjacent to the sun.

**circumsolar radiation** *Solar.* the amount of solar radiation coming from a circle in the sky centered on the sun's disk and having a radius of between 2.5 and 3.5 degrees, depending on the type of instrument being used to measure direct normal irradiance.

**Cistercians** *History.* a European religious order, noted in medieval times for their ingenious use of water resources and water power; in their agricultural practices they also showed an awareness of modern ecological concepts such as sustainability.

**Cisternay du Fay** (Charles-François de) 1698–1739, French chemist who discovered that an electrical charge has both positive and negative values and also was the first to describe the principle of insulating a conductor with some nonconducting substance.

**CitiCar** *Renewable/Alternative.* a pioneering consumer electric vehicle marketed by U.S. entrepreneur Robert Beaumont in the 1970s; several thousand units of this car and an improved version (the **ComutaCar**) were manufactured, due to interest in alternative vehicles spurred by the oil crisis of 1973. However, it was perceived as a glorified golf cart and by the early 1980s it was no longer produced.

**citric acid cycle** another name for the KREBS CYCLE.

**citygate** *Oil & Gas.* a point or measuring station at which a distributing gas utility receives gas from a natural gas pipeline company or transmission system.

**civilian nuclear reactor** *Nuclear.* a nuclear reactor used to generate energy for consumer purposes (e.g., electrical power, steam for district heating), as opposed to a nuclear weapon or warhead for military use.

**cladding** *Materials.* **1.** any of various processes in which two materials are bonded together under high pressure and heat. **2.** the material resulting from such a process. **3.** *Nuclear.* specifically, a material used to construct reactor components and designed to maintain a separation between their contents and the coolant, e.g., the cladding of a fuel pin that separates the fuel pellets from the coolant. Zirconium and zirconium alloys (e.g., Zircaloy) are common cladding materials.

**Clapeyron, Benoit Paul Emile** 1799–1864, French engineer who developed a mathematical reformulation of the CARNOT CYCLE and through this made Carnot's work more generally known.

**Clapeyron equation** *Thermodynamics.* a formula for the heat of vaporization of a liquid as a function of its temperature and volume change upon vaporization.

**Clapeyron relation** *Thermodynamics.* a differential equation that determines the heat of vaporization of a liquid.

**clarifier** *Materials.* a device used to clear a liquid of suspended particles, as through filtration, gravitation, centrifugation, or the addition of an enzyme.

**classical economics** *Economics.* the leading school of economic thought of the 19th century; central to this approach were precepts such as economic freedom, competition, laissez-faire government, a labor theory of value, and free trade. The earlier belief that agriculture was the chief determinant of economic health was rejected in favor of manufacturing development, and the importance of labor productivity was stressed.

**classical mechanics** *Physics.* those aspects of mechanics based on Newton's fundamental laws of motion, applied in a realm where the speeds involved are small compared to the speed of light, and the sizes of systems involved are large compared to their wavelengths. Also, **classical dynamics.**

**classical physics** *Physics.* the discipline of physics from its founding years up until the 20th century, based on Newton's laws of motion and excluding later studies of relativity and quantum mechanics.

**classical thermodynamics** *Thermodynamics.* the discipline of thermodynamics as it developed during the 19th century, dealing with the study of heat, and thus with the collision and interaction of particles in large, near-equilibrium systems; it does not employ the ideas of atoms and molecules. Compare NONEQUILIBRIUM THERMODYNAMICS.

**clastogen** *Health & Safety.* a physical or chemical agent that causes chromosome damage or DNA strand breaks; e.g., certain environmental pollutants. Thus, **clastogenic.**

**clathrate** *Chemistry.* a substance in which the molecules of one compound are encapsulated in lattices or cagelike structures within another compound.

**Claude method** *Physics.* **1.** a process for liquefying air and other gases, in which a gas acts to drive a piston, and then moves to an expansion chamber where it cools by adiabatic expansion. **2.** *Hydrogen.* specifically, the use of this process to produce hydrogen fuel from liquefied natural gas. [Developed by French engineer Georges *Claude,* 1870–1960.]

**Clausius, Rudolf** 1822–1888, German physicist who made major contributions to thermodynamics and the kinetic theory of gases. His statements of the first and second laws of thermodynamics can be considered the modern foundation of this science.

**clausius** *Measurement.* a unit of entropy increment for a closed system that receives 1000 international table calories (4186.8 J) of heat transfer at a temperature of 1 K.

**Clausius inequality** *Thermodynamics.* a principle applying to any real engine cycle; the change in heat over temperature for a thermodynamic cycle is always less than or equal to zero.

**Clausius statement** *Thermodynamics.* an alternative name and description of the second law of thermodynamics: no cyclic process is possible that does nothing but transfer heat from a cold body to a hot one; alternatively, it is not possible to construct a heat pump that will transfer heat from a low-temperature reservoir to a high-temperature reservoir without using external work.

**Clean Air Act (CAA)** *Policy.* a comprehensive U.S. federal law passed in 1970 that regulates air emissions, giving the Environmental Protection Agency (EPA) the authority to develop national ambient air quality standards (NAAQS) for key pollutants, in order to protect public health and the environment.

**clean air market** *Economics.* any market-based regulatory programs designed to improve air quality; e.g., a cap-and-trade program to reduce sulfur dioxide emissions from power plants.

**clean coal technology** *Coal.* any of various processes to reduce the emissions of pollutants from coal combustion or from the conversion of coal to chemicals and other materials; this includes both technologies to remove pollutants and to improve efficiency, which also decreases emissions.

**clean development mechanism (CDM)** *Policy.* an aspect of the Kyoto Protocol intended to promote sustainable development and climate protection. It is designed to allow industrialized countries to proceed in a flexible manner to meet the greenhouse gas (GHG) emission reduction targets that they agreed to achieve under the Protocol, and at the same time provide an opportunity for developing countries, not bound to reduce their emissions under the Protocol, to achieve sustainable development while participating in GHG mitigation.

**clean energy** *Consumption & Efficiency.* a popular term for forms of energy that generate relatively little pollution in their production and consumption; e.g., wind or solar energy as opposed to coal or oil. Also, **clean energy.**

**clean room** *HVAC.* a specially constructed and enclosed area environmentally controlled to eliminate airborne particulates and stabilize temperature, humidity, air pressure and motion, and so on, as for the purpose of manufacturing and assembling highly sensitive equipment.

**Clean Water Act (CWA)** *Policy.* an amendment (1977) to the Federal Water Pollution Control Act of 1972 that regulates discharges of pollutants to waters in the U.S., giving the Environmental Protection Agency (EPA) the authority to establish effluent standards on an industry basis and continuing the requirements to set water quality standards for contaminants in surface waters.

**clearness index** *Solar.* the ratio of the global horizontal solar radiation at a site to the extraterrestrial horizontal solar radiation above that site.

**clear sky index** see CLEARNESS INDEX.

**cleat** *Coal.* the main vertical joint in a coal seam, along which the coal breaks most easily when mined.

**clerestory** *HVAC.* an upper extension of a side wall, built above adjoining roof sections and having a window to admit daylight into a high central room; may also be used for ventilation and solar heat gain. Also spelled **clerestorey.**

**Clermont** *History.* the first commercially successful steamboat, launched by Robert Fulton on the Hudson River in New York (1807).

**climate** *Earth Science.* **1.** average weather as defined by surface variables such as temperature, precipitation, and wind. **2.** a statistical description in terms of the mean and variability of temperature, precipitation, wind, and other meteorological phenomena over a period of time ranging from months to thousands or millions of years.

**climate change** a statistically significant difference noted either in the mean state of the climate or in its variability, persisting for an extended period (decades or longer). See below.

**climate change science** *Climate Change.* the study of the mean and variability of meteorological phenomena, such as temperature, precipitation, and wind, to detect a (possible) change in climate, and then to attribute such a change to natural or anthropogenic (human-influenced) factors.

**climate control** *HVAC.* **1.** the maintenance of air in an enclosed space (e.g., a building, room, motor vehicle, etc.) to control factors such as its temperature, relative humidity, and flow, as well as to remove impurities and contaminants from the air. **2.** specifically, such a system that operates automatically to maintain a pre-set desired temperature.

**climate feedback** *Climate Change.* a change in one component of the climate system, often temperature, that elicits a response in one or more other components. An increase in temperature could cause $CO_2$ to "bubble out" of the ocean, which would reinforce the increase in temperature. This is a positive climate feedback; a negative feedback would offset or reduce an initial increase in temperature.

**climate lag** *Climate Change.* a delay that occurs in climate change as a result of some factor that takes effect very slowly; e.g., the effects of the discharge of carbon dioxide from fossil fuels may not be known for some time because some of the emitted $CO_2$ dissolves in the ocean and is not released to the atmosphere until many years later.

**climate model** *Earth Science.* a mathematical or schematic representation of the climate system based on the physical, chemical, and biological properties of the various components of the system, and their interactions and feedback processes. See also GENERAL CIRCULATION MODEL.

**climate prediction** *Climate Change.* the result of an attempt to produce the most likely description or estimate of the actual evolution of the climate in the future, e.g., at seasonal, interannual, or long-term time scales.

**climate change** In order to define "climate change," one must first define climate. The simplest definition of climate is, the average of the weather. A more rigorous definition, adopted by the World Meteorological Organization in 1966, is that climate is the mean state along with the variability and higher-order statistics of the state of the atmosphere, as computed over a 30-year sampling period. The climate of a given region is defined not only in terms of means or averages, but also in terms of the typical departures from average conditions from year to the next. The atmospheric variables of particular interest are temperature, precipitation, wind speed, relative humidity, and cloudiness. Fluctuations from one year to the next do not constitute a change in climate; rather, this variability is part of what characterizes or defines a given climate. A change of climate occurs only if the means and/or the variability change from one 30-year sampling period to the next. Thus, a change in variability with no change in means constitutes a change in climate, as does as a change in means with no change in variability. The use of 30-year sampling periods is arbitrary. If, for example, climate were defined based on a 100-year sampling period, then changes in the statistics from one 30-year period to the next would be classified as climatic variability (and part of the characterization of the climate) rather than as climatic change. Climate can change due to factors entirely internal to the climate systems (such as slow interactions between the atmosphere and oceans), as well as due to factors external to the climate system (such as changes in solar irradiance, systematic changes in the earth's orbit, or human-induced increases in the concentrations of greenhouse gases).

**L. Danny Harvey**
University of Toronto

**climate projection** *Climate Change.* a projection of the response of the climate system to emission or concentration scenarios of greenhouse gases and aerosols, or radiative forcing scenarios, often based upon simulations by climate models; in technical use this term is distinguished from CLIMATE PREDICTION (see above) to emphasize that climate projections depend upon the scenarios used, which are based on assumptions that may or may not be realized.

**climate protection** *Global Issues.* global efforts to prevent or minimize climate change, generally through efforts to reduce the emissions of greenhouse gases such as carbon dioxide.

**climate scenario** *Climate Change.* a plausible and often simplified representation of future climate, based on an internally consistent set of climatological relationships. Climate scenarios are constructed to study the potential consequences of anthropogenic climate change, and they often serve as input to impact models. A CLIMATE PROJECTION (see above) often serves as the raw material for constructing a climate scenario.

**climate sensitivity** *Climate Change.* an equilibrium change in surface air temperature following a unit change in radiative forcing (°C/Wm$^2$); in practice, it frequently is described as the equilibrium change in global mean surface temperature following a doubling of the preindustrial atmospheric $CO_2$ concentration.

**climate signal** *Climate Change.* a statistically significant difference between the control and disturbed simulations of a climate model.

**climate system** *Earth Science.* a highly complex system consisting of five major components: the atmosphere, the hydrosphere, the cryosphere, the land surface and the biosphere, and the interactions among them.

**climate variability** *Climate Change.* variations in the mean state and other statistics (such as standard deviations, the occurrence of extremes, and so on) of the climate on all temporal and spatial scales beyond that of individual weather events; this may be due to natural internal forcing or to anthropogenic external forcing.

**climatic analog** *Climate Change.* **1.** an area that has a climate closely resembling another area that is distant geographically. **2.** a historic climate situation in which changes similar to the present occurred.

**climatic anomaly** *Climate Change.* a significant deviation of a particular climatic variable (such as temperature) from the mean or normal over a specified length of time.

**climatic optimum** *Climate Change.* a period in history during which surface air temperatures were warmer than at present in nearly all regions of the world. One such period has been identified from about 8000 to about 5000 years ago, when early civilizations began to flourish, and another in the medieval period (approximately the years 900 to 1200).

**climatology** *Earth Science.* the scientific study of climate.

**Clinch River** *Nuclear.* a site in northeastern Tennessee, part of the Oak Ridge portion of the Manhattan Project during World war II; later the location of the **Clinch River Breeder Reactor Project,** an ambitious but controversial plutonium-based breeder reactor program that was initiated in 1972 but suspended before completion in 1983 when the U.S. Congress withdrew funding.

**clinker** *Materials.* **1.** a hard mass of fused ash produced in a coal furnace as a byproduct of combustion. **2.** a term for powdered cement, produced by heating a properly proportioned mixture of finely ground raw materials (calcium carbonate, silica, alumina, and iron oxide) in a kiln to a temperature of about 2700°F. **3.** a general term for a lump of vitrified stony material.

**clipper ship** *History.* a type of sailing vessel of the 19th century, noted for speed and graceful design. [From the idea that such ships could "clip" along at a fast pace.]

**clo** *HVAC.* a unit of thermal insulation, defined as the amount of insulation needed to keep a sitting person comfortable in a normally ventilated room at 70°F (21°C) and 50% relative humidity. [Short for clothing insulation.]

**closed cycle** *Thermodynamics.* **1.** an isolated thermodynamic cycle in which the thermodynamic fluid does not enter or leave the system, but is used repeatedly without introduction of new fluid; typical of a sealed system. **2.** *Nuclear.* a nuclear fuel cycle in which the radioactive material leaving a power plant is reprocessed, thereby reclaiming some of the

*clipper ship* *Clipper ships flourished for a brief period in the mid-19th century, in the final chapter of the commercial sailing ship era.*

plutonium and other radioactive elements for reuse.

**closed-cycle cooling** *HVAC.* a process in which cooling water cycles through a generating facility, absorbs heat from steam passing through a condenser, and then evaporates the heat to the atmosphere through cooling towers before it recycles the remaining water through the facility.

**closed-cycle system** see OPEN-CYCLE SYSTEM.

**closed fuel cycle** see CLOSED CYCLE (def. 2)

**closed-loop biomass** *Biomass.* a process in which biomass feedstocks are grown specifically for the purpose of energy production.

**closed-loop system** *Consumption & Efficiency.* **1.** a system that monitors its own activity and automatically makes adjustments to correct any unwanted deviation from pre-set parameters. **2.** specifically, a heating system (e.g., solar or geothermal) that circulates heat exchange fluids through a series of pipes or panels that are closed off from the external atmosphere.

**closed system** *Thermodynamics.* **1.** a system that exchanges only energy with its surroundings; i.e., no mass can enter or leave the system. **2.** *HVAC.* a cooling and dehumidification device in which the working fluid is enclosed in the machine and interacts with the cooled medium by heat transfer through solid walls.

**cloud chamber** *Nuclear.* an apparatus for delineating the tracks of high-speed particles as they pass through a chamber filled with a saturated vapor that, when supercooled by adiabatic expansion, condenses to form droplets of liquid around ions left in the particle's path.

**cloud condensation nuclei** *Earth Science.* airborne particles that serve as an initial site for the condensation of liquid water and that can lead to the formation of cloud droplets.

**cloud cover** *Earth Science.* a term for the portion of the sky that is covered by clouds, usually measured as tenths of sky covered.

**cloud enhancement** *Solar.* an increase in solar intensity due to reflected light from nearby clouds.

**cloud point** *Measurement.* **1.** the temperature and ionic strength of a solution at the point at which phase separation is induced. **2.** the temperature at which waxy crystals in an oil or fuel form a cloudy appearance; used to measure the ability of a diesel fuel to operate under cold weather conditions.

**Club of Rome** (est. 1968), a non-profit, nongovernmental organization (NGO) that brings together scientists, economists, business leaders, and government officials from various nations to examine global issues such as population, economic development, resource consumption, environmental preservation, and climate change. Known especially for commissioning the best-selling book LIMITS TO GROWTH.

**CMM** coal mine methane.

**CN** cetane number.

**CNG** compressed natural gas.

**CNO** carbon-nitrogen-oxygen (cycle).

**$CO_2$** carbon dioxide.

**Coachella Valley** *Wind.* an area near Palm Springs, California that is the location of a vast wind energy facility supplying the electricity needs of Palm Springs and surrounding communities.

**coal** **1.** a brown to black combustible, carbonaceous sedimentary rock formed by the compaction of partially decomposed plant material. **2.** this substance when burned as fuel. Coal was the main type of fuel in industrial countries before the advent of the petroleum industry, and it is still an important source of energy. See next page.

**Coal Age** *History.* the period in Western history in which coal was the dominant fuel,

**coal** Coal is a carbon-based rock that burns. Its origins are in prehistoric earth where generations of dead vegetation were deposited in an oxygen-depleted swampy environment to form peat. As geological forces compressed these deposits, the resulting high temperature and pressure converted peat to coal and occasionally, coal to diamond. Coal is denser than peat, contains less water, and has higher energy content. The elements found in coal are the same ones found in the vegetation, as well as the soils and waters associated with it. Those elements are most notably carbon, hydrogen, nitrogen, sulfur, and oxygen. Silicon, iron, magnesium, potassium, phosphorus, and virtually the entire Periodic Table of elements are generally present in smaller or trace quantities. Coal is ranked as lignite, bituminous, or anthracite, depending upon energy content; with increasing coalification, moisture content decreases and carbon content increases. Coal provided warmth that enabled the human population to move into northern latitudes as earth emerged from the most recent ice age. Today, in North America, coal is the predominant source of energy used to generate electricity. With advances in clean coal technology, coal is expected to serve as a dominant source of energy well into the future and even beyond the 21st century.

**Mildred B. Perry**
U.S. Department of Energy
National Energy Technology Laboratory

from the latter part of the 18th century roughly up to the middle of the 20th century.

**coal analysis** see PROXIMATE ANALYSIS; ULTIMATE ANALYSIS.

**Coal and Iron Police** *History.* a private police force of the coal and steel industry operating in Pennsylvania from 1865 to 1931. Hired to protect the property of their respective companies, they were used to intimidate coal mine workers and break up strikes, and, if necessary, evict workers and their families from their homes.

**coal ball** *Coal.* a roundish nodule or mass of petrified plant matter found in coal seams or adjacent rocks.

**coalbed** or **coal bed** *Coal.* a layer or stratum of coal thick enough to be mined profitably; a seam.

**coalbed methane (CBM)** *Coal.* an unconventional form of natural gas formed in the coalification process and found on the internal surfaces of the coal. To commercially extract the gas, its partial pressure must be reduced by removing water from the coalbed. The large quantities of water, sometimes saline, produced from coalbed methane wells pose an environmental risk if not disposed of properly.

**coal bench** see BENCH.

**coal bottom ash** *Coal.* the coarse, granular, incombustible byproduct collected from the bottom of furnaces that burn coal.

**coal breccia** *Coal.* naturally occurring angular fragments of coal within a coal bed.

**coal briquette** see BRIQUETTING.

**coal classification** *Coal.* any of various systems for classifying coal by such criteria as heating value, caking or coking property, amount of moisture, volatile matter, or sulfur, and so on.

**coal cleaning** *Coal.* a general term for various procedures employed to prepare mined coal for burning, especially the removal of impurities in order to increase the heat content of the coal and to reduce potential air pollutants, especially sulfur dioxide.

**coal combustion product (CCP)** *Coal.* solid waste left after burning the carbon and hydrocarbons in coal; made up of fly ash, bottom ash, slag, and sulfur dioxide sludge. These materials were disposed of as waste in the past, but are now often used in other industries. Also, **coal combustion byproduct (CCB).**

**coal conversion** *Coal.* the conversion of coal to electricity or into solid, liquid, or gaseous fuels, by means of various processes such as pyrolysis, gasification, liquefaction, maturation, or combustion.

**coal cutter** *Coal.* a power-operated machine that draws itself along the coal face by means of rope haulage and cuts out a thin strip of coal from the bottom of the seam.

**coal dust** *Coal.* minute particles of coal, either airborne or lying on a mine surface; there is no established standard as to how finely coal must be divided to be termed dust. Coal dust is an explosion risk in underground mines and is also a health hazard if ingested into the lungs (as in black lung disease).

**coal face** *Coal.* the working surface of a coal seam, from which coal is actively being mined (or has been) mined.

*coal face Workers at the coal face. As British author George Orwell famously observed, simply reaching the coal face to begin work was a more arduous task in itself than almost any other occupation.*

**coal field** *Coal.* **1.** an area where coal is actively being mined. **2.** more generally, any region or locale in which deposits of coal are found.

**coal fines** *Coal.* very small particles of coal, especially such particles left over as waste after a cleaning or processing operation.

**coal fire** *Coal.* any unwanted or unplanned fire in a coal mine, especially a large, spontaneous fire burning underground for an extended period of time, as for example in the coal fields of eastern Pennsylvania and in northern China.

**coal gas** *Coal.* a gaseous substance produced by the partial oxidation (burning) of coal to make coke, widely used historically as a fuel and for illumination.

**coal gasification** *Coal.* a method of converting coal, coke, or char to a useful gaseous product for heating and illuminating purposes, by reaction with oxygen, steam, carbon dioxide, air, or a mixture of these.

**coal grade** *Coal.* **1.** a quality rating of coal relating to its suitability for use for a particular purpose, based on criteria such as heating value, coking, caking, or petrologic properties, or percentage of carbon, moisture, volatile matter, organic components, and other constituents. **2.** specifically, a grading of various coal types according to their degree of impurity; i.e., the amount of sulfur, ash, and so on.

**coalification** *Coal.* the complex process of chemical and physical changes undergone by plant sediments over geological time to convert them to coal.

**coal leasing** see LEASING.

**coal liquefaction** *Coal.* the conversion of coal to produce synthetic liquid fuels by means of a destructive distillation process.

**coal measures** *Earth Science.* **1.** in Europe, the sequence of rocks corresponding to the Upper Carboniferous period, broadly synchronous with the Pennsylvanian of North America. **2.** any sequence of sedimentary rocks that includes coal beds interstratified with clays, shales, sandstones, and so on.

**coal mine fire** see COAL FIRE.

**coal mine methane (CMM)** *Coal.* a subset of coalbed methane that is released from coal seams as they are fractured and the coal is removed from the mine; post-mining activities such as storing, processing, and transporting coal also emit methane. Capture and use of CMM reduces greenhouse gas emissions and improves mine safety and productivity.

**coal mine reclamation** see RECLAMATION.

**coal oil** *Coal.* a former term for the fuel now known as KEROSENE, based on the fact that historically it was obtained from coal.

**coal pile** *Coal.* **1.** a large mass of coal ready for burning. **2.** a large accumulation of waste matter from coal mining.

**coal planer** *Coal.* a type of continuous coal-mining machine developed especially for longwall mining, in which power equipment drags a heavy steel plow back and forth across the coal face.

**coal preparation** *Coal.* any processing of mined coal to prepare it for market, including crushing and screening, removal of impurities, sizing, and loading for shipment.

**coal-producing region** *Coal.* an official geographic classification of coal-producing states of the U.S., consisting of the Appalachian Region, the Interior (Midwest/Central) Region, and the Western region.

**coal pyrolysis** see PYROLYSIS.

**coal rank** *Coal.* **1.** a system of grading various types of coal according to their quality, based on the degree of physical alteration of the organic matter composing the coal. **2.** specifically, a coal classification system used in the U.S. that includes (ranking from higher to lower) the following types: anthracite, bituminous, sub-bituminous, and lignite or brown coal (and also peat in some classifications), based on various criteria, especially heating value.

**coal refuse** *Coal.* the waste material that remains when raw coal is prepared or cleaned for market; this usually contains a mixture of rock types, but typically with elevated levels of the mineral pyrite, causing it to be a source of acid mine drainage if not handled properly.

**coal reserve** *Coal.* the amount of in situ coal in a defined area that can be recovered by mining at a sustainable profit at the time of determination. The reserve is derived by applying a recovery factor to that component of the identified resources of coal designated as the demonstrated reserve base.

**coal screening** *Coal.* a method to separate coal particles according to size, by passing them through a series of screens with openings of decreasing size.

**coal seam** see SEAM.

**coal seam gas** another term for COALBED METHANE.

**coal sizing** *Coal.* a categorization of coal particles, especially anthracite coal, based on whether or not they will pass through a mesh screen with openings of a given size; sorting is done by a system of overlying screens with successively smaller openings. The various sizes have metaphorical names, such as rice coal, buckwheat coal, pea coal, chestnut coal, stove coal, and egg coal.

**coal slag** see SLAG.

**coal sludge** *Coal.* a coal slurry (see next) that has been partly dewatered by sedimentation, resulting in a more consolidated, molasses-like mass; this can cause significant environmental damage if released or spilled into waterways.

**coal slurry** *Coal.* a mixture of water, fine coal particles, and other wastes resulting from the washing of coal.

**coal spontaneous combustion** see SPONTANEOUS COMBUSTION.

**coals to Newcastle** see NEWCASTLE.

**coal tar** *Coal.* a black, thick liquid formed during the distillation of coal that yields compounds such as benzene and phenol, from which a large number of dyes, drugs, and other compounds are derived, and that finally yields a residual substance which is used in pavements.

**coal washing** *Coal.* the treatment of coal in a liquid medium to reject waste materials such as rock, clay, and other minerals.

**coal workers' pneumoconiosis (CWP)** *Health & Safety.* the technical name for BLACK LUNG DISEASE.

**coarse coal** *Coal.* a size classification for coal particles of more than 12.5 mm.

**Coase theorem** *Economics.* the proposition that the method by which property rights are allocated does not affect economic efficiency (assuming an absence of transaction costs), so long as they are well defined and a free market exists for the exchange of rights between those who have them and those who do not. [Named for British economist Ronald H. *Coase,* born 1910.]

**coastdown** *Nuclear.* an action that permits reactor power level to decrease gradually as the fuel in the core is depleted.

**coaxial cable** *Communication.* a transmission line that is constructed in such a way that an inner conductor carrying the signal is surrounded by, and isolated from, a grounded outer conductor that acts to minimize electrical and radio frequency interference; the primary type of cabling used by the cable television industry and also widely used for computer networks. [So called because both conductors share the same longitudinal axis.]

**Cockroft–Walton accelerator** *Nuclear.* an early particle accelerator consisting of a multistep voltage divider employed to accelerate ions linearly through constant voltage steps; the first device to produce an artificial nuclear disintegration by bombarding lithium with accelerated protons (1932). [Built by English scientists John *Cockroft* and Earnest *Walton*.]

**co-combustion** *Chemistry.* the combined combustion of different fuels (e.g., biomass and coal) in a single furnace.

**COD** chemical oxygen demand.

**code** *Measurement.* governmentally adopted specifications that describe the minimum requirements for a given industrial activity, such as the design and construction of buildings.

**coefficient of coupling** *Electricity.* a numerical rating between 0 and 1 that specifies the degree of magnetic coupling between two circuits.

**coefficient of friction** *Physics.* a constant that, when multiplied by the normal force between two bodies that are in surface contact, indicates the force of friction between them necessary for them to start sliding.

**coefficient of haze (COH)** *Environment.* a measurement of the amount of dust and smoke required to produce a certain reduction of light (haze) in a theoretical 1000 linear feet of air; a coefficient less than 1.0 is considered acceptable air quality and more than 3.0 is considered dirty.

**coefficient of heat transmission** *Solar.* the rate of heat loss in Btu per hour through a square foot wall or other building surface when the difference between indoor and outdoor air temperatures is 1°F.

**coefficient of performance (COP)** *HVAC.* a rating for cooling systems; can be defined as the amount of heat removed from a cooled space divided by the amount of energy supplied to the cooler to accomplish this removal. It can also be a rating for a heating system, as the ratio of heat delivered to a space divided by the energy required for this.

**coefficient of thermal expansion** *Physics.* the amount of increase in size of a solid object that takes place for each degree of increase in its temperature at constant pressure, either an expansion in its length **(coefficient of linear expansion)** or in its volume **(coefficient of cubical expansion)**.

**coenzyme A** *Biological Energetics.* a coenzyme that takes part in fatty acid metabolism; it is found in the cells of all living organisms and is one of the most important substances involved in cellular metabolism. It helps in the conversion of amino acids, steroids, fatty acids, and hemoglobins into energy.

**cofferdam** *Hydropower.* **1.** a temporary watertight barrier that is pumped dry to expose the bottom of a body of water, to allow construction on the foundation of a dam. When the project is completed, the cofferdam can be demolished. **2.** *Oil & Gas.* a heavy double bulkhead used in tankers to separate oil tanks.

**coffin** *Nuclear.* a shipping container for spent fuel rods that is heavily shielded and designed to withstand severe impacts that might occur during transportation.

**co-firing** *Consumption & Efficiency.* the use of a supplemental fuel in a boiler in addition to the primary fuel that it was originally designed to use.

**cofuel** *Renewable/Alternative.* a secondary fuel used in conjunction with a main fuel, especially an alternative fuel (e.g., wood residue, sewage sludge) used to supplement a conventional fuel such as coal.

**cogeneration** *Conversion.* the simultaneous production by means of a single fuel source of both useful energy (usually electricity) and heat (e.g., process steam) that can then be recovered for use as additional energy.

**cogwheel** *History.* a gear wheel with metal teeth; one of the elementary machines of antiquity.

**coil** *Electricity.* **1.** a number of turns of a wire used to introduce inductance into an electric circuit. **2.** *HVAC.* the device in an air conditioner or heat pump through which the refrigerant is circulated.

**coincident demand** *Consumption & Efficiency.* the energy demand required by a given customer or class of customers during a particular time period.

**coke** *Materials.* **1.** a residue of fixed carbon and residual ash left after heating bituminous coal at an extremely high temperature in the absence of air, so that the carbon and ash are fused together; used as a fuel and as a material for the conversion of iron ore into iron and steel. **2.** to convert coal into such a substance. Thus, **coke coal, coke oven, coking coal,** and so on. **3.** a similar residue that is the final product of thermal decomposition in the condensation process in petroleum cracking. This product is described as **marketable coke** or **catalyst coke.**

**coke oven gas** *Coal.* a gas emitted from the production of coke, made up of water vapor, hydrogen, methane, carbon monoxide, and other gases.

**coking** *Oil & Gas.* a process for thermally converting and upgrading heavy petroleum residues into lighter products and byproduct petroleum coke (see COKE above).

**cold cranking amps (CCA)** *Storage.* a performance rating for automobile starting batteries, defined as the current that the battery can deliver for 30 seconds while maintaining a terminal voltage greater than or equal to 1.20 volts per cell, at 0°F (–18°C), when the battery is new and fully charged.

**cold fusion** *Nuclear.* a thermonuclear reaction that can occur at ordinary temperatures, rather than the extremely high temperatures usually required for such fusions; currently being investigated as a possible future source of energy but not yet shown to be effectively possible.

**cold night sky** *Earth Science.* the low effective temperature of the sky on a clear night.

**cold rolling** *Materials.* a processing of metal in which a metal ingot is rolled into a slab through rolling mills, either alone or in a series and below the recrystallization temperature, usually at room temperature. Thus, **cold-rolled steel.**

*cold-rolling Tons of coiled cold-rolled steel stored in preparation for processing into tin-plated steel.*

**cold shutdown** *Nuclear.* the status of a reactor coolant system at atmospheric pressure and at a temperature below 200°F following a reactor cooldown.

**cold start** *Transportation.* difficulty in starting an internal combustion engine in cold weather because of factors such as slower evaporation of gasoline, restricted flow of oil, and slower chemical reactions inside the battery.

**cold stress** *Materials.* stress in materials resulting from low temperatures.

**cold thermal storage** *HVAC.* the storage of thermal energy at low temperature (as in the form of snow, ice, or water) for later use in a cooling procedure; e.g., an ice storage tank employed in an air-conditioning system.

**cold-water pollution** *Environment.* an artificial lowering of the temperature of a body of water; may occur below a large dam that has a valve mechanism for releasing cold water from the bottom of the dam into the river downstream. This can pose a threat to native fish that are acclimated to warmer water temperatures.

**coliform bacteria** *Health & Safety.* any of various fermentative, gram-negative, rod-shaped anaerobic bacteria, typically found in the intestinal tracts of humans and other animals; their presence in waste water is an indicator of pollution and of potentially dangerous contamination.

**collar** *Mining.* the timber or concrete around the top of a shaft or drill hole.

**collecting** *Ecology.* a hunter-gatherer strategy in which human groups move out from their residential base to harvest, process, and transport seasonal concentrations of food and other resources for future use.

**collection and bypass system** see BYPASS SYSTEM.

**collective dose** *Nuclear.* the average effective radiation dose in a population; i.e., the total radiation dose incurred by a population, divided by the number of people in the population.

**collector** *Solar.* any device that absorbs solar radiation and transforms it to thermal energy that can then be used for heating or power generation.

**collector efficiency** *Solar.* the ratio of solar radiation striking a collector to the energy transferred to the collector's heat transfer fluid.

**collector fluid** *Solar.* a liquid or air used to absorb solar energy and transfer it for direct use, indirect heating of interior air or domestic water, or to a heat storage medium.

**collector plate** *Solar.* a metal sheet in a solar collector whose primary function is to absorb incoming solar radiation reaching it through the glazing, while losing as little heat as possible as it transfers the retained heat to the heat transfer medium.

**collector tilt** *Solar.* the angle between the plane of the horizon and the surface of a solar collector, generally used to maximize the collection of solar radiation. Also, **collector angle.**

**collider** *Nuclear.* a particle accelerator in which two beams traveling in opposite directions are steered together to provide high-energy collisions between the particles in one beam and those in the other.

**colliery** *Coal.* a complete coal-mining operation, including the mine, shops, preparation plant, and equipment.

**color** *Lighting.* **1.** the sensation, determined by wavelength, that is generated by light in the visible spectrum. **2.** the property of a given material or object by which it is perceived as resembling the light of a particular wavelength; technically referred to as its *hue* (e.g., red, blue, and so on). Also spelled **colour.**

**color rendering** *Lighting.* the extent to which a light source makes the color of an object appear natural and appealing to the eye and faithful to the true color of the object. Also, **color rendition.**

**color rendering index (CRI)** *Lighting.* a description of the effect of a light source on the color appearance of a given object, compared to a reference source of the same color temperature; expressed on a scale of 0 to 100 (based on a maximum value of 100 for natural daylight), with 80 or above being considered high quality. A typical cool white fluorescent lamp will have a CRI of about 65, while newer types of fluorescent lamps can have a CRI of 80 or more. Incandescent and halide lamps generally have a CRI approaching 100. Also, **color rendition index.**

**color temperature (CT)** *Lighting.* a numerical measurement used to quantify the tone of light being used for viewing, expressed in terms of the Kelvin scale (K). A (theoretical) metal being heated to extremely high temperature will first emit deep red light, then orange, white, and eventually shades of blue; thus the color of the emitted light can be correlated with the temperature of the metal. Typical values that can be assigned according to this scale are: a setting sun 2300 K; a typical incandescent light bulb 2800–2900 K; a summer daylight sky 5500 K; a cloudy overcast sky 6500–8000 K.

**column flotation** *Coal.* a coal-cleaning technique in which coal particles are attached to uniformly sized and evenly dispersed air bubbles rising in a vertical column, so that the coal can be removed at the top of the column.

**combined cycle** *Conversion.* a process in a power plant involving a combination of gas and steam turbines; a combustion turbine converts natural gas into electrical energy, and then exhaust heat from that process is used to produce steam, which powers a steam turbine to generate more energy. Thus, **combined-cycle (power) plant.**

**combined heat and power (CHP)** *Conversion.* another term for COGENERATION; i.e., the process of extracting both useful energy and useful heat from the same process.

**combustible** *Materials.* **1.** describing a material that is able to burn. **2.** specifically, describing a material that is relatively difficult to ignite and slow to burn, as opposed to a *flammable* material that burns relatively easily. Thus, **combustibility.**

**combustion** *Chemistry.* a process of burning, especially the burning of fuel and oxidant to produce heat and/or work. See next page.

**combustion analyzer** *Measurement.* an instrument used to measure the efficiency of a combustion process, as by measuring levels of carbon monoxide, nitrogen dioxide, or sulfur dioxide.

**Combustion Institute** (est. 1954), an educational nonprofit society whose purpose is to promote and disseminate international research in combustion science. The main activity of the Institute is the widely attended International Symposium on Combustion.

**combustion** The burning of fuel and oxidant to produce heat and/or work. It is the major energy release mechanism in the earth and key to humankind. Combustion includes thermal, hydrodynamic, and chemical processes. It starts with the mixing of fuel and oxidant, and sometimes in the presence of other species or catalyst. The fuel can be gaseous, liquid, and solid. The mixture may be ignited with a heat source. When ignited, chemical reactions of fuel and oxidant take place and the heat release from the reactions makes the process self-sustained. The combustion products include heat, light, chemical species, pollutants, mechanical work, and plasma. Sometimes, a low-grade fuel, e.g., coal, biomass, and coke, can be partially burned to produce higher-grade fuel, e.g., methane. The partial burning process is called gasification. Various combustion systems, e.g., furnaces, combustors, boilers, reactors, and engines, are developed to utilize combustion heat, chemical species and work. Advanced measurement and control equipments have been traditionally used as a tool for the analysis of a combustion system. The analysis can lead to the lower energy use and pollutant emissions of the system. Recently, the computational codes, including the kinetics and fluid dynamics codes, have started to be used as an added tool for the combustion analysis. In recent years, concerns over the availability of energy and the impacts of combustion processes on environments have attracted more and more attention. To use energy efficiently and responsively has become a main goal of society.

**Chenn Q. Zhou**
Purdue University, Calumet

**combustion turbine** *Conversion.* an internal combustion engine in which liquid or gaseous fuel is used to generate mechanical energy through a rotating shaft, which then drives an electric generator or other piece of equipment.

**combustor** *Conversion.* **1.** a chamber in which air is mixed with a fuel and ignited. **2.** the system in a gas turbine or jet engine that contains burners, ignitors, and injection devices in addition to the combustion chamber.

**comfort system** *HVAC.* describing an air conditioning system designed to provide a comfortable environment for humans, by controlling both temperature and humidity to acceptable levels. Thus, **comfort cooling** or **conditioning.** Compare PRECISION COOLING.

**comfort zone** *HVAC.* **1.** the conditions of temperature, humidity, and air movement at which the greatest percentage of people will feel comfortable. **2.** based on this, a term for a familiar situation in which a person feels relaxed and comfortable.

**command-and-control** *Policy.* a law or regulation consisting of a "command" that sets a standard such as the maximum level of permissible pollution, and a "control" that monitors and enforces the standard. Examples are the U.S. Clean Water and Clean Air Acts. Thus, **command-and-control instrument.**

**commensalism** *Ecology.* a biological relationship between individuals of two different species in which one benefits and the other is not significantly affected, either negatively or positively.

**commercial** *Consumption & Efficiency.* describing the sector of the economy that is neither residential, industrial, or agricultural; e.g., retail stores, business offices, hotels and motels, health care facilities, educational institutions, and financial services. Thus, **commercial customer, sector, service,** and so on.

**commercial fuel** *Consumption & Efficiency.* **1.** a term for fossil fuel-derived fuels, such as liquefied petroleum gas, kerosene, natural gas, or coal. **2.** fuels that are traded in a formal market. Also, **commercial energy**. Compare TRADITIONAL FUEL.

**comminution** *Mining.* the breaking, crushing, grinding, or pulverizing of coal, ore, or rock.

**committed dose** *Nuclear.* a dose that accounts for continuing exposures over long periods of time (such as 30, 50, or 70 years).

**commodity** *Economics.* **1.** any physical object produced in an economic process. **2.** specifically, a standard agricultural or industrial substance that is marketed in its raw, unprocessed state; e.g., wheat, rice, sugar, cotton, gold, silver, crude oil, natural gas.

**commodity-based** *Economics.* referring to a price for oil or another energy commodity obtained under contracts to buy or sell in a domestic futures market.

**commodity swap** *Economics.* an exchange of streams of cash flows over time between two parties, in which the cash flows are dependent on the price of an underlying commodity. In a typical commodity swap, one side pays a fixed price, while the other pays a floating (varying) price that corresponds to the market price of the underlying commodity. Oil swaps commonly base the variable payment on the average value of an oil index over a period of time, which removes the effects of an unusually volatile single day.

**common carrier** *Transportation.* **1.** a transporter obligated by law to provide service on a regular basis for all interested parties to the limit of its capacity, in return for a fee that is uniformly charged to all users; e.g., a city bus company. **2.** *Communications.* a government-regulated organization that provides telecommunications services to the general public.

**communications satellite** *Communication.* an artificial satellite that relays radio, television, Internet, telephone, and other communication signals around the world; usually follows a geostationary or geosynchronous orbit.

**community energy** another term for DISTRICT ENERGY.

**commutation** *Electricity.* **1.** the repeated reversal of current through the windings of an armature in a direct-current motor so that direct current is provided at the brushes. **2.** the switching of currents between various paths as needed for the operation of a system or device.

**commutator** *Electricity.* a part in a direct-current motor or generator that provides electrical continuity between the rotating armature and the stationary terminal, and also permits reversal of the current in the armature windings.

**commuter-rail** *Transportation.* describing a railway system that provides passenger service between suburban and metropolitan areas, usually having reduced fares for multiple rides and commutation tickets for regular riders.

**compact fluorescent lamp (CFL)** *Lighting.* a lamp employing a smaller fluorescent bulb that typically screws into a standard light socket and produces a color of light similar to an incandescent bulb; CFLs combine the efficiency of fluorescent lighting with the convenience of a standard incandescent lamp. Thus, **compact fluorescent bulb (CFB), compact fluorescent lighting.**

**company town** *Coal.* a historic economic system in coal mining, especially in the U.S., in which the organization owning and operating the mine also provided the miners and their families with housing, maintained a store selling the miners their food, clothing, and other supplies (usually on an exclusive basis), and provided municipal services as a de facto local government.

***company town*** *In a typical company town, the homes of the coal mine workers were built close together next to the rail line used to transport the coal.*

**comparative advantage** *Economics.* a concept formulated by British economist David Ricardo, according to which economic agents (individuals, firms, governments, and other entities) are most efficient when they specialize in the things that they are relatively the best at doing, and then trade with others for other things.

**comparative label** see ENERGY LABEL.

**compass rose** another term for WIND ROSE.

**competent** *Mining.* describing rock that is hard and strong enough to sustain the stable condition of an opening without any structural support or with only minimal support.

**competition** *Economics.* **1.** a market condition in which prices and supply are not established by any particular group; i.e., they are influenced by many market participants and forces, and not determined by a regulatory body. Thus, **competitive. 2.** *Ecology.* the simultaneous demand by two or more organisms or species for a necessary common resource that is in limited or potentially limited supply, resulting in a nonconfrontational struggle among those organisms or species for continued survival.

**competitive equilibrium** *Economics.* the point, defined in prices and quantities, at which the demand for a certain good or service equals the available supply of that item at the time.

**competitive lease** *Economics.* a lease issued in an area believed to contain crude oil or natural gas; auctioned to the highest qualified bidder.

**complementary** *Economics.* describing different goods or services that are consumed at the same time as each other, e.g., the purchase of a motor vehicle and the purchase of fuel to power it. Thus, an increase in the demand for the first will cause an increase in the demand for the second.

**complete mix digester** *Biomass.* a type of anaerobic digester with a mechanical mixing system in which temperature and volume are controlled to maximize the anaerobic digestion process for biological waste treatment, methane production, and odor control.

**complex system** *Measurement.* a physical, biological, or social system with many individual, related parts that interact in such a way as to produce an aggregate behavior that is difficult to predict.

**compliance coal** *Coal.* a low-sulfur coal or blend of coals that when burned will be able to meet sulfur dioxide emission standards for air quality.

**composite** *Materials.* describing a material or material system composed of a combination of two or more smaller constituents that differ in form and chemical composition and that are essentially insoluble in each other.

**compost** *Biomass.* a mixture of decaying organic materials used as a soil amendment or fertilizer, such as manure, hay, dead leaves, cut grass, garden waste, kitchen refuse, and wood ashes.

**compound parabolic collector** *Solar.* a collecting device consisting of two curved reflecting segments that are part of parabolas; a form of solar collector that is stationary (i.e., which does not track the sun). Also, **compound parabolic concentrator.**

**compound parabolic trough** *Solar.* a type of parabolic trough system in which two half-parabolic reflectors with a metal absorber pipe funnel solar radiation to the absorber pipe.

**compressed air** *Storage.* air that is held at a pressure higher (often many times higher) than standard atmospheric pressure, thereby increasing its density; used to inflate automobile tires, operate power tools, and for various other purposes.

**compressed air energy storage (CAES)** *Storage.* a procedure in which off-peak electrical energy is used to compress air into underground storage reservoirs until times of greater electricity demand. To generate power, the compressed air is heated and then passed through turbines.

**compressed gas storage** *Storage.* a storage device for gases (e.g., hydrogen, natural gas, nitrogen) at room temperature under extremely high pressure, typically 20 megapascals. Thus, **compressed hydrogen, nitrogen,** and so on.

**compressed natural gas (CNG)** *Oil & Gas.* natural gas that is comprised primarily of methane, compressed to a pressure at or above 2400 pounds per square inch and stored in special high-pressure containers. It is used as a fuel for natural gas-powered motor vehicles.

**compression** *Physics.* **1.** a condition in which the volume of a substance is reduced as a result of pressure changes. **2.** specifically, the stage in an engine cycle in which the fuel-air mixture is reduced in volume to bring about combustion. **3.** *Communications.* the fact of reducing the number of bits of information required to store or transmit a given data file (without removing any vital data).

**compression ignition (CI)** *Transportation.* the type of ignition that normally initiates combustion in a diesel engine; i.e., rapid compression of the air within the cylinders generates enough heat to ignite the fuel as it is injected (as opposed to the use of spark plugs for ignition). Thus, **compression-ignition engine.**

**compression ratio (CR)** *Transportation.* in a cylinder of an internal combustion engine, the ratio of the volume above the piston at the bottom of its compression stroke (maximum compression) to the volume above the piston at the top of its stroke (minimum compression). In general, the higher the compression ratio of an engine, the greater its efficiency.

**compression stroke** *Transportation.* the regularly occurring phase in a reciprocating engine in which the fuel-air mixture trapped in the cylinder is compressed by piston action.

**compressor** *Chemistry.* any device used to reduce the volume of a substance, especially a device used to increase the pressure of air or another gas, as in a refrigerating or air-conditioning system.

**Compton effect** *Physics.* the elastic scattering of a photon by a massive particle (usually an electron) when the interaction is considered a collision of two otherwise free particles; this applies when the energy of the photon is comparable to or higher than the rest energy of the electron. The collision imparts kinetic energy to the electron, and the quantum energy of the photon is reduced. [Described by U.S. physicist Arthur Holly *Compton,* 1892–1962.]

**computable general equilibrium (CGE)** *Economics.* a top-down model of the economy that includes all of its major components and markets, and the relationships between them.

**computational fluid dynamics (CFD)** *Physics.* the application of computer technology to make quantitative analyses and predictions concerning the behavior of fluids in motion, and their effects on the solids with which they are in contact.

**ComutaCar** see CITICAR.

**concentrating collector** *Solar.* a solar collector that uses reflective surfaces to concentrate sunlight onto a small area, where it is absorbed and converted to heat or, in the case of solar photovoltaic devices, into electricity.

**concentrating system** *Photovoltaic.* a system that uses optical elements to increase the amount of sunlight incident on a photovoltaic (PV) cell. Concentrating PV arrays must track the sun and use only direct sunlight because the diffuse portion cannot be focused onto the cells, and thus are generally limited to very sunny regions.

**concentrator** *Solar.* a solar collector that uses optical methods to concentrate solar radiation before absorption. Thus, **concentrator cell, concentrator system.**

**Concorde** *Transportation.* a British–French supersonic aircraft, the only supersonic aircraft to carry on commercial service for an extended time. It began transcontinental passenger flights in 1976, and it went out of service in 2003 because of safety concerns and the high cost of operation relative to passenger revenue.

**Concordia, Charles** born 1908, U.S. engineer widely known for his work in the analysis and control of electrical power systems.

**condensate** *Oil & Gas.* **1.** hydrocarbons that exist in gaseous form under reservoir conditions, but that condense to a marketable liquid product when brought to the surface, either through natural differences in pressure and temperature or via a production process. **2.** *Materials.* any product of condensation, such as water resulting from the cooling of steam, as in a process of desalination.

**condensate well** *Oil & Gas.* a natural gas well containing a liquid condensate that can be separated from the natural gas either at the wellhead, or during the processing of the gas.

**condensation** *Chemistry.* **1.** the transformation of a gas into a liquid or solid. **2.** a reaction of two or more organic chemicals, one of the products of which is water, ammonia, or a simple alcohol. **3.** *Earth Science.* a process by which water vapor changes to dew, fog, or clouds; brought about either by the cooling of air to its dew point or the addition of enough water vapor to bring the mixture to the point of saturation. **4.** *Physics.* the region of maximum density through which compression waves (such as sound waves) travel.

**condensation trail** see CONTRAIL.

**condenser** *Consumption & Efficiency.* **1.** a chamber enclosing an array of tubes into which the exhaust steam from a steam engine is distributed and condensed by the circulation of cooling water through the tubes. **2.** *Refrigeration.* a heat exchange coil within a mechanical refrigeration system used to reject heat from the system; this is the coil where condensation takes place.

**condensing engine** *Consumption & Efficiency.* a steam engine in which the steam exhaust liquefies in the vacuum space following discharge from the engine cylinder.

**condensing furnace** *HVAC.* a high-efficiency forced-air gas furnace that uses a second condensing heat exchanger to extract the latent heat in the flue gas.

**condensing turbine** *Conversion.* a steam turbine in which the exhaust steam is condensed and the water formed from this process then is used to supply the feedwater for the generator.

**condensing unit** *Refrigeration.* the part of a refrigerating mechanism that pumps vaporized refrigerant from the evaporator, compresses it, liquefies it in the condenser, and returns it to the refrigerant control.

**conditioned** *HVAC.* describing air that has been treated or regulated in some way, as to control its temperature, relative humidity, purity, movement, or pressure. Thus, **conditioned air, conditioned space** or **area.**

**conductance** *Thermodynamics.* **1.** the physical property of an element, device, branch, or system that is the factor by which the mean square voltage must be multiplied to determine the corresponding power lost by dissipation. **2.** the heat transfer through a material divided by the temperature difference across the surfaces of the body.

**conduction** *Physics.* the passage of energy (e.g., heat, sound, or electricity) through a medium (the conductor) while the medium itself experiences no mass movement as a whole.

**conduction band** *Electricity.* a vacant or only partially occupied set of many closely spaced electronic levels resulting from an array of a large number of atoms forming a system in which the electrons can move freely or nearly so; usually used to describe the upper energy band in semiconductor material separated by the energy gap (bandgap) from the valence band.

**conductive** *Materials.* describing a material that has the property of conducting heat or an electric current.

**conductive plastic** *Materials.* a type of polymer consisting alternately of single and double bonds between the carbon atoms, and "doped"; i.e., electrons are removed (through oxidation) or introduced (through reduction). These "holes" or extra electrons can move along the molecule so that it becomes electrically conductive. Conductive plastics have enormous potential, as for solar cells and "smart" windows that can exclude sunlight.

**conductivity** *Electricity.* a measure of the ability of a material to conduct an electrical current, equal to the reciprocal of resistivity.

**conductor** *Electricity.* **1.** a material that is capable of carrying electric current, especially one that is highly suitable for this, such as copper wire. **2.** *Physics.* any material that serves or can serve as a medium for conduction, as of sound, heat, and so on.

**conduit** *Materials.* **1.** a solid or flexible tubing that houses electric wires and that can also serve as a protective shield. **2.** any pipe through which materials may pass. **3.** *Earth Science.* an underground channel that is completely filled with water, and that is always under hydrostatic pressure.

**Conference of the Parties (COP)** *Policy.* an annual meeting of the countries that have signed the Framework Convention on Climate Change. The first COP was held in Berlin in 1995; the noted Kyoto, Japan meeting of 1997 was the third (COP3).

**confinement** *Nuclear.* **1.** the fact of preventing or mitigating the uncontrolled release of radioactive material to the environment by means of a barrier that surrounds the main parts of a facility containing radioactive materials. **2.** the barrier itself.

**confinement time** *Nuclear.* in fusion energy, the length of time that a plasma is maintained at a temperature above the critical ignition temperature. To yield more energy from the fusion than has been invested to heat the plasma, the plasma must be held up to this temperature for some minimum length of time.

**conservation** *Ecology.* **1.** any of various efforts to preserve or restore the earth's natural resources, including protection of wildlife, maintenance of forest or wilderness areas, control of air and water pollution, and prudent use of land and natural resources. **2.** see

ENERGY CONSERVATION. **3.** see CONSERVATION OF ENERGY; CONSERVATION LAWS.

**conservation biology** *Ecology.* the branch of biology concerned with the planning and management of natural resources, and especially with maintenance of natural evolutionary change and the diversity of species and genetic material.

**conservation laws** *Physics.* a general statement that a physical quantity, such as energy, momentum, or mass, is descriptions of the properties of an isolated system (one that does not interact with its surroundings) that have never been observed to change, and thus are said to be "conserved". These laws are among the most fundamental principles of sciences, such as the conservation of energy, conservation of momentum, and conservation of angular momentum.

**conservation of angular momentum** *Physics.* the principle that a system under no external forces will maintain constant total angular momentum.

**conservation of energy** *Physics.* a fundamental law of physics and chemistry stating that the total energy of an isolated system is constant despite internal changes. It is most commonly expressed as "energy can neither be created nor destroyed", and is the basis of the first law of thermodynamics. This conservation principle also applies to other fields; e.g., as expressed in electronics by KIRCHHOFF'S VOLTAGE LAW, or in fluid mechanics by BERNOULLI'S PRINCIPLE.

**conservation of mass** *Physics.* the principle that during an ordinary chemical change, there is no detectable increase or decrease in the quantity of matter; i.e., mass cannot be created or destroyed. This principle generally holds true in larger contexts but can be violated at the microscopic level, as in a nuclear reaction. Also, **conservation of matter.**

**conservation of momentum** *Physics.* the principle that a system under no external forces will maintain constant linear momentum; i.e., the momentum of a closed system does not change.

**conservation supply curve** *Measurement.* a graph depicting the incremental cost of an energy conservation or energy efficiency program (dollars per joule "saved") as a function of the incremental amount of energy saved.

**conservative force** *Physics.* a force for which the work done in displacing a particle from one point to another depends only on the location of those two points, and not on the path taken by the particle in moving from the initial position to the final one; e.g., the force of gravity.

**conservative system** *Physics.* a mechanical system in which there are no losses of energy due to dissipative processes such as friction; i.e., the sum of potential plus kinetic energy is constant.

**conserved** *Physics.* describing a quantity whose value does not change with time during the evolution of a dynamic system (under certain specified conditions).

**constant** *Measurement.* a number, term, quantity, or mathematical object that is assumed to be fixed within the given context.

**constant elasticity of substitution (CES)** *Economics.* a type of production function in which the elasticity of substitution among energy, capital, labor, and other factors is assumed to be constant.

**constant-horsepower load** *Conversion.* a load for which the torque requirement decreases as the speed increases, and vice versa.

**constant-torque load** *Conversion.* a load for which the amount of torque required to drive a machine is constant regardless of the speed at which it is driven.

**constraint** *Physics.* an external restriction on the motion or natural degrees of freedom of a system.

**consumer** *Consumption & Efficiency.* **1.** any individual who purchases or otherwise acquires a good or service for personal use (as opposed to its use in production or resale to others). **2.** more generally, any entity that utilizes energy or resources. **3.** *Ecology.* an organism that obtains energy by feeding on other organisms or on existing organic matter, rather than producing its own energy or obtaining it from inorganic sources.

**consumer price index (CPI)** *Economics.* an aggregate index of the prices of a fundamental "basket" of goods and services, based on the amount paid by consumers in a representative group of urban areas; e.g., food and beverages, housing, transportation, medical care, apparel, and recreation.

**consumer-subsidy equivalent (CSE)** *Economics.* an integrated metric of the amount of monetary support provided to consumers by various government policies.

**consumer surplus** *Economics.* a premium that consumers would be willing to pay for a certain good or service above what they actually do pay for it, according to its market price at a given time.

**consumption** *Consumption & Efficiency.* **1.** the fact of consuming; the use of goods, services, or resources. **2.** specifically, the use of energy resources; e.g., the use of oil or coal.

**consumptive use** *Hydropower.* any withdrawing of water from a given source, in which little or none of this water is later returned to the supply source; can apply to water purposely removed for human use (e.g., irrigation) or to natural loss (e.g., evaporation).

**contact** *Electricity.* **1.** the conducting part of a component, such as a switch or relay, that interacts with another conducting part to make or break a circuit. **2.** *Communications.* the point at which an object, such as an aircraft or ship, is first detected by radar or another detecting device.

**containment** *Nuclear.* the confinement of a hazardous material that is being produced, stored, manipulated, transported or destroyed, in order to prevent or limit its contact with people and the environment.

**contaminant** *Health & Safety.* any undesired radioactive material or residual radioactivity that is deposited in excess of acceptable levels in human tissue or on or in structures, areas, or objects. Thus, **contamination.**

**contango** *Economics.* an energy market condition in which prices for a given commodity are higher in the future delivery months than in the current delivery month, often due to the cost of storing and insuring the commodity. Compare BACKWARDATION.

**continental glacier** *Earth Science.* a thick glacier or sheet of ice of such size that it covers a substantial portion or the entire surface of a continent. Also, **continental ice sheet.**

**continental shelf** *Earth Science.* relatively shallow ocean-bottom lands close to the edge of a continent; the gently sloping section of a continental margin lying between the shoreline and the continental slope. Also, **continental platform.**

**contingent valuation** *Economics.* a survey-based economic method that is used to quantify in monetary terms the benefits (or costs) of an environmental policy; used in contexts in which an environmental good or service does not have a readily definable market price, e.g., the value of recreational fishing.

**continuous accumulation** *Oil & Gas.* a term for oil and natural gas that occurs in an extensive reservoir or reservoirs and that is not necessarily related to conventional structural or stratigraphic traps. Such accumulations lack well-defined petroleum/water contacts and thus are not localized by the buoyancy of petroleum in water.

**continuous miner** *Mining.* a machine designed to remove coal from the working face, break it up mechanically, and load it into cars or conveyors in one continuous operation, without the use of cutting machines, drills, or explosives. Thus, **continuous mining.**

**continuous scale label** *Consumption & Efficiency.* a display on certain energy-using devices (such as household appliances) that shows the energy consumption of the best and worst products of this type currently on the market, as a relative context for the energy consumption of the given product.

**contour mining** *Mining.* a technique of coal mining practiced on hilly or mountainous terrain, in which workers use excavation equipment to cut into the hillside along its contour to remove the overlying rock and then mine the coal, until the overburden becomes uneconomical to remove. Also, **contour stripping.**

**contracts for differences (CFD)** *Electricity.* a type of bilateral contract in which the electric generation seller is paid a fixed amount over time, which is a combination of the short-term market price and an adjustment with the purchaser for the difference.

**contrail** *Earth Science.* condensation trail; a cloudlike streamer that is frequently observed behind jet aircraft flying in clear, cold, humid air; the moving aircraft disturbs particles of ice or water vapor, causing a pressure reduction above the wing surfaces which, when combined with water vapor in the engine exhaust gases, tends to condense and leave a visible trail of condensed water vapor.

**controlled fire** *History.* **1.** the fact of humans having the ability to start fires purposely and control them (as opposed to a natural fire event such as lightning striking a dead tree); a typical estimated date for the first occurrence of this is about 500,000 years ago, though various scholars have proposed dates ranging from 1.5 million to 150,000 years. **2.** *Environment.* a contemporary strategy of purposely setting a fire in a natural setting to achieve some desired effect; e.g., to reduce the threat of uncontrolled wildfires in the future. Also, **controlled burn(ing).**

*controlled fire Controlled burning of grassland, a Native American practice for centuries but generally eliminated by the late 19th century, has been revived in recent years as its ecological value has been recognized.*

**controlled flight** *Transportation.* the historic development of the ability to direct the course of an airborne vehicle, as opposed to simply launching a vehicle into the air.

**controlled fusion** *Nuclear.* the confinement of a sustained fusion reaction in a controlled environment for the purposes of power generation.

**control rod** *Nuclear.* a rod, plate, or tube containing a material such as hafnium or boron, used to control the power of a nuclear reactor. By absorbing neutrons, a control rod prevents the neutrons from causing further fissions.

**control volume** *Thermodynamics.* a fixed region in space chosen for the thermodynamic study of mass and energy balances for flowing systems.

**convection** *Thermodynamics.* **1.** a transfer of heat or mass that occurs when a fluid flows over a solid body or inside a channel while temperatures or concentrations of the fluid and the boundary are different; transfer occurs because of the motion of the fluid. **2.** *Earth Science.* the movement and mixing of ocean water masses, usually caused by temperature differences between them.

**convection cooling** *HVAC.* a natural cooling process in which hot air flows upward from the object or space that is being cooled.

**convector** *HVAC.* a heating unit that has openings for the air to enter, become warm, and then exit. Thus, **convector heater.**

**conventional** *Consumption & Efficiency.* describing energy sources that have been widely used in the industrial world for an extended period of time, such as petroleum, natural gas, or coal, as opposed to alternative sources such as wind or solar energy. Large-scale hydropower and nuclear power generation are usually also considered conventional forms of energy. Thus, **conventional energy, power, farming,** and so on.

**conventional accumulation** *Oil & Gas.* a term for oil or natural gas that occurs in structural or stratigraphic traps, commonly bounded by a water contact and therefore affected by the buoyancy of petroleum in water.

**conventional gas** *Oil & Gas.* a term for natural gas obtained by the traditional method of extraction from deep-lying geologic formations, as opposed to that obtained from other sources; e.g., coalbed methane.

**conventional mining** *Mining.* a term for the older form of room-and-pillar mining, in which the coal seam is cut, drilled, blasted, and then loaded into cars in a series of steps; contrasted with CONTINUOUS MINING which does not involve drilling and blasting.

**conventional oil** *Oil & Gas.* a term for oil obtained by traditional extraction methods (e.g., well drilling), rather than from UNCONVENTIONAL sources such as shale, tar sands, biofuels, and so on.

**conversion** 1. a process of transforming one form of energy to another; e.g., from chemical to thermal energy during combustion, or from thermal to mechanical energy by means of a heat engine. 2. a change from one system of units to another, as from temperature in degrees Celsius to degrees Fahrenheit. 3. *Nuclear.* a process in which radioactive materials, chiefly uranium-238 and thorium-232, are bombarded with neutrons and transformed into plutonium-239 and uranium-233, respectively.

**conversion efficiency** *Conversion.* the physical ratio of desired output to total input in an energy conversion process, such as the conversion of fuel to heat.

**conversion factor** *Measurement.* a numerical factor used to convert a given quantity between one system of units and another; e.g., to convert a metric distance to feet, the quantity in meters is multiplied by a conversion factor of 3.3.

**converter** *Consumption & Efficiency.* 1. a circuit or device that changes signals, frequency, voltage, current, or data from one form or mode to another. 2. specifically, a device that changes alternating current to direct current. 3. *Materials.* a refining furnace in which air is blown through the molten bath in order to oxidize unwanted impurities.

*converter U.S. iron maker William Kelly became known for his Kelly Converter (1857), which supposedly preceded Britain's Bessemer process. However, subsequent tests have shown that his converter could not have been used successfully.*

**converter reactor** *Nuclear.* a nuclear reactor that converts fertile material into fissile material; the fissile material usually is different than the reactor fuel; e.g., a reactor fueled with U-235 that converts U-238 into Pu-239.

**Cook, Earl Ferguson** born 1920, U.S. geologist known for his integrative analysis of the role of energy in society, in particular the relationship of economic growth to fossil fuel consumption.

**cool** *Lighting.* a subjective description of the way in which the human eye perceives a certain light source; colors at the blue end of the spectrum are considered to be "cool" and colors at the red end are considered to be "warm" (based on the traditional association of the color blue with cold or ice, and red with heat or fire). Thus, **cool light(ing), coolness,** and so on. In terms of COLOR TEMPERATURE, cool sources actually have a higher temperature (4000 K and higher) than warm sources.

**coolant** *HVAC.* 1. any material, usually a liquid, that is characterized by its ability to absorb heat from its environment and transfer it away from the heat source. 2. *Nuclear.* a substance circulated through a reactor to remove or transfer heat; common coolants include light or heavy water, various gases such as air and carbon dioxide, liquid sodium, and certain organic compounds.

**cooling capacity** *HVAC.* the rate at which equipment removes heat from the air passing through it under specified conditions of operation, expressed in watts; e.g., the cooling capacity of an air conditioner is frequently measured by its Btu rating (the amount of heat it can remove per hour).

**cooling degree-day** see DEGREE-DAY.

**cooling load** *HVAC.* the amount of cooling per unit time required to maintain the desired temperature for a conditioned space or product.

**cooling tower** *Consumption & Efficiency.* a tall, towerlike structure used for reducing the temperature of water, by bringing it into contact with an airstream where a small portion of the liquid is evaporated and the major portion is cooled; used for large-scale air-conditioning installations and for cooling steam condensers of power plants.

**cooling tower fogging** *Consumption & Efficiency.* a fog condition created at a cooling tower when the exhaust air **(cooling tower plume)** becomes supersaturated, so that part of the water vapor condenses into viable liquid droplets.

**cooling water** *Consumption & Efficiency.* water that is used to remove metabolic or process heat; in a power plant, the water is cycled between a heat exchanger and a cooling tower.

**cooling water intake structure (CWIS)** *Nuclear.* a device designed to extract large volumes of water from a lake, river, ocean, or reservoir for the purpose of cooling at a nuclear power plant or other industrial facility. Also, **cooling water intake system.**

**cooperative** *Economics.* an entity that is owned and operated by those who benefit directly from its products and services; e.g., an electric utility company.

**COP** coefficient of performance.

**COPD** chronic obstructive pulmonary disease.

**Copernicus, Nicolaus** 1473–1543, Latinized version of the name Mikolaj Kopernik, the Polish astronomer who established the "heliocentric" model of the solar system, i.e., the principle that the sun (not the earth) is the central point to which the motions of the planets are to be referred. He was the first person in history to create a complete general system of the solar system (the **Copernican system**), combining mathematics, physics, and cosmology.

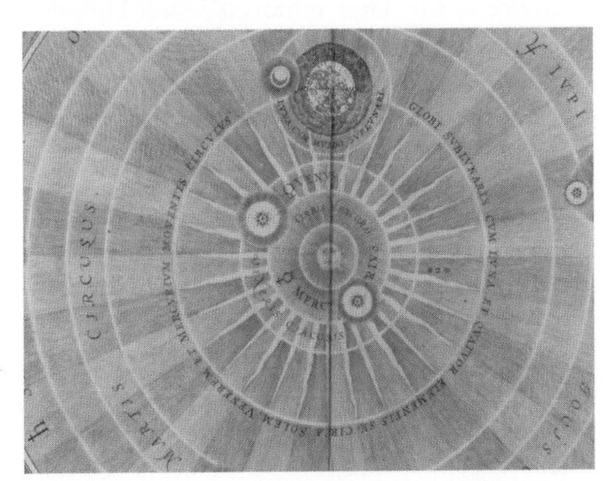

*Copernicus Detail from an artistic representation of the Copernican system, showing the earth (top) and other planets revolving around the sun at the center.*

**copper** *Chemistry.* an element having the symbol Cu, the atomic number 29, an atomic weight of 63.54, a melting point of 1083°C, and a boiling point of 2595°C; a soft, reddish, ductile metal that is an excellent conductor of electricity.

**coppice** *Biomass.* **1.** a method of regenerating a forest stand in which trees in the previous stand are cut and the majority of new growth is associated with vigorous stump sprouting from roots or root suckers. Also, **coppicing.** **2.** to produce new growth in this manner.

**coppice regeneration** *Biomass.* the ability of certain hardwood species to regenerate by producing multiple new shoots from a stump left after harvest.

**coral bleaching** *Environment.* a phenomenon in which coral reef colonies become whitened as a result of higher temperatures or other environmental stresses; e.g., pollution episodes.

**cord** *Biomass.* the common measurement for cut wood, especially fuelwood; one cord measures 4 feet by 4 feet by 8 feet, or 128 cubic feet.

**core** *Earth Science.* **1.** the central region of the earth, originating at a depth of about 2900 km, outside of which are the mantle and the crust. It is thought to consist of a molten outer core and a solid inner core; the temperature of the inner core is not known but has been estimated at 5000–7000° C. **2.** *Nuclear.* the nuclear fuel and fission reaction region in a nuclear reactor.

**core inflation** *Economics.* a measure of inflation that is thought to provide a better indicator of underlying inflationary pressures, because it excludes both food and energy prices, which can be volatile.

**core melt** *Nuclear.* an event or sequence of events resulting in the melting of part of the fuel in the reactor core. Also, **core damage, core melt accident.**

**core sample** *Mining.* a small cylindrical sample taken from an area while drilling, for the purpose of examination or analysis.

**Coriolis force** *Earth Science.* an apparent force exerted on a moving object by the rotation of the earth; an object that is moving horizontally above the earth's surface in the Northern Hemisphere tends to show a rightward deflection, and one in the Southern Hemisphere tends to show a leftward deflection. Also, **Coriolis effect.** [Described by Gaspard de *Coriolis,* 1792–1843, French civil engineer.]

**Corn Laws** *History.* British regulations on the import and export of grain, mainly wheat, intended to control its price. Officially imposed in 1815 and then repealed in 1846, signaling a shift toward free trade.

**corn stover** *Biomass.* residue materials from the corn plant obtained after harvesting, consisting of the cob, leaves, and stalk; used as a source of fuel.

**cornucopia** *Consumption & Efficiency.* **1.** a traditional image of a harvest basket overflowing with food, symbolizing a great abundance of foods. **2.** by extension, the view that the earth's natural resources are unlimited, or, if limited, that any possible shortage can be overcome by technological innovation. Thus, **cornucopian.**

**corona** *Solar.* the extended upper atmosphere of the sun.

**corona hole** *Solar.* a region on the solar disk where surface magnetic fields open out to allow particle streaming and subsequent magnetic storms on the earth.

**corporate average fuel economy (efficiency)** see CAFE.

**corporate environmental strategy (CES)** *Economics.* a functional business strategy that allows a firm to lessen its environmental liabilities, compliance costs, and product development costs in an organized and reportable fashion.

**corporate social responsibility (CSR)** *Social Issues.* the operation of a business so as to generate profit for its owners and shareholders and create long-term value, while at the same time adhering to the ethical, legal, and public standards held by the larger society, such as safeguarding the rights and welfare of workers, offering safe products of reasonable quality, protecting the environment, avoiding bribery and corruption, and contributing to sustainable development.

**corposant** *Earth Science.* a luminous, sporadic electric discharge from tall pointed objects, such as a church steeple or a ship's mast; more commonly known as Saint Elmo's fire.

**correlated color temperature (CCT)** see COLOR TEMPERATURE.

**corrosion** *Materials.* the degradation of metals or alloys due to chemical reactions with their environment, accelerated by the presence of acids or bases; for example, the rusting of metal surfaces exposed to moist air or to impure water. Thus, **corrosive.**

**Cort, Henry** 1740–1800, English inventor who revolutionized the British iron industry with the PUDDLING method, a process of converting cast iron into wrought iron or steel.

**cosmic** *Physics.* characteristic of or relating to the universe as a whole, especially the portion of the universe outside the earth's atmosphere.

**cosmic rays** *Physics.* high energy sub-atomic particles, mostly protons and helium nuclei, that travel across space at close to the speed of light. The lowest energy cosmic rays originate in the sun, and higher energy ones from supernovae and pulsars within the galaxy. Those with the highest energy may be extragalactic in origin, possibly from quasars and active galactic nuclei. Also, **cosmic (background) radiation.**

**cosmological constant** *Physics.* a constant in the relativistic gravitational field equations representing a repulsion that offsets the gravitational attraction which would otherwise make the matter distribution collapse; added by Einstein to his general theory of relativity to account for an apparently non-expanding universe, but later removed when Hubble's observations seemed to indicate that it was not needed.

**cosmology** *Physics.* the study of the origin of the universe, or a theory to account for this.

**cost-benefit analysis** *Economics.* a method of analysis in which the economic impacts of a proposed plan, policy, or project are quantified in dollars (or another monetary unit) for the purpose of comparing the total value of the negative impacts (costs) with the positive impacts (benefits).

**cost, insurance, (and) freight** *Economics.* a type of sale in which the buyer of a product (e.g., coal) agrees to pay a unit price that includes the free on board value of the product at the point of origin, plus all costs of insurance and transportation.

**cost of conserved energy (CCE)** *Economics.* the additional cost that must be invested in order to implement a long-term energy-saving strategy or feature; e.g., the cost to a homeowner to install a green roof on his house or a solar heater for his swimming pool; in these examples CCE may include not only the cost of the installation itself but the interest on money borrowed to pay for it.

**cost-of-service ratemaking** *Economics.* a ratemaking procedure in which prices are set by regulators who seek to match the rates charged to consumers with the costs incurred in providing those consumers with electricity.

**cost-plus** *Economics.* **1.** a method of determining the price of a good or service that includes not only the cost of the item itself but also the seller's expenses in developing or producing it. **2.** specifically, such an arrangement for oil or natural gas in which the consumer price is equal to the wellhead price plus transportation costs, distribution costs, a return to the seller, and taxation.

**Cotella** *Oil & Gas.* (Katalla) a region along the Gulf of Alaska, noted as the site of the first oil discovery in Alaska and thus the earliest evidence of the region's potential as an energy source.

**Cottrell, Frederick** 1877–1948, U.S. inventor of the ELECTROSTATIC PRECIPITATOR, a key technology to reduce air pollution.

**Cottrell, William** born 1903, U.S. sociologist who developed a general theory of social and economic change based on transitions in energy sources and their conversion technologies.

**Coulomb, Charles-Augustin de** 1736–1806, French physicist known for his discoveries in electricity, magnetism, and friction.

**coulomb** *Electricity.* the standard international unit of electric charge, symbol C; equal to the charge that passes through any cross-section of a conductor in one second during a constant current flow of one ampere.

**coulomb barrier** *Nuclear.* the electric repulsion that must be overcome in nuclear fusion in order for particles to become close enough to each other for the attractive nuclear strong force to fuse the particles.

**Coulomb force** *Electricity.* the force between two charged particles, as between electrons and protons, that (by Coulomb's law) is proportional to the product of the charges and inversely proportional to the square of the distance between them and that is either repulsive or attractive, depending on the relationship of the charges.

**Coulomb's law** *Electricity.* the statement that the force between two electric charges is proportional to the product of their magnitudes and inversely proportional to the square of the distance between them; the force between like charges is an attraction, and the force between unlike charges is a repulsion.

**counter-current exchange** *Biological Energetics.* a type of thermoregulation in animals based on blood vessels that run parallel to each other, with the blood flowing in opposite directions; warm blood in the arteries passes very close to cooled blood returning in the veins from the extremities. Heat is transferred from the warmer blood to the cooler as the blood flows through these vessels, thus conserving body heat.

**counterelectromotive force** *Electricity.* a voltage that is induced in an inductive circuit due to a changing current; the polarity of the voltage opposes the polarity of the voltage driving the current.

**cowboy economy** *Economics.* a term for an economic system or philosophy emphasizing free market activities and minimal government regulation; regarded as having a profligate use of nonrenewable resources with little emphasis on issues such as conservation, recycling, and waste disposal. [From the idea that a *cowboy* on the 19th century American frontier conceived of the natural world as having no significant limits or restrictions on the use of resources.]

**cowling** *Transportation.* a streamlined metal covering over all or part of an aircraft engine, usually having hinged or removable panels; designed to protect engine components and to promote airflow cooling.

**CPI** consumer price index.

**cracking** *Oil & Gas.* **1.** any of various refining processes using heat, pressure, and catalysts to decompose and recombine molecules of organic compounds, especially hydrocarbons, in order to form molecules that are suitable for motor fuels and petrochemicals. **2.** *Chemistry.*

any process of breaking up organic compounds into smaller molecules and reassembling the products into other compounds.

**cradle-to-grave** *Consumption & Efficiency.* **1.** a term for all stages in the course of an industrial process, beginning with the material's creation or generation (cradle), extending through its transportation, treatment, storage, and use (lifetime), and culminating in its incineration or disposal (grave). **2.** specifically, a term for hazardous waste control extending over the entire course of an industrial process. The term **cradle-to-factory** is similar to the above but does not include the waste disposal stage.

**crankcase** *Transportation.* a boxlike casing for the crankshaft and connecting rods of certain engines; in an automobile engine, the bottom of the crankcase is a reservoir in which hot motor oil is collected and cooled before being recirculated by a pump.

**crankshaft** *Transportation.* the shaft around which a crank rotates; in most engines or machines, it is the main shaft that transmits power from the crank to the connecting rod.

**creaming curve** *Oil & Gas.* a graphic representation plotting cumulative oil discoveries versus cumulative new field exploratory drilling, meant to illustrate changes over time in the efficiency of drilling.

**creature comfort cooling** see COMFORT SYSTEM.

**creep** *Mining.* **1.** a gradual process in which pillars are forced down into the floor, or up into the roof, of the mine. **2.** a very slow lowering or downhill movement of mining ground.

**creosote** *Materials.* an oily liquid with a burning taste, obtained by distilling coal and wood tar; used as a wood preservative.

**Cretaceous** *Earth Science.* the final geologic period of the Mesozoic era, occurring after the Jurassic and before the Tertiary period (between 136 and 65 million years ago). The amount of Cretaceous coal in North America is greater than that of any other geologic period.

**CRI** color rendering index.

**crib** *Mining.* **1.** a support of horizontally, cross-piled, squared timbers that are cross-piled horizontally in log-cabin style, used to support a structure above. **2.** any structure composed of one or more layers, upon which a load may be spread.

**Crick, Francis** 1916–2004, English biologist who, with James Watson, discovered (1953) the DOUBLE HELIX structure for DNA and its replication scheme.

**Criswell, David** born 1941, U.S. physicist who, with Robert D. Waldron, invented the concept of the LUNAR SOLAR POWER SYSTEM.

**criteria (air) pollutant** *Health & Safety.* any of various air pollutants used by the U.S. Environmental Protection Agency (EPA) as indicators of air quality; the EPA has established a maximum concentration for these, above which adverse effects on human health may occur. Criteria pollutants include ozone, carbon monoxide, lead, nitrogen dioxide, sulfur dioxide, and particulate matter.

**critical** *Nuclear.* describing the state of a reactor when the number of neutrons released by fission is exactly balanced by the neutrons being absorbed (by the fuel and poisons) and escaping the reactor core. A reactor is said to be critical when it achieves a self-sustaining nuclear chain reaction, as when the reactor is operating. Thus, **criticality, critical (nuclear) reactor.**

**critical infrastructure protection (CIP)** *Health & Safety.* a range of activities and procedures designed to protect and maintain essential public services in the event of a deliberate attack or a large-scale natural disaster; e.g., communication networks, electric power systems, economic processes, and so on.

**critical level** *Health & Safety.* the maximum amount of pollutant deposition that a given sector of the environment can tolerate without significant harmful effects. Also, **critical load.**

**critical mass** *Nuclear.* **1.** the smallest mass of fissile material that will support a self-sustaining chain reaction under specified conditions. **2.** in figurative use, the minimal point of development in a process that must be reached in order for significant progress to begin to take place.

**critical point** *Chemistry.* the state at which the properties of the vapor phase of a substance become indistinguishable from those of the liquid phase; i.e., the highest temperature and pressure point at which the liquid and gas phases of a substance can coexist in equilibrium as a single phase.

**critical temperature** *Measurement.* the temperature above which a substance has no transition from the liquid to the gaseous phase; i.e., the critical temperature of a gas is the highest temperature at which it can be liquefied, regardless of the pressure applied.

**Crookes, William** 1832–1919, English chemist and physicist who identified the state of matter known as plasma and discovered the element thallium. He also invented the radiometer and the **Crookes tube,** an early form of cathode-ray tube.

**crop coal** *Coal.* an informal term for inferior coal found on or just beneath the surface of the earth.

**crop residue** *Biomass.* the portion of a harvested crop that remains after the marketable portion of the plant has been removed for use as food or fiber; can be used as a biomass fuel.

**cross (price) elasticity** *Economics.* the change that will occur in the quantity of a demand for a certain good, in response to a change in the price of another, assuming other factors remain constant; e.g., a change in the quantity of oil demanded for power generation as a result of a change in the price of coal; usually expressed as the percentage change that will occur in the latter good in response to a 1% change in the former.

**crossflow turbine** *Conversion.* a water turbine with a drum-shaped runner consisting of two parallel discs connected together near their rims by a series of curved blades; so called because the incoming water hits the blades and then flows across the middle of the wheel to strike the blades again.

**cross subsidy** *Policy.* a policy that reduces costs to particular types of customers, products, or regions by increasing charges to others; the subsidization of one group with the revenues from another. Also, **cross subsidization.**

**CRT** cathode-ray tube.

**crude** *Oil & Gas.* short for CRUDE OIL, especially as used in the oil industry.

**crude assay** *Oil & Gas.* a procedure for determining the general distillation and quality characteristics of crude oil.

**crude oil** *Oil & Gas.* a mixture of hydrocarbons that exists in the liquid phase in natural underground reservoirs and remains liquid at atmospheric pressure after passing through surface separating facilities. It occurs in many varieties, distinguished by specific gravity, density concentrations of other hydrocarbons, volatility, heating value, and sulfur content. Crude oil may contain significant amounts of natural gas. Fuels such as motor gasoline, diesel fuel, and jet fuel are derived from crude oil, as are a variety of materials known as petrochemicals.

**crude oil quality** *Oil & Gas.* one of two properties of crude oil, the sulfur content and API gravity, that significantly affect processing complexity and product characteristics.

**crude oil spill** see OIL SPILL.

**crust** *Earth Science.* the outermost region of the earth, from the surface itself to a depth of about 70 km beneath land surfaces **(continental crust)** and 10 km below ocean surfaces **(oceanic crust).**

**cryo-** *Refrigeration.* a prefix meaning "freezing" or "very cold"; used to describe the effects of low temperatures or activities carried on at low temperature.

**cryoadsorption** *Storage.* the storage of a material (e.g., hydrogen) by means of carbon adsorption at low temperatures.

**cryocondensation** *Refrigeration.* the process of a phase change from a gas to a liquid or solid phase when the gas contacts a surface having a temperature lower than the dew point of the gas. Thus, **cryocondense.**

**cryoelectronics** *Electricity.* a field of engineering that studies the design and functioning of electronic systems, circuits, and devices at temperatures approaching absolute zero (0 K or –270°C), especially as applied to the phenomenon of superconductivity. Also, **cryotronics.**

**cryogenic** *Refrigeration.* **1.** relating to the deep-refrigeration domain involving temperatures below 120 K. **2.** describing a substance (e.g., hydrogen) stored at such a temperature.

**cryogenics** *Physics.* the branch of science that deals with the realm of extremely low temperatures and their effect on the properties of matter.

**cryology** *Earth Science.* **1.** the scientific study of snow and ice. **2.** *Refrigeration.* the study of refrigeration at low temperatures ranging down to absolute zero.

**cryopedometer** *Measurement.* an instrument for measuring the depth to which soil is frozen.

**cryopump** *Materials.* **1.** a device designed to produce an ultrahigh vacuum by the condensation or adsorption of a gas at a very low temperature (usually below 77 K). **2.** to capture a condensable vapor by such a device.

**cryosorption** *Refrigeration.* the process of capturing a gas on or in a material that has a very large surface and that has been cooled to a very low temperature; the weak attractive force between the gas and surface is enhanced by the low temperature. Thus, **cryosorb.**

**cryosphere** *Earth Science.* the portion of the earth's surface that is characteristically covered by snow and ice throughout the year. Studies of the extent and status of the cryosphere provide insights into present and past climate change.

**cryotrap** *Refrigeration.* to capture one type of gas by entrapping it under another gas during the process of condensing the latter one at low temperature.

**cryotron** *Electricity.* a superconductive device in which current in one or more input circuits magnetically controls the superconducting-to-normal transition in one or more output circuits.

**crystal** *Materials.* any homogeneous solid that has a regularly repeating atomic arrangement; this may be a chemical element, a compound, or an isomorphous mixture.

**crystalline** *Materials.* relating to or having a crystal structure.

**crystalline silicon** *Photovoltaic.* c-Si, the dominant light-absorbing semiconductor in photovoltaic (PV) cells; it is a relatively poor absorber of light and requires a considerable thickness (several hundred microns) of material, but it yields stable PV cells with good efficiencies.

**crystallinity** *Materials.* a measure of the crystalline properties and qualities of a substance; i.e., how perfectly ordered the atoms are in a crystal structure.

**crystallization** *Materials.* the formation of crystals from a liquid, a vapor, or an amorphous solid.

**crystallography** *Materials.* the study of the properties and formation of crystals.

**CT** color temperature.

**Ctesibius of Alexandria** c. 285–222 BC, Greek physicist and inventor (in Alexandria) who discovered the elasticity of air and invented several devices using compressed air, including force pumps and an air-powered catapult.

**cubic feet per second** *Hydropower.* a unit of measurement for water flow, equal to one cubic foot of water moving past a given point in a time of one second. One cubic foot per second = 7.48 gallons per second, or 448.8 gallons per minute.

**Cugnot, Nicholas-Joseph** 1725–1804, French inventor who built what many consider to be world's first self-propelled mechanical vehicle or automobile (1769), known as a STEAM WAGON.

**cull** *Materials.* **1.** a popular term for any item of production separated out for rejection because it does not meet certain specifications, especially an unacceptable farm product. **2.** see CULL WOOD. **3.** to remove such an inferior or unmarketable item. Thus, **culling.**

**cull wood** *Biomass.* wood that does not have commercial value as timber (because it is dead, rotten, poor in form or quality, and so on), but that can be removed to use as fuel (and to improve the overall health of the forest). Thus, **cull section, cull tree.**

**culm** *Coal.* solid waste products from a coal-mining operation, especially from anthracite coal mines. A **culm bank** or **pile** is an accumulated mass of such wastes.

**cultivator** *History.* a farm implement that is used to break up the surface of the soil and remove weeds near growing plants; improved cultivators of the mid-19th century contributed greatly to increased agricultural production.

**cultural evolution** *Sustainable Development.* **1.** any fundamental change over time in the

*cultivator* *A contemporary heavy-duty field cultivator.*

social and economic character of a given society. **2.** specifically, an (earlier) theory that cultures will naturally pass through certain progressive stages on the path to industrialization; i.e., from hunter-gatherer societies to nomadism to agricultural and pastoralism and then ultimately to a modern industrial society. This was seen as analogous to biological evolution from invertebrates to lower vertebrates to mammals to humans.

**culturally mediated** *Consumption & Efficiency.* a descriptive term for all forms of energy produced through the application of human technologies.

**Curie Marie** (Marya Sklodowska), 1867–1934 Polish-born scientist active in France; and her husband, French scientist **Pierre Curie,** 1859–1906, the co-discoverers of the elements polonium and radium and pioneers in the study of natural radioactivity. Marie Curie was the first person to receive two Nobel Prizes, one in Physics and the other in Chemistry.

**curie** *Nuclear.* the amount of radioactivity in one gram of the isotope radium-226. One curie is 37 billion radioactive disintegrations per second.

**Curie's law** *Physics.* the statement that the magnetic susceptibilities of most paramagnetic substances are inversely proportional to their absolute temperatures.

**curium** *Chemistry.* a synthetic radioactive element having the symbol Cm, the atomic number 96, atomic weights of isotopes 242–247.07, and a melting point 1340°C; a chemically reactive, silver-white metal used for remote, small-scale power generation. [Named for Marie and Pierre *Curie.*]

**current** *Electricity.* **1.** the rate of transfer of electrons or of positive ions, negative ions, or holes from one point to another; measured in amperes. *Physics.* **2.** the amount of any quantity flowing past a reference point per unit time. **3.** *Earth Science.* a horizontal movement of water in a well-defined, established pattern, as in a river or stream, or in the ocean. **4.** the movement of a definite body of air in a certain direction.

**current collector** *Renewable/Alternative.* the conductive plates or wires in a fuel cell that collect electrons on the negative (anode) side and distribute them on the positive (cathode) side.

**current density** *Electricity.* the current flowing through a given cross-sectional area of a conductor, usually represented by a vector whose direction is in the direction of the current; generally expressed in amperes per square meter.

**current signature** *Electricity.* the unique distortions in a current profile caused by an operating electromechanical device.

**Current War** *History.* a term for the 19th-century rivalry between DC current, promoted

*Marie Curie*

by Thomas Edison, and AC current, promoted by George Westinghouse, with the competition being which one would be adopted as the dominant consumer standard. It would eventually be shown that AC current is more convenient since it can be produced at high enough voltages to be transmitted over large distances.

**curtailable rate** *Electricity.* an option offered by utilities to customers who can accept specified amounts of service reduction in return for reduced energy rates.

**cushion gas** another term for BASE GAS.

**customer class** *Consumption & Efficiency.* a broad category of energy consumers defined by such criteria as their consumption or demand levels and patterns; such a classification may be used to establish rates charged for energy use.

**cut-and-cover method** *History.* a historic advance in the construction of underground (subway) transportation systems; the excavation in which the system operates is roofed over (as opposed to being an open trench), so that surface traffic can use the street.

**cut-in** *Consumption & Efficiency.* **1.** the minimum wind speed at which a wind turbine becomes activated to predict useable power. **2.** the minimum rotational speed at which an alternator or generator has a high enough voltage to make electricity flow in a circuit. Thus, **cut-in speed.**

**cut-out** *Wind.* the wind speed at which a wind generator activates some kind of overspeed mechanism, to either stop the unit's generation of power completely or control the rotational speed to produce constant power, as by changing the blade position, activating spoilers that increase drag, or turning the entire unit sideways to the wind. Thus, **cut-out speed.**

**cutter** see COAL CUTTER.

**Cuyahoga** *Environment.* a river that flows through Cleveland, Ohio; noted because the surface of the river caught fire in 1969 from oil slicks and chemical pollution. The paradoxical image of a burning body of water became a rallying point for passage of the U.S. Clean Water Act.

**CVD** chemical vapor deposition.

**CWG** carbureted water gas.

**CWIS** cooling water intake structure.

**cyanobacteria** *Ecology.* a large and widespread group of bacteria, typically (though not exclusively) having a blueish-green pigmentation and thus also known as blue-green algae, that are capable of photosynthesis and that play an important role in the nitrogen cycle. Current research efforts in renewable energy propose the production of hydrogen as a biofuel through the action of cyanobacteria.

**cybernetics** *Communication.* **1.** the scientific study of communication and control, especially so as compare the communication and control systems of humans and other living organisms with those of complex machines. **2.** this science as it relates specifically to the development and operation of automatic control equipment.

**cybersecurity** *Communication.* the fact of protecting the confidentiality, accuracy, and availability of information stored and transmitted on computer systems, especially sensitive information such as financial records, personal data, and trade and official secrets.

**cycle** *Measurement.* **1.** a process in which the initial and final states of a system are identical; i.e., the initial and final states have the identical values for all respective properties. **2.** *Electricity.* the complete set of values through which an alternating voltage or current passes successively. **3.** a single discharge-and-recharge sequence for secondary cells or batteries. See also CYCLIC PROCESS.

**cycle depth** *Storage.* the degree to which the charge of a secondary battery is drawn from it during discharge, expressed as a percentage of the total battery capacity.

**cycle life** *Storage.* the number of discharge-charge cycles that a battery can tolerate under specified conditions before it fails to meet specified performance criteria.

**cyclic memory** see MEMORY EFFECT.

**cyclic process** *Thermodynamics.* a process in which a system in a given initial state goes through a number of different changes in state and finally returns to its initial values; e.g., steam (water) that circulates through a closed cooling loop.

**cycling loss** *Consumption & Efficiency.* a loss of heat as water circulates through a water heater tank and its system of inlet and outlet pipes.

**cycling unit** *Electricity.* an electric-generating unit that operates with frequent load changes, starts, and stops; these units generally have lower efficiencies and higher operating costs than baseload plants.

**cycloconverter** *Electricity.* an alternation-current converter in which the AC supply from the grid is converted directly into another AC voltage waveform with a different (usually lower) frequency, without an intermediate DC stage.

**cyclone burner** *Consumption & Efficiency.* a furnace in which finely ground fuel is blown in spirals in the combustion chamber to maximize combustion efficiency.

**cyclone (cyclonic) collector** *Environment.* an air pollution abatement device that collects dust particles through centrifugal force. Similarly, **cyclone scrubber.**

**cyclotron** *Nuclear.* a particle accelerator consisting of halves of a hollow cylinder, called **dees** because their shape resembles a letter D, connected to a high-frequency voltage source in a uniform perpendicular magnetic field; charged particles injected into the gap between the dees are propelled in a spiral of increasing radius so that the path's length increases with the particles' speed until they are deflected as a high-energy beam. One of the earliest types of particle accelerators, and still used as the first stage of some large multistage particle accelerators.

**Czochralski, Jan** 1885–1953, Polish chemical engineer who made seminal contributions to the commercialization of photovoltaic cells.

**Czochralski process** *Electricity.* the most widely used technique for making single-crystal silicon, in which a seed of single-crystal silicon contacts the top of molten silicon. As the seed is slowly raised, atoms of the molten silicon solidify in the pattern of the seed and extend the single-crystal structure. Also, **Czochralski method, Czochralski crystal growth.**

**Daimler, Gottlieb** 1834–1900, German engineer and inventor; with his colleague Wilhelm Maybach, Daimler's improvements in the internal-combustion engine, made in the 1880s, contributed largely to the development of the modern automobile industry.

**Daimler-Benz** *Transportation.* a leading contemporary automobile company, formed by the merger in 1926 of two of the pioneering companies of the industry, the Daimler Motor Company founded by Gottlieb Daimler, and Karl Benz's Benz & Cie. Popularly known as Mercedes-Benz from the name of their most famous vehicle, and now officially known as DaimlerChrysler since a 1998 merger with the U.S. auto company Chrysler Corp.

**d'Alembert, Jean le Rond** 1713–1783, French mathematician and physicist who helped to resolve the controversy in mathematical physics over the conservation of kinetic energy by improving Newton's definition of force.

**Dalton, John** 1766–1844, English scientist who developed the atomic theory of matter (1803); he revived the concept of atoms as described by the ancient Greeks (see DEMOCRITUS) and provided an explicit chemical and physical description for it. He also was the first to describe color blindness, a condition from which he suffered and which thus became known as "Daltonism".

**dalton** *Measurement.* an alternate name for the unified atomic mass unit; equivalent to the atomic weight of a hydrogen atom or $1657 \times 10^{-24}$ g.

**Dalton's law** *Chemistry.* the statement that the total vapor pressure of a mixture of nonreacting gases in a closed container is equal to the sum of the individual pressures of the gases in the same container.

**Dalton's theory** *Chemistry.* the atomic theory formulated by John Dalton, specifically the assumptions that: (a) all matter consists of tiny particles (atoms) that are indestructible and unchangeable; (b) chemical elements are made up of these particles, which are identical for a given element in terms of mass, size, and other properties; (c) when such elements react, their atoms combine in simple, whole-number ratios.

**Daly, Herman** born 1938, U.S. economist known as a founder of the field of ecological economics. He applied classical concepts of capital and income to resources and the environment, the laws of thermodynamics, and the insights of ecology, particularly in relation to flows of materials and energy through economic systems.

**dam** *Hydropower.* any barrier that serves to restrict or block the flow of water; the term can apply to even a simple feature in nature (e.g., a beaver dam on a woodland stream), but usually used to describe a large man-made structure intended to generate hydropower, curb flooding, provide an irrigation supply, and so on.

*dam The huge Aswan Dam was built to control the flow of the Nile River (completed 1970); called the High Dam in contrast with the historic Low Dam.*

**damage function** *Climate Change.* a description of the relation between changes in the climate and consequent reductions in economic activity, relative to the rate of activity that would be possible in an unaltered climate.

**Dammam** *Oil & Gas.* the site of a major oil field in Saudi Arabia, the surface expression of a geological formation known as the **Dammam Dome;** its discovery by Aramco (1938) marked the beginning of the commercial oil industry in Saudi Arabia, and enabled Aramco to become the world's largest oil-producing company.

**damp** *Coal.* **1.** a toxic gas present in a mine. **2.** to reduce the fire in a furnace by placing moist coals or ashes on the fire bed, or by restricting the flow of air. **3.** *Physics.* to gradually diminish the amplitude of a wave or oscillation.

**damper** *HVAC.* a shutter or frame used to control or restrict **(damp down)** the flow of air in a furnace, fireplace, or stove.

**Daniell cell** *History.* the first battery (1836) to produce a first reliable and lasting source of direct-current electricity, by providing a barrier between copper and zinc plates in the cell, which, by stopping hydrogen from forming, solved the problem of polarization. [Developed by English chemist John Frederic *Daniell,* 1790–1845.]

**Darby, Abraham** c. 1678–1717, English ironmaker who was the first (1709) to use coke successfully in the smelting of iron and thus revolutionized the production of steel. This alleviated Britain's shortage of charcoal due to wood fuel scarcity, and thus it is seen as a major factor in the future success of the British Industrial Revolution.

**D'Arcy, William Knox** 1849–1917, English businessman who was the principal founder of the Persian (Iranian) oil industry. Although his drilling ventures at first experienced a considerable number of dry holes, oil was finally struck at Masjid-I-Sulaiman in 1908. In 1909 the Anglo-Persian Oil Company was founded with D'Arcy as a director.

**dark energy** *Physics.* the residual energy in empty space that is causing the expansion of the universe to accelerate.

**dark reactions** *Biological Energetics.* the process of using the reducing power and energy produced during light reactions to fix inorganic material into high-energy organic molecules; so called because they do not require light to take place.

**dark respiration** *Biological Energetics.* a three-phase process that occurs in the mitochondria of plant cells, creating energy by oxidizing sugars. It is described as having two components: growth respiration, the biosynthesis of structural compounds, and maintenance respiration, the energy required by the normal activities of cells. Called "dark" in contrast with photorespiration, though it can occur in both dark and light conditions.

**Darrieus turbine** *Wind.* a type of vertical-axis wind turbine that has a characteristic "eggbeater" shape with two or three thin C-shaped blades meeting at the top and bottom of the axis; the first modern vertical-axis wind turbine (patented 1931). [Designed by French engineer Georges *Darrieus.*]

**daughter** *Nuclear.* a nuclide that results from the radioactive decay of another nuclide (the parent nuclide). Also, **daughter product.**

**Davenport, Thomas** 1802–1851, U.S. inventor who developed the first direct-current electrical motor (1834), known as the **Davenport motor.** He later used this device to power the first model electric railway.

**Davy, Humphrey** 1778–1829, English chemist who advanced the field of electrolysis and isolated various elements—potassium, sodium, calcium, strontium, barium, and magnesium. He also developed the innovative DAVY LAMP for use in coal mines.

**Davy (safety) lamp** *History.* an important advance in mining safety technology (1815), reducing the risk of igniting the flammable gas mixtures that are common in mines; noted especially for preventing the lamp temperature from rising above the ignition point of mine gases such as methane.

**day-ahead** *Economics.* describing conditions established in an energy market one day in advance, as opposed to real-time conditions; e.g., suppliers and wholesale customers agree on the price to be paid for electricity for the following day. Thus, **day-ahead price, market, demand,** and so on.

**daylight factor** *Lighting.* the ratio of daylight illumination at a given point on a plane due to the light received from the sky, to the illumination on a horizontal plane due to an unobstructed hemisphere of this sky, expressed as a percentage. Direct sunlight is excluded for both values of illumination.

**daylighting** *Lighting.* **1.** the use of direct, diffuse, or reflected sunlight through architectural design practice and electrical lighting controls, in order to distribute and control natural illumination and reduce electrical energy use. **2.** any use of natural daylight for indoor lighting.

**daylighting control** *Solar.* a device that senses the amount of light provided by daylight and controls electric lighting or shading devices to maintain a specified lighting level.

**daylight simulation** *Lighting.* a measure of the ability of a light source to match its spectral power distribution to the natural spectrum of daylight; the closer the match to true daylight, the less the human eye has to strain and thus the lower the risk of vision-related health problems and loss of productivity. Also, **daylight reproduction.**

**DC** direct current.

**DD** degree-day.

**DDT** *Health & Safety.* dichlorodiphenyl trichloroethane, an insecticide and pesticide. See next column.

**DE** **1.** diesel exhaust; see DIESEL PARTICULATE MATTER. **2.** distributed energy.

**dead air** *HVAC.* **1.** a sealed air space between two material structures, such as the hollow area between two walls. **2.** *Communication.* a temporary, unintentional absence of audio and/or video programming during a broadcast.

**deadband** *HVAC.* an intermediate range of temperature between heating and cooling, in which no conditioning takes place. Thus, **deadband interval.**

**deadband thermostat** *HVAC.* a thermostat that provides two separate control signals, one for the heating equipment and one for the cooling equipment; in the intermediate (deadband) range between the two, no control signal is provided.

**dead state** *Thermodynamics.* the condition of a system when it is in thermal, mechanical, and chemical equilibrium with a conceptual reference environment.

**dead storage** *Coal.* a status of coal that has been mined but is not immediately available for use because it is in a compacted pile to prevent weatherization; i.e., some procedure is needed to reclaim the coal.

**deadweight** *Measurement.* the weight of a truck, railroad car, tanker, or other such carrier, not including any of its load or contents; the weight difference between a completely empty and a fully loaded vessel. Thus, **deadweight ton (DWT).**

**DDT** An abbreviation of dichlorodiphenyl trichloroethane; an insecticide and pesticide in the form of colorless needles or white to slightly off-white powder, is insoluble in water and slightly soluble in alcohol. Because it isn't particularly lethal to humans, it was extensively used in WW II to reduce mosquito populations and thus control malaria to protect U.S. troops. It was also used on civilian populations in Europe to prevent the spread of lice and the diseases they carried. It then became popular as the first modern pesticide, and was hailed as a miraculous advance in pest control (its developer, Paul Müller of Switzerland, won the Nobel Prize in 1948). Peak usage in the U.S. occurred in 1962, when 80 million kilograms of DDT were used and 82 million kilograms produced. However, it was later targeted as harmful to wildlife and human health because of bioaccumulation and biomagnification in food chains. A famous study done in 1967 in the estuary on Long Island Sound showed the biomagnification factor for DDT of more than 200,000x. Rachel Carson sounded the initial alarm against DDT, in her 1962 book *Silent Spring*. DDT was alleged to produce the phenomenon of shell-thinning in birds, particularly raptors (carnivorous birds that hunt and kill other animals), birds that eat carrion (dead animals), and birds that eat fish. The decline in populations of ospreys and bald eagles were attributed to DDT. Many insect pests also may have developed resistance to DDT. It was banned in the U.S. and Sweden in the early 1970s. Other parts of the world continue to use DDT in agricultural practices and in disease-control programs. Some observers maintain that DDT was "demagogued" out of use, with environmentalist groups exaggerating its risk and misrepresenting the scientific research as a means to increase their political power.

**Cutler Cleveland**
Boston University

**deaeration** *Chemistry.* the removal of oxygen, carbon dioxide, or other gases from a liquid or semiliquid substance. Thus, **deaerator.**

**dealer tank wagon (price)** *Economics.* the price, usually of branded gasoline, offered by major refiners and delivered to service stations on a cost, insurance, and freight (CIF) basis.

**deasphalting** *Oil & Gas.* a refining process for removing asphalt compounds from petroleum fractions, such as reduced crude oil. The recovered stream from this process is used to produce fuel products.

**de Broglie, Louis** 1892–1987, French physicist who discovered the wave nature of electrons. He proposed that light was not the only phenomenon to exhibit a wave-particle duality; i.e., ordinary particles such as electrons or protons could also exhibit wave characteristics **(de Broglie waves),** by virtue of their motion.

*Louis de Broglie*

**debt crisis** *Economics.* a prolonged financial crisis in developing economies of the 1980s, especially certain nations of Latin America, resulting from a combination of heavy borrowing and the need to import large quantities of oil in the aftermath of the oil price shocks of the 1970s.

**Debye, Peter** 1884–1966, Dutch-born U.S. physical chemist noted for his investigations of dipole moments, X-rays, and light scattering in gases.

**decay** *Nuclear.* **1.** the gradual decomposition of dead organic matter. **2.** *Physics.* the progressive diminishing of the size or range of a quantity, either spatially or temporally. **3.** *Nuclear.* radioactive decay; i.e., the disintegration of atomic nuclei resulting in the emission of alpha or beta particles.

**decay chain** *Nuclear.* a series of nuclides linked in a chain by radioactive decay, with each nuclide in the chain decaying to the next until a stable nuclide is reached.

**decay constant** *Nuclear.* the ratio between the number of nuclei decaying per second and the total number of nuclei.

**decay cooling** *Nuclear.* the continued circulation of coolant for many days after reactor shutdown, in order to ensure the removal of decay heat and thus prevent damage to fuel elements and possible contamination by fission products.

**decay energy** *Nuclear.* the total energy emitted in radioactive decay.

**decay heat** *Nuclear.* the heat produced by the decay of radioactive fission products after a reactor has been shut down.

**deceleration** *Physics.* a decrease in velocity over time, or the rate of such a decrease. Thus, **decelerate**.

**decentralized energy** *Renewable/Alternative.* a broad term for energy facilities that operate at a local level on a small scale, as opposed to large centralized facilities such as municipal electrical power plants, nuclear plants, and the like; e.g., photovoltaics, wind systems, biomass, and so on.

**deci-** *Measurement.* a metric prefix meaning "one-tenth", as in deciliter.

**deciview** *Environment.* a measurement of visibility; one deciview represents the minimal perceptible change in visibility to the human eye. This is used as a means of describing the presence of particles in the air that obscure vision; e.g., smog.

**declination** *Solar.* the latitude at which the sun's rays are perpendicular to the surface of earth at solar noon.

**decline curve** *Oil & Gas.* a function that describes the rate of oil production from a given reservoir as a function of time, typically starting with a period of rapid production that eventually declines exponentially towards zero.

**declining block rate** *Electricity.* an electricity rate structure in which the per unit price of electricity decreases as the amount of energy purchased increases; usually available only to very large consumers.

**decommissioning** *Nuclear.* the removal of a reactor or other nuclear facility from service, and the subsequent actions of safe storage, dismantling, and so on carried out to make the site available for unrestricted use.

**decomposer** *Biological Energetics.* an organism that feeds by breaking down organic matter from dead organisms; e.g., certain bacteria and fungi.

**decomposition** *Chemistry.* a process in which one or more substances break down into simpler molecular substances, as from the effects of heat, light, chemical or biological activity, and so on.

**decomposition voltage** *Hydrogen.* the minimum voltage applied to an electrolyzer in order to decompose water by electrolysis; this threshold voltage is 1.23 volts.

**decompression** *Physics.* the act of reducing pressure or the effects of pressure.

**decontamination** *Health & Safety.* **1.** the process of making a person, object, or area safe by destroying, neutralizing, or removing toxic matter such as chemical or biological agents. **2.** specifically, the reduction or removal of radioactive material from a surface, as by treating the surface, letting the material stand so that natural decay takes place, or covering or enclosing the contamination.

**deconvolution** *Oil & Gas.* a data processing method that improves the temporal resolution of data from seismic surveys.

**deep-cycle** *Storage.* describing a type of battery with large plates that can be discharged to a low state of charge many times without damaging the battery.

**deep discharge** *Storage.* the process of discharging a battery to a low percentage (20% or less) of its full charge capacity.

**deep gas** *Oil & Gas.* a term for unconventional natural gas deposits that are found at much greater depths than those reached in traditional extraction processes (typically defined as below 15,000 feet).

**Deep Gas Project** *Renewable/Alternative.* a project in Sweden involving the drilling of experimental wells in the Siljan Ring, a crater formed by a meteor impact, to test the theory (see next) that methane migrates upward from the inner regions of the earth to areas where it is feasible for extraction. Two wells were drilled, one to a depth of nearly 5 miles, but they encountered no commercial amounts of hydrocarbons.

**deep gas theory** *Oil & Gas.* a theory that presents an abiogenic (non-living) origin for natural gas deposits, proposing that they originate from primordial materials buried deep within the earth. According to this, hydrocarbons that migrate upward to lesser depths will break up into extractable methane gas.

**deep ocean water applications (DOWA)** *Renewable/Alternative.* a term for various alternative products and processes that can be derived from an ocean thermal energy conversion plant.

**deep saline** *Earth Science.* referring to deeply buried saltwater systems or features that have been proposed as a long-term underground storage option for carbon dioxide captured from fossil fuel use. Thus, **deep saline aquifer, formation,** or **reservoir.**

**deep-sea vent** see HYDROTHERMAL VENT.

**defense in depth** *Nuclear.* an approach to nuclear facility safety that builds in different layers of defense against the release of hazardous materials, so that there is not total reliance on one particular layer by itself.

**deferrable load** *Electricity.* an electrical load that must be supplied with a certain amount of energy over a specified time period in an isolated electrical network, but that has some flexibility as to exactly when the supply takes place.

**de Forest, Lee** 1873–1961, U.S. inventor who devised the first vacuum tube, a device that takes weak electrical signals and amplifies them. His work was instrumental in the development of radio and television broadcasting and other aspects of the modern electronics industry.

**deforestation** *Environment.* the process of clearing land of trees or forest, as for logging or fuelwood, agriculture and livestock raising, settlement, mining, dam building, and so on.

**defrost** *Refrigeration.* **1.** to remove or prevent the accumulation of frost or ice on the coils of a refrigerating unit. **2.** to melt frost, ice, or

*deforestation Deforestation resulting from a timber harvesting process.*

condensation from the windshield of an automobile or other vehicle, typically by means of a series of ducts that deliver warm, dry air from the engine. **3.** to carry out any such process to remove frost or ice from an object or surface. Thus, **defroster.**

**degaussing** *Physics.* the process of neutralizing the strength of the magnetic field of an object.

**DeGolyer, Everette Lee** 1886–1956, U.S. geologist known as a pioneer in applying principles of geophysics to petroleum exploration.

**degradation** *Materials.* **1.** a deleterious change in the chemical structure, physical properties, or appearance of a material; e.g., as caused by exposure to heat, light, oxygen or weathering. **2.** *Thermodynamics.* the transformation of energy into a form that is less available for doing useful work.

**degree** *Measurement.* **1.** any of various units of measure for physical conditions such as temperature or pressure. **2.** specifically, one increment of the Celsius, Fahrenheit, or Kelvin temperature scales. **3.** a unit of angular measure equal to 1/360 of a full rotation.

**degree-day (DD)** *HVAC.* the number of units (degrees) that the average outdoor temperature falls below or exceeds a base value (usually 65°F) in a given period of time. Each degree that the mean daily temperature is above the base value is a **cooling degree-day (CDD)** unit. Each degree that the mean daily temperature is below the base value is a **heating degree-day (HDD)** unit. Degree-days are a reasonably good indicator of the heating and cooling requirements of buildings.

**De Havilland Comet** *Transportation.* a British aircraft that became the first jet aircraft to provide commercial passenger service (1952).

**Dehlsen, James** born 1937, U.S. businessman noted for his central role in the development of the wind power industry in the United States.

**dehumidifier** *HVAC.* any device that removes moisture from its surrounding atmosphere.

**dehumidify** *HVAC.* to remove water vapor from air, especially by cooling the air below the dewpoint. Thus, **dehumidification.**

**dehydrated** *Chemistry.* lacking water; in a state of dehydration.

**dehydration** *Chemistry.* **1.** the removal of water from a compound. **2.** *Biological Energetics.* a condition in which there is an excessive loss of water from the body tissues; this can lead to shock or even death. **3.** *Consumption & Efficiency.* the process of removing water from a food product, especially by an accelerated artificial method as opposed to natural or air drying.

**dehydrogenation** *Chemistry.* the removal of one or more hydrogen atoms from a molecule.

**deka-** *Measurement.* a metric prefix meaning "ten", as in dekaliter (10 liters).

**de Laval** see LAVAL (nozzle, turbine).

**delayed coking** *Oil & Gas.* a process by which heavier crude oil fractions can be thermally decomposed under conditions of elevated temperatures and pressure to produce a mixture of lighter oils and petroleum coke.

**delivered energy** *Consumption & Efficiency.* a term for energy provided at the place where the energy is actually used to produce the desired service; e.g., electricity or fuel entering an end-use consumer device.

**demand** *Economics.* the incentive to purchase a certain good or service at a given price; e.g., gasoline.

**demand elasticity** see ELASTICITY.

**demand ratchet** *Economics.* a minimum average billing for energy use applied to a customer who may have inconsistent or seasonal energy requirements.

**demand-side** *Economics.* **1.** of or relating to the demand for an energy commodity by consumers, as opposed to its supply. **2.** relating to or based on the concept that the proper way for a government to stimulate economic growth is by encouraging increases in aggregate demand. Compare SUPPLY-SIDE.

**demand-side management (DSM)** *Policy.* the planning, implementation, and monitoring of utility activities so as to encourage customers to modify their pattern of energy usage.

**demand water heater** *HVAC.* **1.** any type of water heater that supplies hot water immediately on demand, as opposed to storing heated water on an ongoing basis. **2.** specifically, a TANKLESS WATER HEATER (see).

**dematerialization** *Consumption & Efficiency.* a reduction of energy and material inputs into, and residual outflows from, the production and consumption of goods and services; often measured in terms of intensity, such as the mass of material used to manufacture a good or services.

**Democritus** c. 460–370 BC, Greek philosopher who was an early proponent of atomic theory. He proposed that matter was made of discrete indivisible particles, which he described as *atomos,* meaning "cannot be cut" (since he believed these particles to be indestructible, eternal, and unchanging). In another impressive observation, he stated that the Milky Way galaxy was a conglomeration of stars.

**demographic transition** *Social Issues.* a model that describes the transformation of a country from a pattern of high birth and death rates to one of low birth and death rates; according to this the highest population growth rates will occur in a transitional period when death rates have declined while birth rates remain relatively stable.

**demon of Maxwell** see MAXWELL'S DEMON.

**demonstrated resources** *Coal.* the sum of coal resources including both those for which estimates of the rank, quality, and quantity have been computed to a high degree of geologic assurance and those for which estimates have been computed to a moderate degree of geologic assurance.

**denatured** *Biomass.* describing ethyl alcohol that has had a substance added to it to make it unfit for human consumption as a beverage.

**dendrochronology** *Environment.* the dating of past events and variations in the environment and the climate by studying the annual growth rings of trees.

**dendroclimatology** *Climate Change.* the scientific use of tree growth rings as climate indicators; tree rings provide evidence of a wider range of climatic variables over a larger part of the earth than any other annually dated proxy record.

**denitrification** *Chemistry.* a process of removing nitrogen or nitrogen compounds from a substance.

**densification** *Biomass.* a mechanical process used to compress biomass (usually wood waste) into pellets, briquettes, cubes, or logs that will burn more efficiently.

**density** *Physics.* the mass per unit volume of a substance under specified conditions of pressure and temperature.

**Densmore car** *History.* the first railroad oil tank car, a major breakthrough in bulk transportation compared to the earlier method of stacking individual barrels on a flat car. [Invented by U.S. oil buyer and shipper Amos *Densmore.*]

**denuded zone** *Photovoltaic.* a term for a very thin region in a semiconductor adjacent to its surface cleared from contaminants and defects; devices are built into a denuded zone.

**Department of Energy (DOE)** (est. 1977), the U.S. agency whose mission is to advance the energy security of the nation, as by promoting a diverse supply and delivery of reliable, affordable, and environmentally sound energy; by providing scientific research capacity and advancing knowledge; and by protecting the environment; e.g., providing for the permanent disposal of the nation's high-level radioactive waste.

**dependency ratio** *Social Issues.* the ratio of the population classified as too young or too old to work, relative to the active working-age population; typically defined as those under age 16 or over 65.

**dependency theory** *Global Issues.* the theory that commerce between nations with different levels of economic development will lead to economic growth in the more developed countries but will not have a corresponding benefit for the less developed countries; i.e., they will become overly reliant on the more developed countries.

**depletable** see NONRENEWABLE.

**depleted uranium** *Nuclear.* the byproduct of the uranium enrichment process; uranium containing less than the natural fraction of U-235 (0.71%). This is considerably less radioactive than even natural uranium, though still extremely dense and thus useful for armor and armor-penetrating weapons.

**depletion** *Consumption & Efficiency.* **1.** the consumption of a natural resource at a faster rate than the possible rate of replacement. **2.** see OIL DEPLETION.

**depletion allowance** *Policy.* the reduction of taxes for owners of minerals and fossil fuel resources, to compensate for the exhaustion of an irreplaceable capital asset.

**deposit** *Mining.* any natural accumulation of a useful mineral, or an ore, in an amount sufficient to be worthy of exploitation.

**deposit-refund** *Policy.* a market-based instrument in which the consumer makes a prior payment for potential pollution (the deposit), and is guaranteed a return of the deposit upon proving that the pollution did not take place (the refund).

**deposition** *Earth Science.* a process in which a material is removed from the atmosphere, either by precipitation **(wet deposition)** or by contact with a surface **(dry deposition).**

**depth of discharge (DOD)** *Storage.* the number of ampere-hours removed from a fully charged cell or battery, expressed as a percentage of its rated capacity.

**DER** **1.** discrete emission reduction. **2.** distributed energy resource; see DISTRIBUTED.

**derating hour** *Consumption & Efficiency.* time that a generating unit is forced to operate at less than full power because of a maintenance or operating problem.

**deregulation** *Policy.* a process in which a government agency opens up full retail competition for the supply of a desired energy resource, often down to the household level.

**derrick** *Oil & Gas.* the towerlike framework erected over a borehole to support the drilling tools, casing, and pipe; it also includes apparatus for raising, lowering, and controlling the equipment.

***derrick*** *Oil derrick at a California oil field. The term* "derrick" *was first applied to a structure used to hang criminals, from the name Thomas Derrick, a well-known London hangman.*

**desalination** *Environment.* the removal of salt from a body or system, especially the removal of salt from seawater to make it suitable for drinking.

**Descartes, René** 1596–1650, French philosopher known for his logical and mechanistic approach to scientific inquiry and for his contributions to the development of modern mathematics. He is especially known for the statement *"Cogito, ergo sum"* ("I think, therefore I am"), the idea that although one may doubt all other things, the act of doubting indicates that the doubter must exist.

**desertification** *Environment.* the degradation of land to desert conditions in arid, semi-arid, and dry sub-humid areas, caused primarily by human activities and climatic variations.

**desiccant** *Materials.* a solid or liquid material with an affinity for absorbing water molecules (e.g., moisture from air).

**desiccant cooling** *HVAC.* the cooling of a building space by means of contact between warmer air and a material that absorbs moisture from the air (such as lithium chloride). Similarly, **desiccant dehumidification.**

**design-basis accident** *Nuclear.* a postulated accident that a nuclear facility must be designed and built to withstand without loss to the systems, structures, and components, in order to assure public health and safety. Similarly, **design-basis event.**

**design-basis threat (DBT)** *Nuclear.* a profile of the type, composition, and capabilities of a potential adversary's postulated threat to a nuclear facility, such as acts of sabotage or the theft of nuclear material.

**design day** *Oil & Gas.* a 24-hour period of demand used for planning natural gas capacity requirements.

**design head** *Hydropower.* the precise height of the water surface required to produce a flow (**design flow**) that will provide the maximum efficiency from a turbine (i.e., it will function optimally as designed).

**design heat load** *HVAC.* the total heating requirements of a given building, expressed as the total heat loss from the building that would occur during the most severe winter conditions it is likely to experience.

**design life** *Consumption & Efficiency.* the period of time during which a facility or component is expected to perform satisfactorily according to the technical specifications by which it was produced.

**design month** *Solar.* the month with the lowest mean daily insolation value; used as the basis for the planning of many stand-alone solar energy systems.

**desorption** *Chemistry.* a process by which a substance that had been adsorbed or absorbed by a liquid or solid material is removed from the material.

**destructive distillation** *Chemistry.* a process in which a carbon-containing material such as coal or oil shale is heated in the absence of air, resulting in its decomposition into solids, liquids, and gases, with the solid end product being carbon.

**Destructor** *History.* the first known example of a systematic incineration of urban solid wastes; a system burning mixed waste and producing steam to generate electricity, put into operation in Nottingham, England in 1874.

**desulfurize** *Materials.* to reduce or remove the sulfur content of a material, such as coal, petroleum oil, or molten metals. Thus, **desulfurization.**

**desuperheater** *HVAC.* a heat exchanger installed in a heat pump directly after the compressor, designed to remove a portion of the heat from hot, vaporized refrigerant and transfer it to a domestic hot water tank.

**detection** *Climate Change.* the process of demonstrating that climate has changed in some defined statistical sense, without providing a reason for that change. Compare ATTRIBUTION.

**detection gas** *Nuclear.* a gas that is sensitive to radiation and that produces an electric, electromagnetic, or other such signal that is easily detected.

**Deterding, Henri** 1866–1939, co-founder of the Royal Dutch oil company (1890) and head of the firm that merged with the company, Shell Transport and Trading Company to become the Royal Dutch/Shell Group.

**detonate** *Chemistry.* to carry out a process of detonation; produce an explosion through a chemical change.

**detonation** *Chemistry.* **1.** the explosion of a substance through a chemical change, which may be initiated by heat, friction, or mechanical impact. **2.** *Transportation.* in an internal combustion engine, a premature spontaneous combustion of the air-fuel mixture that can occur when the temperature of compressed air in the cylinder exceeds certain limits; accompanied by loss of power, overheating, and knocking.

**detritus** *Biomass.* dead organic matter.

**Detroit Electric** *History.* the leading electric vehicle in the early years of the U.S. automotive industry, produced by the Anderson Carriage Company from 1907 until the late 1930s.

**deuterium** *Chemistry.* heavy hydrogen; an isotope of hydrogen having the symbol D or $^2H$, with one neutron and one proton in its nucleus and an atomic weight of 2.0144.

**deuterium oxide** *Nuclear.* ($D_2O$) a compound that is chemically the same as normal water, except that the hydrogen atoms are of the heavy isotope deuterium, in which the nucleus contains a neutron in addition to the proton; used as a moderator to slow down neutrons in nuclear reactors. Also known as HEAVY WATER.

**deuteron** *Nuclear.* the nucleus of a deuterium atom, consisting of a proton and a neutron; because of its low binding energy, it is used as a projectile in nuclear bombardment experiments.

**developed** *Sustainable Development.* describing those nation-states with an industrial economy and generally high levels of gross domestic product, literacy, and commercial energy use per capita, though not necessarily evenly distributed throughout the population of the nation. Thus, **developed country, nation,** and so on.

**developing** *Sustainable Development.* describing those nation-states that have not yet made a complete transition to an industrial economy, generally having lower levels of gross domestic product, literacy, and commercial energy use per capita than developed nations such as Japan, Great Britain, and Sweden. Thus, **developing country, nation,** and so on.

**development** *Sustainable Development.* **1.** the process of a nation or other entity making the transition to an industrial economy with relatively high levels of gross domestic product, education, and commercial energy use per capita. **2.** the process of converting a particular area of land from a more natural to a less natural state; e.g., from forest to farmland or from farmland to housing.

**development bank** *Sustainable Development.* a multilateral institution that provides financing for the development incentives of member nations; e.g., the Asian Development Bank, the Islamic Development Bank.

**development drilling** *Oil & Gas.* a term for drilling taking place in an area where wells have been drilled before and geological information has been established. Thus, **development well.**

**devolatilization** *Conversion.* a process of thermal degradation of coal or wood in which the heat is supplied by the combustion of the emerging gases.

**Dewar (vacuum) flask** *Storage.* a double-walled vessel in which the space between the walls is evacuated and the surface bounding the vacuum is silvered, thus providing a high degree of thermal insulation; used to store liquefied gases. The common thermos bottle (a container for storing hot or cold liquids) is an adaptation of the Dewar flask. [Invented by Scottish chemist James *Dewar,* 1842–1923.]

**dewatering** *Materials.* the removal of water from a substance or medium; e.g., the use of a chemical to remove water from a liquid such as oil or gasoline.

**dewaxing** *Materials.* the removal of paraffin wax from lubricating oils to improve low temperature properties.

**dew point** *Chemistry.* **1.** the temperature and pressure at which a gas mixture begins condensing to form a liquid phase. **2.** specifically, the temperature to which a given parcel of air in the atmosphere must cool at constant pressure and water vapor in order for saturation to occur.

**DG** distributed generation; see DISTRIBUTED.

**DHFC** direct hydrogen fuel cell.

**diabatic** *Thermodynamics.* **1.** involving a loss or gain of heat. **2.** specifically, describing a process in which there is a transfer of heat into or out of the system in question. Thus, **diabatic process.** Compare ADIABATIC.

**diagenesis** *Earth Science.* the chemical, physical, and biological changes that sediment undergoes after its deposition but before its metamorphism and consolidation.

**diapause** *Biological Energetics.* a state of energy-conserving, arrested development or growth, accompanied by greatly decreased metabolism, often correlated with the seasons.

**diathermal** see DIABATIC.

**diathermanous** *Materials.* describing a substance or a space that allows the passage of heat, especially one that is highly conducive to heat.

**dieback** *Ecology.* **1.** a progressive death of plant parts or of a growth of vegetation. **2.** specifically, a high incidence of forest decline and individual tree death due to a change in climate conditions (e.g., a prolonged drought) that makes trees vulnerable to disease and insect predators (e.g., bark beetles).

**dielectric** *Electricity.* **1.** describing a substance that is a poor conductor of electricity, but an efficient supporter of electrostatic fields; it can support an electrostatic field while dissipating minimal energy in the form of heat; frequently used in capacitors. **2.** a material of this type; e.g., many ceramics, mica, glass, and plastics, as well as dry air. Thus, **dielectric material.**

**dielectric absorption** *Electricity.* **1.** the energy losses in a dielectric medium when the medium is exposed to a time-varying electric field. **2.** the undesirable tendency of certain dielectrics to retain a portion of an electric charge after removal of the electric field.

**dielectric constant** *Electricity.* the property of a material that determines how much electrostatic energy can be stored per unit volume of the material when unit voltage is applied.

**dielectric strength** *Electricity.* the ability of a dielectric material to withstand high voltages without breaking down; expressed as the highest voltage required per millimeter of material thickness before breakdown occurs.

**Diesel, Rudolf** 1858–1913, German inventor of the diesel engine. (patented 1894), the first engine to operate successfully via self-combustion rather than relying on a separate energy source such as a spark. He was a distinguished thermal engineer who designed many heat engines, including a solar-powered air engine. See also following entries.

**diesel** short for DIESEL ENGINE or DIESEL FUEL.

**Diesel cycle** *Thermodynamics.* **1.** an ideal thermodynamic cycle consisting of four processes: a compression at constant entropy; a constant-pressure heat transfer to the system; an expansion at constant entropy; and a constant-volume heat transfer from the system. **2.** an actual version of this cycle used in Diesel engines: first, air is drawn into the cylinder and compressed; second, energy is added by the injection and combustion of fuel in the cylinder; third, the gases produced by this combustion expand to move the piston downward for the power stroke; fourth, the burned gases in the cylinder are expelled.

**diesel-electric** *Transportation.* describing a vehicle or machine (e.g., a locomotive) powered by an electric motor that receives power from an electric generator, which in turn is powered by a diesel engine.

**diesel engine** *Transportation.* a type of internal combustion engine that burns fuel oil with the ignition being brought about by heat that results from air compression, rather than by an electric spark as it is in a gasoline engine. The diesel engine was historically the first engine in which fuel could be ignited without a spark.

**diesel fuel** *Oil & Gas.* a combustible distillate of petroleum used as fuel for diesel engines; usually the fraction of crude oil that is distilled after kerosene. Used in a variety of on-road transportation applications, as well as off-road uses; e.g., locomotives, ships, tractors, bulldozers, or mining equipment.

**diesel index** *Oil & Gas.* a measure of the ignition quality of a diesel fuel, calculated from a formula involving the gravity of the fuel and its aniline point; the higher the index, the higher the ignition quality of the fuel.

**diesel particulate filter (DPF)** *Transportation.* a device that entraps soot and other diesel particulate matter (see next) in the emissions of diesel-powered equipment.

**diesel particulate matter (DPM)** *Environment.* the solid aerosol component of diesel exhaust, consisting primarily of soot on which organic material is adsorbed, ash, sulfates, and water. This has significant potential to cause adverse health effects; e.g., in coal mines using diesel-powered equipment.

**diesel soot** see DIESEL PARTICULATE MATTER.

**diesel volume equivalent** *Measurement.* the number of standard cubic meters of hydrogen equivalent to a liter of diesel fuel (or standard cubic feet of hydrogen equivalent to a gallon of diesel).

**Dietz, Robert Edwin** 1818–1897, U.S. inventor who patented one of the first practical kerosene lamps (1859). He apparently did this independently of the earlier work of Ignacy LUKASIEWICZ of Poland. The Dietz Company went on to manufacture hundreds of lantern models, and became a pioneer in the automotive electric lighting industry.

**difference engine** *History.* an early computer consisting of a self-controlled device that mechanized the production of the final values in a mathematical table from the pivotal values first computed; designed by George and Edward Sheutz in 1837, based on the model of Charles Babbage.

**differential** *Transportation.* a gear train that allows two shafts to rotate at different speeds while being driven by a third shaft; the sum of the rotational rates remains constant and is equal to twice the rotational speed of the driving shaft; this allows the rear wheels of an automobile to rotate at different speeds when the vehicle is turning.

**differential heating** *Earth Science.* the difference in rate of heat absorption of land and water surfaces. Water is a slower conductor of heat than land and thus a given amount of solar radiation will heat an area of ground to a greater extent than it will a comparable area of ocean.

**differential rotation** *Solar.* a change in solar rotation rate with latitude. Low latitudes rotate at a faster angular rate (approx. 14 degrees per day) than do high latitudes (approx. 12 degrees per day).

**differential thermostat** *Solar.* an electronic difference thermostat for on/off control of the temperature difference between the heating source and storage tank in solar heating units and other types of heat recovery plants.

**diffraction** *Physics.* **1.** a phenomenon observed in the propagation of waves in which the propagation direction is changed when a wave encounters an obstruction or edge, such as an aperture. **2.** specifically, the bending of light as it passes an obstruction.

**diffuse radiation** *Solar.* describing solar radiation that is received indirectly at the surface of the earth, rather than from direct solar rays, as a result of scattering due to clouds, fog, haze, dust, or other obstructions in the atmosphere. Similarly, **diffuse insolation, diffuse irradiance, diffuse illuminance, diffuse sunlight,** and so on. Compare DIRECT RADIATION.

**diffuse reflectance** *Solar.* the ratio of the diffusely reflected part of the whole reflected flux, to the incident flux. Similarly, **diffuse transmittance.**

**diffuse reflection** *Solar.* a reflection process in which the reflected radiation is sent out in many directions, usually bearing no simple relationship to the angle of incidence; characteristic of rough surfaces. Compare SPECULAR REFLECTION.

**diffusion** *Physics.* **1.** a process in which particles disperse, moving from regions of higher density to regions of lower density. **2.** the process by which light reflects off an irregular surface in all directions.

**diffusion furnace** *Electricity.* a furnace used to make junctions in semiconductors by diffusing dopant atoms into the surface of the material.

**Digester** *History.* a device built by French engineer Denis Papin in 1679, regarded as the first atmospheric engine (an early type of steam engine).

**dike** *Environment.* a low ridge or bank of earth used to hold water in place during irrigation or to protect lowland areas against flooding.

*dike A dike in The Netherlands, which literally means "Lower Lands." A system of dikes has greatly contributed to the ability of this low-lying nation to prevent land from being submerged.*

**dilatometer** *Measurement.* an instrument designed to measure incremental changes in the dimensions of materials that are deformed by external influences.

**dimensionless number** *Measurement.* a number with no associated unit of measure such as square feet or pounds; often the ratio of two numbers with the same unit of measure.

**dimer** *Chemistry.* a molecule that is produced by bonding together two molecules of the same chemical structure (monomers). Thus, **dimerization.**

**dimethyl** another term for ETHANE.

**dimethyl ether (DME)** *Renewable/Alternative.* a highly flammable, colorless, compressed gas that is soluble in water and alcohol; can be derived from natural gas or coal. Currently used as a refrigerant, solvent, and catalyst and also considered as a potential fuel for motor vehicles and household cooking and heating use, because it is sulfur and nitrogen free and burns more cleanly than conventional fuels.

**diminishing return(s)** *Consumption & Efficiency.* the principle that, in any production function, as the input of a given factor rises (assuming other factors remain constant), the marginal product of that factor must eventually decline.

**diode** *Electricity.* a two-element active electronic device containing an anode and a cathode, characterized by the ability to pass an electric current more easily from cathode to anode than from anode to cathode; it can be in the form of an electron tube, semiconductor device, or crystal device.

**dioxin** *Health & Safety.* any of a family of hazardous compounds, formed in the burning of chlorine-based compounds with hydrocarbons; dioxin pollution is associated with waste-burning incinerators and certain paper mills, and with the production of chlorinated chemicals (e.g., many pesticides). See also TCDD.

**dipole** *Chemistry.* **1.** a molecule containing both positively and negatively charged groups. **2.** *Electricity.* a localized positive and negative charge distribution that has no net charge, and whose mean positions of positive and negative charges do not correspond.

**dip tube** *HVAC.* a tube of plastic below the cold port of a water heater that carries cold water to the bottom of the tank to be heated, keeping it separated from the already-heated water at the top of the tank.

**Dirac, Paul** 1902–1984, English physicist noted for his contributions to the field of quantum mechanics.

**Dirac equation** *Physics.* a differential equation for an electron in an electromagnetic field, describing the relativistic behavior of the electron in terms of a wave equation that automatically allows for electron spin.

**Dirac theory** *Physics.* a relativistic theory of the electron that predicts the electron's internal angular momentum, according to which a positron may be regarded as an empty negative energy state whose usual occupying electron has been removed.

**direct beam radiation** see DIRECT RADIATION.

**direct combustion** *Conversion.* a term for a biofuel conversion process in which forest and agricultural wastes and residues are burned to produce steam, electricity, or heat; can also be employed for municipal solid wastes following appropriate preparation.

**direct cooling** *Electricity.* the transfer of waste heat from a power plant to a body of water, using a water-to-water heat exchanger; this does not consume significant amounts of water (most is returned) but can impact aquatic species due to contact with screens or pumps.

**direct current (DC)** *Electricity.* a current that flows in only one direction and has an average value that is essentially constant. Compare ALTERNATING CURRENT.

**direct energy** *Consumption & Efficiency.* **1.** the energy required for actual use by consumers, as for transportation, heating and cooling, lighting, cooking, washing, and so on, as opposed to the energy required to allow this to take place, such as energy used in the manufacture of appliances for cooking, washing, and so on. **2.** see DIRECT USE.

**direct expansion** *HVAC.* an operation that is able to absorb desired heat from a heat source at a lower temperature and reject unwanted heat to a heat sink at a higher temperature.

**direct-fired** *HVAC.* describing a fuel-burning device in which the heat from combustion and the products of combustion are transferred directly to the space to be heated.

**direct gain** *Solar.* a type of passive solar heating system in which specially designed south-facing windows (north-facing in the Southern Hemisphere) allow sunlight to enter a home, where it is absorbed and stored by masonry floors and walls (usually dark-colored). When the room cools, the heat stored in the thermal mass is released into the room. Compare INDIRECT GAIN.

**direct hydrogen** *Hydrogen.* describing a technology in which the hydrogen fuel for a transportation vehicle is produced separately and then stored in a fuel tank on the vehicle; contrasted with an ONBOARD HYDROGEN system in which the fuel is produced by a processor within the vehicle. Thus, **direct hydrogen vehicle.**

**direct hydrogen fuel cell (DHFC)** *Hydrogen.* a type of fuel cell that utilizes pure hydrogen as fuel (rather than hydrogen produced from the reformation of hydrocarbon fuels such as gasoline, diesel, propane, or natural gas); currently this technology has the greatest system fuel efficiency and power density among fuel cells.

**direct internal reforming** see INTERNAL REFORMING.

**directional drilling** *Oil & Gas.* a method used in oil and gas drilling in which the direction of the drill string is forced out of a vertical direction, in order to reach target areas laterally displaced from the point where the drill bit had entered the earth.

**directional sailing** *History.* the fact of being able to sail a boat directly on the desired course, as opposed to having to sail only with the wind; a crucial advance in the history of sail power (about 900 AD, probably initiated by Arab merchant ships on the Red Sea).

**direct methanol fuel cell (DMFC)** *Renewable/Alternative.* a type of fuel cell in which the fuel is methanol, in liquid or gaseous form; the methanol is eletrochemically oxidized directly at the anode with no reformation to hydrogen.

**direct radiation** *Solar.* radiation received directly at a point after having traveled a straight path from the sun, with no scattering in the atmosphere; typically measured on a surface that is kept perpendicular to the direction of the center of the sun's disk. Similarly, **direct insolation, direct irradiance, direct illimuminance, direct normal radiation, direct sunlight,** and so on. Compare DIFFUSE RADIATION.

**direct use** *Consumption & Efficiency.* **1.** the utilization of heat as useful energy rather than its conversion to other forms of energy such as electricity. **2.** see DIRECT ENERGY.

**dirigible** *Transportation.* a lighter-than-air craft that is capable of being propelled and steered for controlled flight.

**disaggregation** *Electricity.* the separation of a vertically integrated utility into smaller, individually owned business units (i.e., generation, dispatch and control, transmission, distribution).

**discharge** *Storage.* **1.** the release of stored electrical energy from a source, such as a battery or capacitor. **2.** to release stored electricity in this manner. **3.** the conversion of chemical energy to electrical energy in a battery. **4.** *Electricity.* the flow of electrical current through a gas, usually causing luminescence of the gas.

**discharge factor** *Storage.* a number equivalent to the time in hours during which a battery is discharged at constant current, usually expressed as a percentage of the battery's total capacity.

**discharge lamp** *Electricity.* a type of lamp that produces light when the gas is ionized by an electric current passing through it at low or high pressure; e.g., fluorescent lamps, neon tubing.

**discharge rate** *Storage.* the current that is withdrawn from a battery over time, usually expressed as a percentage of the battery's rated capacity.

**discoveries** *Oil & Gas.* **1.** any new quantities of oil and gas identified through drilling. **2.** a description of this, including the sum of field extensions, new reservoir discoveries in old fields, and new field discoveries occurring during a given reporting year.

**discrete emission reduction (DER)** *Policy.* a discrete amount of emission reduction by a facility, expressed in tons. Unlike an emission reduction credit (ERC), which is a permanent emission reduction expressed in tons per year, DERs do not carry units of time.

**disintegration** *Nuclear.* **1.** a process of breaking or separating into parts; a loss of unity or solidity. **2.** *Nuclear.* a change in the properties of a nucleus, whether spontaneous or induced, in which particles or photons are released.

**disintegration chain** see DECAY CHAIN.

**disintegration energy** *Nuclear.* the energy released or absorbed during a nuclear or particle reaction.

**dispatchable** *Consumption & Efficiency.* a term for an energy system that can be expected to provide a continuous output (given normal conditions), thus offering the ability to furnish power on demand to meet changing loads;

e.g., hydrocarbon-based or nuclear power plants are dispatchable, but solar and wind power are not. Thus, **dispatchable power** or **energy.**

**dispersant** *Environment.* a chemical that aids in the breakup of a conglomerate solid or liquid substance into fine particles or droplets distributed throughout another medium; e.g., used to disperse oil slicks or spills into the water column rather than having them remain floating on the surface.

**displacement** *Transportation.* the total volume displaced by an engine's cylinders, calculated by multiplying the number of pistons times the volume displaced by a cylinder moving from bottom dead center to top dead center; expressed in cubic centimeters

**displacement pump** *Hydropower.* a pump in which valves prevent the return flow of displaced liquid during the retracting phase of the pump cycle, thus creating a pulsing action characterized by alternate filling and emptying of an enclosed volume. A **positive displacement pump** has an expanding cavity on the suction side and a decreasing cavity on the discharge side.

**disposal well** *Oil & Gas.* a well, often a depleted oil or gas well, into which waste fluids can be injected for safe disposal.

**dissipation** *Thermodynamics.* **1.** the conversion of energy into heat by friction in a mechanical process, or by resistive joule heating in an electrical process. **2.** the loss of energy from a system as a result of this.

**dissipative** *Thermodynamics.* **1.** describing an open energy system with a form created by the constant exchange of energy and matter with its surroundings. **2.** describing a loss of energy from a system as a result of this, as from friction in a mechanical process.

**dissipative system** *Thermodynamics.* an open thermodynamic system that maintains or increases its organization through exergy destruction. Also, **dissipative structure.**

**dissociation** *Chemistry.* the separating of a molecule into fragments, such as simpler molecules, atoms, radicals, or ions, because of a change in physical conditions, as when the molecule collides with other material or absorbs electromagnetic radiation. Thus, **dissociation energy.**

**dissolution** *Chemistry.* a process of dissolving; i.e., the dispersal of a solid or gaseous substance into a liquid.

**dissolved gas** *Oil & Gas.* chemicals normally occurring as gases, such as nitrogen and oxygen, that are held in solution in water. See also ASSOCIATED NATURAL GAS.

**distillate fuel (oil)** *Oil & Gas.* a general classification for one of the petroleum fractions produced in conventional distillation operations. Such fuel is used primarily for space heating, on-and-off-highway diesel engine fuel (including railroad engine fuel and fuel for agricultural machinery), and electric power generation. Included are products known as No. 1, No. 2, and No. 4 fuel oils, and No. 1, No. 2, and No. 4 diesel fuels. It is reported in the following sulfur categories: 0.05% sulfur and under, for use in on-highway diesel engines; greater than 0.05% sulfur, for use in all other distillate applications.

**distillation** *Chemistry.* a separation process in which a liquid source is evaporated and the vapor is condensed to a liquid end product, typically for the purpose of purification. Thus, **distillate.**

**distillation curve** *Measurement.* a plot showing the percentages of gasoline that evaporate at various temperatures; an important indicator for fuel standards such as volatility.

**distillers dried grain** *Biomass.* the dried byproduct of the process of fermenting grain (e.g., corn), typically used as a high-protein animal feed.

**distortionary tax** *Economics.* a conventional tax levied on a good or service that imposes an economic burden in excess of the direct revenue generated by the tax.

**distributed** *Electricity.* describing a power-generation technology that allows power to be used and managed in a decentralized and small-scale manner, thereby siting generation close to load, in order to minimize electricity transmission and maximize waste heat utilization. A distributed power unit can be connected directly to the consumer or to a utility's transmission or distribution system. Thus, **distributed generation, distributed power, distributed resource,** and so on.

**distributed collector (system)** *Solar.* a configuration of modular concentrating solar

**distributed energy** The generation of electricity (and heat) at, or close to, the point of demand. Distributed energy (DE) includes fossil technologies—fuel cells, micro-turbines, internal combustion engines and Stirling engines; renewable technologies—photovoltaic cells and wind turbines; and energy storage options. DE technologies come in a range of sizes, dependant on the application—i.e., a residential home versus an industrial facility. Potential DE advantages include lower costs through higher efficiency—waste heat recovery and avoidance of transmission and distribution losses, reduced carbon dioxide ($CO_2$) and sulfur dioxide ($SO_2$) emissions and enhanced flexibility of electricity networks. Potential DE disadvantages include higher costs through loss of generation economies of scale and higher local air pollution (including nitrogen oxides and particulates) near population areas. In many countries, institutional barriers, including lack of interconnection protocols and low electricity buy-back tariffs, have contributed to restricting DE to niche applications, such as emergency back-up power and peaking requirements. However, changes in the relative economics of centralized versus distributed energy, restrictions on new electricity transmission lines, and improved DE control technologies, have resulted in the reconsideration of widespread use of DE. For example, in the Netherlands by year 2000, 6% and 35% of national electricity capacity were DE units of less than 1MWe and less than 50MWe respectively.

**Neil D. Strachan**
Pew Center on Global Climate Change

collectors with an interconnected absorber pipe network to carry the solar heating working fluid to a heat exchanger.

**distributed energy (DE)** *Electricity.* the generation of electricity (and heat) at or close to the point of demand. See above.

**distribution** *Electricity.* a general term for the process of supplying electricity to a customer. See also DISTRIBUTED.

**distribution utility** *Electricity.* the regulated electric utility entity that constructs and maintains the distribution wires connecting the transmission grid to the final customer.

**distributor** *Transportation.* in an automobile or other motor vehicle, a device with a rotating drum that directs the secondary current from the induction coil to spark plugs in the various cylinders of the engine at correct times and in correct sequence.

**district energy** *HVAC.* a term for energy (e.g., steam, hot water, or chilled water) that is produced at a central location and then transmitted to various specific sites in a given area (district) for uses such as space heating and cooling or domestic hot water heating. Such a system can substitute for furnaces, air conditioners, and so on within the district's individual buildings. Thus, **district heat(ing)** or **cooling.**

**disturbance** *Ecology.* a chemical or physical event, often caused by humans, that leads to a sustained disruption in the function and structure of an ecosystem.

**diversion dam** *Hydropower.* a dam used in conjunction with long tunnels, canals, or pipelines to divert water to a powerhouse located at a distance from the dam.

**Divisia index** *Economics.* a method of aggregation that permits variable substitution among material types without imposing any prior restrictions on the degree of substitution. [Developed by French economist François *Divisia,* 1889–1964.]

**DME** dimethyl ether.

**DMFC** direct methanol fuel cell.

**Döberreiner, Johann Wolfgang** 1780–1849, German chemist known for his discovery of similar triads of elements in 1829, a key step in the development of the periodic law and the periodic table of chemical elements.

**Dobson spectrophotometer** *Earth Science.* a ground-based instrument that is used to measure the ozone content of the atmosphere, by comparing the intensity of ultraviolet light at two different wavelengths, one where light is strongly absorbed by ozone and the other where it is only weakly absorbed. [Designed by British meteorologist G. M. B. *Dobson,* 1889–1976.]

**DOE** Department of Energy.

**dog** *Biological Energetics.* the common domesticated animal *Canis familiaris;* larger breeds have traditionally been used as draft animals; e.g., to pull carts or sleds.

**Doheny, Edward** 1856–1935, U.S. entrepreneur who began the development of the oil resources of southern California with his operations in the Los Angeles area in the 1890s, which he later extended to Mexico, South America, and the British Isles. He later was implicated in the notorious TEAPOT DOME scandal of the 1920s.

**Doherty, Henry** 1870–1939, U.S. businessman who formed the holding company Cities Service Co. in 1910, which later became the giant energy company CITGO.

**domestic** *Consumption & Efficiency.* **1.** having to do with private households, as opposed to public buildings or industrial facilities. Thus, **domestic fuel, domestic heating,** and so on. **2.** see DOMESTIC ENERGY.

**domestic energy** *Consumption & Efficiency.* energy from resources existing within the borders of a nation rather than obtained from abroad.

**donkey** *Biological Energetics.* the domesticated ass, *Equus asinus,* used since ancient times as a beast of burden and draft animal.

**donor** *Chemistry.* **1.** a molecule, or part of a molecule's structure, that provides an electron pair to an acceptor. **2.** *Photovoltaics.* a dopant, such as phosphorus, that puts an additional electron into an energy level very near the conduction band; the electron is easily excited into the conduction band where it increases electrical conductivity compared to an undoped semiconductor.

**Donora** *Health & Safety.* a community in Pennsylvania that was the site of a noted environmental incident in 1948, when a "killer fog" created by industrial emissions hovered over the city for several days. Twenty people died and about 7000 others were hospitalized or became ill, about half the population of the town. This has been described as the first incident to make Americans generally aware of the health hazards of air pollution.

**Doomsday Clock** *Global Issues.* a symbolic clock face maintained since 1947 by the Bulletin of the Atomic Scientists. It uses the analogy of the human race being at a time that is a "few minutes to midnight", with midnight representing global destruction by nuclear war. The clock was started at seven minutes to midnight during the Cold War in 1947, and has subsequently been moved forwards or backwards at intervals, depending on the editors' perception of the state of the world.

**dopant** *Materials.* a chemical element (impurity) added in small amounts to an otherwise pure semiconductor material to modify the electrical properties of the material. An **N-dopant** (negative) introduces more electrons. A **P-dopant** (positive) creates electron vacancies (holes). e.g., boron, phosphorous, arsenic, antimony. Thus, **doping, doping agent.**

**Doppler, Christian Andreas** 1803–1853, Austrian physicist who first described (1842) how the observed frequency of light and sound waves is affected by the relative motion of the source and the detector, a phenomenon now known as the DOPPLER EFFECT.

**Doppler effect** *Physics.* an apparent shift in the observed frequency of a wave due to relative motion between the source and the observer; e.g., a train whistle will have higher pitch than normal as the train approaches an observer, and a lower pitch after it passes. The Doppler effect applies to all types of waves, including light. Also, **Doppler shift.**

**Doppler radar** *Earth Science.* a weather radar system that measures the direction and speed of a moving object, such as drops of precipitation, by determining its motion relative to the observer; based on the work of Christian Doppler.

**dose** *Health & Safety.* a specific amount of ionizing radiation or a toxic substance absorbed by a living organism.

**dose commitment** see COMMITTED DOSE.

**dose equivalent** see EQUIVALENT DOSE.

**dosimeter** *Nuclear.* dosage meter; a radiation survey instrument, either fixed in working spaces or worn by personnel (e.g., a film badge), used to measure and record doses accumulated during possible exposure to ionizing radiation.

**dosimetry** *Nuclear.* the principles and techniques involved in the measurement and recording of ionizing radiation doses.

**double dividend** *Economics.* the possibility that an environmental tax can both reduce pollution (the first dividend) and reduce the overall burden of the tax system, by using the revenue generated by this tax to displace other distortionary taxes that tend to slow economic growth (the second dividend).

**double-glazed** *Consumption & Efficiency.* describing a window or a sliding glass door that has two thicknesses of glass with an air space between them; used to provide better insulation. Thus, **double-glazing, double-pane glass.**

**double helix** *Biological Energetics.* the characteristic structure of DNA; i.e., two spiral strands that wind around each other.

**double-rate meter** *Electricity.* a watt-hour meter with two registers that indicate on-peak and off-peak energy consumption on separate dials.

**doubling time** *Social Issues.* the amount of time that it takes for a given entity (e.g., a nation) to double its population from the present time or some earlier date, equal to 70 divided by the annual percentage increase. Thus a nation with a 3% growth rate will double its population in about 23 years.

**DOWA** deep ocean water application.

**downblending** *Nuclear.* the process of blending highly enriched uranium with natural uranium to produce low enriched uranium.

**downflow** *HVAC.* a type of precision air conditioning system that discharges air downward, directly beneath a raised floor; used in computer rooms and other such modern office spaces.

**downhole** *Oil & Gas.* describing equipment, tools, and instruments used in a borehole during drilling. Thus, **downhole drill(ing).**

**downstream** *Consumption & Efficiency.* **1.** referring to the final stage of processing and/or actual use or consumption of energy products. **2.** referring to a latter stage in any process with linear progression, such as a production process. Compare UPSTREAM.

**downwelling** *Earth Science.* a process in which (warmer) surface ocean waters accumulate along a coastline and then are displaced downward to a deeper level.

**downwind** *Wind.* **1.** in the direction toward which the wind is blowing; with the wind. **2.** describing a wind energy device that operates with the hub and blades facing away from the wind direction; this offers the advantage of a lighter and more flexible construction than a similarly rated UPWIND machine.

**DP** distributed power; see DISTRIBUTED.

**DPM** diesel particulate matter.

**Draeger apparatus** *Mining.* a long-service, self-contained oxygen-breathing apparatus with a lung-governed oxygen feed that allows a mine worker to perform normal work for up to seven hours; also used as a defense against poisonous gases or oxygen shortages. [Named for the German company that developed this device.]

**draft** *Physics.* **1.** a movement of air, especially in an enclosed space. **2.** *Hydropower.* the release of water from a storage reservoir, usually measured in feet of reservoir elevation.

**draft animal** *Biological Energetics.* any domesticated animal used for pulling heavy loads, such as a plow or wagon; e.g., a horse, ox, mule, or water buffalo.

***draft animal*** *A traditional use of draft animals is to pull a plow or cultivator to till the soil. The term "draft" in this context comes from an older word meaning "to pull".*

**draft tube** *Hydropower.* a sealed, flared discharge pipe leading vertically from a water turbine to its tailrace; it decreases outlet pressure and increases turbine efficiency.

**drag** *Physics.* **1.** a force on an object parallel with the direction of the relative velocity of the flow; drag can be on an airfoil moving through the air, opposing the direction of the airfoil, or on the surface of an air duct, in the direction of the airflow. **2.** the effect of this force on an aircraft or other body in flight; in combination with available thrust, drag determines the craft's speed.

**dragline** *Coal.* an excavating machine used in surface coal mining, consisting of a large steel scoop bucket that swings on chains or wires from a movable boom; the scoop is cast into the material to be excavated and dragged back to the boom, where it is lifted and dumped. Also, **dragline excavator.**

**drain** *Storage.* the current withdrawn from a battery during discharge.

**drainage basin** *Hydropower.* an area occupied by a closed drainage system, especially a region that collects surface runoff and contributes it to a stream channel, lake, or other body of water.

**drainage windmill** *Wind.* a type of windmill employing a large scoop wheel to carry inland waters out to sea, thus permitting the use of lowland areas for agriculture; used extensively in Holland since medieval times and also elsewhere in western Europe.

**drainback system** *Solar.* a type of solar hot water heating system in which the solar collectors and a storage reservoir are pitched so that when the solar pump turns off, the fluid drains back from the collectors to the storage reservoir. During freezing weather, the fluid is drained back into the heat of the house and the collectors are not subjected to freeze-bursting.

**draindown system** *Solar.* a type of solar hot water heating system in which the collectors in an open loop system are filled with domestic water under house pressure when there is no danger of freezing. Once the system is filled, a differential controller operates a pump to move water from the tank through the collectors.

**Drake, Edwin** 1819–1880, U.S. entrepreneur who drilled the first successful oil well in the United States at a site in Titusville, Pennsylvania (1859); this became known as the **Drake field.** While oil was known prior to this, it was not available in large enough quantities to be considered very useful. Thus, many historians trace the start of the commercial oil industry to Drake's venture (though there are competing claims for other sites).

**draught** another spelling of DRAFT, especially in British usage.

**drawdown** *Hydropower.* **1.** the distance by which the water level of a reservoir or other body of water is lowered as a result of the withdrawing of water. **2.** the act or fact of releasing water in this way. **3.** *Oil and Gas.* the difference between flowing pressure and the static bottom-hole pressure.

**dredge mining** *Mining.* a historic method of recovering coal from rivers or streams by removing large portions of the stream bed.

**dredging** *Hydropower.* the process of digging up and removing solid matter from an underwater area, such as a harbor, river bed, or wetland; e.g., to provide greater depth for navigation or to increase streamflow for hydropower.

**Dresser, Solomon** 1842–1911, U.S. engineer noted for key improvements in the methodology of oil and natural gas recovery. In 1880 he patented a cylindrical packer that sealed crude oil from water and other elements in a well, a problem that had vexed oil producers. The company he founded would grow to be a large commercial venture.

**Dresser coupling** *Oil & Gas.* a type of coupling for pipe joints making it possible to provide a leakproof pipeline over extended distances. This invention (1891) permitted for the first time the long-range transmission of natural gas from the fields where it naturally occurred to urban centers of consumption.

**drift** see SURFACE DRIFT.

**drift mining** *Mining.* the process of working relatively shallow coal seams by tunneling laterally, as when a horizontal seam emerges on the surface at the side of a hill or mountain. Thus, **drift mine.**

**drill** *Consumption & Efficiency.* **1.** a cutting tool with a rotating end that is used to cut or enlarge holes in rock, earth, metal, wood, and other solid materials. **2.** to operate such a device. **3.** a farm implement that plants seeds by making a series of holes or furrows, dropping the seeds into them, and then covering them with soil.

**drill bit** see BIT.

**drill collar** *Oil & Gas.* a heavy, thick-walled tube, usually steel, used between the drill pipe and the bit in a drill stem to put weight on the bit to enhance drilling effectiveness.

**driller's method** *Oil & Gas.* a technique for killing a well by circulating high-density mud after the formation fluids have been circulated out of the well, in order to maintain drill pipe pressure.

**drilling mud** *Oil & Gas.* a mixture of finely divided heavy material consisting of clay, water, and chemical additives that is pumped downhole through a drill pipe; used for such purposes as cooling, lubrication, the transport of cuttings to the surface, and the prevention of foreign fluids entering the wellbore. Also, **drilling fluid.**

**drilling rig** *Oil & Gas.* a structure housing the equipment used to drill for and extract oil and natural gas from underground reservoirs.

**drillpipe** *Oil & Gas.* a heavy steel pipe that is rotated to give motion to a drilling bit and that is used to circulate drilling fluid.

**drill stem** *Oil & Gas.* a term for the entire assembly of a drilling apparatus.

**drill string** *Oil & Gas.* the column (string) of drill pipe with attached tool joints that links the drill bit to a mechanism imparting rotary or reciprocating motion.

**drive** *Physics.* **1.** a source of power or the application of power. **2.** a mechanism that imparts or transfers power to or within a machine.

**driveline retrofit** *Refrigeration.* a conversion of refrigeration equipment to an alternative refrigerant that requires the replacement of the compressor motor.

**drivepower** *Consumption & Efficiency.* a term for energy consumed by motors and motor-driven equipment.

**driver fuel** *Nuclear.* nuclear reactor fuel that contains the fissionable isotopes along with fertile isotopes that are bred into fissionable isotopes.

**driveshaft** *Transportation.* the shaft in a motor vehicle that transfers power from the engine or motor to the unit where this power is applied.

**driving force** *Social Issues.* a term used to describe any of the social forces that are identified as the sources of environmental problems, such as population expansion, economic growth, political and economic institutions, technological developments, and cultural values.

**drought** *Global Issues.* a period of less than average or normal precipitation over an extended period of time in a given area, severe enough to affect the water supply and cause biological damage and/or economic loss.

**drum brake** *Transportation.* a brake system in which a pair of brake shoes are pressed against the inner surface of a shallow metal drum rigidly attached to a wheel; a historic advance (about 1900) in stopping power for motor vehicles.

**dry ash-free (daf)** *Coal.* a theoretical measure of coal (or other organic material), based on a sample in which the moisture and ash are totally eliminated and the remaining constituents are recalculated to total 100 percent.

**dry basis** *Coal.* a theoretical description of coal quality, based on a sample with zero moisture.

**dry-bulb temperature** *Measurement.* air temperature that is not affected by atmospheric humidity; ordinary household thermometers measure ambient temperature in this manner. Thus, **dry-bulb thermometer.**

**dry cask** *Nuclear.* a large container used for storing spent nuclear fuel, using thick layers of materials such as steel, concrete, and lead (instead of water) as a radiation shield. A dry cask can hold multiple fuel assemblies. Thus, **dry-cask storage.**

**dry cell (battery)** *Storage.* a term for cells that use a solid or powdery electrolyte and that rely on ambient moisture in the air to complete the chemical process; the electrolyte

***drought*** *Abandoned Colorado farmhouse submerged by winds that have shifted topsoil from the neighboring arid fields; a depiction of the "Dust Bowl" landscape of the Depression era.*

is immobilized, making the battery portable; e.g., the Leclanche or carbon-zinc cell commonly used in flashlights, toys, and portable electronic devices.

**dry-charged** *Storage.* describing a storage battery in which the electrolyte is drained after the plates are formed; before being placed in service, the battery is filled with electrolyte and charged for a short time.

**dry circuit** *Electricity.* a term for a circuit with extremely low maximum voltages and very small maximum currents, so that there is no arcing to roughen the contacts, and an insulating film develops that prevents closing of the circuit.

**dry deposition** see DEPOSITION.

**dry hole** *Oil & Gas.* **1.** a drill hole that has been abandoned due to a lack of production of marketable petroleum materials. **2.** by extension, any speculative venture that has been unsuccessful.

**dry natural gas** *Oil & Gas.* natural gas that remains after the removal of the liquefiable hydrocarbon portion from the gas stream, or after the removal of any volumes of nonhydrocarbon gases that would render the gas unmarketable.

**dry steam** *Geothermal.* **1.** a term for a geothermal fluid that is primarily very hot steam. **2.** *Consumption and Efficiency.* a vapor with a very low water content, produced at high temperatures; used in various cleaning processes.

**dry steam plant** *Geothermal.* a geothermal power plant in which steam is conveyed directly to a turbine to generate electricity; e.g., Lardarello, Italy or The Geysers in northern California. This is the earliest form of geothermal power generation.

**dry ton** *Measurement.* one ton (2000 pounds) of material dried to a constant weight.

**DTW** dealer tank wagon.

**dual condenser** *HVAC.* a heat pump system that has the capability to switch, usually automatically, between an air and a water heat exchanger.

**dual-fired** *Electricity.* describing a generating unit that can produce electricity by means of two or more input fuels. In some of these units, only the primary fuel can be used continuously; the alternate fuels can be used only for start-up or in emergencies.

**dual flash cycle** *Geothermal.* a second stage of a FLASH STEAM process in which any liquid that remains in the vaporizing tank is flashed again at a different pressure in a second tank, to extract more energy in addition to that provided by the initial stage.

**dual fuel vehicle** *Transportation.* see BI-FUEL VEHICLE.

**duality** *Thermodynamics.* the concept that electrons (and all particles) behave both as classical particles and as waves depending on how they are observed (i.e., they cannot be both at the same time).

**Dubai crude** *Oil & Gas.* one of an array or "basket" of seven crude oil types used by OPEC as reference points for pricing.

**ducted** *HVAC.* describing a heating/cooling system that regulates temperatures throughout a building by means of a series of ducts in the floor or ceiling. The compressor unit is located outside the building.

**dump energy** *Hydropower.* a term for energy that is lost (i.e., dumped) by a hydroelectric system, usually because it is in excess of the capacity or requirements of the system; typically defined as the number of megawatt hours of energy produced in excess of aggregate load.

**dung** *Biomass.* solid animal waste matter; a traditional fuel source in dried form; e.g., cow dung.

**Dunlop, John Boyd** 1840–1921, Scottish inventor credited with the development of the first practical pneumatic tire **(Dunlop tire),** based on experiments with his son's bicycle. The company bearing his name would grow to be a major international corporation.

**Durnin, J. V.** Scottish nutritionist noted for his seminal work in the field of human nutrition and the energetics of human metabolism.

**Dutch disease** *Economics.* **1.** a phenomenon that occurs when a resource boom causes exchange rates to rise and labor and capital to migrate to the booming sector; the result is higher costs and reduced competitiveness for domestic goods and services, effectively crowding out previously productive sectors. **2.** any situation in which an income windfall from an energy resource leads to adverse consequences for the nation's economy. Also, **Dutch curse.** [Named after the negative effects of the recent North Sea oil boom on the *Dutch* nation (The Netherlands).]

**Dutch oven** *Consumption & Efficiency.* an early type of furnace, having a large, rectangular box lined with firebrick on the sides and top, commonly used for burning wood.

**Dutch windmill** *Wind.* a term for the familiar historic type of European windmill; so called because of its early and widespread use by the Dutch.

**duty cycle** *Storage.* **1.** the amount of time it takes to start, operate, stop, and idle a machine when it is being used for intermittent duty. **2.** a percentage that expresses the amount of "on" (working) time as compared to total time for an intermittently operating device or system.

**duty rating** *Consumption & Efficiency.* the amount of time that a unit or system can deliver peak output (e.g., electrical power).

**dye-sensitized solar cell** *Photovoltaic.* an advanced type of photovoltaic cell that uses a dye-impregnated layer of titanium dioxide to generate a voltage, rather than the semiconducting materials used in most solar cells. Also, **dye solar cell.**

**dynamic head** *Hydropower.* the vertical distance from the water level of a pump or reservoir to the point of free discharge of the water; measured during actual flow and taking into account the effects of friction, since resistance to friction is not the same at all flow rates. The **total dynamic head (TDH)** is the total resistance from both friction and gravity that a pump will experience when delivering a given rate of flow.

**dynamic height** *Earth Science.* a measure of the level of a system of ocean water, relative to mean sea level; used to describe the height of a current, eddy, or column of water. The dynamic height can be negative in cases of extreme cold or salinity.

**dynamic pressure** *Hydropower.* an expression of the pressure of a flowing fluid, such as water flowing in a pipeline; equal to one-half the fluid density, multiplied by the fluid velocity squared. It is equivalent to the static pressure minus any actual pressure loss from friction, turbulence, and so on.

**dynamics** *Physics.* **1.** the field of mechanics that deals with the study of motion and of the forces that bring about motion. **2.** see POPULATION DYNAMICS.

**dynamite** *Materials.* a powerful blasting explosive that was originally manufactured by the absorption of nitroglycerine into a porous base material such as charcoal or wood pulp; now generally manufactured with ammonium nitrate or cellulose nitrate.

**dynamo** *Electricity.* a device for converting mechanical energy into electrical energy, especially one that produces direct current.

**dynamoelectric** *Electricity.* relating to or involving the exchange of energy between mechanical and electrical processes.

**dynamometer** *Transportation.* any of various devices used in testing a motor or engine for such characteristics as efficiency and torque, especially a device used to simulate road conditions and loads in stationary settings and to gather data about vehicle performance under those conditions.

**dyne** *Measurement.* a unit equal to the amount of force that will cause a mass of 1 gram to accelerate at a rate of 1 cm per second per second; the basic unit of force in the centimeter-gram-second system.

**E-** *Oil & Gas.* ethanol; a designation of the amount of ethanol contained in a fuel mixture with gasoline; e.g., E-10. 10% ethanol and 90% gasoline (gasohol); E-95. 95% ethanol and 5% gasoline.

**earth** *Earth Science.* **1.** also, **Earth.** the planet inhabited by humans; the third planet from the sun, at a mean distance of about 92.9 million miles. **2.** soil; the loose, fragmented material that composes part of the surface of this planet, especially soil that can be cultivated.

**earth berm** see BERM.

**earth-cooling tube** *Renewable/Alternative.* an underground metal or plastic pipe employed in home cooling; as air travels through the pipe it gives up some of its heat to the soil, and enters the house as cooler air (assuming the earth is significantly cooler than the incoming air). A potential alternative to air conditioning, but at present not sufficiently efficient or economical.

**earth-coupled heat pump** see GROUND-SOURCE HEAT PUMP.

**earth dam** *Hydropower.* a dam whose main section is composed primarily of tightly compacted earth (soil, gravel, sand, silt, clay, and so on), as opposed to concrete or another manufactured material; e.g., Fort Peck Dam on the Missouri River.

**Earth Day** *History.* an annual nationwide demonstration in the U.S. (founded 1970), advocating environmental protection and preservation; originally conceived by Gaylord Nelson, U.S. Senator from Wisconsin, and organized by Denis Hayes.

**earthfill dam** **1.** see EARTH DAM. **2.** see FILL DAM.

**earthmover** *Mining.* any of various large vehicles used to excavate and transport earth at a mine site.

**Earth Summit** *Sustainable Development.* another name for the United Nations Conference on Environment and Development (UNCED), a meeting of people and organizations from 178 nations held in Rio de Janeiro, Brazil in 1992. The meeting produced the Rio Declaration on Environment and Development, a statement of 27 principles upon which the nations agreed to base their actions in dealing with environment and development issues.

**Easter Island** *History.* a remote island in the Pacific Ocean about 2000 miles west of South America, historically settled by Polynesian peoples; the site of a once-flourishing society that began to decline about 1400 AD; now studied as a model of the collapse of a civilization due to overconsumption of resources.

**ebb generation** *Renewable/Alternative.* the simplest mode of operation for a tidal plant, in which the production of electricity occurs during the part of the cycle when water flows through the turbines from the basin to the sea (i.e., during ebb tide).

**ebb tide** *Earth Science.* a falling tide; the phase of the tide between a high water and the following low water.

*ebb tide* *Tides on the north coast of France are among the most extreme in the world. The ebb tide at the historic Mont St. Michel abbey leaves behind an expanse of mud flats.*

**EBR-1** *Nuclear.* Experimental Breeder Reactor 1, the first nuclear reactor in the world to produce useable quantities of electric power, lighting four 100-Watt light bulbs at the National Reactor Testing Station in Arco, Idaho (1951).

**ECCS** emergency core-cooling system.

**ECN** Energy Research Centre of the Netherlands, the largest Dutch research center in the field of energy.

**ecoefficiency** *Economics.* a business strategy to produce goods with lower levels of use of materials and energy, in order to realize the economic benefits of environmental improvements.

**eco-industrial park (EIP)** *Environment.* a community of manufacturing and service businesses located on a common property and sharing resources to improve their environmental and economic performance; e.g., savings from waste recycling, the ability to avoid regulatory penalties, and increased efficiency in material and energy use. These parks employ ecological principles to achieve the least damaging interaction with the environment.

**ecolabel** *Environment.* an official symbol or logo displayed on, or in connection with a certain product to show that the product is "environmentally friendly"; i.e., that it has been designed to do as little harm to the environment as possible. Thus, **ecolabeling** or **ecolabelling.**

**ecological** *Ecology.* having to do with the relationship between organisms and their environment.

**ecological deficit** *Ecology.* the amount by which the ecological footprint (see below) of a human population (e.g., a country or region) exceeds the biocapacity of the space available to that population.

**ecological economics** *Economics.* a transdisciplinary field that examines the relationship between ecological and economic systems from a number of integrated viewpoints, including principles from economics, the natural sciences, and other social sciences; practitioners are interested in the broad question of the use of resources by society and in understanding the nature, behavior, and dynamics of the economy-environment system.

**ecological energetics** *Biological Energetics.* **1.** the branch of ecology that studies ecosystems or communities in terms of the energy flows within them. **2.** specifically, the energy flows associated with a given species; e.g., the ecological energetics of the desert tortoise. See below.

**ecological engineering** *Economics.* the application of basic and applied principles from engineering, ecology, economics, and natural sciences, for the restoration and construction of aquatic and terrestrial ecosystems.

**ecological footprint** *Ecology.* a concept used to describe the impact that a given human entity (e.g., a nation such as Japan, or a large

**ecological energetics** Ecological energetics refers to a suite of approaches in the environmental sciences that attempts to understand ecological structure and especially function from the perspective of energy costs and gains. It has been applied at scales from individual organisms to entire ecosystems. Generally there is an implicit evolutionary backdrop, in other words the question is often how does this or that morphology, physiology, or behavior contribute to selective advantage, that is survival and the leaving of progeny. A particularly good example of this is a study of European chickadees who examined explicitly their reproductive output in terms of the timing of their migration relative to the spring pulse of caterpillars, in turn attuned to the timing of leaf outbreak. Those birds that nested just prior to the outbreak had much food for their young which could be gathered with relatively little effort. Those that were too early or late left far fewer offspring. Climate changes were impacting which behaviors were rewarded and not. Energy studies have also been undertaken at the level of populations and ecosystems, and still one of the best is that of Silver Springs by Howard Odum. Perennial questions of such studies are those of the efficiency of trophic (or food level) processes as energy goes from sunlight to sugars produced by photosynthesis to herbivores to carnivores, and also the role of power in fitness. Finally, some ecologists have used such concepts applied originally to natural ecosystems to look at national economies from an energetic perspective. Such studies are thought to be increasingly useful as humanity appears to be facing an increasingly energy-constrained future.

**Charles A. S. Hall**
College of Environmental Science and Forestry
State University of New York, Syracuse

**ecological footprint** A resource accounting framework for measuring human demand on the biosphere. The Footprint of a population is the biologically productive land and water area that the population requires to produce the resources it consumes and absorb the waste it generates, using prevailing technology. Because people consume resources and ecological services from all over the world, their Footprint is the sum of these areas, wherever they are on the planet. Results are expressed in global hectares, hectare of biologically productive space with world-average productivity. This measurement unit, or 'ecological currency' makes comparisons possible across the world. Governments and organizations to measure and manage sustainability efforts use the Footprint. In consuming nature's products and services, people have an impact on the earth. But since nature has the ability to renew, it can cope with human demand as long as this demand stays within the regenerative capacity of the biosphere. The earth's biologically productive area is approximately 11.3 billion hectares or 1.8 global hectares per person in 2001. The global Ecological Footprint in 2001 is estimated at 13.5 global hectares or 2.2 global hectares per person. This finding implies and suggests that society must develop more sustainable technologies, institutions, and behaviors to avoid eroding the supply of natural resources and environmental services for future generations.

**Mathis Wackernagel**
Global Footprint Network
Oakland, California

city such as São Paulo) has on the biological resources of the earth; a measure of how much biocapacity a population or process requires to produce its resources and absorbs its waste. See above.

**ecological integrity** *Ecology.* the condition of an ecosystem that has its native components intact and is self-sustaining and self-regulating, including abiotic components (e.g., water), biodiversity, and ecosystem processes (e.g., fire, predation); such a system has a self-correcting ability to recover when subjected to a disturbance.

**ecological overshoot** *Ecology.* the overexploitation of resources or accumulation of waste; growth beyond an area's carrying capacity.

**ecological pyramid** *Ecology.* a description of the flow of energy in an ecosystem, expressed in a graphical format resembling a pyramid; the parameter being displayed can be the number of organisms at each level **(number pyramid)**, energy present at each level **(energy pyramid)**, or amount of biomass per level **(biomass pyramid)**.

**ecological risk assessment (ERA)** *Ecology.* a process evaluating the likelihood that adverse ecological effects may occur or are occurring as a result of the exposure of terrestrial, avian, and aquatic receptors to one or more stressors.

**ecological terrorism** see ECOTERRORISM.

**ecology** the scientific study of the relationship between organisms and their environment, including their relationship with other organisms, ranging from the relationship of two individuals to the organismal interactions of the biosphere as a whole.

**econometrics** *Economics.* economic measurement; the application of statistical methods to the analysis of economic data and theories. Thus, **econometric.**

**economic efficiency** *Economics.* a condition in which the organization of production minimizes the ratio of inputs to outputs, and goods are thus produced at minimum cost in money and resources.

**economic geography** *Economics.* the principles and processes associated with the spatial allocation of human and natural resources, and the environmental, economic, and social consequences resulting from such allocations.

**economic globalization** see GLOBALIZATION.

**economic rent** see RENT.

**economiser** a British spelling of ECONOMIZER.

**economizer** *Consumption & Efficiency.* **1.** an apparatus that uses warm flue gases exiting a steam boiler to preheat feedwater entering the system, thus improving boiler efficiency and economy. **2.** a compartment in a continuous-flow oxygen system that collects exhaled oxygen for reuse. **3.** see ECONOMIZER SYSTEM.

**economizer system** *HVAC.* a system that takes advantage of favorable weather conditions to reduce the amount of mechanical cooling required for a building, by introducing naturally cool outside air into the building.

**economy in transition (EIT)** *Economics.* a descriptive term for a nation that is in transition from a centrally-planned economy to a market-based economy; especially applied to nations that were part of the Soviet sphere of influence; i.e., Russia, various Central and East European countries such as Poland and Hungary, and former republics of the USSR.

**ecosphere** *Ecology.* the earth conceived of as a living system; the planet and all the living organisms that inhabit it, along with the environmental factors that affect them.

**ecosystem** *Ecology.* an identifiable entity in nature, consisting of a community of living organisms and their surrounding environment of air, soil, water, mineral cycles, and so on, which they interact with and affect; e.g., the savanna ecosystem of central and southern Africa, the boreal forest ecosystem of northern Canada.

**ecosystem health** *Ecology.* the functional integrity of an ecosystem, as characterized by a state of ongoing self-renewal in the system as a whole and in its particular components (soil, water, animals, plants, and so on).

**ecoterrorism** *Global Issues.* violence carried out to further the political or social objectives of environmental activists, who ostensibly are motivated by concern for the natural environment.

**ecotourism** *Global Issues.* tourist activity that is ecologically sustainable and that focuses on experiencing natural areas in a manner that will encourage environmental appreciation and conservation of resources.

**ecotoxicology** *Health & Safety.* a scientific discipline that studies the adverse effects of chemicals, both synthetic and natural, on biological and ecological systems, especially on a large scale. Thus, **ecotoxicological, ecotoxicologist.**

**ectotherm** *Biological Energetics.* a cold-blooded animal; i.e., an organism that regulates its body temperature largely by exchanging heat with its surroundings, such as a reptile, amphibian, or fish. Thus, **ectothermy.** Contrasted with ENDOTHERM.

**ectothermic** *Biological Energetics.* deriving body heat from the sun and other external sources.

**eddy conductivity** *Thermodynamics.* a fluid-flow quantity describing the heat transfer due to eddies (whirling or circular flow) in a fluid, caused by a temperature difference between the fluid near the boundary and the fluid in the center of the stream, so that hot fluid is carried to the cooler regions and cold fluid to the warmer regions.

**eddy current** *Electricity.* an electric current produced in a conductor as it moves through a magnetic field.

**Edison, Thomas Alva** 1847–1931, prolific U.S. inventor who held more than 1000 patents for his inventions, including innovations such as the incandescent electric lamp (1879) and the phonograph (1877). Edison did not invent the light bulb, but he produced an improved version using a small carbonized filament. In 1882, he opened the first commercial power station in New York City, thus beginning the electric era.

*Thomas Edison*

**Edison battery** another name for a NICKEL-IRON BATTERY.

**Edison effect** *Electricity.* a phenomenon observed by Thomas Edison; the emission of electrons from a heated cathode. Though he did not recognize the importance of this discovery, subsequent scientists used the effect as the basis for the electron tube.

**Edison Electric Institute** (est. 1933), an influential U.S. trade association for shareholder-owned electric companies that represents utility interests in the legislative and regulatory arenas.

**EDP** *Economics.* environmentally adjusted net domestic product; an environmental accounting aggregate obtained by deducting the cost of resource depletion and environmental degradation from net domestic product.

**EEI** Edison Electric Institute.

**EER** energy efficiency ratio.

**effective current** *Electricity.* the amount of alternating current that produces the same power dissipation effect on a load resistor as the corresponding amount of direct current.

**effective dose** *Nuclear.* a measure of toxicity calculated by multiplying equivalent doses by a tissue-specific weighting factor that is intended to take into account differences in sensitivity of different organs and tissues to the effects of ionizing.

**effective dose equivalent** *Nuclear.* the sum of the dose equivalents to various organs or tissues and the weighting factors applicable to each of these that are irradiated.

**effective half-life** *Nuclear.* the time required for the concentration of a radionuclide in a biological system to be reduced to one-half of the original concentration.

**effective internal resistance** *Storage.* the apparent opposition to current within a battery that manifests itself as a drop in battery voltage proportional to the discharge current; depends on factors such as the battery design, state of charge, temperature, and age.

**effective nocturnal energy** *Solar.* the energy transfer required to maintain a horizontal upward-facing blackbody surface at the ambient air temperature, in the absence of solar irradiance.

**effective temperature** *Measurement.* the temperature at which motionless, saturated air would create the same sensation of comfort in a sedentary worker wearing ordinary indoor clothing as that created by actual conditions of temperature, humidity, and air movement.

**efficacy** *Lighting.* a measure used to compare light output to energy consumption, measured in lumens per watt. A 100-watt light source producing 1750 lumens of light has an efficacy (efficiency) of 17.5 lm/W.

**efficiency** *Consumption & Efficiency.* **1.** in general, the relative effectiveness of a system or device, especially in terms of the total resources required to attain a desired output. See also various specific terms such as ECONOMIC EFFICIENCY, ENERGY EFFICIENCY, END-USE EFFICIENCY, and so on. **2.** *Thermodynamics.* a dimensionless quantity that characterizes an energy conversion process based on the relationship of work output to energy input. See also FIRST LAW EFFICIENCY, SECOND LAW EFFICIENCY.

**efficiency gap** *Consumption & Efficiency.* the difference between the actual level of investment in energy efficiency and a higher level that would be cost-beneficial from the standpoint of consumers and society.

**efficiency service company (ESCO)** *Electricity.* a company that offers to reduce a client's electricity consumption with the cost savings being split with the client.

**effluent** *Environment.* **1.** a liquid or gas that flows out or flows away; e.g., a stream that flows out of a larger stream, a lake, or another body of water. **2.** liquid waste matter that results from sewage treatment or industrial processing, especially such waste liquid released into waterways.

**E-fuel** *Oil & Gas.* a fuel consisting of an ethanol/gasoline mixture.

**egg coal** see COAL SIZING.

**eggbeater turbine** *Wind.* a term for a wind turbine whose blades resemble the motion of the common kitchen eggbeater; i.e., it has thin blades that rotate about a vertical axis; e.g., the Darrieus turbine.

**EGS** enhanced geothermal system.

**EIA** **1.** environmental impact assessment. **2.** Energy Information Administration.

**Einstein, Albert** 1879–1955, German-born physicist who published three papers in 1905 that had a profound effect on the development of physics. In one paper, he proposed the theory of special relativity (see RELATIVITY). In a second paper, he explained the PHOTOELECTRIC EFFECT. In the third paper he provided evidence for the physical existence of atom-sized molecules. With the rise of fascism in Germany, Einstein moved to the U.S., where

*Albert Einstein*

his correspondence with President Franklin D. Roosevelt concerning atomic energy contributed to the establishment of the MANHATTAN PROJECT.

**einsteinium** *Chemistry.* a synthetic radioactive chemical element having the symbol Es, the atomic number 99, and an atomic weight (of the most stable isotope) 252.08; produced by bombarding berkelium and californium with helium ions and deuterons.

**Einstein's (relativity) theory** see RELATIVITY.

**Einstein shift** *Physics.* the lengthening of the wavelengths of light that are emitted by bodies with strong gravitational fields, causing a displacement of spectral lines toward the redder part of the spectrum; predicted in Einstein's general theory of relativity.

**EIOLCA** *Economics.* economic input–output life cycle assessment; a method of estimating the overall environmental impacts from producing a certain dollar amount of any of various products, materials, services, or industries, with respect to resource use and emissions.

**EIS** environmental impact statement; see ENVIRONMENTAL IMPACT ASSESSMENT.

**EKC** environmental Kuznets curve.

**Ekman convergence** *Earth Science.* a zone of convergence between warm and cold water masses, caused by the Ekman transport of surface water (see below). Also, **Ekman layer.** [Described by Swedish oceanographer Vagn Wilfrid *Ekman,* 1874–1954.]

**Ekman transport** *Earth Science.* the net mass displacement of water from one place to another, caused by wind blowing steadily over the surface; the net mass transport is 90° to the right (in the Northern Hemisphere) of the wind's direction.

**el** short for ELEVATED RAILWAY.

**elastic** *Materials.* **1.** having the property of elasticity; able to return to its original shape after experiencing strain and removal of deforming stress. **2.** *Economics.* having an economic elasticity that is greater than or equal to one.

**elasticity** *Materials.* **1.** a property of materials in response to stress, indicating the degree to which strain disappears from a material when the stress is removed. **2.** *Economics.* the relative responsiveness of an economic condition to change; generally measured as the percentage change that will occur in a given variable, in response to a 1% change in another variable. Thus, **elasticity of demand, elasticity of supply.** See next page.

**elasticity of substitution (ESUB)** *Economics.* the percentage change in the relative use of inputs (capital, labor, energy, materials) in response to changes in their relative costs. A high ESUB implies an easy ability to switch away from a given input as its relative price increases, and a low ESUB indicates the opposite.

**elastomer** *Materials.* a natural or synthetic material having elastic properties similar to rubber. Thus, **elastomeric.**

**Electra** *History.* an advance in aviation (1937), the first airplane with a fully pressurized cabin, developed by Lockheed Aircraft.

**electric** *Electricity.* **1.** having to do with or involving electricity. **2.** produced by or carrying electricity.

**electrical** *Electricity.* **1.** having to do with electricity; electric. **2.** having to do with the science or technology of electricity. Thus, **electrical engineer(ing).**

**electrical conductance** see CONDUCTANCE.

**electrical energy** *Electricity.* **1.** the energy inherent in an array of charged particles because of their relative positions. **2.** the energy inherent in a circuit because of its position in relation to a magnetic field.

**elasticity** Introduced by Alfred Marshall in the early 1880s, the concept of elasticity has assumed an important role in modern energy analysis. Intuitively, elasticity means responsiveness; in policy circles it is generally measured as the per cent change in one variable that is associated with a 1% change in another (presumably causal) variable. The most common elasticities used in energy analysis are price and income elasticities of demand as well as supply elasticities. Economists typically distinguish short-run elasticities (responses with a fixed capital stock) from long-run elasticities (where the response can include an adjustment of the capital stock) and have found, not surprisingly, that long-run elasticities are typically larger. Elasticities are commonly used in energy forecasting either as the basis for the forecast (in simpler models) or to summarize the demand or supply responsiveness implied by more complicated forecasting models. Elasticities also have behavioral implications for key participants in the energy market. For example, the low price elasticity of demand for oil allows OPEC to increase revenue by raising oil prices, while the higher income elasticity for oil demand subjects the cartel to potentially destabilizing demand volatility. For the energy market, the low short-run price elasticity of demand, combined with the low short-run supply elasticity, imply the potential for quite large swings in the oil prices in response to shocks, such as when production is curtailed by war. For governments, the low price elasticity of oil demand also implies that price-based conservation policies must raise prices by a large amount to have much effect.

**Tom Tietenberg**
Colby College

**electrical fault** see FAULT.

**electrical induction** see INDUCTION.

**electrical power** see ELECTRICAL ENERGY.

**electrical resistance** see RESISTANCE.

**electrical thermometer** *Measurement.* a thermometer using a transducing element whose element properties are a function of its thermal state.

**electrical ground** see GROUND.

**electrical transient** *Electricity.* any voltage or current that deviates from the normal steady-state condition.

**electric baseboard** see BASEBOARD HEAT.

**electric battery** see BATTERY.

**electric car** *Transportation.* a passenger vehicle that is powered exclusively by an electrochemical power source, or partially so powered (hybrid electric).

**electric cell** see CELL (def. 2).

**electric charge** see CHARGE.

**electric current** see CURRENT.

**electric dipole** see DIPOLE.

**electric energy** see ELECTRICAL ENERGY.

**electric field** *Electricity.* a region in space in which lines of force produced by an electric charge exert a force on other electric charges.

**electric force** *Electricity.* a force between two objects such that each have the physical property of charge.

**electric generator** see GENERATOR.

**electric heat(ing)** *HVAC.* a process in which electric energy becomes heat energy by resisting the free flow of electric current; e.g., radiant heating.

**electricity** **1.** a fundamental form of energy, consisting of oppositely charged electrons and protons that produce light, heat, magnetic force, and chemical changes. **2.** the flow of this energy; electric current. **3.** the general phenomenon of charges at rest and in motion.

**electric light(ing)** *Lighting.* **1.** an incandescent lamp, or the light produced by this. **2.** any form of lighting powered by electricity.

**electric–magnetic field** see ELECTROMAGNETIC FIELD.

**electric motor** *Conversion.* a device that converts electrical energy into mechanical energy using forces exerted by magnetic fields on current-carrying conductors.

**electric polarization** *Electricity.* the separation of charges in a material to form electric dipoles, or the alignment of existing electric dipoles in a material when an electric field is applied.

**electric potential** *Electricity.* the potential measured by the energy of a unit positive charge at a point, expressed relative to an equipotential surface that has zero potential, generally the surface of the earth.

**electric power** see ELECTRICAL ENERGY.

**electric radiant heating** see RADIANT HEATING.

**electric railroad** *Transportation.* a railroad having a continuous system of overhead wires or a third rail mounted alongside or between the guide rails throughout the track's length to supply electric power to a transportation system.

**electric resistance heat(ing)** see RESISTANCE HEAT.

**electric space heating** see SPACE HEATING.

**electric utility** see UTILITY.

**electric vehicle (EV)** *Transportation.* a vehicle that is powered solely by an electrochemical power source, such as a battery or fuel cell.

**electrification** *Electricity.* **1.** the process of applying an electric charge to a component or device. **2.** the fact of providing electric power to an area or to specific consumers.

**electro-** a prefix meaning "electric" or "electricity".

**Electrobat** *History.* an early electric car (1894) developed by entrepreneurs Henry Morris and Pedro Salom, who also were the first to operate a motor vehicle service in the U.S. with their Electric Carriage and Wagon Company in New York City.

**electrocatalysis** *Chemistry.* the acceleration of a chemical reaction in the region of an electrode. Thus, **electrocatalyst.**

**electrochemical** *Chemistry.* or relating to electrochemistry; having to do with the relationship of chemical change and electric force.

**electrochemical cell** *Conversion.* a device that converts chemical energy into electrical energy by passing current between a negative electrode and a positive electrode, through an ionically conducting electrolyte phase.

**electrochemistry** *Chemistry.* the scientific study of chemical changes that occur when a chemical reaction produces an electric current, or vice versa.

**electrochromic** *Consumption & Efficiency.* undergoing a change in optical properties upon the passage of an electrical current. Electrochromic materials can be used in window glass to provide energy efficiency in buildings, by electronically varying the level of tint in the window to control the amount of light and solar heat entering the room. Thus, **electrochromics, electrochromism.**

**electrode** *Electricity.* an electronically conductive structure that provides for an electrochemical reaction through the change of oxidation state of a substance; it may contain or support the reactant or act as the site for the reaction. The anode and cathode of an electric cell are electrodes.

**electrode boiler** *Conversion.* a boiler that converts electricity to heat energy, by means of passing a current between electrodes immersed in water to raise the temperature of the water.

**electrodeposition** *Materials.* an electrolytic process in which a metal is deposited at the cathode from a solution of its ions; used in the manufacture of semiconductors.

**electrode potential** *Chemistry.* the potential developed by a metal or other electrode material immersed in an electrolytic solution; usually related to the standard potential of the hydrogen electrode, which is established at zero. Also, **electrode voltage.**

**electrodialysis** *Materials.* a process in which an electric field transports ionized material through a membrane to separate it from other liquids or ions of opposite charge.

**electrodynamics** *Electricity.* the study of the relationships between electromagnetics and mechanical phenomena.

**electrodynamometer** *Measurement.* a device used to measure electric power.

**electrokinetics** *Electricity.* the study of the motion of electric charges.

**electroluminescence** *Materials.* the generation of light by applying electricity to a material such as a semiconductor or phosphor (a substance that can exhibit the phenomenon of fluorescence or phosphorescence).

**electrolysis** *Chemistry.* **1.** the process of splitting water into its components, hydrogen and oxygen, by means of an electrical current. **2.** any process in which the passage of an electric current through a solution or medium produces a chemical reaction. Thus, **electrolytic.**

**electrolyte** *Electricity.* **1.** any liquid or solid substance that while in solution or in its pure state will conduct an electric current by means of the movement of ions; usually it is a solution of water and acids or metal salts. **2.** *Biological Energetics.* any of certain inorganic compounds, such as sodium, potassium, magnesium, and calcium, that dissociate fluids into ions conducting electric currents and that thus constitute the major force in controlling fluid balance **(electrolyte balance)** within the body.

**electrolytic capacitor** *Electricity.* a capacitor in which an electrolyte serves as a plate; the other plate is wound aluminum foil. A thin layer of oxidation on the foil is the dielectric.

**electrolytic cell** *Electricity.* an electrochemical cell in which the reactions are driven by the use of an external potential greater than the thermodynamic or reversible potential of the cell.

**electrolytic cleaning** *Materials.* a process of removing soil, scale, or corrosion products from a metal surface by subjecting it as an electrode to an electric current in an electrolytic bath.

**electrolytic reduction** *Materials.* a process of reducing the oxidation state of a material through the application of a current; commonly used in electroplating.

**electrolyzer** *Hydrogen.* a device or system that can produce a process of electrolysis; i.e., the decomposition of water into hydrogen and oxygen by means of an electrical current.

**electromagnet** *Electricity.* a magnet, consisting of a coil wrapped about a soft iron or steel core, that becomes strongly magnetized when current flows through the coil.

**electromagnetic** *Physics.* **1.** having to do with or caused by electromagnetism; i.e., the interaction of electricity and magnetism. Thus, **electromagnetic energy. 2.** having to do with the science of electromagnetism.

**electromagnetic current** *Physics.* a movement of charged particles in the atmosphere giving rise to electric and magnetic fields, such as those in the ionosphere that transmit radio signals.

**electromagnetic field (EMF)** *Physics.* the combination of electric and magnetic fields that surround moving electrical charges (e.g., electrons), such as those in electric currents. Electromagnetic fields apply a force on other charges and can induce current flows in nearby conductors.

**electromagnetic induction** *Physics.* the generation of an electromotive force by changing the magnetic flux through a closed loop circuit, or by moving a conductor across the magnetic field; this principle is the basis for the electric generator and electric motor.

**electromagnetic interference (EMI)** *Electricity.* the harmful impairment of a transmitted electromagnetic signal by an electromagnetic disturbance.

**electromagnetic pulse (EMP)** *Nuclear.* a broadband, high-intensity, and short-duration burst of electromagnetic energy resulting from a nuclear explosion.

**electromagnetic radiation** *Physics.* the emission and propagation of radiation associated with a periodically varying electric and magnetic field traveling at the speed of light; types include gamma radiation, X-rays, ultraviolet, visible, and infrared radiation, and radar and radio waves.

**electromagnetic spectrum** *Physics.* the ordered sequence of all known forms of electromagnetic radiation, extending from the shortest cosmic rays through gamma rays, X-rays, ultraviolet, visible radiation, and infrared radiation, and including microwaves and all other radio wavelengths.

**electromagnetic theory (of light)** *Physics.* the theory that light consists of electromagnetic radiation and therefore obeys Maxwell's equations; contrasted with earlier concepts such as light as a stream of tiny particles or light as a wave in a medium of ether. See also LIGHT.

**electromagnetic wave** *Physics.* an oscillation of the electric or magnetic field associated with the propagation of electromagnetic radiation.

**electromagnetism** *Physics.* **1.** the interaction between magnetism and electricity, and the phenomena produced by this interaction. **2.** the scientific study and application of such phenomena.

**electromechanical** *Physics.* describing a mechanical device, system, or process that is actuated or controlled by electromagnetic or electrostatic phenomena. Thus, **electromechanics.**

**electrometer** *Measurement.* **1.** an instrument used to measure voltage variation without drawing current from the source. **2.** an instrument used to determine fluctuations in electrostatic potential difference between charged electrodes due to radiation.

**electromotive force (EMF)** *Electricity.* the pressure that causes electrons to move in an electrical circuit; measured as the amount of energy supplied by an electric current passing through a given source, as measured in volts.

**electron** *Physics.* a stable elementary particle that is a primary constituent of ordinary matter, contained in the atoms of all elements and described as having a charge of $-1.602 \times 10^{-19}$ C, a rest mass of $9.11 \times 10^{31}$ kg, and a spin of 1/2. Electrons flowing in a conductor constitute an electric current.

**electron acceptor** see ACCEPTOR.

**electron beam** *Physics.* a stream of electrons emitted from a given source and traveling under the influence of an electric or magnetic field in the same direction and at approximately the same speed.

**electron capture** *Physics.* a process in which an inner shell electron is captured by the nucleus of its own atom; the mass number is unchanged, but the atomic number is decreased by one. This process is accompanied by the emission of a neutrino. Also, **electron attachment.**

**electron donor** see DONOR.

**electronic** *Electricity.* **1.** relating to a device whose operation involves the motion of electrical charge carriers in a vacuum, gas, or semiconductor. **2.** relating to the study of such devices; i.e., electronics.

**electronic ballast** see BALLAST.

**electronics** *Electricity.* the study and application of the conduction of electric charges in various media, including vacuums, gaseous media, and semiconductors.

**electron shell** *Physics.* the arrangement of electrons at various distances from the nucleus of an atom, according to the energy they have. Those with the least energy are in the shell closest to the nucleus, traditionally called the **K shell,** which can hold no more than two electrons. The **Q shell**, farthest from the nucleus, can hold 98 electrons, but is never completely filled.

**electron tube** *Electricity.* a device in which electrons are conducted through a vacuum or gaseous medium within a gas-tight chamber; used to generate, amplify, and rectify electric oscillations and AC currents.

**electron volt** *Measurement.* a unit of energy, symbol eV; defined as the kinetic energy acquired by an electron that is accelerated through a potential difference of 1 V; equivalent to $1.6022 \times 10^{-19}$ joules.

**electroosmotic drag** *Electricity.* the flux of a polar species ($H_2O$) due to its attraction to a proton ($H^+$) that is transported from the anode to the cathode.

**electrophile** *Chemistry.* an ion or molecule that has a partial or complete positive charge, so that it can accept an electron pair or share an electron pair with another atom. Thus, **electrophilic.**

**electrophoresis** *Chemistry.* a process in which electrically charged particles that are suspended in a solution move through the solution under the influence of an applied electric field. Thus, **electrophoretic.**

**electroplating** *Materials.* the process of plating or coating a conducting surface with a metal by a process of electrolysis.

**electrorefining** *Materials.* **1.** a petroleum refinery procedure to aid in separating chemical treating agents from the hydrocarbon phase by use of an electrostatic field. **2.** a method of metal refining in which the metal is dissolved anodically and plated at the cathode of an electrolytic cell.

**electrostatic** *Physics.* having to do with or utilizing the phenomena associated with electric charges at rest. Thus, **electrostatic field, electrostatic generator,** and so on.

**electrostatic precipitator** *Environment.* a device that removes small foreign particles from air, such as ash, dust, and acid by electrically charging and then collecting the particles on a plate that is oppositely charged; used to filter out pollutants (and retrieve valuable materials) from utility and industrial processes before they reach the atmosphere. Also known as a **Cottrell precipitator** or simply a **Cottrell,** after its original developer. U.S. inventor Frederick Cottrell.

**electrostatics** *Physics.* the study of the phenomena associated with electric charges at rest.

**electrothermal** *Physics.* involving both electricity and heat, especially heat produced by electrical current.

**element** *Chemistry.* any of a class of substances that cannot be separated into simpler substances by chemical means; the fundamental units of which all matter is composed, at or above the atomic level. All atoms of a given element have the same nuclear charge and number of protons and electrons, but may vary in mass according to the number of neutrons in the nucleus (isotopes). There are more than 100 known elements, although the number is not precise because some artificially produced elements have been claimed but not confirmed.

**elementary machine** see SIMPLE MACHINE.

**elementary particle** *Physics.* one of the fundamental units of which all matter is composed, including leptons, quarks, and bosons. Elementary particles do not appear to be divisible into smaller units and their size apparently is too small to measure.

**elephant** *Biological Energetics.* the largest living land animal; the Indian elephant, *Elephus maximus,* has traditionally been used as an energy source.

**elevated railway** *Transportation.* a railway that operates on a raised structure rather than on the road surface, in order to permit the passage of vehicles and pedestrians beneath it, especially such a railway is used for urban transportation.

**ellipsoidal reflector (ER)** *Lighting.* a lamp having two conjugate foci; light from one focus passes through the other after reflection. Ellipsoidal reflectors collect a high fraction of total emitted light without greatly increasing heat in the illumination plane and thus are widely used in theatrical lighting. They also are an efficient choice for use in recessed fixtures in household lighting.

**El Niño** *Earth Science.* a large-scale, complex set of changes in the water temperature in the Eastern Pacific equatorial region, producing a warm current; it occurs annually to some degree between October and February, but in some years it intensifies and causes unusually severe storms and some destruction of marine life. [From Spanish for "The Child", meaning the Christ child; because it typically begins around Christmas time.]

**El Niño–Southern Oscillation (ENSO)** *Earth Science.* a periodic warming of surface ocean waters in the eastern tropical Pacific along with a shift in convection in the western Pacific further east than the climatological average; such conditions affect global weather patterns. Occurs roughly every four to five years and can last up to 18 months.

**Elton, Charles** 1900–1991, British scientist, considered one of the founders of ecology; he described many principles central to this field today, including succession, niche, food webs, and the links between communities and ecosystems.

**elution** *Materials.* a process of removing and separating substances adsorbed on a fixed bed by a stream of liquid or gas.

**embankment** *Earth Science.* **1.** a ridge constructed of earth, rock, or other materials, built to carry a highway or railroad track to a higher elevation than the surrounding terrain. **2.** a protective bank to prevent water encroachment or protect against erosion. **3.** a narrow depositional feature built out from a shore by the action of waves and currents.

**embankment dam** *Hydropower.* a common type of dam structure made of fill material, usually earth or rock, constructed with sloping sides and usually with a length greater than its height, often having a reinforced concrete core to provide waterproofing.

**embargo** see OIL EMBARGO.

**embedded cost** *Electricity.* the fixed cost of all facilities in an electric power supply system, including generating plants, substations, and distribution lines.

**embodied energy** *Consumption & Efficiency.* the sum of the energy requirements associated, directly or indirectly, with the delivery of a good or service.

**emdollar (EM$)** *Economics.* a measure of the money that circulates in an economy as the result of some flow or process.

**emergence** *Thermodynamics.* a term for the development of features of whole systems that are not deducible from the features of the constituent subsystems.

**emergency core-cooling system (ECCS)** *Nuclear.* the system in a nuclear power reactor that provides a flow of cooling water, independent of the coolant circuit normally used to transfer heat energy out of the reactor core.

**emergency response planning** *Health & Safety.* the fact or process of developing procedures that will counteract or alleviate the effects of an emergency, especially an environmental event such as a discharge or spill of toxic material; e.g., an oil spill, the release of hazardous gas.

**emerging disease** *Global Issues.* a new disease; an infectious disease known to the international medical community for a relatively short time; e.g., a novel strain of influenza. See also REEMERGING DISEASE.

**emergy** *Measurement.* a statement of all the energy and material resources used in the work processes that have generated a product or service, calculated in units of one particular form of energy, often in solar energy units. See below.

**EMF** **1.** electromagnetic field. **2.** electromotive force.

**EMI** electromagnetic interference.

**emission** see EMISSIONS and following entries (EMISSIONS ALLOWANCE, etc.).

**emissions** *Environment.* **1.** materials that are emitted; e.g., discharged or sent off from a given site. **2.** specifically, the release of primary pollutants directly to the atmosphere by processes such as combustion, and also by natural processes.

**emissions allowance** *Economics.* **1.** a certifiable permit to emit pollutants. **2.** the currency used in cap-and-trade programs.

**emissions banking** see BANKING.

**emissions-based maintenance** *Environment.* an engine maintenance procedure that is based on the engine tailpipe emissions of carbon monoxide and smoke obtained while the engine operates at full throttle against the torque converter load.

**emissions borrowing** *Policy.* the use of emissions reductions from future commitment periods in order to meet current targets.

**emissions cap** *Environment.* a mandated standard that establishes a ceiling on the amount of emissions that can be released into the atmosphere by a given source (e.g., an electrical power plant) within a designated time period.

**emissions coefficient** *Measurement.* a unique value for scaling emissions to activity data in terms of a standard rate of emissions per unit of activity (e.g., pounds of carbon dioxide emitted per Btu of fossil fuel consumed).

**emissions control** *Environment.* a general term for all efforts to control and reduce the amount of material released to the atmosphere by human activities, especially the discharge of harmful pollutants from point

**emergy** An expression of all the energy and material resources used in the work processes that generate a product or service, calculated in units of one form of energy. In the early 1970s Howard T. Odum recognized that traditional methods of measuring energy did not account for the "quality" of different forms of energy (e.g., sunlight, versus fossil fuels or electricity). As a systems scientist with expertise in many fields including: biogeochemistry, ecology, meteorology, and open systems thermodynamics, Odum reasoned that a measure of a product's quality could be obtained from evaluating the total energy and matter directly and indirectly required to make it. Odum and his colleagues began using sunlight as the base to evaluate all other forms of energy, reasoning that, all other forms are nothing more than concentrated sunlight. As the concept matured evaluations were conducted on many processes of the biosphere (ecological and technological) calculating the energy required to transform energy and materials from one form into another. David Scienceman, an Australian colleague of Odum's, coined the term emergy. Odum believed that emergy was a universal measure of the work of nature and society made on a common basis and therefore a measure of the environmental support to any process in the biosphere. The ratio of the total emergy input to the available energy of the product was named *transformity.* Since all energy transformations of the geo-biosphere can be arranged in an ordered series to form an energy hierarchy reinforced and stabilized by web-like and feedback interactions, transformities can be used as indicators of hierarchical position and role of a component in the thermodynamic hierarchy of the whole system.

**Mark T. Brown,**
University of Florida
Sergio Ulgiati, University of Siena

sources such as motor vehicles, power plants, factories, and other industrial facilities.

**emissions factor** *Policy.* the relationship between the amount of pollution produced and the amount of raw material processed; e.g., the emissions factor for a blast furnace making iron would be the number of pounds of particulates per ton of raw materials.

**emissions inventory** *Environment.* an official accounting of the amount of various specified gases (e.g., carbon dioxide, methane, nitrous oxide) released to the atmosphere over a given period of time (e.g., 1 year) in a given area (e.g., the Great Lakes region, the United Kingdom, the earth as a whole), or from a given facility (e.g., a power plant, factory, or university).

**emissions rate** *Environment.* the weight of a given pollutant emitted per unit of time (e.g., tons emitted per year).

**emissions reduction** another term for EMISSIONS CONTROL.

**emissions reduction credit (ERC)** *Policy.* a credit for permanently reducing pollution below a regulatory standard or baseline, expressed in tons per year.

**emissions reduction unit (ERU)** *Policy.* a unit of certified greenhouse gas reduction that is capable of being traded.

**emissions scenario** *Environment.* a projection of the future level of emissions of substances that have the potential to be radiatively active (e.g., greenhouse gases, aerosols), based on assumptions about influential factors such as temperature, population, economic trends, deforestation or reforestation, amount and efficiency of energy use, and so on.

**emissions spectrum** *Physics.* a diagram, graph, plot, or other such display indicating the degree to which a substance emits radiant energy with respect to its wavelength or frequency.

**emissions standard(s)** *Environment.* the highest amount of pollutant allowed by law to be discharged from a given single source; i.e., a specific motor vehicle, machine, power plant, factory, industrial process, and so on. A **performance-based standard** stipulates the emissions limits that each firm is allowed. A **technology-based** standard not only specifies emissions limits, but also dictates the type of technology that must be used to reach the target.

**emissions tax** *Policy.* a type of market-based instrument in the form of tax per unit of pollution released.

**emissions trading** *Economics.* a market-based form of environmental regulation that uses an allowance for emissions (or credit for emission reductions) that can be bought and sold.

**emissivity** *Physics.* the ratio of the energy radiated by an actual substance to the ideal energy radiated by a perfect blackbody at the same temperature; a perfect blackbody has an emissivity of one, and a perfect reflector has an emissivity of zero.

**emittance** *Solar.* for a sample at a given temperature, the ratio of the radiant flux emitted by the sample to that emitted by a blackbody radiator at the same temperature, under the same conditions of measurement.

**emjoule** *Measurement.* emergy joule; the unit of measure of EMERGY; e.g., the solar emergy of wood is expressed in terms of the number of joules of solar energy that were required to produce the wood.

**EMP** electromagnetic pulse.

**Empedocles** c. 495–435 BC, Greek philosopher known for his scientific insights that influenced later thinkers such as Plato and Aristotle. He proposed a theory of matter based on four essential components: air, fire, earth, and water. This foreshadowed later developments in atomic theory.

**emphysema** *Health & Safety.* a pathological accumulation of air in tissues, especially of the lungs, causing excessive inflation of the air sac and destructive changes in the alveolar walls; this results in a decrease in respiratory function and difficulty in breathing. Associated with smoking and also with air pollution and with occupational exposure to irritating dust and fumes.

**empirical** *Measurement.* based on actual observation or experimentation.

**empirical temperature** *Thermodynamics.* a property that is the same for any two systems that are in thermodynamic equilibrium with each other.

**empower** *Measurement.* emergy power; a statement of EMERGY used per unit of time.

**EMS** energy management system.

**emulsification** *Chemistry.* the formation of a mixture of two liquids, such as oil and water, in which one of the liquids is in the form of fine droplets and is dispersed (but not dissolved) in the other. Thus, **emulsify, emulsifier.**

**emulsion** *Chemistry.* any stable mixture of two (or more) immiscible liquids where one liquid (in the form of fine droplets or globules) is dispersed in the other.

**encapsulation** *Photovoltaic.* a method by which photovoltaic cells are protected from the environment, typically by being laminated between a glass superstrate and an ethylene vinyl acetate (EVA) substrate.

**endangered** *Ecology.* a categorization for a species (or other taxonomic group) having such a reduced population that it is in danger of becoming extinct; various specific criteria are employed to establish this classification.

***endangered*** *The whooping crane was one of the first species listed under the U.S. Endangered Species Act (1973). At one point its population level had dwindled to about two dozen known individuals, due mainly to habitat loss.*

**ending stocks** *Oil & Gas.* a classification for primary stocks of crude oil and petroleum products held in storage as of 12 midnight on the last day of the given month.

**end-of-discharge** *Storage.* the point at which the discharge process for a battery terminates; typically at a charge level of one volt per cell for nickel-based batteries.

**endogenous** *Environment.* **1.** originating or occurring within a living system. **2.** *Economics.* determined by factors internal to an economy. Thus, **endogenous growth (model).**

**endotherm** *Biological Energetics.* a warm-blooded animal; i.e., an organism that maintains its body temperature by producing heat within the body. Thus, **endothermy.** Contrasted with ECTOTHERM. See next page.

**endothermic** *Biological Energetics.* describing an organism (i.e., a mammal or bird) that maintains its body temperature at a stable level largely independent of (and typically higher than) the ambient temperature, primarily using generation of heat internal to the animal.

**endowment** *Oil & Gas.* the sum of cumulative oil production, reserves, reserve growth, and undiscovered resources.

**end use** *Consumption & Efficiency.* the desired physical function that is provided to a customer by a given energy service, such as heating, cooling, lighting, or mechanical work.

**end-use efficiency** *Consumption & Efficiency.* the ratio of the amount of energy services provided to the amount of energy consumed.

**energeia** *History.* the source of the modern word *energy,* a metaphysical concept employed by Aristotle in the sense of "activity" or "action toward a goal"; a word formed by combining two root forms meaning "at" and "work".

**energetics** *Physics.* the study of energy and its transformations from one form to another.

**ENERGIA** (est. 1995), an international network whose mission is to link individuals and groups concerned with energy, sustainable development, and gender. Its goal is to contribute to the empowerment of rural and urban poor women through a specific focus on energy issues.

**energy** *Physics.* **1.** a fundamental physical concept, defined classically as the capacity to do work. Can be described more generally as the potential ability of a system to influence changes in other systems by imparting either work (forced directional displacement) or heat (chaotic displacement/motion of system microstructure). **2.** the use of this capacity to perform useful functions for humans, such as

**endotherm** The term endotherm refers to animals (birds, mammals, some fishes and insects, and even some plants) that are capable of generating sufficient amounts of heat energy to maintain a high core temperature (e.g., 37–40°C in birds and mammals) by metabolic means—usually derived from aerobic activity of locomotor muscles in animals and by unique biochemical mechanisms in plants (e.g., skunk cabbage). Endotherms differ from ectotherms because they typically have core temperatures above that of the surrounding environment, whereas the core temperatures of ectotherms depend on external sources of heat—primarily from solar radiation. Endothermic animals birds and mammals that regulate their core body temperature at a relatively constant level are referred to as homeotherms (Greek homeo = similar). To maintain a constant body temperature, a homeotherm must balance heat loss with heat production. Heat loss is minimized in most mammals by having a thick coat of fur or thick layer of subcutaneous fat, whereas heat loss is promoted by sweating, panting or by seeking shelter in cooler environments. Endotherms are sometimes referred to as "warm-blooded," but this term is inaccurate and misleading, as is the term "cold-blooded" for ectotherms. For example the body temperature of a small tropical fish in warm water or desert lizard on hot sand (both of which are considered ectotherms) may have body temperatures higher than birds or mammals in the same environment-largely because of the insulation provided by feathers and fur and associated behavioral and physiological heat dissipating mechanisms that prevent their body temperatures from increasing above critical temperatures. Many ectotherms are able to regulate their body temperature behaviorally, by moving into and out of sunlight. Most endotherms are homeotherms, but by definition, some large reptiles (crocodiles and some of their extinct relatives—dinosaurs), as well as some large fish (tuna) and night-flying moths, are considered endotherms, because of the metabolic activity of skeletal muscles that generate large amounts of heat. However, because these endotherms lack a layer of insulation and do not have a thermostat that regulates either heat production or heat dissipation, they are considered poikilotherms (Greek poikolos = changeable). Some mammals and birds that at times have high and well-regulated body temperatures, but at other times they are more like ectotherms and are referred to as heterotherms (Greek hetero = different). Heterothermy is characteristic of small hibernating rodents and bats.

**Thomas H. Kunz**
Boston University

heating or cooling buildings and enclosures, powering vehicles and machinery, lighting, cooking foods, and so on. See next page.

**energy audit** *Consumption & Efficiency.* a program carried out by a utility company in which an examiner inspects a home and suggests ways that energy use can be reduced.

**energy balance** *Earth Science.* the relationship between the amount of radiant energy that reaches the earth from the sun and the amount of energy that is either absorbed at the surface and in the atmosphere, or transmitted back into space.

**energy balance model** *Measurement.* an analytical technique that attempts to account for all energy coming in and going out of a system. Applied to the earth, this technique tracks incoming solar radiation, how it is absorbed, reflected, or refracted, and ultimately released back to the atmosphere. Called "balance" because all incoming energy equals all outgoing energy, consistent with the first law of thermodynamics.

**energy breakeven** see BREAKEVEN.

**energy budget** *Economics.* **1.** a description of the types and amounts of energy used by a household, firm, or nation for a specified period of time, and the tasks for which that energy is used. **2.** *Earth Science.* a description of the earth's heat transfer on the model of a financial budget; i.e., in terms of a "balance sheet" of gains and losses. The total amount of radiation from the sun represents revenue; a certain amount of this energy is spent by being absorbed at the earth's surface, and another portion is spent by being absorbed in the atmosphere. The remainder is lost by being retransmitted to space. **3.** any similar description for a specific region; e.g., a given ecosystem.

**energy** In the mechanical sense, *energy* is the ability to do work where work is the action of a force acting on an object undergoing a displacement. Matter in motion is said to have kinetic energy because of its ability to change the motion of another object. Matter in a favorable position, such as water atop a dam, is said to have potential energy because of its ability to change the motion of another object once the water flows over the dam. Other important forms of energy include thermal energy, nuclear energy, electromagnetic energy, and gravitational energy. Energy can be transformed from one form to another. For example, a steam turbine converts thermal energy to mechanical energy, a nuclear reactor transforms nuclear energy to thermal energy, and a solar cell converts electromagnetic energy into electric energy. The physics principle of conservation of energy requires that there be no loss of energy in any energy transformation.

**Joseph Priest**
Miami University of Ohio

**energy cane** *Biomass.* a term for a tall grass that has a C4 metabolic system and strong stature, so that it is suitable to be cultivated as a form of biomass energy.

**energy carrier** *Physics.* a form of matter that can transport energy from one point to another; e.g., electricity, hydrogen, or adenosine triphosphate (ATP) in living systems.

**energy cascading** *Consumption & Efficiency.* the use of the residual heat in liquids or steam from one process to provide heating, cooling, or pressure for another process; e.g., the use of steam from an electric power plant in a district heating system.

**energy chain** *Consumption & Efficiency.* **1.** all the successive stages involved with the supply of an energy source to the end user, such as a given fuel product's characterization, exploration, extraction, conversion, processing, and delivery, and the treatment and disposal of its wastes. **2.** *Ecology.* another term for a FOOD CHAIN.

**Energy Charter (Treaty)** *Policy.* an international treaty signed in Lisbon in 1994, intended to facilitate Western investment in the development of energy resources in Eastern Europe and the former Soviet Union.

**energy conservation** *Consumption & Efficiency.* **1.** a collective term for activities that reduce end-use demand for energy by reducing the service demanded, e.g., a reduction in the demand for gasoline by reducing the number of miles driven, or a reduction in the demand for natural gas for space heating by lowering the thermostat. **2.** another term for ENERGY EFFICIENCY. **3.** see CONSERVATION OF ENERGY. See below.

**energy conversion** see CONVERSION (def. 1).

**energy conversion chain** see ENERGY CHAIN.

**energy cost** *Economics.* **1.** the monetary cost of a quantity of purchased energy. **2.** the amount of energy used directly and indirectly to produce a good or service; measured in joules.

**energy cost of living** *Consumption & Efficiency.* the energy used directly and indirectly by a household in its consumption of goods and services.

**energy conservation** Energy conservation has been diversely interpreted as meaning (1) increasing the efficiency or productivity of energy use to maintain or increase economic output with less energy; (2) accommodating behavior by curtailing (rationing or "doing without") energy; and (3) switching to more abundant energy sources. Energy efficiency, in the most common meaning of energy conservation today, involves technological change—the substitution of ingenuity for energy resources—to provide a given amenity level for less. Energy efficiency is sometimes quantified in terms of energy intensity, or the amount of energy required to produce a unit of goods or service. Energy conservation and energy efficiency can be stimulated by information (for example, labeling of appliances), technological progress (research and development), shortages (wartime rationing), price increases (price elasticity of demand), and codes and standards (for example, automotive fuel economy standards). Interest in energy conservation is driven not for its own sake but by economic, security, and environmental concerns, including climate change. Energy conservation, in an economic and policy context, thus means the intelligent use of energy through the reduction of waste.

**Bill Chandler**
Pacific Northwest National Laboratory, USA

**energy crisis** *History.* a general term for economic, political, and social disruptions in a nation or society, caused by an extreme shortage of available energy resources; e.g., in the U.S. in 1973–74 as a result of an Arab oil embargo.

*energy crisis The 1979 energy crisis occurred in the wake of the Iranian Revolution. Faced with mass protests, the Shah of Iran fled the country; the Iranian oil sector was then in disarray, driving up global prices.*

**energy crop** *Biomass.* a crop grown specifically for its value in producing energy; the term usually applies to fast-growing crops used for liquid fuel or electricity, e.g., certain grasses and trees such as poplar, maple, black locust, willow, sycamore, sweetgum, and eucalyptus.

**energy demand** see DEMAND.

**energy density** *Measurement.* a statement of the energy content of a fuel or energy storage device per unit mass or volume.

**energy deregulation** see DEREGULATION.

**energy efficiency** *Consumption & Efficiency.* **1.** a reduction in the quantity of energy used per unit service provided, e.g., a reduction in the quantity of motor gasoline used per kilometer driven. **2.** another term for ENERGY CONSERVATION.

**energy efficiency ratio (EER)** *HVAC.* a measure of the relative efficiency of a heating or cooling appliance; this is equivalent to the unit's output in Btu per hour divided by its consumption of energy, measured in watts.

**energy-efficient diet** *Biological Energetics.* **1.** a human diet emphasizing foods that require a relatively low amount of energy to grow, harvest, and prepare; e.g., locally grown vegetables eaten raw. **2.** in animals, a diet that provides the maximum return for the amount of energy expended in obtaining the food; e.g., a diet based on a prey animal that is relatively easy for the predator to find and kill.

**energy expenditure** see BASAL ENERGY EXPENDITURE.

**energy farming** see ENERGY PLANTATION.

**energy flow** *Social Issues.* **1.** the movement of energy through a society; e.g., from the environment through extraction, processing, transportation, and end use. **2.** *Biological Energetics.* the movement of energy through a biological system; e.g., through the different trophic levels of a food chain.

**energy flux** *Physics.* a quantity measuring the rate of energy flow; the energy per unit time per unit area traveling across a surface element that is perpendicular to the energy flow.

**energy futures** see FUTURES.

**energy/GDP ratio** *Economics.* energy consumption over a given period, divided by the gross domestic product or national income in that period. See next page.

**Energy Information Administration** (est. 1977), the statistical agency of the U.S. Department of Energy; its mission is to provide policy-independent data, forecasts, and analyses regarding energy and its interaction with the economy and the environment.

**energy intensity** *Economics.* the amount of energy required to produce a given economic product or service, e.g., the amount of energy used to heat or cool a certain living space, or to transport a person over a certain distance. The energy/GDP ratio is a common indicator of the energy intensity of an entire economy.

**energy-intensive** *Economics.* **1.** having to do with energy intensity (see above). **2.** describing an industry or process for which energy costs represent a large part of the total production costs; e.g., the manufacture of aluminum, steel, cement, paper, or fertilizers.

**energy/GDP ratio** The energy/GDP ratio is defined as energy consumption over a given period divided by the gross domestic product or national income in that period. This indicator is also sometimes referred to as energy intensity or energy cost as it reflects the energy requirements to produce a given amount of economic output. Some see this indicator as a rough proxy for the environmental impact of economic production as all energy use disturbs the natural environment in some way, though obviously some fuels and activities are more damaging than others. The data shows that the energy/GDP ratio has declined in most developed countries over the last few decades. If energy is measured only in terms of fossil fuels and other modern energy carriers, the energy/GDP ratio has increased in many developing countries and over the last couple of centuries energy intensity follows an inverted U shape path in the developed countries. However, when traditional energy carriers such as wood are included it seems that energy intensities have mostly declined over time. The factors directly affecting energy intensity are the structure of the economy, the mix of energy inputs, the amount of other inputs (like labor) used, and technological change. The shift to greater electricity use over time in most developed countries is probably significant in reducing energy intensities as electricity is more productive than other fuels. But the most important long-term factor seems to be improving technology or total factor productivity.

**David Stern**
Rensselaer Polytechnic Institute

**energy label** *Consumption & Efficiency.* information posted on or affixed to an appliance or other energy-using device, indicating how much energy the appliance uses, its efficiency compared to other models, and how much the energy use will cost the consumer.

**energy ladder** *Social Issues.* the tendency of a population to increase the quality of its energy use as its socioeconomic status improves; biomass fuels such as wood, crop residues, and animal waste appear on the lower "rungs" of the ladder, with coal, kerosene, oil and gas, and electricity above (and with solar, wind, and hydrogen above those in postindustrial societies). Similarly, **energy pyramid.**

**energy level** *Physics.* a stable quantity of energy that a physical system may have, such as the electron configuration of electrons in atoms or molecules.

**energy loss** *Consumption & Efficiency.* the portion of energy input to a process or a device that is not converted into useful work.

**energy management system (EMS)** *HVAC.* a control system capable of monitoring environmental and system loads and adjusting HVAC operations accordingly, in order to conserve energy while maintaining comfort.

**energy metabolism** *Biological Energetics.* the sum of chemical reactions that produce or conserve energy within cells.

**energy option** see OPTION.

**energy payback** see PAYBACK.

**energy plantation** *Biomass.* a system of using land to grow crops, woody or otherwise, to provide fuel rather than food and fiber.

**Energy Policy Act (EPACT)** *Renewable/Alternative.* energy legislation in the U.S. enacted in 1992, whose major provisions include energy efficiency standards, new regulatory options for electricity generation, changes in nuclear power licensing and fuel services, and a variety of tax incentives.

**energy price shock** see PRICE SHOCK.

**energy quality** *Consumption & Efficiency.* the relative economic usefulness per heat equivalent (as measured in joules) of different types of fuels (and electricity).

**energy recovery ventilator (ERV)** *HVAC.* a device that preheats incoming outside air during winter and precools incoming air during summer, in order to reduce the energy required to condition the indoor air.

**Energy Reorganization Act** *Nuclear.* legislation enacted in the U.S. in 1974 to replace the Atomic Energy Commission with two agencies dealing with nuclear power, one for research (ERDA) and the other for regulation (NRC).

**energy reserve(s)** see RESERVE.

**energy return on investment (EROI)** *Economics.* a ratio calculated by dividing the amount of useful energy produced by a given process, divided by the energy used in that process. For example, the energy produced by an oil well compared to the energy used (directly and indirectly) to produce, maintain, and operate the well, or the ratio of the quantity of food energy captured by a living organism to the energy used to capture it (e.g., a predator pursuing prey).

**energy security** *Global Issues.* the various security measures that a given nation, or the global community as a whole, must carry out to maintain an adequate energy supply; this can include a wide range of issues such as developing nonfossil fuel sources, maintaining military forces to protect pipelines and other components of the supply chain, and encouraging the stability of governments in oil-exporting countries.

**energy service company (ESCO)** *Consumption & Efficiency.* a company that specializes in energy efficiency measures under a contractual arrangement in which the company shares the value of energy savings with the customer.

**energy signature** *Ecology.* the total array of external energy forces from which an ecosystem can capture available energy.

**energy source** *Consumption & Efficiency.* a collective term for all resources providing useful energy such as human and animal power, wind, water power, coal, petroleum, natural gas, and nuclear power, as well as alternatives such as geothermal and solar energy.

**Energy Star** *Consumption & Efficiency.* a certification and label given to consumer products (such as household appliances) that meet strict energy efficiency criteria set by the Environmental Protection Agency in the U.S.; e.g., refrigerators qualifying for this label are at least 15% more efficient than the minimum efficiency standard.

**energy storage** *Storage.* any process or state of maintaining energy in a form that permits the energy to be made available in a useful form at a later point in time; five basic methods of storage are chemical, electrochemical, magnetic, mechanical, and thermal.

**energy supply chain** see ENERGY CHAIN.

**energy surplus** *Economics.* a quantity of energy defined as the amount of useful energy produced or obtained by a process, minus the energy used in that process. See also ENERGY RETURN ON INVESTMENT.

**energy theory of value** *Economics.* the concept that the monetary value of a good or service is proportional to and determined by the embodied energy in that good or service.

**energy transfer** see HEAT TRANSFER.

**energy transition** *Social Issues.* a change in the primary form of energy consumption of a given society; e.g., the historic transition from wood to coal and then to oil and gas in industrial Europe; the current shift from biomass fuels to commercial energy in some areas of the developing world. See below.

**energy vector** see ENERGY CARRIER.

**engine** *Conversion.* **1.** any machine in which power is applied to do work, specifically by converting thermal energy into mechanical energy. **2.** another term for a locomotive; i.e., the powered vehicle that pulls or drives a railroad train.

**engine cycle** *Conversion.* any series of thermodynamic processes that occur cyclically and that results in the conversion of heat transfer into work; e.g., a transfer of heat from

**energy transitions** Specific patterns of both energy supply and demand quantities and qualities define a given state of an energy system. Changes from one state to another are called energy transitions. Since the onset of the Industrial Revolution a number of important energy transitions have occurred and continue to unfold. The first important transition was the overcoming of supply limitations imposed by locally available renewable energy flows by high density fossil fuels traded globally. The concomitant expansion of demand and the emergence of high-density energy use in urban areas triggered further transitions in energy supply in a structural change away from direct uses of dirty fossil fuels (especially coal) and increasing reliance on clean, versatile, grid-dependent energy forms (gas and electricity). These transitions are embedded in longer overall secular trends towards more efficient and cleaner energy supply and end-use that will continue to unfold, albeit at different rates across different regions.

**Arnulf Grubler**
International Institute for
Applied Systems Analysis
Laxenburg Austria

a high-temperature reservoir to a device performing useful work on its surroundings.

**engine knock** see KNOCK.

**Englemann, Theodor Wilhelm** 1843–1909, German physiologist who showed that the light reactions of photosynthesis, which capture solar energy and convert it into chemical energy, occur within chloroplasts and respond only to the red and blue hues of natural light.

**enhanced** *Oil & Gas.* describing a system for crude oil that augments the fraction of crude oil recovered from a reservoir, as by the injection of materials not normally present within the reservoir; usually initiated as a secondary or tertiary method. Thus, **enhancement.**

**enhanced geothermal system (EGS)** *Geothermal.* a system used to extract heat from the less productive margins of existing geothermal fields, or from entirely new fields lacking sufficient production capacity under current conditions, employing a combination of hydraulic, thermal, and chemical processes such as rock fracturing, water injection, or water circulation.

**enhanced greenhouse effect** *Climate Change.* the concept that the earth's natural GREENHOUSE EFFECT has recently been enhanced by increased emissions of greenhouse gases relating to human activities, such as the burning of fossil fuels.

**enhanced recovery** *Oil & Gas.* the injection of water, steam, gases or chemicals into underground oil reservoirs to cause oil to flow toward producing wells, permitting more recovery than would have been possible from natural pressure or primary and secondary recovery methods. Also, **enhanced process.**

**ENIAC** *Communication.* a pioneering computer, developed from 1943 to 1946 by J. W. Mauchly and J. P. Eckert at the University of Pennsylvania, using punched cards for input and output data. It contained 17,000 vacuum tubes, weighed over 30 tons, and occupied 1500 square feet of space. ENIAC is often described as the first electronic computer, although a 1973 court decision found that its design was based on the earlier work of John Vincent Atanasoff. [An acronym for Electronic Numerical Integrator and Computer.]

*ENIAC Co-developer J. P. Eckert poses with the ENIAC computer. ENIAC required an entire room of space to perform operations that in current technology could be accommodated by a single chip of about 0.5 mm in size.*

**enriched uranium** *Nuclear.* uranium whose uranium-235 content has been increased through the process of isotope separation.

**enrichment** *Nuclear.* the process of increasing the proportion of a fissile isotope in a nuclear fuel above its natural fraction, typically the isotope uranium-235 to U-238. Thus, **enrichment feed.**

**enrichment tails** *Nuclear.* fissile uranium (uranium-235) remaining in the waste stream from the uranium enrichment process, typically about 0.2–0.3% U-235.

**Enron** *Economics.* a powerful energy trading company that became America's most notorious corporate collapse to date. It had become the seventh-largest U.S. company by buying electricity from generators and selling it to consumers, and was praised in the business media as the model for a new type of energy company. In reality Enron had been losing vast amounts of money and disguising the losses by false reports and accounting tricks.

**ENSO** El Niño–Southern Oscillation.

**enteric fermentation** *Biological Energetics.* a process in the digestive system of an animal in which carbohydrates are broken down by microorganisms into simpler molecules for absorption into the bloodstream.

**entering temperature** *Conversion.* the temperature of air, water, or another working fluid as it enters a heat pump, heat exchanger, chiller, or other such energy conversion system. Thus, **entering air temperature (EAT), entering water temperature (EWT).**

**enthalpy** *Thermodynamics.* heat content; a property expressing the total energy content of a system, defined as $H=U+PV$, where $H$ is enthalpy, $U$ is the internal energy of the system (any kind of energy form of its molecules), $P$ is the pressure exerted on the system by its environment, and $V$ is the volume of the system. If a steady flow process takes place at constant pressure and no work is done (other than mechanical work against the boundaries), then the change in enthalpy is equal to the heat transferred during this process.

**enthalpy of atomization** *Chemistry.* the change in enthalpy that occurs when one mole of a compound is converted into gaseous atoms (atomized) at constant pressure. Similarly, **enthalpy of combustion, fusion, reaction, sublimation, vaporization,** and so on.

**entomb** *Nuclear.* to encase radioactive contaminants in a structurally long-lived material, such as concrete, until such time as the radioactivity decays to a level permitting decommissioning and ultimate unrestricted release of the property. Thus, **entombment.**

**entrained** *Conversion.* describing a type of gasifier in which the feedstock (fuel) is suspended by the movement of gas to move it through the gasifier. Thus, **entrained-flow gasifier** or **entrained-bed gasifier.**

**entrainment** *Ecology.* the process by which an aquatic organism is drawn into the intake of a power plant along with the inflowing water.

**entropy** *Thermodynamics.* **1.** a measure of the disorder or randomness of a closed system; more entropy means less energy is available for doing work. The total entropy of an isolated system cannot decrease when the system undergoes a change; it can remain constant for reversible processes, and will increase for irreversible ones. **2.** in popular use, a condition of disorganization, deterioration, chaos, and so on. See next column.

**entropy generation** *Thermodynamics.* a measure of the irreversibilities or imperfections that occur during a thermodynamic cycle.

**entropy** An index of the amount of unavailable energy in a given system at a given moment. In 1824, Carnot discovered the vital thermodynamic principle that the maximum available energy that can be extracted from a system depends solely on the temperatures of two energy sources. Based on Carnot's pioneering work, Clausius identified in 1865 a quantity that irrevocably increases toward a maximum in an isolated system (second law of thermodynamics) and called that quantity *entropy*, from the Greek word τρ trope, (transformation). In the late 1890s, Boltzmann tried to give a statistical foundation for entropy law. In 1945, Schrödinger added a note to *What is life?* to explain how living things (open systems) constantly strive to compensate for entropic degradation by taking low entropy and giving off high entropy. Starting in the late 1940s, the Prigogine school worked on non-equilibrium thermodynamics, using the concepts self-organization and emergence to explain how open systems employ all available means to resist entropic degradation. From the 1960s, Georgescu-Roegen systematically investigated implications of the entropy law in the economic process and claimed that energy shortage and scarcity of mineral resources limit human survival, i.e., available energy and matter continuously change into unavailable form and disappear. The earth is open with respect to energy but closed materially, so it would be helpful to describe a mechanism by which the earth can avoid entropic degradation. In the late 1970s, Tsuchida showed that the earth can be regarded as a Carnot engine powered by temperature difference between the sun and outer space. Thus, our planet discards thermal entropy through the adiabatic process of water cycles.

**Kozo Mayumi**
University of Tokushima, Japan

**envelope** see BUILDING ENVELOPE.

**environment** **1.** usually, **the environment.** the total of all the surrounding natural conditions that affect the existence of living organisms on earth, including air, water, soil, minerals, climate, and the organisms themselves. **2.** the local complex of such conditions affecting a particular organism and ultimately determines its physiology and survival. **3.** *Physics.* the surroundings of a physical system that can have some effect on the behavior of the system. **4.** the combination of all external conditions that influence the performance of a device or process.

**environmental** *Environment.* **1.** of or relating to the environment or to the particular environment of a given organism. **2.** describing a disease or condition that can be caused or influenced by factors in the environment; e.g., the association of certain toxic chemicals with some forms of cancer.

**environmental accounting** *Economics.* **1.** a method of economic or financial accounting that considers the use of natural resources and the release of pollution when assessing the economic performance of a firm, industry, nation, or other economic entity. **2.** specifically, a system incorporating the depreciation of natural resources and ecosystem services into estimates of net domestic product or net national product.

**environmental degradation** *Environment.* any deterioration in the structure or function of the environment, especially human-caused; e.g., air and water pollution, desertification, deforestation.

**environmental engineering** *Environment.* technological activity with the goal of reducing or preventing the pollution or degradation of the environment; e.g., treatment, management, and control of hazardous wastes; control and monitoring of air pollution and acid deposition; design and management of solid waste facilities.

**environmental equity** *Social Issues.* **1.** the principle that environmental impacts should exist or occur across the spectrum of society, rather than disproportionately affecting any particular segment categorized according to some population characteristic, such as gender, age, ethnic group, place of residence, income level, and so on. **2.** specifically, the equitable incidence of health risks resulting from exposure to toxic substances.

**environmental ethics** *Environment.* the field of applied ethics as concerned with questions of human interaction with the environment, including such issues as conservation and stewardship, future generations, animal rights, land use, development and sustainability, population levels, agricultural ethics and food security, biodiversity, endangered and threatened species, and ecotourism.

**environmental fate** *Environment.* an indication of what happens to a substance once it enters the environment, especially a toxin or pollutant such as a pesticide; this includes how and where it enters the environment, how long it lasts, where it goes, and how its toxicity changes over time.

**environmental gradient** *Environment.* a change in an environmental factor over a given variable, such as time, space, altitude or depth, latitude, type of terrain, and so on; e.g., a change in temperature, soil type, precipitation level, pH, salinity, and so on.

**environmental impact** *Environment.* a general term for any human action that has a significant effect on the natural environment, either negative or positive.

**environmental impact assessment (EIA)** *Environment.* an analysis of the impacts on the natural, social, and economic environment of a proposed project or resource management plan; e.g., the building of a dam, highway, airport, factory, and so on. Also, **environmental impact analysis** or **statement**.

**environmental injustice** *Social Issues.* the fact of different social groups, especially those at different economic levels, being disproportionately impacted by negative environmental effects, such as industrial or agricultural wastes, factory emissions, power plant siting, uranium mining, and other energy-related activities.

**environmentalism** *Social Issues.* an active social movement of the contemporary era focusing on the threats to human health, and to the earth itself, posed by various types of damage to natural systems; includes a broad range of concerns, such as air and water pollution, climate change, loss of biodiversity, threats of extinction for certain species, conversion of wilderness land for commercial purposes, and expanding human population. Thus, **environmentalist.**

**environmental justice** **1.** see ENVIRONMENTAL INJUSTICE. **2.** see ENVIRONMENTAL EQUITY.

**environmental Kuznets curve (EKC)** *Measurement.* a hypothesized inverted U-shape relationship between income (per capita GDP) and indicators of resource use and pollution (e.g., per capita energy use, per capita release of $SO_2$). See next page.

*environmental injustice The concept of environmental injustice indicates that certain segments of the population will be more affected by negative environmental conditions than others. Photo by Arthur Rothstein.*

**environmental movement** see ENVIRONMENTALISM.

**environmental performance bond** see PERFORMANCE BOND.

**environmental pollution cost** *Environment.* the environmental effect of a pollutant, based on such factors as the cost to society for the correction or compensation of environmental damage and the cost incurred to prevent a harmful emission from escaping into the environment.

**Environmental Protection Agency (EPA)** (est. 1970), a U.S. agency whose mission is to protect human health and the environment. Its specific charges include the development and enforcement of regulations that implement environmental laws enacted by Congress; financial support to state environmental programs, non-profit groups, and educational institutions; environmental research and education.

**environmental release** *Environment.* another term for emission; i.e., the discharge of a given material to the environment, especially a pollutant or toxin.

**environmental remediation** see REMEDIATION.

**environmental restoration** see RESTORATION.

**environmental restriction** *Environment.* a land-use regulation that postpones, restricts, or prohibits the development or exploitation of a given area in order to protect environmental resources; e.g., an endangered or threatened species.

**environmental state** *Thermodynamics.* the state of a system when it is in thermal and mechanical equilibrium with the reference environment; i.e., at the pressure and temperature of the reference environment.

**Environmental Sustainability Index (ESI)** *Sustainable Development.* a measure of the overall progress of a given nation towards environmental sustainability, developed for more than 140 nations. A recent ranking (2002) placed Finland, Norway, Sweden, and Canada as highest in rate of progress toward this goal, and Kuwait, United Arab Emirates, North Korea, and Iraq as lowest.

**environmental tax** *Economics.* a tax levied on a pollutant such as carbon dioxide or sulfur dioxide as a means of reducing emissions; the revenue from such a tax may

**environmental Kuznets curve** The environmental Kuznets curve (EKC) hypothesizes that the relationship between per capita income and the use of natural resources and/or the emission of wastes has an inverted U-shape. According to this specification, at relatively low levels of income the use of natural resources and/or the emission of wastes increase with income. Beyond some turning point, the use of the natural resources and/or the emission of wastes decline with income. Reasons for this inverted U-shaped relationship are hypothesized to include income driven changes in: (1) the composition of production and/or consumption; (2) the preference for environmental quality; (3) institutions that are needed to internalize externalities; and/or (4) increasing returns to scale associated with pollution abatement. The term EKC is based on its similarity to the time-series pattern of income inequality described by Simon Kuznets in 1955. A 1992 World Bank Development Report made the notion of an EKC popular by suggesting that environmental degradation can be slowed by policies that protect the environment and promote economic development. Subsequent statistical analysis, however, showed that while the relationship may hold in a few cases, it couldn't be generalized across a wide range of resources and pollutants.

**Amy Richmond**
Boston University

then be used to finance cleanup, prevention, reduction, enforcement or educational efforts intended to reduce pollution and resource depletion.

**environmental terrorism** *Global Issues.* terrorist actions with the purpose (or result) of damaging the environment or destroying natural resources; e.g., an intentional oil spill or oil well fire. See also ECOTERRORISM.

**environmental tobacco smoke (ETS)** *Health & Safety.* smoke emitted from the burning end of a cigarette (cigar, pipe, etc.) or exhaled by the smoker, and thus present in the surrounding air, causing exposure of others nearby to the toxic compounds in the smoke; popularly known as secondhand smoke.

**environmental wacko** *Social Issues.* a derogatory term used by opponents of environmentalism, to characterize environmental activists as being eccentric, silly, unrealistic, irrational, and so on.

**enzyme** *Biological Energetics.* a specific protein molecule that can affect (accelerate) the rate of a chemical reaction without itself being permanently altered. Thus, **enzymatic.**

**EOC voltage** *Storage.* end of charge; the terminal voltage of a secondary cell or battery at the end of charging.

**EOD voltage** *Storage.* end of discharge; the terminal voltage of a secondary cell or battery at the end of discharging.

**eolian** another spelling of AEOLIAN.

**EPA** Environmental Protection Agency.

**EPACT** *Policy.* Energy Policy Act; a comprehensive legislative program in the U.S. that mandates and encourages energy efficiency standards, alternative fuel use, and the development of renewable energy technologies.

**epidemiological** *Health & Safety.* having to do with or indicated by the study of patterns of disease. Also, **epidemiologic.**

**epidemiology** *Health & Safety.* **1.** the branch of medical science concerned with the study of patterns of disease within human populations, including the cause, distribution, prevention, and control of various disease. Thus, **epidemiologist. 2.** the cause and incidence of a specific disease; e.g., the epidemiology of lung cancer. **3.** more generally, the study of patterns of injury or pathological behavior on the model of disease study; e.g., the epidemiology of domestic violence.

**epilimnion** *Earth Science.* the relatively warm and uniformly mixed uppermost layer of water in a lake or reservoir, especially the less dense, oxygen-rich upper layer of a thermally stratified lake; because of its exposure and turbulent mixing, dissolved gases are freely exchanged with the atmosphere.

**epitaxial growth** *Materials.* in semiconductors, the growth of one crystal on the surface of another crystal; the growth of the deposited crystal is oriented by the lattice structure of the original crystal.

**epitaxial layer** *Materials.* a layer or layers of semiconductor material having the same crystalline orientation as the host substrate on which it is grown.

**epithermal neutron** *Physics.* a neutron that has an energy level just above the thermal range, between about 0.02 and 100 eV.

**EPRI** Electric Power Research Institute; an influential U.S. electric power research and development organization.

**equalization** *Storage.* the process of restoring all cells in a battery to an equal state of charge.

**equalization charge** *Storage.* a process of overcharging a battery for a short time to mix the electrolyte solution.

**equation of state** *Chemistry.* an expression that describes the state of a substance in terms of the relationship of its basic physical quantities of volume, pressure, and temperature for a given mass. Used for pure substances and also applied to the relationships among other thermodynamic variables.

**equatorial mount** *Solar.* a sun-tracking mount, usually clock-driven, whose axis of rotation is parallel to that of the earth.

**equilibrium** *Physics.* a balanced state in a system; e.g., a condition in which the energy gained by a system from its surroundings is exactly in balance with the energy that it loses, irrespective of the passage of time.

**equinox** *Solar.* the two occasions of the year when the sun passes over the celestial equator and the length of day and night are almost equal.

**equipartition** *Thermodynamics.* the principle that molecules in thermal equilibrium have the same average energy associated with each independent degree of freedom of their motion.

**equivalence** see MASS-ENERGY EQUIVALENCE.

**equivalence ratio** *Consumption & Efficiency.* the ideal oxidizer to fuel ratio of a particular oxidizer and fuel to the actual ratio at which the unit is operating.

**equivalent carbon dioxide** *Climate Change.* the concentration of carbon dioxide that would cause the same amount of radiation forcing as a given mixture of $CO_2$ and other greenhouse gases.

**equivalent dose** *Nuclear.* the absorbed dose adjusted for the relative biological effect of the type of radiation being measured.

**equivalent temperature** *HVAC.* the air temperature of an imaginary environment in which an occupant would feel the same thermal sensation as in the actual environment.

**ERA** ecological risk assessment.

**ERC** emissions reduction credit.

**ERDA** Energy Research and Development Administration; a U.S. agency created (1974) to serve as a federal research program as a replacement for the research activities of the Atomic Energy Commission. In 1977, ERDA became the U.S. Department of Energy.

**erg** *Measurement.* a unit of energy equivalent to the work done or energy expended by a force of 1 dyne acting through a distance of 1 cm. One erg equals 0.0000001 joule.

**ergonomic** *Health & Safety.* **1.** having to do with the discipline of ergonomics (see below). **2.** describing products or procedures designed so as to provide maximum efficiency and comfort and minimal health risk over time, taking into account human anatomy, physiology, psychology, and so on.

**ergonomics** *Health & Safety.* the study of physical and mental factors that affect people in occupational settings; used in the design of work sites, work processes, and so on; e.g., the design of computer workstations so that users will have minimal strain on posture and vision.

**Ericsson, John** 1803–1889, Swedish engineer noted for his invention of the screw propeller, still the main form of marine propulsion. He designed and built the famous *Monitor* warship for the Union Navy in the Civil War, and was also responsible for engineering advances such as improved transmission of power by compressed air, new types of steam boilers and condensers, and an early steam locomotive.

**Ericsson solar engine** *Solar.* an early solar hot-air engine, a displacer type (Stirling) engine powered by a parabolic reflector, built by John Ericsson in about 1872. He envisioned that Californian farmers would use what he termed a **sun-motor** for irrigation purposes, but that did not come to fruition.

**EROI** energy return on investment.

**erosion** *Earth Science.* a combination of processes in which the materials of the earth's surface are loosened, dissolved, or worn away, and transported from one place to another by natural agents such as wind and rainfall.

***erosion*** *California beachfront homes are among the most expensive in the U.S., but their location can be controversial because of the idea that they contribute to the erosion of the shoreline.*

**ERP** emergency response planning.

**ERR** excess relative risk.

**Erren, Rudolf** German engineer who in the 1920s and 30s developed motor vehicles that were able to operate on either hydrogen or conventional fuels. He devised a fuel injection system that allowed hydrogen to be fed directly into the cylinder, thereby eliminating the carburetor, which was poorly suited to inject a gaseous fuel. Reportedly thousands of vehicles were converted to this system, but none are known to have survived World War II.

**ESCO** *Consumption & Efficiency.* energy services company.

**ESI** Environmental Sustainability Index.

**Esso** a brand name for the Standard Oil Company ("S-O"), now used for various companies of the ExxonMobil Corporation. See EXXON.

**ester** *Chemistry.* an organic compound that is formed by combining an acid with an alcohol, with the elimination of water.

**estimated additional resources (EAR)** *Nuclear.* uranium resources in addition to REASONABLY ASSURED RESOURCES (RAR), including uranium that is expected to occur, mostly on the basis of direct geological evidence, in extensions of well-explored deposits, little explored deposits, and undiscovered deposits believed to exist in association with known deposits.

**estivation** *Biological Energetics.* a dormant state of reduced metabolism, occurring in certain animals in response to higher ambient temperatures. Thus, **estivate.**

**estuarine** *Earth Science.* relating to or occurring in an estuary; i.e., an area where a freshwater river meets the ocean and tidal influences result in fluctuations in the salinity of the intermixed waters.

**estuary** *Earth Science.* a semi-enclosed coastal body of water that has a connection with the open sea and within which fresh water and salt water mix through the influence of currents and tides.

**ESUB** elasticity of substitution.

**ETA** **1.** estimated time of arrival. **2.** event tree analysis.

**ETBE** ethyl tert-butyl ether.

**ETDE** Energy Technology Data Exchange (est. 1987); a program of the International Energy Agency (IEA), whose mission is to provide governments, industry, and the research community with access to the widest range of information on energy and to increase dissemination of this information to developing countries.

**eternal fire** *History.* an ancient term for a continuously burning flame from a natural oil spring, found especially in Middle Eastern sites.

**ethane** *Chemistry.* $C_2H_6$, a flammable, colorless gas, insoluble in water and soluble in alcohol; used in organic synthesis, as a fuel, and in refrigeration. Also, DIMETHYL.

**ethanol** *Chemistry.* one of a group of chemical compounds (alcohols) composed of molecules that contain a hydroxyl group (OH) bonded to a carbon atom, especially, $C_2H_5OH$ **(fuel ethanol),** a transparent, colorless, volatile liquid produced through the fermentation of agricultural products; it is the intoxicant in beverages such as wine, beer, or whiskey. It also is manufactured as a transportation fuel from biological feedstocks such as corn and sugarcane.

**ether** *History.* an invisible medium once presumed to fill all unoccupied space; a hypothesis formerly used to explain the propagation of light, heat, and electromagnetic waves. Attempts to measure the ether in the late 19th century produced no results, and Einstein abandoned the concept in his 1905 paper on relativity.

**ethyl alcohol** another name for ETHANOL.

**ethylene** *Oil & Gas.* $C_2H_4$, an olefinic hydrocarbon recovered from refinery processes or petrochemical processes.

**ethyl gasoline** *Oil & Gas.* gasoline containing the antiknock compound tetraethyl lead; i.e., leaded gasoline.

**ethyl tert-butyl ether (ETBE)** *Oil & Gas.* a (proposed) additive to gasoline to improve engine performance and reduce tailpipe emissions of pollutants; a biofuel derivative.

**ETR** extraterrestrial radiation.

**ETS** environmental tobacco smoke.

**eucalyptus** *Biomass.* any of various tall trees of the genus *Eucalyptus,* native to Australia and now widely grown elsewhere; e.g., in the U.S.; noted as an efficient fuel source because of its fast growth and hardiness.

**eudiometer** *Measurement.* a device used to measure the volume of a gas during combustion.

**eupatheoscope** *HVAC.* an instrument used to measure the operative temperature of a room in terms of the comfort environment that this provides.

**Euratom** European Atomic Energy Community (est. 1958); an intergovernment agency operating to ensure the development of nuclear energy in the European Union, and to ensure that all users in the Union receive a regular and equitable supply of ores and nuclear fuels.

**EUREC** European Renewable Energy Centres Agency (est. 1991); an independent association of more than 40 groups from across Europe seeking to strengthen research and development efforts in renewable energy technologies.

**European Atomic Energy Community** see EURATOM.

**European Energy Charter** see ENERGY CHARTER.

**European Organization for Nuclear Research** see CERN.

**eustatic** *Earth Science.* describing a global fluctuation in sea level caused by changes in water supply (such as continental ice sheets melting) or by large-scale changes in ocean basin capacity. Thus, **eustasy; eustatic sea-level change.**

**eutectic** *Chemistry.* describing a mixture of substances that has a melting or freezing point lower than that of any mixture of the same substances in other proportions.

**eutectic salts** *Solar.* salts used in solar thermal storage applications because they melt at low temperatures. The energy (heat) stored in the molten salts is much higher than a similar volume of solid material; when the salt changes back to a solid, this heat is released.

**euthermy** *Biological Energetics.* the more or less constant regulation of body temperature around a set point. Thus, **eutherm; euthermic.**

**eutrophic** *Ecology.* having to do with eutrophy (a stage in eutrophication), or the overall process of eutrophication.

**eutrophication** *Ecology.* a process that increases the amount of nutrients, especially nitrogen and phosphorus, in a lake or other aquatic ecosystem; occurs naturally over geological time but may be accelerated by human activities, such as waste disposal or land drainage, leading to an increase in algae and a decrease in biodiversity. Eutrophication is divided into four levels of increasing nutrient concentration: oligotrophy, mesotrophy, eutrophy, and hypereutrophy.

**eutrophy** *Ecology.* **1.** one of the phases of eutrophication, the third stage in which the water is extremely rich in nutrients, with high biological productivity. **2.** another term for the entire process of eutrophication.

**evacuated-tube collector** *Solar.* a solar collector in which thermal heat is captured by the use of a collector fluid that flows through an absorber tube contained inside an evacuated glass tube.

**evacuation** *Chemistry.* the process of removing air, water vapor, or other gases from a given medium; e.g., a refrigeration system.

**evaporation** *Physics.* any process in which a liquid is converted to its vapor phase by the addition of heat to the liquid. Thus, **evaporate, evaporative.**

**evaporation pond** *Environment.* a containment pond designed to hold liquid wastes and to concentrate the waste through a process of evaporation.

**evaporative condenser** *Refrigeration.* **1.** an apparatus that utilizes the evaporation of water by air at the condenser surface as a means of heat dissipation. **2.** any cooling device that operates on the basis of a vapor being condensed by the evaporation of water.

**evaporative cooler** *HVAC.* an air-cooling unit that turns ambient air into moist, cooler air by saturating it with water vapor.

**evaporative cooling** *Refrigeration.* **1.** the process of cooling air by evaporating water into it. **2.** the process of lowering the temperature of a large amount of liquid by employing the heat of vaporization of a part of the liquid.

**evaporative emissions** *Transportation.* emissions from the evaporation of fuel from motor vehicle carburetors or fuel systems; can occur while vehicles are operating or refueling, or even while parked.

**evaporator** *Refrigeration.* **1.** the component or element of a refrigeration system where the refrigerant withdraws heat energy as it evaporates (i.e., changes from a liquid to a gas). **2.** any device in which evaporation occurs, especially one designed to concentrate a solution.

**evapotranspiration** *Earth Science.* the total loss of water from a particular area, equal to the sum of the amount of water lost by evaporation from the soil and other surfaces and the amount lost by transpiration from plants.

**event** *Nuclear.* a term for any unintended occurrence, including operating error, equipment failure, or other mishap, having consequences or potential consequences that are significant in terms of protection or safety.

**event tree analysis (ETA)** *Measurement.* a visual representation of all the events that can occur in a complex system; so called because the events are depicted as spreading out from a source, like the branches of a tree. An **event tree** can be used to analyze safety systems, in order to identify the various possible outcomes of the system following an initiating event that is unexpected or unsatisfactory; e.g., a malfunction in a nuclear power plant.

**evolution** *History.* descent with modification; cumulative changes in the characteristics of organisms over generations.

**evolutionary economics** *Economics.* a school of economic thought that views the economy as an evolving system (on the model of the biosphere) and emphasizes dynamics, changing structures, and disequilibrium processes, such as would occur in biological evolution.

**exa-** *Measurement.* a prefix meaning "one quintillion", or $10^{18}$; symbol E.

**exajoule** *Measurement.* a measure of energy, symbol EJ; $10^{18}$ J, equivalent to $10^{15}$ Btu.

**excess air** *Coal.* a term for the air required to burn coal completely, in addition to the theoretical combustion air.

**excess relative risk (ERR)** *Nuclear.* the excess rate of occurrence of a particular health effect associated with exposure to radiation.

**excitation** *Physics.* the process of changing the state of a system from its ground state to a given excited state., especially a process by which the energy state of an atom or molecule is increased above the ground state by radiation or collision.

**excited state** *Physics.* the condition of an atom at a state of higher energy than its ground state (lowest normal energy state); the excess energy is usually released eventually as a gamma ray.

**exciter** *Electricity.* **1.** a small auxiliary generator that provides current for the field structure of a larger generator. **2.** *Physics.* anything that brings a system to an excited state.

**exclusion area** *Nuclear.* a term for the area surrounding a nuclear reactor in which the reactor operator has the authority to determine all activities, including limiting or excluding access to the area by personnel or property.

**excursion** *Nuclear.* a rapid increase in nuclear reactor power, either accidental or deliberate.

**exercise** *Storage.* to fully discharge a nickel-based battery on a periodic basis, in order to dissolve any buildup of crystals that can cause a loss of performance.

**exercise price** see STRIKE PRICE.

**exergoeconomics** *Economics.* **1.** the combination of exergy analysis and economic principles to maximize the thermodynamic and economic energy efficiency of a process. **2.** specifically, the design of industrial activities using both economic and thermodynamic analysis.

**exergy** *Thermodynamics.* the maximum amount of useful work (ordered motion) that a system can perform when it is brought into thermodynamic equilibrium with its surroundings by reversible processes. See next page.

**exergy consumption** *Thermodynamics.* the exergy consumed or destroyed during a process due to irreversibilities within the system boundaries. Also, **exergy destruction, exergy degradation.**

**exfiltration** *HVAC.* the outward flow of air through windows and other openings of a building, especially an uncontrolled outward leakage through cracks and other unintentional openings.

**exhaust** *Consumption & Efficiency.* **1.** a general term for waste gases resulting from combustion or another chemical process. **2.** *Transportation.* specifically, the waste gases emitted by motor vehicles. **3.** the system in a motor vehicle that expels such gases to the environment.

**exhaust gas recirculation (EGR)** *Transportation.* an emissions control technique that reuses engine exhaust gases as part of the intake air supply to help reduce harmful emissions (especially nitrous oxides).

**exhaustible resource** see NONRENEWABLE RESOURCE.

**exhaust manifold** *Transportation.* a system that carries spent gases from the combustion chambers of an internal combustion engine to an exhaust pipe.

**exergy** Exergy is the maximum amount of work that can be done by a subsystem as it approaches thermodynamic equilibrium with its surroundings by a sequence of reversible processes. The term was coined by Zoran Rant (1904–1972) in 1953 from the Greek words *ex* (external) and *ergos* (work). Exergy measurements are made relative to an equilibrium state in which there are no gradients of any kind. This implies uniformity of temperature, pressure, density, chemical composition as well as uniform gravitational and electromagnetic fields. Thus, the exergy of a subsystem is a measure of its 'distance' from equilibrium. Mechanical exergy is known as kinetic energy, while thermal exergy is more familiarly known as heat. These concepts are important in the design of energy-efficient machines. Chemical exergy is used in chemical engineering and thermoeconomics for process optimization, but also in economics and environmental science. Fuel combustion is the spontaneous recombination of hydrocarbons or carbohydrates with atmospheric oxygen, resulting in their mutual chemical equilibrium state. Thus, the heat of combustion (enthalpy) of a fuel is roughly equivalent to its exergy content. For non-fuels, chemical exergy is a measure of distinguishability from the surroundings. A high-grade ore has more embodied exergy than a low-grade ore, and thus needs more energy to be upgraded. Exergy also is an important concept in understanding life processes. Seemingly dead structures in space convert into organized, self reproducing structures as life and life forms by means of converting and partly destroying exergy. Nature creates a state far from thermodynamic equilibrium on the earth by an everlasting redesign of the environment mainly powered by the exergy of the sunlight.

**Göran Wall**
Mölndal, Sweden

**exhaust trail** *Transportation.* a visible condensation stream that forms when water vapor from an aircraft exhaust mixes with and saturates the air in the aircraft's wake.

**exitance** *Lighting.* the density of light reflecting from a surface at a given point, measured in lumens per square foot.

**exogenous** *Environment.* **1.** originating or occurring outside a living system. **2.** *Economics.* determined by factors external to an economy. Thus, **endogenous growth (model).**

**exosomatic evolution** *History.* referring to the evolution of the human species as a whole through the development of tools, language, and various stages of technology.

**exosphere** *Earth Science.* literally, the outer sphere; a region of extremely low air density that is considered to be the outermost limit of earth's atmosphere, from which atmospheric gases can escape into space. Estimated to begin at about 640 km above the surface and to merge with outer space at about 1280 km.

**exothermy** *Biological Energetics.* the regulation of body temperature by means of the management of heat sources that are primarily external to the organism. Thus, **exotherm, exothermic.**

**expanded polystyrene** *Materials.* a type of insulation that is molded or expanded to produce coarse, closed cells containing air, often used to insulate the interior of masonry walls.

**expander cycle** *Thermodynamics.* a thermodynamic cycle in which the working fluid transfers mechanical energy to an external system. Also, **expansion cycle.**

**expansion** *Physics.* **1.** an increase in the volume of a substance while its mass remains constant. **2.** specifically, an increase in the volume of the working fluid in a mechanical process. **3.** *Electricity.* a process that increases the effective gain of a strong electrical signal and decreases that of a weak signal.

**expansion ratio** *Chemistry.* the ratio of the volume at which a gas or liquid is stored, compared to the volume of the same gas or liquid at a standard pressure and temperature.

**expansion work** *Thermodynamics.* the work done when a system expands against an opposing pressure.

**experience curve** *Economics.* the principle that producers tend to become increasingly efficient as they gain experience making their product, thus providing lower unit production costs. See next page.

**expert system** *Communication.* a general term for any computer application in which problems are solved by means of an information base containing rules and data, from which

**experience curve** Experience curves describe the reduction in unit cost of a product that is driven by economies of scale in production and distribution, and through learning by doing when a product moves from test markets into larger and larger markets, and is sold in higher numbers. Unit costs are thought to decay exponentially with the cumulative number of units sold. Experience curves are used to calculate the "progress ratio", which describes the relative cost after a doubling of units sold. A progress ratio of 85% means that costs are reduced by 15% with each doubling of units sold—or in the case of energy, of capacity installed. The progress ratio is assumed to be constant, and independent of the rate at which the experience is accumulated, of the developmental stage of the technology, and of potential qualitative evolution of the product or technology. Typically, progress ratios are assumed to be between 75 and 85%. Experience curves have been estimated for a variety of technologies (e.g., combined cycle gas turbines, wind turbines, solar photovoltaics, infrastructure, non-energy technologies). The results of the estimations are not always consistent, mostly due to problems with the availability and the statistical properties of data. Still, experience curves play an important role in medium- and long-term energy scenario calculations, for example for energy planning or climate policy purposes. They are used for the estimation of future cost of energy and determine the sequence in which technologies in least-cost optimization models are deployed.

**Christine Woerlen**
Global Environment Facility, USA

inferences are drawn on the basis of human experience and previously encountered problems.

**exploration** *Materials.* a general term for any process of searching for materials of value (oil, coal, minerals, ore, and so on), by means of geological surveys, geophysical prospecting, boreholes, tunnels, etc. Thus, **exploratory.**

**explosion** *Chemistry.* an extremely rapid chemical reaction or change of state that generates heat and usually gas. Thus, **explosive.**

**explosive limit** *Chemistry.* the range in which a gas and air are in the proper proportion to each other so that the gas will burn when ignited.

**exponential growth** *Measurement.* **1.** the description of a given quantity as increasing at a rate that is a constant percentage of the existing quantity; e.g., the growth in size of a large city that has increased its population each year from 1960 to the present. **2.** in popular use, any large and rapid rate of growth.

**exposure** see RADIATION EXPOSURE.

**exposure pathway** *Health & Safety.* the route that a toxic substance takes from its source to its end point, during which humans or other organism can be exposed to it; this includes a source of contamination, an environmental transport mechanism, a point and means of exposure, and a receptor population.

**exposure test** *Health & Safety.* a determination of the level, concentration, or uptake of a potentially toxic compound (or its metabolites) in biological samples from an organism, and the interpretation of the results to estimate the absorbed dose or degree of environmental pollution.

**extensions** *Oil & Gas.* the reserves credited to a reservoir because of the enlargement of its proved area by wells drilled in years subsequent to its discovery.

**extensive** *Physics.* describing a property of a system, such as its mass or total volume, that is a function of its extent (size); i.e., that changes with the quantity of material present in the system. Thus, **extensive property, extensive variable.** Compare INTENSIVE.

**external combustion engine** *Transportation.* an engine in which heat transfer from combustion occurs outside the engine and then is transferred across the boundary of the system to the working fluid of the engine (e.g., a steam engine).

**externality** *Economics.* **1.** ancillary effects of production or consumption for which no internal cost is incurred, typically when the actions of firms and individuals have an effect on others than themselves; e.g., pollution of a river that negatively affects the health of people living downstream rather than employees of the firm releasing the harmful material. Thus, **external cost(s). 2.** a similar process having a positive effect; e.g., investment by a nation to reduce its $CO_2$ emissions that provides a more stable climate for other nations.

**external reforming** *Hydrogen.* the production of a desired product (usually hydrogen) from a hydrocarbon fuel (methanol, gasoline, natural gas, propane, and so on) by a method in which the fuel is processed prior to entry to the fuel cell or stack.

**extinction** *Ecology.* the fact of a given species (or other taxonomic group) having no living members; the death of all individuals with no direct descendants.

**extraction loss** *Oil & Gas.* a reduction in the volume of natural gas due to the removal of liquid constituents such as ethane, propane, and butane.

**extraction turbine** *Conversion.* a steam turbine in which a portion of the working fluid is tapped between stages of the expansion process and used for purposes other than generating mechanical power.

**extractive industry** *Mining.* a commercial operation involved in exploring for nonrenewable natural resources and extracting them from the earth, such as mining, oil and gas extraction, forestry, fisheries, or agriculture.

**extraterrestrial** *Earth Science.* originating from sources other than the earth; e.g., the sun.

**extraterrestrial radiation** *Solar.* the ideal amount of global horizontal radiation that a location on earth would receive if there were no intervening atmosphere or clouds; used as the reference amount to which actual solar energy measurements are compared. Similarly, **extraterrestrial (normal) irradiance.**

**extreme weather event** *Earth Science.* a classification for a weather event that is very different from the normal range of weather experienced in the given location, such as a flood, drought, tornado, blizzard, ice storm, heat wave, or cold spell, and so on. An increased incidence of extreme weather events can be an indicator of climate change.

**extruded polystyrene** *Materials.* a type of insulation material with fine, closed cells, containing a mixture of air and refrigerant gas; it has a high R-value, good moisture resistance, and high structural strength.

**exurbanization** *Social Issues.* the fact of becoming exurban; i.e., a population shift from a central city and its nearby suburbs to other (typically newer) communities that are farther from the central city (and that thus require longer commuting distances).

**Exxon** in full ExxonMobil, the world's largest integrated oil company, originating as the Standard Oil Company of John D. Rockefeller. When Standard Oil was broken up by court order in 1911, Rockefeller formed various organizations; two of these, Standard Oil Co. of New Jersey and Standard Oil Co. of New York, were the chief predecessors of Exxon and Mobil, respectively. The two companies merged in 1998.

**Exxon Valdez** see VALDEZ.

**F** Fahrenheit.

**factory farming** *Consumption & Efficiency.* a (negative) term for animal agriculture that is regarded as analogous to the production of manufactured goods in a traditional factory; i.e., it involves high levels of energy use, mechanization, automation, and pollution, as well as the output of large numbers of products (animals) in a relatively small space.

**faculae** *Solar.* bright regions of the photosphere visible in white light near the limb of the sun, brighter than their surroundings because they are higher in temperature and density.

**Fahrenheit, (Daniel) Gabriel** 1686–1736, Polish-born scientist who invented the first accurate thermometer and devised the Fahrenheit temperature scale (see following).

**Fahrenheit** *Measurement.* relating to or expressed by the Fahrenheit temperature scale.

**Fahrenheit (temperature) scale** *Measurement.* a temperature scale based on 32°C for the freezing point of water and 212°C for the boiling point; developed by Gabriel Fahrenheit in about 1724. The Fahrenheit scale was historically used in Europe prior to the adoption of the Celsius (centigrade) scale, and it is still used for common temperature measurements in the U.S. (e.g., weather reports).

**Fall, Albert B.** 1861–1944, U.S. government official who became notorious for his involvement in a widely publicized oil leasing scandal in the 1920s. See TEAPOT DOME.

**fallout** *Nuclear.* radioactive material released into the atmosphere by activities such as the detonation of nuclear weapons, accidental emissions from nuclear power plants, or natural occurrences (e.g., the eruption of Mount St. Helens).

**fallout shelter** *History.* a site intended to give some protection against fallout radiation and other effects of a nuclear explosion, either an existing area such as a basement or tunnel, or a structure specially constructed for this purpose; a widespread feature of U.S. efforts in nuclear civil defense during the Cold War era.

**falls** see WATERFALL.

**fan** *Consumption & Efficiency.* **1.** any of various devices designed to produce a current of air (especially a cooling current) by the movement of one or more broad, often rotating blades or vanes. **2.** *Wind.* a small vane designed to keep the blades of a windmill facing the direction of the wind.

**fantail** *Wind.* a mechanism that automatically rotates the blades of a windmill into the wind; an important historic development in windmill technology (1745) because previously a person had to rotate the blades manually.

**FAO** Food and Agriculture Organization (est. 1945); a UN agency whose mandate is to raise levels of nutrition, improve agricultural productivity, and better the lives of rural populations.

***fallout shelter*** *A model American fallout shelter (c. 1957). An underground facility with a two-week supply of food and water, a battery-operated radio, auxiliary light sources, first aid equipment, and other supplies.*

**farad** *Electricity.* a basic unit of capacitance in the meter-kilogram-second system; equivalent to the capacitance of a capacitor in which a charge of 1 coulomb produces a change of 1 volt in the potential difference between its terminals. [Named for Michael *Faraday.*]

**Faraday, Michael** 1791–1867, English experimental scientist especially noted for his discoveries in electricity, including electromagnetic induction, the battery, the electric arc, and the dynamo. His work on induction formed the basis of modern electromagnetic technology, and his work on electrochemistry laid the basis for this other important modern industry.

**faraday** *Measurement.* a unit of electric charge, $9.64870 \times 10^4$ C, required for 1 gram-equivalent of a substance to be liberated by electrolysis.

**Faraday constant** *Electricity.* a determination of the amount of electricity corresponding to 1 mole of electrons.

**Faraday's law** *Physics.* a law of electromagnetic induction stating that the electromotive force induced in a circuit is proportional to the time rate of magnetic flux change linked with the circuit.

**Farmington** *Health & Safety.* a community in West Virginia that was the site of a coal mine explosion in 1968 that resulted in the death of 78 miners, one of the most serious mine disasters in U.S. history.

**Farnsworth, Philo T.** 1906–1971, U.S. inventor called "the father of television" for his invention of the cathode ray tube and other electronic innovations that removed all moving parts from televisions and made possible today's TV industry. He transmitted his first electronic television picture in 1927.

**fast neutron** *Nuclear.* a term for a free neutron with a kinetic energy level of more than 1 MeV; so called to distinguish these from lower-energy thermal neutrons that have a kinetic energy level similar to the average kinetic energy of a room-temperature gas.

**fast pyrolysis** *Biomass.* the thermal conversion of biomass by rapid heating to a temperature of between 450 and 600°C in the absence of oxygen.

**fast reactor** *Nuclear.* a reactor having little or no moderator and utilizing fast neutrons. It normally burns plutonium while producing fissile isotopes in fertile material such as depleted uranium (or thorium). Thus, **fast fission reactor, fast neutron reactor, fast breeder reactor,** and so on.

**fatigue** *Materials.* the progressive failure of a material due to changes in its properties resulting from repeated stress.

**Fat Man** *Nuclear.* the U.S. code name for the nuclear weapon detonated over Nagasaki, Japan, on August 9, 1945; it had a slightly larger yield than the bomb known as Little Boy dropped on Hiroshima 3 days earlier.

**fault** *Electricity.* **1.** a failure or interruption in an electrical circuit or device, due to an open circuit, short circuit, or ground in a circuit component or line. **2.** *Earth Science.* a rock fracture along which movement or displacement in the plane of the fracture has taken place; earthquakes tend to occur on fault lines.

*fault* *The San Andreas Fault runs roughly north to south through the center of California; it has been associated with many earthquakes in the state.*

**fault trap** *Oil & Gas.* an oil or gas reservoir formed by the presence of one or more fault surfaces underground.

**fault tree analysis (FTA)** *Measurement.* a logical, structured process that can help identify potential causes of system failure before the failures actually occur; it offers the ability to focus on an event of importance, such as a critical safety issue, and work to minimize its occurrence or consequence. So called because it involves a graphical representation resembling the branches of a tree.

**FBC** 1. fluidized bed combustion. 2. fuel-borne catalyst.

**fc** footcandle.

**FCCC** *Policy.* Framework Convention on Climate Change; a binding treaty opened for signature at the UN Conference on Environment and Development (UNCED) in 1992. Convention parties agreed to attempt to limit emissions of greenhouse gases.

**F cell** *Renewable/Alternative.* short for fuel cell, or for a vehicle powered by fuel cell technology.

**feathering** *Wind.* an overspeed protection system in a wind turbine that changes the angle or orientation of the blades to slow them in winds of undesired high velocity.

**Federal Energy Regulatory Commission (FERC)** an independent U.S. federal agency that regulates the interstate transmission of natural gas, oil, and electricity.

**Federal Highway Act** *History.* a program of interstate highway construction beginning in 1956 that greatly expanded the U.S. network of modern roadways. Originally conceived as a system of military transport, its eventual effect was to ensure the demise of the railroad as a major form of interstate transportation.

**Federal Power Commission (FPC)** *History.* a U.S. agency established in 1920 by the **Federal Power Act** to regulate the electric power and natural gas industries, including the licensing of nonFederal hydroelectric projects, the regulation of interstate transmission of electrical energy, and the rates for that electricity's sale. It was replaced by the Federal Energy Regulatory Commission (FERC) in 1977.

**feebate** *Policy.* a revenue-neutral strategy that imposes fees on one type of customer to pay for rebates given to others; e.g., an incentive program with fees on polluting cars and rebates for cars with cleaner technologies.

**feeder (line)** *Electricity.* 1. an electrical supply line, either overhead or underground, which runs from the substation through various paths, ending with the transformers. 2. *Transportation.* a local rail line or highway that connects with a main thoroughfare.

**feeding strategy** *Biological Energetics.* 1. any of various behavior patterns employed by animals to obtain nutrients; e.g., predation. 2. a specific food supply provided to a domestic animal. 3. the method used by a human society to obtain food; e.g., hunting and gathering.

**feedlot** *Consumption & Efficiency.* an enclosed area used to raise (feed) a large number of farm animals (especially beef cattle) at a given site.

**feed material** *Nuclear.* refined uranium or thorium metal or their pure compounds in a form suitable for use in nuclear reactor fuel elements or as feed for uranium enrichment processes.

**feedstock** *Materials.* 1. a raw or processed organic material that is chemically reacted to produce fuel; e.g., trees, grasses, corn, agricultural wastes, wood wastes and residues, aquatic plants, and municipal wastes. 2. more generally, any raw material required for a machine or processing plant whose output is some form of energy product.

**feedwater** *Consumption & Efficiency.* 1. water supplied to a boiler or other heating system for the generation of steam. 2. specifically, water employed in a nuclear reactor to remove heat from the fuel rods by boiling and becoming steam. The steam then becomes the driving force for the turbine generator.

**feller-buncher** *Biomass.* a large, self-propelled caterpillar type vehicle that is used to harvest trees; it has large hydraulic arms and fingers that grasp, lower, cut, and stack the felled trees without debarking or delimbing.

**femto-** *Measurement.* a prefix meaning "one quadrillionth", or $10^{-15}$; symbol f.

**fen** *Earth Science.* a flooded peat marsh having nonacidic soil (in contrast to the acidic soil found in bogs).

**fenestration** *HVAC.* the arrangement and design of windows (and other structural openings) in a building.

**FERC** Federal Energy Regulatory Commission.

**ferment** *Conversion.* to undergo a process of fermentation (natural or induced); decompose into simpler substances.

**fermentation** *Conversion.* the chemical decomposition of a complex substance, especially a carbohydrate, into simpler chemical products, brought about by the action of enzymes, bacteria, yeasts, or molds, generally in the absence of oxygen. May be a natural process, or one brought about or enhanced technically to produce a desired end product; e.g., the fermentation of grape juice to make wine or of corn products to produce ethanol fuel.

**fermentation ethanol** *Biological Energetics.* the conversion of biomass materials to ethanol in a process of enzymatic transformation by microorganisms; e.g., yeasts acting on organic compounds such as sugars.

**fermenter** *Conversion.* **1.** a device employed in a technical process of fermentation; e.g., a fabricated vessel used in brewing beer. **2.** an organism that carries out a process of fermentation.

**Fermi, Enrico** 1901–1954, Italian–American physicist who played a key role in the development of the first atomic bomb. Fermi provided the mathematical statistics required to clarify a large class of subatomic phenomena, discovered neutron-induced radioactivity, and directed the first controlled chain reaction involving nuclear fission (at the University of Chicago on December 2, 1942).

**fermi** *Measurement.* a unit of length in the metric system equal to $10^{-15}$ m; used to express nuclear measurements.

**Fermi–Dirac statistics** *Physics.* the mathematical formulation of the behavior of particles that are subject to the Pauli exclusion principle; i.e., that have completely antisymmetric wave functions and half-integer spin. [Named for Enrico *Fermi* and Paul *Dirac.*]

**Fermi energy** *Thermodynamics.* the energy associated with a particle that is in thermal equilibrium with a system of interest; this energy is strictly associated with the particle and does not consist of heat or work.

**Fermi level** *Physics.* the highest energy level in a substance that can remain populated with electrons at a temperature of absolute zero. In a metal, the Fermi level is the highest occupied molecular orbital in the valence band. In a semiconductor, the Fermi level is located in the band gap.

**fermion** *Physics.* a particle that obeys Fermi–Dirac statistics; a half-integer spin particle.

**fermium** *Chemistry.* a synthetic radioactive element having the symbol Fm, the atomic number 100, and an atomic weight (of the most stable isotope) of 257.10. Made in nuclear reactors, it has properties similar to those of erbium. [Named for Enrico *Fermi.*]

**ferric** *Materials.* describing various compounds of iron, especially those in which the element has a valence of 3.

**ferrite** *Materials.* a compound consisting of the iron oxides $Fe_2O_3$ and FeO; used in computers and other communication devices.

**ferro-** *Materials.* a prefix meaning "iron".

**ferroelectricity** *Electricity.* the property of a substance (such as iron) that exhibits spontaneous electric polarization and hysteresis; analogous to the spontaneous magnetic polarization in ferromagnetism (see next).

**ferromagnetism** *Materials.* the property of certain metals and alloys, especially those of the iron group, rare-earth, and actinide series, that are capable of spontaneous magnetic polarization, resulting in drastic magnetic effects. These materials are strongly attracted to magnets and are used in permanent magnets and various ceramic compounds. Thus, **ferromagnet, ferromagnetic.**

**ferrous** *Materials.* describing various compounds of iron, especially those in which the element has a valence of 2.

**fertile** *Nuclear.* describing a nonfissile element that can be made fissile by the absorption of neutrons; e.g., U-238, Pu-240. Thus, **fertile element.**

**fertilizer** *Consumption & Efficiency.* any substance that improves the plant-producing quality of the soil, such as manure or a mixture of chemicals.

**Fessenden, Reginald Aubrey** 1866–1932, Canadian inventor who was the first person to broadcast words and music over radio waves (1906). He invented the modulation of radio waves, which allowed reception and transmission on the same aerial without interference. This is the basis for amplitude modulation (AM) radio.

**Ffestiniog** *Hydropower.* a large power station in Wales, the UK's first major facility to employ the PUMPED STORAGE technique to provide generating power for electricity.

**FFV** flexible-fuel vehicle.

**fiberglass** *Materials.* a glass in the form of fine, flexible fibers, widely used in the manufacturing of many industrial products such as textiles, filters, and insulation materials, and also used to reinforce or strengthen plastics.

**fiber optics** *Materials.* a branch of optical technology dealing with systems that transmit light signals and images over short and long distances through the use of **optical fibers** (transparent, hair-thin strands of glass or plastic). These fibers have a wide range of applications, as in the transmission of computer data, telephone messages, and other communications.

**Fibonacci** (Leonardo Pisano), c. 1170–1250, Italian mathematician known for his work in the establishment of the decimal system that eventually replaced the Roman numeral system.

**field** *Physics.* **1.** an abstract representation of the idea that matter modifies the space around it, as when a mass sets up a gravitational force in the surrounding space. **2.** thus, any region, volume, or space in which a physical force is operative and influential. **3.** specifically, a region in which an electrical or magnetic force is active. **4.** *Oil & Gas.* an area consisting of a single reservoir or multiple reservoirs grouped on, or related to, the same individual geological structural feature and/or stratigraphic condition.

**field–effect** *Electricity.* describing a device whose operations or characteristics are controlled by the influence of an electric field.

**field gas** *Oil & Gas.* a term for the feedstock gas entering a natural gas processing plant.

**field intensity** *Electricity.* the strength of an electromagnetic field; equal to the vector sum of all the forces exerted by an electrical or magnetic field at a given point in the field.

**field line** *Physics.* a curve that is everywhere parallel to the direction of the prevailing magnetic field; it is part of a family of such lines, one of which passes through each point of space.

**field production** *Oil & Gas.* a collective term for crude oil production on leases, natural gas liquids production at processing plants, new supply of other hydrocarbons/oxygenates and motor gasoline blending components, and fuel ethanol blended into finished motor gasoline.

**field separation facility** *Oil & Gas.* a surface installation designed to recover lease condensate from a produced natural gas stream originating from more than one lease.

**field strength** see FIELD INTENSITY.

**field theory** *Physics.* any theory in which the basic quantities are fields, such as electromagnetic theory, which studies the interaction of electric and magnetic fields. See also UNIFIED FIELD THEORY.

**filament** *Electricity.* **1.** a thread of tungsten, carbon, or similar material that emits light when heated by an electric current. **2.** in an electron tube, a cathode that is heated when an electric current flows through it.

**fill dam** *Hydropower.* any dam constructed of fill; i.e., excavated natural materials such as soil or rocks, or industrial waste materials.

**fill factor** *Photovoltaic.* the ratio of a photovoltaic cell's actual power to its theoretical power if both current and voltage were at their maximum; a standard used to evaluate cell performance.

**film** *Materials.* **1.** a very thin, continuous layer or sheet of a substance, such as a soap bubble or alcohol on water. **2.** any of various synthetic materials in the form of long, very thin sheets, such as cellophane or polyethylene. **3.** *Communications.* a sheet or strip of cellulose-based material covered with a light-sensitive or light-and-sound-sensitive emulsion, used for recording still or moving images.

**final energy** *Consumption & Efficiency.* a collective term for forms of energy sold to or used by the ultimate consumers (e.g., households, industrial facilities).

**fine coal** *Coal.* coal with a maximum particle size usually less than one-sixteenth in. and rarely above one-eighth in.; may pose a significant health threat when inhaled and then lodged in the lungs of miners.

**finished motor gasoline** *Oil & Gas.* a general term for any complex mixture of relatively volatile hydrocarbons, with or without small quantities of additives, blended to form a fuel suitable for use in spark–ignition engines. This includes conventional gasoline, gasohol and other types of oxygenated gasoline, and reformulated gasoline, but not aviation gasoline.

**finite** *Consumption & Efficiency.* describing energy sources whose supply is ultimately limited given existing environmental conditions; i.e., fossil fuels.

**fire** *Chemistry.* **1.** the heat and light produced by combustion. **2.** the process of combustion itself.

**fire air** *History.* an early term for hydrogen, based on its inflammable properties.

**fireball** *Nuclear.* a highly luminous, intensely hot spherical cloud of dust, gas, and vapor generated by a nuclear explosion.

**firedamp** *Mining.* **1.** a combustible gas, primarily methane, formed by the decomposition of coal or other carbonaceous matter. **2.** an airtight stopping used to isolate an underground fire and to prevent the inflow of fresh air and the outflow of foul air.

**fire flooding** *Oil & Gas.* a technique for enhancing secondary oil recovery in a reservoir. A combustion process is begun at an injection well by continually pumping oxygen-containing gas downhole, and as the heat breaks down the crude, the light oil is pushed ahead through the reservoir toward the production well.

**fireproof** *Materials.* **1.** to treat a material or build a structure so that it will be (highly) resistant to combustion. **2.** describing a material or structure having this quality.

**fire-resistant** *Materials.* **1.** describing an artificial material that is resistant to combustion, especially a normally combustible material (e.g., wood, clothing or bedding fabric) that has been specially treated for this purpose. **2.** describing a natural material that is more resistant to combustion than others of its type; e.g., certain plants that retain high moisture levels. Also, **fire-resistive, fire-retardant.**

**fire-tube boiler** *Conversion.* a steam boiler in which hot flue gases pass through tubes surrounded by boiler water before entering a chimney.

**firewall** *Health & Safety.* **1.** a wall erected to prevent or retard the spread of fire from one part of a building to another, utilizing a noncombustible material such as asbestos or stainless steel. **2.** *Communications.* by analogy with this, a protective program or device intended to provide security for a computer network, or an individual computer, by limiting access to the system and preventing certain unwanted types of data from entering.

**firewood** *Biomass.* another term for FUELWOOD.

**firing rate** *Measurement.* **1.** the rate at which fuel is fed to a burner, expressed by volume, heat units, or weight per unit time. **2.** the amount of energy produced by a heating system from the burning of a fuel.

**firm service** *Consumption & Efficiency.* an electricity or natural gas arrangement under which the distributor agrees to provide the buyer with an uninterrupted supply (except during an emergency when continued delivery of power is not possible). This type of contract is usually more expensive, and is used primarily by firms that cannot afford to risk loss of supply for any period of time. Thus, **firm power, firm contract, firm transmission** and so on. Compare INTERRUPTIBLE SERVICE.

**first-hour rating** *Consumption & Efficiency.* the ability of a unit or system to meet peak-hour demands, described in terms of the amount of output (e.g., heat) it can deliver during a busy hour.

**first-law efficiency** *Thermodynamics.* the efficiency of an energy-conversion process as defined by the first law of thermodynamics (see next). It is expressed as the work or energy delivered, divided by the work or energy input.

**first law of thermodynamics** *Thermodynamics.* a fundamental law of nature that applies the conservation of energy principle to heat and thermodynamic processes. It is expressed in various ways, such as (a) the total energy of an isolated system remains constant; (b) the change in internal energy of a system is equal to the heat added to the system minus the work done by the system; (c) energy can neither be created nor destroyed. See also THERMODYNAMICS.

**first-law optimization** *Thermodynamics.* the design of an energy-conversion device or process so as to obtain maximum efficiency as defined by the first law of thermodynamics (see previous).

**first-round effect** *Economics.* the direct, immediate effect that higher energy prices have on consumption and production; e.g., higher prices for crude oil increase the price of gasoline and heating oil charged to consumers with little time lag. Higher oil prices can also have a **second-round effect;** i.e., a nonenergy firm, after having to pay more for its fuel, will then pass on this increased cost in the form of higher prices for its goods or services.

**Fischer assay** *Oil & Gas.* a laboratory method for determining the quantity of synthetic oil that a sample of crushed oil shale will produce.

**Fischer–Tropsch (FT) process** *Conversion.* a process used to convert natural gas or coal to liquid fuels that can be used in conventional vehicles. These processes primarily produce fuel suitable for use in compression ignition (diesel) engines. Such a process can be used to convert biomass into fuel or to utilize associated gas at oil fields. [Developed by chemists Franz *Fischer,* 1852–1932, and Hans *Tropsch,* 1839–1935.]

**fishery** *Ecology.* **1.** a site or area for harvesting or catching fish or other aquatic food sources; modern fisheries are noted as highly energy intensive. **2.** the activities involved with obtaining fish or other aquatic organisms as food (or for sport).

**fish ladder** *Hydropower.* a series of ascending pools, enclosures, steps, or the like, constructed at a dam to enable salmon or other migrating fish to swim upstream around or over the dam.

**fish passage** *Hydropower.* any feature of a dam that enables fish to move around, through, or over it without harm; e.g., a fish ladder.

**fish passage efficiency (FPE)** *Hydropower.* a description of the proportion of juvenile fish passing a dam through the spillway, sluiceway, or a bypass system, as opposed to passing through the turbines.

**fissile** *Nuclear.* describing an isotope that is capable of undergoing fission as a result of interaction with slow neutrons in a nuclear reactor; common examples are uranium-235 and plutonium-239. Thus, **fissile material.**

**fission** *Physics.* the process, either spontaneous or induced, by which a nucleus splits into two or more large fragments of comparable mass, simultaneously producing additional neutrons (on the average) and vast amounts of energy.

**fissionable** *Nuclear.* describing an isotope that readily undergoes fission by thermal neutrons.

**fission bomb** *Nuclear.* another term for an atomic bomb; e.g., a weapon that derives its explosive force from nuclear fission.

**fission fragment** *Nuclear.* a nucleus formed as a direct result of fission (as opposed to fission products formed by the decay of these nuclides).

**fission product** *Nuclear.* a residual nucleus formed in fission, including fission fragments and their decay daughters; e.g., strontium-90, cesium-137.

**fission reactor** *Nuclear.* a nuclear reactor that produces energy by splitting a large nucleus such as uranium into two smaller nuclei.

**fission-track dating** *Nuclear.* a radioisotopic dating method that depends on the tendency of uranium to undergo spontaneous fission as well as the usual decay process.

**Fitch, John** 1743–1798, U.S. inventor who made the first successful trial of a steamboat, on the Delaware River in 1787. However, it was Robert Fulton's *Clermont,* built after Fitch's death, that turned the steamboat into a commercial success.

**fixation** see NITROGEN FIXATION.

**fixed air** *History.* an earlier term for carbon dioxide.

**fixed-bed** *Materials.* describing a procedure in which the additive material, such as a catalyst, remains stationary in the chemical reactor or adsorber bed. Thus, **fixed-bed reactor, fixed-bed gasifier.**

**fixed carbon** *Materials.* the amount of solid combustible material remaining in a sample of coal, coke, or bituminous material after the removal of moisture, volatile matter, and ash; usually expressed as a percentage by weight of the original sample.

**fixed-speed** see VARIABLE SPEED.

**fixed-tilt** *Photovoltaic.* describing a photovoltaic arrangement set in at a fixed angle with respect to the horizontal. With such an arrangement, some portion of direct sunlight will be lost because of oblique sun-angles in relation to the array. Thus, **fixed-tilt array, fixed-tilt system,** and so on.

**fl** footlambert.

**flame** *Chemistry.* the visible burning portion of a material undergoing combustion.

**flame speed** see BURNING SPEED.

**flame temperature** *Chemistry.* the temperature of a flame burning a stoichiometric mixture of fuel and air (i.e., one in which neither fuel nor air is in excess).

**flaming spring** see BURNING SPRING.

**flammability** *Materials.* the fact of being flammable; the extent to which a material will tend to become combustible.

**flammability limit** *Conversion.* either of the limits within which a combustible gas has both enough fuel and enough air for combustion. Below the **lower flammability limit (LFL)** there is not enough fuel to burn, and above the **higher flammability limit (HFL)** there is not enough air to support combustion. Also, **flammability range.**

**flammable** *Materials.* tending to burn; capable of supporting combustion.

**flare** *Solar.* a sudden eruption of energy on the solar disk lasting minutes to hours, from which radiation and particles are emitted.

**flare gas** *Oil & Gas.* waste gas disposed of by burning in flares, usually at the production sites or at gas processing plants.

**flash distillation** *Chemistry.* a method of distilling liquids in large quantities, applied to such diverse processes as petroleum refining and the desalination of seawater. A liquid feed mixture is pumped through a heater to raise the temperature and enthalpy of the mixture. It then flows through a valve and the pressure is reduced, causing the liquid to partially vaporize. Once the mixture enters a large enough volume (the **flash drum**), the liquid and vapor separate.

**flash-freezing** *Refrigeration.* the process of freezing a substance within a very short time; used in food preservation as a means of retaining more flavor and nutritional value than conventional freezing.

**flash gas** *Refrigeration.* gas resulting from the instantaneous evaporation of a refrigerant when its pressure is lowered.

**flashing** *Conversion.* the rapid vaporization of a fluid by either pressure reduction or heat, especially a process of producing steam by discharging water into a region of lower pressure.

**flash point** *Chemistry.* the lowest temperature at which vapor above the surface of a volatile liquid will quickly ignite when the liquid is heated under standard conditions.

**flash steam** *Conversion.* steam that is produced from hot water by a sudden extreme lowering of the pressure of the water, causing some of the hot water to rapidly vaporize, or "flash" into steam.

**flash steam (power) plant** *Geothermal.* a geothermal system employing flash steam to generate electricity. Fluid is sprayed into a tank held at a much lower pressure than the fluid, causing some of the fluid to rapidly vaporize. This vapor drives a turbine which in turn powers a generator. Flash steam plants are the most common type of geothermal power plants in operation today.

**flash vaporization** *Chemistry.* the fast vaporization obtained by passing a liquid through a heat source. See also FLASH DISTILLATION.

**flat-plate** *Photovoltaic.* describing an arrangement of photovoltaic cells or materials mounted on a rigid flat surface with the cells exposed freely to incoming sunlight. Thus, **flat-plate array, flat-plate module,** and so on.

**flat-plate collector** *Solar.* a metal box with a special glazing on top and a dark-colored absorber plate on the bottom; solar radiation passes through the glazing and strikes the absorber plate, which heats up to convert the radiation into thermal energy. The heat is then transferred to a liquid passing through pipes attached to the absorber plate.

**Fletcher, Edward** U.S. engineer noted for his advancement of the field of solar chemistry, in areas such as the production of hydrogen and oxygen from water and the storage of solar energy by thermochemical transformations.

**flexible-fuel vehicle (FFV)** *Transportation.* a vehicle that can switch between gasoline or diesel fuel and an alternative fuel such as alcohol in the same tank.

**flexicoking** *Oil & Gas.* flexible coking; a thermal cracking process that converts heavy hydrocarbons such as crude oil, tar sands bitumen, and distillation residues into light hydrocarbons.

**float** *Hydropower.* a term for the vane, blade, or bucket contacting the water in a water wheel.

**float battery** another term for a STANDBY BATTERY.

**float(ing) charge** *Storage.* a battery charge current that is equal to, or slightly greater than, the self-discharge rate, so as to maintain the battery in a fully charged condition.

**float coal** *Coal.* a small, isolated mass of coal embedded in sandstone or shale, typically formed from eroded peat transported from its original deposit.

**float dust** *Coal.* minute particles of coal dust carried in suspension by air currents.

**float glass** *Materials.* a high quality, transparent flat glass manufactured by floating molten glass on a bath of molten tin at extremely high temperature, and then letting it cool slowly to form a flat sheet. Thus, **float-glass process.**

**float life** *Storage.* the number of years that a battery can keep its stated capacity when it is kept at float charge.

**float-zone process** *Electricity.* a method used to form single crystal semiconductor substrates; an alternative to the CZOCHRALSKI process; this results in very high-purity, single-crystal silicon, but does not produce as large silicon wafers as does the Czochralski process.

**flooded** *Storage.* **1.** describing a form of rechargeable battery in which the plates are completely immersed (i.e., flooded) within a liquid electrolyte. Most passenger cars use this form of battery; they are also the most commonly used type for independent and remote area power supplies. Thus, **flooded (cell) battery. 2.** *Refrigeration.* describing an evaporator containing liquid refrigerant at all times.

**floor** *Mining.* **1.** the upper surface of the stratum underlying a coal seam, or the bottom of the seam itself. **2.** the lower surface of a horizontal mineshaft.

**flotation** *Mining.* a process of separating valuable minerals from waste rock by causing the mineral particles to float to the surface.

**flow** *Physics.* **1.** the continuous motion of water, air, or another fluid. **2.** specifically, the volume of water passing a given point per unit of time. **3.** any fluidlike movement of matter or charges in a system. **4.** *Transportation.* the movement of traffic on a road, especially a high-speed divided highway. See also ENERGY FLOW.

**flow augmentation** *Hydropower.* the release of water from a storage reservoir to supplement downstream flow, especially to enhance stream flows during dry periods or to improve fish migration.

**flow design** *Hydropower.* the ideal flow rate at which a turbine is intended to operate for maximum efficiency.

**flow field plate** *Hydrogen.* a flat plate made of metal, ceramic, graphite, or composite, used to direct hydrogen and oxygen reactants to a membrane electrode assembly, disperse waste heat, and capture the electricity generated by a fuel cell.

**flow oven** see FLÜSOFFEN.

**flow rate** *Measurement.* **1.** a quantity that measures the amount of a fluid moving across a specified unit area in a given amount of time. **2.** *Transportation.* the rate at which vehicles pass by a given point or through a given stretch of highway.

**flue** *Conversion.* a passage for removing the products of combustion from a furnace, boiler, or fireplace, or through a chimney.

**flue gas** *Conversion.* any gaseous combustion product generated in a furnace; i.e., gas that is emitted to the flue.

**flue gas desulfurization** see DESULFURIZATION.

**fluence** *Nuclear.* a measure of the strength of a radiation field, equal to the product (or integral) of particle flux and time, expressed in units of particles per square centimeter.

**fluffing** *HVAC.* a (deceptive) practice of mixing air with loose-fill insulation as it is installed, so that it gives the appearance of fullness but actually has lower density than is required to meet a specified R-value.

**fluid** *Chemistry.* a nonsolid state of matter in which the atoms or molecules are free to move past each other, as in a gas or a liquid.

**fluid bed** see FLUIDIZED BED.

**fluid catalytic cracking** *Oil & Gas.* a widely used method of oil refining in which gas-oil fractions are cracked to form lower molecular weight components in a fluidized catalyst bed.

**fluid coking** *Oil & Gas.* a thermal cracking process utilizing the fluidized-solids technique to remove carbon (coke) for continuous conversion of heavy, low-grade oils into lighter products.

**fluidics** *Physics.* a technology that carries out sensing, control, information processing, and actuation functions with fluid dynamic phenomena rather than mechanical moving parts.

**fluidization** *Materials.* **1.** the fact or process of becoming a fluid, or acting as a fluid. **2.** specifically, a technique in which a finely divided solid is caused to behave in the manner of a fluid by its being suspended in a moving gas or liquid.

**fluidized bed** *Consumption & Efficiency.* a layer of hot air or gas at the bottom of a container upon which a finely divided powdered material floats; used to combust, dry, heat, or cool the material.

**fluidized bed combustion (FBC)** *Consumption & Efficiency.* a process in which pulverized or granulated fuel and air are introduced into a fluidized bed of sand or some other material, where combustion takes place. This is classified as *bubbling* or *circulating,* depending on whether the bed material remains in place or is transported out of the vessel by the fluidizing gas, and then recovered and returned to the bed. Thus, **fluidized bed combustor.**

**fluidized bed process** *Materials.* a process in which finely divided powders act in a fluid-like way when suspended and moved by a rising stream of gas or vapor; primarily used for catalytic cracking of petroleum distillates.

**fluid mechanics** *Physics.* the scientific study of the mechanical properties of fluids (gases and liquids) in motion or at rest, including the observation, description, and mathematical computation of their behavior.

**flume** *Hydropower.* **1.** a channel of steel, reinforced concrete, wood, etc. employed to carry water for industrial purposes; e.g., at a dam. **2.** to divert or direct water in this manner.

**fluorescence** *Physics.* **1.** an effect in which a substance releases electromagnetic radiation while absorbing another form of energy, but ceases to emit the radiation immediately upon the cessation of the input energy. **2.** specifically, the light emission of a given wavelength by a substance that is activated by light of a different wavelength, such as the absorption of ultraviolet light by the coating in a fluorescent tube to give off visible light.

**fluorescent** *Physics.* **1.** having the property of fluoresence (see previous). **2.** see FLUORESCENT LAMP.

**fluorescent lamp** *Lighting.* an electric light consisting of a glass tube containing a small amount of mercury and a chemically inactive gas at low pressure, usually argon; the inside of the tube is coated with phosphors that absorb ultraviolet rays and change them to visible light. Fluorescent lamps are widely used in industrial and institutional sites. Thus, **fluorescent lighting.**

**fluorine** *Chemistry.* a nonmetallic element having the symbol F, the atomic number 9, an atomic weight of 18.998, a melting point of –219°C, and a boiling point of –188°C; a member of the halogen family, the most electronegative element and the strongest oxidizing agent. Used in the enrichment of uranium and manufacture of refrigerants and plastics.

**fluorocarbon** *Materials.* **1.** any of a number of dense, inert compounds analogous to a hydrocarbon in which some or all of the hydrogen atoms have been replaced by fluorine atoms; widely used as solvents, refrigerants, propellants, and lubricants, and for various other purposes. **2.** another term for a chlorofluorocarbon (a fluorocarbon that contains chlorine).

**fluorometer** *Measurement.* **1.** a device used to measure the amount of fluorescent radiation produced by a given sample. **2.** specifically, a device that measures chlorophyll fluorescence as a method for investigating the photosynthetic activity of plants.

**flüssofen** *History.* flow oven; the earliest form of a true blast furnace, developed in the Rhineland in the 1300s.

**flux** *Physics.* **1.** the measure of the flow of some quantity per unit area and unit time. **2.** any substance that will promote the melting of another substance to which it is added. **3.** the electric or magnetic field lines of force that traverse a given cross-sectional area.

**flux concentration** *Solar.* the density of radiation falling on or received by a surface or body.

**flux density** *Electricity.* the electric flux passing through a surface, divided by the area of the surface. Also, **flux displacement.**

**fly ash** *Materials.* **1.** a fine powder formed after combustion of coal, consisting of the noncombustible matter in the coal and a small amount of carbon that remains from incomplete combustion. **2.** any finely divided residue, essentially noncombustible refuse, suspended in the combustion gases from a furnace.

**flying boat** *History.* an earlier term for a seaplane (an aircraft that can land on water).

**flying machine** *History.* an earlier term for an aircraft.

**flywheel** *Consumption & Efficiency.* a heavy wheel that rotates on a shaft so that its momentum imparts uniform rotational velocity to the shaft and attached machinery; also capable of storing mechanical energy.

**FM** frequency modulation.

**foam board** *HVAC.* a plastic foam insulation product, pressed or extruded into board-like forms, typically used for interior basement or crawl space walls or beneath a basement slab; can also be used for exterior applications.

**foam-core panel** *HVAC.* a type of insulation consisting of foam insulation contained between two facings of drywall or wood composition boards such as plywood.

**foam insulation** *HVAC.* a high R-value insulation product usually made from urethane that can be injected into wall cavities, or sprayed onto roofs or floors, where it expands and sets quickly.

**FOB** **1.** free on board. **2.** freight on board.

**focusing collector** *Solar.* a device that focuses solar radiation on a surface or point.

**fog** *Earth Science.* clouds along the ground; i.e., the suspension of a visible aggregation of minute water droplets or ice crystals in the atmosphere near the earth's surface. An international standard defines fog as such an event reducing visibility to below 1 km (0.62 mile).

**food capture** *Biological Energetics.* any of the processes by which living organisms obtains their food.

**food chain** *Ecology.* the transfer of energy in a given ecosystem, moving through various stages as a result of the feeding patterns of a series of organisms. A typical food chain begins with green plants (autotrophs) that derive their energy from sunlight, then continues with organisms that eat these plants (heterotrophs), then other organisms that consume the plant-eaters, then decomposers that break down the dead bodies of those organisms so that they can be used as soil nutrients by plants.

**food pyramid** see ECOLOGICAL PYRAMID.

**food security** *Global Issues.* the various security measures that a given nation, or the world community as a whole, must carry out to maintain an adequate food supply; can include issues such as preventing food-borne diseases, maintaining a sufficient amount of productive agricultural land, and conducting outreach programs to provide food for disadvantaged persons.

**food web** *Ecology.* the interconnected feeding relationships among multiple food chains found in a particular place and time; so called to reflect the complexity of these relationships, such as the proclivity of some organisms to feed on more than one level, or the alterations that may occur from life cycle changes or the availability of food.

**footcandle** or **foot candle** *Lighting.* a unit of illumination, symbol fc, that represents the light intensity over a surface of one square foot located one foot from a standard candle; equal to one lumen per square foot.

**footlambert** *Lighting.* a unit of luminance, symbol fl; equal to 0.3183 candela per square foot, or to the uniform luminance of an ideal diffuse surface emitting or reflecting light at a rate of one lumen per square foot.

**footprint** see ECOLOGICAL FOOTPRINT.

**forage** *Biological Energetics.* **1.** any of various kinds of vegetable matter (such as hay or grain) used as food for domestic animals. **2.** to actively search for food.

**foraging** *Biological Energetics.* **1.** a feeding strategy of preindustrial human societies in which food is obtained by searching out available resources in the surrounding environment, such as suitable wild plants and game animals, as opposed to obtaining food by farming. **2.** the behavior pattern in animals in which they search for suitable food; e.g., plant food for herbivores or prey animals for predators.

**force** *Physics.* the cause of motion; according to Newton's second law of motion, force (f) is equal to the product of mass (m) and acceleration (a), measured in newtons (nt).

**force constant** *Physics.* the restoring force that tends to oppose the displacement of nuclei in a molecule, thus producing vibrations in response to a displacing force.

**forced-air** *HVAC.* a heating system in which the circulation of warm air is generated by means of a blower or fan (i.e., a device that forces the air to circulate). Thus, **forced-air furnace, forced-air unit.**

**forced convection** *Thermodynamics.* a form of convection in which the buoyancy force is negligible and temperature variations are not large enough to alter the flow significantly. Compare FREE CONVECTION.

**forced outage** *Electricity.* **1.** a blackout; an unanticipated loss of electrical power due to the breakdown of a major component, such as a power plant, transformer, or transmission line. **2.** the intentional shutdown of a generating unit or other facility for emergency reasons.

**force field** *Physics.* the generalized region of space in which a specific force has effect and can be described by a vector.

**force-free field** *Physics.* the field in a particular type of magnetostatic equilibrium in which the Lorentz force vanishes and in which field and current are parallel.

**force majeure** *Economics.* a greater force; the effect on economic activities of an extraordinary and unforeseen event such as a flood or storm, war or terrorist activity, large-scale work stoppage, or the like, making it impossible to fulfill a contractual obligation.

**forcing mechanism** *Climate Change.* any process that alters the energy balance of the climate system, i.e., that changes the relationship between the amount of incoming radiation from the sun and outgoing infrared radiation from earth. For example, a volcanic eruption may create a layer of gases that will reduce incoming radiation (cooling effect); conversely emissions of $CO_2$ may create an enhanced greenhouse effect that will absorb outgoing radiation (warming effect).

**Ford, Henry** 1863–1947, U.S. automobile manufacturer who, though he did not build the first American car, became the single individual most associated with the development of the auto industry in the U.S. and the most famous industrialist in the world. He set up the first assembly line to produce the Model T, whose low price made automobile ownership possible for the average citizen. It became the most popular car of its era, with over 15 million units sold (1908–1927).

***Ford*** *Ford Motor Company's famous River Rouge plant is situated on a 2,000-acre site personally selected by Henry Ford in 1915; by the 1920s it had become the largest industrial complex in the world.*

**Fordism** *History.* business practices and theories associated with or derived from the activities of Henry Ford, especially mass production technologies used to produce goods and to provide an improved standard of living for the production workers.

**forebay** *Hydropower.* **1.** the water intake area for a canal, penstock, or turbine, designed to reduce water velocity and turbulence so as to settle suspended material and keep it from entering the system. **2.** the small reservoir of a pipeline that distributes water to the consumer; the last free water surface of a distribution system.

**forebay guidance** *Hydropower.* a large net or curtain placed in the forebay of a dam to intercept fish that would otherwise go through the powerhouse.

**forest ecology** *Ecology.* the scientific study of the relationship between trees and other forest organisms and their surrounding environment.

**forestry** *Ecology.* the practice of growing trees for the commercial production of timber or fuelwood.

*forest ecology* *This large fire (1945) and others of the era led to a Keep America Green program, which was dedicated to eliminating all fires in forest areas. It is now recognized that fire plays a key role in maintaining forest ecosystems.*

**forestry residue** *Biomass.* useful materials remaining after forest harvesting operations; i.e., tree tops, limbs, and other woody material not removed from commercial hardwood and softwood stands, as well as woody material resulting from management operations such as thinnings and removal of dead trees; can be used for fuel or processed into secondary biofuels.

**forest slash** see SLASH.

**forge** *Materials.* **1.** a special furnace in which metal is heated before shaping. **2.** to shape metal in this way.

**formation** *Earth Science.* a layer of rocks capable of being mapped by geological or geophysical methods.

**formation gas** *Oil & Gas.* a term for the original gas yielded underground from a reservoir.

**Forrester, Jay** born 1918, U.S. engineer, a pioneer in the development of computers and also the founder of the field of system dynamics. He and colleague Robert Everett originated (1944) a project called Whirlwind, which produced the first real-time electronic digital computer.

**forward contract** *Economics.* an agreement to purchase or sell a certain amount of energy on a set date at a predetermined price; in most cases forward contracts are fulfilled physically, i.e., crude oil or electricity actually changes hands. Used by producers and consumers to lock in an energy price and hedge against any adverse market changes in the future. Thus, **forward market.** Compare FUTURES.

**forward settlement** *Economics.* the purchase or sale of a specific quantity of emissions reductions, offsets, or allowances at the current or spot price, with delivery and settlement scheduled for a specified future date.

**fossil city** *Consumption & Efficiency.* a term for a city that is heavily dependent on the use of fossil fuel to maintain its economy and infrastructure, especially a city whose transit system relies heavily on private motor vehicles. Compare SOLAR CITY.

**fossil energy (power) plant** *Consumption & Efficiency.* a system of devices for the conversion of fossil energy to mechanical work or electric energy. See next page.

**fossil fuel** *Materials.* a fuel, such as coal, oil, and natural gas, produced by the decomposition of ancient (fossilized) plants and animals. Thus, **fossil energy.**

**Foucault, Jean** 1819–1868, French physicist who in 1850 accurately measured the speed of light. He invented the **Foucault pendulum** that demonstrates the rotation of the earth and discovered the existence of eddy currents **(Foucault currents);** he also invented the gyroscope.

**four-cycle engine** *Transportation.* an engine having a cycle that is completed in four piston strokes: a suction or induction stroke, a compression stroke, an expansion or power stroke, and an exhaust stroke.

**Fourier's law of conduction** *Thermodynamics.* the statement that the rate of heat flow through a homogenous solid is directly proportional to the area of the section at right angles to the direction of flow, and to the temperature difference along the path of flow. [Named for French mathematician Jean Baptiste *Fourier,* 1768–1830.]

**Fourneyron turbine** *Hydropower.* the first efficient water turbine, a reaction turbine developed in France in the 1820s and based on a design of horizontal wheels curved in one direction that rotate around a vertical shaft. Various improvements of this were made in

**fossil energy (power) plants** Systems of devices for the conversion of fossil energy to mechanical work or electric energy. The main systems are the Steam (Rankine) Cycle and the Gas Turbine (Brayton Cycle). In the Steam Cycle, high pressure and high temperature steam raised in a boiler is expanded through a steam turbine that drives an electric generator. The steam gives up its heat of condensation in a condenser to a heat sink such as water from a river or a lake, and the condensate can then be pumped back into the boiler to repeat the cycle. The heat taken up by the cooling water in the condenser is dissipated mostly through cooling towers in the atmosphere. In the Gas Turbine Cycle, air compressed to high pressure, and heated to high temperature by the combustion of natural gas or light fuel oil is the working fluid that expands in the turbine to provide the torque for driving both a compressor and the electric generator. The gas turbine demands clean fuels such as natural gas or light fuel oil. Combustion is the prevailing fuel utilization technology in both the above cycles. Coal is the preferred fuel for the steam cycle; because of its low cost and broad and secure availability worldwide. Combustion generated pollutants, such as oxides of nitrogen ($NO_x$), of sulfur ($SO_x$), and particulates, if uncontrolled and emitted into the atmosphere represent environmental and health hazards. Environmental regulations supported by intensive research and developments have reduced pollutant emissions significantly. Improvements in efficiency and emissions come by increasing steam pressure and temperature in the steam cycle, and by increased turbine inlet temperature in the gas turbine cycle. Coal gasification produces a fuel gas that is capable of being used in the gas turbine. By integrating coal gasification with gas turbine and steam cycles, advantage can be taken of high efficiency and low pollutant emission while using coal, an inexpensive, secure and indigenous fuel in many countries throughout the world. A potential additional advantage of the Integrated Gasification Combined Cycle (IGCC) is the capability of capturing $CO_2$ from the fuel gas and making it ready for high-pressure pipeline transportation to a sequestration site. This will be key to the commercial and clean co-production of electricity and hydrogen from coal.

**János Beér**
Massachusetts Institute of Technology

the ensuing decades and modern versions are still in use today. Invented by French engineer Benoit *Fourneyron*, 1802–1867.]

**four-stroke cycle** *Transportation.* an engine cycle that is completed in four piston strokes: first, a suction or induction stroke in which an air-fuel mixture is drawn into the cylinder; second, a compression stroke in which this mixture is compressed and ignited; third, an expansion or power stroke in which the gases produced by this combustion expand to move the piston downward; fourth, an exhaust stroke in which the burned gases are expelled from the cylinder. This is the standard technology for liquid-fueled automobiles.

**four-stroke engine** *Transportation.* an engine employing the FOUR-STROKE CYCLE.

**fourth dimension** *Physics.* the concept that time represents a fourth context of measurement in addition to the three classical dimensions (length, width, depth); developed to account for the fact that the work of Einstein and others indicates that certain physical events involve variables both of space and time.

**fourth law of thermodynamics** see ONSAGER RECIPROCAL RELATIONS.

**FPC** Federal Power Commission.

**frac sand** *Oil & Gas.* fracturing sand; a pure, high-quality quartz sand pumped at high pressures into subsurface rock strata to crack open fractures created by the pressure; this increases the permeability of the rock, thus improving the rate of oil and gas flow.

**fraction** *Materials.* **1.** any portion of a mixture that exhibits uniform or very similar properties. **2.** the portion of a powder having particle sizes ranging between two established limits. **3.** an individual, recognizable portion of crude oil that is the product of a distillation or refining process.

**fractional distillation** *Chemistry.* a procedure for separating components of different boiling points from a liquid mixture, by vaporizing the liquid and then passing the vapor through a vertical tube containing a packing such as glass beads or metal turnings.

**fractional-horsepower motor** *Conversion.* a common type of small motor with a rated output power of less than 1 horsepower, used for example in computer and office equipment, motor vehicles, or residential heating/cooling systems.

**fractionation** *Chemistry.* **1.** any of various methods of separating the components of a mixture into fractions, such as electrophoresis. **2.** a method used to determine the molecular weight distribution of a polymer, based on the tendency of polymers of high molecular weight to be less soluble than those of low weight.

**fracturing** *Oil & Gas.* a well stimulation technique in which fluids are pumped into a formation under extremely high pressure to create or enlarge fractures for oil and gas to flow through. Proppants such as sand are injected with the liquid to hold the fractures open.

**Francis, James** 1815–1892, British-born hydraulic engineer and inventor of the FRANCIS TURBINE. He also greatly advanced the methods of testing hydraulic machinery and was a founding member of the American Society of Civil Engineers.

**Francis turbine** *Hydropower.* a reaction hydraulic turbine that is widely used in low and medium head applications, especially for large hydroelectric plants; a mixed-flow turbine in which water enters radially to the shaft and exits axially. It was developed in New England in the 1840s as a more efficient successor to the Boyden turbine.

**francium** *Chemistry.* a chemical element having the symbol Fr, the atomic number 87, and an atomic weight 223; the heaviest of the alkali metals, probably existing only as radioactive isotopes, of which **francium-223** is the longest-lived, with a half-life of 21 min.

**Franklin, Benjamin** 1706–1790, American patriot, author, diplomat, and scientist, famous for his leadership in the founding of the United States, for technological achievements such as the invention of bifocals and the FRANKLIN STOVE, and also for his experiments with electricity. In 1752, he flew a kite attached to a silk string in a thunderstorm, and showed that a metal key tied to the thread would charge a Leyden jar. These experiments led to the use of lightning rods.

*Benjamin Franklin*

**Franklin stove** *History.* a type of iron furnace stove conceived by Benjamin Franklin and improved by David Rittenhouse, providing an advance in heating over fireplaces since it could be free-standing in the center of a room; widely used in the U.S. since the late 18th century and still sold today.

**Frasch process** *Mining.* a process for the extraction of sulfur from subsurface deposits, in which superheated steam is piped into the sulfur deposit to melt it; compressed air is then pumped down to force the molten sulfur to the surface. [Developed by German–American engineer Herman *Frasch,* 1851–1914.]

**Fraunhofer Institute** in full, Fraunhofer Institute for Solar Energy Systems (ISE), est. 1981; the first solar research institute in Europe to operate independent of a university, recognized as a leading international research institute in the field of solar energy.

**free convection** *Thermodynamics.* a form of convection produced by buoyancy force; fluid motion due to density differences. Compare FORCED CONVECTION.

**free cooling** *HVAC.* **1.** a cooling effect produced or enhanced by naturally cool outside air. **2.** energy savings achieved in this manner.

**free energy** *Thermodynamics.* a system's capacity to perform useful work; the amount of energy available for work. See also GIBBS FREE ENERGY; HELMHOLTZ FREE ENERGY.

**free enthalpy** *Thermodynamics.* a formula for determining the amount of energy available for work in a system after that system undergoes a change.

**free gas** *Oil & Gas.* **1.** a hydrocarbon that exists at reservoir temperature and pressure in the gaseous phase and remains so when produced under standard conditions. **2.** natural gas produced alone, not with condensate or crude oil. **3.** *Chemistry.* any substance that naturally exists in the gaseous phase.

**free market** *Economics.* an economic system in which market-determined prices guide choices about the production and distribution of goods; in general productive resources are privately owned and managed, and government intervention is kept to a minimum (employed mainly to prevent or correct market failures, such as illegal price-fixing).

**free on board (FOB)** *Economics.* the price paid for a good at the site of operation, excluding any freight or shipping and insurance costs (e.g., coal at the mining site). [So called because of the historic practice in merchant shipping in which a seller was obligated to pay the cost of delivering goods on board a vessel of the buyer's choice.]

**free oxygen** *Earth Science.* oxygen in its molecular forms, either as normal diatomic oxygen ($O_2$) or ozone ($O_3$) not combined with any other elements. Earth's early atmosphere did not have significant levels of free oxygen, which is required for life by higher organisms. The abundance of free oxygen in later geological epochs and up to the present has been largely driven by terrestrial plants, which release oxygen during photosynthesis.

**free trade** *Economics.* the selling of products between countries without tariffs or other such trade barriers.

**free trade area** *Economics.* a group of countries that adopt a policy of free trade (see above) for trade among themselves, while not necessarily changing the barriers that each member country may have for trade with others outside the group.

**freeze** *Chemistry.* to bring about or undergo a phase change in a material from the liquid to the solid state, due to the withdrawing of heat from it; for a given pressure, this occurs at a fixed temperature in a pure substance, and over a range of temperatures in a mixture of substances.

**freeze-drying** *Refrigeration.* the process of dehydrating and preserving a substance by freezing it and maintaining it in the frozen state under high vacuum conditions.

**freezer** *Refrigeration.* **1.** a separate refrigeration unit that is held at a temperature below 32°F (typically 0–5°F), in which perishable foods can be stored in a frozen state. **2.** a compartment or unit of a refrigerator serving this same purpose.

**freezing point** *Chemistry.* the temperature at which a substance in liquid form freezes, equal to the temperature at which its solid form melts; this represents equilibrium between these two phases.

**freight on board (FOB)** *Economics.* an indication of the point at which the buyer becomes responsible for a shipment and its charges; e.g., FOB destination means that the seller owns the goods while in transit and pays charges up to the point of delivery at the destination.

**Freon** *Refrigeration.* the trade name for a series of nonflammable, nonexplosive fluorocarbon or chlorofluorocarbon products. Freon represented a significant advance when first marketed in the 1930s because it did not present the hazards of earlier refrigerants, and it led to much greater consumer use of refrigerators and air conditioners. However, in the 1990s Freon was banned because of concerns over its role in the depletion of the ozone layer.

**frequency** *Physics.* **1.** the number of cycles of a wave that move past a fixed observation point per second; the SI unit of frequency is the hertz (Hz). **2.** specifically, the number of sound waves per second that are produced by a vibrating object, as in radio broadcasting.

**frequency modulation (FM)** *Communication.* the instantaneous variation of the frequency of a carrier wave in response to changes in the amplitude of a modulating signal; provides a more static-free transmission of radio waves than AM broadcasting.

**fresh feed(s)** *Oil & Gas.* a term for crude oil or petroleum distillates that are being fed to processing units for the first time.

**freshet** *Hydropower.* **1.** a sudden rise or overflowing of a small stream as a result of heavy rains or rapidly melting snow. **2.** any small, clear freshwater stream.

**freshness index** *HVAC.* a sensory index indicating a subjective assessment of the purity of indoor air by building occupants.

**Fresnel cooker** *Solar.* a type of solar cooker that uses a parabolic reflector (Fresnel lens) to concentrate and transfer solar radiation onto a baking tray or cooking pot.

**Fresnel lens** *Solar.* an optical device that focuses light like a magnifying glass; concentric rings are faced at slightly different angles so that light falling on any ring is focused to the same point. [Invented by French physicist Augustin *Fresnel,* 1788–1827.]

**friable** *Materials.* easily broken into small pieces. Thus, **friability.**

**friction** *Physics.* a force that opposes the relative motion of two material surfaces that are in contact with one another; the direction of the force on each body is opposite to the direction of its motion relative to the other body.

**friction head** *Physics.* **1.** the amount of reduction in the head (height under pressure) of a fluid, due to the friction between the fluid and its container and also to the intermolecular interaction. **2.** *Hydropower.* the resistance that must be overcome by a pump to offset the friction losses of water moving through a pipe.

**friction layer** *Earth Science.* the thin layer of the atmosphere adjacent to the earth's surface.

**Fritts, Charles** U.S. inventor who built the first genuine solar cell (1883), using junctions formed by coating the semiconductor selenium with an ultrathin, nearly transparent layer of gold. His devices were not efficient, but they proved the viability of light as an energy source.

**front-wheel drive** *Transportation.* describing a vehicle in which only the front wheels receive driving power from the engine (as opposed to the traditional rear-wheel drive).

**Fruehauf, August** 1868–1930, U.S. carriage builder who invented the tractor trailer, a truck with the cab and engine separate from the main cargo-carrying body of the truck.

**FTA** **1.** Federal Transit Administration. **2.** fault tree analysis.

**fuel** *Materials.* **1.** any material that evolves energy in a chemical or nuclear reaction. **2.** specifically, a material that can be used to provide power for an engine, combustor, power plant, nuclear reactor, and so on.

**fuel assembly** see FUEL ELEMENT.

**fuel-borne catalyst (FBC)** *Materials.* a chemical compound of an organic and a metal added to a fuel to make a metal ash that promotes the combustion of soot collected with it in a diesel particulate filter (DPF).

**fuel cell** *Renewable/Alternative.* a device that converts the chemical energy of a fuel directly into electricity and heat without combustion, through a process of oxidation; fuel cells differ from conventional electrical cells in that the active materials, such as hydrogen and oxygen, are not contained within the cell, but are supplied from outside. See next page.

**fuel-cell furnace** *Conversion.* a dual-chamber furnace in which partial combustion takes place in a primary chamber and combustion is then completed in the secondary chamber.

**fuel cell poisoning** *Renewable/Alternative.* a term for the degradation of a fuel cell's performance because of impurities in the fuel binding to the catalyst.

**fuel cell stack** *Renewable/Alternative.* an array of individual fuel cells connected in series, for the purpose of increasing electrical current.

**fuel cell vehicle** *Renewable/Alternative.* an electric-drive vehicle that derives the power for its drive motor(s) from a fuel cell system; a **hybrid fuel cell vehicle** also derives drive motor power from a supplemental battery or ultracapacitor.

**fuel cladding** see CLADDING.

**fuel crisis** see ENERGY CRISIS.

**fuel cycle** *Consumption & Efficiency.* the total life of a given fuel in all of its uses and forms, including its extraction or generation, transportation, combustion, air emission, byproduct removal, and waste transportation and disposal.

**fuel cycle analysis** *Environment.* **1.** an evaluation of environmental impact that considers the effects of obtaining a fuel (e.g., mining coal or harvesting wood), as well as the more commonly assessed impacts of burning the fuels for useful heat or electricity. **2.** another term for WELL-TO-WHEELS ANALYSIS.

**fuel cell** Fuel cells have been known for 150 years. William Grove, British physicist and a justice of Britain's high court, first constructed a "wet-cell" battery in 1838. Grove constructed one of the first fuel cells by immersing two platinum electrodes in a container of sulfuric acid. When the two electrodes are separately sealed in chambers containing oxygen and hydrogen, a constant current would flow between the electrodes. The chemical energy of the fuel oxidation is directly converted to electricity with no intermediate mechanical moving parts. Due to the direct nature of energy conversion, fuel cells are not limited by Carnot cycle considerations. Since the initial construction by Grove, fuel cells have come a long way and are on the verge of large scale commercial applications. The basic unit of a fuel cell consists of three components: cathode, anode, and electrolyte. Ionic transport of one type of species occurs from one electrode chamber to another through the electrolyte with compensating electrical current in the external circuit. Typically, the voltage from a single cell is c. 1.0 Volt. Several cells are connected together in a series-parallel arrangement to form the fuel cell stack. Fuel cells can be classified on the basis of the type of electrolyte: proton exchange membrane fuel cells (PEMFCs), solid oxide fuel cells (SOFCs), alkaline fuel cells (AFCs), molten carbonate fuel cells (MCFCs), and phosphoric acid fuel cells (PAFCs). Fuel cell systems are modular, noiseless, and environmentally friendly with reduced emissions of greenhouse gases and no acid rain causing nitrogen and sulfur oxide emissions. Further, SOFCs and MCFCs can be combined with conventional power generation systems in hybrid cycles resulting in power generation efficiencies as high as 60%. In the coming decades they are expected to find applications in residential, transportation and stationary power generation applications.

**Srikanth Gopalan**
Boston University

**fuel economy** *Transportation.* **1.** a standard measure of the rate of motor vehicle fuel consumption, expressed as the total distance traveled divided by the amount of gasoline fuel consumed in doing this. **2.** a general statement of this based on the average mileage traveled per unit of fuel for a class of vehicles; e.g., a certain car type in a given model year. See next page.

**fuel efficiency** *Transportation.* the efficiency with which a motor vehicle converts energy into movement. Not necessarily equivalent to FUEL ECONOMY, in that one vehicle might have better technology and thus be more efficient than another, but if it is much larger and heavier than the other vehicle, it would have poorer fuel economy.

**fuel element** *Nuclear.* a cluster of fuel pins mounted into a single assembly.

**fuel ethanol** see ETHANOL.

**fuel fabrication** *Nuclear.* the process of making reactor fuel assemblies, usually from sintered uranium oxide pellets that are inserted into zircalloy tubes, comprising the fuel rods or elements.

**fuel injection** *Transportation.* a system for delivering fuel directly under pressure into the cylinder or combustion chamber of a spark–ignition engine, thus eliminating the need for a conventional carburetor; used in diesel engines and in many gasoline engines.

**fuel pellet** *Nuclear.* a small cylinder of enriched uranium, typically uranium oxide; several pellets are assembled into a fuel pin, which then forms a component of a fuel element.

**fuel pin** *Nuclear.* a long slender tube made of a zirconium alloy and containing fuel pellets; several fuel pins bundled together form a fuel element.

**fuel reformer** see REFORMER.

**fuel rod** see FUEL ELEMENT .

**fuel share** *Consumption & Efficiency.* the percentage of total fuel use accounted for by an individual fuel; can be measured in dollars or in physical units such as joules.

**fuel-switching** *Consumption & Efficiency.* the potential ability of a manufacturer or power generator to use a different energy source in place of that actually consumed, by a substitution within a short time and without extensive modifications.

**fuel temperature coefficient** *Nuclear.* the change in reactivity per degree change in nuclear fuel temperature.

**fuel economy** Fuel economy is most often measured by the distance a vehicle can travel on a given volume of fuel (or the inverse). In the U.S. fuel economy is measured in vehicle miles per gallon of fuel. In the European Union liters per 100 kilometers is the preferred measure. Fuel economy is an imprecise measure of energy efficiency as it does not take into account vehicle mass, passenger or cargo capacity, engine power or other attributes that may affect the value of a vehicle's services. In addition, different fuels contain varying amounts of energy per volume. For example, diesel fuel contains approximately 11% more energy per volume than gasoline. Devising comparable volume-based fuel economy measures for liquid and gaseous fuels and electricity raises as yet unresolved problems. For these reasons, some prefer alternative measures of energy efficiency, such as ton-kilometers per megajoule. Fuel economy depends not only on the efficiency with which a vehicle's power train converts the energy in fuel into useful work at the wheels, but also on vehicle mass, aerodynamics and rolling resistance, as well as how the vehicle is driven and the ambient conditions under which it is operated. Speed affects fuel economy through the number of engine revolutions required per distance traveled and because aerodynamic drag increases with the square of velocity. Idling reduces fuel economy because fuel is consumed while the vehicle is not moving. Frequent braking reduces fuel economy by converting the kinetic energy of the vehicle into heat which is then dissipated (regenerative braking allows hybrid vehicles to recapture and store some of this energy in the form of electricity). Cold weather also degrades fuel economy because internal combustion engines are much less efficient before they are fully warmed up and warm up takes longer in cold weather. To permit consistent comparisons among different vehicles, governments and vehicle manufacturers measure passenger car and light truck fuel economy under laboratory conditions over strictly specified driving cycles. A driving cycle is defined by a program of vehicle velocity as a function of time. The U.S., European Union and Japan all use different driving cycles to better approximate typical driving conditions in their regions. The fuel economy individual drivers realize will deviate from these standardized measures depending on the conditions under which they drive and their driving styles.

**David L. Greene**
Oak Ridge National Laboratory

**fuelwood** *Biomass.* firewood; wood and wood products burnt as fuel.

**fuelwood crisis** *Sustainable Development.* a severe shortage of wood for heating and cooking purposes, either historic (e.g., England in the Middle Ages) or contemporary (less developed countries in Asia and sub-Saharan Africa).

**fugacity** *Thermodynamics.* a function that is introduced as an effective substitute for pressure, to allow a real gas system to be considered by the same equations that apply to an ideal gas.

**fugitive air** *Mining.* **1.** air that moves through a ventilating fan but that does not reach the active working face, e.g., that is lost through leaks in the ventilating system. **2.** any unintended release of air during a mining operation.

**fugitive dust** *Mining.* a term for silt, dust, and sand that becomes airborne and is thereby carried away from a mine property by air currents.

**fugitive emission(s)** *Oil & Gas.* an unintended leak of gas from the processing, transmission, or transportation of fossil fuels.

**full-cost pricing** *Economics.* a price for a good or service that would include the cost of externalities; e.g., a price for electricity from a coal-fired power plant that includes the cost of the damage to human health caused by the sulfur dioxide released in coal combustion.

**Fuller, R. Buckminster** 1895–1983, U.S. inventor, architect, engineer, and mathematician, best known for his invention of the GEODESIC DOME. Fuller was an early proponent of renewable energy sources, which he incorporated into his designs.

**fullerene** *Materials.* any of various cage-like molecules that constitute the third form of pure carbon (along with the historically known forms diamond and graphite), whose prototype $C_{60}$ (the buckyball) is the roundest molecule that exists. Fullerenes are a class of discrete molecules, soccer ball-shaped forms of carbon with extraordinary stability. [Named for Buckminster Fuller; their configuration suggests the shape of his famous geodesic dome.]

**full extraction** *Mining.* an underground coal mining technique that removes the coal in large areas, causing the void to be filled in with caved rock.

**full pool** *Hydropower.* the elevation of a reservoir at its maximum normal operating level; i.e., at full height but within holding capacity and not spilling over.

**full sun** *Solar.* the amount of power density of solar radiation received at the earth's surface at noon on a clear day (about 1000 $W/m^2$).

**Fulton, Robert** 1765–1815, U.S. inventor who made steamboating a practical and widespread mode of transportation, with the successful voyage of the *Clermont* on the Hudson River from New York to Albany in 1807.

**fumarole** *Earth Science.* a vent or hole in the earth's surface, usually in a volcanic region, from which steam, gaseous vapors, or hot gases are released.

**fumes** *Materials.* **1.** the smoky particulate matter that emanates from heated materials. **2.** vapors evolved from concentrated acids or solvents. **Fume particles** are solid particles of very small size (under 1 micron in diameter) formed as such vapors condense, or as chemical reactions take place.

**Fumifugium** *History.* the title of a paper published by Englishman John Evelyn in 1661 ("Fumifugium, or the Inconvenience of the Aer and Smoake of London Dissipated"), one of the earliest tracts on air pollution. Evelyn described the poor air quality of London and suggested remedies to deal with this, such as relocating energy-intensive industries (ironmaking, brewing) outside the city, and planting gardens and orchards in their place.

**fundamental particle** see ELEMENTARY PARTICLE.

**fund pollutant** *Environment.* a term for a pollutant substance that can be naturally absorbed by the environment at least to some extent, as opposed to a substance that will remain indefinitely if not actively removed (stock pollutant). However, a fund pollutant will accumulate if its rate of emission exceeds the capacity of the environment to absorb it.

**Fundy, Bay of** *Renewable/Alternative.* a body of water off the coastline of Nova Scotia, Canada; highly suited to use for tidal power because it has the largest tidal changes in the world (a difference between high and low tides of as much as 50 ft).

**furl** *Wind.* **1.** to roll up and fasten a sail so that it will not capture the wind. **2.** to point the blades of a wind turbine out of the wind.

**furling speed** *Wind.* the amount of wind required to produce the maximum power that a wind energy device is capable of generating; any wind in excess of that speed will not generate more than this maximum.

**furling tail** *Wind.* a protection mechanism in a wind energy device that allows the tail of the device to fold up and in during unsuitably high winds; this causes the blades to turn out of the wind, protecting the machine.

**furnace** *HVAC.* an enclosed structure in which heat energy is produced to warm a building space, heat a substance of interest, or create some other physical or chemical effect.

***furnace*** *A contemporary blast furnace used for steelmaking; its temperatures can reach 1650°C (3000°F).*

**fuse** *Electricity.* **1.** a protective device based on a wire or element that melts at low temperature; when the current through the fuse exceeds the fuse rating, the wire melts and opens the circuit. **2.** *Materials.* a combustible substance enclosed in a continuous cord, used for initiating an explosive charge by transmitting fire to it.

**fusion** *Chemistry.* **1.** the process of melting; the conversion of a solid into a liquid by means of heat or pressure. **2.** *Nuclear.* a nuclear process in which two light nuclei combine at extremely high temperatures to form a heavier nucleus and release vast amounts of energy. The explosive force of a hydrogen bomb is an example of uncontrolled fusion, and the energy of the sun and other stars is believed to derive from fusion reactions.

**fusion reactor** *Nuclear.* a nuclear reactor that produces energy by fusing light nuclei together, instead of splitting heavy nuclei apart as in the fission process.

**fusion triple product** *Nuclear.* the three quantities that together define the energy-breakeven point for fusion reactions. The rate of fusion in the fuel, f, is constant for any given amount of fuel in a particular state; the actual net energy released is a function of f (and in turn, the temperature T), the number of particles in a particular area (its density N), and the amount of time they remain together (the confinement time t). The fusion triple product ntT must be above a certain value for a successful confinement scheme.

**futures (contract)** *Economics.* an agreement between a buyer and a seller for purchase or sale of a given quantity of an asset, such as electricity or oil, at a certain time in the future for a certain price; this may or may not result in an actual physical exchange of the asset. Futures often are traded in a **futures market** (e.g., the New York Mercantile Exchange), in which they are used to transfer risk and to generate gains through speculation or arbitrage. Compare FORWARD CONTRACT.

**fuzzy logic** *Communication.* **1.** problem-solving that involves a certain degree of inference and intuition to reach the proper conclusion; regarded as a crucial distinction between human intelligence and machine reasoning. 2. a level of artificial intelligence that can process uncertain or incomplete information; characteristic of many expert systems.

**G** *Measurement.* **1.** a unit of acceleration equal to the standard acceleration of gravity, or about 9.8 meters per second per second. **2.** an abbreviation for gravity or gravitation. **3.** short for giga- (one billion).

**G-7** a group of seven major industrialized countries whose heads of state meet to discuss economic and political issues; the U.S., Canada, Japan, Britain, France, Germany, and Italy. Also known as G-8 with inclusion of Russia.

**G-77** a coalition of developing countries, intended to promote the collective economic interests of its members and enhance their negotiating capacity. Originally with 77 members (1964), it now has more than 130.

**Gadget** *Nuclear.* the code name given to the first nuclear device tested by the U.S. (July 16, 1945) in its efforts to develop an atomic bomb for use as a weapon in World War II.

**Gaia hypothesis** *Earth Science.* the proposal that the earth can be conceived of a single living super-organism, and that the presence of life on earth has played a fundamental role in establishing the physical and chemical conditions that now exist on the planet. [Named for the ancient Greek goddess of the earth.]

**gain** *Electricity.* **1.** the increase in signal power produced by an amplifier. **2.** see HEAT GAIN.

**Galilean** *History.* relating to Galileo or to his theories and discoveries.

**Galilean relativity** *Physics.* a concept based on Galileo's investigations into the performance of falling bodies, reportedly based on an experiment in which he dropped two different weights from the top of the Leaning Tower of Pisa. It states that the acceleration of a given particle will have the same value in all frames of reference that are moving at constant velocity in relation to each other. (This applies at ordinary speeds but not at speeds approaching the speed of light.)

**Galileo Galilei** 1564–1642, Italian astronomer and physicist associated with many of the important advances in science in the late Middle Ages. He made major investigations in mechanics, including experiments on acceleration, friction, inertia, and falling bodies. He improved the telescope and pioneered its use for astronomical observation, discovering mountains on the moon, many new stars, the four satellites of Jupiter, and the composition of the Milky Way. He also supported the theories of Copernicus concerning the motion of the planets.

**gallium** *Chemistry.* a rare metallic element having the symbol Ga, the atomic number 31, an atomic weight of 69.72, a melting point of 29.78°C, and a boiling point of 2403°C; a silver-white metal used for doping semiconductors and producing solar cells and solid-state devices such as transistors.

**gallium arsenide** *Materials.* GaAs, dark gray crystals that are electroluminescent in infrared light; a semiconductor material formed by combining gallium and arsenic, used in transistors, solar cells, and semiconducting lasers.

**gallon** *Measurement.* a traditional unit of volume made up of 4 quarts; the U.S. unit for gasoline and various other liquids. The U.S. **standard gallon** is 128 fluid ounces or 231 cubic inches; equivalent to 3.785 liters. The British **imperial gallon** is 160 fluid ounces or 277.4 cubic inches; equivalent to 4.546 liters.

**Galvani, Luigi** 1737–1798, Italian anatomist who discovered the relationship between electricity and animation (1771), while dissecting a frog. He touched an exposed nerve of the frog with his metal scalpel, and he observed the dead frog's leg kick as if in life. He had discovered the electrical nature of the nerve–muscle function, thus establishing the basis for the study of neurophysiology.

**galvanic** *Electricity.* describing a flow of electricity that results from chemical activity.

**galvanic cell** *Electricity.* a device that generates an electrical current, consisting of dissimilar metals in contact with each other and with an electrolyte. A **galvanic battery** is composed of one or more such cells.

**galvanic corrosion** *Materials.* the corrosion of a metal caused or accelerated by an electrical contact with a more noble metal or nonmetallic conductor in a corrosive electrolyte.

**galvanic series** *Materials.* a listing of metals according to their potential or ease of oxidation in a given environment, with lithium at the negative (least noble) end and platinum at the positive (most noble) end.

**galvanism** *History.* an earlier term for the production of electricity by chemical action; so called because this was studied by Galvani.

**galvanometer** *Measurement.* an instrument that measures a small electric current by measuring the mechanical motion derived from electromagnetic or electrodynamic forces produced by the current.

**game theory** *Economics.* a tool for evaluating optimal behavior strategies, often in economics, that is based on a theory of making the best choice from among available strategies given imperfect information.

**gamma** *Measurement.* a unit of measure for the strength of a magnetic field, equivalent to 0.00001 ($10^{-5}$) oersted.

**gas** *Chemistry.* **1.** one of the three fundamental forms of matter, along with liquids and solids. Unlike a solid (and like a liquid), a gas has no fixed shape and will conform in shape to the space available. Unlike a liquid, it has no fixed volume and will conform in volume to the space available. In comparison with solids and liquids, gases have widely separated molecules, are light in weight, and are easily compressed. **2.** short for GASOLINE.

**gas bubble trauma** *Environment.* a fatal or sublethal fish syndrome that occurs when gas in supersaturated water comes out of solution and reaches chemical equilibrium with atmospheric conditions, leading to the formation of bubbles within the tissue of aquatic organisms.

**gas cap** *Oil & Gas.* a term for the gas that accumulates in the upper portions of a reservoir where the pressure, temperature, and fluid characteristics are conducive to free gas.

**gas centrifuge** *Nuclear.* a system for the enrichment of uranium, in which uranium hexafluoride gas is spun in a centrifuge fast enough for the heavier isotope of uranium to be partially separated from the lighter one.

**gas chromatography** *Chemistry.* a process in which a gas mixture is passed through a cylinder and the different chemicals that make up the gas adhere to the surface of the cylinder at different intervals. This information is then used to determine the chemical makeup of the gas.

**gas condensate** see CONDENSATE.

**gas-cooled fast reactor (GFR)** *Nuclear.* an advanced nuclear reactor design in the development stage featuring a fast-neutron spectrum, helium-cooled reactor and a closed fuel cycle. The high outlet temperature (850°C) of the helium coolant makes it possible to deliver electricity, hydrogen, or process heat with high efficiency.

**gas-cooled graphite-moderated reactor** *Nuclear.* a type of graphite-moderated nuclear reactor distinguished by its use of a gas, usually carbon dioxide or helium, as a coolant, and its ability to use natural uranium as well as enriched uranium as a fuel.

**gas drive** *Oil & Gas.* a primary recovery mechanism for oil wells containing dissolved and free gas, in which the energy of the expanding gas is used to propel the oil from the reservoir into the wellbore.

**gaseous** *Materials.* existing as a gas; having the characteristics of a gas rather than a liquid or solid.

**gas fill** *HVAC.* an inert gas such as argon, used instead of air in a sealed space for greater insulation value; e.g., between dual windowpanes.

**gas guzzler** *Transportation.* a motor vehicle that has a poor fuel consumption performance; currently defined in the U.S. as a fuel economy of less than 22.5 miles per gallon. A **gas-guzzler tax** is a federal tax collected on the original manufacturer's sale of an automobile with fuel economy below this level (or another prescribed level).

**gas hydrate** *Materials.* a solid, crystalline, wax-like substance composed of water, methane, and usually a small amount of other gases, with the gases being trapped in the interstices of a water and ice lattice. Gas hydrates form beneath the permafrost and on the ocean floor under conditions of moderately high pressure and at temperatures near the freezing point of water. Gas hydrates have

***gas guzzler*** *The term "gas guzzler" was coined at the time when most Americans first became aware of the issue of fuel consumption, during the energy crisis of 1973–74. It describes large vehicles such as these, the U.S. norm at the time.*

been proposed as a possible future source of natural gas.

**gasification** *Conversion.* a thermochemical process that converts a solid or liquid fuel source (e.g., coal) to a gaseous fuel.

**gasifier** *Conversion.* any of various devices employed for the production of gas, such as a unit to produce synthesis gas from coal, or to produce gaseous fuel from solid or liquid biomass (e.g., agricultural wastes).

**gaslight** *History.* illumination by coal gas; the dominant form of urban lighting prior to the development of electric light. Thus, **gaslamp.**

**gasohol** *Oil & Gas.* gasoline and alcohol; a mixture typically containing about 90% unleaded gasoline and 10% ethyl alcohol; used as an alternative fuel in some automobile and truck engines.

**gas oil** or **gasoil** *Oil & Gas.* a liquid petroleum distillate having a viscosity intermediate between that of kerosene and lubricating oil. It derives its name from its original use in the manufacture of illuminating gas. It is now used for diesel fuel, heating fuel, and also as a feedstock.

**gas–oil ratio** *Oil & Gas.* the number of cubic feet of natural gas produced with a barrel of oil; an approximation of the composition of oil and gas from a reservoir, expressed in cubic feet of gas per barrel of oil at standard temperature and pressure.

**gasoline** *Oil & Gas.* a volatile liquid mixture of hydrocarbons that is obtained by refining petroleum and that is used as the fuel in most internal-combustion engines.

**gasoline additive** see ADDITIVE.

**gasoline blending** *Oil & Gas.* the technique of mixing up to 15 different hydrocarbon streams to produce a gasoline with specific characteristics, including vapor pressure, boiling points, sulfur content, color, stability, aromatics content, olefin content, octane rating, and also other governmental or market restrictions.

**gasoline equivalent gallon** *Measurement.* the volume of natural gas that equals the energy content of one gallon of gasoline, in order to compare equivalent volumes of the two fuels based on their energy content in Btu (since natural gas is normally measured in cubic feet).

**gasoline grade** *Oil & Gas.* the classification of gasoline by octane ratings; see REGULAR; MIDGRADE; PREMIUM.

**gasometer** *Measurement.* an apparatus used to contain and measure gas, particularly for chemical studies. Thus, **gasometry, gasometric.**

**gas shale** *Oil & Gas.* natural gas stored within an organically rich shale interval (reservoir), and requiring natural fractures and artificial well stimulation for acceptable rates of gas flow to take place.

**gas shift process** *Chemistry.* a process in which carbon monoxide and hydrogen react in the presence of a catalyst to form methane and water.

**gassing** *Storage.* **1.** the evolution of gas from one or more electrodes in the cells of a battery; this commonly results from local self-discharge or from the electrolysis of water in the electrolyte during charging. **2.** any evolution of gases during an event.

**gassing current** *Electricity.* the portion of charge current that goes into electrolytic production of hydrogen and oxygen from an electrolytic liquid; this current increases with increasing voltage and temperature.

**Gassner, Carl** 1839–1882, German scientist who invented the first dry cell battery (1888).

He used zinc as the container for the other elements and for the negative electrode. The electrolyte was absorbed in a porous material and the cell was sealed across the top. This led to the modern carbon–zinc battery.

**gas supersaturation** *Environment.* the overabundance of gases in turbulent water, as at the base of a dam spillway, that can cause a fatal condition in fish known as GAS BUBBLE TRAUMA.

**gassy** *Mining.* describing a situation in which a mine gives off methane or other gas in quantities that must be diluted with pure air to prevent an occurrence of explosive mixtures.

**gas thermometer** *Measurement.* a thermometer that utilizes the thermal properties of gas (as opposed to one that uses a liquid such as mercury), either by maintaining the gas at constant volume and observing changes in pressure, or vice versa.

**gas-to-liquid (GTL)** *Oil & Gas.* a process that combines the carbon and hydrogen elements in natural gas molecules to make synthetic liquid petroleum products, such as diesel fuel.

**gas-to-oil ratio** see GAS–OIL RATIO.

**gas turbine** *Conversion.* a rotary engine in which liquid or gaseous fuel is burned to produce electric power and heat. Hot combustion gases are passed to the turbine, where they expand to drive a generator and are then used to run a compressor.

**gas-turbine engine** *Transportation.* a type of internal combustion engine in which the shaft is spun by the pressure of combustion gases flowing against curved turbine blades located around the shaft.

**gasworks** *Coal.* a facility where coal gas is produced for use in lighting and heating.

**GATT** *Policy.* General Agreement on Tariffs and Trade; a multilateral trade agreement signed in 1948 among autonomous economic entities (not necessarily countries), aimed at expanding international trade as a means of raising global welfare. GATT was superseded by the World Trade Organization (WTO) in 1995.

**Gauss, Karl Friedrich** 1777–1855, German mathematician who proved fundamental theorems of algebra and arithmetic and made major contributions in areas such as geometry, number theory, and probability.

**gauss** *Measurement.* a unit of magnetic induction in the centimeter-gram-second system, equivalent to 1 maxwell per $cm^2$.

**Gauss law** *Electricity.* a description of the relation between the electric flux flowing out from a closed surface and the charge enclosed in the surface; the amount of charge within the surface is proportional to the flux.

**Gauss principle** *Physics.* the principle that a system of interconnected particles, subjected to a certain set of forces, will have nearly the same motion that the particles would have if disconnected from each other but still subjected to the same forces; i.e., the constraints on the overall system are minimal.

**Gay-Lussac, Joseph Louis** 1778–1850, French chemist noted for his original investigations into the behavior of gases. He deduced that gases at constant temperature and pressure combine in very simple numerical proportions by volume.

**GCM** general circulation model.

**GDP** gross domestic product.

**Gedser turbine** *Wind.* a 200 kW wind turbine built in 1957 on the Gedser coast of southern Denmark, considered to be the first wind turbine connected with an (asynchronous) AC generator to an electrical grid. It had huge 24 meter sails and was for many years the largest wind turbine in the world. It operated for 11 years with little maintenance and is now displayed at the Danish Museum of Electricity.

**GEF** Global Environment Facility.

**Geiger counter** *Nuclear.* a device that detects the passage of charged particles by means of the ionization of gas that they cause as they pass through a region; used to detect the particles produced in certain forms of radioactivity. [Invented by German physicist Hans *Geiger,* 1882–1947.]

**Geissler (mercury) pump** *Lighting.* a glass water pump in which a vacuum is created when mercury flows between two reservoirs, one of which is fixed and the other variable; this technology contributed to the success of Edison's first incandescent lamps, being used to remove air from the lightbulb. [Developed by German physicist Johann Heinrich Wilhelm *Geissler,* 1814–1879.]

**Geissler tube** *Electricity.* a gas-filled, dual-electrode discharge tube that emits brilliant and colorful fluorescent light when high voltage is passed through it, a demonstration of the luminous effects of discharges through rarefied gases. The Geissler tube led to the discovery of electrons.

**Gell-Mann, Murray** born 1929, U.S. physicist noted for his classification of subatomic particles and their interactions. In 1961 he and Israeli physicist Yuval Ne'eman independently proposed a scheme for classifying strongly interacting particles into a simple, orderly arrangement of families based on common general characteristics.

**gel-type battery** *Storage.* a lead-acid battery in which the electrolyte is immobilized in a gel; usually used for mobile installations in situations in which the battery will be subject to high levels of shock or vibration.

**GEM lamp** *Lighting.* General Electric Metallized lamp; an improved type of incandescent lamp employing a metal-like carbon filament, introduced in 1904. It provided greater energy efficiency than earlier carbon lamps.

**General Agreement on Tariffs and Trade** see GATT.

**general aviation** *Transportation.* a term for all facets of civil aviation except commercial air carriers.

**general circulation** *Earth Science.* the complete range of large-scale motions of the atmosphere and the ocean as a result of differential heating on a rotating earth, acting to restore the energy balance of the system through the transport of heat and momentum.

**general circulation model (GCM)** *Earth Science.* a computerized model of the dynamics and energetics of atmospheric circulation on a global scale, used to understand historic changes in climate, and to predict future changes due to forces such as the emission of greenhouse gases from human energy use. See below.

**General Electric Company (GE)** a large multinational energy corporation founded by Thomas Alva Edison in 1892 that originally was in the business of generating and selling electricity; it now works in many other areas of the energy industry.

**general equilibrium model** *Economics.* an approach that simultaneously considers all markets in an economy, allowing for feedback effects between individual markets. It is particularly concerned with the conditions that permit simultaneous equilibrium in all markets and with the properties of such an equilibrium. Also, **general equilibrium analysis.**

**general gas law** another term for the IDEAL GAS LAW.

**general circulation models** General circulation models (GCMs) are computer models of the climate system used to quantify climate response to a change in radiative forcing. The models describe the major components of the climate system such as the atmosphere, oceans, terrestrial biosphere, glaciers and ice sheets, and land surface. In order to project the impact of human perturbations on the climate system, GCMs represent the key processes operating in the climate system and the complex interactions and feedbacks among these components. These climate processes are represented in mathematical terms based on physical laws such as the conservation of mass, momentum, and energy. The complexity of the climate system means that the calculations from these mathematical equations can be performed in practice only by using a powerful computer. The equations are solved numerically at a finite resolution using a three-dimensional grid over the globe. Typical resolutions used for simulations are about 250 km in the horizontal and 1 km in the vertical. Many physical processes cannot be properly resolved at such coarse spatial resolutions and one resorts to including their average effect through parametric representations. GCMs form the basis for forecasting future climate change and its possible impacts by calculating how future emissions of greenhouse gases lead to changes into their atmospheric concentrations, how changes in concentration produce changes in global radiative forcing, how that radiative forcing in turn changes global mean temperature, and finally, how changes in temperature generate environmental changes such as global mean sea level rise. A future research challenge is to include chemistry, biology and ecology into the climate system models to improve the representation of the various processes and feedbacks in the system.

**Ranga Myneni**
Boston University

**general theory of relativity** see THEORY OF RELATIVITY.

**generating** *Electricity.* the fact of producing electric power; generation. Thus, **generating unit, generating station,** and so on.

**generation** *Electricity.* **1.** the process of producing electric energy by transforming other forms of energy (mechanical, chemical, or nuclear). **2.** the amount of electric energy produced, expressed in watt-hours (Wh). Thus, **generation unit, company,** and so on.

**generation attribute** *Electricity.* a nonprice characteristic of the electrical energy output of a generation unit, such as the unit's fuel type, emissions, vintage, and so on.

**generation mix** *Electricity.* a term for a diversity of generating units used to produce electricity (e.g., 40% coal, 30% hydropower, and 30% natural gas).

**generator** *Conversion.* any machine that converts mechanical energy into electrical energy.

**genotoxic** *Health & Safety.* causing genetic damage; describing a toxic agent that is capable of damaging DNA; e.g., certain pesticides. Thus, **genotoxin, genotoxicity.**

**genuine progress indicator (GPI)** *Sustainable Development.* an indicator proposed as an alternative to gross domestic product (GDP) as a measure of economic well-being; it is primarily based on the personal consumption data included in GDP, but with adjustments such as income distribution, the value of household work and volunteer work, and the costs of crime, resource depletion, and pollution.

**genuine savings** *Economics.* an indicator developed by the World Bank to assess an economy's sustainability; it aims to represent the value of the net change in the whole range of assets that are important for development, and it deducts the value of depletion of natural resources and pollution damages.

**geocentric** *Earth Science.* using the earth, or a certain location on earth, as a reference point for measurement.

**geocentric theory** *History.* the ancient belief that the sun and other bodies of the solar system revolve around the earth; stated in detail by Ptolemy about 140 AD and later displaced by the HELIOCENTRIC THEORY of Copernicus.

**geochemistry** *Earth Science.* the scientific study of the chemistry of the earth, including the rocks, sediments, and soil that constitute the solid earth and the fluids that compose the ocean, inland waters, and the atmosphere.

**geodesic dome** *Consumption & Efficiency.* a strong prefabricated enclosure constructed of lightweight bars forming a grid of polygons, with no internal supports; it is made of standardized parts that allow quick assembly and dismantling. The dome is energy efficient because it requires less building material and has less surface area, because heat loss due to wind turbulence is lessened, and because its shape minimizes air leakage. [Invented by Buckminster FULLER.]

**geodesy** *Earth Science.* the study of the size and shape of the earth, the measurement of terrestrial gravitational forces, and the location of fixed points on the earth's surface.

**geoengineering** *Earth Science.* an intentional planetary-scale interference in natural systems (e.g., the large-scale storage of $CO_2$ in the ocean interior), with the goal of promoting climate stabilization and reducing undesired anthropogenic change.

**geographic information system (GIS)** *Measurement.* the computer hardware, software, and technical expertise applied to assemble and analyze geographical data, especially the correlation of databases with graphic displays to present information; frequently employed in environmental studies.

**geography of energy** *Global Issues.* the study of energy development, transportation, markets, or use patterns from a geographical perspective.

**geological repository** *Nuclear.* a mined facility for the disposal of radioactive waste, using waste packages and the natural geological formations as barriers to provide waste isolation.

**geologic assurance** *Mining.* the relative degree of certainty with which the existence, abundance, and quantity of a given resource can be ascertained; e.g., coal or oil.

**geologic province** *Earth Science.* an extensive region that is characterized by a similar geologic history, or by particular structural or physiographical features throughout; e.g., a basin or delta.

**geologic storage** *Earth Science.* the long-term accumulation of a substance, such as carbon dioxide or radioactive waste, in a natural geologic formation; e.g., a sedimentary basin, seabed, or underground cavern. Also, **geologic sequestration.**

**geology** *Earth Science.* the study of the earth in terms of its development as a planet since its origin, including the history of its life forms, the materials of which it is made, the processes that affect these materials, and the products that are formed of them.

**geomagnetic** *Earth Science.* of or relating to geomagnetism; having properties of geomagnetism.

**geomagnetic field** *Earth Science.* the magnetic field observed in and around the earth.

**geomagnetic storm** *Earth Science.* a worldwide disruption of the earth's magnetic field, distinct from regular diurnal variations; caused by ionic disturbance from solar events.

**geomagnetism** *Earth Science.* the various magnetic phenomena that are generated by the earth and its atmosphere, and by extension the magnetic phenomena in interplanetary space.

**geometric concentration ratio** *Solar.* the ratio of a solar collector aperture area to the absorber area.

**geophysics** *Earth Science.* the scientific study of the physical characteristics and properties of the solid earth, its air and waters, and its relationship to space phenomena.

**geopolitics** *Global Issues.* **1.** the study of the influence of geography on the course of politics and history. **2.** the influence of geographic factors, population distribution, and natural resources on a nation's foreign policy; e.g., the effort of a nation to control a canal, trade route, oil supply, and so on.

**geopressure** *Earth Science.* pressure beneath the surface of the earth, especially a pressure of greater than usual strength existing in a subsurface formation; e.g., a fluid deposit that is under very high pressure because it bears part of the overburden load.

**geopressured** *Earth Science.* **1.** under very high pressure beneath the earth's surface; usually defined as a fluid existing at pressure greater than 0.465 psi for each vertical foot of depth. **2.** *Geothermal.* describing reservoirs of highly pressurized geothermal fluids trapped underground that offer various energy potentials; e.g., hydraulic energy from the high pressure, or thermal energy from the high temperature of the fluids, or natural gas dissolved in the fluid. Also, **geopressurized.**

**geopressured brine** *Geothermal.* **1.** a reservoir of subsurface saline water existing at very high pressure and offering the potential for geothermal energy. **2.** specifically, such a reservoir that is completely saturated with natural gas and thus can be both a geothermal source and an unconventional gas source.

**Georgescu-Roegen, Nicholas** 1906–1994, Hungarian-born mathematician and economist who described the second law of thermodynamics as playing a central role in production theory, with implications for the sustainability of economic growth. His work became a cornerstone of the fields of ecological and evolutionary economics.

**geosphere** *Earth Science.* the physical earth; a term for the solid mass (lithosphere) of the planet, or for the lithosphere, hydrosphere, and atmosphere as a whole.

**geostationary** *Earth Science.* apparently fixed in position in relation to the earth; describing an artificial satellite that travels above the equator and at the same speed as the earth rotates, so that it constantly appears at the same point in the sky. A **geosynchronous** satellite has an orbit similar to a geostationary one, except that it does not necessarily lie in the earth's equatorial plane.

**geostrophic** *Earth Science.* of or relating to the deflective forces produced by the earth's rotation.

**geostrophic current** *Earth Science.* a wind, ocean current, or other such movement in which the horizontal force is exactly balanced by the Coriolis force. Similarly, **geostrophic flow, geostrophic force.**

**geostrophy** *Earth Science.* the balance between the Coriolis force and the horizontal pressure gradient that determines the first-order circulation patterns of the open ocean.

**geosynchronous** see GEOSTATIONARY.

**geotectonics** see TECTONICS.

**geothermal** **1.** relating to or caused by the heat of the earth's interior. **2.** describing an energy system that makes use of this heat.

**geothermal agriculture** *Geothermal.* the use of geothermal heat in agriculture; e.g., the use of low-temperature geothermal water to warm irrigation water or to sterilize soil.

**geothermal aquaculture** *Geothermal.* the use of geothermal heat in fish farming; e.g., the use of geothermally heated water to provide a controlled environment for the husbandry of marine organisms.

**geothermal cooling** *HVAC.* the use of naturally cooler water or air to lower the temperature of a building, as opposed to air or water that is artificially cooled.

**geothermal direct use** *Geothermal.* the direct use of geothermally heated water; e.g., to heat a home or greenhouse or to provide warmer water for fish farming, as opposed to an indirect use such as driving a turbine to generate electric power.

**geothermal drilling** *Geothermal.* the process of drilling a well to explore for or extract geothermal energy, or to reinject thermal wastewater in the ground after power generation.

**geothermal dry steam** see DRY STEAM.

**geothermal energy** *Geothermal.* energy in the form of natural heat flowing outward from within the earth and contained in rocks, water, brines, or steam. This heat is produced mainly by the decay of naturally occurring radioactive isotopes of thorium, potassium, and uranium in the earth's core.

**geothermal flash steam** see FLASH STEAM.

**geothermal gradient** *Geothermal.* the rate of temperature change in soil and rock from the surface to the interior of the earth; on the average, estimated to be an increase of about 10°C per km.

**geothermal heat(ing)** see GEOTHERMAL ENERGY.

**geothermal heat pump** another term for a GROUND-SOURCE HEAT PUMP.

**geothermal hot spring** see HOT SPRING.

**geothermal mining** *Geothermal.* **1.** the extraction of valuable minerals from geothermal fluids; e.g., market-grade silica from a geothermal brine. **2.** the process of purposely transporting geothermal energy from beneath the earth for human use; e.g., the building of a well and pipeline system to bring heated water to a power plant.

**geothermal plant** *Geothermal.* a power plant utilizing geothermal energy, such as a binary cycle plant or dry steam plant.

**geothermal reservoir** *Geothermal.* a subsurface system consisting of a large volume of hot water and steam trapped in porous and fractured hot rock underneath a layer of impermeable rock; some reservoirs can be commercially developed as an energy source.

**geothermal silica** see GEOTHERMAL MINING.

**geothermal system** *Geothermal.* **1.** a localized geological environment in which circulating steam or hot water carries some of the earth's natural internal heat flow close enough to the surface to be employed for productive human use. **2.** any technological system that makes use of this heat as an energy source; e.g., to power an electrical power plant or to heat or cool a building.

**geothermometry** *Geothermal.* the study of the earth's heat and subsurface temperatures.

**Gerard of Cremona** c. 1114–1187, Italian scholar whose publications in the sciences presented advanced Greek and Arab thought to an intellectually impoverished Europe during the Middle Ages.

**Gesner, Abraham** 1797–1864, Canadian scientist who first devised a method to distill kerosene from petroleum (1849). He also wrote an early standard reference work for petroleum refining.

**gettering** *Electricity.* the process of removing device-degrading impurities or defects from the active circuit regions of a semiconductor wafer.

**geyser** *Earth Science.* a natural hot spring or fountain that periodically discharges a column of hot water and steam into the air.

**Geysers, The** *Geothermal.* an active geothermal area within the Clear Lake volcanic field in northern California, noted as one of the earliest sites to employ geothermal energy to generate electricity (1921) and still one of the world's most productive geothermal fields.

**GFR** gas-cooled fast reactor.

**Ghawar** *Oil & Gas.* a huge oil field in Saudi Arabia, currently the largest conventional oil field in the world. See next page.

**GHE** ground heat exchanger.

*geyser* *Yellowstone National Park in Wyoming is noted for its surface geothermal activity, such as this Riverside Geyser on the Firehole River.*

**ghg** greenhouse gas.

**giant kelp** see KELP.

**Gibbs, Josiah Willard** 1839–1903, American chemist and mathematical physicist who first formulated a concept of thermodynamic equilibrium of a system in terms of energy and entropy. In doing so, he helped found the field of thermochemistry.

**Gibbs free energy** *Thermodynamics.* the minimum thermodynamic work (at constant pressure) needed to drive a chemical reaction (or, if negative, the maximum work that can be done by the reaction). See next page.

**Gibbs–Helmholtz equation** *Thermodynamics.* an equation used to determine the effect that temperature and pressure have on the ability of a chemical reaction to perform useful work.

**Giffard, (Jules) Henri** 1825–1882, French inventor who achieved the first powered and controlled flight. His hydrogen-filled airship had a steam engine that drove a three-bladed propeller and was steered by a sail-like rudder. In 1852 it flew over a distance of about 17 miles (27 km).

**giga-** *Measurement.* a prefix meaning "one billion", or $10^9$; symbol G.

**gigahertz (GHz)** *Electricity.* a unit of frequency equal to 1 billion hertz, or 1 million kilohertz.

**gigantothermy** *Biological Energetics.* the ability of an organism to maintain a constant, relatively high body temperature due to its having a large body and insulation. Larger animals have a relatively low surface area to body volume ratio, so they retain heat better than smaller animals.

**Ghawar** Ghawar is by far the largest and most productive oilfield in the world. It is located in the eastern region of the Kingdom of Saudi Arabia in the Empty Quarter desert (about 200 km from the capital, Riyadh). This super-giant "king of kings" oilfield was discovered by the Arabian–American Oil Company (now Saudi Aramco) in 1948. The northern portion of Ghawar lies about 90 km west of the Arabian Gulf. From its northern extremity, the field extends southward some 230 km as essentially one long continuous anticline, about 40 km across at its widest point. Although Ghawar is a single oilfield, it is divided into six areas, namely Fazran, Ain Dar, Shedgum, Uthmaniyah, Haradh and Hawiyah. Geologically, the field is categorized as a fairly simple underground structure with a complete closure, a typical Middle East reservoir of porous limestone and dolomite. The oil comes almost entirely from a producing zone known as Arab D, about 2100 meters below the surface. Currently, the estimated proven reserves of Ghawar are more than 70–85 billion barrels Arabian Light crude oil (33 API). Commercial production from Ghawar began in 1951 and reached a peak of 5.7 million barrels per day in 1981. This was the highest sustained oil production rate achieved by any single oilfield in world history. Today, Ghawar remains the world's most important oilfield. Its daily production exceeds 5 million barrels per day, which is more than half of Saudi Arabia's cumulative oil production.

**Abdullah M. Aitani**
King Fahd University of Petroleum and Minerals
Saudi Arabia

**Gibbs free energy (G)** is a measure of the maximum available work that can be derived from any system under conditions of constant temperature (T) and pressure (P). G is a thermodynamic "state function", i.e., an equilibrium property that depends *only* upon the conditions—such as T, P and electrical, magnetic and gravitational fields—imposed on the system being considered, and not on that system's past history. Since absolute G values cannot be determined, *changes* in G as a system goes from one state to another become the main focus of attention. These ΔG ("delta-G") values are highly informative. If $\Delta G$ (= $G_{\text{final state}} - G_{\text{initial state}}$) is negative, the process observed liberates energy: it will occur spontaneously and can be harnessed to do useful work. For chemical changes, tabulated standard free energy values can be used to predict the direction and energy yield of a particular reaction. For example, it's easy to calculate that if one burns a mole (114 g) of isooctane to carbon dioxide and water, a total of 5226 kJ (kilojoules) of Gibbs free energy will be released, i.e., $\Delta G = -5226$ kJ/mol. This large negative value predicts a spontaneous process that proceeds completely to products. Performed in an internal combustion engine, about one-third of the ΔG will be recovered. A substantially larger fraction could be extracted by a fuel cell. J. Willard Gibbs first defined the free energy function that bears his name in the landmark theoretical papers of 1876 and 1878 that have led most authorities to rank him as America's greatest native-born scientist.

**Scott C. Mohr**
Boston University

**gigawatt (GW)** *Electricity.* a unit of electric power equal to 1 billion watts, or 1 million kilowatts.

**gigawatt hour (GWhr)** *Electricity.* a unit of energy defined as the amount of energy delivered by a source of one gigawatt over a period of one hour.

**Gilbert, William** 1544–1603, English physician and inventor who published (1600) a standard work on electrical and magnetic phenomena, influencing astronomers such as Johannes Kepler and Galileo.

**Gini coefficient** *Measurement.* a measure of the degree of income inequality in a given society, based on a scale between zero (income is equally distributed among all those in the society) and one (a single person has all the income). [Developed by Italian statistician Corrado *Gini,* 1884–1965.]

**GIS** geographic information system.

**glacial** *Earth Science.* **1.** describing the action, features, movements, and materials produced by or derived from glaciers or ice, or a region covered by glaciers or ice. **2.** describing a geologic time period marked by an ice age.

**glacial epoch** or **period** see ICE AGE.

**glacial maximum** *Earth Science.* the position or time of the greatest advance of a glacier, or the greatest extent of a particular period of glaciation; e.g., the extreme advance of Pleistocene glaciation toward the equator.

**glacial rebound** *Earth Science.* the adjustment of previously glaciated areas after a glacial retreat; e.g., the uplift of Scandinavia after the most recent glaciation.

**glaciation** *Earth Science.* the covering and alteration of the earth's surface by glaciers, including such effects as erosion, deposition, leveling, change of drainage systems, and creation of numerous lakes.

**glacier** *Earth Science.* a large mass of land ice that is formed by the compaction and recrystallization of snow, and which flows slowly down-slope or outward in all directions under its own weight.

***glacier*** *The tidewater face of Muir Glacier, Glacier Bay National Park, Alaska (one of many natural sites in the U.S. named for the famed environmentalist John Muir).*

**glare** *Lighting.* excessive or unpleasant brightness from a direct light source.

**Glaser, Peter** U.S. space scientist known for promoting the idea of generating electricity in space for use on earth, by collecting energy from the sun. Though technically possible, cost considerations have prevented the implementation of such a system.

**glass** *Materials.* a brittle, noncrystalline, usually transparent or translucent material that is generally formed by the fusion of dissolved silica and silicates with soda and lime; a universally produced material for such uses as windows, bottles and containers, lenses and instruments, and many other purposes.

**glass mat battery** see ABSORBED GLASS MAT BATTERY.

**Glauber's salt** *Materials.* $Na_2SO_4 \cdot 10H_2O$, large transparent crystals, needles, or granular powder; soluble in water and insoluble in alcohol; used in solar heat storage and air-conditioning. [First formulated by German chemist Johann Rudolf *Glauber,* 1604–1688.]

**glazing** *Materials.* **1.** the process of fitting a window frame with glass. **2.** a covering of transparent or translucent material (typically glass or plastic) used for admitting light. **3.** *Solar.* a material that admits light and also inhibits the escape of heat generated by the incident light.

**glazing percentage** *HVAC.* the ratio between the window glass area of a home or building and the total area of the space to be heated or cooled.

**glider** *Transportation.* a fixed-wing aircraft designed to glide and sometimes to soar, usually not having any form of onboard power plant.

**Global 2000** *Global Issues.* a report on global environmental trends commissioned by U.S. President Jimmy Carter, published in 1980. The general conclusion of the report was that if present trends were to continue, the world in 2000 would be more crowded, more polluted, less stable ecologically, and more vulnerable to disruption.

**global air pollution** *Global Issues.* increased air pollutant concentrations of long-lived compounds, which affect air quality on a global scale regardless of the location of the emissions.

**global ecology** *Global Issues.* the study of the relationship between organisms and their environment on a global scale, including such issues as climate change, sustainable development, and wildlife population levels.

**Global Environment Facility** (est. 1991), an independent financial organization that provides grants to developing countries for projects that benefit the global environment and promote sustainable livelihoods in local communities.

**globalism** *Global Issues.* **1.** the policy or practice of conducting an activity on an international basis rather than a national or local one; e.g., a corporation that sells its goods to markets throughout the world and that locates its facilities and workforce in various countries according to the need at hand. **2.** another term for GLOBALIZATION.

**globalization** *Global Issues.* **1.** the tendency over time for the nations and citizens of the world to become more closely interconnected, as a result of factors such as increased trade and travel, higher rates of immigration, and the spread of mass media such as film and television. **2.** any specific instance of this trend; e.g., the globalization of the Internet from its beginning as a small network in the U.S.

**global pollutant** *Global Issues.* a pollutant that travels to the upper atmosphere and has effects on a global scale; e.g., ozone-depleting gases.

**global public good** *Global Issues.* a public benefit that is strongly universal in terms of nations, individuals (preferably accruing to all population groups), and generations (both current and future); e.g., the eradication of a formerly widespread disease.

**global radiation** *Solar.* total solar radiation; the sum of direct, diffuse, and ground-reflected radiation (the third of these usually being insignificant compared to the first two). Similarly, **global horizontal radiation, global irradiance, global normal radiation, global solar radiation,** and so on.

**global warming** *Climate Change.* **1.** any general increase in the earth's surface temperatures over an extended period of time. **2.** the proposition that in the current era, global temperatures have been slowly rising. This is attributed to the phenomenon of atmospheric gases such as carbon dioxide absorbing long-wave radiation as it is reradiated off the earth's surface rather than this heat escaping into space, thus making the atmosphere warmer. See below.

**global warming potential (GWP)** *Climate Change.* an index that represents the combined effect of the differing times that greenhouse gases remain in the atmosphere and their relative effectiveness in absorbing outgoing infrared radiation.

**global warming** The theory that global climatic change, surface warming in particular, is increasingly caused by human emissions of greenhouse gases, mainly carbon dioxide from fossil fuel burning. Aerosol particles also emitted from fossil fuel and biomass burning are believed to result in cooling, partly masking global warming. However, the dominant effect is warming the entire troposphere (the atmosphere below about 11 kilometers containing most of the air's mass) by 1.5 to 4.5 degrees Celsius (eventually) if the preindustrial atmospheric $CO_2$ concentration of 270 ppm were to double. Without natural greenhouse gases (water vapor, carbon dioxide and ozone), the earth would be a frozen iceball. As $CO_2$ has climbed above 360 ppm from fossil fuel burning, planetary mean temperatures rose to the highest in millennia; with warming amplified in the Arctic, thawing permafrost, thinning sea ice and creating marked recession of glaciers. Also observed, as predicted, is subsurface warming from heat penetrating the sea. Likewise, satellite-derived troposphere temperatures are rising, and the stratospheric cooling, the latter consistent with a $CO_2$ greenhouse effect. Potential warming of the earth's surface if the bulk of fossil fuel resources are burned could be as much as 10 degrees Celsius, comparable to mid-Cretaceous climates when the poles were deglaciated and dinosaurs roamed Antarctica. Adverse impacts of such dramatic climate shifts on the biosphere and human economy are incalculable. The challenge for energy technology is slowing, stabilizing, and if possible reversing, fossil fuel burning induced global warming, requiring a revolutionary change in the global energy system.

**Marty Hoffert**
New York University

**glow discharge** *Electricity.* a glow inside an electron tube due to the ionization of gas caused by a discharge of electricity. Thus, **glow-discharge tube.**

**glowplug** *Transportation.* an electrically heated wire used in diesel engines to preheat the air in the cylinder to facilitate starting the engine, as in cold conditions when the compression process alone may not raise the air to a sufficient temperature for ignition.

**Gluckauf** *History.* the first oil tanker of modern design, built at Newcastle, England (1866) for a German shipping company; it differed from previous versions in having the liquid cargo contained directly in tankage integral with the hull, rather than in individual barrels.

**glycerol** *Chemistry.* a trihydroxy sugar alcohol that is the basis of many lipids and an important intermediate in carbohydrate and lipid metabolism.

**glycogen** *Biological Energetics.* the chief storage form of glucose within the body, readily available to supply the tissues with an oxidizable energy source.

**glycogenesis** *Biological Energetics.* the synthesis or formation of glycogen.

**glycol-cooled** *HVAC.* describing a type of air-conditioning system that uses freon as a refrigerant and a water/glycol solution as a condensing medium. The glycol keeps the solution from freezing during winter operation.

**glycolysis** *Biological Energetics.* the anaerobic breakdown of carbohydrates, in which a molecule of glucose is converted by a series of steps to two molecules of lactic acid, yielding energy in the form of adenosine triphosphate (ATP).

**GMO** *Health & Safety.* genetically modified organism; a plant or animal whose genetic makeup has been altered by humans in some way, especially by modern DNA technology. Many industries support the development and use of GMOs as foods, while questions of safety have been raised by others.

**GMT** Greenwich Mean Time.

**GNI** gross national income.

**gnomon** *Solar.* the part of a sundial that casts a shadow on the disk, usually a rod or fin pointed at the celestial pole.

**GNP** gross national product.

**gobar gas** *Biomass.* another term for BIOGAS, used especially in India; a term derived from the Hindi word for cow dung.

**Goddard, Robert Hutchings** 1882–1945, U.S. physicist considered one of the founders of modern rocketry, which launched an entirely new field of science and engineering. He constructed and successfully tested the first rocket using liquid fuel (1926), and went on to develop a wide range of technologies that produced more than 200 patents in rocketry. He also made early proposals for transportation by magnetic levitation.

**gold** *Chemistry.* a metallic element having the symbol Au, the atomic number 79, an atomic weight of 196.967, a melting point of 1063°C, and a boiling point of 2800°C. It is a precious, highly malleable, high-density metal that is widely used for jewelry and (formerly) for making coins; it also has many industrial applications.

**Goodenough, John B.** U.S. engineer who produced fundamental advances in battery technology with his research on lithium–ion battery technology. He developed new cathode materials for these batteries that greatly improved their performance.

**Goodyear, Charles** 1800–1860, U.S. inventor who developed a process to produce VULCANIZED RUBBER so that it could be applied to a vast array of industrial uses, especially automobile tires. This process revolutionized the rubber industry.

**Goose Creek** *Oil & Gas.* an oil field on Galveston Bay, the first offshore drilling operation in Texas and the second in the U.S. overall; the site of one of the largest oil refineries in the U.S.

**Gorrie, John** 1803–1855, U.S. physician, a pioneer in refrigeration and air-conditioning who was granted the first U.S. patent for mechanical refrigeration (1851). His apparatus, initially designed to cool yellow fever patients, was an expanding-air cooling machine, similar in concept to the modern refrigerator.

**governor** *Consumption & Efficiency.* **1.** a control device on a machine or engine that is used to regulate speed, pressure, or temperature automatically (as by controlling the fuel supply). **2.** a similar device that prevents a machine or engine from operating at above a certain pre-set limit.

**GPI** Genuine Progress Indicator.

**GPP** gross primary production.

**grade** see COAL GRADE; GASOLINE GRADE.

**gradient solar pond** see SALT GRADIENT POND.

**grain** *Biomass.* **1.** the seed or seed-like fruits of wheat, oats, corn, or other cereal grasses. **2.** any of the plants on which these seeds grow. **3.** *Materials.* any small, hard particle though of as similar to such a seed, as of pollen, sand, salt, and so on. **4.** the orientation or appearance of the composite particles of a solid substance, such as wood. **5.** *Measurement.* the smallest unit of avoirdupois weight, originally based on the typical weight of a grain of wheat, equivalent to 64.8 milligrams. One pound avoirdupois equals 7000 grains.

**gram** *Measurement.* a basic unit of mass in the metric system, equal to the weight of one cubic centimeter of pure water at a standard temperature of 4°C; equivalent to 0.035 avoirdupois ounce.

**Grameen Shakti** a pioneering institution in the field of microenterprise financing for the poor in Bangladesh; originally focused on agricultural, craft-making, and housing loans, it now also finances investments in solar, wind, and biogas power generation.

**gramme** another spelling of GRAM.

**Gramme dynamo** *Electricity.* an early form of continuous-current electrical generator (1869) able to function as an electric motor; a major step in the development of commercial electric power. [Devised by Belgian engineer Zenobe-Theophile *Gramme,* 1826–1901.]

**grams per mile** *Transportation.* a mixed measure for the weight of pollutants emitted into the atmosphere in vehicle exhaust gases; certain pollution control laws set maximum limits for each exhaust pollutant in terms of grams per mile.

**Grandcamp** *Health & Safety.* the ship that was the source of what became known as the TEXAS CITY DISASTER.

**Grand Coulee Dam** *Hydropower.* a dam built in 1942 in Washington State, U.S., to harness the power of the Columbia River. It was the largest hydroelectric dam in the world at that time (in fact the largest concrete structure of any kind), and still remains the largest dam in North America.

*Grand Coulee Dam The Grand Coulee Dam produces up to 6.5 million kilowatts of power and irrigates over half a million acres of farmland in the Columbia River Basin.*

**grandfathering** *Policy.* a method for the initial distribution of permits in an emission trading scheme, in which the government gives away tradable permits to the regulated private entities for free, usually based on their historical emissions. Thus, **grandfather clause.**

**Grandpa's Knob** *Wind.* a hilltop near Rutland, Vermont noted as a site of high winds; during the World War II era a 1.25 MW wind generator was erected here to provide power to the local grid. This is considered the earliest use of wind energy to supply public power on a large scale.

**grand unified theory** see UNIFIED FIELD THEORY.

**Granger causality** *Economics.* a statistical relationship between two variables such that historical observations of one variable will help predict current values of the other. For example, energy use is said to "Granger cause" GDP, if the value for energy use in year $y - 1$ can explain the value for GDP in year $y$ in a statistically significant way. [Described by British economist Clive *Granger,* born 1934.]

**graphite** *Materials.* a black to gray mineral, a crystalline form of carbon found in intensely metamorphosed rocks; widely used as a lubricant and in pencils and paints, and also in very pure form as a moderator in gas-cooled reactors

**graphite-moderated** *Nuclear.* describing a particular type of nuclear reactor distinguished by its use of graphite instead of water as the moderator.

**graphite storage** *Storage.* the use of a material composed of nanosized fibers of graphite to adsorb and store hydrogen; a proposed technology for hydrogen-powered vehicles.

**Grätzel cell** *Photovoltaic.* another name for a DYE-SENSITIZED SOLAR CELL, from its developer, Swiss scientist Michael Grätzel.

**gravitation** *Physics.* a force of attraction between any two bodies having mass, the magnitude of which is dependent on the product of the two masses and the inverse square of the distance between them. Thus, **gravitational energy** or **force.** See also GRAVITY.

**gravitational acceleration** *Physics.* acceleration due to gravity; at or near the earth's surface, denoted by $g$ and approximately equal to 9.8 meters (32.2 feet) per second per second (the value varies slightly with latitude and elevation).

**gravitational constant** *Physics.* the fundamental constant that, when multiplied by the product of the masses of two bodies and divided by the square of their separation distance, will give the gravitational attraction between them.

**gravitational field** *Physics.* the force field created around massive bodies that causes attraction of other massive bodies. Thus, **gravitational force.**

**gravitational potential energy** *Physics.* the energy that an object possesses because of its position in a gravitational field, especially an object near the surface of the earth where the gravitational acceleration can be assumed to be constant, at about 9.8 $m/s^2$.

**gravitational redshift** see EINSTEIN SHIFT.

**gravity** *Physics.* the phenomenon by which massive bodies are attracted to one another. In popular usage, *gravity* is often used to describe the force of attraction itself, but technically *gravitation* is the force and *gravity* is the observed effect of this force.

**gravity convection** *Solar.* the natural movement of heat that occurs when a warm fluid rises and a cool fluid sinks under the influence of gravity.

**gravity dam** *Hydropower.* a large dam constructed of concrete and/or masonry that relies entirely on its own mass for stability against the forces of the water behind it.

**Gray, Elisha** 1835–1901, U.S. educator famous for contesting the invention of the telephone with Alexander Graham Bell. On the same day that Bell filed his application for a telephone patent, Gray applied for a caveat announcing his intention to file a claim for a patent within three months. After years of litigation, Bell was legally named as the inventor of the telephone, although the issue is still debated to this day.

**Gray, Stephen** 1666–1736, British chemist who conducted experiments establishing that some materials conduct electricity better than others and that static electricity travels on the surface of objects, rather than through the interior. This is considered the first discovery of the principle that electricity can flow.

**gray** *Nuclear.* the amount of energy absorbed in a material, symbol Gy; an SI unit equal to an energy absorption of 1 J/kg of absorbing material, i.e., 1 Gy = 1 J/kg. An older unit, the rad, is still often used, particularly in the U.S. (100 rad = 1 Gy). [Named for English scientist L. H. (Hal) *Gray,* 1905–1965.]

**graybody** or **gray body** *Thermodynamics.* a body that has the spectral distribution of a BLACKBODY, except that its spectral emission intensity is a constant fraction less than unity of the blackbody value.

**gray economy** *Economics.* business activities that are not accounted for by official statistics, including both illegal activities (the black market) and activities that are in themselves legal but are not accurately reported to avoid taxation (e.g., barter activities or unrecorded cash transactions).

**gray water** or **graywater** *Consumption & Efficiency.* wastewater generated by water-using fixtures and appliances in a household or building, excluding toilets, urinals, kitchen sinks, and garbage disposals; can be reused for such purposes as landscape irrigation.

**Great Blackout** *History.* a term for an event in North America (Nov, 9–10, 1965) in which an area of about 80,000 square miles of the northeastern U.S. and Canada suffered a loss of electrical power. The failure especially affected the New York City area, with an estimated 800,000 people stranded in the city's subway system.

**Great Smog of London** *Health & Safety.* a toxic event occurring in 1952 in London, after an exceptionally cold winter forced homes and factories to burn large quantities of coal. Consequently a temperature inversion formed, trapping pollutants above the ground. More than 4000 people died from respiratory ailments within the following week, leading to major reforms in air quality controls for Britain.

**Great Western** *History.* a paddle steamer built by English engineer I. K. Brunel, the largest ship of its era and the first transatlantic passenger steamship in regular service (1837).

**Greek fire** *History.* an ancient incendiary weapon ("liquid fire") similar to modern napalm; of unknown composition but thought to consist of lighter elements of petroleum mixed with quicklime; used in warfare in the Middle East about 700 A.D.

**green** *Environment.* **1.** a term for any activity, policy, organization, attitude, and so on that is intended to protect and preserve the natural environment. **2.** specifically, referring to the use of renewable energy sources such as solar and wind power, as opposed to fossil fuels such as oil and coal. [From the association of the color of green plants with a healthy and fertile environment.]

**green accounting** another term for ENVIRONMENTAL ACCOUNTING.

**green algae** *Ecology.* the Chlorophyta, a diverse group of algae capable of photosynthesis, typically living in aquatic habitats and having a greenish color resulting from the presence of chlorophyll pigments. Proposed as a source of biomass energy.

**green car** *Transportation.* a vehicle that uses a form of alternative energy rather than gasoline, or as a supplement to gasoline, such as a hybrid electric vehicle.

**green certificate** *Policy.* a system in which generators of electricity from renewable resources receive a certificate for a predetermined unit of energy produced; such certificates have a market value and can be bought and sold, and thus provide a financial incentive for the use of renewable energy. Thus, **green certificate trading.** See below.

**green certificate** Green certificates are electronic or paper representations of the environmental attributes of electricity generated from approved 'green' or 'renewable' energy power plants. The environmental attributes—for example, reduced emissions of pollutants—are determined by comparing the environmental impact of the green or renewable energy power plant with the environmental impact, on average, of all power plants in the system. Social attributes are occasionally included as well; they could include, for example, reduced displacement of people. In many cases, each green certificate has a face value of one megawatt hour (MWh) of electricity generated at a qualified green or renewable energy power plant. The certificates distinguish the environmental attributes of the electricity from the electrons (that is, the 'power attributes') of the electricity. Each—that is, the green certificates and the electrons—can usually be sold separately in different markets. The purchase of green certificates can help electricity producers and consumers meet their renewable energy obligations (where applicable), even when not sourcing electricity directly from a renewable energy power plant. In jurisdictions without obligatory renewable energy targets, green certificates—marketed to, for example, firms advancing 'corporate social responsibility' goals—can help stimulate voluntary purchases of renewable electricity. Alternative terms used to describe green certificates include green tags, renewable energy certificates or credits, or tradable renewable certificates. Green certificates have been used, for example, in many countries in Europe, as well as Australia and the United States.

**Ian H. Rowlands**
University of Waterloo

**green city** another term for a SOLAR CITY.

**green diesel** *Oil & Gas.* a term for diesel fuel that is mixed with a biodiesel cetane booster.

**greenfield** *Environment.* an area of property that has not previously been used for industrial activity or other commercial development and thus is relatively undisturbed from an environmental perspective. See also BROWNFIELD.

***greenfield*** *The term "greenfield" includes not only land in its natural state but also agricultural sites such as this; farmland near urban areas is often a candidate for development.*

**greenhouse** *Consumption & Efficiency.* **1.** a glass-enclosed, climate-controlled structure in which young or tender plants are cultivated, used especially to provide warmer temperatures for growing plants during cold weather. Thus, **greenhouse heat(ing). 2.** *Earth Science.* the concept of the earth as being similar to such a structure, because of its ability to maintain a higher temperature in its atmosphere than in the surrounding outer space (the greenhouse effect).

**greenhouse effect** *Earth Science.* the trapping of heat by greenhouse gases (see next) that allow incoming solar radiation to pass through the earth's atmosphere but prevent the escape to outer space of a portion of the outgoing infrared radiation from the surface and lower atmosphere. This process has kept the earth's temperature approximately 33°C warmer than it would be otherwise; it occurs naturally but may also be enhanced by certain human activities; e.g., the burning of fossil fuels.

**greenhouse gas (GHG)** *Earth Science.* any of the gaseous constituents of the atmosphere, both natural and anthropogenic, that absorb and emit radiation at specific wavelengths within the spectrum of infrared radiation emitted by the earth's surface, the atmosphere, and clouds. This causes the greenhouse effect (see previous). See below.

**greenhouse gas** A gas that is largely transparent to the visible radiation reaching Earth from the sun but absorbs the thermal, infrared radiation that is radiated outward from the earth's surface. The energy absorbed in the atmosphere is re-radiated in all directions, warming the earth's surface. The greenhouse gases $H_2O$ and $CO_2$ occur naturally in the atmosphere and make earth's surface temperature approximately 33 degrees Celsius warmer than it would otherwise be. Human activities, such as burning fossil fuels, are now releasing additional greenhouse gases to the atmosphere, creating an enhanced greenhouse effect. Greenhouse gases include $H_2O$, $CO_2$, $N_2O$, ozone, the chlorofluorocarbons, and $SF_6$. Greenhouse gases exist in the atmosphere as molecules of three or more atoms. The atoms in the gas molecule vibrate with respect to each other, and the molecules rotate in space. Transitions between different energy levels of vibration and rotation in the molecule are quantized and occur with the absorption or release of infrared radiation. Fourier is generally credited with describing (in 1827) how the earth's surface temperature is controlled by the differential absorption of visible and infrared radiation, but use of the phrase "greenhouse effect" to characterize this phenomena seems to have appeared about 1950. Although the phrase has become entrenched in our vocabulary, it is a bit of a misnomer as the physical processes of the two phenomena are quite different. Greenhouse gases in the atmosphere act to restrict the radiative transfer of energy whereas a glass greenhouse acts primarily by restricting the convective transfer of energy.

**Gregg Marland**
Oak Ridge National Laboratory

**greenhouse skeptic** *Climate Change.* a term applied to scientists and nonscientists (e.g., climatologist Robert Balling, science fiction author Michael Crichton) who do not believe that there is sufficient evidence that global warming is now occurring, or if it is occurring, that there is not sufficient evidence that this is the result of human activity rather than natural processes, or that a warmer climate will negatively affect environmental conditions on the planet.

**Greenpeace** (est. 1971), an international environmental activist group known for its public campaigns to protect the environment. It campaigns to stop climate change, protect ancient forests and the oceans, prevent whaling, and encourage sustainable trade.

**green power** *Renewable/Alternative.* **1.** a term for electricity produced from renewable sources or in such a way that the process results in less adverse impact on the environment than fossil or nuclear sources. **2.** *Social Issues.* the political power of groups, factions, and individuals who favor policies that protect and preserve the environment.

**green pricing** *Renewable/Alternative.* a strategy or formula used by a utility company to recover the cost of producing electricity from renewable sources, often by charging customers premiums or incremental amounts on their utility bills.

**Green Revolution** *Global Issues.* a term for the mid-20th century global movement in agriculture that featured the widespread use of new, high-yielding crop varieties and new technologies such as increased use of irrigation, fertilizer, and pesticides. This vastly increased the nutritional security of the human population, although with some significant negative social and environmental impacts.

**green roof** *Renewable/Alternative.* a rooftop planted with vegetation; these can be classified as **intensive green roofs** that have thick layers of soil and a wide variety of plant or even tree species, or **extensive green roofs** with a shallow soil layer to support turf, grass, or other low-lying groundcover. A green roof offers improved insulation, natural cooling, better water retention and management, and better urban heat island mitigation compared to a conventional roof.

**Greens (Green Party)** *Social Issues.* **1.** a political party established in West Germany in the late 1970s that gradually developed into an important faction with their antinuclear campaign slogan: "Atomkraft? Nein, Danke" (Nuclear Power? No, Thanks), eventually forming part of a coalition that won the German popular vote. **2.** any of various parties subsequently founded in other countries with a similar platform of environmentalism, anti-militarism, and social justice.

**green tax** *Economics.* a tax based on environmental impact, especially on environmental damage from pollution or on consumption of nonrenewable resources; e.g., a gas guzzler tax on vehicles with low fuel efficiency.

**green ton** *Biomass.* a quantity of 2000 pounds of undried biomass material.

**Greenwich Mean Time (GMT)** *Measurement.* the accepted international time system of a 24-hour day, based on a standard time beginning at Greenwich, England, the site of the Royal Observatory, whose calculations were the basis for this method of world time. This site is also the prime meridian of longitude.

**grey** another spelling of GRAY.

**grid** *Electricity.* **1.** the transmission network through which electricity moves from suppliers to customers. **2.** *Storage.* a metal framework in a storage cell or battery, used as a conductor and support for the active material.

**grid-connected** *Renewable/Alternative.* **1.** describing a renewable energy system, such as a photovoltaic array, that acts as an independent generating plant and supplies power to the electrical grid. **2.** more generally, describing any project that is connected to a utility distribution system. Similarly, **grid-tied.**

**gross aperture area** *Solar.* the maximum projected area of a concentrating collector through which the unconcentrated solar radiant energy is admitted. Similarly, **gross collector area.**

**gross domestic product (GDP)** *Economics.* the sum of all the final goods and services produced within the borders of a nation in a given time period (either by domestic or foreign producers); total internal economic activity.

**gross generation** *Electricity.* the total amount of electric energy produced by generating units, measured at the generating terminal in kilowatt-hours (kWh) or megawatt-hours (MWh).

**gross head** *Hydropower.* the vertical distance between the top of a channel or pipeline that conveys water to a pipeline under pressure and the point where the water discharges from the turbine.

**gross heat content** *Measurement.* the total amount of heat energy released when a fuel is burned.

**gross national income (GNI)** *Economics.* the sum of all income received within a given nation by domestic producers of goods and services.

**gross national product (GNP)** *Economics.* the sum of all the final goods and services produced by the companies of a given nation in a given time period (whether produced at home or abroad); the total economic activity of domestically owned producers.

**gross primary production (GPP)** *Environment.* the amount of energy that a plant converts to chemical energy by photosynthesis per unit time; the rate at which new plant biomass is formed by photosynthesis.

**gross ton** *Measurement.* a measure of the internal volume, or enclosed spaces, of a maritime vessel; equivalent to one hundred cubic feet.

**gross world product (GWP)** *Economics.* the total of the gross domestic product (see above) for all nations.

**ground** *Electricity.* **1.** a connection to the earth that conducts electrical current to and from the earth. **2.** to make such a connection. **3.** the voltage reference point in a circuit. **4.** a point in an electrical system with zero voltage.

**ground albedo** see ALBEDO.

**ground-coupled heat pump** another term for a GROUND-SOURCE HEAT PUMP.

**grounded** *Electricity.* describing a conductor system or circuit conductor that is intentionally connected to ground.

**ground fault** *Electricity.* an undesirable connection to ground, or an undesirable loss of ground in a grounded system.

**ground heat** *Geothermal.* the varying natural transfer of heat downward when the surface of the ground is warmer than the subsurface, or upward when the reverse is true; can be employed for the heating and cooling of buildings by means of a system that either adds heat to indoor air, or subtracts it, depending on the need.

**ground heat exchanger (GHE)** *Renewable/Alternative.* a device that exchanges heat with the earth; thermal energy is extracted from or rejected into the ground, depending on mode of operation (to provide heat in winter and cooling in summer).

**grounding** *Electricity.* the process of connecting to ground or to a conductor that is grounded.

**ground-level concentration** *Health & Safety.* the concentration of a chemical species, normally a pollutant, in air; usually measured at a specific height near the ground.

**ground-level ozone** *Environment.* a chemical reaction between volatile organic compounds and oxides of nitrogen in the presence of sunlight. High concentrations of ozone near ground level are harmful to people, animals, crops, and other materials.

**ground loop** *HVAC.* a term for a series of heat exchange pipes employed in a GROUND SOURCE HEAT PUMP; may be a closed or open loop.

**ground-reflected** *Solar.* describing radiation from the sun that is reflected back into the atmosphere after striking the earth.

**ground-source heat pump (GSHP)** *Renewable/Alternative.* a heating and cooling system that relies on the relatively constant temperature of the earth. In a closed-loop system, water or antifreeze solution is circulated through pipes buried in the earth, removing heat from the home in the summer and delivering it in the winter. An open-loop system operates on the same principle but withdraws water from underground or surface sources and then discharges the water back to the environment.

**ground speed** *Transportation.* the actual speed that an aircraft travels over the ground; it combines the craft's air speed and the wind speed relative to the craft's direction of motion.

**groundwater** *Earth Science.* **1.** the portion of subsurface water that is below the water table, in the zone of saturation. **2.** standing water lying temporarily on the surface of solid ground, as after a heavy rainfall.

**Grove, William Robert** 1811–1896, British physicist who reasoned that if electricity could produce hydrogen and oxygen, then combining the two should create electricity. Grove employed this theory to devise the world's first "gas battery" (1839), later renamed the fuel cell.

**Grubb, Willard Thomas** U.S. chemist who made a major advance in fuel cell design in 1955 when he developed the PROTON EXCHANGE MEMBRANE (PEM). This was among the first commercial uses of a fuel cell.

**GSWP** *Measurement.* Global Soil Wetness Project; a modeling activity to provide a global data set of soil moisture, temperature, runoff, and surface flows.

**Gt** gigaton (one billion tons).

**GtC** gigatons of carbon (one billion tons).

**Guericke, Otto von** 1602–1686, German engineer who invented the first air pump and used it to study the phenomenon of a vacuum and the role of air in combustion and respiration. He refuted the long-held notion that it was impossible for a vacuum to exist.

**guide vane** *Hydropower.* a device used in reaction turbines to change water flow direction by 90 degrees, causing the water to whirl and enter all the turbine buckets simultaneously, thus improving efficiency.

**Gulbenkian, Calouste** 1869–1955, Armenian oil magnate who played an important role in the organization of the Royal Dutch–Shell Group and in the evolution of the international oil industry.

**Gulf Oil** *History.* an oil company organized in Texas in 1901, one of the historically dominant companies in the industry; merged with Chevron (Standard Oil of California) in 1984 to form what is now ChevronTexaco.

**Gulf War** *History.* the 1991 conflict between Iraq and the U.S. and allied nations, following Iraq's invasion of Kuwait. More than 600 oil wells were set ablaze in Iraq and Kuwait by retreating Iraqi forces, creating huge oil lakes in the desert and massive regional air and water pollution. In addition, 240 million gallons of crude oil were released to the marine environment, the largest oil spill ever.

**gum bed** *Oil & Gas.* a term for surface oil seeps that have congealed.

**gunite** *Materials.* a construction material composed of cement and sand sprayed onto a surface by means of a high pressure air-water mixture; used to reinforce mine passages.

**gusher** *Oil & Gas.* an informal term for a large, uncontrolled flow of oil from a well, especially such a flow that has just been exposed.

*gusher* *A gusher at the famous Spindletop field of West Texas; this Lucas Well (drilled by Capt. Anthony Lucas) proved the feasibility of drilling on salt domes in flat country and established Texas as a major producing area.*

**Gutenberg, Beno** 1889–1960, U.S. geophysicist noted for his analysis of earthquake waves and the information they furnish about the physical properties of the solid earth. With colleague Charles Richter, he provided some of the basic information used to develop the theory of plate tectonics.

**Gutenberg, Johannes** c. 1400–1468, German inventor who inaugurated the era of movable type with his printing of a Latin Bible, known today as the Gutenberg Bible (1454). Whereas scribes copied manuscripts by hand before Gutenberg's invention, copying became mechanized and much faster after this. His invention enormously increased the number of books in Europe and facilitated greater exchange of ideas, thus spreading the effects of the Renaissance.

**guyed tower** *Wind.* a wind energy tower that is secured by a guy (a supporting chain, wire, rope or rod); the typical configuration for a smaller home energy system, as opposed to the free-standing construction of larger commercial systems.

**GVWR** *Transportation.* gross vehicle weight rating; the maximum weight of a vehicle, including its payload.

**GW** gigawatt.

**GWP** **1.** global warming potential. **2.** gross world product.

**Gy** *Nuclear.* the abbreviation for GRAY (a measure of radiation).

**gypsum** *Materials.* $CaSO_4 \cdot 2H_2O$, a mineral, occurring in long tabular to prismatic crystals; the most common sulfate mineral, found in sedimentary deposits, saline lakes, and deposits associated with vulcanism; used in making cement and plaster.

**gyromagnetics** *Physics.* the study of relationships between the magnetization of bodies and their angular momentum, as in the **gyromagnetic effect** (the mechanical rotation induced in a body due to a change in its magnetization).

**Haber–Bosch process** *Materials.* a widely used industrial process for the synthesis of ammonia by direct combination of nitrogen and hydrogen in the presence of iron catalysts; the basis of modern fertilizer production. [Developed by German chemists Fritz *Haber,* 1868–1934, and Carl *Bosch,* 1874–1940.]

**Hadley circulation** *Earth Science.* a direct, thermally driven, zonally symmetric circulation that consists of an equator-directed movement of air from 30° north or south latitude, with rising wind components near the equator, pole-directed flow aloft, and descending components at 30° latitude again; it was initially proposed as an explanation for trade winds. Also, **Hadley cell.** [Described by English meteorologist George *Hadley,* 1685–1768.]

**hadron** *Physics.* any particle of the largest family of elementary particles, which interact with each other through strong interactions, usually produce additional hadrons in a collision at high energy, and are roughly spherical.

**Hahn, Otto** 1879–1968, German physical chemist who (with Fritz Strassman) discovered the process of fission in uranium and thorium (1938). This immediately contributed to the discovery of chain reactions and the development of nuclear weapons and ultimately nuclear power.

**half-life** *Nuclear.* the time required for one half of a radioactive sample to decay.

**halide lamp** see METAL HALIDE LAMP.

**halide torch** *Health & Safety.* a propane-powered device used to detect gas leaks in refrigerating systems, by means of color changes in the flame caused by the presence of halogen refrigerant leaks in the system.

**Halladay windmill** *Wind.* a 19th-century machine considered to be the first commercially successful self-governing windmill, having four wooden blades and capable of turning automatically to face changing wind directions and controls its own speed of operation. [Invented by New England machinist Daniel *Halladay.*]

**Hall–Héroult process** *Conversion.* a process of electrolytic reduction in a molten bath of natural and synthetic cryolite; the standard industry practice worldwide for the conversion of alumina to aluminum. [Independently developed in the 1880s by Charles Martin *Hall* of the U.S. and Paul *Héroult* of France.]

**Halliburton, Erle P.** 1892–1957, U.S. engineer who perfected the technique of cementing oil wells (1912) to prevent pollution of freshwater aquifers and to make oil and gas production more efficient. The company bearing his name is today one of the largest energy companies in the world.

**Hallidie, Andrew Smith** 1836–1900, English-born builder of the first American municipal cable car system (cable railroad) in San Francisco, California. The first successful trip was made in 1873.

**halocarbon** *Chemistry.* a compound containing carbon and one or more halogens (fluorine, chlorine, and bromine); used as a refrigerant and propellant gas. When polymerized with hydrocarbons, halocarbons yield plastics having extreme chemical resistance and high electrical resistivity.

**halogen** *Chemistry.* a collective term for the electronegative nonmetallic elements of Group VII of the periodic table (chlorine, bromine, iodine, astatine); these exist at room temperature in all three states of matter: iodine and astatine as solids, bromine as a liquid, and fluorine and chlorine as gases. Halogen means "salt-former", and they combine directly with most metals to form salts. Thus, **halogenated.** See also HALOGEN LAMP.

**halogenated chlorofluorocarbon** *Refrigeration.* any of a class of refrigerant substances; see CHLOROFLUOROCARBON.

**halogen lamp** *Lighting.* a type of incandescent lamp with higher energy efficiency than standard lamps. It uses a halogen gas, usually iodine or bromine, that causes the evaporating tungsten to be redeposited on the filament, thus prolonging its life.

**hammermill** *Biomass.* a device consisting of a rotating head with free-swinging hammers, used to reduce chips or hogged fuel to a predetermined particle size through a perforated screen.

**hand crank** *History.* an earlier method of starting an automobile by means of manually turning a crank; supplanted by the electric self-starter after about 1920. Also, **hand starter.**

**handloading** *Mining.* the historic method of removing coal from the working face manually, rather than by means of machinery.

**Hanford** *History.* an area in south-central Washington State that was part of the U.S. effort in World War II to develop nuclear weapons. Plutonium manufactured at the Hanford site was used in constructing the atomic bomb dropped on Nagasaki. Currently, the site is the center of a massive environmental clean-up and restoration effort because of significant releases of radioactive materials.

**HAP** hazardous air pollutant.

**hard coal** *Coal.* another term for ANTHRACITE COAL, because of its hardness.

**hard energy** see SOFT ENERGY.

**hard-rock mining** *Mining.* any of various techniques used to mine valuable minerals, such as gold, that are encased within igneous or metamorphic rock.

**hardware** *Communication.* a collective term for those aspects of computer technology that are actual physical objects, such as a central processing unit and disk storage, as opposed to *software* such as operating systems, applications, and utility programs.

**hardwood** *Biomass.* one of the botanical groups of dicotyledonous trees that have broad leaves, such as the birch, elm, mahogany, maple, or elm, contrasted with softwoods (conifers). The *hardwood/softwood* distinction is a general description and does not literally refer to the hardness of the wood, since some conifers have harder wood than certain hardwoods. Short-rotation, fast-growing hardwood trees are being developed as future energy crops.

**harmonic** *Electricity.* describing electrical voltages or currents with frequencies that are integral multiples of the fundamental frequency, e.g., if 60 Hz is the fundamental frequency, then 120 Hz is the second harmonic and 180 Hz is the third harmonic.

**hartree** *Measurement.* a unit of energy approximately equal to $4.36 \times 10^{-18}$ J or 27.21 eV; used to study atomic structure. [Named for English mathematician Douglas Raynar *Hartree,* 1897–1958.]

**Hartwick rule** *Economics.* a rule describing the conditions under which an economy that relies on a nonrenewable resource can avoid economic decline; it holds that as long as the stock of capital does not decline over time, nondeclining consumption is also possible. The stock could be held constant by investing nonrenewable resource rents in human-made capital (e.g., alternative-energy cars). [Named for Canadian economist John *Hartwick.*]

**Harvey, William** 1578–1657, English physician who first described in detail the complete circulatory system in which blood is pumped around the body by the heart. His ideas were not accepted during his lifetime but are now considered to be the foundation for all modern research on the heart and blood vessels.

**Hauksbee, Francis** c. 1666–1713, English physicist who did some of the first important experiments with electricity. See also INFLUENCE MACHINE.

**haulage** *Mining.* the fact of hauling; the horizontal transport of ore, coal, supplies, waste, and so on.

**haulageway** *Mining.* an underground passageway in a mine that is utilized for the transport of materials, personnel, or equipment, usually by means of a track or conveyor belt.

**HAWT** horizontal axis wind turbine.

**hazard** *Health & Safety.* **1.** a substance for which there is scientifically valid evidence that it has the potential to be dangerous to health, such as a combustible liquid, an explosive, or a toxic chemical. **2.** similarly, a potentially dangerous condition, as at a work site, either a preventable condition or something that is inherent in the locale or the activity taking place there.

**hazardous** *Health & Safety.* having to do with or being a hazard; having the capacity to cause damage to human health or to the environment.

**hazardous air pollutant (HAP)** *Health & Safety.* an official classification for chemicals (e.g., asbestos) that are either known or suspected to cause serious health and environmental effects, such as cancer, birth defects, or neurological disorders.

**hazardous waste** *Health & Safety.* **1.** any waste material that has the potential to damage health or the environment, especially toxic materials resulting from industrial activities such as coal mining, nuclear power generation, chemical processing, steelmaking, paper manufacture, and so on. **2.** more generally, any manufactured item that cannot be safely recycled; e.g., prescription drugs, batteries, television sets and computer monitors, and automobile tires.

**haze** *Earth Science.* fine dust, salt, mist, smoke, or other such solid particles dispersed throughout a portion of the atmosphere, which are not visible to the naked eye as distinct entities, but which diminish horizontal visibility and give the atmosphere an opalescent appearance, rendering distant objects indistinct and subdued in color.

**HCFC** hydrochlorofluorocarbon.

**H-coal process** *Coal.* a direct coal liquefaction process intended to make coal cleaner so that it will produce less ash and less sulfur emissions.

**HDI** Human Development Index.

**head** *Hydropower.* **1.** the pressure of water or another liquid, expressed as the total height of a column of the liquid that would have the same pressure as the base of the column; typically expressed in feet and described as low, medium, or high head. **2.** the height of a reservoir or other body of water above its level at the point of use; i.e., the vertical distance that the water falls. **3.** short for HEADWATER.

**head loss** *Hydropower.* a reduction in the flow of water or another fluid through a pipe or tube; e.g., because of frictional forces or water turbulence.

**headrace** or **head race** *Hydropower.* a waterway that directs water toward a hydropower site, typically running parallel to the water source and diverting water to provide the optimal flow rate.

**headwater(s)** *Hydropower.* **1.** the water upstream from a dam or another such hydropower site. **2.** a stream at a higher elevation that is the ultimate source of a larger river.

**headwater benefit** *Hydropower.* a term for additional electric generation at a downstream hydroelectric site, resulting from flow regulation farther upstream at a headwater project.

**health hazard** see HAZARD.

**health physics** *Health & Safety.* an interdisciplinary science intended to promote radiation safety; i.e., to define and implement proper procedures for protecting radiation workers, the general public, and the environment from radiation hazards.

**heap leaching** *Materials.* **1.** a process in which valuable metals such as gold and silver are leached from a heap, or pad, of crushed ore. **2.** specifically, the separation, or dissolving out from mined rock of soluble uranium constituents by the natural action of percolating a prepared chemical solution through heaped rock material.

**heat** *Thermodynamics.* **1.** a measure of the amount of energy transferred from one body to another because of the temperature difference between those two bodies. **2.** the temperature of a body, substance, or physical environment, especially a relatively high temperature. See below.

**heat** Heat is always the transfer of energy from one substance to another. Heat transfer may occur when two substances at different *temperatures* are brought together. The energy flow will always be from the warmer substance (with a higher temperature) to the cooler substance (at lower temperature). In most cases, the result of heat transfer is warming the cooler body and cooling the hotter body. Under certain conditions, the change in temperature of any substance is the result of: the amount of heat energy flowing into the body, its *heat capacity*, and its mass. However, no change in temperature may occur during heat transfer. If a substance melts or turns to a vapor, as long as the changes of state occur, the substance continues absorbed heat, but its temperature stays constant (until all the material has changed state.) Amounts of heat are expressed in energy units: the *calorie*, the *joule*, and the *BTU*. One calorie will raise one gram of water one degree centigrade. The BTU (British Thermal Unit) is about 252 calories and about 4.2 calories is a joule.

**Fred L. Wilson**
Rochester Institute of Technology

**heat anticipator** *HVAC.* a thermostat feature used to prevent rapid on-off cycling and thus maintain a relatively constant room temperature; it will shut off the heater before the air inside the thermostat actually reaches the set temperature, to allow for the fact that certain parts of the conditioned space will reach this temperature before the specific site of the thermostat does. Thus, **heat anticipation.**

**heat budget** see ENERGY BUDGET.

**heat capacity** *Thermodynamics.* the ratio of the energy or enthalpy absorbed (or released) by a system to the corresponding temperature rise (or fall).

**heat conduction** *Thermodynamics.* the heat transfer of energy through a substance by means of the presence of a temperature difference within the substance; heat will spontaneously flow from a hotter region to a colder region.

**heat conservation** *Thermodynamics.* the observation that the quantity of heat (and other forms of energy) remains constant in any conversion process.

**heat content** *Thermodynamics.* the relative capacity of a system to transfer heat across its boundaries to the surrounding environment; now usually expressed as the ENTHALPY of the system.

**heat convection** see CONVECTION.

**heat death** *Thermodynamics.* a term for the state of a system that has reached its maximum level of entropy (i.e., it has a uniform temperature throughout and a complete lack of order); such a system has no further capacity to do work on its surroundings.

**Heat Economiser** *History.* a heat engine innovation patented by Scottish inventor Robert Stirling in 1816, so called because it stored heat between the hot and cold cycles. See also ECONOMIZER; STIRLING ENGINE.

**heat engine** *Thermodynamics.* any system (not strictly an engine, though it may be) that, with the aid of an external energy source, can operate a cycle of processes to extract useful work.

**heat equator** see THERMAL EQUATOR.

**heat exchange** see HEAT FLOW.

**heat flow** *Thermodynamics.* **1.** the transfer of energy from one body or system to another as a result of a temperature difference between the two. **2.** the amount of energy transferred in such a process, or the rate at which the transfer occurs. **3.** see GLOBAL HEAT FLOW.

**heat flux** *Thermodynamics.* the amount of heat transfer across a surface per unit time and per unit cross-sectional area, measured in joules per second per meter squared.

**heat gain** *HVAC.* any increase in the amount of heat contained in an enclosed space, resulting from such sources as direct solar radiation, heat flow through walls and windows, and the waste heat given off by lights, machinery, and people within the space.

**heat gain factor** *Solar.* an estimate used in calculating cooling loads of the heat gain due to transmitted and absorbed solar energy through clear glass 1/8 inch thick, at a specific latitude, time, and orientation.

**heating degree day** see DEGREE DAY.

**heating intensity** *HVAC.* the ratio of energy used for space heating to the square footage of heated floor space and heating degree days.

**heating season** *HVAC.* the coldest months of the year, in which there is a greater demand for indoor space heating; often defined in North America as those months in which the average daily temperatures falls below 65°F.

**heating seasonal performance factor** *Consumption & Efficiency.* an efficiency rating for heat pumps; a measure of the average number of Btu of heat delivered for every watt-hour of electricity used by the heat pump over the heating season.

**heating value** *Consumption & Efficiency.* the amount of heat produced from the complete combustion of a unit quantity of fuel (e.g., joules per cubic foot of natural gas). **Higher** (gross) **heating value (HHV)** is the amount of energy from the complete combustion of a fuel, including the vaporization heat of the water vapor. **Lower** (net) **heating value (LHV)** is obtained by subtracting the latent heat of vaporization of the water vapor formed by the combustion of the hydrogen in the fuel from the HHV. The differences between LHV and HHV are, depending on the fuel, 5–20%.

**heat island** *Environment.* the phenomenon that on warm summer days, the air in a city can be 6–8°F hotter than its surrounding areas, due to factors such as a lack of trees, shrubs, and other plants to shade buildings, intercept solar radiation, and cool the air, or the fact that buildings and pavement made of dark materials absorb solar radiation.

**heat(ing) load** see LOAD HEAT.

**heat loss** *HVAC.* a decrease in the amount of heat contained in an interior space, resulting from heat flow though walls, windows, and other building surfaces, and from the exfiltration of warm air.

**heat-loss coefficient** *Storage.* the rate at which heat is lost from a storage device, per degree temperature difference between the average temperature of the storage medium and the average temperature of the surrounding air.

**heat of compression** *Thermodynamics.* the increase in temperature brought about when a given amount of gas is compressed by a certain amount.

**heat of condensation** *Thermodynamics.* the heat required per unit mass to change a substance from a vapor to a liquid at its boiling point.

**heat of formation** *Thermodynamics.* the heat evolved or absorbed during the formation of one mole of a substance from its component elements.

**heat of fusion** *Thermodynamics.* the heat absorbed by a unit mass of a solid at its melting point in order to convert the solid into a liquid at the same temperature.

**heat of reaction** *Thermodynamics.* the change in the enthalpy of a system that occurs when a reaction takes place at constant pressure.

**heat of solidification** another term for HEAT OF FUSION.

**heat of subcooling** *Chemistry.* the quantity of heat removed from a liquid to reduce it from its saturation temperature at saturation pressure, to some lower temperature at the same pressure.

**heat of sublimation** *Thermodynamics.* the amount of heat change involved when a certain quantity of a substance is converted from a solid to a gaseous state at a specific pressure and temperature.

**heat of vaporization** *Thermodynamics.* the heat required per unit mass to change a substance from a liquid to a vapor at its boiling point.

**heat pump** *Thermodynamics.* any device used to transfer heat from a lower-temperature resource to a higher-temperature reservoir, thus providing a warmer temperature for space heating. The process can be reversed to provide space cooling.

**heat rate** *Measurement.* an expression of the conversion efficiency of thermal energy to work output; e.g., the total Btu content of a fuel burned for electric generation divided by the resulting net power generation in kilowatt-hours.

**heat recovery** *Consumption & Efficiency.* any process in which what would otherwise be waste heat is recovered for a useful purpose; e.g., heat from the generation of electricity that is then used for heating purposes in the plant.

**heat release** *Thermodynamics.* the liberation of energy through a heat transfer from a system, especially as measured for a certain volume over a given time.

**heat reservoir** *Thermodynamics.* any body that receives or accepts a heat transfer from a heat engine and is sufficiently large to be unaffected by this transfer; a source or sink of heat at constant temperature.

**heat sink** *Thermodynamics.* any body of matter (gaseous, liquid, or solid} that receives a heat transfer from its surrounding environment; an area or medium where heat is deposited.

**heat source** *Thermodynamics.* any body or system that serves to provide energy for another body or system; an area or medium from which heat is removed.

**heat storage** *Storage.* any process or condition in which a device or medium absorbs collected solar heat and stores it for use during periods of cooler temperatures.

**heat stress** *Biological Energetics.* a significant increase (decrease) of body temperature induced by sustained heat (cold) imbalance that cannot be fully compensated for by temperature regulation, or by the activation of physiological regulatory systems.

**heat-to-power ratio (HPR)** *Measurement.* the ratio between heat use and electricity generation in a combined heat and power application.

**heat transfer** *Thermodynamics.* energy transfer across the boundary of a thermodynamic system due to a temperature difference.

**heat transmission** another term for HEAT FLOW.

**heat trap** *Renewable/Alternative.* a device used to retard the natural tendency of heat to rise within a system (and thus be effectively lost to the system); e.g., a valve or piping arrangement that reduces heat loss in a water heater.

**heat wheel** *Consumption & Efficiency.* a device used in industrial heat recovery that transfers heat from one stream of gas to another.

**heavier-than-air** *Transportation.* describing an airborne vehicle having a cumulative weight greater than the air it displaces, and thus acquiring its lift from aerodynamic forces; i.e., a jet or propeller-driven aircraft as opposed to a hot-air balloon.

***heavier-than-air*** *A Wright Brothers military aircraft (1909). Although lighter-than-air (balloon) flights were undertaken in the 1780s, the first successful heavier-than-air flights did not occur until 120 years later.*

**heaving** *Mining.* the rising of the bottom of a mine after removal of the coal.

**heavy crude** *Oil & Gas.* petroleum with a high specific gravity and a low API gravity (20° or less), having a relatively high proportion of heavy hydrocarbon products or fractions; it requires special production procedures to extract from underground formations.

**heavy fuel oil (HFO)** *Transportation.* a group of residual fuels that can be burned in autoignited engines (i.e., diesel fuels). Such fuels typically require heating and filtering for proper combustion.

**heavy ion accelerator** see ACCELERATOR.

**heavy metal** *Chemistry.* **1.** any metal or alloy of high specific gravity, especially one that has a density higher than 5 g/cm$^3$. **2.** *Earth Science.* any of a series of dense metallic elements naturally occurring in the earth's crust, such as chromium, cadmium, lead, nickel, or iron. These can damage living organisms at low concentrations and tend to accumulate in the food chain.

**heavy naphtha** *Materials.* a mixture obtained by the fractional distillation of coal tar: a deep amber to dark red liquid that boils at 160–220°C; used in the manufacture of synthetic resins, as a solvent, and for a variety of other purposes.

**heavy oil** see HEAVY CRUDE.

**heavy rail** *Transportation.* a term for traditional urban transit systems operating in trains of multiple cars on separate rights-of-way from which all other vehicular traffic and pedestrians are excluded; i.e., subways or elevated railways. So called in contrast with a LIGHT RAIL system.

**heavy water** *Chemistry.* **1.** water containing significantly more than the natural proportion (1 in 6500) of heavy hydrogen (deuterium) atoms to ordinary hydrogen atoms. It is heavier than ordinary water by about 10%, and occurs in minute quantities (about one part per 7000 parts water). It is used in thermonuclear weapons and nuclear reactors. **2.** another variety of this in which the hydrogen atoms are in the form of tritium rather than of deuterium.

**heavy-water reactor** *Nuclear.* a reactor that is moderated and cooled by heavy water (deuterium oxide).

**hect-** or **hecto-** *Measurement.* a prefix meaning "one hundred"; symbol h.

**hectare** *Measurement.* a basic metric unit of surface area, symbol ha; equal to 100 ares; equivalent to 10 000 m$^2$ or 2.471 acres.

**hedge** *Economics.* a position in a futures or options market that is intended as a temporary substitute for the sale or purchase of the actual commodity. Thus, **hedging.**

**hedonic** *Economics.* **1.** relating to or based on quality (literally, pleasure or happiness). **2.** specifically, basing the price of a good or service on its characteristics (both intrinsic and external). Thus, **hedonic pricing.**

**Heisenberg, Werner** 1901–1976, German physicist considered to be a founder of quantum mechanics, known especially for his description of the UNCERTAINTY PRINCIPLE. This is also known as **Heisenberg uncertainty.**

**helical** *Physics.* having the pattern of a helix (i.e., a spiral or corkscrew). Thus, **helical flow.**

**helicity** *Physics.* the projection of a particle's spin along its direction of motion. The helicity of a particle is described as being either left-handed or right-handed, depending on whether its spin vector is in the direction of motion or against it.

**helicopter** *Transportation.* a wingless aircraft acquiring its primary motion from engine-driven rotors that accelerate the air downward, providing a reactive lift force, or accelerate the air at an angle to the vertical, providing lift and thrust.

***helicopter*** *The Lynx, a British military helicopter. Fixed-wing aircraft were originally designed in analogy with the flight of birds; a helicopter, with its ability to hover and fly backwards, also has a counterpart in nature, the hummingbird.*

**helimagnetism** *Physics.* a property of certain metals, alloys, or salts that are composed of crystal planes possessing a magnetic moment whose alignment at low temperatures varies uniformly from atomic plane to plane and resembles that of a screw or helix.

**helio-** *Solar.* a prefix meaning "sun".

**heliocentric** *Physics.* **1.** centered on the sun. **2.** having to do with or based on the heliocentric theory (see next).

**heliocentric theory** *History.* the principle that the earth and other planets revolve around the sun; following the work of Copernicus in the 16th century, this replaced the earlier geocentric (earth-centered) system described by Ptolemy.

**heliochemical** *Biological Energetics.* converting energy from the sun to chemical energy, especially by the process in which green plants utilize solar energy through photosynthesis. Thus, **heliochemistry.**

**heliodon** *Solar.* a device used to simulate the angle of the sun; used to assess the shading potentials of building structures or landscape features.

**helioelectrical** *Conversion.* converting energy from the sun to electricity; e.g., by means of photovoltaics (solar cells). Thus, **helioelectricity.**

**heliograph** *Solar.* **1.** a signaling device that uses flashes of sunlight; two mirrors reflect sunlight on a distant point through a shutter so that messages may be transmitted in telegraphic code. **2.** an instrument that records the amount and duration of sunshine, often on a strip of blueprint paper.

**heliostat** *Solar.* a system of plane mirrors that continuously adjust in angle according to the sun's position, so as to reflect a beam of solar radiation to some fixed point in space; used in some forms of concentrating solar power.

**heliothermal** or **heliothermic** *Biological Energetics.* **1.** regulating body temperature primarily by means of heat gain from direct solar radiation, as in reptiles and amphibians. **2.** *Earth Science.* describing a lake whose temperature increases rather than decreases with depth. **3.** *Solar.* utilizing solar radiation to produce useful energy. Thus, **heliothermy.**

**heliothermometer** *History.* an early form of solar collector, invented by Swiss scientist Horace-Benedict de Saussure in about 1780.

**heliotropic** *Biological Energetics.* **1.** oriented toward sunlight; describing a plant that grows or moves toward the sun. **2.** *Solar.* describing a device that follows the sun's apparent movement across the sky.

**helium** *Chemistry.* a gaseous element having the symbol He, the atomic number 2, an atomic weight of 4.002 6 (only hydrogen is lighter), and a boiling point of −268.9°C; it is colorless, odorless, tasteless, and noncombustible. It occurs on earth in natural gas and makes up a small fraction of the atmosphere, and is abundant elsewhere in the universe. Used for filling balloons, as a protective gas in growing silicon, as a coolant for nuclear reactors, in cryogenic applications, and as a rocket fuel.

**helium balloon** *Transportation.* a passenger balloon deriving its lift from an enclosed envelope of helium, which will cause the craft to rise (when it is not tethered), based on the lighter weight of helium relative to air. Early balloon flights employed **hydrogen balloons** operating on a similar principle, but these have been supplanted by helium because of the highly flammable nature of hydrogen.

**helix** *Materials.* **1.** an object or structure having a spiraling or corkscrew shape. **2.** see DOUBLE HELIX.

**Helmholtz, Hermann von** 1821–1894, German physicist who extended Joule's studies of heat and energy to a general principle, now referred to as the law of CONSERVATION OF ENERGY. He described the relationship among mechanics, heat, light, electricity, and magnetism as manifestations of a single force. His use of the word "force" corresponds to what later became known as *energy.*

**Helmholtz free energy** *Thermodynamics.* A thermodynamic property $A$ that describes the capacity of a system to do work, expressed by $A = U - TS$, where $U$ is the internal energy, $T$ the absolute temperature, and $S$ the entropy of the system.

**Helmont, Jan Baptista van** 1580–1644, Belgian chemist who was the first to recognize the existence of gases distinct from atmospheric air, and the first to use the term "gas". He described the gas given off by burning charcoal as *spiritus silvestre* ("wild spirit"); it is now known as carbon dioxide.

**hemicellulose** *Biomass.* $C_6H_{10}O_5$, a high-molecular-weight polysaccharide complex that functions as a structural component of plant cells; it is a polymer of five different sugars.

**hemispheric flux** see RADIANT FLUX.

**hemispheric(al) irradiance** *Solar.* the quantity of solar energy incident on a given surface area in unit time through a unit hemisphere above the surface, commonly measured in watts per square meter.

**Heng, Zhang** Chinese mathematician in the Han Dynasty, thought to have built the first seismoscope or seismometer in AD 132.

**Henry, Joseph** 1797–1878, U.S. scientist who discovered the electromagnetic phenomenon of self-inductance, and also discovered mutual inductance, independently of Faraday. His work on the electromagnetic relay was the basis of the electrical telegraph.

**henry** *Electricity.* a standard unit of inductance in the meter-kilogram-second system, equivalent to that of an induced 1 volt in the presence of a current that is changing at a rate of 1 ampere per second. [Named for Joseph *Henry.*]

**Henry hub** *Oil & Gas.* a pipeline hub on the Louisiana Gulf coast, the delivery point for the natural gas futures contract on the New York Mercantile Exchange (NYMEX).

**Henry's law** *Chemistry.* a law stating that the amount of gas that will dissolve in a given quantity of a liquid is proportional to the partial pressure of the gas above the liquid at constant temperature. [Formulated by English chemist William *Henry,* 1775–1836.]

**heptane** *Chemistry.* $CH_3(CH_2)_5CH_3$, a volatile, irritating, colorless liquid that is soluble in alcohol and insoluble in water; used as a standard for octane ratings of gasoline.

**herbaceous biomass** *Biomass.* **1.** perennial nonwoody crops that are harvested for use as an energy source, such as switchgrass and reed canary grass. **2.** more generally, any plant-derived organic matter available for energy use on a renewable basis, Thus, **herbaceous energy.**

**herbicide** *Ecology.* any agent that is lethal to plants, especially a manufactured substance used to kill unwanted vegetation such as agricultural weeds.

**hermetic** *Materials.* made airtight, as by fusion or sealing.

**hermetic motor** *Refrigeration.* a motor that is sealed within the refrigerant atmosphere inside a chiller, and thus is isolated from the atmosphere outside the chiller. A compressor driven by a hermetic motor has the advantage that the compressor shaft does not have to pass through a seal between the atmosphere and the refrigerant medium inside the chiller.

**Hero (Heron) of Alexandria** first century AD, influential Greek thinker who wrote many books on mathematics, physics, geometry, and mechanics, most notably a description of different types of machines, including devices such as the AEOLIPILE, which is considered to be the first steam-powered engine.

**Hero's fountain** *History.* an apparatus in which gravity and pressure create a continuous fountain of water; an apparent perpetual-motion machine but actually the fountain will gradually subside unless more water is added.

**Hero's principle** *Lighting.* the observation that a ray of light, traveling from one point to another by means of reflection from a plane mirror, will always take the shortest path between the two points.

**herring barrel** *History.* a unit of measure enacted by King Edward IV of England in 1482, establishing 42 gallons as the standard size for a barrel of herring (in order to end cheating in the packing of fish). This 42-gallon measure became the standard unit of measure for various products, and eventually this same standard was adopted for barrels of oil.

**Hertz, Heinrich Rudolf** 1857–1894, German physicist who confirmed Maxwell's electromagnetic theory and was the first to recognize electromagnetic waves (known also as **Hertzian waves).** He also discovered that cathode rays could be produced at a much lower voltage if the cathode were illuminated with ultraviolet light, a phenomenon that became known as the photoelectric effect.

**hertz** *Measurement.* a standard unit of measurement for frequency, equivalent to one wave cycle per second; symbol Hz.

**Hess's Law** *Thermodynamics.* the statement that the amount of heat released or absorbed in a chemical reaction is not affected by the number of steps in the reaction; i.e., the heat change is the same whether the reaction takes place in one step or several. [Described by Germain Henri *Hess,* 1802–1850, Swiss-born Russian chemist.]

**heterodyne** *Electricity.* the combination of two different radio frequencies, one usually being a received external signal and the other a signal generated within the receiving apparatus, in order to produce electrical beats whose frequencies are equal to the sum of and the difference between the original frequencies.

**heterojunction** *Materials.* the junction of two different semiconductor materials (e.g., silicon and silicon germanium); this structure is often chosen for producing cells made of thin-film materials that absorb light better than silicon. Thus, **heterojunction (solar) cell**. Compare HOMOJUNCTION.

**heterosphere** *Earth Science.* the upper of two portions of the atmosphere, as distinguished from each other by the general similarity or lack of same of their composition. It begins at 80–100 km above the earth, closely coinciding with the ionosphere and the thermosphere, and is characterized by variation in composition and the mean molecular weight of constituent gases. Compare HOMOSPHERE.

**heterotroph** *Biological Energetics.* literally, other feeder; an organism that (unlike green plants) cannot obtain its nutrients directly from inorganic sources in the environment, and thus must consume nutrient material from other organisms. Thus, **heterotrophy**. Contrasted with AUTOTROPH.

**heterotrophic** *Biological Energetics.* nourished from outside; describing an organism that is not capable of producing its own nutrients; the condition of most living things except green plants and certain bacteria.

**heterotrophic respiration** *Biological Energetics.* the conversion of organic matter to carbon dioxide by organisms other than plants; e.g., soil bacteria.

**HEU** highly enriched uranium.

**HFC** hydrofluorocarbon.

**HFO** heavy fuel oil.

**HHV** higher heating value; see HEATING VALUE.

**hibernation** *Biological Energetics.* a dormant, sleep-like state characterized by lower body temperature and reduced energy consumption and heart and breathing rates; a physiological condition occurring in some animals in response to cold winter conditions; e.g., bears, some bats and birds, snakes, frogs, turtles. It is presumed that hibernation is triggered (and ended) by certain environmental cues such as day length and circ-annual rhythms, though this is not definitely known.

**Hicksian income** *Economics.* the maximum amount that can be consumed over a given period of time, without depleting the value of the capital stock of natural and man-made resources. [Named for British economist John *Hicks.*]

**HID** *Electricity.* high-intensity discharge (lamp).

**Higgins, Pattillo** 1863–1955, U.S. oil wildcatter responsible for the discovery of the enormous SPINDLETOP oil field in Texas (1901).

**high-energy** *Physics.* describing a particle (such as a cosmic ray) having a mass of several hundred MeV or greater; its physical mass is given by the energy value, in MeV, divided by the square of the speed of light, in $m/s^2$.

**high-energy civilization** *Social Issues.* a term for a contemporary society that requires a high level of energy consumption, especially fossil fuels, in order to maintain its present mode of life.

**high-energy physics** see PARTICLE PHYSICS.

**higher heating value** see HEATING VALUE.

**high explosive** *Materials.* a class of explosives in which the active agent is in chemical combination and is detonated so readily that the chemical reaction is virtually instantaneous.

**high-grading** or **highgrading** *Consumption & Efficiency.* the principle or practice of developing and extracting only resources of the highest quality and lowest cost, known as "best first" or "take the best and leave the rest". An example would be harvesting only certain valuable trees from a forest while those of poor form, health, or potential value are left standing. Compare LOW-GRADING.

**high-income economy (HIE)** *Economics.* in the World Bank's economic classification system, a country with a per capita gross national income (GNI) above $9076 (2002 standard). See also LOW-INCOME ECONOMY, MIDDLE-INCOME ECONOMY.

**high-intensity discharge (HID) lamp** *Lighting.* a lamp that produces light by passing electricity through gas, which causes the gas to glow. Examples of HID lamps are mercury vapor lamps, metal halide lamps, and high-pressure sodium lamps. HID lamps have extremely long life and emit far more lumens per fixture than do fluorescent lights.

**high-level waste (HLW)** *Nuclear.* radioactive waste produced by the use of uranium fuel in a nuclear reactor or nuclear weapons processing. It contains the fission products and transuranic elements generated in the reactor core. It is highly radioactive and hot.

**highly enriched uranium (HEU)** *Nuclear.* a classification for uranium containing more than 20% U-235; used for making nuclear weapons and as fuel for some isotope production, research, and power reactors.

**high-occupancy vehicle** see HOV.

**high-octane** another term for PREMIUM grade fuel.

**high-pressure sodium (HPS) lamp** *Lighting.* a high-intensity discharge light source that produces light by an electrical discharge through sodium vapor operating at relatively high pressure and temperature. A common form of outdoor lighting.

**high-pressure steam** *Geothermal.* a classification for steam having a pressure greatly exceeding that of the atmosphere; used to turn high-speed turbines that drive electrical generators.

**high-sulfur coal** another term for NONCOMPLIANCE COAL.

**high-temperature battery** *Storage.* a term for modern batteries in which the electrolyte is operated at very high temperatures (e.g., 300°C), such as lithium–ion or sodium–sulfur batteries.

**high-temperature collector** *Solar.* a solar thermal collector designed to operate at a temperature of 180°F or higher.

**high-temperature gas-cooled reactor (HTGR)** *Nuclear.* a nuclear reactor design in the development stage that features a fuel mixture of graphite and fuel-bearing microspheres. The coolant consists of helium pressurized to about 100 bars. It can operate at very high temperature because graphite has an extremely high sublimation temperature and helium is completely inert chemically.

**high-temperature superconductor (HTS)** *Electricity.* a term for a superconductor with a critical temperature above the boiling point of nitrogen.

**high-test** another term for PREMIUM grade fuel.

**high-voltage** or **high voltage** *Electricity.* **1.** an imprecise term for voltage that is large enough to present a safety hazard, or that is considerably higher than would normally be expected in a particular application. **2.** relating to or having such a voltage level. Thus, **high-voltage wire, high-voltage generator,** and so on.

**highwall** *Mining.* the unexcavated face of exposed overburden and coal in a surface mine, or in a face or bank on the uphill side of a contour mine.

**highwall mining** *Mining.* a method of coal mining in which a continuous mining machine is driven under remote control into a seam exposed by previous open-cut operations. A continuous haulage system then carries the coal from the miner via conveyors.

**Hill, Archibald** 1886–1977, British physiologist noted for discoveries concerning the production of heat in muscles. His research helped establish the origin of muscular force in the breakdown of carbohydrates with formation of lactic acid in the muscle.

**Hindenburg** *History.* a German dirigible that was the largest aircraft ever to fly, at more than 800 ft long and 135 ft in diameter. After making several transatlantic trips, it caught fire in 1937 while making a landing at Lakehurst, New Jersey. Thirty-five people were killed and this tragedy effectively ended the use of airships for passenger transportation.

**Hipparchus of Rhodes** c. 190–120 BC, Greek scholar who made fundamental contributions to the advancement of astronomy as a mathematical science and to the foundations of trigonometry.

**HLW** high-level waste.

**Hoffman, John** U.S. government official who developed innovative programs for energy efficiency, such as the Energy Star program.

**hog** *Biomass.* **1.** a machine used to process wood materials to produce smaller particles for use as fuel. **2.** to produce fuel material in this way. Materials that are hogged include bark residues; log debris, cull and trim; sawmill wastes; residential yard wastes; industrial packaging; and construction/demolition wood wastes. Thus, **hog (hogged) fuel.**

**holding pond** *Environment.* a pit, reservoir, or structure built to hold a body of liquid for a given purpose, especially polluted runoff or waste water held for later transfer and disposal.

**Holdren, John** U.S. physicist noted for his efforts to reduce the dangers of nuclear conflict, achieve cooperation in energy–technology innovation, and shape new understanding and policies to ensure sustainable development of the earth's resources.

**hole** *Electricity.* the vacancy where an electron would normally exist in a solid. Under the application of an electric field, holes move in the opposite direction from electrons, thus producing an electric current. Holes are induced into a semiconductor by adding small quantities of an acceptor dopant to the host crystal.

**holocellulose** *Materials.* the entire carbohydrate constituent of wood or another fibrous substance (cellulose plus hemicellulose).

**Holocene** *Earth Science.* **1.** the geologic epoch of the Quaternary period extending from the end of the Pleistocene to the present, beginning at the end of the last Ice Age about 11,000 years ago and characterized by the development of human civilizations. **2.** referring to the rocks and deposits formed during this time.

**home heating oil (HHO)** *Oil & Gas.* a classification of distillate fuel oil used in space heating and for other purposes such as off-grid electricity generation, crop drying, and fuel for irrigation pumps.

**homeorhetic** *Thermodynamics.* describing the tendency of a system to maintain constancy or balance in its dynamic properties (e.g., energy flow).

**homeostasis** *Biological Energetics.* **1.** the ability of an organism to maintain a stable and constant internal environment, including such factors as body temperature and fluid pressure and content. **2.** specifically, the maintenance in an organism of energy balance, where input equals output and energy storage is zero (no energy accumulated or depleted over time). Thus, **homeostatic.**

**homeotherm** another term for an ENDOTHERM.

**homeothermic** see ENDOTHERMIC.

**home scrap** *Materials.* a term for materials discarded during a manufacturing or processing operation and directly recycled back to the operation, rather than being sold to others.

**homojunction** *Photovoltaic.* a junction formed of one single semiconductor material; the two parts of the junction may display different properties such as conductivity type or doping concentration. Thus, **homojunction (solar) cell.** Compare HETEROJUNCTION.

**homosphere** *Earth Science.* the lower of two portions of the atmosphere, as distinguished from each other by the general similarity or lack of same of their composition. It extends from the surface to an altitude of 80–100 km above the earth and is characterized by uniformity of composition. Compare HETEROSPHERE.

**Hooke's law** *Physics.* the statement that an object exerts a force proportional to the displacement of the object and in the opposite direction; many elastic solids closely obey this law. [Named for English scientist Robert *Hooke,* 1635–1703.]

**Hoover dam** *Hydropower.* a dam at Black Canyon on the Colorado River, the largest dam in the world at the time of its completion in 1936; serving various purposes and especially the generation of hydroelectric power for the states of the U.S. Southwest. [Named for U.S. president Herbert *Hoover,* one of the leading figures in the initiation of the project; also known as *Boulder Dam* because it was originally planned to be built farther upstream at Boulder Canyon.]

***Hoover dam*** *Hoover dam is situated in one of the harshest deserts of North America; the experiences of workers building the dam provided new insights into human energetics under extreme heat conditions.*

**horizontal axis** *Conversion.* a turbine axis orientation in which the axis of rotation of the power shaft is parallel relative to the ground; a newer technology than vertical axis turbines and more widely used.

**horizontal axis wind turbine (HAWT)** *Wind.* a classification for wind turbines in which the axis of rotation is parallel to the wind stream and the ground; the standard design of modern propeller-type turbines as utilized in commercial wind farms.

**horology** *Measurement.* the science of time; the study of time measurement and instruments used for this purpose.

**horse** *Biological Energetics.* a common domestic mammal, family Equidae, employed as an energy source since ancient times; historically the most widely used source of animal power in Europe and North America.

**horseless carriage** *History.* a historic term for any passenger road vehicle powered by steam, gasoline, electricity, wind, and so on, as opposed to being drawn by horses, the dominant method of transportation in earlier times.

**horsepower** *Measurement.* the traditional unit of power in the English system of measurement, equal to 550 foot-pounds or approximately 746 W. It is equivalent to moving 33 000 pounds over a distance of one foot in 1 min. See also SAE HORSEPOWER. See next page.

**hot-air balloon** *Transportation.* a passenger balloon that acquires its lift from the basic principle that warmer air rises into cooler air; an onboard burner heats air that then flows upward into the balloon envelope, causing the craft to rise.

**hot-air engine** *Consumption & Efficiency.* any engine in which air or another gas such as helium, hydrogen, or nitrogen, serves as the working fluid and is alternately heated and cooled by a furnace and regenerator.

**horsepower** The unit of power in the English system of measurement. The term *horsepower* was coined by James Watt (1736–1819), the Scottish inventor and mechanical engineer renowned for his improvements of the steam engine. In the early 1780s Watt and his partner Matthew Boulton set out to sell their steam engines to the breweries of London, calculating that they would be likely customers because brewing was such an energy-intensive process. In order to convince the breweries of the advantages of the steam engine, Watt needed a method to compare their capabilities relative to horses, the power source they were seeking to replace. The typical brewery horse, attached to a mill that ground the mash for making beer, walked in an endless circle with a 24-foot diameter, pulled with a force of 180 pounds, and traveled at a speed of 180.96 feet per minute. Watt multiplied the speed times the force and came up with 32,580 ft.-lbs./minute. That was rounded off to 33,000 ft.-lbs./minute, the figure used today. A healthy human can sustain about 0.1 horsepower, a car generates several hundred horsepower, while a steam turbine in an electric power plant can produce more than 1.5 million horsepower.

**Cutler Cleveland**
Boston University

**hot blast oven** *History.* a historic advance in ironmaking developed by Scottish engineer James Beaumont Neilson (1828), involving a stream of heated air rather than the previous technique of cold air, thus permitting the use of a smaller amount of coal and lesser grades of coal.

**hot box** see SOLAR HOT BOX.

**hot dry rock (HDR)** *Geothermal.* heated, crystalline rock beneath the surface of the earth; extraction of useful energy from hot dry rock typically involves hydraulic fracturing to create permeability in the rock and then the introduction of water to carry the heat from the rock to the surface for use; e.g., in an electric power plant.

**Hotelling, Harold** 1895–1973, U.S. economist, a pioneer in the field of mathematical statistics and economics. The concepts of HOTELLING RENT, the HOTELLING RULE, and the HOTELLING VALUATION are employed in the economics of natural resources (see following).

**Hotelling rent** *Economics.* a measure of natural resource depletion in environmental accounting; the net return realized from the sale of a natural resource under particular conditions of long-term market equilibrium. Defined as the revenue received minus all marginal costs of resource exploitation, exploration, and development, including a normal return to fixed capital employed.

**Hotelling rule** *Economics.* the principle that the net price of a nonrenewable resource will rise at the rate of interest. Specifically: $(p_t - c_t) = (p_0 - c_0)e^{rt}$, where $p$ is the price, $c$ the marginal cost of extraction, $t$ the time, $r$ the rate of interest, and $e$ the base of natural logarithms. Also, **Hotelling principle.**

**Hotelling valuation principle** *Economics.* the principle that the in situ value ($V$) of a nonrenewable resource depends solely on the current price net of extraction costs. Specifically: $(V_0) = (p_0 - c_0)R_0$, where $p$ is the price, $c$ is the marginal cost of extraction, $t$ is the time, and $R$ the amount of the reserve. Also, **Hotelling valuation.**

**hotspot** or **hot spot** *Earth Science.* **1.** an ancient site of volcanic activity, thought to originate from convection currents of molten mantle material along the core–mantle boundary. **2.** *Ecology.* an area identified as having great

***hot-air balloon*** *The height of a hot-air balloon can be controlled by regulating the flow of gas to the burner; i.e., the craft will rise as the flow increases.*

diversity of endemic species and, at the same time, significant impact on the area by human activities; so called because of the concept that this area should receive high priority in conservation activities. **3.** *Photovoltaic.* a phenomenon in which one or more cells within a PV module or array act as a resistive load, resulting in local overheating or melting of the cells.

**hot spring** *Geothermal.* a site at which geothermal heat influences the temperature of water issuing from the ground, providing a spring or pool whose temperature is significantly above the ambient ground temperature on a regular basis for at least a predictable part of the year.

**hot water pollution** *Environment.* an excessive raising of water temperature, caused by a procedure in which a power plant withdraws cool water from a nearby body of water (such as the ocean) for use as a cooling agent, followed by the return of heated water to the same body.

**Houdry process** *Oil & Gas.* a major advance in the petroleum refining industry (1930s), involving the cracking of larger petroleum molecules into the shorter ones that constitute gasoline by a chemical process, as opposed to earlier processes that relied on heat. The Houdry process doubled the amount of gasoline produced from a given amount of feedstock and also improved octane rating, making it possible to produce today's efficient, high-compression automobile engines. [Developed by French chemist Eugene Jules *Houdry,* 1892–1962.]

**household energy** see DOMESTIC ENERGY (def. 1).

**HOV** *Transportation.* high-occupancy vehicle(s); an **HOV lane** on a highway is restricted to vehicles carrying at least two passengers (and in some cases at least three).

**Hovercraft** *Transportation.* a transportation craft that is designed to travel at a short distance above the surface of the ground or a body of water, moving (hovering) on a cushion of air that is held in a chamber beneath the vehicle.

**HPR** heat-to-power ratio.

**HPS** high-pressure sodium.

**HSPF** heating seasonal performance factor.

**HTGR** high-temperature gas-cooled reactor.

**HTS** high-temperature superconductor.

**hub** *Consumption & Efficiency.* **1.** the part of a wheel or fan to which an axle or blades are attached. **2.** the center of any circular, rotating part, especially the center of a wind generator rotor that holds the blades in place and attaches to the shaft.

**hub-and-spoke** *Transportation.* a current system of air travel based on the analogy of a wheel with a central hub and various spokes radiating from it; passengers reach their ultimate destination by first traveling to one of a limited number of airport hubs (typically central cities with large populations) and then taking another flight to one of a larger number of spokes (smaller cities in the same region).

**Hubbert, (Marion) King** 1903–1989, U.S. geophysicist known for his accurate prediction of the peak in oil production in the U.S., in what is now known as the HUBBERT CURVE. He also made fundamental contributions to geophysics, demonstrating that fluids (e,g., oil and gas) can become entrapped under circumstances previously not thought possible.

**Hubbert curve** *Measurement.* a bell-shaped curve representing annual oil production versus time, employed by M. King Hubbert to predict (correctly) in 1956 that U.S. oil production in the lower 48 states would peak in about 15 years. See next page.

**Hubble, Edwin** 1889–1953, U.S. astronomer often regarded as the most influential in this field since the times of Galileo, Kepler, and Newton, and the founder of the field of cosmology. He proposed a new classification system for nebulae, resolved the question of the nature of galaxies, and developed the notion of the expanding universe and the BIG BANG theory.

**hue** see COLOR.

**Hugoton** *Oil & Gas.* a vast natural gas field including acreage in Kansas, Texas, and Oklahoma; identified as the largest natural gas field in North America and the second largest in the world; discovered 1922.

**Human Development Index (HDI)** *Sustainable Development.* a composite index that measures a country's average status in three basic aspects of human development:

**Hubbert curve** is a method of curve fitting using historical data on oil production and discoveries to forecast future rates developed by the American geophysicist M. King Hubbert. Hubbert's methodology is based on the assumption that cumulative oil discoveries or production follow a logistic curve. Initially, production and consumption rise slowly due to low demand and difficulty finding new deposits. Over time, both supply and discoveries increase rapidly as technological know-how increases. Eventually, demand slows due to resource depletion and higher prices. Finally, cumulative discoveries and production approach an asymptote that represents the ultimate recoverable supply. Hubbert estimated ultimate recoverable supply based on the value that generated the best-fit to the historical data. He used the first derivative of the logistic curve to identify the year that production would peak. Using this methodology, Hubbert correctly forecast the year in which oil production in the lower 48 U.S. would peak about fifteen years before it occurred in 1970. Despite this accuracy, further research indicates that the methodology is flawed because it ignores the technical, economic, and institutional determinants of oil production. These omissions cause the methodology to understate the ultimate recoverable supply in the lower 48 states, as well as the global supply and peak in oil production.

**Robert K. Kaufmann**
Boston University

longevity, knowledge, and an acceptable standard of living. Longevity is measured by life expectancy at birth; knowledge by a combination of the adult literacy rate and school and university enrollment, and standard of living by gross domestic product per capita.

**human disturbance** see DISTURBANCE.

**human ecology** *Ecology.* the study of the interrelationship of human beings and their environment, including such issues as energy use and consumption of resources, population levels, climate change, and global patterns of disease.

**human energetics** *Biological Energetics.* the study of the flow and transformation of energy within humans.

**human power** *Biological Energetics.* human beings as an energy source; e.g., farming or industrial work done by humans.

**humic coal** *Coal.* a commonly occurring category of coal, including the ordinary bituminous varieties, that have been formed from HUMUS deposited at a given site.

**humidifier** *HVAC.* **1.** a device for adding moisture to the air or another gas. **2.** specifically, a home or office appliance of this type, used to provide a desired level of humidity for persons having various conditions associated with hypersensitivity to dry air.

**humidify** *HVAC.* to increase the amount of water vapor in air or another gas. Thus, **humidification.**

**humidistat** *HVAC.* a regulatory device that automatically measures and controls the level of relative humidity in a given space.

**humidity** *Earth Science.* a measure of the amount of water vapor or degree of dampness in a given mass of air. See also ABSOLUTE HUMIDITY; RELATIVE HUMIDITY.

**humidity ratio** *Earth Science.* the ratio of the weight of an actual mass of air to the ideal weight of the dry air in the same sample, with the difference between the two indicating the amount of water vapor in the air.

***human power*** *Laborers carrying baggage aboard a Nile River steamship at the height of the British Empire.*

**humus** *Ecology.* decomposed organic matter in which the remains of specific plants and animals are not identifiable and thus the form and appearance of the material is relatively uniform.

**hunting and gathering** *Sustainable Development.* the practice of obtaining food by hunting wild animals and by collecting wild plants and other naturally occurring foodstuffs, rather than by agriculture and the raising of domestic livestock; a strategy thought to be common to all early human cultures but today found in only a few isolated groups. Thus, **hunter-gatherer, hunting-and-gathering society,** and so on.

***hunting and gathering*** *Although true hunter-gatherer societies are rare today, some of their methods survive, as in this African technique of hut building from materials gathered in the bush.*

**Hütter, Ulrich** 1910–1990, Austrian engineer who first described in mathematical detail the theoretical basis for the construction of modern wind turbines with two- and three-rotor blades.

**Hutton, James** 1726–1797, Scottish scientist considered to be the founder of modern geology. In the rocks of Scotland, Hutton found neatly deposited layers of sedimentary rocks overlaying rock layers that were almost vertical. The lower layers of rock, he concluded, must have been deposited eons before, and then later upturned. He thus was the first to describe the vast expanses of time in earth's history.

**Huygens, Christiaan** 1629–1695, Dutch scientist who invented the pendulum clock and was the first to identify Saturn's rings and largest moon, Titan. He also built several pendulum clocks to determine longitude at sea, and derived the law of centrifugal force for uniform circular motion.

**HVAC** an acronym for heating, ventilation, and air-conditioning.

**HVDC** *Electricity.* high voltage direct-current electrical transmission, the preferred mode in various specilaized contexts such as undersea cables, long-haul bulk power transmission in remote areas, and power transmission between unsynchronized AC distribution systems.

**hybrid** *Renewable/Alternative.* **1.** relating to or involving a hybrid energy system, hybrid vehicle, and so on (see below). **2.** short for HYBRID ELECTRIC VEHICLE.

**hybrid analysis** *Measurement.* a method to calculate the embodied energy of a good or service that combines input–output analysis and process energy analysis.

**hybrid electric vehicle (HEV)** *Renewable/Alternative.* **1.** a vehicle that derives part of its propulsion power from an internal-combustion engine and part of its propulsion power from an electric motor. **2.** a vehicle using an internal combustion engine to power a generator that charges a battery, which will in turn power one or more electric-driven motors.

**hybrid energy system** *Renewable/Alternative.* a system that combines two or more forms of energy or power to provide a particular energy service. This may include an energy storage component; e.g., a home powered by a combination of a diesel generator, wind power, and battery storage.

**hybrid lighting** *Lighting.* a lighting system in which a rooftop collector concentrates and sends sunlight through optical fibers to special lighting fixtures that contain both electric lamps and fiber optics to distribute sunlight directly. When the transmitted sunlight is sufficient to illuminate a room to the desired level, the electric lights do not go on.

**hybrid poplar** *Biomass.* a hybrid tree of the genus *Populus* that has significant potential as a renewable fuel source because of its fast-growth characteristics.

**hybrid propulsion** *Renewable/Alternative.* a combination of two different power sources for vehicle propulsion; one source derives its power from fuel, such as an internal combustion engine or a fuel cell; the second is a device that stores and reuses energy. This can be electrical energy or electromechanical energy, such as a hydraulic accumulator or a flywheel.

**hybrid solar lighting** *Lighting.* a lighting system in which a rooftop collector concentrates and sends sunlight through optical fibers to special fixtures inside a building, which contain both electric lamps and fiber optics to distribute sunlight directly. When the transmitted sunlight completely illuminates each room, the electric lights stay off.

**hybrid system** see HYBRID ENERGY SYSTEM.

**hybrid vehicle** *Renewable/Alternative.* any vehicle that employs two sources of propulsion, especially a vehicle that combines a conventional internal combustion engine with an electric motor (see HYBRID ELECTRIC VEHICLE).

**hydrate** *Materials.* **1.** a substance that contains water combined in the molecular form. **2.** crystalline substance that contains molecules of water of crystallization. **3.** to form or become such a substance.

**hydraulic** *Hydropower.* **1.** of or relating to water or another liquid in motion. **2.** relating to or resulting from the pressure created by forcing water through a relatively small pipe, orifice, or other channel.

**hydraulic fracturing** *Oil & Gas.* a oil-well stimulation technique in which high fluid pressure is applied to the face of an underground well with low permeability to force the strata apart, allowing oil to flow toward the well.

**hydraulic gradient** *Physics.* a pressure head gradient (rate of change per unit distance of flow) measured at a specific point and in a given direction, often due to frictional effects along the flow path.

**hydraulic head** *Hydropower.* the vertical distance between the surface of a dam's reservoir and the surface of the waters immediately downstream from the dam.

**hydraulic mining** *Mining.* the use of directed, high-pressure jets of water to break loose minerals of interest from a hillside or mountainside, stream bed, underground deposit, and so on; a historic method of gold mining that has had severe ecological consequences.

**hydraulic pressure** *Hydropower.* the pressure created by forcing water or another liquid through a relatively small pipe, orifice, or other channel.

**hydraulic turbine** *Consumption & Efficiency.* a turbine that converts the potential energy of falling or fast-flowing water to mechanical energy; most hydraulic turbines today are used to generate electricity in hydropower installations.

**hydride** *Chemistry.* a binary compound of hydrogen with another element or a complex species containing hydrogen bound to another element; common examples are the hydrides of boron, lithium, and sodium.

**hydrocarbon (HC)** *Chemistry.* a compound composed of only carbon and hydrocarbon; these are the most abundant molecules in most crude oils and refined fuels. Hydrocarbons can be solid (asphalt), liquid (crude oil), or gas (natural gas). They are a diverse mixture of linear, branched, and cyclic saturated compounds with 1–80 carbon atoms, together with aromatic compounds, most commonly with more than one ring and with multiple alkyl substituents.

**hydrochlorofluorocarbon (HCFC)** *Environment.* a compound containing hydrogen, fluorine, chlorine, and carbon atoms; they are ozone-depleting substances but are less destructive of the stratospheric ozone layer than chlorofluorocarbons (CFCs) and thus have been introduced as temporary replacements for CFCs in aerosols and in appliances such as refrigerators. The Montreal Protocol calls for the reduction and eventual phasing out of HCFCs.

**hydrocracking** *Oil & Gas.* a catalytic, high-pressure refinery process that involves the cracking of heavy petroleum fractions in the presence of an excess of hydrogen in which special catalysts are used, such as platinum on a solid base of mixed silica and alumina; the process may be considered as a combination of hydrogenation and catalytic cracking. Thus, **hydrocracker.**

**hydrodesulfurization** *Materials.* a catalytic process for the removal of sulfur compounds from hydrocarbons using hydrogen.

**hydrodynamic** *Hydropower.* involving or caused by the motion of a fluid, especially the motion of water through a constricted channel.

**hydrodynamics** *Physics.* the study of the motion of fluids and of interactions at fluid boundaries.

**hydroelectric** *Hydropower.* describing electric current produced from water power, especially the force or pressure of falling water. Thus, **hydroelectricity, hydroelectric power.**

**hydrofinishing** *Materials.* a catalytic treating process carried out in the presence of hydrogen to improve the properties of naphthenic oils.

**hydrofluoric acid** *Health & Safety.* a solution of hydrogen fluoride in water that is released during coal burning; e.g., for electrical power; it is corrosive and highly poisonous, and when concentrated it can pass through the skin and cause serious burns.

**hydrofluorocarbon (HFC)** *Environment.* a compound containing only hydrogen, fluorine, and carbon atoms; HFCs have been promoted as alternatives to ozone-depleting substances such as CFCs and HCFCs for various industrial purposes. Though HFCs do not significantly deplete the stratospheric ozone layer, they are greenhouse gases with major global warming potentials and also have other negative environmental implications.

**hydrofoil** *Transportation.* **1.** an airfoil-shaped plate fitted to the underside of a boat to provide lift at high speeds, reducing hull displacement and associated friction. **2.** a vessel equipped with such a plate.

**hydroforming** *Oil & Gas.* a high-temperature refining method in which naphthas contact a catalyst in the presence of hydrogen to yield high-octane aromatics for motor fuel or chemical manufacture.

**hydrogasification** *Conversion.* the gasification of a fuel by reaction with hydrogen; e.g., the production of methane from pulverized coal at high temperature and pressures (above 1200°F and 500 psi).

**hydrogen** *Chemistry.* a nonmetallic element having the symbol H, the atomic number 1, an atomic weight of 1.007 97, a melting point of about –259°C, and a boiling point of about –253°C. It is the lightest and most abundant element in the universe, comprising about 90% of it by weight, but it is only a minute fraction of dry air on earth. Hydrogen in combination with oxygen as water ($H_2O$) is essential to life and is present in all organic compounds; it also occurs in acids, bases, alcohols, petroleum, and other hydrocarbons. In nature at standard conditions it is a colorless, odorless, and highly flammable gas. See next page.

***hydrofoil*** *Hydrofoil vessels are typically used for passenger ferry service on bays, rivers, lakes, and so on.*

**hydrogen-2** *Chemistry.* another name for DEUTERIUM.

**hydrogen-3** *Chemistry.* another name for TRITIUM.

**hydrogenation** *Chemistry.* **1.** a general reaction in which hydrogen is added to the unsaturated molecules of hydrocarbons or fatty acids, normally by use of a catalyst. **2.** any process in which hydrogen is combined with another substance.

**hydrogen balloon** see HELIUM BALLOON.

**hydrogen bomb** *Nuclear.* an extremely powerful type of nuclear weapon, in which the fusion of deuterium and tritium (heavy isotopes of hydrogen) releases enormous amounts of energy in the form of heat.

**hydrogen economy** *Hydrogen.* the concept of an energy system based primarily on the use of hydrogen as an energy carrier, especially for transportation vehicles.

**hydrogen electrode** *Chemistry.* a standard reference electrode, usually consisting of a platinum surface coated with platinum black

**hydrogen** Hydrogen's atomic structure of one proton, and one electron makes it the lightest of the elements. Identified as an element by Cavendish and named by Lavoisier, hydrogen's historic use is as an industrial chemical. Much of its current use is in improving crude oil in refineries. However, hydrogen was considered for energy even in the 19th Century—by Jules Verne, as well as by contemporary scientists. Hydrogen is like electricity in that it is an energy carrier, not a primary source—it is derived from the conversion of some other form of energy. It can be manufactured from a variety of sources, including natural gas, coal, nuclear energy and all forms of renewable energy. When used as a fuel in combustion processes or in fuel cells, hydrogen has minimal emissions relative to conventional fuels. Although hydrogen has about three times the energy density of gasoline per kilogram, making it ideal as a rocket fuel, it has a very low energy density on a volumetric basis. This poses significant economic and technical challenges to the transmission and storage of hydrogen. Potential end uses of hydrogen include fuel cell vehicle technology as well as stationary power generation. Research and development of hydrogen energy centers on reducing the costs associated with its manufacture and storage. As a potential complement to electricity as one of the two primary long-term energy carriers, hydrogen ultimately could offer a transition from today's energy mix that is as significant as that from wood to coal, or coal to oil.

**David Hart**
Imperial College, UK

that is bathed with a stream of hydrogen gas bubbles and immersed in a solution of hydrogen ions. It has a potential of zero when the activity of all species is unity, and it is used to measure hydrogen–ion concentrations.

**hydrogen passivation** *Photovoltaic.* the injection of hydrogen into a mixture of silicon being made for photovoltaic cells; the hydrogen inactivates crystal defects such as grain boundaries and lattice dislocations to improve overall cell performance.

**hydrogen-rich** *Hydrogen.* describing a type of fuel that contains a significant amount of hydrogen, such as gasoline, diesel fuel, methanol, ethanol, natural gas, or coal.

**hydrogen sulfide** *Chemistry.* HS, a toxic, colorless gas that has an offensive odor and that is soluble in water; a dangerous fire and explosion hazard and a strong irritant. It occurs naturally in crude petroleum, natural gas, volcanic gases, and hot springs, and results from the bacterial breakdown of organic matter and human and animal wastes. It also is generated by industrial activities; e.g., food processing, coke ovens, paper mills, and tanneries.

**hydrogen vehicle** *Hydrogen.* a transportation vehicle powered by hydrogen, either by means of an internal combustion engine **(hydrogen engine)** or electricity-generating fuel cells.

**hydrologic(al) cycle** *Earth Science.* the process of water constantly moving through a vast global cycle in which it evaporates from lakes and oceans, forms clouds, precipitates as rain or snow, and then flows back to the ocean.

**hydrology** *Earth Science.* the science of water; the study of the properties, distribution, and effects of water on the earth's surface, in the soil and underlying rocks, and in the atmosphere. Usually applies to studies of water in relation to land surfaces as opposed to the waters of the open ocean (oceanography).

**hydrolysis** *Chemistry.* **1.** a chemical reaction in which water is chemically combined into a reactant and the reactant is broken down into smaller molecules. **2.** a reaction of water with a salt to create an acid or base. Thus, **hydrolytic.**

**hydromagnetic wave** see ALFVÉN WAVE.

**hydrometer** *Measurement.* an instrument used to measure the specific gravity of a liquid.

**hydronic** *HVAC.* referring a heating or cooling system that transfers heat by circulating a fluid through a closed system of pipes. Thus, **hydronic heating.**

**hydropower** energy derived at a variety of scales from water pressure, especially the force or pressure of falling water used to power a water wheel, turbine, and so on.

**hydropower resettlement** *Hydropower.* the large-scale forced relocation of households and communities as a result of the construction of large hydropower dams and their associated reservoirs; e.g., it is estimated that

*hydropower Historically the main use of water power was for mechanical tasks such as grinding grain, but in modern society it has been used mainly for electrical power.*

the construction of the Three Gorges Dam in China will require about 1.2 million people to be resettled.

**hydroretort** *Oil & Gas.* a retort in which oil shale is thermally decomposed in the presence of hydrogen under pressure.

**hydroscopic** *Chemistry.* capable of absorbing moisture from its surroundings.

**hydrosphere** *Earth Science.* the portion of the earth that is water, especially liquid water, ice, and water vapor on the surface, but also including subsurface and atmospheric water.

**hydrostat** see HUMIDISTAT.

**hydrostatic** *Physics.* having to do with liquids at rest. Thus, **hydrostatic pressure.**

**hydrostatics** *Physics.* the scientific study of liquids at rest, as well as the forces and pressures associated with them.

**hydrostatic test** *Materials.* a strength and tightness for a drain, vessel, pipe, or other such closed hollow object; the item is filled with a test liquid and subjected to pressure.

**hydrothermal** *Earth Science.* relating to or caused by heated water, especially the action of water heated by natural processes rather than by industrial activity.

**hydrothermal vent** *Earth Science.* a geyser on the sea bottom through which superhot aqueous solutions rise from the magma beneath the crust; this creates a surrounding system of mineral-rich water that helps to support a distinctive type of ecosystem not found in typical cold-water environments at the same ocean depth.

**hydrotreating** *Oil & Gas.* an oil refinery catalytic process in which hydrogen is contacted with petroleum product streams to remove impurities.

**hydrous** *Chemistry.* containing water.

**hydrous ethanol** *Transportation.* an alternative transportation fuel consisting of ethanol mixed with some fraction of water.

**hygrometry** *Measurement.* the scientific study and calculation of the relative humidity of the atmosphere. Thus, **hygrometer.**

**Hypercar** *Transportation.* a conceptual motor vehicle that is able to have ultralight weight and ultralow drag by the utilization of advanced composite materials; developed by Hypercar Inc. and the Rocky Mountain Institute.

**hypereutrophy** *Ecology.* the ultimate stage of eutrophication; the "old age" of a lake; i.e., a condition of an excessively large supply of nutrients, resulting in a high production of organic matter by plants and animals and murky, odorous water approaching wetlands status. Thus, **hypereutrophic.**

**hypersonic** *Transportation.* describing an object (e.g., a spacecraft) moving at speeds far above the speed of sound (greater than Mach 5). The space shuttle reenters the atmosphere at **high hypersonic** speeds (at or near Mach 25).

**hypocaust** *History.* an ancient Roman method of heating baths and other buildings by means of circulating heated air beneath the floor; considered the first system of central heating.

**hypolimnion** *Earth Science.* the lowest water layer of a thermally stratified lake or reservoir; it is isolated from wind mixing, often too deep for sunlight penetration and plant photosynthesis, and typically cooler and less oxygenated than the epilimnion layer above.

**hypothetical resources** *Coal.* a classification for undiscovered coal resources that may reasonably be expected to exist in known coal mining areas under known geologic conditions.

**hypsithermal** *Climate Change.* relating to or occurring in a climatic period on earth occurring about 10,000 to 5000 years ago, characterized by generally warmer temperatures and decreased or increased rainfall in various areas.

**hypsometry** *Earth Science.* the science of height; the study or measurement of the elevation or depth of features on the earth, especially natural features. Also, **hypsography.**

**hysteresis** *Physics.* **1.** a condition in which the state of a system depends on its previous history, generally the retardation or lagging of an effect behind the cause of the effect. **2.** specifically, the inclination of a magnetic material to saturate and retain some of its magnetism after the alternating magnetic field to which it is subjected reverses polarity.

**hysteresis coupling** *Electricity.* an electric coupling in which torque is transmitted by hysteresis; i.e., forces from the resistance of magnetic fields within a ferromagnetic material.

**hysteresis loss** *Physics.* an energy loss in magnetic material due to an alternating magnetic field, as elementary magnets within the material align themselves with the reversing magnetic field.

**hythane** *Oil & Gas.* hydrogen–methane; a commercial gas product that contains 20% hydrogen and 80% natural gas.

**Hz** hertz.

**IAEA** International Atomic Energy Agency (est. 1957); an intergovernmental organization that serves as a center of cooperation in the nuclear energy field. It was set up as the world's Atoms for Peace organization within the United Nations.

**IAEE** International Association for Energy Economics (est. 1977), a nonprofit professional organization that provides an interdisciplinary forum for the exchange of ideas, experience, and issues among professionals interested in energy economics.

**ice** *Earth Science.* water in the solid state, produced by the freezing of liquid water, by the condensation of water vapor directly into crystals, or by the compaction and recrystallization of fallen snow.

**ice age** or **Ice Age** *Earth Science.* **1.** an interval of geologic time during which a large portion of the earth's surface was covered by glaciers; characterized by a significantly cold worldwide climate and widespread glacial advance toward the equator. **2.** specifically, a name for the Pleistocene epoch, because of its widespread glacial ice.

**ice-charcoal** *History.* an ancient Chinese term for coal.

**ice core** *Earth Science.* an extensive column of ancient ice, drilled out of an ice sheet or glacier for the purpose of obtaining various types of data, especially indications of past climate trends.

**ice front** *Earth Science.* **1.** a high, steep vertical cliff that forms the boundary of an ice sheet as it meets the sea, beyond which there is open water. **2.** another term for GLACIAL MAXIMUM; i.e., the point of farthest advance of a glacier.

**ice sheet** *Earth Science.* a broad, thick sheet of glacial ice that covers an extensive land area for a long period of time, sufficiently deep to overlie most of the bedrock topography, so that its shape is mainly determined by its internal dynamics.

**ice shelf** *Earth Science.* a continuous and more or less permanent sheet of fast ice, generally level and sometimes extending hundreds of kilometers from the shore, fed by glaciers and by annual accumulation of snow. Although attached to land on one side, an ice shelf's edge is chiefly supported by the water on which it floats.

**ice slurry storage** *Storage.* the use of small ice crystals and a binary solution consisting of water and a freezing point depressant, such as ethylene glycol, ethanol, or sodium chloride in cooling applications; e.g., chilled water systems, air conditioning, or district cooling.

**ICRP** International Commission on Radiological Protection (est. 1928); an independent registered charity established to advance for the public benefit the science of radiological protection.

**ideal gas** *Chemistry.* a theoretical gas whose molecules are infinitely small and exert no force upon each other. This means that their motion is completely chaotic and does not depend on any mutual interaction, but only on completely elastic collisions, which cause a transfer of energy from one molecule to

*ice shelf In the past several decades scientists have observed a series of retreats and collapses by Antarctic ice shelves; this has been attributed to a strong climate warming trend in the region.*

another and a change of their direction and velocity. Also called a PERFECT GAS.

**ideal gas law** *Chemistry.* an expression of the relationship of temperature, pressure, and volume for an ideal gas; $pV = nRT$, in which $p$ is the pressure of the system, $V$ is the volume, $n$ is the number of moles of the sample, $R$ is the gas constant for the gas in question, and $T$ the absolute temperature. Many actual gases will approximately obey this law at sufficiently low pressures or high temperatures.

**ideal rocket** *Physics.* a theoretical rocket that would operate perfectly at a velocity equal to that of its jet gases; used to provide ideal parameters to be compared to practice.

**identified resources** *Coal.* a collective term for the sum of coal resources whose location, quality, and quantity have been established from geologic evidence supported by engineering measurement.

**IDLH** *Health & Safety.* immediately dangerous to life or health; describing the maximum concentration of a hazardous material from which one could escape within 30 minutes, without any symptoms impairing the escape or any irreversible health effects.

**IEA** International Energy Agency (est. 1974), an intergovernmental body that seeks to advance the security of energy supply, economic growth, and environmental sustainability through energy policy cooperation. It was formed in response to the oil crisis in 1973–1974.

**IEA Bioenergy** (est. 1978), an organization of the International Energy Agency (IEA) that seeks to improve cooperation and information exchange between countries that have national programs in bioenergy research, development, and deployment.

**IECC** *Policy.* International Energy Conservation Code; a code of standards developed by the International Code Council (ICC) that facilitates energy conservation through efficiency in envelope design, mechanical systems, lighting systems, and the use of new materials and techniques.

**IEEE** Institute of Electrical and Electronics Engineers (est. 1884), a nonprofit technical professional association recognized as an authority in technical areas ranging from computer engineering, biomedical technology and telecommunications, to electric power, aerospace and consumer electronics, among others.

**IEI** International Energy Initiative (est. 1991), an independent nongovernmental public-purpose organization led by international energy experts, with regional offices in Latin America, Africa, and Asia.

**IETA** International Emissions Trading Association (est. 1999), a nonprofit organization that established the first functional international framework for trading greenhouse gas emission reductions. Its mission is a trading regime that results in real and verifiable greenhouse gas emission reductions, balancing economic efficiency with environmental integrity and social equity.

**IGA** International Geothermal Association (est. 1988), a nongovernmental organization whose objective is to encourage research, development, and utilization of geothermal resources worldwide.

**IGCC** integrated gasification combined cycle.

**igneous** *Earth Science.* of or relating to a rock that was formed by solidification from molten or partly molten material; one of the three principal classifications of rocks along with METAMORPHIC and SEDIMENTARY.

**ignition** *Chemistry.* the point at which a substance begins a process of combustion, or the means by which this process begins.

**ignition energy** *Chemistry.* the amount of external energy that must be applied in order to ignite a combustible fuel mixture.

**ignition system** *Transportation.* a collective term for the components of an internal combustion engine that produce the spark to ignite the mixture of fuel and air; i.e., the battery, ignition coil, spark plugs, distributor, and associated switches and wiring.

**IGY** International Geophysical Year.

**IIASA** International Institute for Applied Systems Analysis (est. 1972); a nongovernmental research organization located near Vienna, Austria that conducts interdisciplinary scientific analyses of environmental, economic, technological, and social issues in the context of human dimensions of global change.

**IIR** International Institute of Refrigeration (est. 1954), an intergovernmental organization that pools international scientific and industrial knowledge in all areas of refrigeration.

**IISD** International Institute for Sustainable Development (est. 1990), a nonprofit group whose mission is to promote change towards sustainable development.

**illuminance** *Lighting.* the photometric equivalent of irradiance used as the standard metric for lighting levels; the amount of luminous flux striking a unit area of a surface, measured in lumens per square meter and expressed in lux (lx) or foot-candles (fc).

**illuminating oil** *History.* an earlier name for KEROSENE.

**illumination** *Lighting.* **1.** lighting; the application and distribution of light to a subject. **2.** another term for ILLUMINANCE.

**ILW** intermediate-level waste.

**image-dissection** *Communication.* a type of television picture tube in which an image is swept past an aperture that dissects it section by section, instead of its being scanned by an electron beam; based on the **Image Dissector** technology developed by U.S. television pioneer Philo T. FARNSWORTH in 1927.

**imbalance energy** *Electricity.* the real-time change in generation output or demand requested by an independent system operator to maintain reliability of the grid.

**Imhoff tank** *Environment.* an underground sedimentation tank used in wastewater treatment; designed to remove settleable solids from the wastewater and then anaerobically digest these solids in the lower portion of the tank.

**immiscible** *Chemistry.* not able to be mixed; describing two liquids that do not mix with each other, such as oil and water; the converse of MISCIBLE. Thus, **immiscibility.**

**immobilized electrolyte** *Storage.* describing a type of lead–acid battery in which the electrolyte (the acid) is held in place against the plates instead of being a free-flowing liquid. The two most common techniques are gel-type and glass mat batteries.

**immunotoxic** *Health & Safety.* harmful to the immune system; describing an agent that suppresses normal immune function. Thus, **immunotoxin, immunotoxicity.**

**impedance** *Electricity.* the effective resistance to the flow of electric current at a given frequency in an alternating current circuit; the reciprocal of ADMITTANCE.

**impeller** *Conversion.* **1.** a rotor for transmitting motion, as in a centrifugal pump, blower, turbine, agitator vessel, or fluid coupling. **2.** a rotating member of a centrifugal flow compressor or supercharger.

**imperial** *Measurement.* describing a system of liquid and dry measure traditionally used in Great Britain and certain Commonwealth countries, in which a gallon is equivalent to 1.2 U.S. gallons and a bushel to 1.03 U.S. bushels.

**impingement** *Ecology.* the process in which an aquatic organism strikes against the intake structure of a power plant after being drawn in along with the inflowing water.

**impoundment** *Environment.* an enclosed wetland that is hydrologically isolated from the surrounding ecosystem; this can be due to a combination of natural features (e.g., beach ridge, natural levee ridge) and anthropogenic ones (e.g., road embankment, spoil bank).

**impulse turbine** *Conversion.* 1. a turbine that produces power when a jet of water from an enclosed diversion pipeline moves with force through a specially shaped nozzle to impact directly onto the turbine blades; e.g., a Pelton turbine. **2.** a similar turbine operating on the same principle but employing pressurized steam rather than water as the high-velocity fluid.

**inactive pool** *Hydropower.* a condition in which the water stored in a reservoir or lake does not provide enough flow to generate hydroelectric power.

**incandescence** *Lighting.* the emission of visible electromagnetic radiation due to the thermal excitation of atoms or molecules.

**incandescent** *Lighting.* giving off a glowing light due to a high temperature; having the property of incandescence.

**incandescent lamp** *Lighting.* a light bulb with a resistive wire filament, usually made of tungsten, that can be heated until it glows white hot; the filament is enclosed in an evacuated bulb to prevent oxidation. This is the original form of electric lighting and historically the most common. Thus, **incandescent lighting.**

**incendiary** *Chemistry.* **1.** a chemical agent designed to cause combustion. **2.** employing such an agent, especially as a weapon.

**incentive-based** *Policy.* describing a regulation that uses the economic behavior of firms and households to attain desired environmental goals. Incentive-based programs involve taxes on emissions or tradable emission permits.

**Inch Lines** *Oil & Gas.* a system of pipelines **(Big Inch and Little Big Inch)** constructed during World War II to carry oil from East Texas to the northeastern U.S., a distance of about 1500 miles, to compensate for the fact that the supply of oil by sea was threatened by German submarines. Following the war these lines were converted to natural gas pipelines.

**inch of mercury** *Measurement.* a unit of measure for atmospheric pressure, equal to the amount of power pressure exerted by a one-inch column of mercury under standard conditions of temperature and gravity; a term derived from the use of mercury barometers to measure air pressure.

**incident** *Nuclear.* a term for any occurrence, or series of occurrences having the same origin, causing nuclear damage or creating a grave and imminent threat of such damage.

**incident (incidence) angle** see ANGLE OF INCIDENCE.

**incident radiation** *Solar.* incoming radiation; the quantity of radiant energy striking a surface per unit time and unit area.

**incinerate** *Conversion.* of a material, to burn or be burned until it is converted to ashes. Thus, **incineration.**

**incinerator** *Conversion.* a furnace that is designed for the destruction of refuse through burning; it may be fired by various means such as gas, oil, or solid fuel.

**inclined plane** *History.* a surface sloped at an angle to the horizontal (or some other reference surface), providing a mechanical advantage for raising loads; one of the simple machines along with the lever, screw, and so on.

**income elasticity** *Economics.* the degree to which the demand for a good or service changes with respect to income; e.g., a change in the demand for gasoline as a result of a change in income.

**incomplete combustion** *Conversion.* the combustion of a fuel in which it is only partially burned, with further burning possible under proper conditions. The incomplete combustion of carbon produces carbon monoxide.

**independent power producer (IPP)** *Electricity.* a generator of electricity for public use other than an electric utility, usually a small capacity plant or industrial facility, e.g., a wind farm.

**independent system operator (ISO)** *Electricity.* an independent, governmentally regulated entity established to coordinate, control, and monitor regional transmission in a nondiscriminatory manner and ensure the safety and reliability of the electric system. Similarly, **independent system planner (ISP).**

**index of sustainable economic welfare (ISEW)** *Sustainable Development.* an alternative measure for assessing the strength of an economy and human well-being. Unlike gross national product and similar indicators, it includes measures for other positive contributions such as the value of household labor, or negative contributions such as income inequality, environmental degradation, and the depletion of nonrenewable resources.

**indicated resources** *Coal.* the sum of coal resources for which estimates of the rank, quality, and quantity have been computed to a moderate degree of geologic assurance.

**indigenous** *Sustainable Development.* **1.** of an organism, native to or occurring naturally in the habitat in which it currently exists. **2.** describing a group of people known to have been living in a given area since ancient times; e.g., the Inuit are described as indigenous peoples of the Arctic region.

**indirect cooling** *Electricity.* the use of a closed loop water system to reject waste heat from a power plant to the atmosphere.

**indirect energy** *Consumption & Efficiency.* all energy requirements not considered to be DIRECT ENERGY; e.g., in the case of automobile transportation, the use of vehicle fuel would be direct energy, while the energy required to manufacture vehicles, build and maintain roads, and so on would be indirect. Also, **indirect use.**

**indirect gain** *Solar.* a type of passive solar heating system having the same basic design and principles as a DIRECT GAIN system, with the distinction being that the sunlight does not directly enter the building through windows to reach the absorbent thermal mass. In an indirect system, radiant energy is collected and trapped in a narrow space between the window and the thermal mass (typically a dark-colored wall) and then distributed to the room through vents above the mass.

**indium** *Chemistry.* a metallic element having the symbol In, the atomic number 49, an atomic weight of 114.82, a melting point of 156°C, and a boiling point of 2075°C; a soft, silver-white metal that occurs in zinc and other ores; used in alloys, semiconductor devices, and liquid crystal displays.

**indium oxide** *Materials.* $In_2O_3$; a wide band gap semiconductor that can be heavily doped with tin to make a highly conductive, transparent thin film; often used as a front contact or one component of a heterojunction solar cell.

**indium phosphide** *Materials.* InP, a brittle, toxic, metallic mass, very soluble in mineral acids; used in semiconductors and solar cells, especially multijunction photovoltaic cells.

**indoor air pollution** *Health & Safety.* a classification for pollutants that occur mainly in an indoor environment or that are considered to be more harmful in such a setting, such as dust, mold, pollen, spores, soot, cigarette or wood smoke, radon, and so on.

**indoor air quality** *Health & Safety.* a description or evaluation of the relative amount of pollutant material present in a given indoor environment. In developing countries the home use of wood fires for cooking and heating has a significant negative impact on indoor air quality.

**induced-draft** *Consumption & Efficiency.* describing a furnace in which a motor-driven fan draws air from the surrounding area or from outdoor air in order to support combustion.

**induced innovation** *Economics.* the theory that the direction and magnitude of innovative activity is shaped by market forces such as prices or supply levels; e.g., higher oil prices that lead appliance manufacturers to produce more energy-efficient refrigerators or air conditioners.

**induced radioactivity** *Nuclear.* radioactivity that is produced by bombarding a substance with radiation.

**inductance** *Electricity.* **1.** the proportionality constant of a coil or other inductive component, equal to the ratio of a generated electromotive force to the rate of change of the current. **2.** the property associated with a circuit **(self-inductance)** or a neighboring circuit **(mutual inductance),** whereby an electromotive force is generated in the presence of a changing current.

**induction** *Electricity.* the generation of voltages, currents, electric fields, or magnetic fields by interactions among these quantities without direct contact.

**induction coil** *Electricity.* a high-voltage step-up transformer that can change direct current into alternating current by periodic interruption of direct current through the primary winding.

**induction generator** *Conversion.* an induction device driven at a higher than synchronous speed by separate external power, used to convert mechanical power to electrical power.

**induction hardening** *Materials.* the localized surface heating of a medium carbon steel by an induction coil, so that the temperature is raised above 900°C.

**induction heating** *Materials.* a method of heating by electrical induction, particularly the process of heating electrically conductive materials, such as steel, by inducing high-frequency currents within the material.

**induction lighting** *Lighting.* light generated when gases are excited directly by radio-frequency waves or microwaves from a coil that creates induced electromagnetic fields; it differs from a conventional discharge that uses electrodes to carry current into the arc.

**induction motor** *Conversion.* an alternating current motor in which the currents in the secondary winding are created by induction; often used because of its simple construction, efficiency, and speed regulation.

**inductive charge** *Electricity.* the charge that is contained on an object as a direct result of its being near another charged object.

**inductive coupling** *Electricity.* the coupling or linkage of two circuits by the changing of magnetic lines of force.

**inductor** *Electricity.* a coil of wire that has the property of inductance (see above).

**Industrial Age** *History.* the period in human history commencing with the INDUSTRIAL REVOLUTION and extending to the present day. Many observers describe the current era as the **Postindustrial Age** because of the reduced role of traditional manufacturing as opposed to the service economy.

*industrial agriculture* *Industrial agriculture is much more energy-intensive than traditional agriculture, and also capable of larger crop yields.*

**industrial agriculture** *Consumption & Efficiency.* a modern method of food production characterized by the use of high-yield varieties of crops and livestock, in conjunction with intensive inputs in the form of machinery, energy, fertilizers, hormones, antibiotics, and other chemicals, and irrigation water. Compare TRADITIONAL AGRICULTURE.

**industrial ecology** *Ecology.* the study of the flows of material and energy resources in industrial and consumer activities, of the effects of these flows on the environment, and of the influences of economic, political, and social factors on the use of these resources. See next column.

**industrial ecosystem** *Ecology.* an industrial system that is in some way similar to or modeled after a natural ecosystem; e.g., a group of industrial organizations that agree to coordinate their inputs and outputs in a web of interdependency, comparable to the manner in which material and energy flows take place in nature.

**industrial fishery** *Ecology.* a fishing operation undertaken by a commercial company in which relatively large quantities of capital and energy are deployed.

**industrial fixation** see NITROGEN FIXATION.

**industrialization** *Sustainable Development.* the process of establishing an industrial society,

**industrial ecology** Industrial ecology is an interdisciplinary field of inquiry based on the idea that industrial economic systems resemble natural ecosystems. By understanding the industrial metabolism, that is, the flows of energy and materials in all or parts of an economy, designers, engineers, business planners, and policy makers should be able to significantly reduce environmental burdens. The term *industrial ecology* was popularized following the publication in 1989 of a seminal article in *Scientific American* by Robert Frosch and Nicholas Gallopoulos. Following this event, the field developed during the 1990s and has spawned academic programs, several scholarly journals, and an international society. Industrial ecology draws on principles from ecology, thermodynamics, and systems theory. Life cycle analysis (LCA) and material flow accounting (MFA) are two primary tools used in the field. Building on the notion of symbiosis in nature, highly interconnected industrial networks using wastes as process inputs (industrial symbioses) should more closely mimic the parsimony of closed-loop natural systems. Going beyond the descriptive or analogical relationship, industrial ecology also draws on the metaphor of ecosystems as examples of sustainability, capable of exhibiting resilience, integrity, and adaptive capability in the face of disturbances. The metaphor extends the view of industrial economies as merely material and energy flow networks to incorporate systemic characteristics such as interconnectedness, cooperation, and community, all of which are thought to underpin sustainable features of living systems. Drawing on the association with ecosystems, industrial ecology has been called "the science of sustainability".

**John R. Ehrenfeld**
Executive Director, International Society for Industrial Ecology

especially the development and widespread use of modern energy technologies such as steam power, coal, oil and natural gas, and electricity as replacements for traditional energy sources such as human and animal power. Thus, **industrial society.**

**industrial metabolism** *Biological Energetics.* a concept that employs a biological systems perspective to describe the flow of materials and energy in industrial activities.

**Industrial Revolution** *History.* the name for a process of technological innovation, mechanized development, and agricultural improvements that first gathered momentum in Great Britain in the late 18th century and then extended to continental Europe, North America, and other areas in the early 19th century and thereafter. See next column.

**industrial society** see INDUSTRIALIZATION.

**industrial spirit** *Oil & Gas.* a term for light oils distilling between 30°C and 200°C.

**industrial symbiosis** *Policy.* a relationship in which at least two industrial facilities exchange materials, energy, or information so as to produce a collective benefit greater than the sum of individual benefits that could be achieved by acting alone. The most famous example of this is the eco-industrial park at Kalundborg, Denmark.

**industrial waste** *Environment.* any of the various categories of waste generated from manufacturing or commercial processes; the term usually implies some type of hazardous or toxic waste.

**inelastic** *Economics.* having an ELASTICITY less than one.

**inelasticity** *Economics.* a lack of ELASTICITY.

**inert** *Chemistry.* not reacting with other elements.

**inert gas** another term for a NOBLE GAS.

**inertia** *Physics.* the property of a material body, due to its mass, by which it resists any change in its motion unless it is overcome by force. Thus if no outside forces are present, a stationary body will tend to remain at rest, and a moving body will tend to continue in the same direction at the same speed. Thus, **inertial.**

**inertial confinement** *Nuclear.* a technique for confining plasma in nuclear fusion that involves imploding a small fuel pellet (usually an equal mixture of deuterium and tritium) at a very high speed, temperature, and density. The inertia of the imploding pellet keeps the reaction confined momentarily. Extremely high densities are required to maintain the initial confinement for any practical period of time.

**Industrial Revolution** This term is a commonly used misnomer for an ill-defined period when traditional economies (based on biomass fuels, muscle power and artisanal manufactures) began to be transformed into more complex systems based on coal and steam power and using mechanized, factory-based processes to produce larger volumes of more affordable goods. There is no consensus about dating this process that stretched across many generations and that began in England but had many important continental and American components. Samuel Lilley favored the most liberal span between 1660 and 1918 and subdivided it into the early (1660–1815) and the mature (1815–1918) period. Walt Rostow defined the most restricted span (1783–1802) which he defined as the period of England's economic take-off. The most common periodizations span the second half of the 18th and the first half of the 19th century: Arnold Toynbee opted for 1760–1840, Thomas Ashton for 1760–1830, Hugh Beales for 1750–1850, but he also argued for no terminal date. Leaving aside the appropriateness of the term 'revolution' for what was an inherently gradual process, it is obvious that because energy transitions and technical and economic take-offs began at different times in different places—and industrialization is yet to commence in most of the sub-Saharan Africa where biomass fuels, animate energies and artisanal work are still so common—there can be no clear consensus on dating this remarkable era of human history.

**Vaclav Smil**
University of Manitobas

**INES** International Nuclear Event Scale.

**infant mortality rate** *Health & Safety.* a measure of reproductive health, typically expressed as the number of infant deaths between birth and one year of age, per 1000 births.

**inferred reserves** *Oil & Gas.* expected cumulative additions to proved reserves in oil and gas fields discovered as of a certain date, where such additions are made by extensions, new pools, or enhancement of flow to the well.

**inferred resources** *Coal.* a term for coal resources of a low degree of assurance, for which estimates of the quality and size are based on geologic evidence and projection.

**infiltration** *Consumption & Efficiency.* **1.** the uncontrolled inward leakage of outdoor air into a building through cracks and gaps in the building envelope, especially around windows and doors; this usually refers to cold air during the winter and hot air during the summer. **2.** *Earth Science.* the movement of water into soil or porous rock.

**infinite** *Consumption & Efficiency.* describing energy sources whose supply is theoretically unlimited given existing environmental conditions; i.e., wind.

**inflammable** *Materials.* tending to burn; likely to catch fire; flammable.

**inflammable air** another term for FIRE AIR.

**inflation** *Economics.* an increase in prices for comparable goods and services across the economy over time. Thus, **inflationary.**

**inflow** *Hydropower.* **1.** any inward flow of water. **2.** the amount of water that flows into a reservoir or other enclosure during a specified period.

**Influence Machine** *History.* a device developed by English physicist Francis Hauksbee (1706) that is considered to be the first static electric or frictional electric machine. It was a mercury-filled glass globe mounted on an axle; when spun it would create a mysterious "luminosity" that crackled like lightning when touched and attracted metal particles.

**informal economy** *Economics.* an exchange of goods and services not accurately recorded in government figures and accounting; this can include transactions such as day care and babysitting, tutoring, house cleaning, yard work, and the like.

**Information Age** *Communication.* a term for the current period in history, based on the idea that the fundamental characteristic of contemporary society in developed countries is the exchange of information, especially by means of computers and other electronic media.

**information processing** *Communication.* the computerized evaluation, analysis, and tabulation of data to generate usable information.

**information technology (IT)** *Communication.* the use of computer and telecommunication systems for the processing and distribution of information in digital, audio, video, and other forms.

**information theory** *Communication.* **1.** a branch of learning that is concerned with the study of information representation and transmission. **2.** specifically, an aspect of mathematics postulating that information can be measured and described if it were a physical quantity, such as mass or energy, in terms of the degree of order (or randomness) of the communication system.

**infrared** *Physics.* **1.** the portion of the invisible spectrum consisting of electromagnetic radiation with wavelengths in the range from about 750 nanometers to 1 millimeter; i.e., longer than those of visible light, but shorter than those of radio waves. **2.** relating to or involving such rays. Thus, **infrared energy, infrared radiation.** [Literally, "below the red"; because it is beyond the redder end of the visible spectrum.]

**infrared background radiation** see BACKGROUND RADIATION.

**Ingenhousz, Jan** 1730–1799, Dutch physician who established that in sunlight, plants absorb carbon dioxide and give off oxygen, the first indication of light's role in the photosynthetic process. He also discovered that only the light of the sun (and not the heat it generates) is necessary for photosynthesis, and that only the green parts of plants carry out photosynthesis.

**inhalation unit risk** *Health & Safety.* the excess lifetime cancer risk estimated to result from continuous lifetime exposure to a carcinogenic substance at a concentration of 1 $mg/m^3$ in the air.

**injection molding** *Materials.* a process of molding shapes out of softened, usually thermoplastic material that has been forced or squirted out of a heated chamber by a ram or screw into a water-cooled metal mold; used in metallurgy.

**injection well** *Oil & Gas.* a well in which fluids are injected rather than produced, as to augment hydrocarbon production or maintain reservoir pressure. Thus, **injection fluid.**

**injectivity** *Measurement.* the rate at which natural gas can be injected into gas storage; usually measured in volume expressed at 15.6°C (60°F) and standard atmospheric pressure per unit day.

**injector** *Transportation.* the tube or nozzle in a fuel injection system through which fuel is introduced into the intake airstream or the combustion chamber.

**inorganic** *Chemistry.* **1.** not an organic substance; not a hydrocarbon or a derivative of hydrocarbon. **2.** relating to or involving inorganic chemistry. **3.** *Materials.* not composed of living or formerly living organisms, i.e., plants and animals.

**inorganic chemistry** *Chemistry.* **1.** the scientific study of all chemical substances except hydrocarbons and their derivatives; in general this excludes all substances that contain carbon. **2.** formerly, the study of elements and compounds that are not directly associated with living organisms.

**input** *Electricity.* **1.** the electrical energy that is fed to a device, or the point at which this energy enters the device. **2.** to convey or transfer electrical energy to a device. **3.** *Communications.* data that is transferred from some external source to a computer's main memory; e.g., by typing on a keyboard. **4.** to introduce data into a computer. **5.** *Materials.* any substance or form of energy introduced to a system.

**input-intensive** *Consumption & Efficiency.* describing agricultural production that relies heavily on energy inputs, including fossil fuels, machinery, artificial fertilizers, pesticides, and irrigation systems.

**input-output (I-O) analysis** *Economics.* a technique used to track resources and products within an economy, in order to examine how a change in one economic sector affects other economic sectors. See below.

**insecticide** *Ecology.* any agent that is lethal to insects, especially a manufactured substance used to kill insect pests in agricultural areas, yards and gardens, households, and so on.

**insensible heat** *Thermodynamics.* heat transfer to a material that does not result in a change in temperature of this material; e.g., that is lost through evaporation.

**Insight** *Renewable/Alternative.* the brand name of the first modern hybrid-electric vehicle sold in the United States (introduced to the market by Honda in 1999).

**in situ combustion** *Oil & Gas.* an enhanced oil recovery method for recovering oil of low API gravity and high viscosity from a field; it involves igniting the oil downhole in the formation. The heat from this breaks down the

**input-output analysis** A framework for the quantitative description and analysis of the interdependence of the different sectors of an economy. It was created by Wassily Leontief, who described the basic input-output model and database in publications starting in the 1930s and received the Nobel Prize in Economics for these contributions in 1973. Input-output economics is a systems approach intended as an operational system of general equilibrium. However, it makes different assumptions and different simplifications from what became the dominant interpretation of general equilibrium among academic economists. The field is best known for the input-output tables, which have for half a century been a central part of the National Accounts prepared by Statistical Offices around the world, and for the basic input-output model, a system of linear equations that are used by researchers in a variety of fields. Its popularity for the analysis of energy flows and material flows, and for environmental analysis more generally, is due to two distinguishing characteristics: it maintains a substantial amount of interindustry detail while comprising a macro perspective, and its mathematical formulation facilitates the measurement of resources and goods in both physical and price units. Input-output economists have developed nonlinear dynamic input-output models and nonlinear input-output models of the world economy that rely on the familiar input-output tables plus other types of data, especially data in physical units that begin to be compiled and analyzed on a more systematic basis. The field has an international professional society with a scholarly journal, biannual conferences, and a membership that includes industrial ecologists and ecological economists.

**Faye Duchin**
Rensselaer Polytechnic Institute

oil into coke and light oil, and while the coke burns, the lighter, less-viscous oil is forced ahead to the wellbores of the producing wells.

**in situ gasification** *Coal.* the process of converting coal into synthetic gas directly at the site where the coal was located and mined.

**in situ leach mining** *Mining.* solution mining; a mining process that involves leaving the ore where it is in the ground rather than extracting it, and pumping liquids through it to recover the minerals out of the ore by chemical leaching. As a result there is little surface disturbance and no generation of tailings or waste rock.

**insolation** *Solar.* solar radiation that is received at the earth's surface.

**instantaneous irradiance** *Solar.* the quantity of solar radiation incident on a given surface area over a specified time period, measured in watts per square meter.

**instantaneous water heater** *HVAC.* another term for a DEMAND WATER HEATER, so called because it supplies water immediately on demand.

**instream flow** *Hydropower.* the amount of water flowing in a given stream.

**insulate** *Materials.* **1.** to cover or line a space or substance with a material that prevents or retards the passage of heat energy, electricity, or sound. **2.** to insert an intermediate nonconducting material that will provide electric isolation of two or more conductors.

**insulation** *Materials.* **1.** the act or fact of insulating. **2.** any material employed to reduce or prevent the transfer of electricity, heat, cold, or sound; used in walls, ceilings, floors, and other building spaces. **3.** a nonconducting material used to provide electric isolation of two or more conductors.

**insulator** *Materials.* **1.** any material used for insulation, especially building insulation. **2.** a material that conducts almost no current; used to provide voltage isolation.

**Insull, Samuel** 1859–1938, English-born financier who formed a huge interlocking directorate of American public utility companies in the U.S., including hundreds of steam and hydroelectric plants and numerous other power plants. He helped form the basis for electric utility regulation by states.

**intake** *Consumption & Efficiency.* an opening for air, water, fuel, or other fluid to enter a system; e.g., for fuel to enter the combustion chamber of an engine, or for water to enter a hydropower system.

**intake manifold** see MANIFOLD.

**integral collector system** *Solar.* another term for a BATCH HEATING system.

**integrated assessment** *Measurement.* a method of analysis that combines results and models from the physical, biological, economic and social sciences, and the interactions among them to evaluate the status and consequences of environmental changes and the policy responses to them. This is a common method of studying climate change and its impacts.

**integrated biosystem** *Ecology.* an industrial enterprise in which wastes or unused resources from one aspect of the enterprise are employed as resources in another aspect, so as to reduce overall consumption levels and waste emissions; this can be a traditional system such as a farm using crop wastes to feed its livestock, and the livestock waste to fertilize crops, or a more complex industrial practice such as a brewery converting its organic wastes to biofuel.

**integrated circuit** *Electricity.* a miniaturized electronic circuit whose components are formed on a single semiconductor substrate; a key element in the development of the personal computer and other modern electronic devices.

**integrated gasification combined cycle (IGCC)** *Conversion.* a conversion process in which coal is first gasified to form a synthesis gas, which is then used as fuel for a combined-cycle power plant.

**integrated irradiance** *Solar.* the quantity of solar radiation incident on a given surface area during a specified time, divided by the duration of that time period.

**integrated resource planning (IRP)** *Policy.* **1.** a planning framework that compares the costs and benefits of alternative electricity supply mechanisms, including demand-side savings and nontraditional sources, across an entire electric power system. IRP typically includes consideration of environmental and social externalities. **2.** the management of two or more resources in the same general area, such as water, soil, timber, grazing land, wildlife, and recreation.

**intelligent transportation system (ITS)** *Transportation.* a range of technologies aimed at improving the safety, efficiency, and convenience of a surface transportation network, including electronics, information processing, wireless communications, and control systems.

**intensive** *Thermodynamics.* describing a property of a system, such as its pressure, temperature, or density, that is not dependent on the quantity of material present in the system. Thus, **intensive property, intensive variable.** Compare EXTENSIVE.

**intensive farming** *Ecology.* the practice of farming in which the yield is sufficient for the farmer to feed his family and also have a surplus that can be sold to others for a profit.

**interburden** *Coal.* rock and other material that occurs in between coal seams, as opposed to the overburden that overlies the entire deposit.

**interconnection** *Electricity.* **1.** the link between a distributed energy generator and the load being served by the utility electricity network. **2.** the linking of two or more power generators in phase to produce more energy than a single one. **3.** a conductor within a photovoltaic module that provides an electrical connection between the solar cells.

**interface** *Chemistry.* an area or surface that represents the boundary between two separate phases of a chemical or physical process. The three different states of matter have five possible types of interface: liquid/liquid, liquid/gas, solid/solid, solid/liquid, and solid/gas. (Gas/gas has no definable interface because gases are completely soluble in one another.)

**interferometer** *Measurement.* an instrument that splits a light beam into two or more beams, which then recombine to form interference; used for determining the spectral distribution of light.

**interfuel substitution** *Economics.* the substitution of one fuel for another following a change in their relative prices.

**intergenerational equity** *Social Issues.* an evaluation of the fairness of a given policy or practice on the basis of costs and benefits to different generations; e.g., the present use of a depletable resource that will therefore not be available in the future, or the present lack of a policy to ameliorate future impacts such as climate change.

**interim retrievable storage** *Nuclear.* the storage of radioactive waste materials in a temporary facility, with the premise that the materials can be retrieved from storage when a satisfactory treatment method for them has been developed.

**intermediate-level waste (ILW)** *Nuclear.* radioactive waste typically made up of resins, chemical sludges, and metal fuel cladding, as well as contaminated materials from reactor decommissioning. It can be solidified in concrete or bitumen for disposal. ILW contains high amounts of radioactivity and may require shielding.

**intermittent** *Renewable/Alternative.* describing energy sources whose availability is not constant, such as wind or solar energy.

**intermittent ignition** *Consumption & Efficiency.* a system in which the ignition source is automatically shut off when the equipment is in an off or standby condition

**intermolecular forces** *Physics.* the forces that act on an individual molecule to cause it to attract or repel other adjacent molecules; the strength of these forces in a given molecular substance at a certain temperature determines whether the substance exists as a gas, liquid, or solid. See also INTRAMOLECULAR FORCES.

**internal combustion engine** *Transportation.* an engine in which the process of combustion takes place in a cylinder or cylinders within the engine; the working fluid is a fuel and air mixture, which reacts to form combustion products and is then exhausted; e.g., a gasoline or diesel engine. See next page.

**internal energy** *Thermodynamics.* a property of a system that includes the kinetic energies of the individual particles of the system, the interaction energies between particles, and the intrinsic energies of individual particles, but that does not include the kinetic and potential energies of the system as a whole.

**internal gain** *Solar.* the energy produced as heat inside an enclosed space by its occupants (i.e., body heat) and appliances (such as lighting and electrical equipment).

**internal combustion engine** The internal combustion engine, "the" motor of the early 20th century economy, has brought far-reaching changes to society that enabled convenient and affordable individual transportation. The conversion of chemical energy to mechanical work is accomplished via combustion of mostly hydrocarbon fuels under high pressure conditions. A reciprocating piston-in-cylinder arrangement is used to compress air and fuel and then this mixture is ignited. Combustion rapidly produces high pressure and high temperature gases in the enclosed volume. This gas rapidly accelerates the piston and mechanical work can be transferred to a rotating shaft. The late 19th century designs of Nicolaus Otto and Rudolf Diesel are still the most commonly used engine types; the spark-ignition Otto engine and the compression-ignition Diesel engine. While substantial improvements in fuel economy and pollutant release have been made, internal combustion engines are still a major source of air pollution. The complex interaction of turbulent flow and chemical reactions that occur at high but finite rates, make it difficult to find the overall best approach for efficient and clean combustion of fuel. However, advanced methods, like fuel direct-injection, have pushed thermal efficiencies above 40%. The use of catalytic converters and filters has helped to remove most of the unburned hydrocarbons and nitric oxide from the exhaust—the sources of the infamous Los Angeles smog. More and more, internal combustion engines benefit from sophisticated electronic controls that are used to optimize performance based on actual power demands during operation.

**Volker Sick**
University of Michigan

**internal mass** *Materials.* a term for materials with high thermal energy storage capacity, contained in a building's walls, floors, or free-standing elements.

**internal reforming** *Hydrogen.* the production of a desired product (usually hydrogen) within a fuel cell from a hydrocarbon fuel (methanol, gasoline, natural gas, propane, and so on) that has been fed to the fuel cell or stack; generally a less expensive and more efficient process than external reforming.

**internal resistance loss** *Renewable/Alternative.* a reduction in fuel cell performance because of resistance losses caused by internal structures that create resistance to electron or ion flow.

**International** For names of organizations beginning with the word "International", see the acronym; e.g., for International Energy Agency, see IEA; for International Geothermal Association; see IGA; for International Petroleum Exchange, see IPE.

**International Commission on Illumination** see CIE.

**International Geophysical Year (IGY)** *Earth Science.* by international agreement, a period during which increased observation of worldwide geophysical phenomena is undertaken through the cooperative effort of the various participating nations. July 1957 to December 1958 was the first such year.

**International Nuclear Event Scale (INES)** *Nuclear.* a scale designed for prompt public communication of the significance of an event at a nuclear facility; a rating of 0 is an event with no safety significance, while 7 is an accident involving a major release of radioactive material with widespread health and environmental effects.

**international poverty line** *Global Issues.* an income level established by the World Bank to determine which people in the world can be classified as poor; the level was set at $1 a day per person in 1985, adjusted to local currencies using purchasing power parities.

**International System of Units** *Measurement.* a globally accepted system of measurement employing as its fundamental units: the meter (m) for length, the kilogram (kg) for mass, the second (s) for time, the Kelvin (K) for temperature, the ampere (a) for electric current, the candela (cd) for illumination, and the mole (mol) for the amount of a substance. Abbreviated SI (from the French *Système International*).

**international temperature scale** *Measurement.* a standard temperature scale defined on the basis of certain fixed and easily reproducible points that are assigned definite temperatures; the primary fixed-point temperatures in degrees Celsius include the triple point of water (0.01); the normal boiling points of water (100) and oxygen (–182.692), and the normal freezing points of gold (1064.43) and silver (961.93).

**Internet** *Communication.* a publicly available worldwide system of interconnected computer networks that has revolutionized communications and methods of commerce throughout the world.

**interpollutant trading** *Environment.* an arrangement in which a business or other entity makes reductions of one type of pollutant in order to offset the increases of another pollutant.

**interruptible service** *Consumption & Efficiency.* an electricity or natural gas arrangement under which, in return for lower rates, the customer must either reduce energy demand on short notice or allow the utility to cut off the energy supply temporarily for the utility to maintain service for higher priority users, as during periods of high demand. Thus, **interruptible power, interruptible contract, interruptible transmission,** and so on. Compare FIRM SERVICE.

**intertie** *Electricity.* **1.** a high voltage line that carries electric power long distances. **2.** a circuit connecting two or more control areas of an electric system.

**interventionism** *Economics.* a direct process of involvement by the government in the economy or a sector of the economy, as by acquiring property rights and undertaking productive activities.

**intramolecular forces** *Physics.* those forces that act to cause individual atoms to attract other adjacent atoms within the same molecule, and which are relatively strong in comparison to the forces that act between different molecules (INTERMOLECULAR FORCES).

**intrinsic** *Materials.* describing a layer of semiconductor material (especially as used in a photovoltaic device) whose properties are essentially those of the pure, undoped material. Thus, **intrinsic layer, intrinsic semiconductor.**

**inverter** *Electricity.* a device that converts something to its opposite; e.g., DC current to AC current, a positive signal to a negative one, a computer input pulse to an output pulse, and so on.

**investor-owned utility (IOU)** *Electricity.* a privately-owned electric utility whose stock is publicly traded; it is rate regulated and authorized to achieve a certain allowed rate of return.

**ion** *Chemistry.* an atom, radical, or molecule that has gained or lost one or more electrons and thus acquired a net negative or positive charge. In electrolysis, positive ions (cations) travel to the cathode, while negative ions (anions) travel to the anode.

**ion beam fusion** *Nuclear.* a method of internal confinement fusion in which an energy beam of electrons or other particles is directed onto a tiny pellet of a deuterium–tritium mixture, causing it to explode like a miniature hydrogen bomb, fusing the deuterium and tritium nuclei within a time span too short for them to repel each other.

**ion exchange** *Chemistry.* **1.** a chemical reaction in which ions are interchanged between one substance and another, usually by means of passing a liquid through a porous, granular solid that is relatively insoluble. **2.** the use of such a process to replace certain selected anions or cations in a solution; e.g., to remove undesirable substances, as in water softening, or to recover desirable ones, as in the separation of valuable metals from wastes.

**ionic** *Chemistry.* having to do with or involving ions; occurring in the form of an ion.

**ionic bond** *Chemistry.* a type of chemical bond in which atoms of different elements join by transferring one or more electrons from one to the other to form an ionizing or polar compound.

**ionic conduction** *Electricity.* the movement of charges within a semiconductor due to the displacement of ions within the crystal lattice; an external source of energy is required to maintain this movement.

**ionic semiconductor** *Electricity.* a solid in which the electrical conductivity from the flow of ions predominates over that from the movement of electrons or holes.

**ionization** *Chemistry.* the process of adding an electron to, or removing an electron from, an atom or molecule so as to give a negative or positive net charge; the atom is then called an ion.

**ionization chamber** *Nuclear.* a device that measures the intensity of ionizing radiation.

**ionization energy** *Physics.* the amount of energy required to remove an electron from a specific atom or ion to an infinite point, generally expressed in electron volts and numerically equal to the **ionization potential.**

**ionization potential** see IONIZATION ENERGY.

**ionizing radiation** *Nuclear.* photons of high-energy electromagnetic radiation and particle forms of radiation that have sufficient energy to produce ions by removing electrons from atoms or molecules.

**ionosphere** *Earth Science.* a region in the earth's atmosphere, beginning at an altitude of 70–80 km and extending to an indefinite height, in which free electrons and ions produced by solar radiation are abundant and affect certain radio waves that propagate through this region

**IOU** investor-owned utility.

**IPAT equation** *Global Issues.* an equation that describes human impact on the environment as Impact = Population × Affluence × Technology; i.e., the impact is a function of population size, affluence (wealth), and available technology. Thus a large population of wealthy people in a modern consumer society will have a greater impact on the environment than a small indigenous group in a preindustrial economy. See next column.

**IPCC** Intergovernmental Panel on Climate Change (est. 1988), the most authoritative and influential scientific body on climate change. Its main activity is to provide in regular intervals an assessment of the state of knowledge on climate change.

**IPE** International Petroleum Exchange (est. 1980), Europe's leading energy futures and options exchange and the second largest in the world after the New York Mercantile Exchange (NYMEX). The IPE lists three main energy contracts: Brent Crude oil futures and options, heating oil futures and options, and natural gas futures.

**IPP** independent power producer.

**IRAC** *Economics.* imported refiner acquisition cost; a volume-weighted average price of all crude oils imported into the U.S. over a specified period.

**iridescence** *Materials.* the appearance of rainbowlike colors due to the interference of light reflected from the front and rear surfaces of a thin film or thinly layered material, such as is exhibited in nature in soap bubbles.

**IPAT Equation** The IPAT equation, though phrased mathematically, is a simple conceptual expression of the factors that create environmental impact. IPAT is an accounting identity stating that environmental impact (I) is the product of three terms: (1) population (P); (2) affluence (A); and (3) technology (T). It is stated I = P x A x T or I = PAT. Generally credited to ecologist Paul Ehrlich, the IPAT formulation arose from a dispute in the early 1970s among the most prominent environmental thinkers of the day about the sources of environmental impact. Ehrlich and John Holdren identified population size and growth as the most urgent IPAT factor, whereas Barry Commoner argued that Post WWII production technologies were the dominant reason for environmental degradation. Passing over more complex models, IPAT has been chosen by many scholars in both the social and natural sciences as a starting point for investigating interactions of population, economic growth, and technological development. Operationalizing IPAT terms has figured in work on consumption, agriculture, and energy decomposition analysis. IPAT variants have been used by the International Panel on Climate Change in energy-related carbon emission studies. IPAT variants since the 1990s have reversed the original 1970s view of the harmful role of technology. World Resources Institute studies and founders of industrial ecology reinterpreted IPAT to suggest that given increases in population and affluence, the T term of the IPAT equation then becomes an essential counterweight to P and A requiring environmentally effective technological choices to reduce environmental impact per unit of economic activity.

**Marian Chertow**
Yale University

**IRN** International Rivers Network (est. 1985), a nonprofit, volunteer organization that promotes the work of local organizations around the world striving for the wise use of rivers and fresh water.

**iron** *Chemistry.* a heavy, malleable, silver-white metallic element having the symbol Fe, the atomic number 26, an atomic weight of 55.847, a melting point of 1536°C, and a boiling point of about 3000°C; a widely distributed substance that has been universally used since ancient times as a base for materials such as steel, cast iron, and wrought iron.

**Iron Age** *History.* the final technological and cultural stage in the archaeological sequence known as the three-age system (Stone Age, Bronze Age, and Iron Age) in which the working of iron came into general use, replacing bronze as the basic material for implements and weapons. It began in the Middle East and southeastern Europe about 1200 BC and in China about 600 BC.

**iron hypothesis** *Ecology.* the concept that iron plays a major regulatory role in phytoplankton productivity; it has been proposed that fertilizing the ocean with iron would enhance photosynthesis and thereby reduce atmospheric concentrations of carbon dioxide.

**iron ore** *Materials.* a rock or deposit containing iron-bearing compounds from which metallic iron can be economically extracted.

**iron oxide** *Materials.* a term for any of various iron–oxygen compounds such as ferric oxide or ferrous oxide.

**iron sponge** *Hydrogen.* a proposed form of hydrogen storage in which hydrogen and rust ($Fe_3O_4$) are converted into pure iron, which is then transported to the point of end use where the reverse reaction (oxidation) liberates the hydrogen.

**irradiance** *Solar.* a radiometric term for the rate at which radiant energy is transferred across a unit area of a surface, commonly measured in watts per square meter.

**irradiated** *Nuclear.* subject to irradiation; exposed to ionizing radiation or other radioactivity. Thus, **irradiated (nuclear) fuel.**

**irradiation** *Solar.* **1.** the integration of IRRADIANCE over a specific time period, commonly measured in joules per square meter. **2.** *Nuclear.* exposure to radiation in any form, either intentional (for therapeutic purposes) or unintentional.

**irreversible** *Thermodynamics.* describing a process that cannot return both the systems and its surroundings to their original conditions if the process were reversed.

**irrigation** *Consumption & Efficiency.* the use of water for agricultural purposes in any manner that directs the water artificially to the point of use (as opposed to water that is naturally provided by rainfall or streamflow); used especially in areas that lack sufficient rainfall during the growing season.

***irrigation*** *Huge irrigation pipes scale the walls of the Columbia River Gorge at Grand Coulee Dam, carrying water from the reservoir behind the dam.*

**isenthalpic** *Thermodynamics.* at constant enthalpy; not involving a change in enthalpy. Thus, **isenthalpic compression, expansion,** or **process.**

**isentropic** *Thermodynamics.* at constant entropy; not involving a change in entropy. Thus, **isentropic compression, expansion,** or **process.**

**ISES** International Solar Energy Society (est. 1954), a nongovernmental organization whose principal mission is to encourage the use and acceptance of renewable energy technologies.

**ISFSI** *Storage.* independent spent fuel storage installation; a complex designed and constructed for the interim storage of spent nuclear fuel and other radioactive materials. The typical design for an ISFSI is a concrete pad with dry casks containing spent fuel bundles.

**ISO** independent system operator.

**iso-** *Measurement.* a prefix meaning "equal".

**isobar** *Physics.* **1.** a line or curve that represents a locus of equal pressure points in a fluid. **2.** any of a group of atomic species having the same mass number but different chemical or physical properties. **3.** *Earth Science.* a line on a map or chart connecting points of equal or constant pressure.

**isobaric** *Thermodynamics.* describing a thermodynamic locus that exists at constant pressure. Thus, **isobaric process.**

**isoelectric** *Chemistry.* having a net electrical charge of zero.

**isofootcandle** another term for ISOLUX.

**isohel** *Solar.* a line on a weather map or chart connecting points that receive equal amounts of sunshine during a given interval of time, usually one year.

**isoilluminance curve** *Solar.* a locus of points on a surface where the illuminance has the same value.

**isolated gain** *Solar.* a type of passive solar heating system in which the principal components are physically separated from the main part of the building to be heated, such as a greenhouse or sun room that conveys heat through vents or pipes to the living area of a house.

**isolated system** *Thermodynamics.* an entity that exchanges neither matter nor energy with its surroundings.

**isolation diode** *Photovoltaic.* a diode that prevents one segment of a photovoltaic array from interacting with another array segment; usually used to prevent array energy from flowing backwards through a subvoltage series string.

**isolation transformer** *Electricity.* a transformer with physically separated primary and secondary windings, thus preventing primary circuit voltage from being impressed onto the secondary circuits.

**isolux** *Lighting.* a curve, line, or surface whose points all have equal light intensity. Thus, **isolux line, isolux chart,** and so on.

**isomer** *Chemistry.* **1.** one of two or more substances that have the same chemical composition but differ in structural form; e.g., glucose and fructose. **2.** *Nuclear.* one of several nuclides with the same number of neutrons and protons capable of existing for a measurable time in different nuclear energy states.

**isomerate** *Chemistry.* a product made in an isomerization process; e.g., a gasoline blendstock.

**isomerism** *Chemistry.* **1.** a condition in which two or more chemical compounds possess the same number of atoms of the same elements, and thus have the same molecular formula, but differ in the arrangement of the atoms, and thus have different chemical properties. **2.** *Nuclear.* the existence of excited states in two or more isomers.

**isomerization** *Chemistry.* any process that changes a substance into an isomer, such as butane into isobutane.

**isometric** *Thermodynamics.* **1.** having a constant volume. **2.** describing a process in which the volume of a substance remains constant.

**isometric transition** *Nuclear.* a mode of radioactive decay in which a nucleus goes from a higher to a lower energy state while the mass number and the atomic number remain unchanged.

**isopleth** *Measurement.* a line connecting points of equal value for some quantity, e.g., temperature, rainfall, or solar radiation.

**isotherm** *Earth Science.* a contour line on a weather map connecting points having equal or constant air temperature.

**isothermal** *Thermodynamics.* taking place at or having a constant temperature.

**isotope** *Chemistry.* a member of a chemical element family that has two or more nuclides with the same number of protons but a different number of neutrons, so that while they have the same chemical attributes, they often display different physical attributes; e.g., carbon-12, which is stable, and carbon-14, which is radioactive. Thus, **isotopic.**

**isotopic enrichment** *Nuclear.* the process of increasing the concentration of one isotope in a sample relative to that of the other isotopes of that element

**isotropic** *Physics.* having physical properties that do not vary with direction. Thus, **isotropic energy, isotropic fluid, isotropic radiation,** and so on.

**IT** information technology.

**Itaipú Dam** *Hydropower.* a joint project of Brazil and Paraguay on the Parana river, which began operation of its first unit in 1983. Itaipu eventually became the largest hydroelectric power facility in the world, supplying about one-fourth of Brazil's electrical power and nearly all of Paraguay's.

**ITDG** Intermediate Technology Development Group (est. 1966), a nonprofit group whose aim is to demonstrate and advocate sustainable use of technology in order to reduce poverty in developing countries.

**ITER** *Nuclear.* International Thermonuclear Experimental Reactor; a large-scale experiment conducted by an international consortium to study burning plasmas.

**ITS** intelligent transportation system.

**I-type** *Materials.* intrinsic type; describing a semiconductor material that is left intrinsic (i.e., that has essentially the same properties as the pure, undoped material), so that the concentration of charge carriers is characteristic of the material itself rather than of added impurities. Thus, **I-type layer, I-type semiconductor.**

**I-V curve** *Photovoltaic.* a graphical presentation of the current versus the voltage from a photovoltaic device as the load is increased from the short circuit (no load) condition to the open circuit (maximum voltage) condition; the shape of the curve depicts cell performance.

**Ixtoc** *Environment.* an experimental oil well off the coast of Mexico that experienced a blowout in 1979, releasing about 140 million gallons of crude oil. This is the largest nonwar related oil spill to date (though its environmental impact was relatively small for a spill of that magnitude).

**J** joule.

**jackup drilling** *Oil & Gas.* a process of offshore drilling using a self-contained combination drilling rig **(jackup rig)** and floating barge, fitted with long support legs that can be raised or lowered independently of each other.

**Jahre Viking** *Oil & Gas.* T.T. Jahre Viking, the world's largest oil supertanker (in fact the largest floating object on earth), over 1500 feet long and able to carry more than 4 million barrels of crude oil.

**Jessica** *Environment.* an Ecuadorean tanker that ran aground on rocks off the Galapagos Islands in 2001, releasing about two-thirds of its cargo of 240,000 gallons of oil. The accident occurred in the prophetically named Shipwreck Bay, outside the harbor of San Cristobal Island.

**jet** *Physics.* **1.** a continuous stream of concentrated and well-defined incompressible or compressible fluid (e.g., water, fuel), emitted by a nozzle or other such opening of a small size that compresses (and thus accelerates) the flow of the fluid. **2.** any device that is powered by such a stream of fluid. **3.** *Transportation.* specifically, a jet aircraft or jet engine (see below).

**jet aircraft** *Transportation.* an aircraft, especially a fixed-wing airplane, that attains its thrust from one or more jet engines.

**jet coal** *Coal.* another term for CANNEL COAL, from its black color (jet black).

**jet engine** *Transportation.* any of a class of reaction engines that take in outside air for use as the fuel oxidizer and eject a jet of hot exhaust gases backward for thrust in the opposite direction.

**jet fuel** *Transportation.* any fuel produced specifically for jet engines; usually a kerosene-like petroleum distillate.

**jetstream** *Earth Science.* **1.** a concentration of relatively strong winds within a narrow stream in the atmosphere, especially a quasihorizontal concentration of strong westerly winds in the high troposphere. **2.** *Transportation.* the exhaust stream of a jet or rocket.

**Jevons, W. Stanley** 1835–1882, British economist who was one of the first to examine the question of depletion of energy resources. See JEVONS PARADOX.

**Jevons paradox** *Consumption & Efficiency.* the concept that increased efficiency in the use of a natural resource, which might reasonably be expected to reduce consumption, will actually result over time in greater consumption. See next page.

**JNOC** Japan National Oil Corporation (est. 1967), a corporation formed by the Japanese government to invest in oil and gas exploration and production projects overseas. It uses public money to stockpile petroleum in order to reduce Japan's dependence on foreign oil.

**Joiner, Dad** (Columbus Marion Joiner) 1860–1947, Texan oil wildcatter who discovered the East Texas oilfield (1930), the largest field in the world at that time.

**joint implementation (JI)** *Sustainable Development.* an agreement made between two or more nations under the auspices of the Kyoto Protocol, in which a developed country can receive emissions reduction units (ERUs) when it helps to finance projects that reduce net emissions in another developed country (including countries with economies in transition).

*jet engine Closeup view of the turbine that powers a jet engine.*

**Jevons paradox** Jevons paradox (often referred to as 'rebound' or 'take-back' effect) is the observation that greater energy efficiency, while in the short run producing energy savings, may in the long run result in higher energy use. It was first noted by the British economist Stanley Jevons, in his book *The Coal Question* published in 1865, where he argued that "it is a confusion of ideas to suppose that the economical use of fuel is equivalent to diminished consumption. The very contrary is the truth". The Jevons paradox is an observation based on economic theory and long-term historical studies, and its magnitude is a matter of considerable dispute: if it is small (less than 100%) then energy efficiency improvements will lead to lower energy consumption, if it is large (greater than 100%) then energy consumption will be higher. A key problem in resolving the two positions is that it is not possible to run 'control' experiment to see whether energy use is higher or lower than if there had been no efficiency improvements—there is after all only one future. A further problem is that the rebound effect has differing impacts at all levels of the economy, from the micro-economic (the consumer) to the macro-economic (the national economy), and it has not yet been able to determine its magnitude at all levels of the economy.

**Horace Herring**
The Open University, UK

**joint production** *Consumption & Efficiency.* a process that yields two useful outputs from the same source; e.g., oil and natural gas from the same well or electricity and heat from the same power plant.

**Joskow, Paul** U.S. economist who made important contributions to energy and environmental economics, particularly in the areas of markets for tradable air pollution permits and competitive electricity markets.

**Joule, James Prescott** 1818–1889, English physicist who was a pioneer in the study of heat and energy. Joule's work contradicted the existing CALORIC THEORY, which described heat as a fluid that could neither be created nor destroyed; he concluded that heat was only one of many forms of energy, and only the sum of all the forms was conserved (see CONSERVATION OF ENERGY).

**joule** *Measurement.* the basic unit of energy in the meter-kilogram-second system, symbol J; the amount of work done by a force of one newton acting through a distance of one meter in the direction of the force. Equivalent to $10^7$ ergs and one watt-second.

**Joule equivalent** *Measurement.* another term for the mechanical equivalent of heat; i.e., the relationship between thermal energy and mechanical energy.

**Joule's law** *Electricity.* **1.** the relationship between the steady flow of current through a resistance and the heat dissipated in this process; the heat energy per unit time equals the resistance of the circuit multiplied by the square of the current. **2.** *Thermodynamics.* the principle that the internal energy of a sample of an ideal gas depends only on its temperature, and is independent of its pressure and volume.

**Joule–Thomson effect** *Chemistry.* the change in temperature of a sample of gas as it experiences a Joule–Thomson expansion (see next).

**Joule–Thomson expansion** *Chemistry.* a gas-expansion process in which an enclosed sample of gas is allowed to escape through a small aperture or porous plug into a lower pressure region; generally it is observed that the final temperature of the sample differs from the original temperature. Also, **Joule–Thomson process.**

**Joy, Joseph Francis** 1883–1957, U.S. inventor who pioneered many new concepts in mining. His contributions provided an increase in productivity and safety that enabled coal to be a major energy source in the U.S.

**jumbo jet** *Transportation.* a term for larger passenger jets such as those that went into service in the 1970s; i.e., the Boeing 747, Lockheed L-1011, and McDonnell Douglas DC-10.

**junction** *Electricity.* **1.** the connection between conductors or between sections of a transmission line. **2.** a point or place at which two or more semiconducting materials of opposite polarity meet, such as a P–N junction. **3.** *Transportation.* a major joining point for railroad lines or other thoroughfares.

**junction diode** *Photovoltaic.* a semiconductor device with a junction and a built-in potential that passes current better in one direction than the other; all photovoltaic cells are junction diodes.

**Juul, Johannes** Dutch wind energy pioneer who built the innovative GEDSER TURBINE in Denmark (1956–1957), which for many years was the largest in the world.

**K** Kelvin (scale).

**Kalundborg** *Policy.* a small coastal town in Denmark that is noted for its innovative industrial practices. Six entities (an oil refinery, power station, gypsum board facility, pharmaceutical plant, and the city itself) share ground, surface, and waste water, steam and electricity, and also exchange a variety of residues that become feedstocks in other processes. Thus, **Kalundborg model.**

**Kamen, Martin** 1913–2002, Canadian biochemist who discovered carbon-14, the radioactive isotope of carbon used to trace biochemical pathways and mechanisms and to date archeological and anthropological objects.

**Kamerlingh-Onnes, Heike** 1853–1926, Dutch physicist who discovered the phenomenon of SUPERCONDUCTIVITY (1911), which would lead to many modern applications of superconducting materials.

**Kaplan turbine** *Conversion.* an improved type of propeller-type water turbine developed in the early 20th century, having blades that are adjusted according to different conditions in order to provide for greater efficiency; widely used in various current applications, especially hydroelectric plants. [Developed by Austrian engineer Victor *Kaplan,* 1876–1934.]

**Karcher, John C.** 1894–1978, U.S. engineer who developed a method of oil exploration by means of a seismograph, which would become a revolutionary tool in detecting subsurface structures capable of holding oil.

**katabatic** *Earth Science.* of air, moving downward.

**Katalla** another spelling of COTELLA.

**Kay, John** c. 1704–1780, English inventor of the flying shuttle (1733), a machine that dramatically improved the productivity of textile manufacture.

**Kaya identity** *Measurement.* a method used to analyze the driving forces behind the emission of a pollutant. It defines total emissions as the product of population growth, per capita value added, energy consumption per unit value added, and emissions per unit energy. For example, the Kaya identity for $CO_2$ emissions is: $CO_2$=(Population) × (GDP/Population) × (Energy Use/GDP) × ($CO_2$/Energy Use). [Developed by Japanese environmentalist Yoichi *Kaya,* born 1934.]

**K-capture** *Nuclear.* a form of radioactive decay in which an atomic nucleus first absorbs an electron from among the atom's orbiting electrons and subsequently radiates a neutrino.

**Keeling curve** another name for the MAUNA LOA CURVE. [from its developer, U.S. oceanographer Charles D. *Keeling.*]

**Keenan, Joseph Henry** 1900–1977, U.S. engineer, a pioneer in engineering thermodynamics and the author of classic books on this subject.

**kelp** *Ecology.* the common name for any large, brown cold-water seaweed, especially of the orders Laminariales and Fucales, noted for growing in vast underwater masses **(kelp forests)**. Certain types of giant kelp are proposed as sources of biomass energy; e.g., *Macrocystis pyrifera, M. integrifolia.*

**Kelvin, Lord** William Thomson, Baron Kelvin, 1824–1907, Scottish mathematical physicist who helped lay the foundation for modern physics. He played a major role in the development of the second law of thermodynamics, the absolute temperature scale, the dynamical theory of heat, the mathematical analysis of electricity and magnetism, the geophysical determination of the age of the earth, and the principles of hydrodynamics.

**kelvin** or **Kelvin** *Measurement.* **1.** the fundamental SI unit of temperature, equal to 1/273.16 of the absolute temperature of the triple point of pure water. A temperature increment of one degree kelvin (1 K) is equal to one degree Celsius (1°C). **2.** relating to or expressed by the Kelvin temperature scale.

**Kelvin equation** *Chemistry.* a principal equation of surface chemistry that calculates the free energy of a small particle as a function of its size and pressure. Minute droplets of a liquid have a higher vapor pressure than the bulk liquid and thus a greater rate of evaporation; larger drops decrease in free energy with their increase in size and should grow spontaneously.

**Kelvin (temperature) scale** *Measurement.* a temperature scale that is based on the absolute zero temperature (0 K) and the triple point of water (273.16 K); defined so that the ratio of the temperatures of two heat reservoirs is equal to the ratio of heat transfer from the hotter reservoir to the colder one.

**Kelvin statement** *Thermodynamics.* the statement that there is no cyclic process in which heat can be converted completely into work. This is equivalent to the second law of thermodynamics.

**kenaf** *Biomass.* a hardy herbaceous annual plant, *Hibiscus cannabinu,* widely cultivated as a fiber crop. It has potential as an alternative fiber material to wood; e.g., for papermaking, because it requires less energy to process and has a higher yield per acre.

**Kepler, Johannes** 1571–1630, German astronomer who was the first to correctly explain planetary motion. The three laws of planetary motion bearing his name were published in 1609 and 1619; he described the orbits of the planets not as the circles as described by Aristotle, but instead "flattened circles" (ellipses). He also did important work in optics, especially the study of human vision.

**kerf** *Coal.* a horizontal section cut out of a seam or block of coal, often to facilitate its fall.

**kerogen** *Oil & Gas.* a fossilized, highly insoluble material found in sedimentary rocks, especially oil shales; it may be transformed into petroleum products by distillation. Typical organic constituents of kerogen are algae and woody plant material. Bitumen forms from kerogen during petroleum generation.

**kerosene** *Oil & Gas.* a combustible, water-white, oily liquid with a strong odor that boils at 180–300°C; it is distilled from petroleum and is used as a fuel, as a cleaning solvent, and in insecticides.

**kerosene clipper** *History.* a tall-masted, square-rigged sailing vessel employed in the late 1800s for the international shipping of kerosene (an early enterprise of the John D. Rockefeller company that became Standard Oil).

**kerosene lamp** *Lighting.* a historic lighting device using kerosene (paraffin) as the fuel; a successor to the whale oil lamp and predecessor to the electric light. Modern versions of this are still in use as a portable outdoor lighting source.

**kerosene-type jet fuel** *Transportation.* a kerosene product with a relatively low-freezing point distillate of the kerosene type, used primarily for turbojet and turboprop aircraft engines.

**ketone** *Materials.* any of a class of organic compounds, such as acetone, that have the carbonyl group C=O attached to two alkyl groups; obtained by the oxidation of secondary alcohols.

**Kettering, Charles** 1876–1958, U.S. inventor who designed the first electrical ignition system for the automobile (1911). This eliminated the need for the hand crank, an arduous and sometimes dangerous task, thus helping to make cars more appealing to drive. Kettering became the founder of Delco (Dayton Engineering Laboratories Company), a major automotive company.

**kettle bottom** *Mining.* a large, rocklike object that protrudes through or drops out of the roof of a mine, described as the fossilized root or stump of a tree.

**Keys, Ancel** 1904–2004, U.S. public health scientist noted for his achievements in human nutrition and physiology. He was among the first to emphasize the relation among energy intake, energy expenditure, and resting metabolic rate. He also described the relationship of diet, cholesterol, and cardiovascular disease. During World War II, he developed the famous K rations consumed by millions of soldiers.

**kgU** kilogram of contained uranium.

**Khazzoom–Brookes (KB) postulate** *Consumption & Efficiency.* a description by economists J. Daniel Khazzoom and Leonard Brookes of the REBOUND EFFECT (the concept that improvements in energy efficiency will lead to greater rather than lower energy consumption levels).

**killer fog** *Health & Safety.* a term for an environmental event in which fog combines with airborne pollutants (e.g., smoke from coal fires, steel mill emissions) to cause deaths and acute health effects. Used especially to refer to 1952s disastrous GREAT SMOG OF LONDON.

**kiln** *Consumption & Efficiency.* an oven-like compartment built to contain the heat around the item to be cooked, baked, burned, fired, and so on.

**kilo** *Measurement.* a prefix or word meaning "one thousand"; symbol k.

**kilocalorie (kcal)** *Biological Energetics.* the amount of heat required to raise the temperature of one kilogram (or one liter) of water by one degree Celsius. This is the unit most commonly used to measure the energy content of food and human energy requirements, and the term "calories" as it appears on a food label actually refers to kilocalories.

**kiloelectronvolt** *Electricity.* a unit of energy equal to one thousand electron volts.

**kilowatt (kW)** *Electricity.* a standard unit of electric power equal to 1000 watts or approximately 1.34 hp; e.g., the amount of electric energy required to light ten 100-watt light bulbs.

**kilowatt-hour (kWh)** *Electricity.* a standard unit of electric consumption, corresponding to a usage of 1000 watts for 1 hour; e.g., a 100-watt light bulb burning for 10 hours consumes 1 kWh.

**kinematics** *Thermodynamics.* the branch of mechanics concerned with the study of motion without reference to the effects of force or mass.

**kinematic viscosity** *Measurement.* the absolute viscosity (resistance to flow) of a fluid divided by its density.

**kinetic** *Physics.* involving or causing motion; having to do with motion and its effects.

**kinetic energy** *Physics.* the energy that a moving body possesses because of its motion, dependent on its mass and the rate at which it is moving; equal to $1/2\ mv^2$, where $m$ is mass and $v$ velocity.

**kinetic–molecular theory** *Thermodynamics.* a physical theory that explains the behavior of gases on the assumption that any gas is composed of a very large number of very tiny particles called molecules, which are very far apart compared to their size and thus can be considered as points, and that these molecules exert no forces on one another except during rare collisions which are perfectly elastic. An ideal gas corresponds to these assumptions.

**kinetic pressure** *Physics.* the kinetic energy per unit volume in a fluid, equal to $(1/2)rv^2$, where $r$ is the fluid density and $v$ is the fluid velocity.

**kinetics** *Physics.* the branch of mechanics concerned with the relationship between motion and force and mass.

**King Coal** *Coal.* a characterization of the crucial role that coal played in the Industrial Revolution, and its dominance of the fuel budgets of the United States, the United Kingdom, and other nations in the early 20th century.

**Kirchhoff, Gustav** 1824–1887, German physicist noted for his formulation of laws related to the conduction of electricity (see following entries). He also made major contributions in the study of spectroscopy and advanced research into blackbody radiation.

**Kirchhoff's current law** *Electricity.* the statement that the sum of the currents into a given node equals the sum of the currents out of that node.

**Kirchhoff's radiation law** *Electricity.* one of the fundamental laws of radiation, equating the absorptivity of matter to its emissivity at the same wavelength; it implies that good absorbers of radiation at a given wavelength are also good emitters at that wavelength.

**Kirchhoff's voltage law** *Electricity.* the statement that the sum of all currents flowing into

***Kitty Hawk*** *One century later, Pilot Kevin Kochersberger attempts to duplicate the Wright brothers' original 1903 flight; his attempt failed, mainly due to adverse weather conditions.*

a node is zero; conversely, the sum of all currents leaving a node must be zero.

**Kitty Hawk** *History.* a beach in North Carolina that was the site of the first controlled heavier-than-air flight (Wright Brothers, December 17, 1903).

**Klaproth, Martin Heinrich** 1743–1817, German chemist who discovered or helped verify many important chemical elements, including uranium, zirconium, strontium, titanium, chromium, tellurium, and cerium. His experimental methods also added precision to the field of chemistry.

**Kleiber's law** *Biological Energetics.* a basis for predicting the basal metabolic rate of animals and for comparing nutrient requirements among animals of different size, given as: $MR = aM^{0.75}$, where $MR$ is metabolic rate (in watts), $M$ is body mass (in kg), and $a$ is a constant. This has been confirmed in a broad spectrum of animals spanning more than 20 orders of magnitude in size. [Described by Swiss-born U.S. scientist Max *Kleiber,* 1893–1976.]

**Kleist, Ewald Jurgens von** 1700–1748, German scientist who discovered the LEYDEN JAR (1745). Pieter van Musschenbroek independently discovered a similar device at about the same time.

**knapping** *Mining.* the process of chipping away manually at an ore deposit to remove material that has no value.

**knock** *Transportation.* a metallic, rattling sound in an internal combustion engine, resulting from imperfect fuel combustion; a common problem in early gasoline engines prior to the development of additives to prevent this (originally tetraethyl lead).

**known recovery** *Oil & Gas.* the sum of an oil or gas field's past production and its current estimate of proved reserves.

**Kocs** *History.* a town in Hungary that was the source of an innovative type of passenger vehicle (1457), featuring light spoked wheels and a springed undercarriage. The modern English word *coach* is derived from the expression "cart of Kocs".

**Kopp's law** *Chemistry.* the statement that the heat capacity of a solid compound, at standard temperature and pressure, is approximately equal to the sum of the heat capacities of the elements in this compound; there are exceptions to this for certain lighter elements. [Named for Hermann *Kopp,* 1817–1892, German chemist.]

**kraft** *Materials.* a sturdy tan paper made from unbleached sulfate pulp, often used in combination with other fibers to make container board, wrapping, and grocery bag papers, and the like. [From a German word meaning "strong".]

**kraft process** *Materials.* the predominant method for converting wood chips into wood pulp; it involves the use of caustic sodium hydroxide and sodium sulfide to extract the lignin from the wood fiber in large vats called digesters. Also, **kraft pulping.**

**Krebs, Hans Adolf** 1900–1981, German-born English biochemist who described the cycle of chemical reactions that proved to be the major source of energy in living organisms, known as the citric acid cycle or Krebs cycle.

*Hans Adolf Krebs*

**Krebs cycle** the cycle of chemical reactions that are the major source of energy in living organisms. [Described by Hans Adolf *Krebs.*] See next page.

**Krosno** *History.* a town in Poland that is regarded as the earliest site to have street lighting from oil lamps (late 1500s).

**Krebs cycle** The Krebs cycle, also known as the citric acid cycle or the tricarboxylic acid cycle, is one of the most important reaction sequences in biochemistry. Not only is this series of reactions responsible for most of the energy needs in complex organisms, the molecules that are produced in these reactions can be used as building blocks for a large number of important processes, including the synthesis of fatty acids, steroids, cholesterol, amino acids for building proteins, and the purines and pyrimidines used in the synthesis of DNA. Fuel for the Krebs cycle comes from lipids (fats) and carbohydrates, which both produce the molecule acetyl coenzyme-A (acetyl-CoA). This acetyl-CoA reacts in the first step of the eight step sequence of reactions that comprise the Krebs cycle, all of which occur inside mitochondria of eukaryotic cells. While the Krebs cycle does produce carbon dioxide, this cycle does not produce significant chemical energy in the form of adenosine triphosphate (ATP) directly, and this reaction sequence does not require any oxygen. Instead, this cycle produces NADH and $FADH_2$, which feed into the respiratory cycle, also located inside of the mitochondria. It is the respiratory cycle that is responsible for production of large quantities of ATP and consumption of oxygen. In addition, the respiratory cycle converts NADH and $FADH_2$ into reactants that the Krebs cycle requires to function. Thus, if oxygen is not present, the respiratory cycle cannot function, which shuts down the Krebs cycle. For this reason, the Krebs cycle is considered an aerobic pathway for energy production.

**Clarke Earley**
Kent State University

**krypton** *Chemistry.* a gaseous element that has the symbol Kr, the atomic number 36, an atomic weight of 83.80, a melting point of –156.6°C, and a boiling point of –152.9°C; a colorless and odorless noble gas used in lasers and fluorescent lamps. The earth's atmosphere is about one-millionth part krypton.

**kt** kiloton (1000 tons).

**Kurchatov, Igor** 1903–1960, Russian physicist who was the chief scientist chosen by dictator Joseph Stalin to develop a Soviet nuclear weapon. His efforts produced the USSR's first detonation of a nuclear device in 1949 (a plutonium implosion bomb).

**Kuznets, Simon** 1905–1985, Russian–American economist who pioneered the field of development economics, focusing on an analysis of the economic experiences of modern underdeveloped countries. See ENVIRONMENTAL KUZNETS CURVE.

**kW** kilowatt.

**kWh** kilowatt-hour.

**kymatology** *e.* the scientific study of (ocean) waves and wave motion.

**Kyoto forest** *Policy.* a forest that industrial countries can use to discount their greenhouse gas emissions by certain land-use change and forestry activities under the provisions of the Kyoto Protocol. In particular, growing forest areas may be credited as carbon sinks in some circumstances.

**Kyoto mechanism** *Policy.* a measure included in the Kyoto Protocol to help countries achieve their Kyoto targets by allowing them to participate in international cooperative activities that benefit the global environment; e.g., the clean development mechanism (CDM).

**Kyoto Protocol** *Policy.* an international agreement signed in 1997 at a convention in Kyoto, Japan; it sets binding emissions reductions of greenhouse gases with an average 5.2% reduction below 1990 levels for industrial countries. See next page.

**Kyoto Protocol** Kyoto Protocol, named after the city where it was agreed by negotiators in December 1997, is the daughter Treaty intended to implement the objectives and principles agreed in the 1992 UN Framework Convention on Climate Change (UNFCCC). The core idea is that stabilising the atmosphere (the UNFCCC's "ultimate objective") will require governments to agree quantified limits on their greenhouse gas emissions, through sequential rounds of negotiations for successive 'commitment periods'. The Protocol sets out the general architecture of sequential negotiations, and defines specific first period commitments for industrialised countries (almost synonymous with those in Annex I to the UNFCCC), intended to fulfil the UNFCCC requirement for rich country 'leadership'. These national 'assigned amounts' limit emissions of the six main anthropogenic greenhouse gases (of which $CO_2$ accounts for about 80%) during 2008–12, embedded in mechanisms to increase the flexibility, reach and efficiency of the commitments. Emissions trading allow countries to exchange emission allowances, whilst Joint Implementation and the Clean Development Mechanism allow commitments to be offset against investment in emission-reducing projects worldwide. The Protocol also contains a range of other, more limited, provisions including in relation to technology standards, development and transfer, and stipulates that negotiations on future commitments should start by 2005. Whilst the UNFCCC easily secured almost global participation including all major countries, the Kyoto Protocol was highly charged in its negotiation and aftermath. Although the U.S. government under Clinton signed (and indeed designed much of) the Treaty, it was repudiated by President Bush. Following a long drawn-out ratification decision by Russia, Kyoto entered into force on 16 February 2005, with about 130 countries having ratified. Australia has also refused to ratify but says it intends to fulfil its emissions target and will participate in negotiations on subsequent commitments.

**Michael Grubb**
The Carbon Trust and The University of Cambridge

**La Cour, Poul** 1846–1908, Danish engineer, a pioneer of modern aerodynamics and electricity-generating wind turbines who built his own wind tunnel for experiments. La Cour founded a society of wind electricians in 1905, and also founded the world's first journal of wind electricity.

**lacustrine** *Earth Science.* having to do with a lake; relating to or occurring in lakes.

**Lagrange, Joseph-Louis** 1736–1813, French mathematician regarded as the greatest in this field in the 18th century. He developed the calculus of variations, established the theory of differential equations, and provided many new solutions and theorems in number theory.

**Lagrangian** *Physics.* for a discrete mechanical system, a quantity equal to the system's kinetic energy minus its potential energy, regarded as a function of the generalized coordinates, generalized velocities, and time.

**laissez-faire** *Economics.* literally, leave alone; a concept developed in the late 18th century based on the principle that government should generally refrain from involvement in the economic affairs of the nation.

**Lakeview gusher** *History.* an oil well between the towns of Taft and Maricopa in the Central Valley of California, noted for having one of the largest outputs of any single well in the world, an estimated nine million barrels of oil in about 18 months beginning in March 1910, before the well caved in and was abandoned.

**Lambert** *Lighting.* a unit of luminance in the centimeter-gram-second system, symbol La (or Lb or L); equal to the luminance of a surface that emits or reflects one lumen per square centimeter. [Named for German mathematician and astronomer Johannn Heinrich *Lambert,* 1728–1777.]

**Lambertian reflection** See DIFFUSE REFLECTION.

**Lambert's law** *Physics.* the observation that the intensity of radiation passing through a material decreases exponentially with path length. [Named for J. H. *Lambert.*]

**laminar flow** *Physics.* the streamline flow of a viscous, incompressible fluid in which fluid particles travel along well-defined separate lines in substantially parallel paths.

**Lamm, August Uno** 1904–1989, Swedish electrical engineer known as the father of HVDC (high voltage direct-current electrical transmission).

**lamp** *Lighting.* **1.** any device that provides an electrically energized source of light. **2.** another term for an electric light bulb or tube. **3.** a light source powered by some means other than electricity, e.g., coal gas, whale oil, and kerosene. [From the Greek word *lampas* meaning torch.]

**lamp efficacy** see EFFICACY.

**lamp efficiency** see LUMINAIRE EFFICIENCY.

**lamp life** *Lighting.* an indication of the expected life of an electric lamp, based on the average number of hours at which half of a large group of lamps have failed when operated at standard lamp voltage and current.

**lamp oil** see COAL OIL.

**land degradation** see DEGRADATION.

**land disturbance** *Environment.* a general term for any human removal of soil, rock, or vegetation, especially on a large-scale basis.

**landed cost** *Oil & Gas.* the price of crude oil at the port of discharge.

**landfill** *Environment.* **1.** the practice of filling in land, especially a low-lying wetland area, with garbage, ash, and dirt, as a means of waste disposal. **2.** the materials used in this process.

**landfill gas (LFG)** *Biomass.* a medium-energy fuel gas high in methane and carbon dioxide, produced by natural decomposition of organic material at municipal landfills that contain solid wastes and other waste biomass.

**land reclamation** see RECLAMATION.

**land use** *Global Issues.* the manner in which humans make use of a given area, as for housing, commerce, agriculture, and so on; this also includes land that is not explicitly used; i.e., left in its natural state.

**Langley, Samuel Pierpont** 1834–1906, U.S. astronomer who became a pioneer in heavier-than-air flight. Beginning in the 1880s he produced various versions of a flying machine he called the **Aerodrome.** One model with a gas-powered engine attempted a manned flight in 1903, but crashed immediately after being launched, being too nose-heavy to fly. Shortly after this the Wright Brothers successfully flew the first powered manned flight.

**langley** *Solar.* a measure of solar energy flux per area; equal to one calorie per square centimeter. [Named for S. P. *Langley.*]

**Langmuir, Irving** 1881–1957, U.S. chemist who developed the modern field of surface chemistry and the theory of adsorption catalysis. He promoted understanding of plasmas, heat transfer, and thermionic phenomena, and also invented a high-vacuum electron tube and gas-filled incandescent lamp.

**La Niña** *Earth Science.* a large-scale, complex set of changes in the water temperature in the Eastern Pacific equatorial region, producing a cold current. The colder ocean temperatures associated with La Niña tend to bring weather effects in North America opposite to those produced by EL NIÑO.

**Laplace, Pierre-Simon** 1749–1827, French physicist who published a monumental work on celestial mechanics (1799–1825). He transformed the geometrical study of mechanics used by Newton to one based on calculus, known as physical mechanics.

**lapse rate** *Earth Science.* the rate of change of an atmospheric variable with increasing altitude, especially the rate at which temperature decreases with altitude. The normal lapse rate is defined to be 3.6°F per 1000 ft change in altitude.

**La Rance** *Renewable/Alternative.* the world's largest tidal power plant, at the mouth of the La Rance river in northern France. The tide is channeled through the narrow entrance to the river, across which is a dam with turbines and generators.

**Lardarello** *Geothermal.* a city in Tuscany, Italy, noted as a pioneering site for the use of geothermal power and still a major source of this form of energy.

**large calorie** see KILOCALORIE.

**laser** *Consumption & Efficiency.* a device that emits a high-intensity, highly directional beam of light with a narrow spectral width; capable of producing intense light and heat when focused at close range, and having many applications in industry, medicine, and science. [An acronym for light amplification by stimulated emission of radiation.]

***laser*** *Laser beam employed in the process of manufacturing a jet engine part.*

**laser fusion** *Nuclear.* a method of internal confinement fusion in which the fusion process is induced in small pellets of a deuterium–tritium mixture by striking them with a high energy density that fuses them before they repel each other.

**Laspeyres index** *Economics.* a technique for describing the change in cost of purchasing a given array (basket) of goods and services in the current period, as opposed to a specified base period. The prices are weighted by quantities in the base period. [Proposed by German economist Étienne *Laspeyres* 1834–1913.]

**lateen sail** *History.* a triangular sail providing a historic advance in navigation (about AD 900); it made directional sailing possible, whereas earlier square-sail vessels were limited to sailing with the wind. Thought to have originated with Arab merchant ships on the Red Sea.

**latent heat** *Physics.* the amount of thermal energy that is absorbed or released by a unit amount (usually one mole) of a substance, in the process of a phase change (e.g., from liquid to gas) under conditions of constant pressure and temperature. Compare SPECIFIC HEAT.

**latent heat flux** *Earth Science.* the global movement of latent heat energy through circulations of air and water.

**latent heat of condensation** see HEAT OF CONDENSATION.

**latent heat of sublimation** see HEAT OF SUBLIMATION.

**latent heat of vaporization** see HEAT OF VAPORIZATION.

**latent heat storage** *Storage.* a heat storage system that operates on the basis of a change in the state of the storage medium (e.g., ice to water).

**latent load** *HVAC.* the load (energy requirement to cool a building space) created by moisture in the air, from outside air infiltration and also from indoor sources such as occupants, plants, cooking, showering, and so on. Also, **latent cooling load.**

**latex** *Materials.* **1.** a milky, generally white substance that is excreted by rubber trees such as *Hevea brasiliensis,* used in making natural rubber and similar elastic materials. It is highly unstable and is preserved by the addition of a small percentage of ammonia. **2.** a similar synthetic material, a colloidal suspension of fine particles of plastic or rubber like material in water.

**latitude** *Earth Science.* a statement of the location of a given site on earth, expressed in terms of angular distance north or south of the equator, measured from 0° latitude at the equator to 90° north or south at the poles.

**lattice energy** *Physics.* the energy associated with the construction of a crystal lattice, relative to the energy of all its constituent atoms separated by infinite distances.

**lava** *Earth Science.* **1.** a molten mass of rock material that is extruded by a volcano or through a fissure in the earth. **2.** a rock that is formed by the cooling and solidifying of such molten material.

**Laval nozzle** *Conversion.* a nozzle that is able to generate high-speed gas flows by means of a converging–diverging operation; originally used in steam turbines and then widely used in rocket propulsion. [Invented by Swedish engineer Carl de *Laval,* 1845–1913.]

**Laval turbine** *Conversion.* an improved turbine design of the late 19th century; a high-speed steam turbine having special reduction gearing to allow a turbine rotating at high speed to drive a propeller at a comparatively slow speed, an advance in marine engineering.

***lava*** *Lava flows provide an ideal context for ecologists to study how a new natural community is formed, since the lava will produce a completely sterile habitat that will then be recolonized.*

**Lave, Lester** U.S. economist noted for important research on the costs, benefits, and health risks associated with air pollution, automobile safety standards, and carcinogenic chemicals. He also did important work on the life cycle of energy and transportation systems, the hidden costs of environmental regulation, and the risks of climate change.

**Lavoisier, Antoine** 1743–1794, French scientist who demonstrated the true nature of combustion, overturning the PHLOGISTON theory. His experiments with combustion also provided evidence for the law of conservation of matter. He named the components of water as oxygen and hydrogen, and helped to devise a chemical nomenclature that serves as the basis of the modern system. He worked as a tax collector, and when tax collectors were branded as traitors during the Reign of Terror, he was executed by guillotine.

**law of capture** *Oil & Gas.* **1.** the principle that migratory material assets existing in nature (e.g., wild game and groundwater) belong to the party who obtains (captures) them on his property, as long as this is done legally. **2.** specifically, the application of this principle to oil and gas exploration; historically based on the belief that the position of oil or gas deposits was transitory, in the sense that they flowed in underground streams and thus might be under one tract of land at present but under another tract in the future.

**Law drill** *History.* an advanced form of rock drill in which an air-driven piston operates a hammer tool; considered to be the first practical application of compressed air to drive a motor (1865). [Developed by British inventor George *Law.*]

**law of conservation of energy** see CONSERVATION OF ENERGY.

**law of conservation of mass** see CONSERVATION OF MASS.

**law of gravitation** see NEWTON'S LAW OF GRAVITATION.

**law of stable equilibrium** see UNIFIED PRINCIPLE.

**law of the minimum** *Biological Energetics.* the principle that plant growth is limited by the amount of the most limiting nutrient, whichever nutrient that may be. Thus, if one crop nutrient is missing or deficient, plant growth will be poor, even if the other required elements are abundant. Also know as Liebig's law after its formulator, Justus von Liebig.

**law of universal gravitation** see NEWTON'S LAW OF GRAVITATION.

**Lawrence, Ernest O.** 1901–1958, U.S. physicist noted for his invention of the CYCLOTRON (1929). During World War II he was in charge of the Manhattan Project's process of separating uranium-235 for the atomic bomb. **Lawrencium** (element 103) is named in his honor.

**laws of motion** see NEWTON'S LAWS.

**laws of thermodynamics** see THERMODYNAMICS.

**Lawson criterion** *Nuclear.* a general measure of a system that defines the conditions needed for a fusion reactor to generate more energy in the fusion reactions than is lost in thermal and other radiation out of the fuel; it combines the fuel temperature (which defines the reaction rate), fuel density (how many reactions per unit volume), and the time that the energy of the reactions remains confined by the reaction. Similarly, **Lawson number** or **condition.** [Described by British physicist J. D. *Lawson.*]

**lazy battery effect** another term for MEMORY EFFECT.

**LC50** lethal concentration 50; an abbreviation for MEDIAN LETHAL CONCENTRATION.

**LCA** life cycle assessment.

**LCD** liquid crystal display.

**LCLUC** *Environment.* land cover and land use change; a general term for any process that changes the land cover or land use of a portion of the earth's surface; e.g., the conversion of forest to cropland via deforestation.

**LD** lethal dose.

**LD50** lethal dose 50; an abbreviation for MEDIAN LETHAL DOSE.

**LDCs** less developed countries.

**leachate** *Materials.* the solution or soluble material that results from a leaching process.

**leaching** *Materials.* **1.** the process of separating a soluble substance from a solid by washing or by the percolation of water or another liquid through the substance. **2.** *Earth Science.* the natural or artificial removal of soluble substances from rock, ore, or layers of soil by the action of percolating substances.

**lead** *Chemistry.* **1.** a metallic element having the symbol Pb, the atomic number 82, an atomic weight of 207.2, a melting point of 327.4°C, and a boiling point of 1755°C; a soft, heavy solid occurring naturally in galena. **2.** this substance in the form of a heavy, corrosion-resistant metal, used for such purposes as storage batteries, cable sheathing, and low-melting alloys such as solders. Lead presents a health hazard, and its use is restricted or prohibited in certain applications.

**lead–acid battery** *Storage.* a general term for batteries with plates made of a pure lead, lead–antimony, or lead–calcium material immersed in an acid electrolyte. These were the first commercially practical batteries and remain widely used today because they are relatively inexpensive to manufacture. Most automobile batteries are of this type.

**lead-cooled fast reactor (LFR)** *Nuclear.* an advanced nuclear reactor design that features a fast-spectrum lead or lead–bismuth eutectic liquid metal-cooled reactor and a closed fuel cycle. The fuel is metal or nitride-based, containing fertile uranium and transuranics. The LFR is cooled by natural convection with a reactor outlet coolant temperature of 550°C, possibly ranging up to 800°C with advanced materials. The higher temperature enables the production of hydrogen by thermochemical processes.

**leaded gasoline** *Oil & Gas.* gasoline to which tetraethyl lead has been added to increase the octane number, typically defined as gasoline containing more than 0.05 grams of lead per gallon. Its use has now been curtailed or banned in many areas because it is a source of air pollution. New vehicles manufactured for sale in the U.S. are now required to use unleaded gasoline.

**leaf area index (LAI)** *Ecology.* a measure of the vegetation surface in a given area, defined as the total one-sided green leaf area per unit of ground surface area.

**leakage** *Policy.* **1.** a term for the indirect effect of emission reduction policies or activities, that leads to a rise in emissions elsewhere (e.g., fossil fuel substitution in the developed world, leading to a decline in fuel prices and thus a rise in fuel use in other countries). **2.** *Climate Change.* the unexpected loss of estimated net carbon sequestered. **3.** *Materials.* any unwanted and slow escape or entrance of particles or material. **4.** see NEUTRON LEAKAGE.

**leakage current** *Electricity.* **1.** unwanted stray current that flows across the surface of an insulator or an insulating material. **2.** current flowing between electrodes in a tube that does not flow across the interelectrode space.

**leakage inductance** *Electricity.* a small inductance associated with flux lines of a transformer winding that are not magnetically coupled to the other windings of the transformer.

**lean combustion limit** *Oil & Gas.* the point at which combustion cannot occur because the fuel–air mixture contains too little fuel, resulting in unstable combustion and increased hydrocarbon emissions from misfires.

**lean mixture** *Oil & Gas.* an air–fuel mixture that has a relatively low proportion of fuel in relation to air; engine operation with a lean mixture (using as little fuel as possible) reduces harmful emissions and improves fuel efficiency.

**lease** *Economics.* **1.** an area of surface land that overlays a mineral or fuel (oil, coal) on which exploration or production activity occurs. **2.** a contract that grants the rights to explore and produce from the owner of the mineral rights **(lessor)** to a tenant **(lessee),** usually for a fee and with a specified duration. In oil and gas production, a lease often includes a provision for sharing revenue from production.

**lease and plant fuel** *Oil & Gas.* a classification for natural gas used in well, field, and lease operations, and as fuel in natural gas processing plants.

**least conservation principle** another term for GAUSS PRINCIPLE.

**least developed countries** see LESS DEVELOPED COUNTRIES

**least developed regions** see LESS DEVELOPED REGIONS

**Lebon, Philippe** 1769–1804, French engineer and chemist who invented the Thermolampe ("heat lamp"), an early lamp powered by an illuminating gas, which he patented and exhibited in 1799.

**Le Châtelier's principle** *Thermodynamics.* the principle that if a system in equilibrium is disturbed by some external influence, the system will react in such a way as to alleviate the disturbance. [Formulated by French chemist Henri *Le Châtelier* 1850–1936.

**Leclanche battery** *Storage.* an early form of zinc–carbon battery considered to be the first successful battery (1866). It was known as a "wet cell" because both the anode and cathode were submerged in a liquid solution of ammonium chloride that acted as the electrolyte. In a slightly modified contemporary form, the Leclanche battery (now called a dry cell or zinc–carbon battery) is produced in great quantities and is widely used in devices, such as flashlights and portable electronic devices.

**LED** light-emitting diode.

**Leeghwater, Jan Adriaanzoon** 1575–1650, Dutch hydraulic engineer who masterminded vast land reclamation program along the coast of The Netherlands. His techniques included a large-scale use of windmills to drain and pump water. This reclamation led to an unprecedented prosperity known as the Dutch Golden Age.

**Leeuwenhoek, Anton von** 1632–1723, Dutch scientist whose observations laid the foundations for the sciences of bacteriology and microbiology. His skill at grinding lenses enabled him to build microscopes that magnified over 200 times. Using these microscopes, he was the first to observe sperm, algae, bacteria, mineral crystals, fossils, and red blood cells.

**leeward** *Wind.* away from the direction in which the wind blows; opposite of WINDWARD.

**legacy waste** *Nuclear.* a term for the backlog of stored waste remaining from the development and production of U.S. nuclear weapons, about which a permanent disposal determination remains to be made.

**Leibniz, Gottfried Wilhelm von** 1646–1716, German mathematician who invented the differential and integral calculus (independently of Newton). He introduced several notations used in calculus to this day, and he also developed the binary numeral system now used in digital computers.

**Leiden jar** see LEYDEN JAR.

**leisure energy** *Social Issues.* the human consumption of energy during free-time activities, such as driving to a vacation spot, as opposed to business and household energy use. In the broad sense this also includes energy required to construct and maintain leisure-time facilities, such as swimming pools, golf courses, and ski resorts.

**Lenard, Phillip** 1862–1947, Hungarian–German physicist noted for his research on cathode rays and for the discovery of many of their properties. His work formed the basis for Braun's development of the cathode ray tube.

**Lenoir, Jean** 1822–1900, Belgian-born French engineer and inventor who produced the first practical internal-combustion engine (see next) and a vehicle powered by it.

**Lenoir engine** *History.* an early transportation engine that first powered a road vehicle in 1863, consisting of a single cylinder with a storage battery (accumulator) for the electric ignition system. Its two-stroke cycle was provided by slide valves, and it was fueled by coal gas. Because of improved designs by Nikolaus Otto and others, the Lenoir engine became obsolete and only about 500 such engines were ever built.

**Lenz's law** *Physics.* a law stating that an induced electromotive force in a conductor is always polarized in a direction so as to oppose the change that causes the induced electromotive force. [Formulated by Russian physicist Heinrich *Lenz* 1804–1865.]

**Leonardo da Vinci** 1452–1519, Italian Renaissance figure known for achievements in various fields, including descriptions or depictions of many technological innovations that would come to fruition in later eras; e.g., a flying machine, a bicycle, a windmill, a parabolic mirror, and a parachute.

**Leontief, Wassily** 1906–1999, Russian-born U.S. economist noted for developing the economic methodology known as INPUT–OUTPUT ANALYSIS.

**Leontief technology** *Consumption & Efficiency.* a technology model in which production is constrained by the input of the item with the lowest availability; in this situation, substitution for the item is not possible and thus increasing the availability of other production factors will not allow an increase in production.

**lepton** *Physics.* a particle that does not experience the strong nuclear force; specifically, electrons, muons, and tau leptons and their respective neutrinos and antiparticles.

**leptoquark** *Physics.* a hypothetical elementary particle having a mass of approximately 100–300 GeV that would decay into a lepton and a quark.

**less developed countries (LDCs)** *Sustainable Development.* a classification including those nation-states with low levels of industrial development and consumption of commercial energy. Most have high rates of population growth.

**less developed regions** *Sustainable Development.* the United Nations designation for those areas of the world that contain less developed countries (see above); e.g., sub-Saharan Africa.

**LET** linear energy transfer.

**lethal dose (LD)** *Health & Safety.* the amount of a toxic substance that will cause death in a given subject.

**LEU** low enriched uranium.

**leucaena** *Biomass.* a perennial legume tree or shrub, *Leucaena leucocephala,* widely grown in tropical regions; has multiple uses including for forage and as a shade tree, wind break, and energy source.

**LEV** low emission vehicle.

**levee** *Earth Science.* **1.** a naturally occurring, broad and low ridge of sand and coarse silt deposited along the banks of a stream or river during flooding. **2.** a similar man-made ridge or bank of earth, used to contain flood waters, control irrigation flow, and so on.

**levelized cost** *Economics.* the total cost of generating electricity from a new power plant. See below.

**lever** *History.* one of the simple machines of antiquity, a device consisting of a rigid bar that pivots about a stationary or fulcrum point; used to transmit and enhance power or motion, as in prying, raising, or dislodging an object.

**leverage** *Physics.* **1.** the mechanical advantage of a lever, equal to the length of the arm to which force is applied divided by the length supporting the load. **2.** *Economics.* the use of borrowed capital to increase the potential value of an investment or operation. **3.** the amount of debt incurred in such a process. **4.** to carry out such a strategy; e.g., a company with significantly more debt than equity is said to be highly leveraged.

**levelized cost** This is a term most frequently used to characterize the total cost of generating electricity from a new power plant, though it can be applied to any other technology as well. It is similar to an annual home mortgage payment, except that it covers not only the cost of purchasing or building the plant (i.e., the capital cost, or initial cost), but also the costs of maintaining and operating the plant over its useful life. Because capital and operating expenses occur in different amounts at different points in time, economists employ a "time value of money"—expressed as an interest rate, discount rate, or rate of return—to determine the "net present value" (NPV) of all lifetime costs. This is the total amount of money needed at the start of the project to acquire, operate and maintain the plant over its useful life, as well as pay all interest (or make a profit) on funds used to finance the project. The levelized cost is the uniform annual amount (dollars per year), paid over the life of the plant, which produces the same total NPV (dollars) as the actual project. The levelized cost can also be expressed as a cost per unit of electricity generated (e.g., cents per kilowatt-hour) by dividing the total annual cost by the total annual kWh generated. Technologies with the lowest levelized cost are generally preferred in economic decision-making.

**Edward S. Rubin**
Carnegie Mellon University

**Lewis, Gilbert Newton** 1875–1946, U.S. physical chemist who was the first (1933) to produce a pure sample of HEAVY WATER (deuterium oxide). He also coined the term "photon" for the smallest unit of radiant energy.

**Lewis, John L.** 1880–1969, legendary president of the United Mine Workers of America (UMWA) from 1919 to 1960, noted for successfully organizing previously nonunion coal regions and greatly improving working conditions for coal miners. During the 1940s he led controversial coal mine strikes that alienated much of the public and led to a public falling out with President Franklin D. Roosevelt.

**Leyden (Leiden) jar** *Electricity.* a notable early form of the capacitor, an electricity-storage device that was the first practical way to store static electricity; so called because of its use of a glass jar as the dielectric. [From *Leiden* University, where it was developed.]

**LFR** lead-cooled fast reactor.

**LH2** liquid hydrogen.

**LHV** lower heating value; see HEATING VALUE.

**Libby, Willard Frank** 1908–1980, U.S. chemist noted for developing the widely used technique of RADIOCARBON DATING.

**libertarianism** *Social Issues.* a theory for the organization of society emphasizing private property rights and individual freedom. The basic premise is that individual freedom prevails, except in contexts where other individuals might be harmed by this exercise of freedom.

**Liebig, Justus von** 1803–1873, German chemist who was one of the founder of organic and agricultural chemistry. He revised chemical notation and discovered chloroform (independently of Eugène Soubeiran of France and Samuel Guthrie of the U.S.). He formulated the LAW OF THE MINIMUM which is also called **Liebig's law** in honor of him.

**life cycle** *Consumption & Efficiency.* the consecutive and interlinked stages of a product system, from raw material acquisition or generation of natural resources through all subsequent steps, including ultimate waste disposal.

**life cycle assessment (LCA)** *Consumption & Efficiency.* the process of evaluating energy, water, and other natural resource use, and the emissions associated with a good or service from "cradle to grave". This includes raw material extraction, material fabrication, and product manufacture, use, recycling, and/or disposal. Typically used for consumer products. Also, **life cycle analysis.**

**life-cycle cost** *Economics.* the amortized annual cost of a product, including capital costs, installation costs, operating costs, maintenance costs, and disposal costs discounted over the lifetime of the product.

**life expectancy at birth** *Health & Safety.* within a given population, a statement of the number of years that a representative newborn infant would be expected to live, if the patterns of mortality existing at the time of birth were to stay the same throughout the infant's life.

**life history trait** *Ecology.* a trait that defines the life cycle of an organism (e.g., size at birth, growth pattern, mode of reproduction, age and size at maturity, size of offspring, reproductive life span, and so on).

**life-support systems** *Environment.* a term for the components and processes of the environment that provide the physiological necessities of life; i.e., food and other energies, mineral nutrients, oxygen, carbon dioxide, and water.

**lifetime dose** *Health & Safety.* the total amount of a substance or agent, especially a toxic one, received by an individual during his/her lifetime; e.g., radiation exposure, a carcinogen.

**life zone** *Ecology.* one of the various divisions of the earth's surface into distinct regions that have relatively uniform conditions of climate, altitude, soil, and plant and animal life; e.g., tropical wet forest, cool temperate desert.

**lift** *Physics.* **1.** the total amount of force that acts on a body perpendicular to the undisturbed airflow as the body moves through a fluid medium. **2.** the upward force exerted on a balloon or airship by the lighter-than-air gas within it. **3.** in a vapor compression cycle, the difference between high side (condenser) and low side (cooler) conditions, expressed as a temperature or pressure difference. **4.** any device used to raise materials or objects to a higher level.

**lifting cost** *Mining.* a term for the costs involved with the extraction of a mineral reserve from the producing property.

**lift-to-drag (L/D)** *Transportation.* a measure of aerodynamic efficiency; the ratio for an aircraft of lift force generated to drag experienced.

**light** *Physics.* **1.** the form of electromagnetic radiation perceived by the eyes, ranging from about 380–770 nm in wavelength and propagated at about 186,000 miles per second. **2.** a similar type of radiation that is not visible, including X-rays, ultraviolet, infrared, radio waves, and radar sources. See below.

**light** Light is a propagating electromagnetic field. In a vacuum space, it propagates straight with a speed of $c$ = 2.99792458 x $10^8$ m/sec to any direction. It can have any frequency $\nu$ and wavelength $\lambda$, where $\nu\lambda = c$. But only that light with a wavelength between 380 μm and 770 μm is called "visible" light because we can only see such light. An electrically charged particle (electron, proton, and so on), emits a light quantum, called a photon, of energy $h\nu = E_b - E_a$, where $E_a$ and $E_b$ are the energies of the particle before and after the emission process, respectively, and $h$ is Planck's constant. The light thus created can be reflected, or change its propagation direction without changing its frequency, when it hits systematically arranged atoms (e.g., a crystal). In a transparent crystal, it can be refracted, or propagate with the same frequency but with smaller wavelength (thus, with slower speed). The light quantum can be absorbed, or annihilated, when it hits an atomic system that has states with energies $E_a$ and $E_b$ satisfying the relation $h\nu = E_b - E_a$. Thus, the energy of an emitter is transported by light to an absorber. There are other "selection rules" the atomic states should satisfy to allow the above-stated transitions to take place. The refraction and reflection are the second-order effects that take place when the absorber re-emits another photon. A "laser" can be built when the re-emitted photon is the same photon.

**Masataka Mizushima**
University of Colorado

**light crude** *Oil & Gas.* a crude oil of light color that flows freely at atmospheric temperatures, having predominantly hydrocarbons with low viscosity and low molecular weight; the source of high-value light products such as gasoline, jet fuel, and light-distillate fuel oil.

**light distillate** *Oil & Gas.* a classification for petroleum fractions that have a relatively low boiling point range, the source of gasolines and other high-value products.

**light-duty vehicle** *Transportation.* a general classification for cars, passenger vans, sport utility vehicles, and pickup trucks commonly used for personal transport.

**light-emitting diode (LED)** *Lighting.* a semiconductor diode that converts electrical energy into light when an alternating current is applied; LED displays are a widely used light source in consumer electronics, traffic signals, public information messages, and many other applications.

**lighter-than-air** *Transportation.* describing an airborne vehicle having a cumulative weight less than the air it displaces, and thus acquiring its lift from aerostatic buoyancy; e.g., a dirigible.

***lighter-than-air*** *Lighter-than-air (LTA) craft employ a gas, typically helium, contained in the vehicle's envelope that has less weight than the surrounding air and thus provides buoyancy.*

**light fuel oil (LFO)** *Oil & Gas.* lighter petroleum product distilled off during a refining process; suitable for use in liquid-fuel burning equipment without preheating, Virtually all petroleum used in internal combustion and gas-turbine engines is light oil; the category also includes kerosene and jet fuel.

**light hydrogen** *Hydrogen.* another term for protium (the ordinary hydrogen isotope), so called in contrast with heavy hydrogen (deuterium).

**lighting efficacy** see EFFICACY.

***lightning*** *Each cloud-to-ground lightning flash involves an amount of energy approximately equal to the energy required to operate five 100-watt light bulbs continuously for one month.*

**lightning** *Earth Science.* a luminous, high-current, electric discharge with a long path that flows between a cloud and the ground, between two clouds, or between two parts of the same cloud.

**lightning arrestor (arrester)** *Electricity.* a protective device that allows voltages higher than a certain value to be discharged to ground; used to protect lines, transformers, and equipment from lightning surges.

**lightning rod** *Electricity.* a grounded metal electrode placed on the roof of a building to protect the structure from lightning damage by intercepting the lightning and carrying it to ground.

**light oil** **1.** see LIGHT CRUDE. **2.** see LIGHT FUEL OIL.

**light pollution** *Lighting.* **1.** light that enters the sky from an outdoor lighting system by indirect light reflected from atmospheric particles such as fog, dust, or smog. It often surrounds highly urban areas and reduces the ability to view the night sky. **2.** any outdoor source of light that is considered excessive, intrusive, overly bright or harsh, or otherwise offensive.

**light quality** *Lighting.* a description of the extent to which a given interior space provides the proper lighting for people to perform visual tasks efficiently and feel visually comfortable.

**light rail** *Transportation.* a metropolitan transit system using self-propelled electric rail cars, either on a conventional railway line or on its own right of way. Generally a passenger system, but may carry freight. *Light rail* is a newer term for what traditionally would have been known as a *streetcar* or *trolley* system, and the cars are not necessarily lighter than on other rail systems. Thus, **light rail transit (LRT); light rail vehicle (LRV).**

**light truck** *Transportation.* a category of personal vehicle distinct from cars, including single-unit vehicles with two axles and four tires, such as pickup trucks, sports utility vehicles, and passenger vans; defined in one context as all trucks weighing 8500 pounds or less.

**light water** *Hydrogen.* **1.** ordinary water, as distinguished from heavy water (deuterium oxide or tritium oxide). **2.** *Materials.* a fire-fighting agent consisting of a water solution of perfluorocarbon compounds mixed with a water-soluble thickener of the polyoxyethylene type.

**light water reactor (LWR)** *Nuclear.* a reactor that is moderated and cooled by ordinary (light) water rather than heavy water; LWRs account for most of the world's installed nuclear generating capacity.

**light watt** *Lighting.* a unit measuring the relative power output of a light source. One light watt is the power required to produce a perceived brightness equal to that of light at a wavelength of 550 nm and 680 lumen.

**lightweighting** *Consumption & Efficiency.* a policy of reducing vehicle weight (and thus fuel consumption) through the use of lighter materials or more efficient design.

**light year** *Physics.* the distance that light would travel in a vacuum over a period of 1 year, stated as 9.46 trillion kilometers or 5.8 trillion miles; used in measuring astronomical distances.

**lignin** *Biomass.* **1.** a complex polymer that forms an extensive network within the cell walls of certain plants and that confers strength and rigidity to the cell wall; one of the chief substances found in wood. Lignin represents about 15–20% of the dry weight of cellulosic biomass and is not fermentable at an appreciable rate. **2.** *Materials.* a brown to transparent crystalline form of this substance, derived from paper-pulp sulfate liquor; used for various industrial purposes.

**lignite** *Coal.* the lowest rank of coal, intermediate between bituminous coals and peat; often referred to as brown coal because of its color. It is high in moisture and volatile matter content and relatively low in heating value, producing less than 8300 Btu per pound on a moist, mineral-matter-free basis. Lignite is used mainly as fuel to generate electricity, with smaller percentages of use for the production of synthetic natural gas and fertilizer products.

**Li-Ion** lithium-ion (battery).

**Likens, Eugene** born 1935, U.S. ecologist, leader of the team of scientists whose observations at Hubbard Brook, New Hampshire provided the first documented link between fossil fuel combustion and an increase in the acidity of precipitation in North America ("acid rain"). This prompted the first international symposium on acid rain and greatly increased public awareness of human-accelerated environmental change.

**Lilienthal, Otto** 1848–1896, German inventor who made fundamental contributions to the development of heavier-than-air flight. He made over 2000 flights in gliders of his own design from 1891 to 1896. His book of aerodynamic data (1889) emphasized the curvature of a bird's wings as the secret of lift, and greatly influenced the early designs of the Wright Brothers and other aviation pioneers.

**limb darkening** *Solar.* the dimming of the solar disk toward its edge; an indicator of a photospheric temperature gradient that diminishes with height.

**lime** *Materials.* an industrial product composed of the chemical substance calcium oxide, CaO, a solid in the form of white to grayish pebbles, produced by heating crushed limestone to a temperature of about 900°C. It is one of the most commonly produced materials, and is widely used in construction, agriculture, and metal refining, and in water and waste treatment.

**limestone** *Materials.* a sedimentary rock composed primarily of calcium carbonate, formed from the skeletons of marine microorganisms and coral; used in construction and in the manufacture of lime.

**limiting factor** *Ecology.* the single condition or circumstance that is most influential in making it difficult for a species to live and grow, or to reproduce in its environment; e.g., food, water, light, and temperature.

**Limits to Growth** *History.* **1.** the title of a famous book published in 1972 by the Club of Rome, in which computer simulation models were used to forecast the future of world population growth, resource depletion, pollution, human health, and economic well-being. The results were extremely pessimistic, suggesting a near collapse of society within a short period of time. **2.** the general notion that there are biophysical limits to economic growth.

**Lindbergh, Charles A.** 1902–1974, U.S. aviator who became one of the most famous people in the world when he made the first solo nonstop flight across the Atlantic Ocean. He flew from Roosevelt Field (Long Island, New York) to Paris on 20–21 May 1927 in his one-motor airplane, *Spirit of St. Louis.*

*linear accelerator*

**Lindeman, Raymond** 1915–1942, U.S. ecologist who established the concept of TROPHIC DYNAMICS from his studies of Cedar Creek Bog in Minnesota. He presented a theoretical model of nutrient cycling expressed in terms of energy flow symbolized by mathematical equations. From this he introduced the concept of the efficiency with which energy is transferred from lower to higher trophic levels. He died just months before his seminal work was published.

**linear accelerator** *Nuclear.* a type of particle accelerator in which charged particles are accelerated in a straight line, either by a steady electrical field or by means of radiofrequency electric fields.

**linear energy transfer (LET)** *Nuclear.* energy dissipation of ionizing radiation over a given linear distance; gamma rays have low LET, while alpha rays have high LET.

**linear motor** *Electricity.* **1.** a type of alternating-current induction motor in which the motion between the stator and rotor is linear rather than rotary. **2.** an electric motor that has in effect been split so that the stator is laid out flat; the stator can become the track of a magnetically levitated (maglev) motor or vehicle.

**linear wave** *Earth Science.* a theoretical wave form of sinusoidal shape; can describe offshore wave conditions with satisfactory accuracy.

**line fault** *Electricity.* an open or short circuit, or other adverse condition, in a transmission line that causes partial or total loss of the power or signal at the output end.

**line-focusing collector** or **concentrator** see PARABOLIC COLLECTOR.

**line loss** *Electricity.* electrical energy lost during a transmission or distribution of electricity; much of the loss is thermal in nature.

**line of force** *Physics.* a field line in a force field, in which a tangent to the line will indicate the direction of the force.

**lines** *Electricity.* a general term for any system of poles, wires, cables, transformers, and accessory equipment used to distribute electricity to the public.

**Linnaeus, Carolus** (Carl von Linné) 1707–1778, Swedish botanist who devised a system of classifying organisms that is still in wide use today. He established the modern binomial (two-name) system of nomenclature for plants and animals; i.e., genus and species. He classified more than 4000 species of animals in all, including human beings (which he was the first to designate as *Homo sapiens*).

**Lipmann, Fritz Albert** 1899–1986, U.S. biochemist noted for his description of the manner in which cells convert food into energy. He discovered the role of COENZYME A in cellular metabolism.

**liquefied hydrocarbon gas** another term for LIQUEFIED PETROLEUM GAS.

**liquefied hydrogen** see LIQUID HYDROGEN.

**liquefied natural gas (LNG)** *Oil & Gas.* a liquefied product derived from natural gas (primarily methane) that has been reduced to a temperature of about –260°F at atmospheric pressure; it can be stored under low temperature and high pressure to reduce fuel volume for shipping and storage.

**liquefied petroleum gas (LPG)** *Oil & Gas.* a liquefied or compressed gas consisting of flammable hydrocarbons such as propane or butane, obtained as a byproduct from petroleum refining or from natural gas; widely used as a heating and cooking fuel, as in mobile homes or in areas not connected to gas pipelines. Thus, **liquefied propane gas.**

**liquefied refinery gas (LRG)** *Oil & Gas.* a term for liquefied petroleum gases fractionated from refinery or still gases. Through compression and/or refrigeration, they are retained in the liquid state.

**liquefy** *Chemistry.* to make or become liquid; convert to the liquid phase. Thus, **liquefaction.**

**liquid** *Chemistry.* one of the three fundamental states of matter, along with solids and gases. Unlike a solid (and like a gas), a liquid has constituent particles that are free to move past each other rather than being fixed in a given shape or position. Unlike a gas, it is relatively difficult to compress, and it lacks the capability to expand without limit to fill the space available. Most liquids will assume the shape of a container in which they are confined and will seek the lowest level available.

**liquid biofuel** *Biomass.* any of various fuels obtained from biomass in a liquid form; e.g., bioalcohols, such as methanol or ethanol.

**liquid-cooled** *Transportation.* describing an engine that is cooled by a liquid (e.g., water, glycol, or oil) that circulates through internal passages to a heat exchanger or liquid source.

**liquid crystal** *Materials.* a substance that displays the optical properties characteristic of crystals but other properties characteristic of liquids; commonly seen as an intermediate stage in crystalline forms that do not melt directly into a liquid, and in living organisms, such as the refracting portion of muscle fiber.

**liquid crystal display (LCD)** *Communication.* a screen in which a liquid crystal material placed between two sheets of glass exhibits data when a voltage passes across leads that are connected to character-forming segments etched onto its inner glass surface; a widely used technology for various electronic applications, such as watches and clocks, computer displays, and television screens.

**liquid-dominated** see VAPOR-DOMINATED.

**liquid electrolyte battery** *Storage.* any battery containing a liquid solution of acid and water. Distilled water may be added to the battery to replenish the electrolyte as necessary.

**liquid helium** *Chemistry.* helium gas that has been cooled to a temperature of at least 4.2 K to become liquefied; if cooled to below 2.9 K, it exhibits superconductivity and near-zero viscosity.

**liquid hydrogen** *Hydrogen.* $LH_2$, hydrogen cooled to the point where it changes from a gaseous to a liquid state. This requires extremely low temperatures; liquid hydrogen typically has to be stored at 253°C (423°F), which requires specially insulated and reinforced storage tanks. Also, **liquefied hydrogen.**

**liquid metal fast breeder reactor (LMFBR)** *Nuclear.* a type of fast reactor that produces more fissile material than it consumes. It converts uranium-238 into the fissionable isotope plutonium-239 by means of artificial radioactive decay. The most common form uses liquid sodium as a coolant that transfers the thermal energy to the power generation cycle.

**liquid oxygen** *Transportation.* oxygen gas that has been cooled to a temperature of at least 160.2 K or below, thus becoming liquefied; used in rocket propellants and other high-performance fuels.

**liquid petroleum** *Oil & Gas.* **1.** a term used to describe crude oil combined with natural gas liquids. **2.** another term for MINERAL OIL.

**liquid-to-gas ratio** *Oil & Gas.* the ratio of total petroleum liquids (including oil, condensate, and natural gas liquids) to gas in a gas field (usually expressed in barrels per million cubic feet).

**liquify** another spelling of LIQUEFY.

**listed hazardous waste** *Health & Safety.* an official category of hazardous wastes developed by the U.S. Environmental Protection Agency (EPA), classified in one of the following categories: wastes from non-specific sources, such as solvent wastes, electroplating wastes, or metal heat-treating wastes; wastes from specifically identified industries, such as wood preserving, petroleum refining, or organic chemical manufacturing; certain discarded commercial chemicals, either toxic or highly toxic, such as benzene, parathion, formaldehyde, or vinyl chloride.

**lithium** *Chemistry.* a very soft, silvery metallic element having the symbol Li, the atomic number 3, an atomic weight of 6.941, a melting point of 179°C, and a boiling point of 1317° C. It is an alkali metal and the lightest of all solid elements; used in the production of tritium and alloys.

**lithium bromide** *Materials.* LiBr, white crystals or a white to pinkish granular powder; very soluble in water, alcohol, and ether; used as a refrigerant, drying agent, and for various industrial purposes.

**lithium-ion battery** *Storage.* a battery that uses lithium metallic oxide in its positive electrode and a carbon material in its negative electrode; the lithium ions inside the battery transfer between the positive and negative electrodes during charge or discharge. It has greater capacity and efficiency than the earlier nickel cadmium version and is now widely used for applications such as laptop computers, cell phones, video cameras, and other mobile communication devices.

**lithium-ion-polymer battery** *Storage.* a modification of the lithium-polymer battery that employs hybrid cells containing some gelled electrolyte, to compensate for the relatively low conductivity of a completely dry lithium polymer electrolyte.

**lithium-manganese battery** *Storage.* a type of primary battery that uses manganese dioxide and carbon as the positive electrode, lithium metal foil as the negative electrode, and lithium perchlorate dissolved in propylene carbonate as the electrolyte; such batteries have very high energy storage density.

**lithium-polymer battery** *Storage.* a type of lithium battery that uses a dry plastic-like ultrathin film as the electrolyte rather than the traditional liquid or gelled electrolyte. This offers certain advantages in terms of size, design, and power.

**lithium-sulfur battery** *Storage.* a type of battery that uses lithium in the negative electrode and a metal sulfide in the positive electrode; it can store large amounts of energy per unit weight and is an emerging technology for such applications as laptop computers and electric powered/hybrid vehicles.

**lithology** *Earth Science.* **1.** the study and description of the general, gross physical characteristics of a rock, including its color, structure, grain size, mineral composition, and arrangement of its component parts. **2.** specifically, a characterization of the structural aspects of a coal bed and coal texture obtained from the macroscropic and microscopic observation of the coal.

**lithosphere** *Earth Science.* the strong outer shell of the earth, consisting of the crust and upper mantle; approximately 100 km thick and composed of rigid plates.

**Little Big Inch** see INCH LINES.

**Little Boy** *History.* a nickname for the atomic bomb that was the first nuclear weapon used in warfare, detonated over Hiroshima, Japan, on 6 August 1945.

**Little Ice Age** *Climate Change.* a cold period in Asia, Europe, and North America (1550–1850 AD). Its latter stages coincided with the advent of the Industrial Revolution and an increase of greenhouse emissions.

**littoral** *Earth Science.* **1.** relating to or involving the shore of a lake, sea, or other body of water. **2.** see LITTORAL ZONE.

**littoral zone** *Earth Science.* **1.** the area of shallow fresh water in which light penetrates to the bottom and nurtures rooted plants. **2.** the shoreline where a normal cycle of exposure and submersion by the tides occurs.

**live storage** *Coal.* a term for stored coal that is immediately accessible for use, as opposed to coal that is in long-term inactive storage **(dead storage)** and coal that is being sent directly into use rather than stored. Typically live storage coal would be placed in an enclosed silo or bunker while dead storage coal would be left in outdoor piles.

**llama** *Biological Energetics.* a herbivorous mammal of the Camelidae family, used in the Andes Mountains as a draft animal.

**LLW** low-level waste.

**lm** lumen.

**LMFBR** liquid metal fast breeder reactor.

**LNG** liquefied natural gas.

**load** *Physics.* **1.** the imposed force of pressure, weight, and so on that is supported by a structure or body. **2.** the forces that must be overcome to accelerate a vehicle and maintain its speed. **3.** any device that consumes energy; e.g., an electric motor, a refrigerator. **4.** the amount of energy consumed by such a device. **5.** *Refrigeration.* a term for the space within a refrigerating device from which heat must be withdrawn in the cooling process, or for the amount of energy required to do this.

**load center** *Electricity.* a geographical area in which large amounts of power are drawn by end users.

**load curve** *Electricity.* a graph describing the amount of power supplied as a function of time, especially one illustrating the varying magnitude of the load during a 24-hour period.

**load demand** *Electricity.* an electrical load on generating units, especially a sudden load.

**load factor** *Electricity.* **1.** the ratio of the average load in kilowatts supplied during a designated period to the peak or maximum load in kilowatts occurring in that same period. **2.** *Transportation.* in air travel, the ratio of the number of passengers actually traveling to the total number of available seats.

**load following** *Electricity.* the process by which a utility meets variations in electricity demand by preparing generating units for operation under unit commitment schedules, which reflect forecasted load changes over daily, weekly, and seasonal cycles, plus an allowance for random variations.

**load heat** *Refrigeration.* the heat energy that must be removed from an enclosed space (load) in a refrigeration process.

**load heat exchanger** *Refrigeration.* another term for an evaporator; i.e., a vessel in which a liquid refrigerant vaporizes to a gaseous state to withdraw heat energy from the space to be cooled.

**load leveling** *Electricity.* any load control technique that dampens cyclical daily load flows and increases baseload generation, as by peak load pricing or time-of-day charges.

**load rejection** *Electricity.* a sudden cessation of electrical load on the generating units.

**load shaving** **1.** see LOAD SHEDDING. **2.** see PEAK SHAVING.

**load shedding** *Electricity.* a policy of turning off or disconnecting certain loads (typically nonessential ones) to limit peak demand.

**load shifting** *Electricity.* a load management objective that moves loads from on-peak to off-peak periods.

**local air pollution** *Environment.* increased air pollutant concentrations, and the effect caused by this, relatively near to the source of the emissions, up to distances of a few tens of kilometers.

**location theory** *Economics.* **1.** the study of the relationship between geographic location and economic activity; e.g., the siting choices made by firms in relation to the location of their customers, or the residential choice made by households in relation to their workplace. **2.** the principles involved in selecting the optimal zone or location for a business operation, weighing such factors as transportation costs, production and labor costs, and location of competitors.

*lock* Locks are employed on the Panama Canal to allow for the fact that the water level is higher at the Pacific Ocean end of the canal system.

**lock** *Hydropower.* a section of a waterway at which the level of the water can be artificially raised or lowered.

**locomotion** *Biological Energetics.* **1.** the process of moving from one place to another. **2.** specifically, the fact of an animal moving from place to place, and the energy costs involved with this.

**locomotive** *Transportation.* a self-propelled vehicle, powered by fuel, electricity, or compressed air, used for moving loads on railroad tracks.

*locomotive* The Hercules, built in Pennsylvania in 1837; an advance in locomotive engineering for the time, noted especially for its large engine.

**lodestone** *Materials.* the historic term for a naturally occurring magnetic iron oxide, ($Fe_3O_4$, the mineral magnetite), that exhibits polarity.

**LOE** lowest observed effect.

**log** see WELL LOG.

**logging top** *Biomass.* the unmerchantable upper portion of the stem of a tree; typically a stick about 16–18 feet in length with a diameter of about 7 inches at the base.

**logistics fuel** *Transportation.* a term for transportation fuels employed for military rather than civilian use.

**lognormal distribution** *Measurement.* a statistical distribution function that is closely related to the normal distribution; it applies only when the quantity of interest must be positive, since $lnX$ exists only when the random variable $X$ is positive. It has a wide range of applications, including failure analysis in engineering, the distribution of income in economics, the distribution of oil reservoirs in the earth's crust, and survival times after diagnosis of cancer.

**London forces** *Physics.* a general term for the universal forces of attraction that occur between fluctuating dipoles in atoms and molecules that are very close together. [Described by German physicist Fritz *London* 1900–1954.]

**long** *Economics.* the market position of a futures contract buyer whose purchase obligates him to accept delivery, unless he liquidates his contract with an offsetting sale. Compare SHORT.

**long string** *Oil & Gas.* the last string of casing in a well, set just above or through the producing zone; it is the longest casing in a well and the smallest in diameter.

**long ton** *Measurement.* a British unit of weight equal to 2240 pounds or 1.016 metric tons.

**longwall** *Mining.* an extended horizontal section of coal isolated as the working area in longwall mining (see below).

**longwall advance** see ADVANCE MINING.

**longwall cutter** *Mining.* a compact machine driven by electricity or compressed air, with a jib at right angles to its body, used for cutting into the coal on relatively long faces.

**longwall mining** Longwall mining is an underground method of excavating coal from tabular deposits, as well as soft mineral deposits such as potash. Large rectangular blocks of coal are defined during the development stage of the mine and are then extracted in a single continuous operation. Generally each defined block of coal, known as a panel, is created by driving a set of tunnels, known as headings, from main or trunk roadways in the mine, some distance into the panel. These roadways are then joined to form the starting face of longwall face. As the coal is cut, the longwall face is supported with hydraullically operated supports. The function of these supports is to provide a safe working environment by supporting the roof as coal is extracted as well as advancing the longwall equipment. There are two methods of longwall mining. In retreat mining, the longwall face is installed and as mining continues into the panel the entries are allowed to collapse behind the face line to form part of the gob. In advance mining, the longwall face is set up a short distance from the main development tunnels or headings. The gate entries of the longwall face are formed as the coal is mined, with roadways formed adjacent to the gob. Each coal panel is separate from the adjacent workings with a solid barrier pillar. Longwall mining is an efficient and a high production mining method; a single longwall face can produce more than 25 000 tons of coal per day. Health and safety measures often result in lower rates of production.

**Naj Aziz**
University of Wollongong

**longwall mining** *Mining.* a mining method in which very long rectangular blocks of coal are defined during the development stage of the mine and then extracted in a single continuous operation by an automated cutting head moving parallel to the coal face. When the coal is cut, the working area is protected by a movable, powered roof support system. As the face progresses, the immediate roof above the coal is allowed to collapse behind the line of supports. Also, **longwall system.** See above.

**longwall retreat** see RETREAT MINING.

**longwave radiation** *Physics.* radiant energy emitted by the earth and atmosphere at relatively long wavelengths (between about 5 and 25 micrometer) in the electromagnetic spectrum.

**loop** *Electricity.* **1.** a complete electrical circuit. **2.** a set of branches that form a closed current path, where the omission of any branch breaks up the closed path. **3.** an electrical circuit providing two sources of power to a load or substation, so that if one source is interrupted the other continues to provide power.

**loop flow** *Electricity.* **1.** the movement of electric power from generator to load by dividing along multiple parallel paths. **2.** power flow along an unintended path that loops away from the most direct geographic path or contract path.

**loose fill** *HVAC.* insulation made from products such as rockwool fibers, fiberglass, cellulose fiber, vermiculite or perlite minerals, and composed of loose fibers or granules; it can be put in place by pouring the material directly from the bag or applying it with a blower.

**Lorentz, Hendrik Antoon** 1853–1928, Dutch physicist, a pioneer in formulating the relations among electricity, magnetism, and light. See following entries.

**Lorentz contraction** *Physics.* an effect of special relativity in which all moving objects contract in the dimension parallel to their motion; thus the length of a body contracts as its speed increases. Also known as **Lorentz-Fitzgerald contraction.** [Described by Hendrik *Lorentz* as an extension of the hypothesis of G. F. *Fitzgerald.*]

**Lorentz force** *Physics.* the force per unit volume on a fluid conductor that carries an electric current (density $J$) and lying in a magnetic field $B$; mathematically, it is the vector product of $J$ and $B$ and can also be represented by magnetic stresses.

**Lorentz-Lorenz formula** *Physics.* the relationship between the velocity of light in a medium and the density and composition of the medium. [Proposed independently by Hendrik *Lorentz,* and Danish physicist Ludwig *Lorenz,* 1829–1891.]

**Lorentz transformation** *Physics.* a relationship in which the space and time coordinates of one moving system can be correlated with the known space and time coordinates of any other system. This concept influenced (and was confirmed by) Einstein's special theory of relativity.

**Lorenz, Edward** born 1917, U.S. meteorologist noted for his pioneering descriptions of atmospheric general circulation, including the equations governing atmospheric energetics. See also BUTTERFLY EFFECT.

**loss of coolant** *Nuclear.* in a reactor, the inadvertent escape of water from the primary coolant system at a rate in excess of the capability of the reactor makeup system. Thus, **loss of coolant accident.**

**loss of load probability (LOLP)** *Measurement.* a measure of the reliability of an electrical grid; the probability that there is insufficient generating supply to support electrical demand.

**lost hole** *Oil & Gas.* a term for a well that has suffered a serious incident, such as a blowout, and can no longer be worked.

**lost work** *Thermodynamics.* an effect associated with every irreversible process, representing the difference between the heat transfer to the system and the amount of actual work done. In a reversible process, the lost work is zero.

**Lotka, Alfred J.** 1880–1949, U.S. biophysicist known for his energetic perspective on evolution; i.e., the idea that natural selection is fundamentally a struggle among organisms for available energy. Extending this to human societies, he suggested that the historic transition to nonrenewable energy would pose fundamental challenges to society.

**Lotka-Volterra model** *Ecology.* a predator–prey model proposed independently by Alfred J. Lotka and Vita Volterra, the basis of many models used in the analysis of population dynamics.

**Lovins, Amory** born 1947, U.S. physicist noted for his influential advocacy of renewable energy and energy efficiency. He was among the first to articulate the social and environmental imperatives for a shift from fossil and nuclear fuels to renewable energy sources. He formed the ROCKY MOUNTAIN INSTITUTE and led the design of the HYPERCAR.

**low-Btu gas** *Oil & Gas.* fuel gas with a heating value of 90–200 British thermal units per cubic foot. In contrast, the heating value of natural gas is 1030 Btu per standard cubic foot. Low-Btu gas is often derived from biomass, especially solid wastes.

**low-E** *HVAC.* describing a special window coating that helps to prevent the warmth inside a building from escaping through the glass in the winter, or to block the heat from the summer sun. Thus, **low-E coating, low-E glazing.**

**low emission(s) vehicle (LEV)** *Transportation.* an improved vehicle technology that produces only minimal harmful emissions, by using an alternative power source (such as electricity) as opposed to petroleum.

**low enriched uranium (LEU)** *Nuclear.* a classification for uranium containing less than 20% U-235.

**lower heating value** see HEATING VALUE.

**lower mantle** see MESOSPHERE (def. 3).

**lowest observed (observable) effect** *Health & Safety.* a measure of the point of exposure to a toxic substance at which some evidence of biological effect becomes apparent.

**low-grade heat(ing)** *HVAC.* a term for heat supplied from a source (e.g., a solar or geothermal source) whose temperature is only slightly higher than the space to be heated. Similarly, **low-grade cooling.**

**low-grading** or **lowgrading** *Consumption & Efficiency.* the principle or practice of extracting resources (especially forest resources) of the lowest quality first, on the basis that this will gradually improve the overall quality of the resource; i.e., an effort to maximize the process of "survival of the fittest", as opposed to the HIGH-GRADING approach.

**lowhead dam** *Hydropower.* a dam at which the water in the reservoir is maintained at a relatively small vertical distance above the turbine units.

**low-income economy (LIE)** *Economics.* in the World Bank's economic classification system, a country with a per capita gross national income (GNI) of $735 or less (2002 standard). See also MIDDLE-INCOME ECONOMY, HIGH-INCOME ECONOMY.

**low-level waste (LLW)** *Nuclear.* radioactive waste that is generated from hospitals and industry, as well as the nuclear fuel cycle, typically made up of paper, rags, tools, clothing, filters, and so on that contain small amounts of mostly short-lived radioactivity. LLW does not require shielding during handling and transport and is suitable for shallow land burial.

**low-pressure sodium lamp** *Lighting.* a type of lamp that produces light from sodium gas contained in a bulb operating at a partial pressure of 0.13–1.3 pascal. These lamps have very high efficiency and a long service life; they produce a characteristic yellow light and thus are well suited to applications where color rendition is not important, such as illumination for streets, highways, parking lots, and tunnels.

**low-pressure steam** *Geothermal.* a classification for steam having a pressure less than, equal to, or not greatly above that of the atmosphere.

**low-sulfur coal** another term for COMPLIANCE COAL.

**low-temperature collector** *Solar.* a solar thermal collector that generally operates at temperatures below 110°F and uses pumped liquid or air as the heat transfer medium.

**low-voltage** *Electricity.* describing voltage that is too low to cause an electric shock.

**low-voltage cutoff (LVC)** *Electricity.* the voltage level at which a charge controller will disconnect the load from the battery. Also, **low-voltage disconnect.**

**low-voltage lighting** *Lighting.* lighting equipment that can be operated at a voltage of 30 volts or less; e.g., a residential landscape lighting system that operates on 12-volts current by means of a step-down transformer from a standard 120-volts source.

**LOX** liquid oxygen.

**LPG** liquefied petroleum gas.

**LPS** low-pressure sodium.

**LPW** lumens per watt.

**LRG** liquefied refinery gas.

**LSP** lunar-solar power.

**lubricant** *Materials.* any substance, such as grease, oil, silicone, or graphite, that reduces friction between two interacting surfaces; manufactured predominantly from crude oil. Thus, **lubricate, lubrication.**

**lubricity** *Materials.* a measure of the ability of an oil or other compound to act as a lubricant (reduce friction) between two surfaces in contact.

**Luddite** *History.* **1.** a member of any of various groups of British workers who rioted and attempted to destroy textile machinery in the early period of the Industrial Revolution. **2.** in contemporary use, a person who is regarded as hostile to technological and scientific innovation; e.g., someone who pointedly refuses to use a computer. [Said to be from Ned *Ludd*, a workman who was an originator of this movement.] Thus, **Luddism.**

**luff** *Wind.* to turn a sail, windmill or turbine blade, and so on into the wind so that it will capture the force of the wind.

**Lukasiewicz, Ignacy** 1822–1882, Polish inventor regarded as one of the founders of the modern petroleum industry. He is said to be responsible for the world's first oil well, in Bobrka near Krosno, Poland (1854), and he reportedly built the world's first oil refineries near Gorlice and Jaslo, Poland. He also developed one of the first kerosene lamps, powered by seep oil. (It is not clear whether his kerosene lamp pre-dated the 1859 patent of Robert Dietz of the U.S.)

**LUKOIL** (est. 1993), the largest Russian oil corporation. It originally was an entirely state-owned enterprise composed of three separate production and refining operations; in 2004 the Russian government announced that it planned to divest its remaining stake in LUKOIL.

**lumber mill residue** see MILL RESIDUE.

**lumen** *Lighting.* the SI unit of measurement for the flux of light produced by a light source or received by a surface. It is based on the spectral sensitivity of the photosensors in the human eye under high (daytime) light levels. Photometrically, it is the luminous flux emitted with a solid angle (1 steradian) by a point source having a uniform luminous intensity of 1 candela.

**lumen depreciation** *Lighting.* the gradual loss of efficacy (lumen output) of a lamp over time, typically caused by the loss of chemical additives in the lamp and wall darkening in the arc tube. This can be indicated by a **lumen depreciation curve** depicting the pattern of decreasing light output.

**lumen hour** *Lighting.* a unit of quantity of light, equal to one lumen of light flux continued for one hour. Thus, **lumen second.**

**lumens per watt (LPW)** *Lighting.* a measure of the efficiency of a light source, expressed as the number of lumens of light the source generates per watt of electricity that it requires to do this. The higher the ratio, the more efficient the lighting system.

**luminaire** *Lighting.* a complete lighting unit consisting of a lamp (or lamps), along with the component parts that distribute the light, hold and protect the lamp, and connect the unit to a power supply. [French for "light" or "lighting".]

**luminaire efficacy** *Lighting.* a description of the performance of a luminaire, expressed as its total light output (in lumens) divided by the input power required to produce this. Thus, **luminaire efficacy rating (LER).**

**luminaire efficiency** *Lighting.* the ratio of the total light output from a luminaire to the total output from the bare lamp(s) fitting into it. Luminaire efficiency will always be lower than **lamp efficiency** because the luminaire's component materials will absorb some light.

**luminance** *Lighting.* **1.** the photometric equivalent of radiance. **2.** a measure of the luminous intensity of a surface, equal to the luminous flux leaving, arriving at, or passing through a unit area of that surface as viewed from the same direction. **3.** more generally, the quality or state of emitting light; brightness.

**luminescence** *Lighting.* any radiation of light from a body produced by some means other than heat; any emission of light at temperatures below that required for incandescence. Luminescence can be subdivided into fluorescence and phosphorescence.

**luminosity** *Physics.* **1.** the fact or process of giving off light; the quality of an object that produces light. **2.** the amount of energy radiated into space per second by a star. **3.** the number of particles, per square-centimeter per second, generated in the beams of high energy particle experiments.

**luminous** *Solar.* emitting light; describing the emission of visible radiation.

**luminous efficiency** *Lighting.* the ratio of the radiant energy sensed by the average human eye at a particular wavelength to that received.

**luminous exitance** see EXITANCE.

**luminous flux** *Lighting.* **1.** the flow of light. **2.** a measurement of this, expressed as the time rate of light emission within a unit solid angle of one steradian by a point source having a uniform luminous intensity of one candela.

**lump coal** *Coal.* any bituminous coal material remaining in the form of large lumps after an initial screening, often designated by the size of the mesh used for screening.

**lunar-solar power (LPS)** *Renewable/Alternative.* a proposed new energy source that would capture a portion of the solar energy that is incident on the moon and deliver it to earth for the purpose of electrical power. See next page.

**lung cancer** *Health & Safety.* any of several diseases characterized by malignant (cancerous) growth in an area of the lungs; one of the most common forms of cancer. Strongly associated with smoking and also associated with exposure to airborne toxic chemicals; e.g., occupational exposure to asbestos.

**Lunik II** *History.* a Soviet rocket that was the first man-made object to land on the moon, in 1959.

**Lurgi process** *Coal.* a high-pressure coal gasification process using pure oxygen as the combustion gas, developed in Germany in the 1930s. Coal is introduced into a gasifier, where steam and oxygen are added through a rotating grate. [Developed by the *Lurgi* company of Frankfurt, Germany.]

*Charles Lyell*

**lunar solar power** The system in which a portion of the 13,000 terawatts ($1.3 \cdot 10^{16}$ W) of solar power incident on Earth's Moon is converted at pairs of lunar power bases into beams of microwaves that illuminate, through all atmospheric conditions, receivers on Earth directly or indirectly via relay satellites in orbit about the Earth. Power base pairs are located on opposite east and west sides of the Moon as seen from Earth. Each power base consists of many thousands of small power plots. A power plot consists of solar arrays, wiring to gather the power, microwave generators and control circuitry, and a microwave sub-aperture (reflectors or lenses). Power plots are arranged within a power base so that the sub-apertures appear from Earth to overlap into one large, filled circular aperture that, for 10 cm microwaves, is 30 to 100 km in diameter. Plot components are constructed on the Moon of local lunar materials by machinery brought from Earth. Many thousands of separate receivers on Earth, termed rectennas, absorb their load-following beam (<20% the intensity of sunlight at Earth), rectify their beam into electrical power, and then deliver electric power to their local and regional electrical grids. The LSP System can safely and sustainably deliver many 10s of terawatts (or many kilowatts per person) of non-polluting electric power to Earth without significant use of Earth's resources. LSP can also power facilities and transports in cis-lunar and deep space. Drs David R. Criswell and Robert D. Waldron invented the LSP System.

**David R. Criswell**
University of Houston

**lux** *Lighting.* the SI derived unit of illuminance, symbol lx; equal to a luminous flux of one lumen uniformly distributed over a surface of one square meter.

**LWR** light water reactor.

**lx** lux.

**Lyell, Charles** 1797–1875, British geologist noted for establishing UNIFORMITARIANISM, the principle that features of the earth's surface were produced by natural forces operating for extremely long times. Prior to Lyell, most scientists supported CATASTROPHISM, the idea that these features were due to specific shorter-term events.

**lysimeter** Measurement. an instrument that measures the amount of water-soluble matter in soil

**M** an abbreviation for mega- (one million).

**m** or **m.** an abbreviation for meter.

**Mabro, Robert Emile** Egyptian-born economist noted for his economic analysis of the geopolitics of the Middle East, particularly the role of OPEC in world oil markets.

**MAC** maximum allowable concentration.

**macadam** *Transportation.* a traditional road-making material utilizing uniformly sized stones rollered into layers and finished with asphalt. [Developed by Scottish engineer John *McAdam,* 1756–1836.]

**MacCready, Paul** born 1925, U.S. engineer and founder of AeroVironment, a company noted for its development of alternate energy sources in aviation and other modes of transportation; e.g., a human-powered airplane, solar-powered and battery-powered cars.

**maceral** *Coal.* any of the microscopic, organic constituents of coal, as distinguished from the macroscopic constituents seen in hand specimens.

**Mach, Ernst** 1838–1916, Austrian physicist who established that one of the most important variables affecting aerodynamic behavior is the speed of the airflow over a body relative to the speed of sound (the MACH NUMBER).

**Mach** *Measurement.* an expression of the speed at which an object is moving through a fluid medium according to the MACH NUMBER. An aircraft traveling at Mach 1 is moving at the speed of sound, Mach 2 is twice the speed of sound, and so on.

**machine** *Consumption & Efficiency.* **1.** an assembly of interrelated parts, each with a definite motion and separate function, used to perform a specific task. **2.** any device that transmits or modifies energy. **3.** relating to or produced by such a device; mechanical. See also SIMPLE MACHINE.

**machine drive** *Conversion.* any direct process of end use in which thermal or electric energy is converted into mechanical energy.

**Mach number** *Measurement.* a dimensionless quantity given by the ratio of the velocity of a source in a fluid medium (e.g., an aircraft in flight, or a high-speed fluid flow in a channel) to the velocity of sound in the same medium. *Subsonic* conditions occur for Mach numbers less than 1.0; *transonic* conditions when the Mach number is at or near 1.0; *supersonic* conditions occur above Mach 1.0; and *hypersonic* conditions above Mach 5.

**MacMillan, Kirkpatrick** 1812–1878, Irish inventor who developed the first rear-wheel driven safety bicycle (1842). It was propelled by a horizontal reciprocating movement of the rider's feet on the pedals; this movement was transmitted to cranks on the rear wheel by connecting rods.

**macroalgae** *Ecology.* larger forms of algae; see ALGAE.

**macroclimate** *Earth Science.* the climate of a large area of the earth's surface, especially as measured over an extended period of time.

**macroeconomics** *Economics.* the study of the economy as an aggregated whole, generally ignoring the behavior of individual industries or consumers and instead concentrating on economy-wide magnitudes such as unemployment, inflation, interest rates, money supply, the trade deficit/surplus, and growth of gross national product. Thus, **macroeconomic.**

**macromolecule** *Chemistry.* a complex molecule of relatively large size, having masses in the thousands to millions of daltons; e.g., a polymer or protein. Thus, **macromolecular.**

**macronutrient** *Biological Energetics.* a chemical component of food that is necessary in large amounts as a source of energy; these include carbohydrates, fats, and proteins.

**macroscopic** *Physics.* **1.** having to do with or considering the state of a system as a whole or on a large scale. **2.** specifically, involving an analysis or study in which the individual molecules and atoms are not measured or explicitly considered in the description. **3.** more generally, relating to any large-scale perspective or context. Contrasted with MICROSCOPIC.

**MAD** maximum allowable dose.

**Madrid Protocol** *Policy.* an international agreement in which Antarctica is designated as a natural reserve dedicated to peace and the environment, and environmental principles are laid down for planning and conducting all activities there; e.g., any activity relating to mineral or energy extraction, other than scientific research, is banned.

**mafic** *Materials.* describing an igneous rock that is composed primarily of dark-colored ferromagnesian minerals.

**maglev** *Transportation.* **1.** short for MAGNETIC LEVITATION, as applied in the context of transportation vehicles. **2.** a rail system in which the cars are suspended above or below a guide rail by means of magnetic levitation. **3.** a train that operates on this basis.

**magma** *Earth Science.* the naturally occurring, mobile mass of molten rock material generated within or beneath the earth's surface, and from which igneous rocks are derived.

**magma heating** *Geothermal.* the (experimental) effort to obtain geothermal energy by sinking a well that is adjacent to a magma body and bringing heat from the magma to the surface.

**magnesium** *Chemistry.* a silvery alkaline-earth metallic element having the symbol Mg, the atomic number 12, an atomic weight of 24.305, a melting point of 650°C, and a boiling point of 1107°C. The lightest of all structural metals, used in alloys for structural parts, in pyrotechnics, and in batteries.

**magnet** *Physics.* a substance or object having the property of MAGNETISM, composed of certain materials whose domains are aligned in such a way as to produce a net magnetic field outside the material, or to experience a torque when placed in an external magnetic field.

**magnetic** *Physics.* having to do with or produced by a magnet, or the property of magnetism.

**magnetic bias** *Physics.* a steady magnetic field used in a magnetic circuit.

**magnetic compression** *Nuclear.* a method to provide initial heat for a tokamak fusion reactor, in which the plasma is heated through a rapid compression, which is made possible by increasing the magnetic field.

**magnetic confinement** *Nuclear.* a method now being researched for the containment of plasma in a fusion reaction; it prevents contact between the hot plasma and the walls of its container by keeping it moving in circular or helical paths through magnetic force on charged particles. Thus, **magnetic confinement fusion.**

**magnetic domain** *Materials.* the regions in the microstructure of a material with magnetic properties where atoms are aligned to produce a magnetic field in one direction. When no magnetic flux is present outside the specimen, the magnetic domains aligned in one direction are balanced by domains of opposite alignment.

**magnetic field** *Physics.* a vector field occupying physical space in which a magnetic force can be detected, typically in the presence of a permanent magnet, current-carrying conductor, or electromagnetic wave. Also, **magnetic force.**

**magnetic field line** *Physics.* an imaginary line in a magnetic field along which a compass needle will tend to align.

**magnetic flux** *Physics.* the lines of magnetic force arising from and found in the vicinity of a magnetized body.

**magnetic flux surface** *Physics.* a surface composed of field lines so that the magnetic field, $B$ is everywhere tangential to it. The flux of $B$, contained within the surface is the same everywhere along its length.

**magnetic force** see MAGNETIC FIELD.

**magnetic helicity** *Physics.* a property of a magnetic field preserved in the motion of a fluid that is a perfect electrical conductor.

**magnetic induction** *Physics.* the process of establishing a condition of magnetism in an object that is placed in a magnetic field.

**magnetic levitation** *Physics.* **1.** the noncontact support of an object above a surface by means of magnetic forces. **2.** *Transportation.* specifically, the support of a railway train above the rails by means of this.

**magnetic line** see MAGNETIC FIELD LINE.

**magnetic permeability** *Materials.* the ratio of change in the magnetic flux density in a material to the change in magnetizing force applied to it.

**magnetic pole** *Earth Science.* either of two points on the earth's surface at which the magnetic dip is exactly 90°; the point where the meridians join; i.e., where the magnetic field is vertical.

**magnetic pressure** *Physics.* the pressure associated with a magnetic field, equal to the square of the magnetic induction divided by twice the magnetic permeability.

**magnetic relaxation** *Physics.* the process by which a magnetofluid relaxes to its minimum energy state over time, subject to certain constraints.

**magnetic resonance** *Physics.* a frequency-matching phenomenon in which the spin systems of certain atoms will readily absorb energy from an externally applied, alternating magnetic field, when the frequency of the magnetic field is the same as a frequency characteristic of the spin system.

**magnetic shear** *Measurement.* a measure of the extent to which magnetic field lines vary in direction as the structure of the magnetic field becomes twisted (sheared).

**magnetic susceptibility** *Physics.* the ratio of the magnetization of a substance to the applied magnetic field strength.

**magnetism** *Physics.* the phenomenon by which materials will have either an attracting or repelling force with respect to other materials. This occurs when electrically charged particles are in motion, either from their movement in an electric current or from their presence in a permanent magnet (a substance or object that retains its own magnetic properties). Certain metals are observed to have a strong property of magnetism.

**magneto** *Electricity.* an electric generator containing one or more permanent magnets, which provide the magnetic flux.

**magnetofluid** *Materials.* any liquid that becomes less fluid under the influence of a magnetic field.

**magnetogram** *Measurement.* a graphic representation of magnetic field strengths and polarity. Similarly, **magnetograph.**

**magnetohydrodynamics (MHD)** *Physics.* the study of the motions of electrically conducting fluids, such as plasmas and liquid metals, and their interactions with magnetic fields.

**magnetomotive force** *Physics.* the force that sets up a magnetic field within and around an object, expressed as the work required to carry a magnetic pole of unit strength once around a magnetic circuit.

**magnetopause** *Earth Science.* the outer layer of the earth's magnetosphere, bordering the solar wind.

**magnetosphere** *Earth Science.* a region of space surrounding the earth in which charged particles are controlled by the planetary magnetic field.

**magnetostatics** *Physics.* the branch of physics dealing with magnetic phenomena that are constant in time.

**maintenance-free** *Storage.* describing a sealed battery to which water is not added to maintain electrolyte level.

**majority carrier** *Photovoltaic.* a current carrier (either a free electron or hole) that is in excess in a specific layer of a semiconductor material (electrons in the n-layer, holes in the p-layer) of a cell. Compare MINORITY CARRIER.

**makeup** or **make-up** *Consumption & Efficiency.* water or air added to a system to replace a loss; e.g., water added to boiler feed to compensate for leakage, or air brought into a building from outside to replace exhaust air.

**malaise speech** *Policy.* the name applied to a speech by President Jimmy Carter in 1979, a time of rising energy prices and fuel shortages in the U.S. He called for import quotas, a tax on oil profits, development of synthetic fuels, and individual energy conservation. The speech was poorly received and is said to have contributed to Carter's loss in the next election. [So called because it referred to a "crisis" of spirit and confidence in America, though the specific word "malaise" was not actually used.]

**Malthus, Thomas** 1766–1834, English economist regarded as a founder of the field of population study. He argued that human societies will always be afflicted by poverty and famine, because population is bound to grow at a faster rate than the food supply. His work has led to many subsequent analyses of the relationship between population and available resources, in particular Darwin's ideas about natural selection.

**Malthusian** *Global Issues.* relating to or supporting the theories of Thomas Malthus, in particular the argument that the human population has a natural tendency to grow at a rapid enough rate to offset and perhaps even overshoot productivity gains in food production.

**Malthusian trap** *Consumption & Efficiency.* the principle that human societies will descend over time to a minimum subsistence level, as a result of the tendency of population to increase at a much greater rate than productivity. Also, **Malthusian equilibrium.**

**Manhattan Project** *History.* the code name for the large-scale secret effort by the United States to develop an atomic weapon during World War II, involving various sites, principally Oak Ridge, Tennessee; Richland, Washington; and Los Alamos, New Mexico. [So called because the initial work for the project was done by the Manhattan Engineer District in New York City.]

**manifold** *Consumption & Efficiency.* a pipe or chamber having multiple openings to allow the passage of a fluid; e.g., the conduit of a heating appliance that supplies gas to the individual burners.

**manioc** another name for CASSAVA.

**manometer** *Measurement.* a device for measuring pressure differences, usually by the difference in height of two liquid columns.

**mantle** *Earth Science.* the part of the earth's interior lying below the crust and above the core, extending to a depth of about 3500 km below the surface.

**mantle convection** *Earth Science.* an assumed process deep within the earth in which hotter materials move toward the surface while cooler materials are sent to the interior.

**manufactured coal gas** *Coal.* a fuel gas used for illuminating and heating purposes, produced by heating coal in a retort and capturing the resulting vapors for distribution.

**manufacturing** *Consumption & Efficiency.* any activity that involves the mechanical, physical, or chemical transformation of substances into new products with economic value.

**manway** *Mining.* a small passageway in a mine used for miners to move through; in modern mines also used to convey electrical, water, and air lines.

**Marchetti, Cesare** born 1927, Italian physicist who was among the first to discuss producing hydrogen as an energy carrier and also among the first to propose pumping $CO_2$ into the ocean to ameliorate climate change.

**Marconi, Guglielmo** 1874–1937, Italian physicist, noted for his development of wireless telegraphy. He patented his system in England (1896) and organized a wireless telegraph company to develop its commercial applications. In 1899 he established wireless communication between France and England across the English Channel.

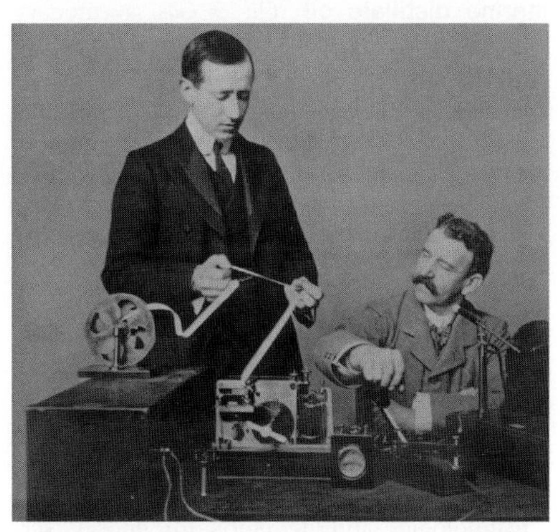

***Guglielmo Marconi*** *(standing left; colleague G. S. Kent seated right)*

**Marco Polo** 1254?–1324?, Florentine author noted for writings on his travels to China and other lands of Asia. Among his descriptions are reports of people collecting oil from seeps on the shores of the Caspian Sea, and the use of "stones that burn like logs" (coal) in China, both practices that were not well known to the West at that time.

**marginal abatement cost** *Economics.* the incremental change in abatement cost as a function of an incremental change in the amount of a pollutant emitted.

**marginal cost** *Economics.* the cost of the additional inputs (labor, capital, energy, and so on) required to produce one additional unit of output, assuming other factors remain constant.

**marginal extraction cost** *Economics.* the cost of the additional inputs (labor, capital, energy, and so on) required to extract one additional unit of a natural resource; e.g., a barrel of oil, assuming other factors remain constant.

**mariculture** *Ecology.* aquaculture in a marine setting; the cultivation of marine (saltwater) organisms under controlled conditions for the purpose of human food or other uses.

**marine biomass** *Biomass.* a collective term for all organic substances of marine origin that can be used for energy production, such as kelp.

**marine distillate oil** *Oil & Gas.* a category of distillate oils that can be burned in auto-ignited marine engines (diesel engines). It requires very little pretreatment for proper combustion. Also, **marine diesel oil.**

**marine gas oil** *Oil & Gas.* a category of distillate oils that can require spark ignition for combustion in marine engines (i.e., gasoline engines).

**marine riser** *Oil & Gas.* in offshore drilling, a large-diameter steel casing with buoy sections that serves as an extension of a wellbore from the sea bottom to the rig surface area.

**MARKAL** *Economics.* market allocation model; a linear programming model that simulates a region's energy system by representing its technologies and demands for energy services.; it calculates the least-cost way to meet a given set of demands for energy services (such as heating and cooling or transportation), within the limits of a user's available technologies and any environmental constraints.

**marker crude (oil)** *Oil & Gas.* a type of crude oil that is sold in sufficient volumes to provide liquidity (many buyers and sellers) in the market, as well as having similar physical qualities of alternative crudes; e.g., West Texas Intermediate (U.S.), Brent (Europe and Africa), Dubai and Oman (Middle East), and Tapis and Dubai (Asia). Thus, **marker crude price.**

**market** *Economics.* any setting or situation in which buyers and sellers interact to exchange goods and services for an agreed-upon price.

**market allocation model** see MARKAL.

**market-based** *Policy.* describing the use of taxes, subsidies, tradable instruments (e.g., emission reduction credits), or other such market factors to achieve desired goals (e.g., lower pollution levels) instead of, or as a complement, to government regulation. Thus, **market-based instrument, incentive, regulation,** and so on. See next column.

**market-based** There are many market-based instruments that can be used to manage behaviors to achieve environmental objectives. These include economic, technological, legal and moral suasion mechanisms, all of which establish positive or negative incentives for environmental related behaviors. Market-based instruments are those that utilize economic incentives through processes that either explicitly mimic markets or explicitly alter market conditions as means of achieving environmental objectives and allow for some flexibility in how those objectives are met. Markets rely on clearly established property rights and clear terms for trade. Market mimicking instruments would include tradable permit systems, or other methods that establish tradable property rights or remove barriers for trading. Instruments that alter market conditions directly would include imposition or removal of taxes or subsidies that change cost or demand conditions, or "Green Product" designations that change demand conditions. While other instruments also have economic incentive effects, such as expected liabilities or fines, market-based instruments are designed to have more straightforward effects on the demand and supply conditions facing individuals and enterprises, and may allow more flexibility in achieving environmental objectives than mandatory or restrictive instruments, such as technological mandates.

**Stephen Farber**
University of Pittsburgh

**market-based environmentalism (MBE)** *Environment.* strategies, policies, or regulations intended to preserve and protect the quality of the environment by means of market forces; e.g., pollution taxes, cap and trade allowances, emission reduction credits.

**market-clearing price** *Economics.* the price at which supply equals demand for future energy delivery; i.e., the point at which buyers and sellers converge in a free market.

**market economy** *Economics.* an economy in which the allocation of resources is based primarily on the voluntary interactions of many buyers and sellers, and prices are established mainly by forces of supply and demand rather than government control.

**market equilibrium** *Economics.* a situation in which supply equals demand in the market for a good or service at a prevailing price.

**market failure** *Economics.* any situation in which a market economy fails to attain economic efficiency and in which there is a potential role for government; i.e., the self-interested actions of market participants have not led to an efficient allocation of resources. See below.

**market forces** *Economics.* the dynamic occurring when competition among firms determines the outcome in a given situation, and a supply-and-demand equilibrium is reached without government intervention.

**market valuation** *Economics.* a process in which characteristics of the market generate a value. For example, the value of a barrel of oil in the ground to its owner, as a function of its current market price less extraction costs.

**Marsh, George Perkins** 1801–1882, U.S. naturalist whose ecological insights pioneered awareness of human impacts on the earth. His *Man and Nature* (1864, revised 1874) is widely regarded as the first modern discussion of environmental problems.

**marsh gas** *Biomass.* a colorless, odorless, tasteless gas (mainly methane, $CH_4$) that is produced by decaying vegetation and other organic materials in wetlands.

**Marxism** *Economics.* **1.** the theories of the German philosopher Karl Marx (1818–1883), especially the proposition that the chief means of production in a society should be publicly owned. **2.** the (nominal) enactment of these theories from 1917 onward in various states with government-controlled economies; e.g., the Soviet Union, East Germany, China, North Korea, Cuba.

**maser** *Electricity.* a device that can produce coherent electromagnetic radiation in the microwave frequency band; a highly precise timing device used in atomic clocks and also to amplify weak radio signals. [An acronym for microwave amplification by stimulated emission of radiation.]

**masonry** *Materials.* a general term for stone or stonework of any type, usually cast or formed, including ceramic brick, tile, concrete, glass, mud, adobe, and the like.

**masonry dam** *Hydropower.* a dam constructed of brick or stone masonry laid and fitted with mortar; may be of arch or gravity design.

**masonry stove** *HVAC.* a type of heating appliance similar to a fireplace, but more efficient and cleaner burning. Heat from the combustion of a fuel is transferred to the heavy masonry mass of the stove, which releases the heat into the room.

**mass** *Physics.* a fundamental property of an object that makes it resist acceleration, and

**market failure** Market failure comes in two varieties. The first type occurs within the framework of neoclassical economic theory, when unregulated markets do not produce the "Pareto optimal" outcome (in which no agent can improve his or her position without adversely affecting at least one other agent) that the theory asserts will ensue if the requirements of perfect competition, complete information, and an absence of external effects are met. These requirements are very stringent, suggesting that market failure is likely to be quite common. Monopolies arise naturally if there are economies of scale in production or distribution; the informational requirements for fully efficient markets are immense; and the absence of property rights in all the goods and services affecting the well-being of individuals means that not all socially relevant costs and benefits will be taken into account by private economic decision-makers. The second type of market failure arises because real individuals and organizations do not have the properties of the ideal types that populate economic models. People, business firms, and public institutions are boundedly rational rather than completely rational, they operate under internal and external rules and procedures that necessarily result in less-than-perfect outcomes, and they respond to incentives that, while pushing them in the direction of improved economic performance, nevertheless do not guarantee optimality. While these real-world departures from the neoclassical model are commonly labeled "market failure", they might more properly be described as psychological, institutional, or organizational deviations from the abstractions of neoclassical economics.

**Stephen J. DeCanio**
University of California, Santa Barbara

that determines its gravitational attraction. This property can be generally regarded as equivalent to the amount of matter in the object. *Mass* and *weight* are not strictly synonymous; mass is not affected by the forces acting on an object, but weight is a relative property that can change according to the gravitational force exerted on the object.

**mass balance** *Physics.* **1.** the principle that the total mass of materials entering a process is equal to the amount leaving the process. See also CONSERVATION OF MASS. **2.** *Earth Science.* the total volume of a glacier, which may increase due to accumulation of snow and ice, or decrease due to melt.

**mass burn** *Environment.* the practice of combusting municipal solid waste (MSW) in a furnace where the material is fed in with little or no prescreening or separation; within the furnace, the combustible content is burned away and the residue of ash, glass, and metal is removed as a bottom ash. Thus, **mass burn facility.**

**mass conservation** see CONSERVATION OF MASS.

**mass-energy equivalence** *Physics.* the principle of the interconversion of mass and energy, as given by Einstein's famous equation $E = mc^2$, where $E$ is the energy in ergs, $m$ is the mass in grams, and $c$ is the velocity of light in centimeters per second.

**mass number** *Physics.* the total number of protons and neutrons in the nucleus of an atom.

**mass production** *History.* a type of technology in which identical products are manufactured in very large quantities, as opposed to a series of single items being fashioned or constructed individually.

**mass transit** *Transportation.* a general term for all forms of passenger transportation other than private motor vehicles; i.e., subways, buses, light rail, and other such public conveyances.

**master metering** *Electricity.* a method of measuring the use of electricity in which multiple customers are all metered cumulatively on the same meter.

**material flow account** *Materials.* a component of a nation's economic accounts that represents inputs of materials, their accumulation,

***mass transit*** *A combination of mass transit (a Swedish ferry) and private transportation, since the ferry is used to transport individual motor vehicles.*

and their outflow to the natural environment and other economies. Similarly, **material flow analysis.**

**material intensity** *Materials.* **1.** the relation between primary material demand and the gross domestic product of a country in a given year; e.g., tons of copper used per dollar of gross domestic product. **2.** a similar relation between material demand and population.

**material recycling** see RECYCLING.

**maternal mortality rate** *Health & Safety.* a statement of the annual number of deaths of women from pregnancy-related or childbirth-related causes, usually expressed as the number of deaths per 100,000 live births (or per 1000 or 10,000 births).

**matrix** *Renewable/Alternative.* the material framework within a fuel cell that supports or contains an electrolyte.

**matter** *Physics.* an aggregate of physical particles possessing inertia and capable of occupying a position in space. See also MASS.

**mature forest** *Ecology.* an intermediate classification for a forest area that is not truly old-growth (i.e., about 80 to 150 years old) but that typically contains many, of the same characteristics as an old-growth forest, such as large mature trees, variety of tree age and species, and the presence of a diversity of plant and animal life.

**Mauna Loa curve** *Climate Change.* a representation of carbon dioxide concentrations in the atmosphere, as measured on the peak of Mauna Loa volcano in Hawaii. Measurements at this location have shown that atmospheric concentrations of $CO_2$ are increasing. Also known as the **Keeling curve** for its description by oceanographer Charles D. *Keeling.* See below.

**maximization principle (hypothesis)** *Ecology.* as formulated by Alfred Lotka, the principle that natural selection will operate among living organisms so as to maximize energy flow through the system, so far as is possible given the existing constraints on available energy.

**maximum allowable concentration (MAC)** *Health & Safety.* a measure used in standards of workplace safety that represents the largest possible airborne concentration of a toxic substance that will not produce significant adverse effects. Similarly, **maximum permissible concentration.**

**maximum allowable dose (MAD)** *Health & Safety.* a measure that represents the largest possible amount of an agent or substance to which an individual can be exposed without significant adverse effects; e.g., radiation. Also, **maximum allowable dose level (MADL).**

**maximum power principle** see MAXIMIZATION PRINCIPLE.

**maximum residue level (MRL)** *Health & Safety.* the highest concentration of a toxic substance remaining in or on a food product (e.g., pesticide residue on a fruit or vegetable, a veterinary medicinal product in meat) that can safely be ingested by humans in eating the food.

**maximum tolerated (tolerable) dose** *Health & Safety.* **1.** the largest possible dosage of a drug or agent that will not produce significant adverse effects. **2.** a measure used in tests of chronic toxicity that represents the largest possible dose that will not produce either death or a 10% decrease in body weight in the subject animal.

**Maxwell, James Clerk** 1831–1879, Scottish physicist renowned for his path-breaking work in electricity and magnetism. He developed the theory of the electromagnetic field on a mathematical basis, and made possible a much greater understanding of the phenomena in this field. He discovered that electric and magnetic energy travel in transverse waves that propagate at a speed equal to that of light; light is thus only one type of electromagnetic radiation.

**maxwell** *Measurement.* a unit of magnetic flux in the centimeter-gram-second system, equivalent to the flux that produces one abvolt in a one-turn circuit when the flux is reduced to zero at a uniform rate in one second.

**Maxwell's demon** *Thermodynamics.* a tiny hypothetical creature that can see and sort individual hot and cold molecules and thus violate the second law of thermodynamics. The demon controls a tiny door between two gas-filled vessels of the same temperature; it opens and closes the door so as to concentrate hotter molecules in one vessel and colder ones in the other, thus decreasing the total

**Mauna Loa curve** In the context of energy consumption, a diagram representing the rate of change of the concentration of atmospheric carbon dioxide as a function of time since 1958 as measured at the Mauna Loa Observatory, a station situated at 3500 meters above sea level on the island of Hawaii in a lava field almost devoid of vegetation. During a part of almost every diurnal interval the air carried to station by wind currents has a chemical composition representative of a large fraction of the northern hemisphere. The Mauna Loa curve shows a strikingly regular annual oscillation, reflecting seasonally varying plant growth and decay, on which is superimposed a steady year to year increase from a seasonally adjusted concentration of 315 ppm (parts per million of carbon by volume) in 1958 to 377 ppm in 2004. Because the earth's atmosphere mixes globally within about a year's time, and another such record at the South Pole shows almost the same rate of rise, this rise is clearly global; it closely tracks the rate of combustion of fossil fuels, evidence that human activity is steadily changing the chemical composition of the earth's atmosphere.

**Charles D. Keeling**
Scripps Institution of Oceanography

*James Clerk Maxwell*

entropy of the system. [Devised by James Clerk *Maxwell.*]

**Maxwell–Boltzmann distribution** *Thermodynamics.* the classical distribution function for the distribution of an amount of energy between identical but distinguishable particles.

**Maxwell's (field) equations** *Physics.* a set of four equations presented by James Clerk Maxwell to describe the electric and magnetic fields as functions of position and time.

**Mayer, Julius Robert von** 1814–1878, German physicist who was the first person to describe the law of the conservation of energy (1842). He did so independently of Joule, who received credit for the discovery a few years later. Mayer also is thought to be the first to describe oxidation as the primary source of energy for living organisms.

**Mazda** or **MAZDA (lamp)** *Lighting.* **1.** the brand name for a historic improvement in the incandescent lamp (1909), utilizing a filament of drawn tungsten, thus providing a higher-quality light output and greater efficiency than earlier carbon-filament lamps. **2. mazdas.** a popular name for electric lights in general. [Named for the Persian god of light.]

**MBE** market-based environmentalism.

**McAfee cracking process** *Oil & Gas.* the petroleum industry's first commercially viable catalytic cracking process, developed in 1923; a method that ultimately would double or even triple gasoline yield from crude oil by then-standard distillation methods. [Developed by U.S. chemist Almer McDuffie *McAfee,* 1886–1972.]

**McCormick, Cyrus Hall** 1809–1884, U.S. inventor who first patented the mechanical reaper (1834), which revolutionized the harvest of key agricultural crops by combining all the steps that earlier harvesting machines had performed separately. This time-saving **McCormick reaper** allowed farmers to more than double their crop size and spurred innovations in farm machinery.

*Cyrus McCormick*

**MCFC** molten carbonate fuel cell.

**McKelvey box** *Measurement.* a two-dimensional scheme that combines criteria of increasing geologic assurance (undiscovered/possible/probable/proved reserves) with those of increasing economic feasibility (subeconomic resources as compared with economic reserves, depending on price and cost levels). [Named for U.S. geologist Vincent *McKelvey.*]

**MDO** marine distillate oil.

**MEA** membrane electrode assembly.

**Meadows, Donella** 1941–2001, U.S. environmental scientist noted for reporting on long-term global trends in population, economics, and the environment, especially as an author of the book LIMITS TO GROWTH.

**mean daily insolation** *Solar.* a statement of the average solar energy available per square meter per day for a given month.

**mean sea level** see SEA LEVEL.

**measured resources** *Coal.* coal deposits for which estimates of the rank, quality, and quantity have actually been computed, within a high degree of geologic assurance, from sample analyses and measurements from well-known sample sites.

**mechanical** *Physics.* **1.** relating to or caused by a physical force. **2.** having to do with or produced by a machine or tool.

**mechanical advantage** *Physics.* the ratio of the output force to the input force for a machine that transmits mechanical energy; i.e., the work produced by the machine, divided by the force applied to it. Actual machines can provide a mechanical advantage that is greater than unity; however, the greater the mechanical advantage, the greater the distance that the input force must move in relation to the output force.

**mechanical efficiency** *Physics.* the ratio between the brake (useful) horsepower and indicated horsepower of an engine.

**mechanical energy** *Physics.* the sum of the kinetic energy and potential energy of an object.

**mechanical engineering** *Consumption & Efficiency.* the branch of engineering concerned with the efficient design, operation, and maintenance of machines.

**mechanical equivalent of heat** *Thermodynamics.* a constant that expresses the number of units of heat in terms of a unit of work, typically expressed as the amount of heat transfer required to raise the temperature of 1 gram of water from 14.5 to 15.5°C.

**mechanical property** *Materials.* any property of a material that influences its behavior when it is exposed to external forces, such as its hardness or elasticity.

**mechanical refrigeration** *Refrigeration.* a standard process of refrigeration, in which a motor-driven compressor drives the circulation of a refrigerant through a closed loop, so that the refrigerant withdraws heat energy from the load (space to be cooled) as it changes from a liquid to a gaseous state.

**mechanical theory of heat** *Thermodynamics.* the principle that heat consists of motions of the particles that make up a substance.

**mechanical weathering** *Earth Science.* a collective term for all the natural physical processes that break rock into smaller fragments without chemical change; e.g., wind, abrasion, temperature change, frost action, biological effects such as plant root extension or animal burrowing.

**mechanics** *Physics.* the branch of physics that deals with motion and with the reaction of physical systems to internal and external forces.

**mechanization** *History.* a process in which machines come into use to replace human and animal power, especially a transitional process of this kind in an entire society.

**MECS** Manufacturing Energy Consumption Survey; a description of the amount of energy consumed by manufacturing firms in the U.S.; conducted by the Department of Energy.

**MECs** *Consumption & Efficiency.* microtechnology-based energy and chemical systems; a class of miniaturized devices integrating heat and mass transfer components along with chemical reactor technology in a single integrated system, with typical feature sizes in the 50 mm to 1 cm size.

**median lethal concentration** *Health & Safety.* the concentration level of a toxic substance that is sufficient to have a lethal effect in 50% of the subject group of organisms under specified conditions; written as $LC_{50}$.

**median lethal dose** *Health & Safety.* a commonly used measure of the toxicity of a substance, expressed as the amount of the substance that will produce a lethal result in 50% of a group of test organisms under specified conditions. written as $LD_{50}$.

**medium-temperature collector** *Solar.* a solar thermal collector designed to operate in the temperature range of 140–180 °F.

**medium-Btu gas** *Oil & Gas.* a gas having a heating value approximately half that of natural gas; i.e., about 500 Btu; a classification for products of biomass gasification.

**mega-** *Measurement.* a prefix meaning "one million" ($10^6$); symbol M.

**megaohmmeter** *Electricity.* an instrument for measuring extremely high resistance.

**megaton** *Measurement.* a unit of explosive power or energy equivalent to the yield of $10^6$ metric tons of trinitrotoluene (TNT); used in describing the output of a nuclear weapon. Thus, **megatonnage.**

**megavolt (MV)** *Electricity.* a unit of electricity equivalent to one million volts.

**megawatt (MW)** *Electricity.* a unit of power equivalent to one million watts (1000 kilowatts); e.g., the amount of electric energy required to light 10,000 100-watt bulbs.

**megawatt-hour (MWh)** *Electricity.* a unit of electricity equivalent to the amount of energy produced by one megawatt of power over a period of one hour. Similarly, **megawatt-month, megawatt-year.**

**megawatt-mile rate** *Electricity.* an electric transmission rate based on distance; contrasted with POSTAGE STAMP RATES, which are based on zones.

**megohm** *Electricity.* a unit of electricity equivalent to one million ohms.

**Meissner effect** *Electricity.* the property of superconductors that allows them to expel magnetic flux from their interiors; responsible for the levitation of permanent magnets above chilled superconductor samples. The operation of maglev (magnetic levitation) trains is based on this property. [Named for German physicist Walther *Meissner,* 1882–1974.]

**melatonin** *Health & Safety.* a hormone secreted by the pineal gland that is identified as playing a key role in the sleep-wake process and also affecting numerous endocrine functions; e.g., secretion of reproductive hormones, inhibition of certain cancer cells. Recent studies indicate that exposure to electromagnetic fields may lower melatonin levels.

**meltdown** *Nuclear.* the melting of the core of a nuclear reactor due to the overheating of nuclear fuel.

**melting energy** *Chemistry.* the energy involved in the transition of a substance from solid to liquid form, by means of heat or the application of heat.

**melting point** *Chemistry.* the temperature at which a solid begins to change to a liquid state. The melting point of most substances is slightly dependent on the pressure; pure metals will melt at a constant temperature, whereas alloys tend to melt over a range.

**melt refining** *Nuclear.* the partial decontamination of metallic nuclear fuel by a process of melting and allowing fission products to volatilize or react with a substrate.

**membrane** *Materials.* **1.** a very thin separating or covering layer, either natural or artificial; e.g., the outer boundary of a living cell. **2.** specifically, the separating layer in a fuel cell that acts as an electrolyte (ion-exchanger) as well as a barrier film separating the gases in the anode and cathode compartments.

**membrane electrode assembly (MEA)** *Renewable/Alternative.* a core component of a fuel cell structure, consisting of an electrolyte membrane, anode and cathode electrodes, and two microporous conductive layers that function as gas diffusion layers and current collectors. Electrochemical reactions occur when a fuel (e.g., hydrogen) and an oxidant are applied to the anode and cathode side of the MEA.

**memory effect** *Storage.* an effect observed in some rechargeable nickel-based batteries that causes them to hold less charge; this occurs when the battery is repeatedly discharged to a particular level that is above a full discharge, that is, only partially used, and then recharged to an equally precise full level. Also known as **cyclic memory** or simply **memory.**

**MEMs** *Consumption & Efficiency.* microelectromechanical systems, a class of miniaturized devices typically made from silicon or employing it in the fabrication process. These are robotic systems that integrate electronics and mechanical devices onto a single substrate, with typical feature sizes in the 1 mm to 50 mm range.

**Mendeleev, Dmitri** 1834–1907, Russian chemist who formulated the periodic law of chemical elements and devised the periodic table.

**mendelevium** *Chemistry.* a synthetic radioactive element having the symbol Md, the atomic number 101, and an atomic weight of 256; produced by bombarding einsteinium with alpha particles in a cyclotron, it has a half-life of 1.5 hour, decaying by spontaneous fission. [Named for Dmitri *Mendeleev.*]

**MEPS** *Consumption & Efficiency.* minimum energy performance standards; a policy tool intended to prevent the least energy-efficient products from participating in the market, by either banning the sale of products that fail to meet specified minimum energy efficiency levels or restricting the availability of low-efficiency products.

**Mercator, Gerardus** 1512–1594, Flemish geographer noted for important innovations in mapmaking, through the use of the MERCATOR PROJECTION. In 1568 he published a noted book of maps (for which he introduced the word *atlas*), and he also produced accurate terrestrial and celestial globes.

*Mercator According to Greek mythology the god Atlas supported the earth on his shoulders; Mercator employed a similar image of Atlas in his first book of maps.*

**Mercator projection** *Measurement.* a map projection in which the surface of the earth is developed on a cylinder that is tangent along the equator. Geographic parallels and meridians are rendered as straight lines spaced so as to produce an accurate ratio of latitude to longitude at any point.

**mercuric oxide battery** *Storage.* a type of alkaline primary battery with a positive electrode of mercuric oxide (often with manganese dioxide), a negative electrode of metallic zinc and either potassium or sodium hydroxide as an electrolyte.

**mercury** *Chemistry.* a silvery-white metallic element having the symbol Hg, the atomic number 80, an atomic weight of 200.59, a melting point of –38.87°C, and a boiling point of 356.58°C; an extremely heavy liquid (the only common metal to exist in this state at ordinary temperatures) with extremely high surface tension; used in thermometers, mercury vapor lamps, and as a neutron absorber. [Named for the Greek god *Mercury,* a symbol of speed, because of its unusual property of flowing easily and quickly.]

**mercury barometer** *Measurement.* an instrument used to measure atmospheric pressure by noting the height to which a column of mercury will rise in an open-ended glass tube inverted into a dish of mercury.

**mercury lamp** see MERCURY VAPOR LAMP.

**mercury poisoning** *Health & Safety.* poisoning due to ingestion of or excessive exposure to mercury or its salts; studies have indicated that certain large predatory fish such as swordfish and sharks may contained elevated levels of methyl mercury and thus could be a risk for human consumption.

**mercury thermometer** *Measurement.* a liquid-in-glass or liquid-in-metal thermometer using mercury as the reference liquid.

**mercury vapor lamp** *Lighting.* a type of high-intensity discharge (HID) lamp source in which light is produced by the passage of a current through a small amount of mercury vapor. Used mainly for street lighting and for large indoor spaces such as gymnasiums and sports arenas.

**mesoclimate** *Earth Science.* the climate of a relatively small and defined region, such as a valley, lake, plantation, or city.

**meson** *Physics.* a generic name for a group of strongly interacting particles having baryon number zero and ranging in mass from 140 MeV to near 10 GeV; all are unstable and decay to the lowest mass states that are accessible.

**mesopause** *Earth Science.* the upper boundary of the earth's mesosphere, corresponding to the level of minimum temperature at 70–80 km.

**mesophile** *Ecology.* an organism that is best suited to living in conditions of moderate temperature (25–40°C). Thus, **mesophily.**

**mesophilic** *Ecology.* growing best or thriving in conditions of moderate temperature.

**mesoscale** *Earth Science.* a definition or model for the study of weather processes over a relatively small area (less than 200 km).

**mesosphere** *Earth Science.* **1.** the atmospheric shell above the earth that extends from the top of the stratosphere to the mesopause. **2.** the atmospheric shell between the top of the ionosphere and the bottom of the exosphere. **3.** the part of the earth's mantle that extends about 1000 km below the surface.

**mesotrophy** *Ecology.* an intermediate stage in EUTROPHICATION; the condition of a lake or other body of water having an intermediate amount of plant nutrients and therefore moderately productive. Thus, **mesotrophic.**

**met** *Measurement.* an approximate unit of heat produced by a person at rest, equal to about 18.5 Btu per square foot per hour; used as a basis for measuring levels of physical activity. [Short for *metabolic.*]

**metabolic** *Biological Energetics.* relating to or involved with metabolism.

**metabolic energy** *Biological Energetics.* **1.** the energy required by an organism to carry on the biochemical processes necessary for life. **2.** the amount of this necessary for some specific use of energy; e.g., for locomotion.

**metabolic heat** *Biological Energetics.* heat produced by the oxidation of food elements (i.e., metabolism) in humans or animals.

**metabolism** *Biological Energetics.* **1.** the sum of chemical and physical processes that take place within an organism to maintain life. **2.** specifically, the digestion of food and its conversion into useful molecules to meet the body's energy needs.

**metabolizable energy** *Biological Energetics.* energy obtained by an organism from food; the basic amount of energy available for all activities.

**metal** *Chemistry.* **1.** any of a class of elements that generally are solid at ordinary temperatures, have a grayish color and a shiny surface, and will conduct heat and electricity well. In a pure electrolytic solution, a metal will form positive ions. Metals constitute about three-fourths of the known elements and can form alloys with each other and with nonmetals. Common metals include copper, gold, silver, tin, iron, lead, aluminum, and magnesium. **2.** an alloy or mixture composed of such substances. Thus, **metallic.**

**metal (air) fuel** *Renewable/Alternative.* a fuel cell technology that uses metals such as zinc, aluminum, and magnesium (in place of hydrogen) to provide electrical power, so as to overcome certain disadvantages associated with hydrogen as a fuel (e.g., cost, flammability). Thus, **metal fuel cell, metal fuel technology.**

**metal halide lamp** *Lighting.* a type of high-intensity discharge (HID) lamp in which most of the light is produced by radiation of metal halide and mercury vapors in the arc tube, similar in construction and appearance to mercury vapor lamps. The addition of metal halide gases results in higher light output, more lumens per watt, and better color rendition than from mercury gas alone.

**metal hydride storage** *Hydrogen.* a device that can store hydrogen by means of a metal carrier; considered a relatively safe and compact method of hydrogen storage. The hydrogen can either be stored in the cavities of a grid of metal, such as magnesium or titanium, or it can enter into an ionic bond with the metal.

**metalimnion** *Earth Science.* the middle or transitional layer of water in a reservoir, between the epilimnion and hypolimnion, in which the temperature exhibits the greatest difference in a vertical direction.

**metallic bond(ing)** *Chemistry.* a force that holds atoms together in a metal, formed by the attraction between positively charged metal ions and mobile free electrons surrounding them; this accounts for many of the characteristic properties of metals and alloys.

**metallic element** another term for a METAL.

**metalloid** *Chemistry.* **1.** an element having both metallic and nonmetallic properties, such as boron or silicon. **2.** a nonmetallic element that can combine with a metal to produce an alloy.

**metallurgical coal** *Coal.* a term for coking coal and pulverized coal used in making metals and alloys.

**metallurgical coke** *Coal.* a low-sulfur coke, suitable for the smelting of iron ores.

**metallurgical grade silicon** *Materials.* silicon prepared by the heating of high-purity silica in an electric arc furnace at temperatures over 1900°C using carbon electrodes, producing a material that is at least 99% pure.

**metallurgy** *Materials.* the science and technology of metals and metal processing. Thus, **metallurgical.**

**metal-organic chemical vapor deposition (MOCVD)** *Materials.* a process of producing materials for semiconductors, including photovoltaic materials, in which a surface layer is produced by the deposition on a substrate of a volatile organo-metallic compound (e.g., methyl and $CH_3$) that is transported to the surface through the gas phase at elevated temperatures.

**metamorphic** *Earth Science.* describing any rock changed (morphed) from a pre-existing solid rock by extreme changes in temperature, pressure, shearing stress, or chemical environment.

**metaphysics** *History.* a systematic investigation of the ultimate nature of being or reality; originally considered an activity of science but later regarded as unverifiable and classified as philosophy. Thus, **metaphysical.**

**metastable equilibrium** *Physics.* **1.** a state of a system that is in pseudo-equilibrium, such that a small disturbance may not disrupt the system but a larger one would render it unstable; a practical example is a ball at rest in a slight depression at the top of a hill. **2.** a condition in which a substance appears to be stable but can undergo a spontaneous change, as when a supercooled liquid suddenly transforms into a solid.

**meteorology** *Earth Science.* **1.** the scientific study of chemical and physical processes in the earth's atmosphere, especially as they relate to weather and climate. **2.** the weather conditions or patterns of a specific area.

**meter** *Measurement.* a basic unit of length in the metric system, roughly equivalent to 39.37 inches. It is currently defined as the distance that light will travel through a vacuum in a period of 1/299,792,458 of a second. Previously, it was defined as 1,650,763.73 wavelengths of the orange-red light from the isotope krypton 86. Originally, it was supposed to represent one ten-millionth of the distance from the North Pole to the equator along a given meridian.

**meter-kilogram-second** see MKS.

**methane** *Chemistry.* $CH_4$, the simplest hydrocarbon and the primary constituent of natural gas; a flammable, explosive gas that is slightly soluble in water and colorless, odorless, and tasteless. It is an important source of hydrogen and some organic chemicals and the principal feedstock in the manufacture of ammonia for fertilizers and explosives. It is a greenhouse gas, released to the atmosphere by anaerobic decomposition of waste in landfills, animal digestion and decomposition of animal wastes, production and distribution of natural gas and oil, coal production, and incomplete fossil fuel combustion.

**methane monitor** *Mining.* an electronic instrument used to detect and measure the methane content of ambient air in a mine area, typically having a digital readout and emitting a warning signal if the methane level rises above a certain point.

**methanogenesis** *Biological Energetics.* the biochemical process by which certain microorganisms produce methane; e.g., bacteria in the ruminant digestive system; archaea in ocean sediments.

**methanogenic** *Biological Energetics.* describing certain microorganisms **(methogens)** that produce methane.

**methanol** *Materials.* $CH_3OH$, a toxic alcohol derived largely from natural gas or coal, sometimes referred to as the simplest alcohol; it can be used as an alternative fuel or as a gasoline additive, where it is less volatile than gasoline. When blended with gasoline it lowers carbon monoxide emissions but increases hydrocarbon emissions. It is also used as an antifreeze and solvent.

**methanol blend** *Oil & Gas.* a mixture containing a high percentage by volume of methanol with gasoline (typically 70% to 85%).

**methanotrophic** *Biological Energetics.* describing bacteria that have the distinctive ability to use methane as a sole source of energy and carbon and oxidize it into carbon dioxide. Thus, **methantroph, methantrophic.**

**methyl** *Chemistry.* a member of the alkyl group $CH_3$–, which is derived from methane.

**methyl alcohol** another name for METHANOL.

**methyl bromide** *Environment.* $CH_3Br$, a highly effective fumigant used to control insects, nematodes, weeds, and pathogens in many crops, in forest and ornamental nurseries, and in wood products. Because it contains bromine, it depletes the ozone layer when released to the atmosphere and thus its use is now being phased out in developed countries.

**methyl chloride** *Materials.* $CH_3Cl$, a narcotic, colorless compressed gas or liquid with a faintly sweet odor; slightly soluble in water and soluble in alcohol; used as a refrigerant, solvent, catalyst carrier, and methylating agent.

**methyl chloroform** *Materials.* $CH_3CCl_3$ (trichloroethane), a nonflammable, colorless liquid, insoluble in water and soluble in alcohol and ether; used as a solvent, aerosol propellant, pesticide, and for metal degreasing.

**methyl hydride** another term for METHANE.

**methyl tertiary-butyl ether (MTBE)** *Materials.* $C_5H_{12}O$, a volatile, flammable, colorless liquid, a hydrocarbon containing 18.15% oxygen; widely used as an octane booster and oxygenate for gasoline blending, with the latter use leading to lower vehicle emissions.

**metre** another spelling of METER.

**metric** *Measurement.* **1.** of or relating to measurement, especially the metric system of measurement. **2.** any standard or value used in measurement. Thus, **metrics.**

**metric system** *Measurement.* a standard system of measurement using decimal units, in which the meter is the basic unit of length, the gram is the basic unit of mass, and the liter is the basic unit of volume or capacity; used in science and technology and as the common measuring system of most nations (the U.S. being a notable exception).

**metric ton** *Measurement.* a unit of mass that is equal to 1000 kilograms, equivalent to 2204.62 pounds.

**metrology** *Physics.* the science of measurement, especially the measurement of length or distance, mass, and volume.

**Meucci, Antonio** 1808–1889, Italian inventor who developed several working telephone models prior to the patent granted to Alexander Graham Bell in 1876. In 1871 Meucci obtained a caveat stating his intent to patent his "Talking Telegraph". However, his application lapsed and Bell became known as the inventor of the telephone.

**MGO** marine gas oil.

**MHD** magnetohydrodynamics.

**MHz** megahertz.

**micelle** *Chemistry.* a spherical arrangement formed by a group of lipid molecules in an aqueous environment.

**Michaux Velocipede** *History.* an advance in bicycle design developed by Pierre and Ernst Michaux of Paris (1863), having the pedals directly connected to the front wheel via crankshafts.

**Michell-Banki turbine** another name for a CROSSFLOW TURBINE.

**micro-** *Measurement.* **1.** a prefix meaning "one millionth" ($10^{-6}$). **2.** occurring or existing at the microscopic level. **3.** very small; tiny; minute.

**microalgae** *Ecology.* smaller, typically unicellular forms of algae; see ALGAE.

**microchannel array** *Consumption & Efficiency.* an array of channels with characteristic dimensions of less than 1 mm, designed for conveying a heat or mass transfer fluid; when used as the basis for heat exchangers, boilers, and condensers, high rates of heat transfer result.

**microchip** *Communication.* a tiny piece of superconductor material on which the components of an integrated circuit are placed.

**microcircuit** *Communication.* an integrated circuit; i.e., a minute electrical circuit on silicon.

**microclimate** *Earth Science.* the climate of a highly specific area (typically with a radius of less than one kilometer), in which conditions can be distinguished from the overall climate; it extends from the surface up to a point at which the general local conditions prevail. The term often refers to a specific plot of farm land for growing a particular crop (e.g., wine grapes).

**microcomputer** *Communication.* a personal computer or other such device with a microprocessor (see below) as its central processing unit.

**microcredit** *Economics.* small amounts of credit made available to households or small businesses for investments in energy supplies or services, usually within a rural development context.

**microcrystalline silicon** another term for NANOCRYSTALLINE SILICON.

**microeconomics** *Economics.* the study of the behavior of one or a few consumers, households, or firms in a larger economy. Thus, **microeconomic.** Compare MACROECONOMICS.

**microelectromechanical systems** see MEMs.

**microelectronics** *Electricity.* the special methods and techniques used in producing miniature circuits.

**microemulsion** *Materials.* a thermodynamically stable type of emulsion in which the dispersed droplets are extremely small (less than 100 nm).

**microgroove** *Photovoltaic.* a small groove scribed into the surface of a solar photovoltaic cell that is filled with metal for contacts.

**microhydro(power)** *Hydropower.* **1.** hydropower systems with less than a rated capacity of approximately 100 kW; the actual size designation varies among countries. **2.** the use of hydropower by means of miniaturized devices, in contexts where a conventionally sized device is not practicable; e.g., in a subsurface mine.

**microlog** *Oil & Gas.* a drill-hole electric log utilizing electrodes mounted at short distances from one another; used to determine the permeability of a selected formation.

**micronutrient** *Biological Energetics.* chemical component of food that is necessary in small amounts for the promotion of health; e.g., specific vitamins and minerals.

**microprocessor** *Communication.* a type of computer architecture based on a general-purpose central processing unit (CPU) contained on a single microchip (chip). Thus, **microcomputer.**

**microscopic** *Physics.* **1.** having to do with or considering the state of a system in terms of its individual components or on a small scale. **2.** specifically, involving an analysis or study in which the individual molecules and atoms (or smaller entities) are the context for the description. **3.** more generally, relating to any small-scale perspective or context. Contrasted with MACROSCOPIC.

**microtechnology** *Consumption & Efficiency.* the application of technology at a highly miniaturized level, especially operating at dimensions of around one micron (one millionth of a meter).

**microtechnology-based systems** see MECS.

**microturbine** *Conversion.* a category of small combustion turbines that produce between 25 and 500 kilowatts of power.

**microvolt** *Electricity.* a unit of electricity equivalent to one millionth of a volt.

**microwatt** *Electricity.* a unit of power equivalent to one millionth of a watt.

**microwave** *Physics.* **1.** electromagnetic radiation having a free-space wavelength between 0.3 and 30 centimeters, corresponding to frequencies of 1–100 gigahertz. **2.** see MICROWAVE OVEN.

**microwave background radiation** see BACKGROUND RADIATION.

**microwave oven** *Consumption & Efficiency.* an electrically operated oven utilizing high-frequency electromagnetic waves to vibrate molecules of food, thus generating the heat required to warm or cook the food in a shorter period of time than conventional electric or gas ovens.

**microwave sounding unit (MSU)** *Measurement.* a sensor carried aboard an earth-orbiting satellite to monitor tropospheric temperatures.

**middle distillate** *Oil & Gas.* a classification for petroleum fractions that have an intermediate boiling point range; light products that are heavier than gasoline, primarily used as diesel fuel, kerosene, and heating oil.

**middle oil** *Materials.* the fraction of oil distilled from coal tar at 200–250°C, yielding primarily naphthalene.

**middle-income economy (MIE)** *Economics.* in the World Bank's economic classification system, a country with a per capita gross national income (GNI) from $736 to $9075 (2002 standard). See also LOW-INCOME ECONOMY, HIGH-INCOME ECONOMY.

**middling** *Coal.* **1.** a composite coal particle that contains both organic and inorganic material. **2.** any of various similar products that are intermediate in quality or grade, such as an ore product that is intermediate between a valuable concentrate and a tailing.

**Midgley, Thomas** 1889–1944, U.S. chemist who developed the TETRAETHYL LEAD additive to gasoline (1921) and CHLOROFLUOROCARBONS (CFCs) for refrigeration and air conditioning (1930), two products that contributed greatly to industrial efficiency but would eventually be identified as harmful to the environment. Thus, his legacy is mixed, and one historian said that "he had more impact on the atmosphere than any other single organism in earth history".

**midgrade** or **mid-grade** *Oil & Gas.* a classification for gasoline having an intermediate octane rating (i.e., the relative ability to resist engine knock); the classification varies by region but typically indicates an octane rating of 88–90. Thus **midgrade gas(oline)** or **fuel.**

**Mie scattering** *Solar.* the scattering of solar radiation by (mathematically spherical) particles in the atmosphere which have an approximate size of the wavelength of light. [Described by German physicist Gustav *Mie* 1868–1957.]

**migration** *Biological Energetics.* a predictable, recurring group movement that is characteristic of the members of a given species, and that occurs regularly in response to seasonal changes in temperature, precipitation, food availability, and so on. The process usually involves a round-trip movement between two areas, to seek a more suitable breeding place, a greater food and water supply, or other more favorable environmental conditions.

***migration*** *Migration involves significant levels of energy expenditure for certain birds; these snow geese make an annual round trip of more than 5000 miles at speeds of 80 km or more.*

**migratory** *Biological Energetics.* characterized by or involved in migration.

**Milankovitch theory** *Climate Change.* the proposal that changes in climate are associated with fluctuations in the seasonal and latitudinal distribution of solar radiation received by the earth, determined by periodic variations in three aspects of the geometry of earth's orbit (eccentricity, obliquity, and precession). These three orbital variations are now known as the **Milankovitch cycles.** [Formulated by Serbian mathematician Milutin *Milankovitch,* 1879–1958.]

**miles per gallon** *Transportation.* the standard U.S. measure for the fuel consumption of a motor vehicle; i.e., miles traveled divided by gallons of fuel consumed.

**miles per gallon equivalent (MPGE)** *Transportation.* the average number of miles that a vehicle can travel on a gallon equivalent of an alternative fuel, or an amount of fuel equal in energy content to one gallon of gasoline; often used to compare the energy content of various fuels or to compare vehicle miles per quantity of fuel.

**military nuclear reactor** *Nuclear.* a facility used to create materials that are used in nuclear weapons, as opposed to a nuclear reactor used to generate energy for consumer purposes such as electrical power.

**Mill, John Stuart** 1806–1873, British philosopher, a key figure in the classical school of economics. He suggested the need for a stationary or steady-state economy; this laid the foundation for ecological economists of the present day to argue that population growth, resource depletion, environmental degradation, and socioeconomic inequalities demand a steady state (as opposed to a growth economy).

**mill** *Mining.* **1.** a machine or device for grinding grain into flour and other cereal products. **2.** the building in which such a grinding operation takes place. See also WINDMILL. **3.** any machine or device for grinding or crushing solid substances. **4.** a factory in which paper, steel, or textiles are manufactured. **5.** *Materials.* a place where ore is crushed or concentrated, or the machinery used to perform these functions.

**mill** *Economics.* a monetary cost and billing unit used by utilities; equal to 1/1000 of the U.S. dollar (1/10 of 1 cent).

*mill* *A historic mill in Georgia (U.S.), producing corn meal and flour through the use of water power from the flow of the Chattahoochee River.*

**mill feed** *Nuclear.* uranium ore supplied to a crusher or grinding mill.

**milli-** *Measurement.* a prefix meaning "one thousandth" ($10^{-3}$), as in *millimeter.*

**Millikan, Robert Andrews** 1868–1953, U.S. physicist who accurately measured the charge carried by an electron, using the elegant "falling-drop method". He proved that this quantity was a constant for all electrons, thus demonstrating the atomic structure of electricity. He also made important studies of cosmic rays (which he named), X-rays, and physical and electric constants.

**millimeter** *Measurement.* a unit of length equal to one-thousandth of a meter; equivalent to 0.039 inches.

**milling** *Materials.* **1.** the mechanical treating of solid materials, such as grain, to produce a powder. **2.** the process of removing valueless material and harmful constituents from ore. **3.** the process of producing plane or shaped surfaces by the use of a cutting machine.

**millpond** *Hydropower.* a pond formed by damming a stream to create a head of water that will provide the power to turn a mill wheel.

**mill residue** *Biomass.* wood materials and bark generated at manufacturing plants when logs are processed into primary wood products (lumber, plywood, and paper); can include slabs, edgings, trimmings, sawdust, and pulp screenings.

**millstone** *Materials.* in traditional grain mills, a flat, circular stone used for the grinding process.

**mill tailings** see URANIUM TAILINGS.

**minable** *Mining.* able to be mined; capable of producing material of value given current technology and environmental and legal restrictions.

**Minas** *Oil & Gas.* **1.** a major oil field in Indonesia. **2.** one of the group or "basket" of crude oil types used by OPEC as a reference point for pricing, representing Far East heavy oil.

**mine** *Mining.* **1.** any opening or excavation in the earth that is used to extract minerals, coal, ore, or other such material of value. **2.** to extract material in this manner.

**mine-mouth (power) plant** *Coal.* a large electrical power plant built directly adjacent to the coal mine that is the source of its fuel; often used with lignite and other such coals that are expensive to ship by train. A strategy intended to minimize transportation costs and provide a more reliable supply flow, though in practical use these plants have faced challenges such as environmental regulations or the depletion of the resources at the site.

**miner's lamp** *Mining.* **1.** an attachment worn on a miner's helmet to provide illumination in a mine; now powered by electricity but formerly employing a gas flame or candle. **2.** the entire helmet itself. **3.** a safety device used to detect methane gas concentrations and oxygen deficiency.

**mineral** *Materials.* a general term for any naturally occurring inorganic substance, of inorganic or possibly organic origin, that has a definite chemical composition and an orderly internal structure, a crystal form, and characteristic chemical and physical properties.

**mineralization** *Materials.* **1.** the fact of becoming a mineral; the conversion of organic compounds to inorganic ones. Thus, **mineralize, mineralizer. 2.** *Biological Energetics.* a process in vertebrates in which the mineral component of bone tissue increases in content and/or density.

**mineral matter** *Coal.* the material in coal from which ash is formed; material in coal that is not formed from decomposed plant products; e.g., minerals that were present in the original plant materials or that were assimilated from extraneous sources, such as sediments and mineralized water. Clay, pyrite, and calcite are minerals often present in coal.

**mineral-matter-free basis** *Coal.* a standard for evaluating coal quality, assuming that all mineral matter (see above) has been removed from it, leaving "pure" coal.

**mineralogy** *Materials.* the scientific study of minerals. Thus, **mineralogical.**

**mineraloid** *Materials.* **1.** a naturally occurring, usually inorganic, mineral-like material that is amorphous and lacks a highly ordered atomic arrangement and characteristic external form; e.g., opal or volcanic glass. **2.** a metallic element in liquid form; e.g., mercury at ordinary temperatures.

**mineral oil** *Materials.* a colorless liquid petroleum derivative with little discernible odor or taste, widely used as a lubricant and for various other purposes.

**Miner's Friend** *History.* a name for the early steam engine of Thomas Savery of England; so called because it was intended to pump water out of coal mines.

**Mines Act** *Mining.* a milestone in mine legislation in England (1842), decreeing that women and girls, and boys under the age of 10, could not work underground in the nation's coal mines. This was in response to a report describing "cruel slaving revolting to humanity", such as women and children chained to carts and working 15-hour days in the mines.

**mineshaft** *Mining.* a primary opening, usually vertical, made through mine strata to connect the surface with underground workings; used for ventilation and drainage and for transporting personnel, materials, and equipment to or from the working area.

**minigrid** *Electricity.* a separate AC electricity grid that is not (always) part of the regular grid; can be used to power small regions or villages or to maintain the power supply in a region when the main grid fails.

**minihydro(power)** *Hydropower.* a hydropower system with a rated capacity less than 500 kW; the actual size designation varies among countries.

**minimum dissipation** *Thermodynamics.* the principle that if certain boundary conditions prevent a system from reaching thermodynamic equilibrium, the system will eventually reach a state in which internal dissipation is at a minimum (assuming input flow remains steady).

**minimum energy standards** see MEPS.

**minimum flow level** *Hydropower.* **1.** the level of stream flow sufficient for a hydroelectric plant to operate at the desired efficiency. **2.** *Environment.* the instream flow required to support fish and other aquatic life, to minimize pollution, or to maintain other uses such as recreation and navigation; a standard employed to regulate withdrawals of water from a river or stream.

**mining** the act or business of discovering, extracting, and using or marketing materials of value from the earth.

**mining engineering** *Mining.* the branch of engineering that deals with the discovery, development, processing, and exploitation of ores and minerals.

**mining geology** *Mining.* the branch of geology dealing with the study of the structures, modes of formation, and occurrence of mineral deposits, as well as the geologic considerations of mine planning.

**minivan** *Transportation.* a type of passenger vehicle, larger than a station wagon but smaller than a conventional van, typically with seats for six to eight people.

**minority carrier** *Photovoltaic.* a current carrier, either an electron or a hole, that is in the minority in a specific layer of a semiconductor material; the diffusion of minority carriers under the action of the cell junction voltage is the current in a photovoltaic device. Compare MAJORITY CARRIER.

**Mintrop, Ludger** 1880–1956, German scientist who was one of the founders of refraction seismic exploration for oil and natural gas (1919). His work led to the discovery of the first salt dome to be found in the U.S. by the seismic method.

**mire** *Biomass.* a peatland where peat is currently being formed and accumulating.

**miscible** *Materials.* able to be mixed; describing two or more liquids that are able to mix with or dissolve into each other in various proportions. Thus, **miscibility.** Compare IMMISCIBLE.

**mitigation** *Environment.* **1.** any direct human effort to alleviate or minimize (mitigate) environmental damage. **2.** specifically, an anthropogenic intervention to reduce the emissions of greenhouse gases or enhance the sinks of such gases.

**mitochondria** *Biological Energetics.* singular, **mitochondrion.** self-replicating organelles, bounded by two membranes, that are found in the cytoplasm of all eukaryotic cells and produce cellular energy in the form of ATP (adenosine triphosphate) via the oxidative phosphorylation reactions.

**mitochondrial respiration** see DARK RESPIRATION.

**mixed (mixing) layer** *Earth Science.* **1.** a surface layer of water having roughly uniform temperature that often exists above the thermocline; the water in this layer is mixed by wind stirring or convective overturning. **2.** a near-surface area of the earth's atmosphere where air is well mixed from turbulence caused by the interaction of the surface and atmosphere; usually located at the base of a temperature inversion.

**mixed oxides fuel (MOX)** *Nuclear.* a mixture of uranium and plutonium oxides that can be used as fuel in nuclear reactors; the plutonium typically is extracted from reprocessed nuclear fuel.

**mixer-settler** *Materials.* a liquid–liquid extraction device that mixes phases and then allows the liquids to settle and separate. In its simplest form, it consists of an agitation tank (the mixer) in which the solutions are contacted, followed by a shallow gravity basin (the settler) where the solutions disengage into individual layers for separate discharge.

**MKS system** *Measurement.* meter-kilogram-second system; a system of measurement having as its fundamental units the meter for length, the kilogram for mass, and the second for time. The MKS system gradually replaced the CGS (centimeter-gram-second) system during the 20th century, and its essential elements were codified in the adoption of the International System of Units (SI) in 1960.

**Mobil** the brand name for a former large oil company, formed by the merger (1931) of Standard Oil of New York (Socony) and the Vacuum Oil Company. Merged with Exxon (originally Standard Oil of New York) to form ExxonMobil in 1998.

**mobile source** *Environment.* a collective term for moving objects that release pollution, especially motorized forms of transportation such as cars, trucks, motorcycles, buses, boats, aircraft, and so on. Thus, **mobile-source pollution.**

**MOCVD** metal–organic chemical vapor deposition.

**Model Energy Code** *Consumption & Efficiency.* a U.S. standard for minimum levels of energy efficiency in residential buildings (three stories or less) for insulation, windows, heating and cooling equipment, air infiltration, and so on.

**Model T** *History.* the name for an inexpensive automobile model manufactured by the Ford Motor Company from 1908 to 1927; its introduction made car ownership possible for the average American and it became the largest-selling car model of its era.

**model year** *Transportation.* a designation of the "age" of a motor vehicle based on the assumption that a new year begins when the model is first sold rather than with an actual calendar year; e.g., a new vehicle introduced in September 2006 would be considered a 2007 model.

**moderator** *Nuclear.* **1.** a material, such as ordinary water, heavy water, or graphite that is used in a reactor to slow down high-velocity neutrons, thus increasing the likelihood of fission. **2.** also, **moderate.** to employ such a material in a reactor.

**modern energy** *Consumption & Efficiency.* a term for energy sources that have come into use at a relatively recent time in human history; e.g., fossil fuels such as coal and oil as opposed to traditional sources, such as human and animal power and firewood.

**modern physics** *Physics.* a term for the science of physics as it has developed since about 1900, including such studies as relativity and quantum mechanics.

**modular** *Consumption & Efficiency.* **1.** describing the use of complete sub-assemblies or sub-systems to produce a larger system. **2.** *Solar.* describing an energy system in which individual collectors can be connected in series to increase energy output; e.g. a photovoltaic system. **3.** *Communication.* describing a computer program that is composed of a collection of well-defined, logically self-contained segments or subroutines. Thus, **modularity.**

**modulation** *Physics.* the fact of changing in some characteristic way the amplitude, frequency, or phase of a wave, or the velocity of the electrons in an electron beam.

**module** see PHOTOVOLTAIC MODULE.

**Mohs scale** *Measurement.* a standard scale for determining the hardness of a rock or mineral, in terms of its resistance to abrasion. It is defined by 10 fairly common minerals ranked from the softest to the hardest: talc (1), gypsum (2), calcite (3), fluorite (4), apatite (5), orthoclase (6), quartz (7), topaz (8), corundum (9), and diamond (10). It is based on the simple fact that a harder material will scratch a softer one. [Devised by German mineralogist Frederick *Mohs,* 1773–1839.]

**moist (coal) basis** *Coal.* a standard for evaluating coal quality, assuming that the coal has not acquired or lost any moisture beyond what exists in its natural state of deposition; i.e., it has no additional water adhering to its surface, nor has it lost any moisture due to drying.

**moisture content** *Materials.* the water content of a solid fuel as measured under specified conditions; this can be on a dry basis (as a fraction of the oven-dry weight of the fuel) or on a wet basis (as a fraction of the weight of the fuel as received).

**molar** *Chemistry.* **1.** of a solution, containing enough solvent so that 1 mole of a solute will dissolve to make one liter of the solution. **2.** describing a physical quantity of some substance in terms of one mole of the substance.

**molar heat capacity** *Chemistry.* the quantity of heat needed to raise the temperature of one mole of a substance by one degree centigrade.

**moldboard** *History.* a curved plate above the share of a plow that lifts and turns over the soil, creating a true furrow; an advance in agriculture associated with China, about 100 BC.

**mole** *Chemistry.* a standard measurement of the amount of a substance, equal to the amount of a substance containing the same number of given elementary entities (such as atoms, molecules, ions, or electrons) as there are atoms of carbon in 12 grams of carbon-12. One mole of atoms contains Avogadro's number of atoms ($6.02 \times 10^{23}$).

**molecular** *Chemistry.* **1.** relating to or consisting of molecules. **2.** another term for MOLAR.

**molecular theory** see KINETIC-MOLECULAR THEORY.

**molecule** *Chemistry.* the smallest unit of matter of a substance that retains all the physical and chemical properties of that substance, consisting of a single atom or a group of atoms bonded together.

**mole fraction** *Chemistry.* in a system of mixed constituents, the ratio of the number of moles of a single constituent in a given volume, to the total number of moles of all constituents in that volume.

**Molly Maguires** *History.* a secret society of Irish–American miners in the coal mining region of Pennsylvania in the 1870s; a sensational murder trial resulted in the execution of 20 members of the group, though the historical record is unclear as to what extent this violence had been incited, or even committed, by private detectives hired by mine owners to infiltrate the miners' union.

**molten** *Materials.* reduced to a liquid by heating; melted.

**molten carbonate fuel cell (MCFC)** *Renewable/Alternative.* a type of fuel cell that uses a molten carbonate salt mixture as the electrolyte, typically lithium carbonate ($Li_2CO_3$) and sodium carbonate ($Na_2CO_3$). At the operating temperature of about 650°C (1200°F), the salt mixture is liquid and a good ionic conductor.

**molten salt reactor (MSR)** *Nuclear.* an advanced nuclear reactor design that features a circulating molten salt fuel mixture with an epithermal-spectrum reactor and a closed fuel cycle. The heat generated in the molten salt is transferred to a secondary coolant system through an intermediate heat exchanger, and then through a tertiary heat exchanger to the power conversion system.

**molybdenum** *Chemistry.* a heavy metallic element having the symbol Mo, the atomic number 42, an atomic weight of 95.94, a melting point of 2617°C, and a boiling point of 5560°C;

it is highly conductive and resistant to heat and is used in high-temperature alloys and resistors.

**moment of inertia** *thermodynamics.* the rotational analog of mass, in units of mass × length$^2$.

**momentum** *Thermodynamics.* a measure of the motion of an object, equal to the product of its mass and its velocity.

**monochromatic** *Physics.* **1.** having a single color; relating to or describing one color or hue. **2.** relating to or involving electromagnetic radiation with one wavelength, or a narrow range of wavelengths. Thus, **monochromatic radiation.**

**monocoque** *Transportation.* **1.** a type of vehicle design or structure in which all or most of the stresses are carried by the outer shell of the vehicle, which is integrated into a single unit. **2.** describing a structure built according to this principle; e.g., an aircraft fuselage or the body of a racing car. [From a French word meaning "one shell".]

**monocrystalline silicon** *Photovoltaic.* a form of crystalline silicon in which the atoms making up the framework of the crystal are repeated in a very regular, orderly manner from layer to layer.

**monoculture** *Ecology.* the cultivation of a single crop or product, to the exclusion of other possible uses of the land.

**monoenergetic** *Physics.* **1.** having a single energy. **2.** describing radiation of a given type in which all particles or photons originate with and have the same energy.

**monomer** *Materials.* a relatively simple compound, usually containing carbon and of low molecular weight, that is able to combine in long chains with other like or unlike molecules to produce very large polymers. Thus, **monomeric.**

**monoplane** *Transportation.* an airplane of the standard contemporary design, with a single main set of wings, usually divided by the fuselage; contrasted with the earlier BIPLANE type with two sets of wings.

**monopoly** *Economics.* a market in which there are many buyers but only one seller; characterized by the seller's power to fix prices or exclude competition, coupled with policies designed to use or preserve that power.

**monopropellant** *Transportation.* a propellant that combines fuel and oxidizer in a single substance, as in a liquid fuel for rockets.

**monorail** *Transportation.* a railroad system that uses a single overhead rail or a single large rail, over which trains are held in place by side wheels.

***monorail*** *Monorail technology is often employed for short routes where laying a conventional two-rail track is not feasible, as for transportation within airports.*

**Montgolfier** **1.** the brothers Joseph Michel (1740–1810) and Jacques Étienne Montgolfier (1745–1799), inventors of the first practical balloon. In 1782 they succeeded in launching a balloon that reached an altitude of 250 meters. **2.** a balloon designed by these brothers; the first free flight by humans was made in a Montgolfier balloon in 1783.

**Montreal Protocol** *Policy.* a landmark international agreement designed to protect the stratospheric ozone layer, originally signed in Montreal, Canada in 1987. It stipulates that the production and consumption of ozone-depleting compounds (chlorofluorocarbons, halons, carbon tetrachloride, and methyl chloroform, among others) are to be phased out over time.

**more developed countries (MDCs)** *Sustainable Development.* a classification including those nation-states with a higher level of industrial development and usually a higher consumption of commercial energy; e.g., France.

**more developed regions** *Sustainable Development.* the United Nations designation for those areas of the world that contain more developed countries (see above); e.g., western Europe.

**Morse code** *Communication.* the standard communication system used in telegraphy, consisting of a prescribed pattern of short and long sounds, clicks and spaces, or flashes of light that represent the letters of the alphabet, numbers, or phrases. [Named for U.S. inventor Samuel F. B. *Morse,* 1791–1872.]

**Mossadegh, Mohammad** 1882–1967, nationalist Iranian political leader who was displaced as head of government (1953) in a coup backed by the U.S. and Britain, who feared he would align Iran with the USSR and thus end the flow of Iranian oil to the West. After his ouster Reza Shah PAHLAVI maintained a pro-Western stance until he too was ousted in 1979.

**motion** *Physics.* a change in the position of a physical system over time.

**motive energy** *History.* an earlier term for MOTION.

**motor** *Conversion.* **1.** any device that converts electrical energy into mechanical energy. **2.** another term for an engine, especially a relatively small engine, such as the internal-combustion engine in an automobile. **3.** *Bioenergetics.* describing structures, such as muscles or nerves, that are involved in or cause movement.

**motor oil** *Oil & Gas.* a petroleum product used as the lubricant in an internal-combustion engine.

**Mouchout, Augustin** 1823–1912, French inventor of the first known device that directly converted solar energy to power (1865). He succeeded in using his apparatus to operate a small conventional steam engine.

**Mount Pinatubo** *Climate Change.* a volcano in the Philippine Islands that erupted in 1991, ejecting huge amounts of particulate and aerosol matter (an estimated 20 million tons of sulfur dioxide entered the stratosphere) and creating a dense cloud layer that spread around the earth and attained global coverage after about one year. This caused significant decreases in the amount of net radiation reaching the earth's surface, which effectively cooled the planet from 1992 to 1994.

**mountaintop mining** *Mining.* coal-mining activities taking place at a coalbed that underlies the top of a mountain; this involves the complete removal of the mountaintop to access the coal and thus it is a controversial technique because of its environmental impact. Also, **mountaintop removal.**

**movable insulation** *HVAC.* a term for an object that can be adjusted according to conditions to provide an insulating effect, such as a window shade, shutter panel, or curtain, as opposed to conventional stationary insulating materials such as foam insulation sprayed into wall spaces.

**movistor** *Electricity.* metal oxide varistor; a device used to protect electronic circuits from surge currents such as those produced by lightning.

**MOX** mixed oxides fuel.

**MPG** or **mpg** miles per gallon.

**MPH** or **mph** miles per hour.

**MRL** maximum residue level.

**MSR** molten salt reactor.

**MSW** municipal solid waste.

**MTBE** methyl tertiary-butyl ether.

**MTD** maximum tolerated dose.

**mud pot** *Geothermal.* a surface feature found in a geothermal area where there is not enough water to support a geyser or hot spring; steam and gas vapors bubble up through a turbulent layer of mud formed by the interaction of gases with rock.

**Muir, John** 1838–1914, U.S. naturalist and political activist known for his pioneering conservation efforts. He worked to preserve the Yosemite area, served as the first president of the Sierra Club, played a key role in the creation of several national parks, and wrote hundreds of articles and several books on the virtues of conservation and the natural world.

**multicriteria analysis** *Policy.* a decision-making process in which various criteria are taken into account simultaneously to provide a comparative assessment of alternative measures or strategies; used in studies of energy issues; e.g., the siting of a power plant.

**multicrystalline silicon** another term for POLYCRYSTALLINE SILICON.

**multijunction (solar) cell** *Photovoltaic.* a device in which individual cells with different band gaps are stacked on top of one another so that sunlight falls first on the material having the largest band gap. Photons not absorbed in the first cell are transmitted to the second cell, which then absorbs the higher-energy portion of the remaining radiation while remaining transparent to the lower-energy portion. These selective processes continue to the final cell, thus achieving greater total conversion efficiency of the incident light.

**multilateral drilling** *Oil & Gas.* a standard drilling technique in which several lateral extensions are branched out from the main trunk hole.

**multimedia exposure** *Health & Safety.* the fact of being exposed to the same toxic substance from multiple sources, such as air, water, soil, and food.

**multinational** *Global Issues.* describing a corporation or other entity that conducts its operations in many nations rather than only in the country in which it originated or is headquartered.

**multiple-arch dam** see ARCH DAM.

**multiple-completion well** *Oil & Gas.* a well equipped to produce oil or gas separately from more than one reservoir.

**multiple-effect** *Materials.* describing a process in which water or another fluid passes through a series of vessels (effects) of successively lower pressure; this permits the fluid to undergo multiple stages of evaporation and condensation without the application of additional heat after the first effect; used for example to desalinate seawater. Thus, **multiple-effect evaporation** or **distillation.**

**multiple junction cell** see MULTIJUNCTION CELL.

**multiplying chain reaction** *Nuclear.* a chain reaction in which an average of more than one fission is produced by the neutrons released by previous fission.

**municipal solid waste (MSW)** *Environment.* a general term for the solid materials that result from the accumulation of all residential, commercial, and light industrial waste; normally handled by local municipalities and typically deposited in a landfill. MSW consists of metal, glass, paper, plastic, dirt, vegetative matter, and the like, either recyclable or nonrecyclable in nature.

**Munk, Walter Heinrich** born 1917, U.S. oceanographer who performed pioneering work in the energetics of wind-driven ocean circulation, vertical mixing in the ocean, wave propagation, and tidal dissipation. He also performed important research on the effects of global warming on changes in sea level.

**Munters, Carl** 1897–1989, Swedish inventor who with Baltzar von Platen developed an improved absorption technology based on the single pressure ammonia/water hydrogen cycle (1922); this would become the basis for modern refrigeration technology.

**muon** *Physics.* a subatomic particle larger than an electron but smaller than a proton.

**Murdoch, William** 1754–1839, Scottish engineer and inventor who was a pioneer of gas lighting. He discovered the properties of coal gas as an illuminant and used this to light his house and offices (1792).

**Muskat, Morris** 1906–1998, U.S. petroleum engineer who first described the fundamental aspects of reservoir dynamics and provided the analytical foundation for reservoir engineering that is used today.

**Muskie** see BIG MUSKIE.

**Musschenbroek, Pieter van** 1692–1761, Dutch physicist who invented the LEYDEN JAR (1745). His experiments provided one of the first scientific studies of electrical charge and its properties. He is also said to have coined the word "physics".

**must-take** *Economics.* describing energy resources that must be taken by the customer and consumed before any other resources are made available.

**mutual inductance** *Electricity.* the ability of one conductor to induce an electromotive force in a nearby conductor when the current in the first conductor changes.

**MV** megavolt.

**MW** megawatt.

**MWh** megawatt-hour.

**Mysterious Island, The** *History.* a novel by the famous science fiction author and futurist Jules Verne (1874), predicting that hydrogen will one day be a major form of energy.

**n** nano- (one billionth).

**NAAQS** *Policy.* National Ambient Air Quality Standards; regulations established by the U.S. Clean Air Act prescribing levels of pollution that may not be exceeded during a specified time in a defined area.

**nacelle** *Transportation.* **1.** a streamlined enclosure designed to house and protect some component of an aircraft, such as the crew, engine, or landing gear; e.g., an airplane cockpit or an airship gondola. **2.** *Wind.* the main body of a horizontal axis wind turbine that contains the gearbox, generator, controller, brake, blade hub, and other parts.

**NAD** *Biological Energetics.* nicotinamide adenine dinucleotide; a coenzyme found widely in nature and involved in numerous enzymatic reactions in which it serves as an electron carrier by being alternately oxidized ($NAD^+$) and reduced (NADH); required in the glycolytic and other metabolic pathways.

**NADPH** *Biological Energetics.* nicotinamide adenine dinucleotide phosphate; a coenzyme found in all living cells and composed of NAD (see above) with an extra phosphate group attached. The energy of NADPH is an important source of reducing power in the cell.

**Naflon** *Materials.* the brand name used by DuPont for a series of fluorinated sulfonic acid copolymers, the first synthetic ionic polymer. It is resistant to chemical breakdown, making it useful for membranes in proton exchange membrane fuel cells.

**Naft Khana** *History.* the site of the first oil field discovered in Iraq (1923).

**nameplate** *Consumption & Efficiency.* a metal tag attached to a machine or appliance by the manufacturer, providing various types of information about the device; in this context referring specifically to the statement of the device's maximum power output.

**nameplate capacity** *Consumption & Efficiency.* the peak ability of a power generation facility (such as a hydropower dam or a fossil-fuel power plant) to provide electricity under specified conditions, measured in units of power such as watts.

**nano-** *Measurement.* a prefix meaning "one billionth" ($10^{-9}$); symbol n.

**nanocrystalline silicon** *Photovoltaic.* nc-Si, a form of silicon used as a light-absorbing semiconductor in photovoltaic cells; similar to amorphous silicon (a-Si), in that it has an amorphous phase, but nc-Si has small grains of crystalline silicon within the amorphous phase.

**nanometer** *Measurement.* a unit of length that is equal to one-billionth of a meter.

**nanoscale** *Measurement.* referring to extremely small dimensions in the range of 100 nanometers or less, or to materials or objects existing at this level.

**nanoscience** *Consumption & Efficiency.* the study of materials and phenomena existing at a highly miniaturized level, especially at dimensions of around one nanometer. Similarly, **nanotechnology.**

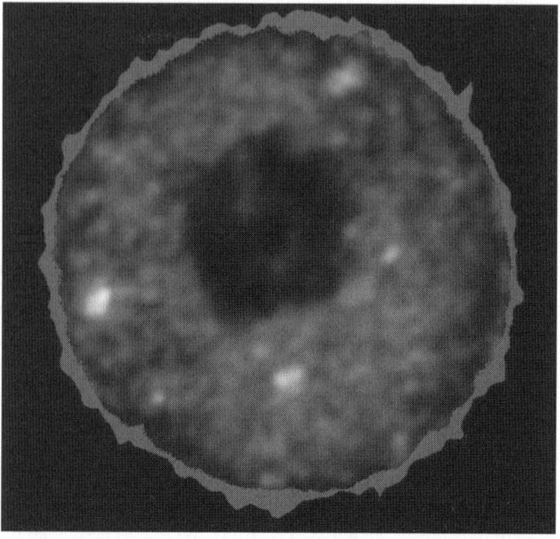

***nanoscience*** *Nanoscience permits technology to operate within a single human cell. Here a laser is used to identify cancer-affected cell mitochondria by the fact that they are dispersed throughout the cell.*

**nanosecond** *Measurement.* **1.** a unit of time that is equal to one-billionth of a second. **2.** a popular term for an extremely short time.

**naphtha** *Oil & Gas.* **1.** any of various products of petroleum, not less that 10% of which distill below 175°C and not less than 95% of which distill below 240°C; flammable and a dangerous fire risk. Used as a gasoline ingredient, a source of synthetic natural gas, a paint and varnish thinner, a dry cleaning fluid, and as a raw material for plastics. **2.** another term for CONDENSATE (def. 1). **3.** see HEAVY NAPHTHA; SOLVENT NAPHTHA.

**naphthalene** *Materials.* $C_{10}H_8$, a white, volatile, monoclinic or platelike solid having a strong coal-tar odor; insoluble in water and soluble in alcohol, ether, and benzene; melts at 80.55°C and boils at 218°C. Consisting of two benzene rings, it serves as the starting point for many organic syntheses, including plastics, dyes, solvents, antiseptics, and insecticides, especially mothballs.

**Napier grass** or **napiergrass** *Biomass.* a robust perennial bunchgrass native to Africa, *Pennisetum purpureum,* that is cultivated for livestock forage and as a form of biomass energy; it can reach a height of 6 meters, grows on a wide range of well-drained soils and is drought tolerant.

**NAS battery** see SODIUM–SULFUR BATTERY.

**natural assets** *Environment.* resources of the natural environment consisting of biological assets (produced or wild), land and water areas with their ecosystems, subsoil assets (minerals and fuels), and air.

**natural attenuation** *Environment.* any natural process taking place in soil or groundwater to evaporate, degrade, dilute, disperse, or absorb toxic contaminants; e.g., digestion by bacteria.

**natural bitumen** *Oil & Gas.* bitumen or other petroleum with very high viscosity, contained in bituminous sands, oil sands, or tar sands, which is not recoverable by conventional means.

**natural capital** *Environment.* components of the natural environment that can be defined as having value for human activities, especially economic activity, such as timber, mineral assets, water resources, clean air, and ecological systems. **Natural capitalism** is a business strategy based on valuation of these assets. See below.

**natural convection** *Thermodynamics.* heat transfer by fluid motion resulting from density gradients in the fluid.

**natural draft** *Consumption & Efficiency.* the natural, unenhanced flow of gases into and out of a combustion chamber, caused by differences in temperature; e.g., when hot air in a fireplace is drawn up the chimney by the presence of cooler air above it.

**natural-draft cooling** *HVAC.* a cooling process in which circulating air produced by natural convection comes in contact with the medium to be cooled. Thus, **natural-draft cooler, natural-draft cooling tower.**

**natural-draft furnace** *HVAC.* a furnace in which the natural flow of gases from around the furnace provides the air to support combustion.

**natural fate** *Environment.* the process of breakdown of a pollutant that has been released into the atmosphere (e.g., a crude oil spill in the ocean), assuming that no cleanup or other such human intervention takes place.

**natural fire event** *Environment.* a description for a fire that occurs because of some condition in the environment rather than its having been ignited by humans; e.g., lightning striking a dry tree stand.

**natural capital** An extension of the economic notion of capital (manufactured means of production) to environmental goods and services. A functional definition of capital in general is: "a stock that yields a flow of valuable goods or services into the future". Natural capital is thus the stock of natural ecosystems that yields a flow of valuable ecosystem goods or services into the future. For example, a stock of trees or fish provides a flow of new trees or fish, a flow which can be sustainable indefinitely. Natural capital may also provide services like recycling wastes or water catchment and erosion control. Since the flow of services from ecosystems requires that they function as whole systems, the structure and diversity of the system are important components of natural capital.

**Robert Costanza**
University of Vermont

**natural flow** *Oil & Gas.* a condition in a well in which the formation pressure is sufficient to produce oil at a commercial rate without its having to be pumped. Thus, **naturally flowing well.**

**natural gas** *Oil & Gas.* a mixture of hydrocarbon compounds and small quantities of various nonhydrocarbons, widely used as a fuel throughout the industrialized world; it exists in the gaseous phase or in solution with crude oil in natural underground reservoirs. See below.

**natural gas liquids** *Oil & Gas.* the hydrocarbons in natural gas that are separated from the gas through the processes of absorption, condensation, adsorption, or other methods in gas processing or cycling plants. Generally such liquids consist of propane and heavier hydrocarbons and are commonly referred to as condensate, natural gasoline, or liquefied petroleum gases (LPG).

**natural gasoline** *Oil & Gas.* a natural gas liquid product that consists primarily of pentanes and higher alkane hydrocarbons ($C5^+$); it is a liquid at ambient temperatures and atmospheric pressure and has a low octane number.

**naturally aspirated** *Transportation.* describing a non-supercharged internal combustion engine that relies solely on valve timing and piston movement to import the air–fuel mixture into the cylinder (or import the atmospheric air in a diesel engine).

**natural monopoly** *Economics.* a situation in which one firm in an industry can serve the entire market at a lower cost than would be possible if the industry were composed of many smaller firms. Gas and water utilities are two classic examples of natural monopolies.

**natural pressure** *Oil & Gas.* energy within an oil or gas reservoir that is sufficient to cause the oil or gas to rise to the surface unassisted by other forces, when the reservoir is penetrated by a well.

**natural resource** *Environment.* a broad term for any feature of the natural environment that can be useful or beneficial to humans, including exhaustible or nonrenewable resources, such as coal and copper, and renewable resources such as trees, fish, and water.

**natural resource accounting** another term for ENVIRONMENTAL ACCOUNTING.

**natural selection** *Ecology.* a theory of evolutionary change asserting that (a) every organism naturally displays slight variations from others of its kind; (b) organisms collectively produce more offspring than can survive; (c) a struggle for limited natural resources consequently ensues, as a result of which the individuals that survive and reproduce will be those whose natural variations adapt them better to their specific environment; and (d) the favorable variations will accumulate in subsequent generations, resulting in divergence and eventual speciation.

**natural gas** Natural gas is a mixture of naturally occurring compounds that are in a gaseous state at the conditions of temperature and pressure at the earth's surface. These compounds are mostly simple molecules formed of hydrogen and carbon (thus hydrocarbon). Methane, with molecules formed from four atoms of hydrogen and one of carbon, is the simplest form and generally is the most common compound in the mixture. Ethane, propane, butane, and pentane are slightly more complex compounds with two, three, four, and five carbon atoms respectively in each molecule. Other more complex hydrocarbon compounds can also be part of the mixture. Oil, by contrast, is an even more complex mixture of larger hydrocarbon molecules. Other non-hydrocarbon compounds are also present in natural gas but generally in lesser amounts. These non-hydrocarbon compounds include carbon dioxide, nitrogen, and hydrogen sulfide. Some natural gas is present in fields that produce little or no volume of liquids (non-associated gas). Natural gas is also present in oil fields. In an oil field, natural gas can occur in separate pools or as a gas cap floating on the denser oil (associated gas). It can also be dissolved in the oil (dissolved gas). Some simple hydrocarbon molecules, such as methane, can be formed by inorganic processes; however, most of the natural gas that occurs in commercial gas or oil fields is likely of organic origin. Buried plant or algal material forms natural gas when subjected to sufficient burial and heating (thermogenic gas) or by the action of microbes (biogenic gas).

**Ronald R. Charpentier**
United States Geological Survey

**natural uranium** *Nuclear.* a term for uranium as it is mined from the earth. The composition of this is 99.3% U-238 and 0.7% U-235.

**navigation** *Transportation.* **1.** any or all of the various processes used in determining position and directing movement from one place to another, especially the movement of a craft in water or air. **2.** the science and technology of guiding such movements.

**ncad** nickel–cadmium (battery).

**NCI** Nuclear Control Institute.

**nc-Si** nanocrystalline silicon.

**NEA** **1.** Nuclear Energy Agency. **2.** net energy analysis.

**neap tide** *Earth Science.* a twice-monthly tide of minimal range that occurs when the earth, sun, and moon are at right angles to each other, decreasing the total tidal force exerted on the earth so that the differences between high and low tides are unusually small, with both the high tide lower and the low tide higher than usual.

**near-zero vehicle (near-ZEV)** see ZERO EMISSION VEHICLE.

**neat** *Materials.* pure; unadulterated; describing a substance that is free from mixture or dilution with other substances; e.g., a fuel that is 100% ethanol would be a **neat fuel** and one that is 95% ethanol would be a **near neat fuel.**

**nebular hypothesis** *History.* the proposal by French scientist Pierre-Simon Laplace that our solar system was formed by the condensation of a nebula of dust and gas, an early (1796) statement of a theory that is still widely accepted today.

**negative** *Electricity.* the opposite of POSITIVE; specific uses include: **1.** having the type of electric charge in which there is an excess of electrons in relation to protons, as in a body of resin that has been rubbed with silk. **2.** describing the part of an electric cell from which the current flows into the cell. **3.** describing an element or compound that tends to gain electrons.

**negative radiative forcing** see RADIATIVE FORCING.

**negative temperature** *Thermodynamics.* a nonequilibrium state of a thermodynamic system in which there are more particles having higher energies than particles having lower energies.

**negawatt** *Consumption & Efficiency.* **1.** negative watt; a term for a unit of electricity "produced" (in fact, saved rather than generated) by electricity-saving measures such as the use of more efficient lamps. **2.** a measure of energy efficiency based on this concept.

**negentropy** *Thermodynamics.* negative entropy; the difference between the entropy of a system at thermodynamic equilibrium and the entropy of the present state.

**NEMS** *Measurement.* National Energy Modeling System; a computer modeling system maintained by the U.S. Department of Energy that produces a general equilibrium solution for energy supply and demand in U.S. energy markets. It reflects market economics, industry structure, and energy policies and regulations that influence market behavior.

**neoclassical economics** *Economics.* a school of economic thought postulating that consumers act to maximize their well-being or utility (subject to their income and other such constraints), that firms act to maximize their profits (subject to the technological constraints of production), and in general that economic agents act to maximize an objective function which can be mathematically described (subject to constraints that also can in principle be so described).

**neoclassical growth (model)** *Economics.* economic growth in which the long-run growth rate of output per worker is determined by an exogenous rate of technological progress.

**neoclassical theory of value** *Economics.* the theory that market prices are determined by the intersection of demand curves and supply curves, which in turn arise from the behavior of individual, self-interested economic agents.

**neodymium** *Chemistry.* a metallic element having the symbol Nd, the atomic number 60, an atomic weight of 144.24, a melting point of 1024°C, and a boiling point of about 3030°C; a soft, malleable, yellow rare-earth element of the lanthanide series that is highly flammable and easily tarnished, and that has high electrical resistivity; used in electronics, alloys, and astronomical lenses and lasers.

**neon** *Chemistry.* an inert element having the symbol Ne, the atomic number 10, an atomic weight of 20.179, a melting point of –249°C, and a boiling point of –246°C; a colorless, odorless, tasteless noble gas that ionizes in electric discharge tubes; used in fluorescent lighting, electronics, and lasers.

**neon lamp** *Lighting.* a tube that generates a bright red glow (or, if treated with mercury, bright blue) when the neon gas inside it is ionized by an electric current; commonly used in outdoor signs and as an indicator light. Thus, **neon bulb** or **tube, neon light(ing).**

***neon lamp*** *Because of its ability to provide bright coloring, neon lighting is often used for displays and advertising, as in this classic example of a Wurlitzer jukebox.*

**NEP** net ecosystem production.

**NEPA** *Policy.* National Environmental Policy Act; a U.S. federal law (1969) that requires all federal agencies to file statements for any action that might have a significant environmental impact; sometimes referred to as "the environmental Magna Carta".

**neptunium** *Chemistry.* a radioactive element having the symbol Np, the atomic number 93, an atomic weight of 237.048, and a melting point of 640°C; a silvery-white metal found naturally in traces, with weighable amounts produced as a byproduct in plutonium production; isotope 237 of neptunium is used in neutron detection instruments.

**NERC** North American Electric Reliability Council (est. 1968), an organization that regulates the bulk electric system in the U.S., Canada, and a portion of Mexico.

**Nernst, Walther Hermann** 1864–1941, German chemist who was awarded the Nobel Prize for his description of the THIRD LAW OF THERMODYNAMICS. He also invented the NERNST LAMP (1897).

**Nernst equation** *Chemistry.* an equation demonstrating that the voltage developed in an electrochemical cell is determined by the activities of the reacting species, the reaction temperature, and the standard free-energy change of the overall reaction.

**Nernst lamp** *Lighting.* an electric lamp whose active element is a wire or rod that is formed of magnesium oxide mixed with oxides of rare metals such as zirconium oxide, heated to a brilliant white incandescence by the application of a current through it.

**Nernst potential** *Chemistry.* the potential energy that is equivalent to the reversible equilibrium in an electrolytic system between charged hydrogen ions and hydrogen gas at a pressure of one standard atmosphere.

**Nernst theory** *Chemistry.* the principle that in an electrolytic cell an equilibrium will develop between the tendency of the electrode to dissolve and form ions in the solution, and the tendency of the ions in the solution to build up on the electrode.

**netback** *Economics.* a method of determining the wellhead price of oil or natural gas that shares the end market value of the resource with all parties in the supply chain (production, transportation, distribution). For example, in a contract for the purchase of liquefied natural gas, the wellhead value of natural gas (the netback price to the producer) is the residual amount after subtracting from the final market price the cost of liquefaction, transportation and storage, and regasification. Thus, **netback pricing, netback purchase.**

**net biome production (NBP)** *Environment.* the net gain or loss of carbon or organic matter from a given area at the biome level; equal to the net ecosystem production minus the carbon lost due to a disturbance such as a forest fire or a logging harvest. Also, **net biome productivity.**

**net ecosystem production (NEP)** *Environment.* the net gain or loss of carbon or organic matter from a given area at the ecosystem level, equal to the NET PRIMARY PRODUCTION minus the carbon lost through heterotrophic respiration. Also, **net ecosystem productivity.**

**net energy** *Consumption & Efficiency.* the amount of energy available from an energy transformation system, after deduction of the energy expended in creating the capital equipment and the energy used in operating the system. See below.

**net energy** Net energy is the difference between the energy produced by an energy facility or technology and the energy required to develop and operate it. Required energy includes the energy to produce all inputs such as steel, concrete, etc. Positive net energy is equivalent to (energy produced/energy required) greater than 1. The latter is called the net energy ratio or energy return on investment. Net energy analysis allows comparison of different energy types, and changes over time as we reach to more distant, dilute, and complicated sources. While readily understandable in broad terms, net energy analysis is complicated in its conceptual and bookkeeping details (e.g., how to compare different types of energy such as liquid fuels and electricity, or whether to include the energy needed to support research and development?). In spite of some controversy, three conclusions have emerged. First, coal, natural gas, and oil continue to produce positive net energy. Second, for these sources the net energy ratio has declined over the past 75 years (e.g., 100 to 20 for crude oil). Third, oil from oil shale or coal has a net energy ratio no greater than 10. Ethanol from grain in the U.S. has an energy ratio of approximately 1.5.

**Robert Herendeen**
Illinois Natural History Survey

**net energy analysis (NEA)** *Measurement.* a technique that compares the quantity of energy delivered by an energy system to the energy used directly and indirectly in the process of delivery.

**net energy balance** *Measurement.* the relationship between the energy provided by a source and the energy required for its production; e.g., the energy of ethanol fuel compared to the fossil energy used for corn farming and ethanol production.

**net generation** *Electricity.* the amount of gross generation from a generating station, less the electrical energy consumed at the station itself for station service or auxiliaries.

**net head** *Hydropower.* the vertical distance between the top of a channel or pipeline that conveys water to a pipeline under pressure and the point where the water discharges from the turbine, minus any pressure losses from friction and turbulence.

**net metering** *Consumption & Efficiency.* the use of a single meter to measure how much power is consumed and produced by a house with its own power source (such as a solar or wind system). Any electricity produced by the resident power system can be transmitted back to the utility; when more power is being produced than used, the meter spins backwards (in effect recording a credit for the customer). Similarly, **net billing.**

**net primary production (NPP)** *Biological Energetics.* a measure of plant growth, expressed as the difference between the total amount of carbon taken up by plants via photosynthesis (gross primary production) and the amount of carbon lost by living plants via autotrophic respiration. usually measured in units of carbon produced per unit area per unit time. Also, **net primary productivity.**

**net radiation** *Solar.* the total solar irradiance on a surface, plus the long-wave radiation per unit area of that surface, minus the upward radiation from the surface.

**net price** *Economics.* a valuation used in natural resource economics to estimate the economic value of a natural resource and its depletion; defined as the actual market price of a natural resource output minus all marginal extraction costs, including what would represent a normally expected return to capital.

**netting** *Policy.* regulations that allow an entity to use emissions reductions achieved at a permitted facility to avoid some of the preconstruction review requirements that would normally apply to a proposed major modification at that same facility.

**neural network** see ARTIFICIAL NEURAL NETWORK.

**neutral** *Electricity.* **1.** having no net electric charge; i.e., an equal number of electrons and protons are present. **2.** *Chemistry.* describing a substance that is neither acid nor alkaline. **3.** *Lighting.* describing a light source that is perceived as being neither warm nor cold; i.e., that has a white appearance as opposed to reddish or bluish.

**neutral-beam injector** *Nuclear.* a method to provide initial heat for a tokamak fusion reactor in which high-energy neutral atoms are shot into the plasma and are immediately ionized. These ions are then trapped by the magnetic fields, and transfer some of their energy to the surrounding plasma particles through collisions, thus raising the overall temperature.

**neutrino** *Nuclear.* a small particle that has no charge and is thought to have very little mass. Neutrinos are created in massive numbers by the nuclear reactions in stars; they are very hard to detect because the vast majority of them pass completely through the earth's atmosphere without interacting.

**neutron** *Physics.* an elementary particle having zero charge, a rest mass of 1.00894 amu (about the same as that of a proton), and a spin quantum number of 1/2. Neutrons are found to be naturally present in all nuclei with a mass number greater than 1 (i.e., all known nuclei except the lightest isotope of hydrogen). They are used to produce fission and other nuclear reactions that release atomic power.

**neutron balance** *Nuclear.* the balance in a reactor between the rate of production of both prompt and delayed neutrons, and their rate of loss due to both absorption and leakage from the reactor; necessary for a constant power level in the reactor.

**neutron bomb** *Nuclear.* a type of nuclear weapon specifically designed to release a relatively large portion of its energy as energetic neutron radiation, and thus destroy life but spare property.

**neutron bombardment** see BOMBARDMENT.

**neutron chain reaction** *Nuclear.* a process in which some of the neutrons released in one fission event cause other fissions to occur.

**neutron economy** *Nuclear.* a term for the degree to which neutrons in a reactor are used for desired ends, instead of being lost by leakage or nonproductive absorption

**neutron flux** *Nuclear.* a measure of the intensity of neutron radiation, expressed as the number of neutrons passing through 1 square centimeter of a given target in 1 second.

**neutron gun** *Nuclear.* a block of moderating material with a channel through it, used for producing a beam of fast neutrons.

**neutron hardening** *Nuclear.* a increase in the average energy of a beam of neutrons brought about by passing them through a medium that selectively absorbs slow neutrons.

**neutron leakage** *Nuclear.* neutrons that escape from the vicinity of the fissionable material in a reactor core; these neutrons are no longer available to cause fission and must be absorbed by shielding placed around the reactor pressure vessel.

**neutron poison** see POISON.

**neutron scattering** *Nuclear.* the process by which neutrons collide with atomic nuclei and either gain or lose kinetic energy.

**neutron shield** see RADIATION SHIELD.

**neutrosphere** *Earth Science.* the atmospheric shell from the earth's surface upward in which the atmospheric constituents are, for the most part, electrically neutral.

**Newcastle** *History.* **1.** a historic industrial city on the Tyne River in northeast England (in full **Newcastle on (upon) Tyne**), the early transport center for the British coal industry. **2. carry (take, bring) coals to Newcastle.** to perform an activity that is a waste of time, because it supplies something already present in abundance (as coal was abundant in Newcastle).

**Newcomen, Thomas** 1663–1729, English inventor of an early atmospheric steam engine (about 1712). It was an improvement over an earlier engine patented (1698) by Thomas Savery, who shared the patent for the later engine with Newcomen.

**new field discoveries** *Oil & Gas.* the discovery of oil or natural gas in an area outside of that containing already identified reserves.

**new-growth** *Ecology.* describing a forest without older trees (i.e., not old-growth); a managed forest, typically having a collective of trees of the same species and age specifically planted for commercial purposes.

**newly industrializing country (NIC)** *Economics.* an intermediate classification for countries that have a high level of economic growth and export expansion, outpacing the less developed countries but not as industrialized as the developed countries. Examples are Mexico, Brazil, and Portugal, as well as the four "Asian Tigers" (Hong Kong, Singapore, South Korea, and Taiwan).

**Newton, Isaac** 1642–1727, English mathematician and natural philosopher considered by many to be the most influential scientist who ever lived. He summarized his discoveries in terrestrial and celestial mechanics in his 1687 work *Philosophiae naturalis principia mathematica* (Mathematical Principles of Natural Philosophy). Newton surpassed all scientists who came before him, concisely stating simple yet elegant scientific principles and methods that applied to every branch of science.

***Newton*** *Newton performed a famous experiment with a prism to show that the different colors of light we observe are actually the components of white light, rather than alterations of it.*

**Newtonian** *Physics.* **1.** having to do with Sir Isaac Newton or his work and theories. **2.** relating to the study of mass, energy, and motion on the basis of the theories of Newton, including especially the assumptions that time is an absolute quantity, and that distance can be absolutely measured rather than being dependent on the time and point of observation. Thus, **Newtonian mechanics, Newtonian physics.**

**Newton's law of cooling** *Thermodynamics.* the statement that for a body cooling in a draft (i.e., by forced convection), the rate of heat loss is proportional to the difference in temperature between the body and its surroundings.

**Newton's laws of motion** *Physics.* the three laws proposed by Newton governing mechanics: first, that a body at rest will continue at rest, and a body in motion will continue in uniform motion along a straight line, unless the body is acted upon by some outside force; second, that a body acted upon by a net force will accelerate, and that it will do so in the direction of the force; third, that every action will produce an equal and opposite reaction.

**New York Mercantile Exchange** see NYMEX.

**NG** natural gas.

**NGL** natural gas liquids.

**NGO** nongovernmental organization.

**NGPA** *Policy.* Natural Gas Policy Act; a U.S. federal law (1978) that effectively ended decades of natural gas price controls by the federal government. Its goal is to deregulate natural gas prices over time, to encourage exploration, and to reduce the price differentials between interstate and intrastate markets.

**niche vehicle** *Transportation.* a term for a motor vehicle that has a relatively small production output and that is sold to a specific component of the overall market; e.g., an expensive sports car.

**nickel** *Chemistry.* **1.** a metallic element having the symbol Ni, the atomic number 28, an atomic weight of 58.70, a melting point of 1455°C, and a boiling point of 2900°C; a malleable, silver-white transition metal having excellent resistivity to corrosion and tarnish. **2.** this element in the form of a metal extensively used for electroplating and as an alloying element and a base for specialty alloys.

**nickel–cadmium battery** *Storage.* a sealed storage battery having a nickel anode, a cadmium cathode, and an alkaline electrolyte; the dominant battery type of the 20th century because of its durability, long life, and relatively low cost, but recently supplanted for some applications by batteries with higher energy densities and less toxic metals (e.g., lithium-based batteries).

**nickel–hydrogen battery** *Storage.* a hybrid battery type that combines the technology of batteries and fuel cells; it has a nickel oxide positive electrode similar to a nickel–cadmium cell, and it is like a hydrogen–oxygen fuel cell in that it has a hydrogen negative electrode. It is used in various spacecraft applications.

**nickel–iron battery** *Storage.* an alternative to the more popular nickel–cadmium battery, less expensive to build and to dispose of than the nickel–cadmium battery; very reliable but does not efficiently recharge.

**nickel–metal hydride battery** *Storage.* a battery type that employs a metal hydride alloy as the alloy, rather than the traditional cadmium; it provides higher energy density than the nickel–cadmium battery and thus has replaced it in some contemporary applications, such as wireless communications and mobile computing.

**nickel–zinc battery** *Storage.* an alternative to the more popular nickel–cadmium battery, having a useful lifetime of only 200 or so charging cycles.

**night sky radiation** *HVAC.* a method of cooling a building through radiant energy exchange; relatively warm surfaces are exposed directly to the colder night sky so that they reject the heat they collected during the day.

**NIMBY** *Social Issues.* not in my backyard; an expression used to describe a common public attitude in which people will acknowledge the theoretical need for a certain form of development (e.g., nuclear power plants, electric power lines, wind farms, low-income housing), as long as the actual facility does not affect the quality of life in their own neighborhood. Thus, **NIMBYism.** See next column.

**NIMBYism** The acronym NIMBY: Not In My Backyard has become widely used to describe the attitude and reaction among the local population protesting the introduction of something unwanted in their community. It is a reaction or attitude towards any project, such as the siting of an actual or perceived hazardous enterprise, such as a power plant, or affordable housing projects that are perceived to pose a threat to health or safety, status or reputation of a neighborhood or geographical area. NIMBYism can take the form of a protest against authorities or industry by the formation of action groups comprised by local residents. This response can be viewed as a reaction from the local population based on a variety of reasons. They may stem from a sense of being overrun by the authorities or industry to a genuine concern for the health and safety of the community. The context of the situation needs to be examined to determine the causes for NIMBYism.

**Misse Wester-Herber**
Örebro University, Sweden

**NiMH** nickel–metal hydride (battery).

**N-I-P** *Photovoltaic.* negative–intrinsic–positive; a structure typically used in CdTe (cadmium telluride) solar cells, employing the same principles as the P-I-N structure but a different arrangement. See also P-I-N.

**Nipkow disc** *Communication.* a pioneering electromechanical image scanning system, consisting of a rapidly rotating disc with a pattern of holes; an important early step in the development of television that was employed in the first transmission of true television pictures. [Invented by German physicist Paul *Nipkow,* 1860–1940.]

**nitrate** *Chemistry.* **1.** any salt or ester of nitric acid, such as sodium nitrate, $NaNO_3$. **2.** *Materials.* a fertilizer consisting of sodium nitrate, $NaNO_3$, or potassium nitrate, $KNO_3$.

**nitric** *Chemistry.* of or relating to nitrogen or to various compounds of nitrogen, especially those in which the element has a valence of five.

**nitric acid** *Materials.* $HNO_3$, a toxic, corrosive, hygroscopic, colorless liquid; completely miscible with water; it emits suffocating fumes and attacks most metals and is a dangerous fire hazard in contact with organic substances. Widely used in the manufacture of fertilizers and explosives, in metallurgy and plastic manufacture, and for various other industrial purposes.

**nitrification** *Biological Energetics.* the process by which nitrogen in ammonia and organic compounds is oxidized to nitrites and nitrates by soil bacteria of the family Nitrobacteraceae.

**nitrogen** *Chemistry.* a gaseous element having the symbol N, the atomic number 7, an atomic weight of 14.0067, a melting point of –209.9°C, and a boiling point of –195.5°C; a colorless, odorless, tasteless gas that makes up about four-fifths of the atmosphere. Nitrogen is an essential ingredient of proteins, which are the principal metabolic entities of organic life. It is used in ammonia synthesis and as an inert gas, refrigerant, and fertilizer component.

**nitrogen balance** *Biological Energetics.* **1.** the relationship between the gain and loss of nitrogen in a living organism. **2.** a similar ratio in an ecosystem, in soil or water, and so on.

**nitrogen cycle** *Earth Science.* the continuous process by which nitrogen circulates among the air, soil, water, plants, and animals of the earth. In a specific cycle, nitrogen in the atmosphere is converted by bacteria into substances that green plants can absorb from the soil; animals then eat these plants (or eat other animals that feed on the plants); the animals and plants then die and decay; the nitrogenous substances in the decomposed organic matter then return to the atmosphere and the soil.

**nitrogen dioxide** *Chemistry.* $NO_2$, a compound of nitrogen and oxygen formed by the oxidation of nitric oxide (NO), produced by the combustion of solid fuels. It is a major air pollutant, playing a key role in atmospheric reactions that produce ground-level ozone, a major component of smog. It is also a precursor to nitrates, which contribute to increased respirable particle levels in the atmosphere.

**nitrogen fixation** *Earth Science.* the conversion of atmospheric nitrogen into chemical compounds, such as ammonia, that can be used by green plants in the formation of proteins. One form of this process is **biological fixation,** which is carried out by certain bacteria that are present on the nodules of the plants. Two other forms are **atmospheric fixation** (the energy of lightning) and **industrial fixation** (human activities such as the burning of fossil fuels and fertilizer production and use).

**nitrogenous** *Chemistry.* containing or utilizing nitrogen.

**nitrogen oxide** *Chemistry.* $NO_x$, the generic term for a group of highly reactive gases, all of which contain nitrogen and oxygen in varying amounts; many are colorless and odorless, although one common pollutant, nitrogen dioxide ($NO_2$) can be seen along with particles in the air as a reddish-brown layer over many urban areas. Nitrogen oxides form when fuel is burned at high temperatures, in motor vehicles, electric utilities, and other industrial, commercial, and residential sources. The components of $NO_x$ are acid rain precursors and also participate in atmospheric ozone chemistry.

**nitroglycerine** *Materials.* a toxic, yellow, viscous liquid that is slightly soluble in water and soluble in alcohol; it freezes at 13.1°C and explodes at 218°C; used as an explosive and rocket fuel.

**nitrous** *Chemistry.* of or relating to nitrogen, or to various compounds of nitrogen, especially those in which the element has a valence of three.

**nitrous oxide** *Chemistry.* $N_2O$, a colorless gas, naturally occurring in the atmosphere; narcotic in high concentrations. Used as an aerosol propellant, leak detector, and anesthetic; in the latter context it is popularly known as **laughing gas.**

**N-layer** see N-TYPE.

**NMR** nuclear magnetic resonance.

**no. 2 fuel oil** *Oil & Gas.* a classification of fuel oil that does not require any preheating and is fired at an ambient temperature, widely used for general purpose domestic heating.

**Nobel, Alfred Bernhard** 1833–1896, Swedish chemist who invented dynamite (1867) and used his wealth to establish the **Nobel Prizes,** awarded annually for achievement in the sciences and literature and efforts in the cause of peace.

**noble gas** *Chemistry.* a gas that is unreactive (inert), or reactive only to a limited extent with other elements; a collective term for six gases that make up a group on the periodic table: helium, neon, argon, krypton, xenon, and radon. [So called because they do not mix readily with other elements, a trait thought of as similar to the attitude of those of noble (aristocratic) birth toward the common people.]

**noble metal** *Materials.* a metallic substance having an electrochemical potential that is much more positive than the potential of the standard hydrogen electrode; a substance that has high resistance to corrosion and oxidation; e.g., gold, silver, or platinum.

**NOCT** normal operating cell temperature.

**nodal pricing** *Economics.* a transmission pricing method resulting in an energy price at every major transmission node (generator, transmission line junction, or substation) equal to the marginal cost of meeting an increase in demand at that node.

**noise pollution** *Health & Safety.* by analogy with air or water pollution, a term for sounds in the environment that are considered to be unwanted, annoying. or harmful, because of their loudness, persistence, unpleasant character, and so on.

**noise thermometry** *Measurement.* the measurement of temperature on the basis of the random electrical noise generated by a resistor, with the power of the noise signal being proportional to the temperature of the material.

**no-load loss** *Electricity.* energy losses resulting from a system that is powered but not in use; virtually all electrical service suffers from at least some no-load loss.

**nominal voltage** *Storage.* the average terminal voltage of a cell or battery during its discharge.

**nonassociated** *Oil & Gas.* describing natural gas that is not in contact with significant quantities of crude oil in the reservoir. Thus, **nonassociated (natural) gas.**

**nonattainment area** *Environment.* a geographic area in the U.S. identified as having air quality that fails to meet national standards for clear air, due to unacceptably high concentrations of criteria air pollutants.

**nonbiodegradable** *Materials.* not biodegradable; describing a substance that cannot be broken down in the environment by natural processes.

**non-caking coal** *Coal.* a description for coal that does not fuse together or solidify as coke when heated, but burns freely and leaves only a slight residue of ash.

**noncommercial fuel** *Biomass.* a classification for traditional fuels, such as fuelwood and dried cow dung, that are collected and used by energy consumers directly, without involving market transactions or energy conversions to processed fuels.

**non-compliance coal** *Coal.* a high-sulfur coal or blend of coals that will not be able to meet sulfur dioxide emission standards for air quality, unless it is specially treated. Officially defined in the U.S. as any coal that emits greater than 3.0 pounds of sulfur dioxide per million Btu when burned.

**non-condensing turbine** *Conversion.* a steam turbine in which the exhaust steam leaving the turbine is used either as co-generated process steam or released into the atmosphere.

**nonconservative force** *Thermodynamics.* a force for which the work done in displacing a particle from one point to another is dependent on the path taken by the particle in moving from the initial position to the final one; e.g., the force of friction.

**nonconvective** *Solar.* not allowing convection; describing the middle layer of water in a solar salt pond that blocks transfer of heat to the surface from the hot bottom layer.

**non-conventional** see UNCONVENTIONAL.

**noncrystalline silicon** another term for AMORPHOUS SILICON.

**non-dispatchable** *Consumption & Efficiency.* a term for an energy system that cannot be expected to provide a continuous output to furnish power on demand, because production cannot be correlated to load. Hydrocarbon-based or nuclear power plants are dispatchable, but solar and wind power are non-dispatchable (without some added component for storage), since the supply of sunlight or wind is periodic and cannot be predicted and controlled. **Thus, non-dispatchable power** or **energy.** Compare DISPATCHABLE.

**non-equilibrium thermodynamics** *Thermodynamics.* a branch of thermodynamics concerned with the study of time-dependent thermodynamic systems, irreversible transformations, and open systems. Compare CLASSICAL THERMODYNAMICS.

**nonexcludable** *Social Issues.* describing a social benefit that is available to all people once the benefit is provided; e.g., pollution control devices, public street lighting. Thus, **nonexcludability.**

**non-Fourier equation** *Thermodynamics.* a constitutive equation of heat conduction for special phenomena (e.g., low temperatures).

**nongovernmental organization (NGO)** a general term for any of various international, national, and local membership organizations that represent the public interest (or the interests of groups of citizens). See below.

**nongovernmental organization**
There is no commonly-agreed definition of 'nongovernmental organizations' (NGOs), although the term generally refers to non-profit organizations that are not instruments of government and that generally aim to represent, protect, and advance public interest and values (or the interest and values of specific groups of citizens). It can also include local membership organizations such as grassroots-based or community-based organizations. Over the last few decades, there has been a huge proliferation of NGOs in both industrialized and the developing countries, and NGOs now come in all shapes and sizes—some may work exclusively at the local scale, and on the other end, there are a number of global NGOs. They can have a wide variety of goals including community development, environmental and human rights protection, providing relief, and alleviating suffering. NGOs engage in a range of activities, including provision of goods and services, advocacy, research, and involvement in policy development and implementation (and often an NGO may engage in multiple activities). In the energy area, the industrialized-country focus of NGOs is mainly (although not exclusively) directed towards reducing the environmental impacts of energy production, conversion, and use, as well as promoting energy efficiency. In the developing-country context, in addition to these goals, NGOs often also help enhance the provision of modern energy services to rural and other poor communities.

**Ambuj Sagar**
Harvard University

**nonhydrocarbon** *Oil & Gas.* describing gases other than hydrocarbons that may be present in natural gas reservoirs; e.g., carbon dioxide, helium, hydrogen sulfide, or nitrogen.

**nonideal** or **non-ideal** *Chemistry.* describing an actual chemical system that does not conform to the behavior established for an ideal system, such as the equation of state for perfect gases. Thus, **nonideal gas, nonideal behavior, nonideal solution,** and so on.

**non-imaging optics** *Lighting.* a field of optics that focuses on transferring light efficiently and controlling its distribution. Non-imaging optic devices act as light funnels that in some cases collect and intensify radiation more effectively than lenses and mirrors. Potential applications include collection of solar energy, radiant heating, laser pumping, and illumination systems.

**non-market valuation** *Economics.* a valuation method used to place a monetary value on goods or services that are not traded in markets, such as biodiversity, pest control services, and so on.

**nonmetal** or **non-metal** *Chemistry.* any chemical element that is not a metal or a metalloid; an element that is not capable of forming positive ions in an electrolytic solution. In the solid state such elements are brittle and nonductile, and are not able to conduct electricity or heat very well; e.g., hydrogen, carbon, nitrogen, oxygen. Thus, **nonmetallic.**

**non-Newtonian** *Physics.* describing systems or processes that do not conform to the theories of Sir Isaac Newton; i.e., the speeds involved are too high, and/or the particles are too small, for Newton's laws of motion to apply.

**non-point source** *Environment.* pollution that cannot be attributed to a single, identifiable facility such as, a power plant, factory, or mine; e.g., runoff pollutants originating with various diffuse sources, accumulating when rainfall moves over the ground. Thus, **non-point source pollution.**

**nonproliferation** *Global Issues.* **1.** a collective term for efforts to prevent or slow the spread of nuclear weapons and the materials and technologies used to produce them. **2.** a situation in which nuclear weapon capability is not extended to states that do not currently possess this ability.

**nonrecoverable** *Consumption & Efficiency.* describing a resource asset (e.g., an offshore oil deposit) that cannot be obtained on an economically sound basis, given present physical, economic, environmental, political, or technological constraints.

**nonrenewable** *Renewable/Alternative.* not able to be renewed at a rate comparable with human use; likely to be effectively depleted or exhausted at some future date. Thus, **nonrenewable resource.**

**nonrenewable energy** *Renewable/Alternative.* an energy source formed and accumulated over a very long period of time in the past, such as a fossil fuel, whose rate of formation is many orders of magnitude slower than the rate of its use, so that it will be depleted in a finite time period at the current rate of consumption. Contrasted with RENEWABLE ENERGY sources such as solar energy and wind power.

**non-shivering thermogenesis** *Biological Energetics.* a process of thermogenesis by an organism involving metabolic heat production by non-muscle tissues; found in newborn and hibernating mammals when cold causes adrenergic stimulation, which in turn causes the heat-releasing metabolism of brown adipose tissue (brown fat).

**non-state function** see STATE FUNCTION.

**nonsustaining chain reaction** *Nuclear.* a chain reaction in which an average of less than one fission is produced by the neutrons released by each previous fission.

**nonthreshold** *Health & Safety.* describing a toxic substance that is assumed to carry some risk of harmful effect even at the lowest levels, and that thus cannot be assigned a threshold point below which exposure is acceptable.

**nontraditional** *History.* **1.** describing an energy source that has a relatively short period of use in human history, such as oil or nuclear power, as opposed to historic sources such as human and animal power. **2.** specifically, describing an energy source that has come into widespread use only within the past few decades, such as solar power. Thus, **nontraditional energy.**

**nonutility** *Electricity.* describing an entity that owns or operates facilities for electric generation but is not an electric utility; e.g., a cogenerator or small power producer. Thus, **nonutility generator.**

**no regrets** *Policy.* a term for technology intended to reduce greenhouse-gas emissions, whose other benefits (in terms of efficiency or reduced energy costs) are so extensive that the investment is worth it for those reasons alone. Also sometimes known as "measures worth doing anyway".

**NORM** *Materials.* naturally occurring radioactive materials; i.e., the low-level radioactivity that is present in the natural environment (and in organisms), such as minerals that contain uranium, thorium, or radioactive potassium, as opposed to radioactivity induced by artificial processes, such as the detonation of a nuclear weapon.

**normal distribution** *Measurement.* a statistical distribution function that is a two-parameter family of curves; the most important distribution for probability and statistics. The first parameter, $\mu$, is the mean. The second, $s$, is the standard deviation. The standard normal distribution sets $\mu$ to 0 and $s$ to 1. The shape of the normal distribution resembles that of a bell, so it sometimes is referred to as the "bell curve".

**normal operating cell temperature (NOCT)** *Photovoltaic.* the estimated temperature of a photovoltaic module when operating under reference conditions specified as 800 watts per square meter irradiance, 20°C ambient temperature, and wind speed of 1 meter per second.

**normal radiation** *Solar.* a term for radiation striking a surface that is facing the sun. See also DIRECT RADIATION; GLOBAL RADIATION.

**normal transmittance** *Solar.* the ratio of the transmitted radiant or luminous flux to the incident flux striking a material in a direction perpendicular to its surface.

**not in my backyard** see NIMBY.

**nozzle** *Consumption & Efficiency.* a contracting, tapering tube or vent used at the end of a pipe, tube, or hose to accelerate and/or direct the flow of a liquid.

**NPP** net primary production.

**NPS (pollution)** see NON-POINT SOURCE.

**NPT** Nuclear Non-Proliferation Treaty.

**NRC** Nuclear Regulatory Commission.

**NREL** National Renewable Energy Laboratory (est. 1974), a U.S. federal agency for research and development in renewable energy and energy efficiency; it is home to the National Center for Photovoltaics, National Bioenergy Center, and National Wind Technology Center.

**N-type** *Electricity.* negative type; a semiconductor to which an impurity has been added so that the concentration of electrons is much higher than the concentration of holes; the electrical current is carried chiefly by these electrons. Thus, **N-type layer, N-type semiconductor, N-type silicon.** Compare P-TYPE.

**nuclear** **1.** of or relating to the nucleus of an atom. **2.** relating to or powered by nuclear energy. **3.** relating to nuclear weapons and their use in warfare. **4.** *Bioenergetics.* of or relating to the nucleus of a cell.

***nuclear*** *The U.S. nuclear-powered submarine Memphis.*

**nuclear absorption** *Physics.* the process by which a nucleus acquires energy; any interaction of ionizing radiation with matter that produces a loss of energy of the incident radiation.

**nuclear accident** see ACCIDENT.

**Nuclear Age** *History.* a term for the current era, as characterized by the discovery, technological applications, and sociopolitical consequences of nuclear energy.

**nuclear battery** *Nuclear.* a battery in which the electric current is produced from the energy of radioactive decay, either directly by collecting beta particles or indirectly, as by using the heat liberated to operate a thermojunction.

**nuclear binding energy** see BINDING ENERGY.

**nuclear bomb** see NUCLEAR WEAPON.

**nuclear bombardment** see BOMBARDMENT.

**Nuclear Control Institute** (est. 1981), an independent research and advocacy center specializing in problems of nuclear proliferation.

**nuclear disintegration** see DISINTEGRATION.

**nuclear energy** *Nuclear.* energy released by radioactive decay, through a nuclear reaction, or in the course of fission (splitting) or fusion (fusing) of atomic nuclei.

**Nuclear Energy Agency** an agency within the Organisation for Economic Co-operation and Development (OECD) that assists its member countries in developing the scientific, technological, and legal capabilities required for the safe, peaceful, and economical use of nuclear energy.

**nuclear event** see EVENT.

**nuclear excursion** *Nuclear.* the rapid release of nuclear reactor power above established operation levels, either deliberately for an experiment or by accident.

**nuclear fission** see FISSION.

**nuclear force** *Physics.* a powerful short-ranged attractive force that binds together the particles inside an atomic nucleus.

**nuclear fuel** *Nuclear.* **1.** light elements such as hydrogen, deuterium, tritium, lithium, boron, or beryllium that are capable of fusing or combining to form a nucleus of higher mass number. **2.** special nuclear material capable of sustaining a chain reaction, such as uranium-235.

**nuclear fuel cycle** *Nuclear.* the totality of operations necessary to fuel nuclear reactors, including the mining and milling of uranium ores, enrichment of uranium, fabrication and use of nuclear fuel, reprocessing of used nuclear fuel, and disposal or long-term management of radioactive wastes or spent fuel.

**nuclear fuel element** *Nuclear.* see FUEL ELEMENT.

**nuclear fuel reprocessing** *Nuclear.* the chemical treatment of spent reactor fuel to separate uranium and plutonium from the small quantity of fission product waste products and transuranic elements, leaving a much reduced quantity of high-level waste; the recovered uranium and plutonium may be recycled into new fuel elements.

**nuclear fusion** see FUSION.

**nuclear incident** see INCIDENT.

**nuclear magnetic resonance (NMR)** *Nuclear.* a phenomenon in which atomic nuclei spin around the axis of a strong magnetic field. The spinning nuclei create oscillating magnetic fields and emit a detectable amount of electromagnetic radiation; this can be used to visualize body structures and analyze their function.

**nuclear magneton** *Physics.* a unit of measure of the strength of the magnetic moment of baryons.

**nuclear material** *Nuclear.* any solid, liquid, or gas that emits radiation spontaneously.

**Nuclear Non-Proliferation Treaty (NPT)** *Policy.* an international treaty whose objective is to prevent the spread of nuclear weapons and weapons technology, promote cooperation in the peaceful uses of nuclear energy, and further the goal of achieving nuclear disarmament.

**nuclear physics** *Physics.* the branch of physics that studies the properties and behavior of atomic nuclei.

**nuclear power** *Nuclear.* **1.** see NUCLEAR ENERGY. **2.** electricity generated by a nuclear fission or nuclear fusion reactor; e.g., the use of the energy from fission to produce steam that will drive a turboelectric generator. Thus, **nuclear power plant.**

**nuclear proliferation** *Global Issues.* an increase in the world's arsenal of nuclear weapons, either through their development or acquisition by states and subnational groups that do not now possess them, or through further development by existing nuclear powers.

**nuclear radiation** *Nuclear.* particulate and electromagnetic radiation emitted from atomic nuclei in various nuclear processes; e.g., alpha and beta particles, gamma rays, and neutrons.

**nuclear reaction** *Nuclear.* **1.** a change in the nucleus of an atom brought about by subjecting it to an impact with other nuclei, or an impact with subatomic particles or high-energy electromagnetic radiation. **2.** the decay of a radioactive material.

**nuclear reactor** see REACTOR.

**Nuclear Regulatory Commission** (est. 1974), an independent U.S. agency to regulate civilian use of nuclear materials. NRC's primary mission is to safeguard public health and protect the environment from the effects of radiation from nuclear reactors, materials, and waste facilities.

**nuclear shield(ing)** see RADIATION SHIELD.

**nuclear steam supply system (NSSS)** *Nuclear.* a nuclear reactor system consisting of the reactor itself and the equipment required to transform heat from the reactor into steam to drive a turbogenerator.

**nuclear threat** *Nuclear.* **1.** the possible danger to society from the use of nuclear weapons. **2.** the possible danger to society and to the environment from the release of radiation from a nuclear power plant.

**nuclear waste** *Nuclear.* liquid or solid material remaining as products of commercial nuclear energy; this exhibits varying amounts of radiation and can be harmful to humans and the environment.

**Nuclear Waste Policy Act (NWPA)** *Nuclear.* the first comprehensive U.S. nuclear waste legislation (1982); it established a program for the disposal of high-level radioactive waste, including spent fuel from nuclear power plants.

**nuclear weapon** *Nuclear.* a weapon whose destructive power is released through nuclear fission or, in thermonuclear weapons, through fusion.

**nuclear weapons state (NWS)** *Nuclear.* one of the five nations recognized by the United Nations as having the right to own nuclear weapons: China, France, Russia, the United Kingdom, and the United States.

**nuclear winter** *Global Issues.* a proposed state of the environment theorized to result from a large-scale nuclear conflict, in which the smoke, dust, and other materials added to the atmosphere by nuclear explosions would drastically lower global temperature by reducing incoming solar energy.

**nucleon** *Nuclear.* a proton or a neutron, particularly as a component of the nucleus of an atom; a general term for either a proton or neutron.

**nucleonics** *Physics.* the branch of physics that studies the internal structure and behavior of atomic nuclei and the practical applications of this.

**nucleosynthesis** *Physics.* the creation of atomic nuclei in nuclear reactions at extremely high temperatures and pressures.

**nucleus** *Physics.* **1.** the positively charged central region of an atom, composed of neutrons and protons and containing nearly all the mass of the atom. **2.** *Bioenergetics.* a central cellular body found in nearly all living cells, involved in fundamental processes such as growth, metabolism, and reproduction.

**nuclide** *Physics.* a species of atom existing for a measurable length of time and distinguishable by the number of neutrons or protons it contains, and by the amount of energy in its nucleus.

**nutrient** *Biological Energetics.* food; any substance that provides the specific materials necessary for the growth, maintenance, and reproduction of living organisms.

**nutritional value** *Biological Energetics.* a measure of the amount of energy and nutrients that become available to the body from a given food product after ingestion.

**NWPA** Nuclear Waste Policy Act.

**NWS** nuclear weapons state.

**NYMEX** New York Mercantile Exchange (est. 1882), the largest physical commodity exchange in the world. The prices quoted for transactions on the Exchange are the basis for the prices that people in many countries pay for commodities such as crude oil, heating oil, gasoline, natural gas, propane, electricity, coal, gold, silver, and aluminum.

**NYMEX futures (price)** *Economics.* the market-determined value of a futures contract to either buy or sell 1000 barrels of West Texas Intermediate (WTI), or some other light, sweet crude oil, at a specified time. The NYMEX futures price is a marker crude oil price because it forms the benchmark for many different crude oils, especially in the Americas.

**Oberth, Hermann** 1894–1989, German scientist considered to be one of the founders of modern rocketry; he was the first to describe the basic principles of spaceflight (1923). During World War II he joined his former student Werner von Braun at the V-2 rocket complex, and then joined him again after the war at the U.S. Army Ballistic Missile Agency.

**obligation to serve** *Electricity.* the responsibility of an electric utility to provide electric service to any customer who seeks that service, and is willing to pay the set rates.

**occupancy sensor** *Lighting.* a device used in a building to turn off lights when a given area is unoccupied. Infrared occupancy sensors detect heat given off by people present in the room, while ultrasonic sensors detect movement in the room; in the absence of heat or motion, the lights are turned off.

**occupational disease** *Health & Safety.* any disease resulting from a specific job or workplace, usually caused by the effects of long-term exposure to certain toxic substances or by continuous or repetitive physical acts.

**occupational health** *Health & Safety.* a branch of medicine concerned with the prevention of disease and injury among people at work; e.g., the identification and control of on-the-job hazards. Also, **industrial health.**

**occupational risk** *Health & Safety.* a risk or danger that is inherent in a particular occupation; e.g., inhalation of coal dust in coal mining, exposure to radiation in a nuclear plant, exposure to hydrogen sulfide in the petroleum industry. Also, **occupational hazard.**

**ocean energy** *Renewable/Alternative.* a collective term for any form of energy that is extracted from the ocean, including thermal energy from the difference between warmer surface waters and cooler deep waters, or mechanical energy from tides, waves, and currents. See below.

**ocean fertilization** *Climate Change.* the process of seeding waters of the open ocean with iron or other nutrients, with the goal of increasing phytoplankton growth so as to increase the uptake of carbon dioxide from the atmosphere into ocean waters.

**ocean mixing** *Climate Change.* the various processes (such as, winds, currents, and

**ocean energy** Oceans, constituting about 70% of the world's surface area, receive the majority of solar radiation, generating thermal gradients that drive currents and winds, and ultimately, waves. Evaporation maintains their overall salinity and a chemical energy differential between fresh and saline water. Gravitational pulls of the moon and the sun impart energy to the ocean in the form of tidal currents, which are significant only where basin geometries result in resonance near a 12 hour period. Oceans are potentially the largest source of renewable energy, but with imposing engineering and economic problems. During the early industrial period, small waterwheel mills used tidal elevation changes. In the 20th century, tidal plants utilizing large dams and turbine-driven generators were constructed for electrical power in France, Russia, and China. Non-tidal ocean currents are generally too slow to be economically attractive. Because wave energy is maximized at high latitudes and on the western continental coasts, research on wave energy extraction has been carried on for several decades by Norway and the United Kingdom. The thermal gradient between warm tropical surface waters and cold deep waters below has been shown to support a low pressure thermal power plant. Basic research has been conducted on the salinity gradient source but no useful systems have yet emerged. Wind energy farms well offshore have been installed in Scandinavia and elsewhere, solving some of the environmental problems associated with land installations. The ocean is the largest source of deuterium and tritium, potential fuels for fusion power generation.

**Richard Seymour**
University of California, San Diego

surface waves) operating in the ocean to equalize the distribution of heat, salt, and various chemicals (including pollutants) that enter ocean waters at various rates and different locations; the rate of ocean mixing affects the extent to which carbon dioxide is exchanged between the atmosphere and the oceans.

**oceanography** *Earth Science.* the scientific study of the oceans, including geology, chemistry, life forms, and physical processes such as, the motion of ocean waters.

**ocean thermal energy conversion (OTEC)** *Renewable/Alternative.* any of various techniques for extracting energy from the vertical temperature difference in the oceans; in principle, OTEC can be used to generate electricity, desalinate water, support deep-water mariculture, and provide refrigeration and air-conditioning.

**ocean thermal gradient (OTG)** *Geothermal.* the temperature difference between warmer surface waters of the ocean and colder deep waters, with deeper water likely to be 20–50°F colder; in principle, this temperature gradient can be utilized with various types of ocean thermal energy conversion systems (see previous).

**octane** *Oil & Gas.* **1.** any of a number of chemical compounds that have the general formula $C_8H_{18}$; colorless, highly flammable liquids, some of which are found in petroleum. **2.** an expression of the octane number (see next) of a fuel; e.g., a rating of 93 *octane* would be a high-octane fuel.

**octane number** *Oil & Gas.* an index of the ability of a fuel to resist preignition (knock) in an internal combustion engine, by giving the percentage of isooctane in a blend of this mixture and normal heptane that manifests the same antiknock characteristics as the fuel sample in question. Gasoline grades (such as premium) are based on the octane number of the fuel.

**Odeillo** *Solar.* the world's largest solar furnace (a facility that uses a series of concave mirrors to concentrate sunlight into an intense, focused beam), built at Font-Romeu in the eastern Pyrenees region of France. It generates temperatures in the range of 800–2500°C, and is used for research on solar materials and solar energy conversion.

**odorization** *Oil & Gas.* a process of adding a distinctive odor to natural gas so that its presence can be more easily detected.

**ODS** ozone-depleting substance.

**Odum, Eugene** 1913–2002, U.S. scientist, called the father of modern ecology for elevating the ecosystem to the dominant concept of this field. He made the relationship between human activity and "natural processes" a cornerstone of this ecosystem concept. Prior to Odum, ecology focused on small-scale studies of individual ponds, forests, lakes, marshes, and other systems that were examined in isolation from each other and from society.

**Odum, Howard** 1924–2002, U.S. ecologist and the brother of Eugene Odum, noted for his pioneering studies of energy flows in ecosystems, and for the application of these principles to energy use in society; i.e., the argument that society faces many of the same energetic constraints that confront other organisms and systems. He introduced EMERGY, a key concept in modern energy studies.

**Oersted, Hans Christian** 1777–1851, Danish physicist who discovered (1819) that a magnetic needle is deflected at right angles to a conductor carrying an electric current; this established a relationship between magnetism and electricity and initiated the study of electromagnetism.

**oersted** *Measurement.* a unit that represents magnetic field intensity, equivalent to the intensity at the center of a single-turn circular (1 cm radius) coil with a current of $1/2\pi$ abamp.

**off-grid** *Electricity.* describing an electrical system that is not connected to the main electrical grid; e.g., a stand-alone photovoltaic system.

**Office of Surface Mining (OSM)** *Mining.* a bureau of the U.S. Department of the Interior whose mission is to ensure that coal mines are operated to protect citizens and the environment during mining, to assure that land is restored to beneficial use following mining, and to mitigate the effects of past mining by pursuing reclamation of abandoned mine lands.

**official development assistance (ODA)** *Sustainable Development.* loans, grants, technical assistance and other forms of cooperation extended by governments of developed nations to developing countries, aimed especially at promoting sustainable development in these countries, particularly through natural resource conservation, environmental protection, and population programs.

**off-road** *Transportation.* describing a recreation vehicle designed for use on natural terrain rather than (or as well as) on conventional highways; e.g., on unpaved roads, dirt trails, and so on.

**offset** *Policy.* a form of credit-based emissions trading where a source makes voluntary, permanent emissions reductions that are surplus to any required reductions. Existing sources that create offsets can trade them to new sources to cover growth or relocation.

*offshore Offshore drilling rig in the Gulf of Mexico, including a helicopter port and platforms that can be raised or lowered.*

**offset ratio** *Environment.* the amount of offsite emissions reduction that must be secured by a polluting facility relative to an on-site emission increase.

**offshore** *Oil & Gas.* a general term for oil and gas industry operations taking place along a coastline (e.g., in Louisiana) or in open ocean waters (e.g., the North Sea field). Thus, **offshore drilling, offshore lease,** and so on.

**Ohain, Hans von** 1911–1998, German physicist who developed an operational turbojet engine that powered the first jet aircraft flight in August of 1939. Frank Whittle in England performed simultaneous and parallel work on jet airplanes, and both men are recognized as co-inventors of the jet engine.

**Ohl, Russell** 1898–1987, U.S. physicist who while working at Bell Labs discovered the P–N barrier in silicon (1940). The P–N junction was to play a crucial role in the ultimate development of the transistor and the solar cell.

**Ohm, Georg Simon** 1787–1854, German physicist who first described the fundamental relationship between voltage, current, and resistance (1827). See OHM'S LAW.

**ohm** *Electricity.* a standard unit of electrical resistance, equal to the resistance of a circuit in which an electromotive force of one volt will maintain a current of one ampere; it has been defined as the resistance to a flow of steady current that will be offered by a thread of mercury 14.4521 grams in mass with a length of 1.063 meters and an unvarying cross-sectional area of one square millimeter, at a constant temperature of 0 °C.

**ohmic heating** *Nuclear.* a method to provide initial heat for a tokamak fusion reactor in which the plasma can be heated to temperatures up to 20–30 million K through the current passing through the plasma. The heat generated depends on the resistance between the plasma and current.

**ohmic polarization** *Renewable/Alternative.* a loss of potential voltage in a fuel cell because of resistance to the flow of ions in the electrolyte and resistance to the flow of electrons through the electrode.

**ohmmeter** *Electricity.* a meter that measures electrical resistance in a circuit, as expressed in ohms; some types also measure current flow.

**Ohm's law** *Electricity.* a law stating that the voltage across an element of a DC circuit is equal to the current through the element multiplied by the resistance of the element; it can be written as: $I = E/R$, where $I$ is the current in amperes, $E$ is the applied voltage in volts, and $R$ is the resistance in ohms.

**oil** *Materials.* **1.** any of a wide variety of greasy, viscous, combustible substances that are liquid at room temperature or when slightly warmed, and insoluble in water; they may be derived from animal, vegetable, or mineral sources. **2.** a mixture of hydrocarbons that is the source of various fuels. See CRUDE OIL. **3.** see MOTOR OIL.

*oil* *Historic photo of an oil boom in Wichita Falls, Texas, at a time (1919) when this area was one of the leading oil-producing regions in the world.*

**oil barge** see BARGE.

**oil-base mud** *Oil & Gas.* a drilling mud containing oil as the liquid component; used to drill through clay formations or in very deep wells.

**oil blowout** see BLOWOUT (def. 1).

**oil crisis** *Health & Safety.* an increase in oil prices large enough to cause a worldwide recession or a significant reduction in global real gross domestic product (GDP). See next column.

**oil curse** see RESOURCE CURSE.

**oil dependence** *Social Issues.* **1.** the fact of a oil-exporting nation's economy and social system being overly dependent on this one source of revenue. **2.** conversely, the fact of an oil-importing nation being overly dependent on this one source of energy.

**oil embargo** *Policy.* the prohibition of oil commerce and trade with a certain country, usually declared by a group of nations; e.g., the 1973 OPEC embargo on oil shipments to Western countries, in particular the U.S. and the Netherlands, in retaliation for their support of Israel in the Yom Kippur War.

**oilfield** see FIELD (def. 4).

**oil fingerprinting** see FINGERPRINTING.

**Oil for Food** *Policy.* a UN Security Council resolution (1995) that allowed partial resumption of Iraq oil exports to finance the purchase of humanitarian goods (food, medicine, and so on); enacted to ameliorate an embargo on Iraqi oil exports that had been in effect as a result of Iraq's invasion of Kuwait.

**oil lake** *Oil & Gas.* a large quantity of oil spreading over a flat surface of the earth, as from seepage, a spill, or the dumping of waste oil. Huge oil lakes resulted from the destruction of Kuwaiti oil wells by retreating Iraqi forces during the Persian Gulf War.

**oil lamp** *Lighting.* a historic form of lamp employing some type of oil as the fuel; the earliest versions are thought to have used vegetable oils such as olive oil.

**oil-led development** *Sustainable Development.* a developing nation's extreme dependence on revenues derived from the export (and not the internal consumption) of petroleum, as

**oil crisis** For economic purposes, an oil crisis is defined as an increase in oil prices large enough to cause a worldwide recession or a significant reduction in global real gross domestic product (GDP) below projected rates by two to three percentage points. The 1973 and 1979 oil episodes both qualify as oil crises by this definition. The 1973 oil crisis caused a decline in GDP of 4.7% in the U.S., 2.5% in Europe, and 7% in Japan whilst the 1979 crisis caused world GDP to drop by 3%. An oil crisis could be precipitated by a rapid expansion in the global economy fueling greater consumption of oil or by a lack of spare production capacity causing demand to outstrip supply, or a combination of both. It could also be caused by a sudden disruption in oil supplies triggered by political events or acts of sabotage against oil installations in major oil-producing countries or even by a deliberate political decision by a major producer or a group of producers to cut supplies.

**Mamdouh Salameh**
Oil Market Consultancy Service
Haslemere, Surrey, UK

measured by the oil and gas sector's share of gross domestic product and total exports, and its contribution to central government revenues.

**oil patch** *Oil & Gas.* an oil industry term for specific regions of the U.S. noted for oil production and refining, such as Texas, Oklahoma, Louisiana, California, and Alaska.

**oil pit** *Oil & Gas.* a hole or cavity in the earth containing waste oil and other residue from drilling and/or refining operations.

**oil pond** another term for an OIL LAKE.

**oil pool** *Oil & Gas.* an economically useful accumulation of petroleum locally confined by subsurface geologic features.

**oil price shock** see PRICE SHOCK.

**oil refinery** see REFINERY.

**oil reserves** see PROVED RESERVES (def. 2).

**oil reservoir** see RESERVOIR (def. 3).

**oil sand** *Oil & Gas.* any porous stratum, unconsolidated porous sand formation, or sandstone rock containing bitumen that can be mined and converted to a liquid fuel. See below.

**oilseed crop** *Biomass.* any of various varieties of plants whose seeds **(oilseeds)** can be used as biomass energy feedstock, such as rapeseed, corn, or safflower.

**oil seep** *Oil & Gas.* a site at which petroleum spontaneously flows to the surface. Also, **oil spring.**

**oil shale** *Oil & Gas.* a general term applied to a group of fine black to dark brown shales rich enough in bituminous material (kerogen) to yield oil upon heating in a retort. Such a fuel is called SHALE OIL, a form of unconventional oil.

**oil spill** *Environment.* an accidental or unwanted release of oil into the environment from the receptacle in which it is being transported or stored (such as, a tanker at sea), especially such a release on a large scale that causes damage to a marine ecosystem. See next page.

**oil state** *Social Issues.* an oil-exporting nation, especially one whose economy is dominated by the business of petroleum extraction and export; e.g., Kuwait.

**oil sand** Oil sands are sand grains coated by water and clay, with bitumen, an especially heavy, viscous crude oil, filling intervening pore spaces. Oil sands are found in 16 major deposits around the world, the two greatest being Canada's Athabasca deposit and Venezuela's Orinoco deposit. The Athabasca deposit is in the province of Alberta, and with the Peace River and Cold Lake deposits, comprises one-third of the known oil reserves on earth, covering 141,000 km$^2$ and containing 1.7–2.5 trillion barrels of bitumen. The Albertan oil sands are the most commercially important on the planet, and their development has major socio-economic impacts. In 2003, over 1300 Aboriginal people were employed by oil sands developers, oil sands industries' revenues were $11 billion (CAD), and Alberta government royalties on them totaled $183 million. By June 2004, $28 billion had been cumulatively invested in oil sands development, creating 33,000 jobs. There are two methods for extracting bitumen from sand: traditional surface mining, and in situ recovery. The former employs giant shovels to excavate bituminous sands within 75 m depth of the surface. The bitumen is separated or "cracked" from its sedimentary matrix by the addition of hot water, mechanical agitation, and skimming. In situ recovery techniques have been pioneered to extract the 80–90% of the Athabasca bitumen lying deeper than 75 m. The most salient in situ recovery technique is Steam Assisted Gravity Drainage, wherein a massive injection of steam cracks the bitumen underground, facilitating its gravity-driven drainage to extraction points. Oil sands production is energy-intensive and environmentally upsetting. Surface mining strips vegetation and topsoil from the surface, and leaves cavernous mine pits. Oil sands mining creates mountains of tailings sands high in sodium, chlorides, and sulfates, which may leak to the environs. Oil sands production is invariably water-consumptive and may reduce the level of local water tables, lakes, and rivers. Hot wastewater released during the cracking process may thermally disrupt ecosystems. Particulate and gaseous emissions, including greenhouse gases such as carbon dioxide, and acidifying compounds such as sulfur dioxide, may be the most deleterious by-products from oil sands production.

**David Chanasyk and Maine McEachern**
University of Alberta

**oil spill** An accidental petroleum release into the environment. On land, oil spills are usually localized and thus their impact can be eliminated relatively easily. In contrast, marine oil spills may result in oil pollution over large areas and present serious environmental hazards. The major accidental oil input into the sea is associated with oil transportation by tankers and pipelines (about 70%), whereas the contribution of offshore drilling and production activities is minimal (less than 1%). Large and catastrophic spills releasing more than 30,000 tons of oil belong to the category of relatively rare events and their frequency in recent decades has decreased perceptibly. Yet, such episodes could pose the serious ecological risk (primarily for sea birds and mammals) and result in long-term environmental disturbances (mainly in littoral zones) and economic impact on coastal activities (especially on fisheries and mariculture). The public concern about oil spills in the sea have been clearly augmented since the tanker *Torrey Canyon* accident off the UK coast in 1967 when 100,000 tons of spilled oil caused heavy pollution of the French and British shores with serious ecological and fisheries consequences. Since then, an impressive technical, political and legal experience in managing the problem has been accumulated in many countries and at the international level, mainly through a number of Conventions initiated by the International Maritime Organization (IMO).

**Stanislav Patin**
Russian Federal Institute of Fisheries and Oceanography (VNIRO)

**oil swap** see COMMODITY SWAP.

**oil tanker** see TANKER.

**oil well** see WELL.

**old-growth** *Ecology.* describing a forest having some trees at least several hundred years old; a forest with large mature trees in the overstory, standing dead trees (snags), dead and decaying logs on the ground, and a multilevel canopy with trees of different ages (i.e., young as well as old) and usually different types.

**OLED** organic light-emitting diode.

**olefin** *Chemistry.* any of a class of unsaturated open-chain hydrocarbons containing one or more double bonds, obtained by cracking petroleum fractions at high temperatures.

**oleo-** *Materials.* a prefix meaning "oil" or "containing an oil".

**oligopoly** *Economics.* a market with few competing producers.

**oligotrophy** *Ecology.* the "youth" or first stage in the process of EUTROPHICATION; the condition of a lake or other body of water that is deficient in nutrients and therefore generally biologically unproductive. Thus, **oligotrophic.**

**omnibus** *History.* the original term for a bus, literally meaning "a vehicle for all".

**onboard hydrogen** *Hydrogen.* describing a technology in which the hydrogen fuel for a transportation vehicle is produced by a processor within the vehicle; contrasted with a DIRECT HYDROGEN system in which the fuel is produced separately and then stored in a fuel tank on the vehicle. Thus, **onboard hydrogen vehicle.**

**onboard reformer** *Hydrogen.* a fuel-reforming device directly installed in a vehicle to converts a fuel into hydrogen within the vehicle itself.

**once-through** *Consumption & Efficiency.* describing a heating process in which water is pumped from a source through the heat exchanger and then discharged; i.e., the water passes through only once rather than being recirculated. Thus, **once-through boiler, once-through system,** and so on.

**once-through cooling** *Consumption & Efficiency.* the process of disposing of waste heat from a power plant by a single process of drawing cold water (as from a nearby river or bay) into the system and then returning it to the same body of water.

**oncogenic** *Health & Safety.* causing cancer; capable of inducing a tumor or other neoplasm.

**oncology** *Health & Safety.* the branch of medical science concerned with the disease of cancer; the study of the physical, chemical, biological, and genetic properties of tumors, including their causes, pathology, and treatment.

**one-axis tracking** *Solar.* a solar energy system capable of rotating about one axis and thus tracking the sun from east to west. Compare TWO-AXIS TRACKING.

**one-fluid theory** *History.* an 18th-century conception of electricity (as described by Benjamin Franklin), according to which electrical charges are contained in a single natural fluid that is invisible and omnipresent. Matter having less than a normal amount of this fluid would be negatively charged and matter having an excess would be positively charged. Compare TWO-FLUID THEORY.

**Onsager reciprocal relations** *Thermodynamics.* an expression of the equality of certain relations between flows and forces in thermodynamic systems that are out of equilibrium, but where a notion of local equilibrium exists. This has been termed the "fourth law of thermodynamics" because it is a general mathematical description of irreversible chemical processes. [Developed by Norwegian chemist Lars *Onsager,* 1903–1976.]

**opacity** *Environment.* **1.** the fact of being opaque; i.e., of being less than fully transparent or translucent. **2.** specifically, the degree to which light is obscured by pollutant particles in the air.

**OPEC** Organization of Petroleum Exporting Countries (est. 1960), the eleven-nation organization that provides a major share of the global oil supply and collectively agrees on production and price targets. See below.

**OPEC basket** *Economics.* an aggregation or "basket" of seven crude oils sold by the Organization of Petroleum Exporting Countries (OPEC): Algeria's Saharan Blend, Indonesia Minas, Nigeria Bonny Light, Saudi Arabia Arab Light, Dubai Fateh, Venezuela Tia Juana, and Mexico Isthmus (a non-OPEC oil). The **OPEC basket price** is an arithmetic average of these oils.

**OPEC price band** *Economics.* the desired price range for the OPEC basket price of crude oil (see above) established in 2000 . According to this, OPEC basket prices above $28 per barrel for 20 consecutive trading days, or below $22 per barrel for 10 consecutive trading days, would result in production adjustments aimed to move the basket price back within the band.

**open-access** *Consumption & Efficiency.* describing a resource to which virtually any potential user can have access to gain benefits from it; e.g., an ocean fishery.

**open-cycle system** *Renewable/Alternative.* a system of ocean thermal energy conversion that directly boils warm surface seawater to create steam to drive a turbine, as contrasted with a **closed-cycle system** that uses the warm seawater to boil a separate working fluid (e.g., ammonia).

**OPEC** Organization of Petroleum Exporting Countries (est. 1960) the eleven-nation organization that supplies a major share of global oil production that collectively agrees on global oil production and/or price targets. The five original members were Iran, Iraq, Kuwait, Saudi Arabia and Venezuela, who were later joined by Algeria, Indonesia, Libya, Nigeria, Qatar, United Arab Emirates, Ecuador and Gabon. Ecuador and Gabon withdrew from the organization in 1992 and 1994, respectively. The Oil and Energy Ministers of the OPEC Members meet regularly to decide on the Organization's oil output level, and consider whether any action to adjust output is necessary to meet oil production and/or price targets in light of current oil market conditions. Member countries hold about 75% of the world's oil reserves, and supply almost 40% of the world's demand. As a result, OPEC's decisions play an important role in the determination of world oil prices. For example, on October 17, 1973, OPEC refused to ship oil to the U.S. who directly supported Israel in its conflict with Egypt, the Yom Kippur War (the embargo was soon expanded to the Netherlands, Portugal, South Africa, and Rhodesia). This embargo caused a fourfold increase in oil prices, which in turn caused inflation and an economic slowdown in oil-importing nations around the world. OPEC's power waxes and wanes. A higher oil price increases OPEC revenue in the short run, but also curtails demand, increases non-OPEC production, and encourages improved energy efficiency, all of which reduce demand for OPEC oil. However, with the majority of the world's remaining oil reserves, OPEC will continue to play a dominant role in the international oil market.

**Cutler Cleveland**
Boston University

**open fuel cycle** *Nuclear.* a nuclear fuel cycle in which there is no reprocessing of spent nuclear fuel; the spent fuel is treated as waste and disposed of rather than being used to produce fresh fuel as in a closed fuel cycle.

**open-hearth** *Consumption & Efficiency.* describing a steelmaking process in which the charge is laid in a furnace on a shallow hearth and heated directly by preheated gas with flat temperatures up to 1750°C, and also radiatively by the furnace walls; a widely used process from the late 19th through most of the 20th century, but now generally replaced by the BASIC OXYGEN PROCESS. Thus, **open-hearth furnace, open-hearth process,** and so on.

*open-hearth The open-hearth process employed at a plant in Gary, Indiana, one of the traditional centers of the U.S. Steel industry.*

**open pit** *Mining.* a term for a broad, massive, often funnel-shaped excavation that is open to the air.

**open-pit mining** *Mining.* a method of mining in which the surface excavation is open for the duration of mining activity, employed to remove ores and minerals near the surface by first removing the waste or overburden and then breaking and loading the ore. Used to mine thick, steeply inclined coalbeds and also used extensively in hard-rock mining for metal ores, copper, gold, iron, aluminum, and other minerals. Thus, **open-pit mine.**

*open-pit Various nations prohibit open-pit mining in certain protected areas, because of its significant negative impacts on the environment.*

**open system** *Thermodynamics.* any system that exchanges mass and energy with its surroundings.

**Operation Ajax** *History.* a secret British–U.S. plan (1953) to oust nationalistic (and supposedly pro-Soviet) Iranian Premier Mohammed Mossadegh, resulting in the restoration to power of pro-U.S. Reza Shah Pahlavi and the establishment of a new oil consortium channeling Iran's oil to the West. The U.S. then replaced Britain as the major foreign player in the Iranian oil industry.

**Oppenheimer, J. Robert** 1904–1967, U.S. physicist who served as scientific director of the Manhattan Project, the U.S. effort to develop an atomic bomb during World War II. After the War, Oppenheimer opposed developing the hydrogen bomb, and in 1953 he was accused of having Communist sympathies. Although found not guilty of treason, he lost his security clearance and his role with the Atomic Energy Commission was terminated.

**Oppenheimer–Phillips reaction** *Nuclear.* a nuclear reaction in which low energy deuteron approaches sufficiently close to a nucleus for the nucleus to strip the neutron from the deuteron but still repel the remaining proton.

**opportunity cost** *Ecology.* **1.** the opportunities or alternative resource uses given up by an organism as a result of a particular use of a resource. **2.** *Economics.* the cost of something in terms of foregone alternative opportunities (and the benefits that could be received from such opportunities).

**optical** *Lighting.* **1.** relating to light or vision, or to techniques, devices, or systems that are designed to utilize light and/or assist vision. **2.** relating to the science of optics.

**optical concentrator** *Photovoltaic.* a lens or mirror system that concentrates radiant solar energy in a concentrating photovoltaic system in order to increase electrical output.

**optical fiber** see FIBER OPTICS.

**optical pyrometer** *Measurement.* an instrument used to measure the temperature of a heated material by comparing the incandescent brightness of its surface with that of a test wire heated to a known temperature.

**optical thickness** *Measurement.* **1.** the product of the physical thickness of a transparent optical medium multiplied by its refractive index. **2.** *Solar.* the degree to which a cloud prevents light from passing through it, which depends on factors such as its physical composition, concentration, and vertical extent. Also, **optical depth.**

**optics** *Lighting.* the scientific study of light; the study of the generation and detection of light, the interaction of light with matter, and the use of light for various industrial, commercial, and scientific purposes.

**optimal depletion** *Economics.* according to neoclassical economic theory, the rate of extraction of a nonrenewable resource (such as coal) that is economically efficient over time. Also, **optimal extraction path.**

**optimal foraging theory (OFT)** *Ecology.* **1.** the prediction that animals will seek out food in a manner that leads to the highest energy return on investment (EROI); i.e., the amount of useful energy obtained relative to energy used. **2.** a similar model for human systems that explains various behaviors as the effort to obtain maximum benefit per unit cost; e.g., the pattern of a hunter-gatherer group trying to obtain food, or of a researcher trying to obtain information on the Internet.

**optimality** see OPTIMALIZATION.

**optimal price** *Economics.* a price on an energy good or service that incorporates all of the marginal costs to society of using that good or service.

**option** *Economics.* a contract that gives the holder the right, but not the obligation, to buy (or sell) a financial instrument or commodity at a predetermined price by or on a certain future date.

**optoelectronics** *Electricity.* the branch of electronics that deals with the relationship between electricity and optical power.

**orbit** *Physics.* **1.** a closed or nearly closed path that is the trajectory of a body acted upon by a force. **2.** the path followed by objects in space moving under gravity, such as the path of one celestial body around another (e.g., the earth around the sun) or the path of an artificial object around a celestial body (e.g., a communication satellite around the earth). **3.** the circular to elliptical path followed by a water particle affected by wave motion.

**orbital theory** *Physics.* a description of the nature and behavior of electrons in an atom or molecule, based on the propositions that each individual electron moves in a definite wave pattern around a nucleus and that it is possible to predict accurately the region in which an electron with a specific energy level will be located.

**ore** *Mining.* **1.** a general term for any naturally occurring material in the earth that will yield minerals of economic value. **2.** such minerals themselves.

**orebody** *Mining.* a well-defined, continuous mass of solid material containing enough of a valuable substance to make extraction economically feasible.

**organ dose** *Health & Safety.* the mean absorbed dose of a toxic agent or substance in a specified organ of the human body; e.g., the liver, spleen, or lungs.

**organic** *Chemistry.* **1.** relating to or involving chemical compounds containing carbon. Thus, **organic compound.** **2.** *Biomass.* having to do with or produced by living organisms, as opposed to inanimate matter. **3.** short for ORGANIC FARMING.

**organic agriculture** see ORGANIC FARMING.

**organic chemistry** *Chemistry.* the scientific study of substances that are compounds of carbon, such as foods or fuels.

**organic farming** *Consumption & Efficiency.* a system of agriculture in which organic products and techniques are used; e.g., the use of natural animal and plant products, such as manure, compost, or bone meal, instead of chemical fertilizers, or the control of plant pests by means of insects or birds that prey on them rather than by chemical pesticides. Thus, **organic foods, organic fertilizers,** and so on.

**organic light-emitting diode (OLED)** *Lighting.* a type of light-emitting diode (LED), made with organic thin films rather than silicon and other traditional semiconductor materials; provides a thinner and more compact display than LCD technology and requires much less power to operate.

**organic waste** *Environment.* a collective term for any plant or animal matter that is discarded or dispelled into the environment; some components of this can be recycled (e.g., as compost) or utilized as a source of biomass energy.

**Organization of Petroleum Exporting Countries** see OPEC.

**organochlorine** *Environment.* any of a class of chemicals with biocidal properties developed in the 1940s; the best known is DDT. Lacking natural breakdown pathways, they have accumulated in ecosystems.

**orientation** *Physics.* the position of any object with respect to some reference point; e.g., the angle that a solar energy collector has with respect to the horizontal or that it has to North.

**orifice** *Consumption & Efficiency.* a small opening or hole that controls the flow rate of a fluid into a cavity, especially the flow of fuel gas to a burner.

**orphaned** *Mining.* describing a mine that has been abandoned and for which no current owner or responsible party can be identified.

**orphan site** *Environment.* a term for a site contaminated by a release of hazardous substances, where the parties responsible for the contamination are either unknown, or are unwilling or unable (e.g., bankrupt) to carry out the needed remedial actions.

**Orsat apparatus** *Measurement.* a device used to determine the composition of a mixture of gases by passing the mixture through a series of solvents, each of which absorbs only one of the gases. Also, **Orsat analyzer.**

**oscillation** *Physics.* a periodic variation in the value of a physical quantity, especially a regular variation above and below some mean value; e.g., the voltage of an alternating current or the pressure of a sound wave.

**oscillator** *Conversion.* a circuit or device that produces oscillation, especially such a device used to convert DC energy into AC energy.

**oscilloscope** *Electricity.* an instrument that uses a cathode-ray tube or similar instrument to display fluctuating electrical signals on a fluorescent screen; applications include testing electronic equipment and monitoring electrical impulses from the heart or brain.

**osmosis** *Chemistry.* **1.** the passage of a pure solvent through a semipermeable membrane, from a solution in which it has higher concentration to one in which it has lower concentration; this tends to equalize the concentration on either side of the membrane. **2.** the manifestation of this process in living systems, such as the movement of water molecules in and out of cells past the cell wall.

**osmotic** *Chemistry.* relating to or involving a process of osmosis.

**osmotic pressure** *Chemistry.* the differential pressure that must be applied to a fluid on one side of a semipermeable membrane in order to prevent its diffusion through the membrane.

**Ossberger turbine** another name for a CROSS-FLOW TURBINE.

**Ostwald, Wilhelm** 1853–1932, German chemist noted for his work on catalysis and his investigations into the fundamental principles governing equilibrium and rates of reaction. He also incorporated thermodynamics into a general theory of economic development.

**OTEC** ocean thermal energy conversion.

**OTG** ocean thermal gradient.

**Otis, Elisha Graves** 1811–1861, U.S. inventor who installed the first passenger elevator (safety elevator) in New York City in 1857. His sons Charles and Norton established Otis Brothers, which became the leading firm in elevator manufacture.

**Otis shovel** *History.* an early type of steam shovel that became the foundation for mechanized construction. Technological descendants of the Otis shovel have been important in large-scale surface mining of coal, uranium, and oil sands. [Invented by U.S. engineer William S. *Otis,* 1813–1839.]

**Otto, Nikolaus August** 1832–1891, German engineer who was the coinventor of an internal-combustion engine, and the inventor of the four-stroke OTTO CYCLE, which provided the first practical alternative to the steam engine.

**Otto cycle** *Thermodynamics.* **1.** an ideal thermodynamic combustion cycle, as follows: a compression at constant entropy; a constant-volume heat transfer to the system; an expansion at constant entropy; and a constant-volume heat transfer from the system. The thermal efficiency of the ideal Otto cycle increases with an increasing compression ratio. **2.** an actual version of this cycle used in internal combustion engines. ☼ See below.

**Otto engine** *Transportation.* an engine employing the Otto cycle (see above).

**oustee** *Social Issues.* a term for people forced to leave their homes and/or livelihoods because they are displaced by large energy-related projects, such as the building of a hydroelectric, nuclear, or thermal power plant.

**outage** *Electricity.* **1.** a temporary suspension of electric service; a loss of electric power. **2.** specifically, such a loss that is unplanned and unintended.

**outcrop** *Coal.* the part of a coal deposit that is exposed at the earth's surface or that is covered only by minimal surface deposits.

**Outer Continental Shelf** *Earth Science.* the shallow portion of the continental margin that extends from the coastline to the more steeply inclined continental slope.

**output** *Electricity.* **1.** the electrical energy that emanates from a device, or the point at which this energy leaves the device. **2.** to convey or transfer electrical energy from a device. **3.** *Communications.* data that is transferred from a computer's main memory to some external source; e.g., a printer. **4.** to provide data from a computer. **5.** *Materials.* any substance or form of energy produced by a system.

**oven** *Consumption & Efficiency.* a general term for any enclosed or semi-enclosed compartment in which substances are artificially heated for such purposes as cooking, drying, or annealing.

**oven dry** see BONE DRY.

**overburden** *Mining.* any barren rock material or soil that overlies a useful mineral deposit or coal seam, and that must be removed before mining can commence.

☼ **Otto cycle** The Otto Cycle is named after Nikolaus Otto (1832–1891, German) who is credited as the first creator of a petroleum fuel based internal combustion engine operating under a four stroke cycle. Otto's engine work during the eve of the industrial revolution was a critical development leading to the dawn of the automotive industry in the late 19th century. The Otto Cycle involves four engine strokes. A stroke is defined as a process in which the cylinder's piston moves up to the top of the engine or down to the bottom of the engine. The cycle begins with the intake stroke in which fresh reactants (fuel and air) are drawn into the engine's cylinder by a downward-expanding piston motion. The second stroke is compression in which the piston moves from the bottom to top of the engine thereby reducing the volume of the fuel and air. Compressing the reactants before burning them allows for greater work and power to be extracted from the following combustion process. When the piston is near the top of its stroke combustion begins and expansion of the burned gases continues into the expansion stroke. The last of the four strokes is the exhaust stroke in which the burned products of combustion are pushed out of the engine in order to make space for the next cycle's intake stroke.

Today the Otto Cycle is frequently associated with an ideal cycle analysis of automotive spark ignition gasoline internal combustion. This ideal cycle assumes that the combustion process occurs instantaneously at constant volume. To a first order this Otto Cycle analysis can provide insights into how engine efficiency is affected by changing operating and design parameters. Engines operating under the Otto Cycle are the principal power-plants used in automobiles worldwide at the present time, although in some regions of the world the Diesel Cycle is a very close second in automotive applications. Engines operating under the Otto cycle are relatively inexpensive to manufacture and provide reasonable power to weight ratios with good efficiencies. In recent years engine design improvements coupled with exhaust after-treatment have resulted in vehicles that produce very low levels of pollutants.

**Chih Wu and Jim S. Cowart**
United States Naval Academy

**overburden ratio** *Coal.* the amount of overburden that must be removed, in relation to the amount of coal that will be obtained from the bed below.

**overcast** *Mining.* **1.** an enclosed mine airway made of concrete, tile, stone, or other incombustible material, allowing one air current to pass over another without interruption. **2.** in surface mining, to shift the overburden from an active site to another site from which the coal has already been mined.

**overcharging** *Storage.* the fact of attempting to store more charge into a secondary cell or battery than its electrochemical system can safely absorb; repeated overcharging can cause physical or chemical damage, often reducing the effective life of the battery. Undercharging a battery can produce the same problems.

**overdischarging** *Storage.* the fact of discharging a cell or battery beyond its cutoff voltage and possibly into voltage reversal, which can shorten its working life or in extreme cases cause irreparable damage.

**overdrive** *Transportation.* **1.** a method of reducing engine revolutions per minute in relation to vehicle speed, using a separate epicyclic gear unit. **2.** the gear unit used in this process.

**overfire** *Consumption & Efficiency.* air introduced into a furnace above its combustion zone with the objective of completing combustion.

**overload** *Electricity.* **1.** a load greater than an electrical system can accommodate or that it is designed to carry. **2.** to provide such a load to a system.

**overpotential** another term for OVERVOLTAGE.

**overshot wheel** *Hydropower.* a vertical water wheel turned on a horizontal shaft by the force of water pouring from above onto buckets attached to its circumference; more efficient than an UNDERSHOT wheel but more difficult to build and less suited to a variety of sites.

**overspeed** *Wind.* a protective feature of a wind energy device that is activated by a wind velocity that is higher than a predetermined maximum, operating to deactivate, realign, or regulate the device so as to prevent physical damage and electrical overload. Thus, **overspeed protection.**

**overvoltage** *Electricity.* a voltage that is higher than the normal or predetermined limiting value for a system.

**ovonic** *Renewable/Alternative.* relating to or involving a device that converts heat or sunlight directly to electricity, by means of a unique glass composition that changes from an electrically non-conducting state to a semiconducting state. [Named for U.S. inventor Stan *Ovshinsky.*]

**Ovshinsky, Stan(ford)** born 1922. U.S. inventor noted for many innovations in various areas, such as batteries, solar cells, superconductors, and hydrogen-powered vehicles.

**Ovshinsky effect** *Materials.* an effect by which a specific glassy thin film switches from a nonconductor to a semiconductor upon application of a minimum voltage; this has led to important advances in the engineering of semiconductors, solar energy, and electric cars.

**own-price elasticity** *Economics.* changes in the quantity demanded for a good due to changes in the price of that good, holding other things constant; e.g., changes in the quantity of motor gasoline demanded as a result of a change in its price.

**ox** *Biological Energetics.* plural, **oxen.** an adult castrated bovine male, widely used since ancient times as a draft animal.

**oxidant** *Chemistry.* **1.** any agent that carries out a process of oxidation or converts a substance to an oxide. **2.** specifically, a substance, usually containing oxygen that supports the combustion reaction of a rocket fuel. Together, the fuel and the oxidizer constitute a propellant. **3.** *Health & Safety.* a collective term for products of the reactions of volatile organic compounds with nitrogen oxides, which impact human health, agriculture, and ecosystems. The most important oxidant is ozone; hydrogen peroxide and peroxy acetyl nitrate are other examples.

**oxidation** *Chemistry.* any reaction resulting in the removal of one or more electrons from a species, thus increasing its valence **(oxidation state)**; oxidation always occurs simultaneously with reduction, in which another species gains the electrons lost from the oxidized species. In an electrochemical cell, oxidation occurs at the anode

**oxidation state** see OXIDATION.

**oxidation-reduction** *Chemistry.* the release of energy in a chemical reaction in which electrons are transferred from one species, which is thereby oxidized (i.e., it loses electrons) to another species which is thereby reduced (i.e., it gains electrons).

**oxidative metabolism** *Biological Energetics.* the cellular production by organisms of energy-containing adenosine triphosphate (ATP) from foodstuffs by means of an oxygen-using process.

**oxide fuel** *Nuclear.* enriched or natural uranium in the form of the oxide $UO_2$, used in many types of reactor.

**oxidize** *Chemistry.* **1.** to convert a substance into an oxide; combine with oxygen. **2.** to undergo or cause to undergo oxidation; lose or remove electrons.

**oxidizer** another term for an OXIDANT.

**oxyfuel combustion** *Renewable/Alternative.* a method of carbon dioxide sequestration from power plants in which the fuel is burned with oxygen, leaving a high-purity $CO_2$ flue gas for capture.

**oxygen** *Chemistry.* a gaseous element having the symbol O, the atomic number 8, an atomic weight of 15.9994, a melting point of –218.4°C, and a boiling point of –182.962°C; a colorless, odorless, tasteless gas that is the most abundant element on earth, about 20% by volume of the atmosphere at sea level, about 50% of the material of the earth's surface, and about 90% of water. Oxygen readily forms compounds with nearly all other elements except inert gases, and it is used for many industrial purposes. See also FREE OXYGEN.

**oxygenate** *Oil & Gas.* **1.** any of various chemicals containing oxygen that are added to fuels, especially gasoline, to make them burn more efficiently and thus produce less pollution. The most common oxygenates are methyl tertiary-butyl ether (MTBE) and ethanol (alcohol). **2.** to combine or treat a substance with oxygen. Thus, **oxygenated fuel.**

**oxygenated gasoline** *Oil & Gas.* conventional gasoline to which chemicals that are rich in oxygen have been added to increase the octane number and/or to meet clean air regulations by reducing carbon monoxide emissions.

**oxygen steelmaking** *Materials.* **1.** any steelmaking process in which oxygen gas rather than nitrogen or hydrogen is used to remove carbon and phosphorus from pig iron. Thus, **oxygen steel.** **2.** see BASIC OXYGEN PROCESS.

**ozone** *Chemistry.* $O_3$, a triatomic form of oxygen; a pungent, unstable blue gas that is formed naturally in the upper atmosphere; it constitutes a protective layer against excess ultraviolet radiation. Ozone in the troposphere (ground level ozone) is a harmful pollutant formed primarily by reactions of hydrocarbons and nitrogen oxides; it is a major ingredient of photochemical smog.

**ozone-depleting substance (ODS)** *Environment.* a classification for chemical compounds that contribute to the loss of ozone in the upper atmosphere; major examples include CFCs, HCFCs, halons, methyl bromide, carbon tetrachloride, and methyl chloroform. They are generally stable in the troposphere and only degrade under intense ultraviolet light in the stratosphere, at which time they release chlorine or bromine atoms, which in turn deplete ozone. Thus, **ozone depletion.**

**ozone-depletion potential (ODP)** *Environment.* a description of the potential of a given substance to contribute to the loss of ozone in the stratosphere, measured against the impact of a similar mass of CFC-11 (a classic ozone-depleting substance) as a standard of 1.0; e.g., the ODP of carbon tetrachloride is 1.2.

**ozone hole** *Earth Science.* a term for the recently observed thinning of the ozone layer (see below) over the Antarctic region during the spring months. The ozone hole has grown in size and annual length of existence over the past two decades.

**ozone layer** *Earth Science.* a layer of the upper atmosphere where most atmospheric ozone is concentrated, about 10–50 km above the earth with a maximum concentration in the stratosphere at an altitude of about 25 km. This layer performs the essential task of filtering out most of the sun's biologically harmful ultraviolet (UV-B) radiation; it is now regarded as subject to depletion by certain industrial products, especially chlorofluorocarbons (CFCs) from aerosol sprays.

**ozone precursor** *Environment.* a chemical compound, such as carbon monoxide, methane, non-methane hydrocarbons, and nitrogen oxides, that in the presence of solar radiation will react with other chemical compounds to form ozone.

**packaged air conditioner** *HVAC.* a complete, self-contained air-conditioning unit assembled into one casing.

**PAD** Petroleum Administration for Defense.

**PADD** Petroleum Administration for Defense District.

**paddlewheel** *Transportation.* a steam vessel propelled by rotating, horizontal-axis wheels that enter the water more or less perpendicularly to propel the ship.

*paddlewheel The historic paddlewheel packet boat* Gordon Greene, *seen on the Mississippi River near Baton Rouge, Louisiana.*

**PAFC** phosphoric acid fuel cell.

**PAH** polycyclic aromatic hydrocarbon.

**Pahlavi, Mohammed Reza** 1919–1980, Shah of Iran from 1941 to 1979, including a period of tumultuous change in the world oil market. In 1979 religious opposition, led by Ayatollah Ruhollah Khomeini, drove the Shah into exile, setting off a chain of events that contributed to a sharp rise in oil prices and eventually triggered a global recession.

**paint pot** another term for MUD POT.

**pair production** *Nuclear.* a process in which a high-energy photon, generally interacting with an atomic nucleus, produces a particle and an antiparticle; this is the chief method by which energy from gamma rays is observed in condensed matter.

**paludification** *Biomass.* the process of development of a peat bog; the decaying of plant matter to form a bed of peat.

**palynology** *Climate Change.* the scientific study of microscopic fossils (spores, pollen grains, and other such organic materials); e.g., as a means of obtaining data relating to past flora and past climate.

**Panamax** *Oil & Gas.* Panama maximum; the largest size oil tanker that can safely navigate the Panama Canal, given the limitations of the canal locks.

**pancaking** *Economics.* the payment of multiple electricity transmission rates for a single transaction across multiple systems. For example, to deliver energy to a load that is located far from the remote site where the power was generated, a wind farm may have to use the transmission systems of multiple operators, each of whom can collect access charges.

**panel** *Mining.* the term for a large rectangular block of coal that is isolated in the process of panel mining (see below).

**panel heat(ing)** *HVAC.* a heating system utilizing floor, wall, or roof panels containing electric coils or pipes that conduct hot air, water, or steam. Similarly, **panel cooling.**

**panel mining** *Mining.* a coal mining technique used in longwall mining, in which a moveable roof support is used to hold up the overburden while the coal is removed; the support is then moved to allow the overburden to collapse.

**panemone** *Wind.* a classic type of windmill design, involving a vertical axis and blades that can react to the wind's direction to maintain a relatively constant angle of attack into the wind.

**Panhandle** *Oil & Gas.* a vast oil field north of Amarillo, Texas that began operation in 1918, historically one of the most productive oil and gas fields in the world.

**paper barrel** *Economics.* a term for trade in non-physical oil markets (futures, forwards, swaps), giving a buyer or seller the right to a certain quantity and quality of crude oil or refined products at a future date, but not to any specific physical lot.

**Papin, Denis** 1647–1712, French physicist whose 1679 steam digester (pressure cooker) demonstrated the influence of atmospheric pressure on boiling points and contributed to the development of the steam engine.

**PAR** parabolic aluminized reflector.

**parabola** *Physics.* a curve formed by a set of points equally distant from a given line and a given point not on the line. Thus, **parabolic.**

**parabolic aluminized reflector (PAR)** *Lighting.* a precision reflector lamp having a lens of heavy durable glass; it may utilize either an incandescent filament, a halogen filament tube, or HID arc tube. PAR lamps rely on both the internal reflector and prisms in the lens for the control of the light beam and are typically used for outdoor floodlighting.

**parabolic collector** *Solar.* a solar energy device having a parabola-shaped reflector (trough) that focuses the sun's direct beam radiation onto a linear receiver located at the focus of the parabola. A heat transfer fluid is heated as it circulates through the receiver, with the captured energy used to generate steam for power generation or for process heat. Also, **parabolic trough, paraboloidal collector.**

**parabolic cooker** *Solar.* a type of solar cooker that uses a parabola-shaped reflector to concentrate solar energy and a control arm that holds a pot or baking dish at the focal point; a stand holds the two components together and allows the cooker to rotate to follow the sun across the sky. Also, **paraboloid cooker.**

**parabolic dish** *Solar.* a solar thermal technology that uses a modular mirror system that approximates a parabola and incorporates two-axis tracking to focus the sunlight onto receivers located at the focal point of each dish. The concentrated sunlight may be used directly by a heat engine at the focal point of the receiver, or used to heat a working fluid that is carried to a central engine.

**parabolic mirror** *Solar.* **1.** a mirror whose reflecting surface is in the shape of a parabola; all the rays striking the mirror are reflected to pass through the same point (the focus of the parabola). Also, **paraboloidal mirror.** **2.** a form of this reportedly used as a weapon in ancient times. See BURNING MIRROR.

**parabolic reflector (PR)** *Lighting.* **1.** any reflector type light source whose reflecting surface has the shape of a parabola to focus and project a light beam. **2.** see PARABOLIC COLLECTOR.

**paraboloid** *Physics.* a reflecting surface that is formed by the revolution of a PARABOLA about its axis of revolution. Thus, **paraboloidal.**

**paraffin** *Materials.* any of a class of saturated aliphatic hydrocarbons having a straight or branched carbon chain and the general formula $C_nH_{2n+2}$, ranging in physical form from methane gases to waxy solids; found in some petroleum deposits. Heavy paraffins occur as wax-like substances that may build up on producing oil wells and may, if severe, restrict production.

**paraffinic naphtha** *Oil & Gas.* a less dense naphtha with a higher paraffin content, used as a feedstock for petrochemical production.

**paraffin oil** *Materials.* a light-colored, combustible wax-free oil that is either pressed or dry-distilled from paraffin distillate; used in floor treatment and as a lubricant.

**paraffin wax** *Materials.* ordinary paraffin in its solid state, a combustible, translucent, tasteless and odorless, white hydrocarbon that is obtained from paraffin distillate; used for various industrial purposes.

**parallel** *Electricity.* describing a side-by-side connection in which the same voltage is applied to all components. Thus, **parallel circuit.**

**parallel connection** *Electricity.* a way of joining solar cells or photovoltaic modules by connecting positive leads together and negative leads together; this configuration increases the current, but not the voltage.

**parallel hybrid** *Renewable/Alternative.* a type of hybrid electric vehicle in which the alternative power unit is capable of producing motive force and is mechanically linked to the drivetrain.

**parallel path flow** *Electricity.* a flow of electrical power on an electric system's transmission facilities, resulting from scheduled power transfers between two other systems.

**parallel series** *Electricity.* a circuit in which two or more parallel circuits are connected in series.

**paramagnetism** *Physics.* the property of a substance having a magnetic permeability greater than that of a vacuum but significantly less than that of ferromagnetic materials such as iron; the permeability of such a substance is considered to be independent of an external magnetizing force. Thus, **paramagnet, paramagnetic.**

**parasitic load** *Storage.* the natural self-discharge of a battery; the constant electrical load that is present even when no power demand is being made upon it by a user.

**parasitism** *Ecology.* a biological relationship between individuals of two different species in which one (the parasite) benefits and the other (the host) does not and typically is harmed.

**Pareto efficiency** *Economics.* an allocation of goods and services in such a manner that there is no feasible alternative that would make some other individual better off and no other individual worse off. Also, **Pareto optimality.** [Named for Italian economist Vilfredo *Pareto,* 1848–1923.]

**PAR (Par) lamp** *Lighting.* short for PARABOLIC ALUMINIZED REFLECTOR lamp.

**Parsons, Charles Algernon** 1854–1931, British mechanical engineer who invented the steam turbine, which greatly increased the efficiency of converting steam into power. His turbine-driven marine engines provided much higher speeds for British warships.

**partial extraction** *Mining.* a type of underground coal mining in which pillars are left in place over time to support the surface, so as to avoid surface subsidence.

**partial oxidation** *Consumption & Efficiency.* a combustion process in which the fuel is partially oxidized to carbon monoxide and hydrogen rather than fully oxidized to carbon dioxide and water; a common process for the production of hydrogen since it can be dissociated from a fuel without being completely consumed.

**partial pressure** *Chemistry.* the portion of the total gas pressure of a mixture that is attributable to one component of the mixture.

**particle** *Physics.* **1.** any finite object that may be considered to have mass and an observable position in space. **2.** specifically, a minute subdivision of matter constituting a fundamental entity so small that it cannot be further subdivided.

**particle accelerator** *Physics.* a device used to produce high-energy, high-speed beams of charged particles, such as electrons, protons, or heavy ions, for research in high-energy and nuclear physics, synchrotron radiation research, medical therapies, and some industrial applications. The world's largest is at CERN (see).

**particle flux** *Earth Science.* **1.** a measure of the rate of settling of particles in the ocean. **2.** a stream of high-speed protons and electrons leaking away from the sun's outer layers.

**particle physics** *Physics.* the branch of physics concerned with the subatomic particles that represent the smallest known constituents of matter.

**particulate matter (PM)** *Materials.* **1.** dust, dirt, soot, smoke and liquid droplets directly emitted into the air by sources such as factories, power plants, cars, engines, construction activity, fires, and natural windblown dust. PM is considered a significant respiratory health risk. **2.** a similar composition of particles formed in the atmosphere by condensation or transformation of emitted gases. **3.** any matter composed of particles that are not superficially bound together, such as sand.

**particulate trap** *Transportation.* an emission control device for diesel vehicles that traps and incinerates diesel particulate emissions after they are exhausted from the engine, but before they are expelled into the atmosphere.

**partitioning** *Nuclear.* the separation of nuclear waste into its constituents.

**pascal** *Measurement.* a standard unit of pressure in the meter-kilogram-second system, equal to the pressure of a force of one newton per square meter. [Named for French mathematician Blaise *Pascal,* 1623–1662, who conducted important early studies of pressure and vacuum conditions.]

**Pascal's law** *Physics.* a law stating that a fluid in equilibrium contained in a vessel externally exerts a uniform isotropic pressure, i.e., equal intensity in all directions.

**passenger-mile** *Transportation.* a measure of traffic carried, based on a unit of one passenger transported one mile; e.g., a bus carrying 50 passengers on a 20-mile trip provides 1000 passenger miles of service. Used to compare different modes of transportation; e.g., fatalities per million passenger miles, or energy use per passenger-mile.

**passivation** *Electricity.* **1.** the practice of growing a thin oxide film on the surface of a semiconductor to protect exposed elements from environmental contaminates, thus ensuring the electrical stability of the device. **2.** a major decrease in the reactivity of a surface; e.g., a solar cell.

**passive** *Solar.* describing systems in which solar energy is somehow captured, stored and distributed by the building; contrasted with ACTIVE systems using collectors, tanks, and pumps.

**passive cooling** *HVAC.* a process of space cooling achieved without mechanical devices, through the use of design features such as shading, unassisted ventilation, convection, or evaporation.

**passive heating** *HVAC.* a process of space heating without mechanical devices (such as gas or electric heat) that typically involves the capture and use of heat from the sun. See also PASSIVE SOLAR HEAT.

**passive regeneration** *Consumption & Efficiency.* a process in which the soot in a diesel particulate filter spontaneously burns off during the normal work cycle because of the heat of the exhaust temperatures.

**passive safety** *Nuclear.* a system that provides nuclear core or containment cooling by natural processes such as buoyancy, condensation, and evaporation, or by a system of prepressurized tanks filled with coolant.

**passive solar cooling** *Solar.* the converse of passive solar heating; a solar energy system that provides cooling without the use of a separate mechanical or storage facility; e.g., a pond of water on a flat roof that cools by means of evaporation.

**passive solar design** *Solar.* structural design and construction techniques that enable a building to utilize solar energy for heating, cooling, and lighting by nonmechanical means.

**passive solar energy** *Solar.* the use of solar energy by PASSIVE means to reduce the heating demand of a building. See below.

**passive solar heat(ing)** *Solar.* a solar energy system that utilizes the structure of a building and its operable components to collect, store, and distribute thermal (heat) energy as appropriate without any mechanical redistribution or storage of the energy. Thus, **passive solar energy, passive solar heater, passive solar system,** and so on.

**passive ventilation** *HVAC.* a process of ventilation without mechanical devices such as blowers or fans; typically created by

**passive solar energy** The use of solar energy by passive means to reduce the heating demand of a building. This term was popularized by the 1st National Passive Solar Conference in Albuquerque, NM, US May 18–19, 1976. Solar energy is captured, stored and distributed by the building in contrast to active systems using collectors, tanks and pumps. Three passive strategies were defined: The most common strategy is direct solar gain through south-facing windows. The building interior is typically massive to store daytime solar heat for nights. Roof overhangs cut off the high summer sun. In indirect gain systems, solar radiation is absorbed outside the heated space. An attached sunspace is an example. Another example is an uninsulated massive wall, painted black and protected from the ambient by a glass facade. Heat from the sun-warmed wall radiates to the interior space with a time lag, ideally shifting the gain to night hours. An improvement is to use transparent insulation to reduce night losses by the wall. In the third strategy, isolated gain, solar absorption occurs outside the insulated envelope. An example is solar air collectors below a heated space so sun heated air rises naturally up into the building interior. By each strategy success depends on useful gains exceeding losses. Overdimensioning can lead to overheating, even in winter, net heat loss, and comfort problems. New glass technologies enable positive energy balances for orientations towards east and west as well as due south. Numerous design tools are available today to optimize such systems.

**Robert Hastings**
Architecture, Energy & Environment Ltd., Switzerland

differences in the distribution of air pressures around a building as the air moves from areas of high pressure to areas of lower pressure, with gravity and wind affecting this airflow.

**pastoral** *Sustainable Development.* describing a society whose mode of life is based on maintaining herds of domesticated animals that are pastured over large areas, with seasonal migrations that track the availability of water and animal food; e.g., the traditional Mongol society of central Asia. Thus, **pastoralism.**

**patent fuel** *Coal.* a composition fuel manufactured from small coal particles compressed into a solid block (briquette) with the addition of a binding agent.

**path dependency** *Sustainable Development.* **1.** the tendency for technological change to continue to occur on the basis of existing patterns and acquired knowledge; i.e., along the same path. **2.** specifically, persistent differences in development patterns among nations, resulting from variation in initial conditions and determining factors (e.g., economic, institutional, and technical).

**Pauli, Wolfgang** 1900–1958, Austrian–American physicist noted for proposing (1924) a new quantum degree of freedom to resolve inconsistencies between observed molecular spectra and the developing theory of quantum mechanics. He was also the first to recognize the existence of the NEUTRINO.

**Pauli exclusion principle** *Physics.* the statement that no two identical fermions may occupy the same quantum state simultaneously.

**PAW** Petroleum Administration for War.

**payback** *Economics.* **1.** the time required for an energy-using device (e.g., an appliance) to recover the initial cost of its purchase, or an energy-producing facility (e.g., a power plant) to recover the cost of its construction. **2.** also, **energy payback (time).** the time required for an energy-producing system to produce as much energy as has been consumed directly and indirectly in its manufacture, operation, and maintenance. Thus, **payback time, payback period.**

**payload** *Transportation.* the maximum allowable weight of cargo (or passengers and cargo) that a vehicle is rated to carry.

**PBL** planetary boundary layer.

**PCBs** *Materials.* polychlorinated biphenyls; any of a group of toxic, chlorinated aromatic hydrocarbons once widely used in a variety of commercial applications, including fire retardants, paints, electrical condensers, batteries, and lubricants. Now discontinued in many countries because of their toxic effects (e.g., association with skin diseases, birth defects, and cancer) and persistence in the environment.

**PCM** phase-change material.

**$pCO_2$** *Earth Science.* partial carbon dioxide; a measurement of the partial pressure of $CO_2$ in the atmosphere as compared to that of the ocean. The partial pressure of $CO_2$ in the atmosphere is the pressure that the $CO_2$ would exert if all other gases were removed; the partial pressure of $CO_2$ in the ocean is the amount of dissolved $CO_2$ and $H_2CO_3$.

**PCR** polymerase chain reaction.

**pea coal** see COAL SIZING.

**peak** *Consumption & Efficiency.* **1.** the highest point of production of a given energy resource; e.g., oil. **2.** the maximum demand for power from a system during a given time period. Thus, **peaking, peak power,** and so on. See next page.

**peaking unit** *Electricity.* a power generator used by a utility to produce additional electricity during peak load times.

**peak load** *Electricity.* the greatest amount of power given out or taken in by a machine or power distribution system during a given time period.

**peak load plant** *Electricity.* a plant normally used during peak-load periods; typically housing older, lower-efficiency steam units, gas turbines, diesels, or pumped-storage hydroelectric equipment.

**peak-load pricing** *Economics.* a system of charging higher rates for power used during high-use periods, to discourage use at those times and reduce a system's peak demand.

**peak shaving** *Consumption & Efficiency.* any of various strategies to reduce the amount of electricity (or another energy source) purchased during those periods when demand is greatest; e.g., by purchasing additional capacity when demand is low and storing it to be released when demand is high, or by shedding nonessential loads during peak periods.

**peak oil** Peak Oil was formed but rarely in time and place in the earth's geological past, meaning that it is a finite resource subject to depletion. It has to be found before it can be produced. The peak of discovery in the 1960s therefore heralds a corresponding peak of production. The larger fields were found first in most areas. Production is also constrained by the physics of the reservoir. The production profile in a country or region with a large population of fields is normally symmetrical, with peak coming when half the total has been produced. Gas follows a different trajectory with a steep terminal decline, with the World peak coming a few years after oil. Public data on oil and gas reserves are grossly unreliable, subject to both over-reporting and under-reporting in different countries, which allows economists to argue that production is simply a function of investment and technology. The true state of affairs would otherwise be almost self-evident. The World is in fact now very close to Peak, spelling the End of the First Half of the Oil Age. It lasted 150 years and saw the growth of industry, transport, trade, agriculture, and financial capital, made possible by an abundant flow of cheap oil-based energy. The Second Half now dawns, and will be characterised by the decline of oil and all that depends upon it. Peak Oil is accordingly an unprecedented historic discontinuity with grave consequences.

**Colin Campbell**
Association for the Study of Peak Oil

**peak shifting** *Electricity.* the process of moving existing loads to off-peak periods.

**peak sun hours** *Solar.* the equivalent number of hours per day when solar irradiance averages 1000 watts per square meter; e.g., six peak sun hours means that the energy received during total daylight hours equals the energy that would have been received had the irradiance for 6 hours been 1000 watts per square meter.

**peak watt** *Photovoltaic.* the maximum nominal output of a photovoltaic device, under standardized test conditions, symbol Wp; usually expressed as 1000 watts per square meter of sunlight with other conditions specified, such as a temperature of 25°C. A unit used to rate the performance of solar cells, modules, or arrays.

**Pearson, Gerald** 1905–1986, U.S. physicist who, with colleagues Daryl Chapin and Calvin Fuller at Bell Labs, produced the first photovoltaic cell that could provide a useful amount of electric power (1954).

**peat** *Biomass.* partially decayed plant material that is extracted, compressed, and dried for use as fuel, typically formed in an anaerobic water-saturated environment such as a bog. Peat is considered an early stage in the formation of coal; it develops when decaying plant material is accumulated and deposited at a greater rate than it is decomposed. See next page.

**peat bog** *Biomass.* a wet, spongy swamp area in which a layer of peat has formed from the decay of the existing plant matter under acidic conditions. Also, **peatland.**

**peat briquette** *Biomass.* a quantity of peat compressed into a small block for use as fuel. Similarly, **peat brick, peat pellet,** and so on.

**peat coal** *Biomass.* **1.** a form of coal that is intermediate between true peat and brown coal or lignite. **2.** peat that has been artificially carbonized for use as fuel.

**peatification** *Biomass.* the biochemical conversion of plant material to peat by means of anaerobic bacterial action.

**pebble bed reactor** *Nuclear.* an advanced nuclear reactor design that uses helium as the coolant, at very high temperature, to drive a turbine directly, producing very high efficiencies. The uranium, thorium, or plutonium fuels are in oxides contained within spherical pebbles made of pyrolitic graphite. This has potential safety advantages with its passive cooling mechanism and a negative temperature coefficient.

**Peccei, Aurelio** 1908–1984, Italian businessman who provide the initial impetus for the founding of the think tank CLUB OF ROME.

**Pechelbronn** *History.* a city in France that was the site of the first known wells dug specifically to search for oil (1745).

**pedestrianism** *Transportation.* the act or fact of walking; transportation on foot rather than in a vehicle.

**pelagic** or **pelagial** *Ecology.* **1.** of or relating to aquatic organisms that live in the open ocean, without direct dependence on the shore or bottom or on deep-sea sediment.

**peat** Peat is vegetable matter that has been accumulated and preserved in anaerobic waterlogged areas (wetlands). The former extent of tropical and non-tropical peatlands is c. 4 million km$^2$, largely in North America, Asia and Europe. In boreal and sub-boreal peatlands alone 270–370 x $10^{15}$ g of carbon is stored, about 1/3 of the total global soil carbon pool. The current rate of carbon sequestering is 40–70 x $10^{12}$ g/a. Outside the tropics about 500,000 km$^2$ has stopped accumulating peat due to human exploitation (drainage, mainly for agriculture, forestry and peat extraction). Dried peat has been used for domestic energy purposes since prehistoric times, sometimes with locally large environmental consequences. For example, in the last 800 years about 9 km$^3$ of peat have been used or lost in the Netherlands, a principal cause for one-third of the country being below sea level. In Finland, Ireland, and Sweden peat is currently used for the commercial production of electricity. In Finland, for example, 7.6% (2003) of the electricity production, or 6.4 TWh; in the other two countries smaller but significant amounts are used. The annual production varies considerably as the drying of peat depends strongly on the weather. The sustainability of energy use of peat is a subject of debate. Between fossil fuels that are hundreds of millions of years old and renewable fuels like trees or annual crops, peat from boreal peatland may be up to 8000 years old and so is neither a fossil nor a renewable fuel. It is sometimes named a "slowly renewable fuel". The sustainability question and whether peat for energy is carbon neutral is largely a matter of definition and the time scale used.

**Anne Jelle Schilstra**
University of Groningen

Thus, **pelagic zone.** **2.** relating to the deeper regions of a lake that are characterized by the absence of aquatic vegetation.

**pellet** *Biomass.* a quantity of biomass fuel (e.g., wood sawdust, peat) that has been compacted under high pressure to form a small object suitable for use as fuel.

**pellet stove** *Biomass.* a space-heating device specifically intended to burn wood pellets rather than conventional cord wood.

**Peltier effect** *Electricity.* the evolution or absorption of heat produced by an electric current passing across junctions of two suitable dissimilar metals, alloys, or semiconductors. [Described by French physicist Jean C. A. *Peltier,* 1785–1845.]

**Pelton turbine** *Conversion.* the earliest form of impulse water turbine (1870s), a key to the commercial development of hydroelectric power. In this design, water pressure is increased with a nozzle, and the resulting water jet impacts curved turbine blades, reversing the water's flow; this causes a runner (rotating part) to spin, producing mechanical energy. The runner is fixed on a shaft, and the rotational motion of the turbine is transmitted by the shaft to an electrical generator. Also, **Pelton wheel.** [Developed by U.S. inventor Lester Allan *Pelton,*1829–1908.]

**PEM** **1.** proton-exchange membrane. **2.** polymer electrolyte membrane.

**PEMFC** **1.** proton-exchange membrane fuel cell. **2.** polymer electrolyte membrane fuel cell.

**PEN** positive electrolyte negative.

**pendulum** *Physics.* any body that is free to swing under the influence of gravity; a simple practical example is a small weight suspended from the end of a string.

**penetrometer** *Measurement.* a device used to measure the penetrating power of a beam of electromagnetic radiation.

**penstock** *Hydropower.* **1.** a gate or valve used to control the flow of water in a hydropower system. **2.** a pressurized pipe or tube used to bring a rapid flow of water to a waterwheel or turbine. [From an older sense of "pen" as an enclosure to confine water for use by a mill or water wheel.]

**pentanes plus** *Oil & Gas.* a mixture of hydrocarbons, mostly pentanes and heavier fractions, extracted from natural gas; this includes isopentane, natural gasoline, and plant condensate.

**pentose** *Chemistry.* a monosaccharide, such as ribose, containing five carbon atoms in a molecule.

**pentose phosphate pathway** *Biological Energetics.* a major metabolic pathway through which animal tissues catabolize glucose to form NADPH and ribose 5-phosphate and to produce pentoses, especially the D-ribose.

**Penydarren** *History.* the world's first steam locomotive designed to run on iron rails, developed by Cornish engineer Richard Trevithick (1804) to operate on a tramway running to the village of Penydarren in south Wales. Though it worked, it was not financially successful because it was too heavy for the rails that were designed for horse-drawn trains.

**percolation** *Materials.* **1.** the gradual movement of a liquid through a porous medium. **2.** *Earth Science.* the movement of water through the interstices in rock, soil, or other porous material, under hydrostatic pressure or by the force of gravity. Thus, **percolate.**

**percussion** *Physics.* the striking of one object against another, usually with significant force.

**percussion drill** *Mining.* a pneumatic drill in which a piston delivers hammer blows rapidly on a drill shank; the development of efficient percussion drills in the 19th century made it possible to excavate large amounts of earth and rock in tunneling and mining.

**perfect gas** another term for an IDEAL GAS.

**perfluorocarbon (PFC)** *Environment.* a group of human-made chemicals composed of carbon and fluorine only: these chemicals, specifically $CF_4$ and $C_2F_6$, have been promoted along with hydrofluorocarbons (HFCs) as alternatives to ozone-depleting substances such as CFCs and HCFCs for various industrial purposes. Though HFCs do not significantly deplete the stratospheric ozone layer, they are powerful greenhouse gases with significant global warming potentials and also have other negative environmental implications.

**performance-based** *Consumption & Efficiency.* describing a mechanism or procedure that is based on the results to be achieved rather than the specific activities to be conducted; e.g., goals such as a certain increase in energy inputs or a certain reduction in waste output. Thus, **performance-based contracting, ratemaking,** and so on.

**performance standard** see PERFORMANCE-BASED.

**perhydrous** *Coal.* describing a coal that is relatively rich in hydrogen.

**perihelion** *Earth Science.* the point at which the earth or another body orbiting the sun is closest to it.

**periodic law** *Chemistry.* **1.** a law stating that all properties of the elements that depend on atomic structure tend to change with increasing atomic number in a periodic manner. **2.** an early statement of this principle that governs the classification of elements according to the repetition of physical and chemical properties at definite intervals (periods).

**periodic table** *Chemistry.* a schematic arrangement of the chemical elements in order of increasing atomic numbers, according to a pattern that represents the periodic law; basic vertical groups consist of elements having similar properties, and elements having properties between those of adjacent neighbors form horizontal "periods".

**Perkins, Jacob** 1766–1849, U.S. inventor living in England who built a practical vapor compression machine and used it to produce ice (1834). This is thought to be the first patent for mechanical refrigeration.

**perm** *Measurement.* short for permeance; a unit of measure for the ability of a material to retard the diffusion of water vapor at a standard temperature, in terms of the number of grains of water vapor that will pass through one square foot of the material per hour at a differential vapor pressure equal to one inch of mercury.

**permafrost** *Earth Science.* perennially frozen ground; typically identified as soil, subsoil, or other ground deposit that is at or below 0°C for at least 2 years. Characteristic of Arctic regions, it can be composed of soil, bedrock, ice, or any combination of these.

**permanent cultivation** *Ecology.* the replacement of a preexisting natural ecosystem with a permanently maintained agricultural system, in order to produce crops as frequently as climate and soil conditions allow.

**permanent magnet** *Materials.* a ferromagnetic substance that has been subjected to a magnetic field strong enough to cause the material to retain its own magnetization indefinitely.

**permeable** *Materials.* allowing the passage of a fluid; describing the (relative) ability of a porous medium to allow a liquid or gas to filter through it. Thus, **permeability.**

**Permian** *Earth Science.* the last geologic period in the Paleozoic era, occurring from 286 to 245 million years ago; a period that

*permafrost Caribou are adapted to a permafrost environment; e.g., their hoofs have specialized dew-claws that can penetrate the frozen surface.*

included the mass extinction of many corals, brachiopods, and trilobites, as well as the diversification and growing dominance of reptiles.

**Permian Basin** *Oil & Gas.* a major oil and gas producing region in the United States, located in West Texas and the adjoining area of southeastern New Mexico.

**permitted explosive** *Coal.* a term for an industrial explosive specially produced and tested for safe use in coal mines where firedamp and/or dangerous coal dust occur; i.e., an explosive whose use is permissible in such situations.

**peroxyacetal nitrate (PAN)** *Health & Safety.* any of a group of compounds formed from the photochemical reactions of nitrogen and organic compounds. PANS are components of smog and known to cause eye and lung irritation.

**perpetual motion** *Thermodynamics.* the (undemonstrated) theory that a machine producing work can continue to operate for an indefinite time solely by the use of its own energy; various supposed perpetual motion machines were described in earlier times prior to an understanding of the laws of thermodynamics. See next column.

**Persian Gulf** *Oil & Gas.* a shallow arm of the Arabian Sea between Iran and the Arabian peninsula; the Persian Gulf oil fields are among the most productive in the world. Gulf nations include Bahrain, Iran, Iraq, Kuwait, Qatar, Saudi Arabia, and the United Arab Emirates.

**Persian Gulf War** see GULF WAR.

**persistence** *Environment.* the length of time that a compound stays in the environment in its original state, once it has been introduced.

**persistent** *Environment.* describing a pollutant with a complex molecular structure, so that it is not effectively broken down but remains in its toxic form over time in the environment; e.g., mercury in the marine food chain.

**perpetual motion** The action of a moving device that requires no input of energy to maintain it, and therefore can continue for ever. This is impossible if the laws of thermodynamics hold, and despite many attempts since antiquity, such a device has yet to be demonstrated. A great number of designs have been published that appear to allow work produced during the operation of a device to both maintain the motion of the device and provide excess work outside the system. Examples include springs, overbalancing wheels, and magnetic and electrical machines. These are considered impossible because the first law teaches that the energy that is consumed by the device to cause its motion, is the maximum that could be delivered by that motion. Practically, because of frictional and other heat losses, the available energy would be less, causing the machine to eventually come to a halt. Another kind of perpetual motion violates the second law, which requires useful work always to be accompanied by an increase in the entropy of the universe. Engines that are driven by the spontaneous evaporation of condensed vapors such as liquid ammonia are such devices, because the energy to condense the vapor before the next cycle is greater than that obtained by harnessing its evaporation. Famous names in perpetual motion include Anthony Zimara (self-blowing windmill, 13th century), Robert Fludde (self-powered water wheel, 15th century), Robert Boyle (self-filling cup, 17th century), J. W. Keely (fraudulent motor, 19th century), J Gamgee (liquid ammonia ship engine, 19th century).

**D. Brynn Hibbert**
University of New South Wales

**personal computer** *Communication.* a general term for any self-contained and relatively small computer system that is designed for use by a single individual for operations such as electronic mail, Internet access, word processing, and financial calculations.

**perturbation** *Physics.* **1.** an influence on a system that disturbs the system only slightly. **2.** a deviation in the motion of a celestial object as a result of the gravitational pull of other bodies.

**perylene** *Photovoltaic.* a semiconducting organic material with potential applications in both light-emitting diodes and photovoltaics.

**pesticide** *Ecology.* any agent that is lethal to pests (unwanted organisms such as, insects and weeds), especially a manufactured substance used to kill agricultural pests.

**PET** *Materials.* polyethylene terephthalate; a thermoplastic, combustion resistant polyester derived from ethylene glycol; widely used for beverage containers, films, fibers, and recording tapes.

**peta-** *Measurement.* a prefix meaning "one quadrillionth" ($10^{-15}$); symbol P.

**petrochemical** *Oil & Gas.* short for petroleum chemical; any of a wide range of chemicals or materials derived directly or indirectly from petroleum or natural gas.

**petrochemical feedstock** *Oil & Gas.* any chemical derived from petroleum that is used for the manufacture of other products, such as synthetic rubber and plastics.

**petrodollars** *Economics.* the U.S. dollars earned by a petroleum producing and exporting country from the sale of oil.

**petro-electric car** *Transportation.* a car considered to be the first hybrid-electric vehicle (1897), using a gasoline engine as the primary drive and an electric motor to assist when more power is required. Developed by Justus B. Entz, chief engineer of the Electric Storage Battery Company of Philadelphia.

**petrography** *Materials.* the systematic classification of igneous and metamorphic rocks on the basis of their mineralogical and textural relationships, especially by means of microscopic study.

**petrol** *Oil & Gas.* **1.** short for petroleum. **2.** another word for gasoline, especially in British use.

**petroleum** *Oil & Gas.* a collective term for crude oil, natural gas, natural gas liquids, and other related products, usually found in deposits beneath the earth's surface and thought to have originated from plant and animal remains of the geologic past. Petroleum is a naturally occurring mixture of hydrocarbon and nonhydrocarbon compounds that may be found in a gaseous, liquid, or solid state, depending on the nature of the compounds and conditions of temperature and pressure. It is the most widely used fuel source in the industrialized world and is also used in many industrial products, such as plastics, synthetic fibers, and fertilizers.

**Petroleum Administration for Defense (PAD)** *Policy.* a U.S. agency operating during the Korean War as the successor to the Petroleum Administration for War (see below). The PAD was established in 1950 and terminated in 1954.

**Petroleum Administration for Defense District (PADD)** *Policy.* one of five geographic aggregations of the United States made by the Petroleum Administration for Defense in 1950; e.g., District IV (Rocky Mountain) includes the states of Colorado, Idaho, Montana, Utah, and Wyoming. Also, **PAD District.**

**Petroleum Administration for War (PAW)** *History.* a U.S. agency operating during World War II to assure the most effective development and utilization of petroleum in the U.S.; it rationed civilian petroleum use to ensure adequate military supplies. The PAW was established in 1942 and terminated in 1946.

**petroleum coke** *Oil & Gas.* a residue from the final product of the condensation process in cracking.

**petroleum cracking** see CRACKING.

**petroleum engineering** *Oil & Gas.* the branch of engineering that includes the recovery, production, distribution, and storage of oil, gas, and liquefiable hydrocarbons, as well as exploration for these products.

**petroleum geology** *Oil & Gas.* the branch of economic geology that is concerned with the origin, migration, accumulation, and occurrence of hydrocarbon fuels, and with exploration for these fuels.

**petroleum isomerization** *Oil & Gas.* a petroleum-refinery process that converts straight chain compounds into branched isomers with the use of a catalyst, in order to obtain higher octane gasolines and other products.

**petroleum microbiology** *Oil & Gas.* a branch of microbiology that deals with the relationship between microorganisms and the formation and industrial uses of petroleum, including both negative microbial effects such as souring and corrosion, and positive effects that can be harnessed to enhance oil recovery.

**petroleum trap** see TRAP.

**petrology** *Earth Science.* **1.** the scientific study of rocks, including their origin, history, composition, structure, and technological applications. **2.** a branch of this discipline studying organic materials such as coals and cokes.

**petroviolence** *Global Issues.* **1.** conflicts or wars that arise because of competition for oil resources. **2.** more generally, any form of violence that can be associated with or attributed to the extraction of oil in a certain locality, or with the relative distribution of wealth generated by that extraction.

**Peukert number** *Storage.* a value that indicates how well a lead–acid battery performs under heavy currents. A value close to 1 indicates that the battery performs well; the higher the number, the more capacity is lost when the battery is discharged at high currents.

**Peukert's equation** *Storage.* a description of how the available capacity of a lead–acid battery changes according to the rate of discharge. Formally, $C = I^nT$, where $C$ is the theoretical capacity of the battery, $I$ is the current, $t$ is time, and $n$ is the Peukert number. The equation describes the fact that at higher currents, there is less available energy in the battery.

**PFC** perfluorocarbon.

**pH** *Chemistry.* a measure of the relative degree of acidity or alkalinity of a substance, based on a scale in which a pH of 7 represents neutrality at standard temperature (e.g., pure water). Acidic substances have lower pH numbers than this and alkaline substances have higher ones. [An abbreviation for potential of hydrogen.]

**phantom** *Nuclear.* a term for a volume of material approximating as closely as possible the density and effective atomic number of living tissue, used in biological experiments involving radiation.

**phantom load** *Consumption & Efficiency.* continuous electricity use by appliances, television sets, computers, and other such devices even when they are ostensibly not operating; e.g., a computer or printer in "sleep" mode.

**phase** *Chemistry.* **1.** the distinct condition of a substance, either as a solid, a liquid, or a gas; a quantity of matter that is homogeneous throughout its environment. **2.** a portion of a system that is uniform both in its chemical composition and physical properties, and that is separated from other homogeneous portions of the same system by boundary surfaces. **3.** *Physics.* a particular stage or point in a sequence through which time has advanced, measured from some arbitrary starting point.

**phase-change material (PCM)** *Materials.* a substance that undergoes a change of state (as by melting, freezing, boiling, or condensing) while absorbing or rejecting thermal energy, normally at a constant temperature.

**phlogiston** *History.* in early chemistry, a term for a supposed volatile component of combustible substances captured from sunlight by plants and released in vital processes, combustion, and reduction. The loss of weight in the burning of wood and other common materials was thought to be a result of the phlogiston in the substance escaping into the air.

**phosphate** *Chemistry.* **1.** a salt or ester of phosphoric acid. **2.** a fertilizer containing phosphorus compounds.

**phosphor** *Materials.* a material that is capable of phosphorescence; i.e., that can absorb any of various types of energy and emit visible light after the energy source is removed.

**phosphorescence** *Lighting.* **1.** an emission of light that follows the absorption of energy by a substance from sources such as visible light, infrared light, ultraviolet light, electricity, cathode rays, or X-rays, and that continues for a relatively long time after the energy supply has ceased. **2.** any continuing emission of light that occurs in the absence of significant heat, such as the bioluminescence of fireflies. Thus, **phosphorescent.**

**phosphoric acid** *Materials.* $H_3PO_4$, a clear colorless liquid that is used in fertilizers, detergents, food flavorings, and pharmaceuticals.

**phosphoric acid fuel cell (PAFC)** *Renewable/Alternative.* an established type of fuel cell technology that uses concentrated phosphoric acid ($H_3PO_4$) as the electrolyte. The operating temperature range is generally 150–220°C (300–400°F), since at lower temperatures the phosphoric acid tends to be a poor conductor and at higher temperatures it tends to decompose.

**phosphorus** *Chemistry.* a widely occurring nonmetallic element having the symbol P, the atomic number 15, an atomic weight of 30.9738; it occurs in three main allotropic forms: white (or yellow), red, and black, with a melting point of 44.1°C and a boiling point of 280°C (white form). It is used in matches, fertilizers, pesticides, semiconductors, and various other industrial products. Phosphorus is an essential element of the human diet and is the main component of bones and teeth, and is involved in some form in virtually all processes of metabolism. Also spelled **phosphorous.**

**photobiological (hydrogen) production** *Hydrogen.* a proposed process to produce commercial hydrogen, based on the fact that certain photosynthetic microbes produce hydrogen from water in their metabolic activities.

**photobiology** *Biological Energetics.* the branch of biology that deals with the effects of radiant energy on living organisms. Thus, **photobiological.**

**photocatalysis** *Chemistry.* a change in the rate of a chemical reaction due to the action of light.

**photocatalyst** *Chemistry.* a substance that is able to produce, by absorption of light, chemical transformations of the reaction participants; e.g., chlorophyll in a process of photosynthesis.

**photocell** *Electricity.* a device that generates electrical energy from light energy, usually as a voltage or current.

**photochemistry** *Chemistry.* the scientific study of chemical changes caused or influenced by the effect of light. Thus, **photochemical.**

**photochromism** *Materials.* the fact of changing color due to exposure to incident radiation; a property of certain organisms when exposed to light of a particular wavelength. Thus, **photochromic.**

**photoconductive** *Materials.* having the property of photoconductivity; i.e., describing certain materials in which exposure to light results in an increase in electrical conductivity. Thus, **photoconductive cell, photoconductive device.**

**photoconductivity** *Electricity.* an effect, observed in many substances, in which the electrical conductivity is increased when the substance is exposed to electromagnetic radiation, usually in the visible region of the spectrum. Also, **photoconduction.**

**photoconversion** *Conversion.* the conversion of a substance from one form to another using the energy supplied by light

**photocurrent** *Electricity.* an electric current produced when a form of radiant energy, such as ultraviolet light, strikes an electrode.

**photodegradation** *Materials.* the decomposition of a substance due to the action of light.

**photoelectric** *Physics.* relating to or produced by the electrical effects of light, including the emission of electrons, the generation of a voltage, or a change in resistance.

**photoelectric cell** see PHOTOCELL.

**photoelectric effect** *Physics.* a phenomenon in which electrons are emitted from a material (generally a metal) when it is exposed to light with a given frequency. The classical concept of light as a continuous wave could not account for this, and it was then explained by Einstein (1905) on the basis that light is a stream of separate particles. This interpretation, and his subsequent elaboration of it, formed the basis for much of quantum mechanics.

**photoelectricity** *Electricity.* the electricity produced when light or other forms of electromagnetic radiation strike certain materials, such as cesium, selenium, and silicon.

**photoelectrochemical cell** *Electricity.* a galvanic cell in which usable current and voltage are simultaneously produced upon absorption of light by at least one of the electrodes.

**photoelectrochemistry** *Electricity.* the study of electrochemical reactions that are affected or promoted by light.

**photoelectrode** *Materials.* a material that absorbs incident light to produce current through an external circuit.

**photoelectrolysis** *Chemistry.* a process that uses light energy rather than electrical energy to produce a chemical change in an electrolytic solution. Thus, **photoelectrolytic.**

**photoelectrolytic (hydrogen) production** *Hydrogen.* a proposed process to produce commercial hydrogen, employing the electrolysis (splitting) of water by means of light energy.

**photoemission** *Physics.* the ejection of electrons from a material, usually a solid, when it is exposed to electromagnetic radiation.

**photofission** *Nuclear.* nuclear fission induced by high energy photons.

**photoluminescence** *Physics.* the luminescent emission of electromagnetic radiation from a material after it has absorbed another form of electromagnetic radiation, particularly infrared or ultraviolet radiation.

**photolysis** *Chemistry.* **1.** the decomposition of a substance into simpler units as a result of its absorbing light; e.g., the separation of hydrogen from hydrogen sulfide in water. **2.** any process in which radiant energy produces a chemical change.

**photometer** *Measurement.* an instrument that measures characteristics of visible radiation (light), such as its luminous intensity or illuminance, relative to a standard reference source. Thus, **photometric.**

**photometry** *Measurement.* the measurement of the visible portion of radiant energy (light). Compare RADIOMETRY.

**photon** *Physics.* the quantum of the electromagnetic field that manifests itself by absorption or emission only in multiple quantum units of energy; a unique massless particle that carries the electromagnetic force.

**photooxidation** *Chemistry.* a process of oxidation driven by radiant energy (light).

**photoperiod** *Biological Energetics.* the length of a day; the amount of time that an organism is exposed to daylight.

**photoperiodism** *Biological Energetics.* any of various behavioral and physiological changes in an organism in response to the amount of daylight to which the organism is exposed; i.e., the relative length of day and night on a seasonal or daily basis.

**photophoresis** *Physics.* a phenomenon in which unidirectional motion is imparted to a system of very fine particles of a solid or liquid, while it is either suspended in a gas or falling through a vacuum when an intense beam of light is fired at it.

**photopic** *Lighting.* relating to vision by the normal eye in bright light; i.e., day vision. Thus, **photopia, photopic vision.**

**photoreactive** *Conversion.* describing a molecule that gathers light and converts it into energy. Thus, **photoreactor, photoreaction.**

**photorespiration** *Biological Energetics.* a process in which an organism takes in oxygen and releases carbon dioxide in the presence of light, occurring during photosynthesis in conditions in which there is a low concentration of carbon dioxide and intensive levels of light.

**photosphere** *Solar.* the outermost visible layer of the sun; the layer of the sun that corresponds to the solar surface viewed in white light. Sunspots and faculae are observed in the photosphere.

**photosynthesis** *Biological Energetics.* the fundamental chemical process in which green plants (and certain other organisms) utilize the energy of sunlight or other light to convert carbon dioxide and water into organic chemical energy. Photosynthesis is often described as the most important chemical reaction on earth. See next page.

**photosynthetic** *Biological Energetics.* relating to or taking part in a process of photosynthesis. Thus, **photosynthetic bacteria, photosynthetic products.**

**photosynthetic efficiency** *Biological Energetics.* the percentage of the total available light captured by plants that is subsequently converted into chemical energy. Also, **photovoltaic conversion efficiency.**

**photovoltaic (PV)** **1.** having to do with or employing the photoelectric effect; i.e., the production of electric power from electromagnetic radiation. **2.** relating to or designating devices that absorb solar energy and transform it directly into electricity.

**photovoltaic array** *Photovoltaic.* an interconnected system of photovoltaic modules that functions as a single electricity-producing unit. The modules are assembled as a discrete structure, with common support or mounting;

*photosynthetic* The energy of sunlight (arrows above) is captured by green plants and utilized to convert carbon dioxide into common sugars and other photosynthetic products (arrows below); arrows to right indicate input of carbon dioxide and output of oxygen and water vapor.

in smaller systems, an array can consist of a single module.

**photovoltaic cell** *Photovoltaic.* a single semiconducting element of small size (for example, 1 square centimeter) that absorbs light or other bands of the electromagnetic spectrum and emits electricity.

**photovoltaic device** *Photovoltaic.* a solid-state device that converts light directly into electricity, with voltage–current characteristics that are a function of the characteristics of the light source and the design and materials of the device. These devices are made of various semiconductor materials including silicon, cadmium sulfide, cadmium telluride, and gallium arsenide, in single crystalline, polycrystalline, or amorphous forms.

**photovoltaic effect** *Photovoltaic.* the phenomenon that occurs when photons (the "particles" in a beam of light) strike electrons and free them from their bound position. When this property of light is combined with the properties of semiconductors, electrons flow in one direction across a junction, setting up a voltage. With the addition of circuitry, current flows and electric power is available.

**photovoltaic efficiency** *Photovoltaic.* the ratio of the electric power produced by a photovoltaic device to the power of the sunlight incident on the device.

**photovoltaic energy** *Photovoltaic.* energy emanating from the sun as electromagnetic radiation that is converted into electricity by means of solar (photovoltaic) cells or concentrating (focusing) collectors.

**photosynthesis** The process of converting light energy to organic sugars. It is found in bacteria, cyanobacteria, algae and most plants. A Dutch physician, Jan Ingenhousz (1730–1799), first suggested that, if placed in the sunshine, a plant might absorb the carbon from carbon dioxide, "throwing out at that time the oxygen alone, and keeping the carbon to itself for nourishment". Photosynthesis involves the capture of the energy of light, the formation of an excited electron, the use of the excited electron to reduce an acceptor substance, and the formation of energy-rich molecules. This process occurs in two stages. During the first stage, the light reactions, light harvesting pigments capture light from the sun forming excitation energy. This excitation energy moves through several pigment molecules to a reaction center where an electron is excited and lost. The electron lost from the reaction center is then rapidly replaced. In plants, the electron donor is water and oxygen is produced during the process. In other photosynthetic organisms (forms of bacteria), compounds other than water are used for electron donation and oxygen may or may not be produced. In both cases, the energy of the excited electrons is used to produce energy-rich molecules for use in the second stage of photosynthesis, the dark reactions. During this process, the energy-rich molecules are used to convert carbon dioxide into common sugars and other organic molecules. This coupling of light energy to chemical energy is directly or indirectly the basis for nearly all life on earth.

**Jed P. Sparks**
Cornell University

**photovoltaic generator** *Photovoltaic.* the total of all interconnected strings of a photovoltaic power supply system.

**photovoltaic hybrid (system)** *Photovoltaic.* a photovoltaic system combined with some other complementary energy system, such as wind turbines or diesel generators.

**photovoltaic module** *Photovoltaic.* a panel assembled from a number of individual photovoltaic cells electrically interconnected in series and parallel so as to provide a specific useful voltage and current.

**photovoltaic panel** *Photovoltaic.* another term for a PHOTOVOLTAIC MODULE, or for a physically connected collection of modules.

**photovoltaic process** another term for PHOTOVOLTAIC EFFECT.

**photovoltaic system** *Photovoltaic.* a complete set of components for converting solar energy into electricity by the photovoltaic effect, including the array and balance of system constituents.

**photovoltaic-thermal** *Photovoltaic.* describing a photovoltaic system that, in addition to converting sunlight into electricity, collects the residual heat energy and delivers both heat and electricity in usable form.

**photovoltaics** *Photovoltaic.* the technology or activity of employing solar radiation as a direct energy source for electrical power.

**phurnacite** *Coal.* a manufactured fuel, not now in widespread use, in the form of carbonized, egg-shaped briquettes; a product that is smokeless and virtually dust free and that can be competitive with anthracite for uses such as cooking and space heating.

**physical accounting** *Economics.* the practice of natural resource and environmental accounting of stocks and changes in stocks in physical (nonmonetary) units; e.g., for the weight of oil, volume of natural gas, and so on.

**physical activity level (PAL)** *Biological Energetics.* the amount of physical activity that a person engages in during a day, including all forms of activity other than resting or sleeping; used as a means of describing the person's daily energy expenditure through exercise and other physical activities.

**physical constant** see CONSTANT.

**physical vapor deposition** *Photovoltaic.* a method of depositing thin-film semiconductor materials in which physical processes, such as thermal evaporation or bombardment of ions, are used to deposit elemental semiconductor material on a substrate.

**physics** the study of mass and energy and the interactions that they are observed to have throughout the universe, including such subfields as mechanics, astrophysics, nuclear physics, particle physics, and quantum mechanics.

**Physiocracy** *History.* a school of thought in France in the 18th century, arguing that agriculture, and more generally, all the extractive sectors of the economy (grasslands, pastures, forests, mines, and fishing) are the basis of economic production. Physiocracy provided the foundation for Adam Smith and the classical economists. Thus, **Physiocrat.**

*phytomass The royal palm, Roystonea regia, a tree of tropical America used as a phytomass fuel source because of its relatively rapid growth rate.*

**phytomass** *Biomass.* living or dead organic matter of plants, especially that of trees and crop residues used for fuel.

**phytoplankton** *Ecology.* any of a wide variety of species of microscopic plants growing abundantly in ocean waters around the world and constituting the foundation of the marine food chain. Phytoplankton exert a global-scale influence on climate because the larger the world's phytoplankton population, the more carbon dioxide gets pulled from the atmosphere, and thus the lower the average temperature due to lower volumes of this greenhouse gas.

**Piccard, Auguste** 1884–1962, Swiss physicist who, with his twin brother **Jean Felix Piccard** (1884–1963) made famous balloon ascents into the stratosphere, ultimately reaching an altitude of 23,000 m (72,177 ft). He and his son **Jacques Piccard** also made notable descents to extreme ocean depths.

**picking yard** *Renewable/Alternative.* a site used to separate out waste materials that have value for reuse, such as paper, metal, glass, or rubber.

**pico-** *Measurement.* a prefix meaning "one trillionth" ($10^{-12}$); symbol p.

**piezo-** *Physics.* having to do with or involving pressure.

**piezochemistry** *Chemistry.* the study of chemical reactions that occur at very high pressures, such as under the earth's crust.

**piezoelectric** *Electricity.* describing the ability of a solid to generate a voltage when subjected to a mechanical stress, or the ability to generate a mechanical force when subjected to a voltage. When compressed, some crystalline materials will produce a voltage proportional to the applied pressure; when an electric field is applied across the material, there is a corresponding change of shape. Thus, **piezoelectricity.**

**piezometer** *Measurement.* a device used to measure fluid pressure, or to measure the compressibility of a material that is subjected to hydrostatic pressure.

**pig** *Nuclear.* **1.** a heavily shielded container used to ship or store radioactive materials. **2.** *Oil and Gas.* a robotic device sent through buried oil and gas pipelines to inspect interior walls for corrosion and defects, measure pipeline interior diameters, and remove accumulated debris; pigs carry a small computer to collect and transmit data for analysis.

**pig iron** *Materials.* a term for crude, high-carbon iron produced by the reduction of iron ore in a blast furnace. [So called because the shape of such an iron casting was thought to resemble a pig.]

**Pigouvian tax** *Economics.* a tax levied on an agent causing an environmental externality (especially pollution damage), enacted as an incentive to avert or mitigate such an external activity. [Named for British economist Arthur C. *Pigou,* 1877–1958.]

**pile** *Nuclear.* another term for a nuclear reactor, especially a reactor constructed by stacking graphite blocks and uranium.

**piledriver** *Consumption & Efficiency.* a framed construction for driving columns of steel, concrete, or timber into the ground, by means of a heavy weight or hammer that is raised and dropped on the head of the column.

**pillar** *Mining.* an area of coal left to support the overlying strata or hanging wall in a mine; sometimes left permanently in place to support surface structures.

**pilot light** *Consumption & Efficiency.* a small permanent flame used to ignite a gas burner.

**P–I–N** *Photovoltaic.* positive–intrinsic–negative; a semiconductor (photovoltaic) device structure that layers an I-type (undoped) semiconductor between a P-type and an N-type; this geometry sets up an electric field between the P- and N-type regions that stretches across the middle I-type region. Light generates free electrons and holes in this region, which are then separated by the electric field. This structure is most often used with amorphous silicon PV devices. Thus, **P-I-N (solar) cell.**

**Pinatubo** see MOUNT PINATUBO.

**pinch effect** *Electricity.* the self-constriction that occurs in a plasma as a result of the passage of a current that tends to constrict (pinch) the plasma.

**Pines, Herman** 1902–1996, U.S. chemist who, with Vladimir Ipatieff, discovered the catalytic alkylation and isomerization of hydrocarbons (1938), a major breakthrough in the production of gasoline.

**pion** *Physics.* a particle with mass between that of electrons and protons, that interacts with neutrons and protons in the atomic nucleus.

**pipeline** *Oil & Gas.* an extended length of pipe with pumping or compressing machinery and apparatus for conveying natural gas, crude oil, or other fluids.

*pipeline The Trans-Alaska Oil Pipeline was established in 1977 to transport oil southward from the North Slope of Alaska to the Valdez terminal on Prudhoe Bay.*

**pipeline fuel** *Oil & Gas.* a term for gas consumed in the operation of a natural gas pipeline.

**pipeline quality** *Oil & Gas.* describing gas that meets the delivery requirements of a pipeline with regard to impurities, water content, Btu content, and other such physical attributes.

**piston** *Consumption & Efficiency.* a movable part fitted into a cylinder, that can receive or transmit motion as a result of pressure changes in a fluid; e.g., in an internal combustion engine.

**pit** *Mining.* **1.** a general term for a coal mine. **2.** any mine, quarry, or excavation site that is worked by open cutting.

**pitch** *Physics.* **1.** the angular displacement of a body (such as a propeller or turbine blade) about a transverse horizontal axis parallel to the lateral axis of the body. **2.** *Coal Mining.* the inclination or rise of a seam of coal.

**pitch** *Materials.* **1.** a historic term for asphalt or bitumen in a liquid state. **2.** a black, viscous residue resulting from the distillation of various tars (e.g., coal tar, wood tar) or petroleum, used for purposes such as caulking, paving, roofing, waterproofing, and so on.

**pitchblende** *Materials.* a black mineral that is one of the most important ore minerals of uranium.

**pitch control** *Wind.* the process of controlling a wind turbine's speed by varying the orientation (pitch) of the blades, and thus altering its aerodynamics and efficiency.

**pit coal** *Coal.* **1.** coal obtained by open-pit mining. **2.** historically, coal obtained from a bell pit (shallow underground mine), as opposed to sea coal which was gathered on beaches as it washed up on shore.

**pitting** *Mining.* **1.** the process of digging or sinking a pit. **2.** the sinking of small pits or shafts to expose material for evaluation and testing.

**pixel** *Communication.* the smallest discrete element of a computer display image, corresponding to a single displayed spot or color value on a display, or a single input spot from a camera.

**placer** *Mining.* a mineral deposit of value at the surface, formed by a sedimentary concentration of heavy mineral particles from weathered debris. Thus, **placer deposit, placer mine, placer mining,** and so on.

**Planck, Max** 1858–1947, German physicist who played a vital role in the development of quantum theory. His study of the distribution of energy in the electromagnetic spectrum led him to deduce the relationship between this and the frequency of radiation. His work influenced a wide range of subsequent research, including Einstein's explanation of the photoelectric effect.

**planck** *Measurement.* a unit of action (energy expended over time), given as the product of an energy of one joule over a period of one second.

**Planck's constant** *Measurement.* a universal constant that is fundamental to quantum theory, having the dimensions of angular momentum and action and the value $6.6260755 \times 10^{-34}$ joule-second in SI units.

**Planck's (radiation) law** *Measurement.* a fundamental law stating that radiation is composed of indivisible units of energy, or quanta, whose energies are proportional to the frequency of the radiation: $E = hv$, where $h$ is Planck's constant and $v$ is the frequency of the resonator absorbing or emitting energy.

**Planck unit** *Measurement.* any of the physical units of measurement originally proposed by Max Planck; a system of natural units defined exclusively in terms of fundamental physical constants.

**planetary albedo** *Solar.* the fraction (approximately 30%) of incident solar radiation that is reflected by the earth–atmosphere system and returned to space, mostly by scattering from clouds in the atmosphere.

**planetary boundary layer (PBL)** *Earth Science.* the lower part of the earth's atmosphere that is directly influenced by conditions on the surface, such as evapotranspiration; its height is variable and ranges from 100 to 3000 m.

**plankton** *Ecology.* a collective term for various animals and plants, usually microscopic in size, that float or drift freely in aquatic systems and form a fundamental part of the system's food web. They can be categorized as bacterioplankton (bacteria), phytoplankton (plants), or zooplankton (animals).

**plant condensate** *Oil & Gas.* one of the natural gas liquids, mostly pentanes and heavier hydrocarbons, recovered and separated as liquids at gas inlet separators or scrubbers in processing plants.

**plant heat rate** *Consumption & Efficiency.* the efficiency of a thermal power plant, typically expressed in British thermal units per kilowatt-hour (Btu/kWh). The average heat rate for U.S. thermal power plants is about 10,000 Btu/kWh.

**plant-use** *Electricity.* referring to electric energy generated by a plant that is then used in the operation of the plant, as opposed to its being transmitted to end users.

**plasma** *Physics.* a gas-like state of matter consisting of positively charged ions, free electrons, and neutral particles. Found in stars, the sun, the solar wind, lightning, and fire, and considered to be a fourth basic state of matter (along with solid, liquid, and gas), making up about 99% of the visible universe. The behavior of most plasma systems is dominated by the electromagnetic interaction between the charged particles.

**plasma-assisted catalysis** *Transportation.* a diesel exhaust treatment system that uses a small amount of electricity from the engine to generate a nonthermal plasma on a ceramic surface.

**plasma diode** *Electricity.* **1.** a tube that generates energy when heat is applied to a space between two electrodes which is filled with cesium vapor. **2.** a tube in which an ionized gas generates electricity that conducts in only one direction.

**plasma mantle** *Earth Science.* a layer of plasma found immediately below the magnetopause, having a tailward flow of speeds from 100 to 200 km/s and showing a decrease of density, temperature, and speed corresponding to the increase of depth inside the magnetosphere.

**plasmatron** *Hydrogen.* **1.** a compact plasma conversion device that generates hydrogen-rich gas mixtures from hydrocarbon fuels such as gasoline and diesel fuel; it can be employed as an onboard fuel-reforming device for spark ignition engines. **2.** *Electricity.* a gas-discharge tube in which a neutral gas, usually helium, is ionized so that it will conduct electricity across electrodes and, under certain conditions, amplify microwave frequencies.

**plastic** *Materials.* **1.** describing a substance or material that is capable of being shaped or molded, with or without the application of heat. **2.** any of various synthetic or organic materials that can be molded or shaped, generally when heated, and then hardened into a desired form; for example, polymers, resins, and cellulose derivatives. Plastics are used in virtually all areas of modern industrial technology.

**plasticity** *Materials.* the fact of being plastic; the ability to be shaped or molded.

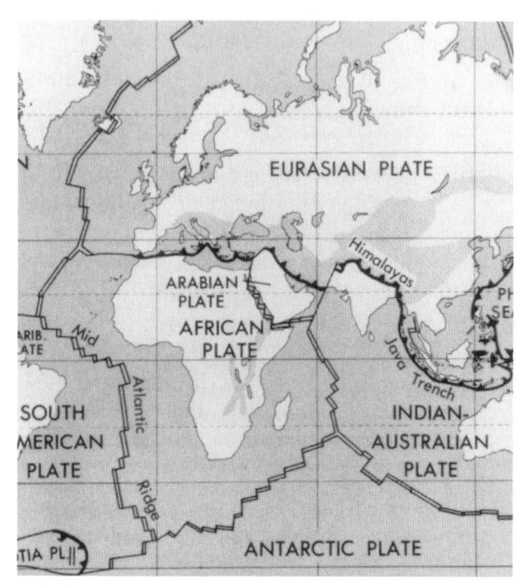


***plate tectonics*** *Depiction of the plates surrounding the continent of Africa.*

**plate** *Earth Science.* **1.** a large coherent portion of the earth's lithosphere that moves as a single rigid entity. **2.** *Electricity.* another term for an anode; i.e., the electrode toward which current flows or to which electrons are attracted.

**plate tectonics** *Earth Science.* a modern geological theory according to which the earth's crust is divided into a limited number of large, rigid plates whose independent movements relative to one another cause intense geologic activity along their margins, such as deformation, volcanism, earthquakes, and mountain building.

**platform** *Oil & Gas.* an immobile offshore structure from which wells are drilled and produced.

***platform*** *Offshore oil platform.*

**platinum** *Chemistry.* a metallic element having the symbol Pt, the atomic number 78, an atomic weight of 195.09, a melting point of 1769°C, and a boiling point of 3827°C; a heavy, silvery-white, highly ductile and malleable substance that does not tarnish or corrode. A very valuable metal used in various industrial applications and for catalysts, notably those that decrease automotive emissions.

**play** *Oil & Gas.* a set of known or postulated oil and gas accumulations sharing similar geologic, geographic, and temporal properties, such as source rock, migration path, trapping mechanism, and hydrocarbon type.

**P-layer** see P-TYPE.

**playtime** or **play time** see RUNTIME.

**Pleistocene** *Earth Science.* a period of the earth's history that began about 2.5 million years ago, characterized by repeated glacial/interglacial fluctuations in climate.

**plenum** *HVAC.* a system of ventilation in which air is forced into an enclosed space so that the outward air pressure is greater than the inward air pressure.

**Ploesti** *History.* a city in Romania that Sir Winston Churchill called "the taproot of German might" during World War II because it was where the Nazi war machine obtained much of its oil. Its refineries were the target of intensive Allied bombing until they were finally rendered inoperative in late 1944.

**plug and abandon** *Oil & Gas.* the process of preparing a well to be permanently closed and isolated.

**plug flow digester** *Biomass.* a type of anaerobic digester that has a horizontal tank to which a constant volume of material is added, in order to force the material already present to move through the tank and be digested.

**plume** *Earth Science.* **1.** a deep-seated upwelling of volcanic magma within the earth's mantle. **2.** a subsurface zone containing predominantly dissolved and sorbed contaminants that originate from a contaminant source area.

**plutonium** *Chemistry.* a metallic element with the symbol Pu, the atomic number 94, an atomic weight (for its most stable isotope) of 244, a melting point of 640°C, and a boiling point of 3230°C; it occurs as fifteen radioactive isotopes (having mass numbers from 232 to 246) and in six allotropic forms. Plutonium and all elements of higher atomic number are radiological poisons because of their high rate of alpha emission and their specific absorption in bone marrow.

**plutonium-238** *Chemistry.* a plentiful but nonfissile isotope of plutonium with a half-life of 87.74 years; used in radioisotopic thermoelectric generators to provide electricity for space probes that are too far from the sun to use solar power.

**plutonium-239** *Chemistry.* the most important isotope of plutonium, because it is fissionable, has a relatively long half-life (24,360 years), and can be readily produced in large quantities in breeder reactors by neutron irradiation of uranium-238. Used as a preferred fuel in nuclear weapons and fast reactors (over one third of the energy produced in most nuclear plants comes from Pu-239). One kilogram is equivalent to about 22 million kilowatt-hours of heat energy; the complete detonation of one kg produces an explosion equal to about 20,000 tons of chemical explosive.

**PM** particulate matter.

**P–N** or **P/N** *Photovoltaic.* positive–negative; describing an area where the surfaces of P-type and N-type semiconductors meet and create an electrical field. Thus, **P–N junction.**

**pneumatic** *Physics.* **1.** activated or set in motion by air. **2.** operated by air pressure or by compressed air.

**Pneumatica** *History.* a book of about AD 60, published by Hero (Heron) of Alexandria, in which he described over 100 different machines, including the first working steam engine.

**pneumatic tire** *History.* an air-filled tire; an advance in vehicle technology made known by Scottish inventor John Boyd Dunlop in the late 1800s, allowing tires of much lighter weight with less vibration and greater traction.

**pneumoconiosis** *Health & Safety.* a respiratory disease characterized by permanent deposition of substantial amounts of particulate matter in the lungs, caused by the chronic inhalation of irritating chemical or mineral substances, including dusts from coal, silica, talc, or rock iron oxides.

**pod** *Oil & Gas.* a term for a defined entity of petroleum source rock.

**Podolinsky, Sergei Andreyevich** 1850–1891, Ukrainian socialist who was among the first to examine the economic process from a thermodynamic perspective. He concluded that the ultimate limits to economic growth are not constraints on production, but physical and ecological laws. His work thus foreshadowed by nearly a century concepts now widely used by biophysical analysts.

**poikilotherm** *Biological Energetics.* an organism such as a fish, amphibian, or reptile that has a body temperature that varies with the temperature of its surroundings. Thus, **poikilothermy.**

**poikilothermic** *Biological Energetics.* having a body temperature that varies according to external conditions.

**Poincaré electron** *Physics.* a classical model of an electron in which the electron is held together by nonelectromagnetic forces and therefore has no self-stress; the model is unstable and thus has infinite self-energy as a point electron. [Named for French mathematician Henri *Poincaré,* 1854–1912.]

**point-contact cell** *Photovoltaic.* a highly efficient silicon concentrator cell that employs light-trapping techniques and point-diffused contacts on the rear surface for current collection.

**point-focus(ing) concentrator** *Solar.* any solar power system that focuses solar energy onto a single central receiver, such as a boiler, engine, or photovoltaic array; a concentrator that directs the solar flux to a specific point. Also, **point-focus(ing) collector.**

**point source** *Environment.* a stationary location or fixed facility that is a single identifiable source of pollution, such as a power plant, factory, mine, or municipal facility that discharges pollutants into air or surface water. Thus, **point-source pollution.**

**poison** *Nuclear.* a term for a material in the vicinity of a reactor core, other than fissionable material, that will absorb neutrons. The addition into the reactor of these materials, such as control rods or boron, is described as an addition of negative reactivity.

**polar** *Physics.* having to do with a pole or with the quality of polarity.

**polarimeter** *Lighting.* an instrument for determining the amount of polarization of light.

**polarity** *Physics.* **1.** a physical property of some systems by which there exists two points having opposite characteristics, such as an electric dipole. **2.** a characteristic of the poles of a magnet (north or south), or of the terminals of a battery (positive or negative). **3.** the direction in which a direct current flows.

**polarization** *Physics.* **1.** a state in which rays of light or other radiation traveling in different directions exhibit different properties. **2.** the separation of the positive and negative charges of an atom or molecule by an external force, typically an electric field. **3.** *Storage.* an increase in the internal resistance of a battery cell, which shortens the cell's useful life as a result of an active chemical change occurring within it. Thus, **polarize.**

**polder** *Wind.* **1.** an area of land reclaimed from the sea by means of drainage. **2.** the use of wind power to accomplish such a reclamation.

**pole** *Physics.* **1.** one of two points, parts, or regions that have opposing qualities or tendencies, such as the ends of a magnet, the electrodes of an electrolytic cell, or the terminals of a battery. **2.** see MAGNETIC POLE.

**poleward flux** *Earth Science.* a process caused by the fact that more heat is incipient on and absorbed at lower than higher latitudes; this excess heat then moves from the tropics to the poles in both hemispheres.

**polishing treatment** *Environment.* a secondary or final level of treatment for domestic sewage to further reduce suspended solids and other pollutants before discharge.

**pollutant** *Environment.* a substance or agent that causes pollution; any harmful human-produced substance present in the environment.

**polluter pays** *Policy.* the principle that those causing pollution should bear the cost of preventing such pollution, or the cost of mitigating its impacts.

**pollution** *Environment.* any alteration of the natural environment producing a condition that is harmful to living organisms. Pollution may occur naturally, as when an erupting volcano emits sulfur dioxide, but the term usually refers to the negative effect of human activities; e.g., automobile exhaust emissions, oil spills, the dumping of industrial wastes in the water supply, the overuse of pesticides and chemical fertilizers, improper disposal of solid wastes, and so on.

**pollution control** *Environment.* a general term for any process of preventing or reducing the generation and release of pollutants, contaminants, hazardous substances, or wastes. Also, **pollution prevention.**

**pollution credit** *Policy.* **1.** a unit of pollutant emission allotted to a given company or operation, which can be sold or traded to others if it is not used; i.e., transferred from a firm that has reduced its emissions below the level required to another entity that has failed to meet the required standard. **2.** informally, a government incentive or subsidy that is thought to lead to increased pollution; e.g., taxpayer support for the construction of new highways.

***pollution control*** *West Virginia scene near a Union Carbide manufacturing plant, prior to pollution control efforts that reduced emissions from the plant by more than 90%.*

**pollution haven** *Global Issues.* a country or state that has weaker environmental standards than others, with the hypothesis that companies engaging in highly polluting activities will choose to relocate to such a site to avoid the restrictions enforced elsewhere.

**pollution tax** see EMISSIONS TAX.

**polonium** *Chemistry.* a radioactive metallic element having the symbol Po, the atomic number 84, an atomic weight of 210, a melting point of 250°C, and a boiling point of 962°C; a very hazardous member of the uranium decay series that has no stable isotopes, occurs naturally in uranium ores, and is prepared artificially by neutron bombardment of bismuth; used as a source of alpha particles and neutrons.

**polychlorinated biphenyls** see PCBs.

**polycrystalline silicon** *Photovoltaic.* a form of crystalline silicon used as a light-absorbing semiconductor in photovoltaic cells, composed of many variously oriented individual crystals.

**polycyclic aromatic hydrocarbon (PAH)** *Chemistry.* a major class of unsaturated hydrocarbon compounds characterized by the presence of variable numbers of rings, formed in the incomplete burning of fossil fuels and vegetable matter. Many PAHs are identified as toxic and particularly as carcinogenic.

**polyethylene** *Materials.* a ductile, easily molded thermoplastic that is chemical resistant and has good insulating properties; it has various industrial uses, primarily in the form of packaged film and in pipe, electrical insulation, and molded products.

**polyforming** *Oil & Gas.* a petroleum refinery process to manufacture gasoline from petroleum gases, in which the gases containing unsaturated hydrocarbons are heated and injected into the middle of the cracking furnace.

**polymer** *Materials.* **1.** a large molecule (macromolecule) formed by the union of simpler units that are identical to each other (monomers); it may be natural, such as cellulose or DNA, or synthetic, such as nylon or polyethylene; polymers usually contain many more than five monomers, and some may contain hundreds or thousands of monomers in each chain. **2.** an industrial material or product composed of such molecules.

**polymerase chain reaction (PCR)** *Materials.* a technique in which repeated cycles of DNA synthesis are carried out to produce a large number of a specific DNA sequence; a technology to rapidly multiply fragments of DNA for application in fields such as molecular biology, gene analysis, biomedical research, and forensic science.

**polymer electrolyte membrane (PEM)** *Renewable/Alternative.* a solid, aqueous membrane impregnated with an appropriate acid, used as an electrolyte in fuel cells. The membrane allows positively charged ions to pass through it, but blocks electrons. It typically operates at relatively low temperatures (80–100°C). Thus, **proton-exchange membrane fuel cell (PEMFC).**

**polymerization** *Materials.* the formation of a polymer; a chemical process in which a series of simpler structural units combine to form large, chain-like molecules.

**polymer photovoltaic cell** *Photovoltaic.* an advanced cell design based on semiconducting polymers that have potential applications in both light-emitting diodes and photovoltaics.

**poly-S** polycrystalline silicon.

**polysaccharide** *Chemistry.* any of a group of carbohydrates composed of long chains of simple sugars; e.g., starch, cellulose, insulin, or glycogen.

**polystyrene** *Materials.* $(-C_6H_5CHCH_2-)_n$, a combustible, transparent polymerized styrene of high strength and impact resistance that is an excellent electrical and thermal insulator, widely used for packaging, moldings, insulation, and lamination.

**polysulfide bromide battery** *Storage.* a regenerative fuel cell technology that provides a reversible electrochemical reaction between two salt solution electrolytes (sodium bromide and sodium polysulfide).

**polytropic** *Thermodynamics.* describing a process in which a sample of gas is compressed or expanded while keeping the quantity $pV^n$ held at a constant value; if the process is at constant pressure, then $n = 0$; if the process is at constant temperature for an ideal gas, then $n = 1$; if the process is at constant entropy for an ideal gas, then $n$ is given by the ratio of specific heat at constant pressure to that at constant volume; and if it is at constant volume, then $n$ is infinitely large.

**polyurethane foam** *Materials.* a flexible or rigid cellular substance that is created through the reaction of a polyester with a substance that contains two –N=C=O groups; used for insulation or padding.

**polyvinyl chloride** see PVC.

**pondage** *Hydropower.* a body of water held in a pond or reservoir behind a dam, especially a small amount held for storage to compensate for temporary water shortages.

**poolco** *Electricity.* a utility-independent, privately owned business that can act as the intermediary between all power users and sellers in a given region.

**population dynamics** *Ecology.* **1.** the changing number of individuals of a particular species in a certain habitat over time, as well as the environmental factors that determine this. **2.** specifically, the level of human population in a given area as it changes (grows or in some cases decreases) over time.

**population momentum** *Social Issues.* the tendency for population growth to continue beyond the time that a replacement-level birth rate has been achieved, because of a relatively high concentration of people in the childbearing years.

**Population Reference Bureau** (est. 1929), a non-governmental organization whose mission is to inform people about the population dimensions of important social, economic, environmental and political issues.

**porosity** *Earth Science.* **1.** the fact of being porous; i.e., containing small voids or cavities that can be permeated by water, air, gas, oil, and so on. **2.** the ratio of the total amount of void space in a material to the overall bulk volume of the material.

**Porter hypothesis** *Economics.* the principle that firms facing stringent environmental regulations, rather than being hurt by higher costs of production, will derive a competitive advantage because they are forced to innovate in response to the regulation. [Named for economist Michael E. *Porter.*]

**portfolio standard** see RENEWABLE PORTFOLIO STANDARD.

**Portland cement** *Materials.* a hydraulic cement made of pulverized limestone and clay or shale; historically one of the fundamental materials of industrial society as a basic ingredient in concrete and mortar.

**positive** *Electricity.* the opposite of NEGATIVE; specific uses include: **1.** having the type of electric charge in which electrons are minimally present, as that developed on a glass object that has been rubbed with silk. **2.** describing the part of an electric cell toward which the current flows. **3.** describing an element or compound that tends to lose electrons.

**positive displacement** *Consumption & Efficiency.* a term for mechanical equipment characterized by a reduction of the internal volume of a chamber, usually by a piston. Thus, **positive-displacement compressor, pump,** and so on.

**positive electrolyte negative (PEN)** *Renewable/Alternative.* the assembled cathode, electrolyte, and anode of a solid oxide fuel cell.

**positive radiative forcing** see RADIATIVE FORCING.

**positron** *Physics.* the antiparticle of an electron, having the same mass and spin as the electron, but opposite charge and magnetic moment.

**possible sun duration** *Solar.* the sum of the time intervals within a given period during which the sun is above the real horizon, which may be obscured by mountains, trees, buildings, and so on.

**post-aeration** *Environment.* describing a wastewater facility or basin into which oxygen is introduced at a later stage of treatment to further reduce biological and chemical oxygen demand.

**postage stamp rate** *Electricity.* a flat rate charged for transmission service without regard to distance; so called because of the analogy with government postage, for which the rate is based on the level of service and the distance over which the posted item is transmitted is not a factor. Compare MEGAWATT-MILE RATE.

**posted price** *Oil & Gas.* a stipulated price of a barrel of oil that was historically used by exporting nations as a base to collect taxes from oil companies. In most cases, posted prices were significantly higher than actual market prices.

**post-industrial** or **postindustrial** *Social Issues.* describing the premise that certain societies (e.g., the U.S.) have moved from a conventional industrial economy to a newer model based primarily on the provision of services and the exchange of knowledge, rather than on the marketing of finite goods.

**post mill** *Wind.* a traditional European mechanical windmill used for grinding grain and other materials (e.g., tobacco, paper), in which the tower body is supported on a vertical post resting on a wooden trestle.

**potable** *Health & Safety.* suitable to drink; describing water that is satisfactory for drinking and culinary purposes, from a health and taste standpoint.

**potash** *Materials.* **1.** a commercial form of potassium carbonate, $K_2CO_3$. **2.** a popular name for various other compounds that contain potassium, such as potassium hydroxide, KOH, or potassium chloride, KCl.

**potassium** *Chemistry.* a metallic element having the symbol K, the atomic number 19, an atomic weight of 39.098, a melting point of 63°C, and a boiling point of 770°C. It is a soft, silver-white, extremely reactive alkali metal that is fairly abundant in the earth's crust; it is essential for plant growth and for human and animal nutrition and is used extensively in its compound form as a fertilizer component.

**potential** *Physics.* the amount of work per unit mass (or charge) that is required to move a mass (or charge) through a gravitational (or electrostatic) field, from an infinite distance to the point at which this force is to be evaluated. See also ELECTRIC POTENTIAL.

**potential difference** *Electricity.* the work required to move an amount of charge between two positions in a circuit, divided by the strength of the charge; measured in volts.

**potential energy** *Thermodynamics.* the energy stored in a body or system as a consequence of its position, composition, shape or state; e.g., gravitational energy, electrical energy, nuclear energy, or chemical energy.

**potential of hydrogen** see PH.

**potential temperature** *Thermodynamics.* the temperature of a compressible fluid that ideally would result if it were compressed or expanded to a specific pressure (typically standard atmospheric pressure), with no gain or loss of heat.

**potentiometer** *Electricity.* an instrument that measures electromotive force or potential difference by comparing a part of the voltage to be measured against a known electromotive force.

**pound** *Measurement.* **1.** a unit of avoirdupois weight equal to 16 ounces, equivalent to 0.45 kilograms. **2.** a unit of troy weight equal to 12 ounces, or 0.37 kilograms.

**pound force** *Measurement.* a force that will accelerate a mass of one pound at a rate of 32.2 feet per second (i.e., an acceleration equal to that of gravity).

**pound of steam** *Measurement.* one pound mass of water converted to steam.

**pour point** *Materials.* the lowest temperature at which a liquid will flow under prescribed conditions.

**poverty** *Social Issues.* the fact of being poor; the absence of wealth. A term with a wide range of interpretations depending on which markers of income and material possessions are employed, and also construed by some observers to include not only economic factors but also cultural, social, and geographic ones.

**powdered coal** *Coal.* the finely crushed form of coal produced by pulverization.

**power** *Thermodynamics.* **1.** the rate in which energy is transferred or converted per unit of time. **2.** the rate at which work is done. The SI unit of power is the watt, which is equal to one joule per second.

**power ascension** *Nuclear.* the time between a nuclear plant's initial fuel loading date and its date of first commercial operation.

**power brake** *Transportation.* an automotive brake operated by pressure from a power source, such as a compressed-air reservoir, proportionately to a smaller amount of pressure applied to the brake pedal.

**power coefficient** *Wind.* a measure of the technical efficiency with which a wind turbine converts wind energy to electricity, expressed as the power produced by the rotating turbine as a percentage of the total energy of the wind passing through the area of rotation. It does not necessarily follow that the higher the coefficient the better, because above a certain wind speed excess energy must be wasted to avoid undue stress on the system.

**power conditioner** *Electricity.* a device designed to ensure that continuous, usable current is supplied from available power, as by the regulation of line voltage (e.g., to prevent surges) or the prevention of energy loss to electromagnetic fields and heat. Thus, **power conditioning.**

**power density** *Thermodynamics.* the amount of energy harnessed, transformed, or used per unit area per unit time.

**power exchange** *Electricity.* an entity providing a competitive spot market for electric power through the auction of generation and demand bids.

**power factor** *Measurement.* the ratio between actual electrical power and apparent power (the product of voltage times current).

**power generation** see GENERATION.

**power grid** see GRID.

**powerhouse** *Hydropower.* **1.** the main structure of a hydropower facility, housing the generating units and related equipment. **2.** any large structure employed to generate electrical power.

**power outage** see OUTAGE.

**power plant** or **powerplant** *Consumption & Efficiency.* a general term for any facility in which some other form of energy (e.g., steam, hydropower) is converted into electrical energy.

**power pool** *Electricity.* an association of two or more interconnected electric systems having an agreement to coordinate operations and planning.

**power surge** see SURGE.

**power system dynamics** *Electricity.* the planning, operation, and security of power systems, including issues such as turbine and generator design and performance, load behavior under normal and disturbance conditions, system control and protection, and the analysis and prevention of power outages.

**powertrain** *Transportation.* **1.** all the moving parts that connect an engine to the point at which work is accomplished. **2.** specifically, the interconnected set of components driving the wheels to propel a motor vehicle, including the engine, transmission, driveshaft, and differential.

**POx** partial oxidation.

**PP** proton-proton (chain reaction).

**ppb** *Measurement.* parts per billion. See PPM.

**ppbv** *Measurement.* parts per billion by volume. See PPMV.

**pphpd** *Transportation.* passengers per hour per direction; a measure of capacity of a rail transit system, in terms of the maximum number of passengers that the system can move in one direction.

**ppm** *Measurement.* parts per million; the number of constituent parts by weight of a specific substance contained within one million parts of a larger medium; used to measure extremely small concentrations in water, gas, soil, rock, and so on.

**ppmv** *Measurement.* parts per million by volume; the fraction of the total volume of a given medium occupied by a specific component of the medium; e.g., carbon dioxide in air.

**ppt** *Measurement.* parts per trillion. See PPM.

**pptv** *Measurement.* parts per trillion by volume. See PPMV.

**prairie plow** *History.* an innovative steel-bladed plow developed by John Lane in 1833, much more efficient in plowing the thick prairie sod of the American Midwest than existing wooden plows.

**PRB** Population Reference Bureau.

**precautionary principle** *Social Issues.* **1.** the ethical principle that if the consequences of a given action are not definitively known, but have a certain potential to be harmful, it is better not to proceed than to risk this negative outcome. **2.** *Environment.* specifically, an approach to environmental management in which actions are avoided if there is a probability that they will lead to environmental damage.

**precession** *Earth Science.* a slow, periodic movement of the earth's rotation axis, caused by gravitational pulls from other celestial objects in the solar system. The earth moves through one complete precession cycle over a period of about 26,000 years; this is one of the factors causing the planet to receive different amounts of solar energy over time.

**precision cooling** *HVAC.* describing a cooling system designed to provide a properly conditioned environment for modern computers and other sensitive electronic equipment; in comparison with a conventional human-occupied environment, heat densities are much higher in an electronic facility and there is a narrower tolerance range for temperature and humidity.

**precocial** *Ecology.* describing organisms in which the offspring are hatched or born at a relatively advanced stage of development; a term applied to certain birds and mammals.

**precombustion capture** *Conversion.* a method of capturing carbon dioxide from fossil fuels prior to their conversion; this often occurs in gasification systems, where carbon in the coal, oil, or gas is transformed into $CO_2$ or CO (carbon monoxide) at high pressures. This process also results in the production of hydrogen, which could then be used as fuel.

**precombustion chamber** *Transportation.* a small chamber adjacent to the cylinder head of some compression ignition engines into which the fuel is injected at the end of the compression stroke.

**precursor** *Environment.* a term for atmospheric compounds that themselves are not greenhouse gases or aerosols, but that have an effect on greenhouse gas or aerosol concentrations by taking part in physical or chemical processes that regulate the production or destruction rates of these substances.

**predation** see PREDATOR–PREY SYSTEM.

**predator–prey system** *Ecology.* an ecological system in which one species (the prey) forms an important part of the food consumption regimen for another (the predator); e.g., the North American lynx and the snowshoe rabbit. Extensive studies have been made of the relationship in population levels of such a pair of species.

*predator–prey This prairie dog* (Cynomys spp) *is the primary prey animal for the black-footed ferret* (Mustela nigripes)*; as prairie dogs have experienced severe loss of habitat in recent years, this predator's population has virtually disappeared.*

**predatory pricing** *Economics.* the practice of charging particular customers artificially low prices in order to underprice or eliminate a competitor; electricity and natural gas markets are frequently investigated for the possibility of this behavior, although market complexities can make it difficult to determine if it actually exists.

**preferential oxidation (PrOx)** *Hydrogen.* a catalyzed chemical reaction in which species other than hydrogen are oxidized to yield a product stream having maximum amounts of hydrogen with minimum amounts of carbon monoxide and other contaminants.

**pregnant** *Materials.* a term for a solution containing valuable minerals that have not yet been extracted.

**pre-ignition** *Transportation.* the ignition of the fuel mixture in a gasoline engine cylinder before normal ignition by the spark plug, typically caused by overheated plug points. This results in an inefficient, rough-running engine.

**preindustrial** *Social Issues.* **1.** preceding industrialization; describing an event, process, society, and so on that does not involve the use of industrial methodology such as fossil energy, mass production, and capitalism. **2.** specifically, describing the world (or a specific society) prior to the Industrial Revolution of the early 19th century, that relied principally on renewable energy (wood) and animate energy converters (humans and draft animals).

**premium** *Oil & Gas.* a classification for gasoline having a higher octane rating; the classification varies by region but typically premium has an octane rating of 92–93. Thus, **premium gas(oline)** or **fuel.**

**preparation plant** *Coal.* a facility at which coal is made ready to be transported to the end user, including crushing, screening, and mechanical cleaning.

**prescribed burning** *Ecology.* **1.** among traditional peoples, the deliberate setting of surface fires in woodland or grassland; e.g., to clear away dead plants after a dry season, to provide greater access to forests, to improve feeding conditions for game animals, and so on. **2.** a similar contemporary strategy, to help prevent more destructive unplanned fires, to kill off unwanted plants that compete with commercial species, and so on.

**pressure** *Physics.* **1.** the amount of force that is exerted per unit area. **2.** specifically, the condition of a fluid, such as air or water, in which it exerts a perpendicular force per unit area on a surface in contact with it that has the same value at this point, regardless of the orientation of the surface. **3.** a measurement of this condition.

**pressure vessel** *Nuclear.* **1.** a strong-walled container surrounding the core of a nuclear reactor. It usually also contains the moderator, neutron reflector, thermal shield, and control rods. **2.** *Physics.* any structure designed to contain a fluid at a different pressure from the pressure surrounding the structure without changing the fluid volume.

**pressurize** *Physics.* **1.** to maintain normal atmospheric pressure within a system that has a higher or lower external pressure. **2.** conversely, to maintain a higher pressure within a system that would otherwise have normal atmospheric pressure. **3.** to control the pressure of a system or vessel in any manner. Thus, **pressurized, pressurizer, pressurization.**

**pressurized blast furnace** *Consumption & Efficiency.* a smelting furnace operated under pressure to reduce iron ore to pig iron, in which the off-gas line is constricted to allow more air to pass through the furnace at a faster rate, increasing the smelting rate.

**pressurized fluidized bed combustion** see FLUIDIZED BED COMBUSTION.

**pressurized solid oxide fuel cell (PSOFC)** *Hydrogen.* a type of fuel cell system with a tubular ceramic stack; typically used in stationary power plants, in a hybrid power system with a gasoline-fueled microturbine.

**pressurized water reactor (PWR)** *Nuclear.* a type of light water nuclear reactor distinguished by a primary cooling circuit that flows through the core of the reactor under very high pressure, and a secondary circuit in which steam is generated to drive the turbine; this is the most common type of reactor in use.

**Prestige** *Environment.* an oil tanker that broke in half and sank off Spain's northwest coast in 2002, while carrying nearly 20 million gallons of heavy fuel oil. The tanker's slow breakup released thousands of tons of oil into the waters of the Atlantic Ocean, causing one of Europe's worst environmental disasters.

**Prévost's theory** *Thermodynamics.* the principle that a body continuously exchanges heat with its environment, radiating energy which is independent of the environment, and increases or decreases its temperature depending upon whether it absorbs more radiation than it emits. [Described by Swiss physicist Pierre *Prévost,* 1751–1839.]

**prey** see PREDATOR–PREY SYSTEM.

**Price–Anderson Act** *Policy.* U.S. federal legislation (1957) that limits the amount of insurance that nuclear power plant owners must carry and caps their liability in the event of a catastrophic accident or attack. The amount of liability is considered to be far short of the actual financial consequences that could be incurred, and in effect the Act thus serves as a taxpayer subsidy of the nuclear power industry. [Sponsored by Senator Clinton *Anderson* and Congressman Melvin *Price.*]

**price band** *Oil & Gas.* the minimum and maximum price range for the average OPEC basket price of oil. OPEC members attempt to adjust their production relative to demand so as to keep prices within this band.

**price elasticity** *Economics.* the ratio of the percentage change in energy demand or supply, relative to the percentage change in energy price. Thus, **price elasticity of demand, price elasticity of supply.**

**price indexation** *Economics.* a method of determining the price of natural gas in which price is typically set at a level that is competitive with the market's other fuel options; the indexation formula typically ties the price of gas to a basket of alternate fuel prices, such as fuel oils, light oils, and coal.

**price realization** *Economics.* the unit price of oil or natural gas realized by a producer for certain volumes; individual price realizations typically differ from benchmark pricing because of factors such as variation in energy content or quality, distance from major markets, and fixed-price contracts.

**price shock** *Economics.* a large, sudden increase (or, less often, a decrease) in energy prices, especially the price of crude oil in the world market. Such shocks often have significant large-scale effects on economic growth, productivity, and inflation.

**price volatility** see VOLATILITY.

**Priestley, Joseph** 1733–1804, English chemist whose research helped laid the foundation for modern chemistry. He is credited with the discovery of oxygen in 1774, and he also performed important experiments with many other substances such as carbon dioxide, nitrous oxide, and ammonia.

**Prigogine, Ilya** 1917–2003, Russian-born Belgian chemist noted for his development of mathematical models of irreversible thermodynamics. He extended the second law of thermodynamics to systems that are far from equilibrium, and demonstrated that new forms of ordered structures (dissipative structures) could exist under such conditions.

**primary battery** *Storage.* an electrochemical cell or battery that contains a fixed amount of stored energy when manufactured, and that cannot be recharged after that energy is withdrawn.

**primary cell** *Electricity.* an electrochemical cell that is self-initiating in that the chemical energy of its constituents is changed to electrical energy when current is permitted to flow; it cannot be recharged electrically because of the irreversibility of the chemical reaction that occurs within it.

**primary energy** *Consumption & Efficiency.* **1.** the energy directly embodied in natural resources, prior to its being converted or transformed for use. **2.** all energy consumed by end users, usually also including the additional energy required for delivery of this energy to the end user. See below.

**primary energy consumption** *Consumption & Efficiency.* the total amount of energy consumed by end users, plus any losses that occur in the generation, transmission, and distribution of this energy.

**primary energy requirement** *Consumption & Efficiency.* the amount of primary energy needed to deliver a product to a consumer, including the energy needed for the production, use, and disposal of the product.

**primary fiber** *Biomass.* pulp made from trees or from other plant material in its natural state, rather than from recovered materials.

**primary gain** *Solar.* solar radiation transmitted directly through a window or other glazed aperture to the interior of a building. Compare SECONDARY GAIN.

**primary productivity** *Biological Energetics.* **1.** the process in which certain organisms (autotrophs; i.e., green plants and certain bacteria) utilize sunlight or chemical nutrients as a source of energy. **2.** the biomass tissue created in this manner. Also, **primary production.**

**primary energy** Primary energy is the energy embodied in natural resources prior to undergoing any human-made conversions or transformations. Examples of primary energy resources include coal, crude oil, sunlight, wind, running rivers, vegetation, and uranium. When primary energy is converted to a different form like the conversion of moving air molecules to rotational energy by the rotor of a wind turbine, which in turn may be converted to electrical energy by the wind turbine generator, part of the energy from the source is converted into *unusable* heat energy or, loosely speaking, is "lost". Rotors, gearboxes, or generators are never 100% efficient because of heat losses due to friction in the bearings, or friction between air molecules. These losses are attributable to the second law of thermodynamics. The example of wind underscores an important issue with respect to measuring the primary energy content of wind and solar since these are not combustion technologies. The primary energy of wind and solar can be measured as the kinetic energy of wind and the solar energy received on a surface, which depends on the surface's angle to the sun and the distance of the sun from the earth. These measurements, however, are not helpful for use in integrated energy systems models where all energy forms compete to satisfy end-use demand. At least two conventions for measuring non-fossil fuel primary energy have been adopted by the energy forecasting community for renewable energy and nuclear generation: (a) the *output* of the conversion technology is assumed to be the primary energy, which implicitly assumes a conversion efficiency of 100%, or (b) an average fossil fuel conversion factor is assumed and used to back calculate an equivalent fossil energy primary equivalent (e.g., kWh or Btu). Each method has shortcomings, but the convention does make it convenient to incorporate non-fossil sources in energy system models.

**Andy S. Kydes**
Energy Information Administration
U.S. Department of Energy

**primary products** *Economics.* unprocessed or partially processed goods, often used to produce other goods; e.g., agricultural commodities such as grain or vegetables, raw materials such as iron ore, and fuels such as crude oil.

**primary radiation** see SECONDARY RADIATION.

**primary recovery** *Oil & Gas.* any oil recovery process that relies only on the natural energy available in the reservoir and adjacent aquifers, such as the use of natural water drive.

**primary refrigerant** *Refrigeration.* the main refrigerant in a circulating system that changes its state from a liquid to a vapor and back again, such as Freon or ammonia.

**prime mover** *Conversion.* **1.** an engine or device by which a natural source of energy is converted into mechanical power. **2.** in general, the initial agent in a process that activates or energizes all later stages of the process.

**principle of equivalence** see MASS-ENERGY EQUIVALENCE.

**prism** *Lighting.* an element bounded by two polished plane surfaces that deviates or disperses a beam of light; typically a wedge-shaped piece of glass.

**Prius** *Renewable/Alternative.* the first modern hybrid-electric vehicle, produced by Toyota and initially marketed in Japan in 1997.

**private good** *Consumption & Efficiency.* a good that is typically transferred from buyer to seller in a market; its consumption is not freely available to all and it cannot be consumed by one person without reducing the amount left for others; e.g., a piece of bread or a gallon of gasoline. Compare PUBLIC GOOD.

**privatization** *Economics.* the process of moving from a state-controlled economic system to a privately owned system operating for profit.

**probabilistic risk analysis (PRA)** *Measurement.* a process of systems analysis that calculates the likelihood of undesirable events and their uncertain consequences; used to quantify the risks caused by high-technology installations and other such complex systems for which classical statistical analysis is difficult or inadequate.

**probability of causation** *Health & Safety.* the estimated probability that a toxic or pathological event (such as the incidence of cancer) was caused by a postulated action (such as radiation exposure).

**process energy analysis** *Measurement.* a method to calculate the embodied energy of a good or service, by assessing the energy used directly in each successive step of the production of that good or service.

**process heating** *HVAC.* a procedure that conveys heat through the circulation of combustion gases.

**process integration** *Consumption & Efficiency.* a broad term for various integrated approaches to industrial processes, with the goal of increasing energy efficiency and reducing the environmental impact from process industries.

**processing gain** *Oil & Gas.* the volumetric amount by which total output from an oil refinery is greater than input for a given period of time, due to the processing of crude oil into products that, in total, have a lower specific gravity than the crude oil processed. Thus, **processing loss.**

**produced water** *Oil & Gas.* water that is brought to the surface along with petroleum in an oil production process; it often contains high levels of salt, heavy metals, and hydrocarbons.

**producer gas** *Chemistry.* **1.** a highly flammable, toxic gas obtained by burning coal or coke with a restricted supply of air, or by passing air and steam through a bed of incandescent fuel so that the carbon dioxide formed is converted into carbon monoxide. **2.** *Biomass.* a similar gaseous fuel obtained from wood or other biomass.

**producer-subsidy equivalent** *Economics.* an integrated metric of the aggregate support provided to producers by various government policies; policies that act as taxes rather than subsidies are incorporated by using the opposite sign.

**production function** *Economics.* a mathematical representation of the quantitative relationship between productive inputs and the level of output of some good or service, expressing the maximum quantity of output that can be obtained from any particular amounts of inputs.

**production tax** *Economics.* **1.** a tax levied per unit of energy produced. **2.** a tax per unit of pollution generated.

**productivity** *Economics.* **1.** the output of any production process, per unit of input. **2.** the marginal relationship of inputs (land, labor, capital) to outputs (GDP), often used to measure the economic efficiency of production. Thus, **labor productivity, land productivity, capital productivity. 3.** *Ecology.* a measurement of the potential rate of biological production of a given area; defined as the amount of light energy converted to chemical energy by autotrophs (self-feeders) per unit time and/or unit space.

**product recycling** see RECYCLING.

**programmable** *Consumption & Efficiency.* describing a device or system (e.g., a thermostat) that can be adjusted by the end user to operate in a certain desired manner.

**progress option** *Consumption & Efficiency.* a long-term strategy for industrial societies based on the premise that future technological advances and forces of the marketplace will be sufficient to overcome any problems caused by resource shortages (especially fossil fuels). Compare PRUDENCE OPTION.

**projected dose** *Nuclear.* the radiation dose that can be expected to be incurred if a specified countermeasure or set of measures (or no countermeasures) were to be taken.

**projectile** *Physics.* **1.** any object that is projected by an external force and that continues to travel by its own inertia. **2.** such an object fired as a weapon of war; e.g., a bullet, artillery shell, or grenade. **3.** a rocket or guided missile.

**proliferation** **1.** see NUCLEAR PROLIFERATION. **2.** see CELL PROLIFERATION.

**Prometheus** *History.* in Greek mythology, the Titan who stole fire from Zeus and gave it to mortal humans, which unleashed a flood of inventiveness and productivity on earth. (The term **Promethean** thus refers to great originality or creativity.) For this offense, Zeus punished Prometheus by having him chained to a rock with an eagle tearing at his liver.

**promethium** *Chemistry.* a radioactive rare-earth (lanthanide) element having the symbol Pm, the atomic number 61, an atomic weight of 147, and a melting point of 1160°C; an artificial element produced by nuclear fission.

**prompt neutron** *Nuclear.* a neutron emitted during fission in the formation of fission fragments.

**propane** *Chemistry.* $CH_3CH_2CH_3$, a colorless gas with a natural gas odor; soluble in alcohol and slightly soluble in water; boils at –42.5°C and freezes at –187.7°C; a dangerous fire risk. It is widely used as an industrial and household fuel, and also as a solvent, refrigerant, and aerosol propellant.

**propellant** *Materials.* an agent that propels; i.e., that moves something forward with force, such as a fuel mixture that provides the thrust for a rocket, or a compressed gas that dispenses the contents of an aerosol container.

**propeller** *Transportation.* **1.** an assembly of radiating blades around a revolving hub that produces thrust or power to move an aircraft. **2.** *Wind.* a similar device that employs the force of wind to provide mechanical energy, as in a windmill.

**propeller turbine** *Wind.* a collective term for wind turbine blades that are analogous in external design to an airplane propeller; common types may have either two or three blades and the blades may be fixed or self-adjusting.

**proppant** *Oil & Gas.* a material consisting of comparably sized particles, used to hold fractures open after a hydraulic fracturing treatment to a producing oil or gas well; may be naturally occurring sand grains or specially engineered materials such as high-strength ceramics.

**propulsion** *Physics.* the fact of moving an object and maintaining this motion.

**protectionism** *Policy.* an economic policy of promoting domestic industries through the use of high tariffs and other regulations to discourage imports. Thus, **protectionist.**

**protein biosynthesis** *Biological Energetics.* the complex mechanism by which proteins are synthesized, amino acid by amino acid, in accord with a sequence of nucleic acid bases in a messenger RNA molecule that has been transcribed from a DNA gene. Many proteins and RNA molecules are involved in the apparatus that performs this function.

**protium** *Hydrogen.* the ordinary light hydrogen isotope, consisting of one proton and one electron.

**protocol** *Policy.* an agreement linked to an existing international convention as a separate, additional agreement that must be signed and ratified by the parties to the convention; e.g., the Kyoto Protocol.

**proto-industry** *History.* an early phase of industrialization involving the decentralized production of handcrafted work in households for sale in relatively distant markets; e.g., textiles. Thus, **proto-industrialization.**

**proton** *Physics.* a stable elementary particle found in the nucleus of ordinary matter.

**proton–exchange membrane (PEM)** *Hydrogen.* a type of fuel cell technology featuring a membrane through which protons but not electrons can pass, forcing the latter to move along an electrode and generate a current. Thus, **proton-exchange membrane fuel cell (PEMFC).**

**proton–proton (PP) chain reaction** *Nuclear.* one of the two fusion reactions by which stars convert hydrogen to helium, thought to be the dominant source of energy in stars the size of the sun or less. The reaction turns four hydrogen nuclei (four protons) into one helium nucleus (an alpha particle). Since the alpha particle has less mass than the four protons individually, the reaction releases energy, and the excess energy heats the surrounding gas. Compare CARBON–NITROGEN–OXYGEN CYCLE.

**proved (proven) reserves** *Consumption & Efficiency.* the estimated quantities of energy sources, such as crude oil or natural gas, that can be demonstrated with reasonable certainty to be recoverable under existing economic and operating conditions. The location, quantity, and grade of the energy source are usually considered to be well established in such reserves. See also specific entries, such as COAL RESERVES.

**provider of last resort** *Electricity.* a legal obligation traditionally given to utilities, to provide service to a customer after other competitors have decided they do not want to provide that customer's services.

**PrOx** or **PROX** preferential oxidation.

**proximate analysis** *Coal.* an analysis of the proportionate amounts of various constituents of coal, made by classifying the coal as moisture (water), volatile matter, fixed carbon, or ash.

**proxy data** *Climate Change.* an indicator used to examine variations in climate over time; e.g., tree ring records, crop yields and harvest dates, glacier movements, ice cores, snow lines, coral characteristics, seasonal sediment levels, insect and pollen remains, marine microfauna, and so on. Also, **proxy (climate) indicator.**

**prudence option** *Consumption & Efficiency.* a long-term strategy for industrial societies based on the premise that significant changes in lifestyle and values will be necessary in the future to deal with energy and resource realities, because technological advances and forces of the marketplace will not be sufficient to overcome shortages. Compare PROGRESS OPTION.

**prudent investment (test)** *Policy.* a regulatory standard requiring that costs recovered from customers by a regulated public utility must have been reasonable, given what was known or knowable at the time the costs were incurred.

**psi** *Measurement.* pounds (per) square inch; a unit of measure equal to the amount of pressure applied by a mass of one pound acting on an area of one square inch.

**psia** *Measurement.* pounds (per) square inch absolute; a unit of measure for pressure that excludes atmospheric pressure, with the assumption of a partial to full vacuum (i.e., a pressure below 14.7 psi).

**psig** *Measurement.* pounds (per) square inch gauge; a unit of measure for pressure relative to ambient atmospheric pressure; i.e., a psig of 0 for a pressure gauge at sea level indicates an absolute pressure (psia) of 14.7 psi, since the gauge exists in an environment with that pressure.

**PSOFC** pressurized solid oxide fuel cell.

**psychro-** *Measurement.* a prefix meaning "cold".

**psychrometer** *Measurement.* an instrument used to measure the moisture content of the atmosphere, composed of two thermometers, one having a wet bulb and the other a dry bulb.

**psychrometry** *Thermodynamics.* the study of the thermodynamic properties of moist air and the use of these properties to analyze conditions and processes involving moist air. Also, **psychometrics.**

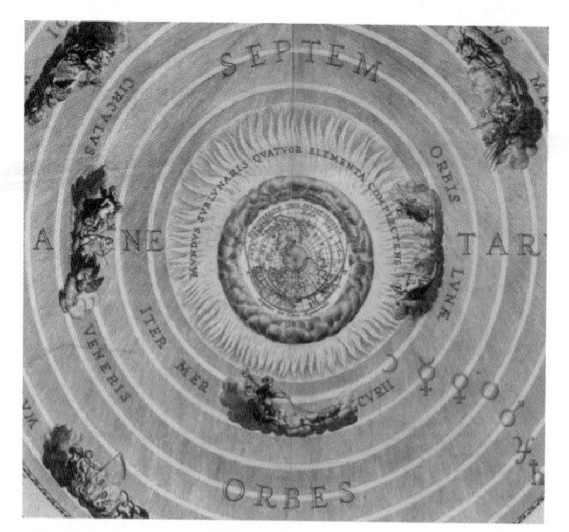


***Ptolematic*** *Detail from an artistic representation of the Ptolemaic model of the solar system, showing the earth at the center.*

**Ptolemaic system** *History.* a theory developed by Ptolemy about 150 AD holding that a motionless earth is the center of the universe with the sun, moon, and planets revolving around it, while the fixed stars are attached to an outer sphere concentric with the earth; this model was generally accepted in the West until the establishment of the COPERNICAN theory about 1500 years later.

**Ptolemy** c. AD 100–170 Greek philosopher who presented a widely accepted model of the solar system (the PTOLEMAIC SYSTEM). He also made important contributions to geography and cartography.

**P-type** *Electricity.* positive type; a semiconductor to which an impurity has been added so that the concentration of holes is much higher than the concentration of electrons; the electrical current is carried chiefly by electron flow into these holes. Thus, **P-type layer, P-type semiconductor, P-type silicon.** Compare N-TYPE.

**public bad** *Consumption & Efficiency.* an item whose consumption cannot be avoided and whose presence makes one worse off (or whose absence would make one better off); e.g., crime, disease, air pollution, offensive street noise. The converse of a PUBLIC GOOD.

**public domain** *Social Issues.* the realm of assets of knowledge and creativity that are freely available to all citizens, rather than having their access restricted by a copyright, patent, or the like held by some individual or organization.

**public exposure** *Nuclear.* a term for radiation exposure incurred by members of the public at large, as opposed to specific exposure of individuals through medical treatment or occupational contact.

**public good** *Consumption & Efficiency.* a relatively nondepletable good that is generally available to all and that can be consumed by one person without reducing the amount available for others; e.g., clean air, a public parade, a lighthouse. Compare PRIVATE GOOD.

***public good*** *The lighthouse is often cited as an example of a public good, on the grounds that all passing ships can benefit from its presence while no ship's use of it detracts from that of others.*

**public health** *Health & Safety.* the branch of medicine concerning with improving health and preventing disease at the population level, including such measures as: the control of communicable diseases, both traditional and emerging; the prevention of noncommunicable diseases (e.g., cancer, heart disease); health education and the dissemination of health information; the provision of community services, outreach, and relief efforts.

**public monopoly** *Economics.* the exclusive right of the state to undertake an economic activity such as oil or natural gas extraction.

**public utility** see UTILITY.

**puddling** *Materials.* a process of converting cast iron into wrought iron or steel by subjecting it to intense heat and frequent stirring in a reverberatory furnace in the presence of oxidizing substances, by which it is freed from a portion of its carbon and other impurities. A historic advance in the quality and efficiency of steel production, developed by English ironmaster Henry Cort in 1784. Thus, **puddled steel.**

**PUHCA** *Electricity.* Public Utility Holding Company Act; U.S. legislation enacted in 1935 to protect utility stockholders and consumers from unfair financial and economic practices of utility holding companies.

**pulley** *History.* one of the simple machines of antiquity, typically consisting of a metal wheel fixed to a shaft, with a rope, belt, band, or the like passing over the wheel to transmit motion or energy.

**Pullman car** *History.* the first successful railroad sleeping car, developed by U.S. entrepreneur George Pullman (1857).

**pulp** *Materials.* fibrous material prepared from wood, cotton, grasses, and other such natural materials by chemical or mechanical processes, for use in making paper or cellulose products.

**pulp substitute** *Biomass.* a combination of paper and paperboard trimmings from paper mills and converting plants.

**pulpwood** *Biomass.* a collective term for forest products that are used to produce pulp, such as woodchips and logs.

**pulse** *Biological Energetics.* **1.** the rhythmic pressure of the blood against the walls of a vessel, especially an artery. **2.** *Measurement.* an abrupt change in a quantity, characterized by a rise and a decline, typically occurring over a short time interval; e.g., a sharp variation of a current or voltage having a normally constant value.

**pulse amplitude** *Measurement.* the magnitude of a pulse, measured with respect to the normally constant value.

**pulsed power technology (PPT)** *Consumption & Efficiency.* a technology used to generate and apply energetic beams and high-power energy pulses for a wide variety of applications.

**pulse duration** *Electricity.* the time interval between the first and last instants at which the instantaneous value is a specified fraction, often 50%, of the peak pulse amplitude.

**pulse duration modulation** *Electricity.* a type of modulation by which the duration of a pulse varies in accordance with a characteristic of the modulating signal.

**pulse width** see PULSE DURATION.

**pulsing paradigm** *Ecology.* the principle that complex systems and processes in nature (and human society) operate in a pulsing manner; i.e., alternating between different states rather than existing in a particular steady state.

**pulverization** *Coal.* the technique of breaking down coal into a fine powder for burning, which provides advantages such as a higher combustion temperature, improved thermal efficiency, and a lower air requirement for combustion.

**pulverized coal** *Coal.* coal that has been crushed to a fine dust in a grinding mill, and then blown into the combustion zone of a furnace where it burns rapidly and efficiently.

**pump** *Consumption & Efficiency.* **1.** a device that converts mechanical energy into fluid energy, typically by suction or compression; used to move water, air, or other fluids into, through, or out of a system. **2.** to move a fluid by means of such a device. **3.** *Bioenergetics.* any process in nature that is comparable to this fluid movement; e.g., the movement of blood in the body by the heart, or the transport of carbon to ocean depths. **4.** *Physics.* an energy source that increases the number of highly energized electrons in an electron stream.

**pumped storage** *Hydropower.* a method of storing and producing electricity to supply high peak demand. During periods of low demand, excess generation capacity is used to pump water into an elevated reservoir. When demand rises, water is released back into the lower reservoir through a turbine, generating electricity. Between 70 and 85% of the electrical energy used to pump the water into the elevated reservoir can be recaptured. Also, **pumped hydrostorage.**

**pump-to-wheels** *Transportation.* describing the aspect of vehicle fuel use that occurs when the fuel is consumed by the end user; i.e., after the fuel has actually gone into the vehicle from the gas pump. Compare WELL-TO-WHEELS.

**punch mining** *Mining.* a relatively low-cost technique used to recover coal from deposits not economically accessible by other methods, in which a series of narrow entries are "punched" into the outcrop of a coal bed and then a series of rooms and crosscuts are developed further within the coal. Thus, **punch mine.**

**purchasing power parity (PPP)** *Economics.* a rate of exchange that accounts for price differences across countries, allowing international comparisons of real output and incomes. At the U.S. PPP rate, the PPP of $1 has the same purchasing power in the given domestic economy as $1 has in the United States.

**purge** *Chemistry.* to employ a gas to remove residual gases or liquids from a container; e.g., to remove air and water vapor from the refrigerant inside a chiller.

**PURPA** *Policy.* Public Utility Regulatory Policies Act; a U.S. law (1978) intended to encourage greater use of renewable and energy-efficient technologies.

**Putnam, Palmer Coslett** 1900–1984, U.S. wind energy pioneer who conducted the nation's first great experiment in converting energy from the wind atop a peak in Vermont called GRANDPA'S KNOB (1941).

**putty-clay model** *Economics.* an approach to technology based on the assumption that a high degree of flexibility ("putty") is available in the choice of production equipment before the process begins, but that once it is initiated flexibility is extremely limited ("clay"). Thus new investment would be required to improve productivity.

**PV** the abbreviation for photovoltaic; see PHOTOVOLTAIC entries.

**P-V** pressure-volume.

**PVC** *Materials.* polyvinyl chloride $(-H_2CCHCl-)_x$; a white powder or colorless granules available as film, sheet, fiber, or foam, resistant to weathering and moisture and having good dielectric properties; a polymer of vinyl chloride that is widely used for many industrial purposes, especially for construction uses such as piping, conduits, and siding.

**PV diagram** *Thermodynamics.* pressure–volume diagram; a tool for the study of heat engines that graphically illustrates the amount of work done during a cyclic heat engine process.

**PV/T** *Photovoltaic.* photovoltaic-thermal.

**P–V work** *Thermodynamics.* a form of energy that arises from the pressure ($P$) and the volume ($V$) of a fluid. It is numerically equal to $PV$, the product of pressure and volume. Also, ***P–V* energy.**

**PWR** pressurized water reactor.

**pygas** short for PYROLYSIS GASOLINE.

**pyranometer** *Solar.* an instrument that measures the intensity of total solar radiation.

**pyrgeometer** *Solar.* an instrument that measures net atmospheric irradiance on a horizontal, upward-facing black surface at the ambient air temperature.

**pyrheliometer** *Solar.* an instrument that measures the intensity of direct solar radiation.

**pyro-** *Chemistry.* a prefix meaning "fire" or "heat."

**pyrochemical** *Materials.* describing a high-temperature reprocessing technique that uses selective reduction and oxidation in fused salts or metals to recover nuclear materials.

**pyrolysis** *Chemistry.* **1.** the process of decomposition by heat. **2.** *Biomass.* specifically, the thermal decomposition of biomass at high temperatures in the absence of air; the end product is a mixture of solids (char), liquids (oxygenated oils), and gases (methane, carbon monoxide, and carbon dioxide) with the proportions determined by operating temperature, pressure, oxygen content, and other conditions. Thus, **pyrolytic.**

**pyrolysis gasoline** *Oil & Gas.* a byproduct from the manufacture of ethylene by steam cracking of hydrocarbon fractions such as naphtha or gas oil, used either for gasoline blending or as a feedstock.

**pyrometallurgical** *Materials.* describing a high-temperature reprocessing technique that processes materials in metal form.

**pyrometer** *Measurement.* an instrument for measuring high temperature, usually those above the range of mercury thermometers. Thus, **pryometry.**

**pyrophoric** *Materials.* describing a material that can spontaneously ignite in air.

**pyrotechnics** *Materials.* **1.** the use of chemicals to produce smoke or light, as for signaling, illumination, or screening. **2.** another term for fireworks or fireworks displays.

***pyrotechnics*** *Fireworks display at the Statue of Liberty. The word "pyrotechnics" derives from a Greek term meaning literally "the art of fire".*

**quad** *Measurement.* a unit of energy equivalent to $10^{15}$ British thermal units, or approximately $1.055 \times 10^{18}$ joules.

**Quadricycle** *History.* an early self-propelled car developed by Henry Ford (1896); a four-wheeled vehicle built on a steel frame with no body, steered by a tiller. [So named because it ran on four bicycle type wheels.]

**quadrupole** *Electricity.* a characteristic distribution of charges or magnetic poles formed by placing two electric or magnetic dipoles in close proximity to each other, usually in an alternating-pole arrangement.

**qualifying facility** *Electricity.* a cogenerator or small power producer that has the legal right to sell its excess power output to a public utility.

**quantitative risk analysis** another term for PROBABILISTIC RISK ANALYSIS.

**quantity-based** *Policy.* describing a policy that specifies a quantity outcome (as opposed to a price-based policy such as an environmental tax or subsidy); e.g., the renewable portfolio standard policy, which requires production of a specific quantity of electricity.

**quantize** *Physics.* to restrict the magnitude of a given physical quantity to that of a single member of a discrete set of values; these are typically integral multiples of some fundamental unit, or quantum. Thus, **quantization.**

**quantum** *Physics.* plural, **quanta;** the fundamental unit of electromagnetic energy that is absorbed in integral multiples of $E = h\nu$, where $h$ is Planck's constant.

**quantum number** *Physics.* a number indicating which member of an allowed set of magnitudes a quantized physical quantity possesses, usually an integer or half-integer.

**quantum theory** *Physics.* a modern branch of physics based on the premise that energy and momentum exist in discrete amounts called quanta (see QUANTUM) and that, at the atomic and subatomic levels, the effects of this quantization are significant. This field extends or supersedes classical mechanics in certain respects, especially in descriptions of very small systems (atom size or less), and of certain phenomena (e.g., superfluidity). Also, **quantum mechanics.** See below.

**quantum theory** Developed in the early 1900s by scientists such as Planck, Einstein, Bohr, Heisenberg, and Schroedinger, quantum theory is a term used to describe a physical theory that applies to systems normally at very small length scales (such as the level of the atom). Two primary motivations for this theory were the phenomena of the photoelectric effect and the spectra of light emitted by the Hydrogen atom, which before 1900 were not understood in terms of "classical" mechanics. Quantum theory was able to provide an explanation for these, but at the expense of violating many of our (classical) intuitive beliefs. The word "quantum" itself refers to a system which can only be in certain discrete states, so in order to evolve from one state to another, the system must make a "quantum leap". There are a number of surprising effects associated with quantum systems: tunneling, whereby a particle "tunnels" through a (classically) impenetrable barrier; the uncertainty principle, which states that it is theoretically impossible to know both the position and speed of a particle at one time to arbitrarily small accuracy; quantum teleportation, in which, for example, a packet of light is destroyed at one point in space and an identical packet is made to appear at another point. Such effects may appear counter-intuitive (if not impossible!), but they are very real (and measurable experimentally) in the quantum world of the very small. The modern development of devices such as superconductors, lasers, MRI scanners, and semiconductors all required applications of quantum theory, and with things like quantum computers on the horizon, there is little doubt that quantum theory will play an increasingly important role in the technology of the future.

**Randy Kobes**
University of Winnipeg

**quark** *Physics.* any of a group of elementary particles that make up the hadrons; quarks are acted on by the strong, electroweak, and gravitational forces. Three main types (flavors) of quarks are identified: up (u), down (d) (these are the two types found in ordinary matter), and strange (s). Other types are charm or charmed (c); bottom or beauty (b); and top or truth (t). [A term coined by U.S. physicist Murray Gell-Mann; said to be from a word used by the Irish author James Joyce in his unconventional novel *Finnegan's Wake.*]

**quartz** *Materials.* $SiO_2$, a transparent to translucent trigonal mineral with a vitreous luster, commonly white or colorless but also occurring in a variety of colors; the most abundant and widely distributed of all minerals.

**quasar** *Physics.* quasi-stellar radio source; a compact radio source or visible object, characterized by spectra with high redshifts, implying great distances and consequent high rates of energy production. Quasars are thought to be the most distant and most luminous objects in the universe.

**quenching** *Materials.* **1.** the process of quickly cooling a material, such as steel, by immersion in a cold liquid or gas. **2.** any process by which a heated element is rapidly cooled.

**quern** *History.* a rotary hand mill used to grind grain, likely developed by the Romans about 150 AD and in widespread use in Europe over the next several centuries through Roman influence there.

***quartz*** *Magnified view of quartz crystals.*

**Quesnay, François** 1694–1774, French economist and founder of PHYSIOCRACY, a school of economic thought based on the concept that all wealth originated with the land and that agriculture alone could increase wealth.

**quintessence** *Physics.* a time-varying, hypothetical form of energy postulated to exist as a possible explanation of observations of an accelerating universe.

**race** *Hydropower.* **1.** a channel that conducts water from upstream to the place where it performs work (headrace), or away from that site downstream (tailrace). **2.** a very rapid current, or the narrow channel through which it flows.

**rack railway** *Transportation.* a railway used for additional traction in mountain regions, having an intermediate cogged rail set between the running rails that engages with cogwheels on the locomotive.

**rad** *Nuclear.* a measure of the amount of energy absorbed from ionizing radiation, equal to 100 ergs per gram or 0.01 joules per kilogram of irradiated material; it has been replaced as a standard scientific unit by the gray.

**radar** *Communication.* a system that uses reflected electromagnetic radiation to determine the velocity and location of a targeted object; widely used in such applications as aircraft and ship navigation, military reconnaissance, automobile speed checks, and weather observations. [An acronym for radio detection and ranging.]

**radial** *Electricity.* **1.** an electric transmission or distribution system that is not networked and does not provide sources of power. **2.** see RADIAL TIRE.

**radial engine** *Transportation.* an internal-combustion engine in which the cylinders are arranged like spokes of a wheel at regular intervals around the crankshaft; used formerly as an air-cooled aircraft engine.

**radial tire** *Transportation.* an automotive tire whose cords run across its width, with an extra layer of fabric laid up around the circumference between the plies and the thread.

**radial velocity** *Physics.* a type of velocity that expresses motion toward or away from a given central location or point.

**radiance** *Solar.* the rate at which radiant energy, in a set of directions confined to a unit solid angle around a particular direction, is transferred across the unit area of a surface projected onto this direction.

**radiant** *Physics.* **1.** describing the emission or the measurement of electromagnetic radiation. **2.** specifically, transmitting visible light.

**radiant air temperature** *Solar.* a hypothetical temperature at an exposed surface, calculated to account for the solar absorptance of shortwave radiation and the exchange of long wave emission with the sky and ground, as well as forced convection due to wind at a given ambient air temperature.

**radiant asymmetry** *HVAC.* a situation in which human heat loss occurs more on one side of the body than the other and contributes to thermal discomfort, as when a person sits in front of a campfire on a cold winter night.

**radiant barrier** *Solar.* thin sheets of highly reflective material, such as aluminum, that reduce heat transfer from thermal radiation across the air space between the roof and attic floor of a building.

**radiant density** *Solar.* the radiant energy passing through a unit volume from all directions; expressed in units of joules per cubic meter.

**radiant energy** *Solar.* energy in the form of electromagnetic waves; radiation.

**radiant exposure** *Solar.* the radiant energy incident on a unit surface over some specified time period; expressed in units of joules per square meter.

**radiant flux** *Solar.* radiant energy per unit time passing some specified area from one side; expressed in watts.

**radiant flux density** *Solar.* another term for IRRADIANCE; i.e., the rate at which radiant energy is transferred across a unit area of a surface.

**radiant heat(ing)** *HVAC.* **1.** a heating system that radiates heat from a surface to its surroundings. **2.** any of various means of heating objects or persons by radiation without heating the intervening air. **3.** another term for RADIATION.

**radiant intensity** *Physics.* the amount of electromagnetic radiation striking a given unit area per unit time; expressed in units of watts per steradian.

**radiant panel** *HVAC.* **1.** a panel containing electric resistance heating elements to provide radiant heat to a room. **2.** a panel used in passive solar heating, having integral passages for the flow of warm fluids, either air or liquids. Heat from the fluid is conducted through the metal and transferred to a room by thermal radiation.

**radiation** *Physics.* **1.** energy transferred through space or other media in the form of particles or waves, especially electromagnetic waves; e.g., visible light, ultraviolet light, and infrared light. Certain radiation types are capable of breaking up atoms or molecules. **2.** the complete process by which waves or particles are emitted, pass through a medium, and are absorbed by another body. See below.

**radiation area** *Nuclear.* a classification for any area in which radiation levels could result in an individual receiving a deep dose equivalent in excess of 0.005 rem in one hour at 30 centimeters from the source or from a surface that the radiation penetrates.

**radiation biology** another term for RADIOBIOLOGY.

**radiation burn** *Health & Safety.* damage to the skin resulting from exposure to some form of radiant energy (sunlight, X-rays, radium, and so on).

**radiation chemistry** another term for RADIOCHEMISTRY.

**radiation damage** *Health & Safety.* a general term for any destructive effect brought about by radiation exposure, especially harm to human health from radiation.

**radiation detriment** *Nuclear.* the total harm that would eventually be experienced by an exposed group and its descendants as a result of the group's exposure to radiation from a given source.

**radiation dose** *Nuclear.* the quantity of radiation energy deposited in a material.

**radiation dosimetry** see DOSIMETRY.

**radiation exposure** *Nuclear.* **1.** the fact or state of being irradiated by ionizing radiation. **2.** the time integral of the radiation intensity incident at a given position.

**radiation law** *Solar.* a physical law or equation commonly used to explain the emission of radiation, such as Kirchhoff's law, Planck's law, Stefan–Boltzmann law, or Wien's displacement law.

**radiation** Radiation (from Latin *radiare*, 'to emit beams') is energy transmitted through space as particles or electromagnetic waves or the process of their emission. The most familiar forms of radiation are sunshine and alpha-rays, beta-rays, and gamma-rays emitted by radioactive substances. Particle radiation refers to the radiation of energy by means of small fast moving particles that have energy and mass. Electromagnetic radiation is emitted in discrete units known as photons and are classified (by increasing energy or decreasing wavelength) into radio waves, microwaves, infrared, visible light, ultraviolet, X-rays and gamma-rays. Radiation is separated into two categories, ionizing and non-ionizing, to denote the energy and danger of the radiation. Ionizing radiation is radiation in which individual particle carries enough energy to ionize an atom or molecule. Corpuscular ionizing radiation consists of fast-moving charged particles such as electrons, positrons, or small atomic nuclei. Thermal, epithermal and fast neutrons interact with atomic nuclei creating secondary ionizing radiation and are called indirectly ionizing radiation. Electromagnetic ionizing radiation includes X-rays and gamma-rays. Ultraviolet light also can ionize atom or molecule, but refers usually as non-ionizing radiation. The amount of ionizing radiation, or 'absorbed dose' is measured by the gray. One gray (Gy) is one joule of the energy deposited per kilogram of mass. Some types of radiation, such as neutrons or alpha particles, are more biologically damaging than photons or fast electrons when the absorbed dose from both is equal. To estimate this, dose equivalent, a unit called the sievert (Sv) is used. Regardless of the type of radiation, one sievert of radiation produces the same biological effect. High radiation doses tend to kill cells, while low doses tend to damage or alter the genetic code of irradiated cells. The effect of very low doses is a subject of current debate.

**Valery Chernov**
Universidad de Sonora, Mexico

**radiation pressure** *Solar.* the pressure (or force) on a body illuminated by electromagnetic radiation; the radiation pressure on an object exposed to intense sunlight is about $10^{11}$ times smaller than normal (sea level) atmospheric pressure.

**radiation shield** *Physics.* **1.** any material or obstruction that absorbs radiation and thus tends to protect personnel, materials, or work areas from the effects of ionizing radiation. **2.** a device used on certain instruments to prevent unwanted radiation from biasing the measurement of a quantity. Thus, **radiation shielding.**

**radiation sickness** *Health & Safety.* an illness resulting from excessive exposure to ionizing radiation. The earliest symptoms are nausea, vomiting, and diarrhea, which may be followed by loss of hair, hemorrhage, inflammation of the mouth and throat, and general lethargy.

**radiation thermometer** *Measurement.* an instrument that determines the temperature of a body by measuring the thermal radiation it emits.

**radiation view factor** another term for SHAPE FACTOR.

**radiative capture** *Nuclear.* a nuclear capture process whose immediate result is the emission of electromagnetic radiation only, as when a nucleus captures a neutron and emits gamma-rays.

**radiative-convective model** *Earth Science.* a thermodynamic model that determines the equilibrium temperature distribution for an atmospheric column and the underlying surface, subject to prescribed solar radiation at the top of the atmosphere and prescribed atmospheric composition and surface albedo.

**radiative cooling** *Solar.* any process by which temperature decreases due to an excess of emitted radiation over absorbed radiation.

**radiative damping** *Climate Change.* an imposed positive radiative forcing (see below) on the earth-atmosphere system that represents an energy surplus, as through the addition of greenhouse gases. The temperature of the surface and lower atmosphere will then increase and in turn increase the amount of infrared radiation being emitted into space, thus establishing a new energy balance.

**radiative flux (density)** *Solar.* the radiative energy per unit time and unit area.

**radiative forcing** *Climate Change.* a response to change in the energy balance of the earth-atmosphere system following, for example, a change in the atmospheric concentration of carbon dioxide or a change in the output of the sun; typically measured in watts per meter squared. A **positive radiative forcing** tends to warm the surface (i.e., incoming solar radiation exceeds outgoing infrared radiation) and a **negative radiative forcing** tends to cool the surface.

**radiative heat(ing)** *Solar.* any process by which temperature increases due to an excess of absorbed radiation over emitted radiation.

**radiatively active** *Climate Change.* describing a gas that absorbs incoming solar radiation or outgoing infrared radiation, thus affecting the vertical temperature profile of the atmosphere; e.g., water vapor, carbon dioxide, methane, nitrous oxide, chlorofluorocarbons, and ozone.

**radiator** *HVAC.* **1.** a heating apparatus, typically a coil of tubes through which hot water or steam passes. **2.** an apparatus used to cool circulating fluids, usually a series or coil of thin-walled tubes that are exposed to air or another fluid; commonly employed in water-cooled engines. **3.** *Physics.* any body that emits radiation (as of energy, particles, or waves). **4.** specifically, a body that is capable of emitting electromagnetic radiation, either by inherent properties or by stimulation from an external source.

**radio** *Communication.* a system of communicating sounds over a short or long distance by means of modulating and radiating electromagnetic waves.

**radio-** *Nuclear.* **1.** a prefix describing a relationship to radiant energy, rays, or ionizing radiation. **2.** a prefix describing a relationship to a radioactive isotope of the element to which it is affixed, e.g., radiocarbon.

**radioactinium** *Nuclear.* a member of the actinium series that exists as the thorium isotope with a half-life of 18.5 days; it is both radioactive and toxic, and is found in a number of minerals.

**radioactivation** see ACTIVATION.

**radioactive** *Nuclear.* displaying the property of radioactivity; describing a material having an unstable nucleus that decomposes and emits radiation.

**radioactive cell** *Conversion.* a battery that converts nuclear energy to electrical energy.

**radioactive cloud** *Nuclear.* a cloud of hot gases, smoke, dust, and other matter carried aloft to the atmosphere after the explosion of a nuclear weapon or a nuclear accident.

**radioactive dating** *Measurement.* a technique for measuring the age of an object or a sample of material by determining the ratios of various radioisotopes or products of radioactive decay that it contains.

**radioactive decay** *Nuclear.* **1.** the disintegration of atomic nuclei resulting in the emission of alpha or beta particles (usually with gamma radiation). **2.** the exponential decrease in radioactivity of a material as nuclear disintegrations take place and more stable nuclei are formed.

**radioactive equilibrium** *Nuclear.* a condition of radioactive products in which the rate of formation equals the rate of decay.

**radioactive series** *Nuclear.* a succession of nuclides, each of which transforms by radioactive disintegration into the next until a stable nuclide results. The first member is called the parent, the intermediate members are called daughters, and the final stable member is called the end product. Similarly, **radioactive (decay) chain.**

**radioactive standard** *Nuclear.* a radiation source for calibrating radiation measurement equipment, usually having a long half-life and decaying in such a way that the number and type of radioactive atoms at a given reference time are known.

**radioactive waste** *Nuclear.* the materials, both liquid and solid, that result from using radioactive materials for purposes such as electricity generation by nuclear power plants, defense activities and nuclear weapons manufacture, medical treatment, nuclear research, industrial processes, and the mining and milling of uranium ore. See also specific entries; e.g., HIGH-LEVEL WASTE.

**radioactivity** *Nuclear.* the spontaneous disintegration of atomic nuclei accompanied by the emission of small particles (alpha or beta radiation) or rays (gamma radiation); the natural property of all chemical elements having an atomic number above 83, and possible by induction in all other known elements.

**radio astronomy** *Physics.* the study of the universe in the radio part of the electromagnetic spectrum (approximately 1 millimeter to 30 meters).

**radiobiology** *Ecology.* the study of the relationship of radiation and biological systems, especially the effects of light and ultraviolet and ionizing radiations on living tissue or organisms.

**radiocarbon** *Nuclear.* a radioactive isotope of carbon, especially carbon-14.

**radiocarbon dating** *Measurement.* a precise method of dating ancient organic artifacts, an important tool in modern archaeology and other branches of science. Carbon-14 is an unstable radioactive isotope that decays at a measurable rate upon the death of an organism, allowing a determination of the time at which the organism lived, based on the amount of C-14 remaining.

**radiochemistry** *Chemistry.* the branch of chemistry that studies the properties, applications, and relationships of radioactive matter.

**radiodense** another term for RADIOPAQUE.

**radioecology** *Ecology.* the science dealing with the effects of radiation on species of plants and animals in natural communities or ecosystems.

**radioelement** *Nuclear.* short for radioactive element; i.e., any element having isotopes that emit various forms of radiation while spontaneously decomposing into different nuclides.

**radio emission** *Solar.* emissions from the sun in radio wavelengths from centimeters to dekameters, under both quiet and disturbed conditions.

**radiofrequency heating** *Nuclear.* a method to provide initial heat for a tokamak fusion reactor, in which high-frequency radio waves are launched into the plasma through the use of oscillators.

**radiogenic** *Nuclear.* **1.** of radioactive origin. **2.** describing a disease condition, especially cancer, that is associated with or thought to be induced by exposure to radiation.

**radiography** *Nuclear.* the use of ionizing radiation for the production of shadow images on a photographic emulsion. Thus, **radiograph, radiographing**.

**radioisotope** *Physics.* short for radioactive isotope; an unstable isotope of an element that decays or disintegrates spontaneously, emitting radiation.

**radioisotope thermoelectric generator (RTG)** *Renewable/Alternative.* a self-contained power source that converts the heat generated by radioactive decay into electrical energy; used for various independent types of equipment such as spacecraft, satellites, navigation beacons, and seamarks. Most RTGs employ plutonium-238 as the radioactive fuel, but some versions use strontium-90.

**radiology** *Nuclear.* a branch of the health sciences dealing with radioactive substances and radiant energy and with the diagnosis and treatment of disease by means of both ionizing (e.g., X-rays) and nonionizing (e.g., ultrasound) radiations. Thus, **radiological, radiologist.**

**radiolucent** *Nuclear.* **1.** permitting the passage of X-rays or other forms of radiant energy with little attenuation. **2.** having the property of being transparent to radio waves (and waves of higher frequencies). Thus, **radiolucency.**

**radioluminescence** *Nuclear.* luminescence produced by the bombardment of radiant energy such as X-rays, radioactive waves, or alpha particles. Thus, **radioluminescent.**

**radiolysis** *Nuclear.* **1.** the chemical decomposition of materials stimulated by ionizing radiation. **2.** specifically, the breakdown of molecules that occurs when the water that is used to cool a nuclear reactor core separates into hydrogen and oxygen.

**radiometer** *Nuclear.* an instrument used to detect and measure radiant energy, such as X-rays or microwaves.

**radiometric dating** another term for RADIOACTIVE DATING.

**radiometry** *Measurement.* the process or technique of measuring the characteristics of radiant energy, such as its intensity, in the portion of the electromagnetic spectrum lying adjacent to the visible region. Thus, **radiometer, radiometric.** Compare PHOTOMETRY.

**radiomimetic** *Chemistry.* describing chemical substances that can cause biological effects similar to those caused by ionizing radiation.

**radionuclide** *Nuclear.* a radioactive species of an atom; e.g., tritium and strontium-90 are radionuclides of elements of hydrogen and strontium, respectively.

**radiopaque** *Nuclear.* unable to be penetrated by radiation, especially X-rays. Thus, **radiopacity.**

**radiophotoluminescence** *Nuclear.* luminescence generated by exposing to light a material that has previously been irradiated.

**radioresistance** *Health & Safety.* a relative resistance of cells, tissues, organs, or organisms to the harmful effects of radiation.

**radioresistant** *Nuclear.* able to withstand considerable radiation doses without injury. Thus, **radioresistance.**

**radiosensitive** *Nuclear.* easily injured or affected by radiation. Thus, **radiosensitivity.**

**radiosity** *Solar.* the total radiant energy that leaves a unit area of a surface, or the rate at which this occurs.

**radiosonde** *Measurement.* a balloon-borne instrument for the simultaneous measurement and transmission of meteorological data up to a height of about 30,000 m.

**radiothermoluminescence** *Nuclear.* luminescence released upon heating a substance previously exposed to radiation.

**radiotoxic** *Health & Safety.* describing the potential of a radioactive substance to cause harm to living tissue by radiation after its introduction into the body. Thus, **radiotoxicity.**

**radio wave** *Communication.* an electromagnetic wave whose frequency is in the radio-frequency range, usually between 10 kHz and 300,000 MHz.

**radium** *Chemistry.* a radioactive element having the symbol Ra, the atomic number 88, an atomic weight of 226.0, a melting point of 700°C, and 14 radioactive isotopes. Its only usable isotope, **radium-226,** is a highly toxic, luminescent, white solid derived from uranium ores and used in medical treatment, as a source of neutrons, and in industrial radiography. Commonly used forms include **radium chloride** and **radium bromide,** which are used primarily in the treatment of cancer.

**radon** *Chemistry.* a gaseous radioactive element having the symbol Rn, the atomic number 86, an atomic weight of 222, a melting point of –71°C and a boiling point of –62°C, with 18 radioactive isotopes; an extremely toxic, colorless gas that can be condensed to a transparent liquid and to an opaque, glowing solid. It is derived from the radioactive decay of radium and is used in cancer treatment, leak detection, and radiography. Radon can accumulate in soil and rock beneath homes, in well water, and in building materials, where it can pose a health risk because it is a human carcinogen.

**raffinate** *Materials.* a chemical residue or remainder; e.g., the portion of an oil that is not dissolved during refining, or the residue of contaminants removed from a product stream.

**railroad** *Transportation.* **1.** a permanent road laid with iron or steel tracks forming a continuous line, used to move linked cars from one place to another for the transportation of passengers and goods. **2.** the entire system of transportation using such a road.

**railway** *Transportation.* another term for a railroad, especially in British usage.

**rainforest** or **rain forest** *Ecology.* an area characterized by tall, broad-leaved, densely growing, primarily evergreen trees and a moist climate, having a dry season that is brief or absent; commonly found in the tropics, subtropics, and some sections of the temperate zone.

***rainforest*** *Aerial view of the rainforest surrounding the Amazon River, the largest ecosystem of this type and the most biologically diverse area on earth.*

**Raman effect** *Physics.* the scattering of light as it passes through a transparent medium, caused by the light's interaction with the vibrational or rotational energy of the medium's molecules. This causes the frequency of some of the scattered light to change; the amount of change is a function of the scattering particles and the wavelengths of light. [Identified by Indian physicist Chandrasekhara Venkata *Raman,* 1888–1970.] Thus, **Raman scattering.**

**Raman spectroscopy** *Physics.* the study of radiant energy scattered inelastically when a sample is irradiated with an intense beam of monochromatic light, usually from a laser; a technique used in condensed matter physics and chemistry to study vibrational, rotational, and other low-frequency modes in a system.

**ramjet** *Transportation.* a jet engine that is essentially a tube or duct with both ends open. Air is admitted at one end, compressed by the forward movement of the flight vehicle, heated by the combustion of fuels, and expelled at the other end, producing thrust.

**rammed earth** *Materials.* a traditional construction material made by compressing a soil mixture in a form; also an alternative material in current use.

**ramp metering** *Transportation.* a system of traffic signals employed at highway on-ramps to control the rate of vehicles entering the highway; metering rates are set to optimize traffic flow and to minimize congestion on the highway.

**Ramsey pricing** *Economics.* a system in which a producer of multiple goods apportions costs to the final product in inverse proportion to the price elasticity of demand; e.g., charging more for admission to popular movies than for unpopular ones. A system used in energy pricing, such as the cost of electricity in congested markets. [Named for British economist Frank *Ramsey,* 1903–1930.]

**Ramsey tax** *Economics.* a set of taxes on goods and services designed to raise a given sum of money for the government while minimizing the welfare loss to consumers from the higher prices caused by the taxes.

**rank** see COAL RANK.

**Rankine, William J. M.** 1820–1872, Scottish engineer and physicist, a founding contributor to the science of thermodynamics. He developed a complete theory of the steam engine, and he stated the law of conservation of energy as "all different kinds of physical energy in the universe are mutually convertible".

**Rankine cycle** *Thermodynamics.* an ideal thermodynamic cycle that consists of four processes: heat transfer to the system at constant pressure; an expansion at constant entropy; a constant-pressure heat transfer from the system; and a compression at constant entropy; used as a standard of efficiency **(Rankine efficiency).**

**Rankine (cycle) engine** *Thermodynamics.* an actual device or system that approximates the ideal Rankine cycle (see above); it uses a liquid that evaporates when heated and expands to produce work, such as turning a turbine. The traditional steam locomotive is a common form of Rankine engine. Also, **Rankine (cycle) system.**

**Rankine (temperature) scale** *Measurement.* a temperature scale devised by William Rankine, based on a temperature of 0° for absolute zero and 491.67° for the triple point of water; degree intervals in this scale correspond to degrees Fahrenheit. Another such scale, the KELVIN SCALE, is more commonly used for scientific measurements.

**rapeseed** *Biomass.* an annual plant, *Brassica napus,* of the family Cruciferae (mustard family), growing as a tall and spindly stem topped with green leaves and bright yellow flowers. Its seeds are used as animal feed and as the source of rapeseed oil (see next).

**rapeseed oil** *Biomass.* an agriculturally valuable oil obtained from the rapeseed plant, widely used in food products and also harvested commercially as feedstock for biofuels. Also known as **Canola oil.**

**Rappaport, Roy** 1926–1997, U.S. anthropologist noted for his descriptions of cultural phenomena in terms of energy and material factors among people and the surrounding natural environment.

**rare earth** *Chemistry.* a collective term for a series of fifteen related metallic elements having atomic numbers ranging from 57 to 71, placed in a special row of the periodic table; the group consists of lanthanum, cerium, praseodymium, neodymium, promethium, samarium, europium, gadolinium, terbium, dysprosium, holmium, erbium, thulium, ytterbium, and lutetium. [Though these are not earths and are not literally rare, they are so called because they were associated with more familiar substances known as *common earths.*]

**Rasmussen Report** *Nuclear.* an influential report published by the U.S. Nuclear Regulatory Commission in 1975 that evaluated the probability of accident, human damage, and asset damage from accidents at nuclear power plants, using the technique of fault tree analysis. Its chief author was nuclear scientist Norman *Rasmussen,* (1927–2003).

**rate base** *Electricity.* a measure of a utility's total property value, including plant, equipment, materials, supplies, and cash; the value upon which a utility is permitted to earn a specified rate of return as established by a regulatory authority.

**rate class** *Economics.* a group of customers identified as a class and subject to a rate different from the rates of other groups; e.g., residential consumers of natural gas.

**rated life** *Consumption & Efficiency.* the length of time that a product or appliance is expected to meet a certain level of performance under typical operating conditions.

**rated power** *Consumption & Efficiency.* the power output of a given device under specific or typical operating conditions.

**ratemaking** *Economics.* any of various procedures by which retail energy prices are set, especially for electricity; e.g., cost-of-service ratemaking; performance-based ratemaking; value-of-service ratemaking. Also, **rate design.**

**rate-of-return** *Electricity.* describing rates set to the average cost of electricity, as an incentive for regulated utilities to operate more efficiently at lower rates where costs are minimized.

**rate pancaking** see PANCAKING.

**ratepayer** *Electricity.* a retail consumer of the electricity distributed by an electric utility, either a residential, commercial, or industrial user.

**raw fuel** *Consumption & Efficiency.* a fuel source that is used for energy in essentially the same form in which it is found in nature, with little or no processing; e.g., coal, natural gas, wood.

**raw in place tons** *Coal.* a measurement for the total weight of coal contained within the coal seams of a reserve, including in-seam impurities.

**raw recoverable tons** *Coal.* a measurement for the total weight of coal, including in-seam impurities, that would be mined by a specific underground or surface mine plan applied to a coal reserve.

**ray** *Physics.* **1.** a line or part that emanates from a central point and continues in one direction. **2.** specifically, a stream of light or other radiant energy proceeding in a specific direction.

**Rayleigh, Baron** (John William Strutt) 1842–1919, English physicist who developed theories of sound and light and also discovered the element argon.

**Rayleigh atmosphere** *Solar.* an ideal form of the earth's atmosphere containing only air molecules (the actual atmosphere also contains aerosols, cloud droplets, and ice crystals).

**Rayleigh frequency** *Wind.* a statistical distribution function that is often used to describe the time that the wind spends blowing at a particular speed at a particular location, a key factor in siting wind turbines.

**Rayleigh scattering** *Solar.* the selective scattering of light in the atmosphere by particles that are small compared to the wavelength of light (having a radius less than approximately 1/10 the wavelength of this radiation); this effect produces the appearance of a blue sky during the day.

**RBE** relative biological effectiveness.

**RBOB** *Transportation.* reformulated (gasoline) blendstock for oxygenate blending; a type of reformulated gasoline supply.

**RCRA** *Policy.* Resource Conservation and Recovery Act; U.S. legislation (1976) that directed the Environmental Protection Agency to develop and enforce regulations governing the disposal of hazardous and non-hazardous wastes, and the monitoring and control of air emissions at waste treatment, storage, and disposal facilities.

**RDF** refuse-derived fuel.

**REA** Rural Electrification Act.

**reactance** *Electricity.* opposition to the flow of alternating current by pure inductance or capacitance of the circuit; expressed in ohms.

**reaction** *Physics.* **1.** the equal force or torque offered by a body in opposition to a force or torque applied to it, according to Newton's third law of motion. **2.** see NUCLEAR REACTION. **3.** see CHEMICAL REACTION.

**reaction turbine** *Conversion.* a collective term for turbines in which the water or steam jet enters the runner (rotating part) under a pressure exceeding the atmospheric value. The fluid flowing to the runner still has potential energy, in the form of pressure, that is converted into mechanical power along the runner blades. [So called because motion is created mainly by a *reaction* to the pressure difference.] Compare IMPULSE TURBINE.

**reactive power** *Electricity.* the part of instantaneous power that does no real work; it is absorbed by the inductive or capacitive components of the power system; measured in volt-amperes reactive. The function of reactive power is to establish and sustain the electric and magnetic fields required to perform useful work.

**reactivity** *Chemistry.* **1.** the fact of reacting or taking part in a reaction; the ability to combine chemically with another atom, molecule, or radical. **2.** *Nuclear.* a description of the departure of the state of a nuclear reactor from the critical state; positive values of the reactivity correspond to supercritical states, and negative values to subcritical states.

**reactor** *Nuclear.* a structure inside which an induced nuclear reaction is confined, manipulated, or controlled, as for the production of electrical power, artificial elements, or materials for the construction of nuclear weapons.

**reactor criticality** see CRITICALITY.

**reactor-grade** *Nuclear.* **1.** describing plutonium produced in the normal operation of light water reactors; typically this has less than 60% of the isotope plutonium-239, the isotope best suited for weapons production. **2.** uranium that is enriched to about 3.5% uranium-235; weapons-grade uranium is usually more than 90% U-235. Compare WEAPONS-GRADE.

**reactor trip** see TRIP.

**reactor vessel** *Nuclear.* a large steel container that houses the nuclear reactor core, control rods, moderator and coolant, and other control systems to maintain safe operation.

*reactor* *Nuclear reactors such as this emit huge columns of steam from the water that becomes heated in the process of cooling the reactor.*

**real (fluid) flow** *Physics.* a flow that takes into consideration the energy lost by the flowing fluid through friction with the boundaries restricting its motion.

**real force** *Physics.* a force that can be traced to the effect of its actual physical origin, as distinguished from a theoretical force that is postulated to account for some observed effect.

**real price** *Economics.* the relative price of a commodity in terms of another basket of commodities. GDP (gross domestic product) is often measured in real prices to eliminate the effects of inflation.

**real-time pricing** *Electricity.* the instantaneous pricing of electricity, based on the cost of the electricity available for use at the precise time it is consumed by the customer.

**reaper** *History.* a historic device (19th century) used for harvesting grain by mechanical means.

**reasonably assured resources (RAR)** *Nuclear.* uranium that occurs in known mineral deposits of such size, grade, and configuration that it could be recovered within given production cost ranges, with currently proven mining and processing technology.

**rebound effect** *Consumption & Efficiency.* the principle that an increase in the energy efficiency of one technology, or in one sector of the economy, can actually produce an overall increase in energy use at the macroeconomic level. For example, an increase in motor vehicle fuel efficiency might appear as a cost reduction to consumers, and thus cause an increase in demand (more miles driven). Alternatively, the cost reduction in this sector may result in increased spending for other goods or services that require significant amounts of energy. See also JEVONS PARADOX.

**REC** **1.** rural electric cooperative. **2.** renewable energy credit.

**receiver** *Solar.* **1.** a device that intercepts concentrated solar radiation and transforms it into another form of energy. **2.** *Refrigeration.* a vessel permanently connected to a refrigeration system for the storage of liquid ammonia.

**rechargeable** *Storage.* describing a battery that can be charged again with electricity; i.e., a secondary battery.

**recharging** *Storage.* the process of restoring electrical energy to a secondary battery after it has been discharged, by driving a current back into it from an external source.

*reaper* *The original reaper of Cyrus Hall McCormick (1830s); this machine required only two people to cut as much grain as four to five men could cut previously.*

**reciprocating engine** *Conversion.* another term for a piston engine; i.e., an engine that employs one or more pistons to convert pressure into a rotating motion.

**recirculation system** *Solar.* a type of solar heating system that circulates warm water from storage through the collectors and exposed piping in cold weather conditions to prevent freezing.

**RECLAIM** *Environment.* a current U.S. program to reduce air pollution from oxides of nitrogen and sulfur by providing financial incentives beyond what clean air laws and traditional regulations require. [An acronym for Regional Clean Air Incentives Market.]

**reclamation** *Environment.* the act or fact of returning a given site to a previous environmental state, especially a more natural or more productive state; e.g., the return of a surface coal mine site to its approximate original appearance, as by restoring topsoil and planting native grasses, or the process of draining areas of submerged or marshy lands for agriculture.

**recombination** *Electricity.* in a semiconductor, the action of a free electron falling back into a hole. Recombination processes are either *radiative,* in which the energy of recombination results in the emission of a photon, or *nonradiative,* in which the energy of recombination is given to a second electron, which then relaxes back to its original energy by emitting phonons.

**recoverable** *Consumption & Efficiency.* **1.** available for recovery; describing a resource asset (e.g., coal or natural gas) that is physically and technologically capable of being obtained on an economically sound basis. **2.** describing materials that are available for recycling. Thus, **recoverability.**

**recoverable reserves** *Consumption & Efficiency.* reserve estimates based on a demonstrated reserve base, adjusted for assumed accessibility factors and recovery factors; the portion of the reserve base estimated to be recoverable at a given time.

**recovery** *Consumption & Efficiency.* **1.** the obtaining of an energy resource for practical use. **2.** the process of recovering materials for recycling. **3.** the process of converting municipal solid waste to energy.

**recovery boiler** *Conversion.* a pulp mill boiler in which lignin and spent cooking liquor (black liquor) are burned to generate steam.

**recovery factor** *Coal.* a classification for the percentage of total tons of coal estimated to be recoverable from a given area, in relation to the total tonnage estimated to be available in the reserve base.

**rectenna** *Conversion.* a system in which specialized radio antennas and diodes are linked together into a large plane of rectifying antennas to receive microwave power and convert it into electrical power.

**rectifier** *Electricity.* a device through which current can flow only in one direction; often used alone or in sets to convert AC current into DC pulsating current.

**recuperative heater** *Consumption & Efficiency.* an air heater in which the heat-transferring metal parts remain stationary to form a boundary between the heating and cooling fluids.

**recuperator** *Consumption & Efficiency.* a continuous heat exchanger that preheats combustion air with the heat from the exhaust gases.

**recyclable** *Materials.* able to be recycled; useful for additional purposes after its initial use.

***recyclable*** *Ferrous metals are here separated by magnet from other manufacturing byproducts that do not have recyclable value.*

**recycle** *Materials.* to carry out a process of recycling (see next).

**recycling** *Renewable/Alternative.* **1.** any process of recovering or extracting valuable or useful materials from waste or scrap. **2.** specifically, the reuse of specific consumer or industrial items in order to conserve scarce materials, reduce pollution and littering, and so on. ☼ See below.

**Reddy, Amulya K. N.** Indian electrochemist known for his studies and leadership role with respect to the energy needs of the rural and urban poor in India and other nations. He was founder of the International Energy Initiative (1991).

☼ **recycling** Recycling is the process of turning used products into raw materials that can be used to make new products. Its purpose is to conserve natural resources and reduce pollution. Recycling reduces energy consumption, since it generally takes less energy to recycle a product than to make a new one. Similarly, recycling causes less pollution than manufacturing a new product, and conserves raw materials. It also decreases the amount of waste sent to landfills or incinerators. Although people have always reused things, recycling as we know it today emerged in the 1970s with the environmental movement. Non-profit recycling centers began opening around the country, followed by municipal recycling programs. Today, most U.S. communities have such programs. A typical program asks people to separate their recyclables from their trash before placing them at the curb for collection. To encourage recycling, some communities also charge residents for the quantity of trash put out for collection. The most commonly recycled household items are paper and cardboard; metal, glass, and plastic containers and packaging; and yard waste. Recycling the recovered materials is simple for metals and glass; they can be melted down, reformed, and reused. Yard waste can be composted with little or no equipment. Paper, the most important recycled material, must be mixed with water, and sometimes de-inked, to form a pulp that can be used in papermaking. Plastics recycling requires an expensive process of separation of different resins; without that, it yields only a low-quality product, such as "plastic lumber".

**Frank Ackerman**
Tufts University

**Redfield ratio** *Earth Science.* the principle that the atomic ratios of the chemical components of marine plankton, specifically nitrogen, phosphorus, carbon, and oxygen, are generally identical with their relative proportions in the open ocean; i.e., relatively constant proportions exist during the synthesis and export of organic matter and its subsequent remineralization. Typically used ratios are N:P:C:O = 16:1:106:138. [Described by U.S. oceanographer Alfred *Redfield* 1890–1983.]

**Red Line Agreement** *History.* an agreement (1928) on oil production in most of the former Ottoman Empire by U.S. and European major international oil companies; this established the structure of the Iraq Petroleum Company and was an important foundation of the international oil company cartel.

**redox** short for REDUCTION-OXIDATION.

**redox battery** *Storage.* a rechargeable battery consisting of a semipermeable membrane having different liquids on either side. Electrical contact is made through inert conductors in the liquids, and as the ions flow across the membrane, an electric current is induced in the conductors. These batteries have low volumetric efficiency, but are reliable and long-lived.

**red shift** or **redshift** *Physics.* **1.** the phenomenon that the wavelength of light emitted by an object moving away from the observer is shifted toward longer wavelengths (i.e., the redder end of the visible spectrum), due to a DOPPLER EFFECT. **2.** specifically, this phenomenon as observed with respect to celestial bodies, believed to occur because of their movement outward at increasing rates of speed; this provides the basis for theories suggesting that the universe is expanding.

**red tide** *Environment.* a term for an explosive population growth of dinoflagellates, a minute, single-celled group of algae whose presence may give a reddish tint to water; a red tide may be caused by excessive runoff of nutrients from land or by the discharge of untreated sewage into coastal waters, both of which serve as fertilizers for the algae. Dinoflagellates produce a neurotoxin that can cause death in affected fish and shellfish populations.

**reduction** *Chemistry.* a chemical change that involves a gain of electrons, either by a removal of oxygen or an addition of hydrogen, or simply by the addition of electrons from a donor molecule to an acceptor molecule.

**reduction-oxidation** *Chemistry.* another term for oxidation-reduction; i.e., a chemical change in which one species is oxidized (loses electrons) while another species is reduced (gains electrons).

**redundancy** *Nuclear.* a term for the duplication of safety and control systems to ensure safe and reliable operation of a reactor or other device.

**REE** resting energy expenditure.

**reference cell** *Photovoltaic.* a photovoltaic cell whose short-circuit current is calibrated against the total irradiance of a reference spectral irradiance distribution.

**reference dose (RD)** *Health & Safety.* an estimate of the daily exposure to a given toxic chemical by the human population (including sensitive subgroups) that is likely to be without an appreciable risk of adverse effects during a lifetime.

**reference environment** *Environment.* an idealized version of the natural environment that is characterized by a perfect state of equilibrium; i.e., the absence of any gradients or differences involving pressure, temperature, chemical potential, kinetic energy, and potential energy.

**reference man** *Nuclear.* a hypothetical individual (male or female) postulated to have the physical and physiological characteristics of an average human; used by researchers and public health workers in calculations of the biological effects of radiation.

**reference state** *Thermodynamics.* the most stable state of an element under the prevailing conditions.

**refill** *Hydropower.* the point at which a hydropower reservoir is considered to be full after the seasonal runoff from rain and snowmelt.

**refine** *Materials.* **1.** to remove impurities from a substance, or free the substance from foreign matter; remove unwanted components. **2.** specifically, to transform crude oil into useful commercial products, as by fractional distillation, cracking, purifying, and treating.

**refinery** *Materials.* an industrial facility that carries out processes of refining, as for petroleum or metals.

***refinery*** *The Richmond, California oil refinery on San Francisco Bay, one of the largest and oldest on the West Coast of the U.S.*

**refinery gas** another term for STILL GAS.

**refinery yield** *Oil & Gas.* the percentage of finished product obtained from the input of crude oil and net input of unfinished oils, calculated by dividing the sum of input into the individual net production of finished products.

**reflect** *Physics.* to turn back a ray; be involved in a process of reflection.

**reflectance** another term for REFLECTIVITY.

**reflected radiation** *Solar.* the fraction of incident solar radiation on a surface reflected at a specific wavelength.

**reflection** *Physics.* **1.** a process in which a ray or wave strikes a surface that it does not penetrate. **2.** specifically, the return of a light ray from a surface with no change in the ray's wavelength. **3.** an image produced in such a process, as by a mirror.

**reflective** *HVAC.* describing window glass to which a thin layer of material has been applied to increase its reflective properties and thus reduce heat gain and loss through the window. Thus, **reflective glass, reflective coating** or **film.**

**reflectivity** *Physics.* **1.** the ratio of the energy carried by a reflected wave to that of the incoming wave before reflection; the portion of incident radiation reflected by a surface of discontinuity. **2.** specifically, the ratio of the amount of solar radiation striking a surface to the reflected flux from that surface.

**reflector** *Consumption & Efficiency.* **1.** any surface or element whose purpose is to reflect radiation in a desired direction; e.g., an object that reflects the light of a vehicle's headlights as an aid to locate in the dark, or a device used to reflect and focus solar radiation. **2.** *Nuclear.* a layer of water, beryllium, or graphite that surrounds the core of a reactor to reduce the amount of escaping neutrons.

**reflector lamp** *Lighting.* (Type R lamp) a lamp with a built-in reflecting surface, designed to spread light over specific areas; this may be an incandescent, compact fluorescent, or HID lamp. Used mainly indoors to maximize illumination for theatrical or retail applications.

**reflux** *Materials.* a process by which a portion of the distillate from the top of a distillation column is returned to the column to assist in attaining better separation into desired fractions.

**reforestation** *Ecology.* the restoration of significant tree growth in an area that had previously been forest land; especially an intentional human effort to do this.

*Reforestation of ponderosa pine in the Lassen National Forest of northern California.*

**reformate** *Materials.* the product of a reforming process (see below).

**reformer** *Hydrogen.* **1.** a device that converts a hydrocarbon fuel into a hydrogen-rich gaseous stream and removes impurities, so that the resulting gas can supply the anode of a fuel cell to power a transportation vehicle. An alternative to storing existing hydrogen fuel in a fuel-cell powered vehicle. **2.** more generally, any device that carries out a process of reforming (see below).

**reforming** *Materials.* **1.** a chemical process using heat in the presence of a catalyst to break down a substance into desired components. **2.** an oil refinery process that enhances gasoline quality by changing chemical characteristics rather than breaking up molecules, as in cracking. **3.** *Hydrogen.* in a fuel cell, the production of hydrogen gas from natural gas by first heating the gas with high temperature steam, producing carbon dioxide and carbon monoxide, and then treating the carbon monoxide with high-temperature steam to produce hydrogen.

**reformulated gasoline (RFG)** *Oil & Gas.* gasoline treated so as to be in closer compliance to air quality standards; the usual blending agent is MTBE.

**refraction** *Physics.* a change in the direction of propagation when a wave passes from one medium to another medium of different density.

**refraction method** *Oil & Gas.* a traditional method of oil prospecting involving the use of elastic earth waves initiated by a concussive force to travel down to a dense or high velocity bed, and then move along that bed until being re-refracted up to seismic detector locations on the surface. This method aids petroleum explorers in locating salt domes that transmit waves at high rates of speed.

**refractory** *Materials.* describing a material that has a high softening point and a very high melting point.

**refrigerant** *Refrigeration.* something that refrigerates; a substance that transfers to its surroundings the heat removed from a space to be cooled; can be any material whose properties (e.g., low vaporization temperature) make it suitable for this, either a solid (such as ice), a liquid (such as Freon), or a gas (such as a halogenated hydrocarbon).

**refrigerant charge** *Refrigeration.* the required amount of refrigerant in a system.

**refrigerate** *Refrigeration.* to carry out a process of refrigeration; remove unwanted heat.

**refrigeration** **1.** the process of lowering the temperature of an area or object to a level that is significantly lower than the temperature of its surrounding environment. **2.** specifically, the use of this process to preserve foods. See below.

**refrigeration system** *Refrigeration.* a general term for any system that, when operating between a heat source and a heat sink at two different temperatures, is able to absorb heat from the source and reject it to the sink. Thus, **refrigeration cycle, refrigeration unit.**

**refrigeration** The benefit of refrigeration to preserve food was known since prehistoric times, and refrigeration using natural ice was the practical method for a very long time. In fact, even in the late 19th century natural ice supply became an industry unto itself. By 1909, there were 2000 commercial ice plants in the U.S.; no pond was safe from scraping for ice production. From some big ponds, more than 1000 tons of ice were extracted each day. One of the most significant and beneficial contributions to mankind in the 20th century is the development of efficient mechanical refrigeration systems, which not only produce ice artificially but also provide any desired low-temperature environments. Its significance is clearly understood when one realizes the wide-ranging applications of refrigeration: meatpacking, brewing, food transportation, medical applications, and almost all industries including textile mills, oil refineries, chemical plants, paper, drugs, soap, glue, shoe polish, perfume, celluloid, photographic materials, among many others. We cannot imagine our modern human life without this technology. The mechanical refrigeration system is based on the thermodynamic principle, a so-called vapor compression cycle. Gas (called a refrigerant) is compressed by a mechanical compressor, condensed into liquid, then expanded into a low pressure/low temperature state, where the low temperature liquid evaporates by taking heat from the environment (refrigeration effect here), and the evaporated gas is compressed again as a cycle. One of the best refrigerant gases was Freon-12 (dichlorodifluoromethane) in the past, but now hydrofluorocarbons and some hydrocarbons are being used for refrigerants due to the environmental concerns such as the depletion of stratospheric ozone.

**Akimichi Yokozek**
Dupont Corporation

**refrigeration ton** *Refrigeration.* a measure of refrigerating or air conditioning capacity, expressed as the amount of heat energy absorbed when one ton of ice melts over a period of one day; this is the transfer in a cooling operation of 288,000 Btu in 24 hours (12,000 Btu per hour).

**refrigerator** *Refrigeration.* **1.** an enclosed, insulated appliance or area that is used to maintain food or other perishable items at a low temperature. **2.** any device that carries out a process of refrigeration.

**refuse bank** *Coal.* a large accumulation of waste material generated by a coal-cleaning process.

**refuse-derived fuel (RDF)** *Renewable/Alternative.* the combustible portion of municipal solid wastes (MSW), remaining after processing, separating, and sizing of the MSW to remove any recyclable components and noncombustible materials; typically this consists of paper, plastics, vegetation, and wood along with minimal glass and metal.

**refuse mining** *Mining.* the process of obtaining coal residue that can be economically burned from the waste material left over from previous mining operations.

**regeneration** *Consumption & Efficiency.* **1.** a process of burning off the soot accumulated on a diesel particulate filter. **2.** the treatment of used nuclear fuel elements so that they may be used again in the reactor. **3.** any process of collecting and/or purifying energy or materials for reuse in another process. Thus, **regenerative.**

**regenerative braking** *Transportation.* the process of slowing a vehicle by the application of power to a motor so that the motor acts as an electric generator. This converts the rotational energy of the motor into electrical energy, thus braking the vehicle and also producing electric power. Hybrid-electric vehicles can employ regenerative braking to reclaim much of the energy that would be normally be lost in braking.

**regenerative cooling** *HVAC.* a type of cooling system in which a part of the air supply is used to cool water through direct evaporation, and then this cold water is used to further cool the supply air indirectly through a heat exchanger.

**regenerative cycle** *Conversion.* a cycle in a steam engine using heat that would ordinarily be lost, by extracting exhaust steam from the turbine in stages and using it to heat the feedwater.

**regenerative fuel cell (RFC)** *Renewable/Alternative.* **1.** a type of fuel cell in which the water produced by the electrochemical process is then separated into its constituent components of hydrogen and oxygen, which are fed back to the fuel cell for further use. **2.** any fuel cell in which the reaction product can then be recycled to provide more reactants.

**regenerative furnace** *Consumption & Efficiency.* a furnace in which fuel gas and the air for combustion pass through a heated chamber (regenerator) before combustion, thus improving efficiency; a large portion of the heat of combustion is then recycled back to the regenerator to be stored there; a 19th century improvement on traditional furnaces that lost much of the heat of combustion in the hot gases passing up the chimney.

**regenerative heating** *Thermodynamics.* the process of using heat that is expelled in one part of a cycle for another function, or in another part of the cycle,

**regenerative pump** *Conversion.* a rotating-vane instrument that creates high liquid heads at low volumes with the use of mechanical impulse and centrifugal force.

**regenerative thermal oxidizer (RTO)** *Consumption & Efficiency.* a device used for the destruction of volatile organic compounds (VOCs) and other air pollutants that are discharged in industrial processes, by means of a process of high-temperature thermal oxidation.

**regenerator** *Consumption & Efficiency.* any device that is employed in a process of regeneration (see above), especially one that absorbs heat from exhaust gases in a regenerative furnace.

**regional air pollution** *Environment.* a term for increased air pollutant concentrations, and the effects of this, caused by emissions at distances up to 1000 km.

**regular** *Oil & Gas.* a classification for gasoline having a standard octane rating (i.e., the relative ability to resist engine knock); the classification varies by region but typically indicates an octane rating of 85–87. Thus **regular gas(oline)** or **fuel.**

**regulated streamflow** *Hydropower.* the controlled rate of flow past a given point after a purposeful release of water from a reservoir.

**Reid vapor pressure (RVP)** *Measurement.* a standard measurement of a liquid's vapor pressure in pounds per square inch at 100°F; it is an indication of the tendency of the liquid to evaporate.

**Reitwagen** *History.* a two-wheeled vehicle built (1885) by automotive pioneer Gottlieb Daimler, employing what is considered to be the first successful version of the modern gas engine. [A German term for "riding carriage".]

**relamping** *Lighting.* the substitution of one type of lamp for another to conserve energy; e.g., the replacement of a traditional incandescent lamp with a reflector lamp or compact fluorescent lamp.

**relative biological effectiveness (RBE)** *Nuclear.* the ratio of the number of rads of gamma (or X-ray) radiation of a certain energy that will produce a specified biological effect, to the number of rads of another radiation required to produce the same effect.

**relative humidity** *Earth Science.* the ratio of the amount of water vapor in a given mass of air to the saturation point (i.e., the maximum amount of vapor that the same volume of air could hold at the given temperature).

**relative sea level** *Earth Science.* sea level in comparison to the level of the land; a rise in relative sea level can result from a combination of eustatic sea level rise (e.g., worldwide sea level rise associated with global warming) and sinking of the land due to subsidence.

**relativity** *Physics.* **1.** the fact of being relative; dependence of a quantity or condition on the status of something else. **2.** a theory introduced by Albert Einstein, concerned with physical laws (such as those that describe mass, space, motion, and time) as they are formulated by different observers in uniform relative motion with respect to each other.

The **special relativity theory** postulates that all the laws of physics have the same form in all inertial frames of reference. It also postulates that the speed of light is a constant, independent of any relative motion between the light source and the observer. The **general relativity theory** postulates that in the neighborhood of a given point, it is not possible to distinguish between a gravitational field and an accelerated frame of reference; i.e., gravitational mass and inertial mass are completely equivalent.

**relay** *Electricity.* **1.** a device that is activated by variations in an electric circuit and that when activated makes or breaks a connection in this or another circuit. **2.** any transmission system that passes on a signal from one communication link to another.

**reluctance** *Materials.* the resistance of a material or combination of materials to the magnetic flux in a magnetic circuit.

**rem** *Nuclear.* a unit of ionizing radiation equivalent to the dosage that produces the same amount of damage from radiation exposure as does one roentgen of high-voltage X-rays; 100 rem = 1 sievert (the sievert is the SI unit of absorbed dose). [An acronym for roentgen equivalent man.]

**remaining resources** *Consumption & Efficiency.* **1.** a classification for coal that is still in the ground after a certain amount of mining activity, excluding material that has purposely been left in place because it is not useable or not economically recoverable. **2.** a classification for recoverable oil and natural gas liquid volumes that have not yet been produced.

**remanufacturing** *Renewable/Alternative.* the fact of manufacturing again; a process of bringing large amounts of similar products together for purposes of disassembly, evaluation, renovation, and eventual reuse.

**remediation** *Environment.* **1.** research, design, and engineering efforts focused on dealing with ongoing environmental problems that exist across a range of locations, especially issues of contaminated industrial sites and the treatment of stored waste and contaminated waters. **2.** specifically, cleanup or other methods used to treat, contain, or remove a toxic spill or hazardous materials from a given site.

***remediation*** *Workers carry out a process of remediation in a wetlands area damaged by an oil spill.*

**remining** *Mining.* the reclamation of previously mined land by companies that extract the coal or ore left behind by earlier mine operators.

**remote sensing** *Communication.* the gathering and analysis of data from an object physically removed from the sensing equipment, as in satellite and aerial photography or subsurface detection instruments.

**removal unit (RMU)** *Policy.* under the Kyoto Protocol, the technical term for a SINK CREDIT generated in Annex B countries.

**Renault, Louis** 1877–1944, French automotive pioneer noted for advances such as an expanding drum brake, the first direct drive (gearbox) system, and an improved four-cylinder engine, as well as for supplying tanks and airplane engines to the Allies during World War I. His Renault company became the leading French car manufacturer.

**renewable** **1.** able to be renewed, regenerated, or replenished by natural processes of the earth. Thus, **renewable resource.** Renewable resources can be converted to nonrenewable resources if they are depleted or degraded by humans faster than natural processes renew them. **2.** an item or material of this type.

**renewable energy** *Renewable/Alternative.* any energy resource that is naturally regenerated over a short time scale and either derived directly from solar energy (solar thermal,

***renewable*** *This windmill operates by means of a renewable form of energy, the wind. Energy from the flow of water in the nearby stream would also be considered a renewable form.*

photochemical, and photoelectric), indirectly from the sun (wind, hydropower, and photosynthetic energy stored in biomass), or from other natural energy flows (geothermal, tidal, wave, and current energy). Contrasted with NONRENEWABLE ENERGY forms such as oil, coal, and uranium.

**renewable energy credit (REC)** *Renewable/Alternative.* an indication that a unit of electricity has been generated from renewable energy with low net greenhouse gas emissions. Also, **renewable energy certificate.**

**renewable portfolio standard (RPS)** *Renewable/Alternative.* a market-based policy requiring that a minimum percentage of all electricity must be provided from renewable energy sources such as wind, wood and other biomass, or geothermal. See next column.

**rent** *Economics.* **1.** any income gained from the natural wealth of the earth, such as ownership of land or natural resources. **2.** the surplus generated as the difference between the price of a certain good using a natural resource and the unit costs of converting that natural resource into the good.

**renter state** *Global Issues.* a nation that derives its public revenue from externally generated rents (i.e., profits from natural resource sales), rather than from the surplus production of its own population. In oil-exporting states, this is measured by the percentage of total government revenues represented by natural resource rents.

**rent-seeking** *Sustainable Development.* efforts, both legal and illegal, to acquire access to or control over opportunities for earning economic rents (see above); e.g., activities in oil-dependent countries aimed at capturing oil revenues through economically inefficient means.

**replacement rate** *Sustainable Development.* a condition of population equilibrium in which the number of people born equals the number of people who die, resulting in a constant population size.

**repowering** *Electricity.* the process of rebuilding, upgrading, or replacing major components of a power plant as an alternative to building a new plant.

**renewable portfolio standard (RPS)** A regulation that requires a minimum market share for electricity generated from energy sources that have desirable environmental characteristics. While these are normally understood to be sunlight, wind, geothermal, waves, tides, hydropower, and organic matter, the definition varies somewhat between jurisdictions. Large hydropower is sometimes excluded while municipal solid waste, even if low in organic matter, is sometimes included. Some jurisdictions even include cogeneration of heat and power produced from natural gas combustion, as a way of valuing the much higher energy efficiency. The RPS may allow participants to trade among themselves in achieving the minimum market share requirement at the market-wide level, which in Europe sometimes involves the trading of green energy certificates. Because it specifies a physical requirement, the RPS is referred to as a quantity-based policy. This is in contrast to price-based policies such as environmental taxes or subsidies (in the form of above-market feed-in tariffs), both of which improve the relative economic advantage of renewables relative to fuels associated with emissions or extra risks of damage.

**Mark Jaccard**
Simon Fraser University

**repressuring** *Oil & Gas.* the process of injecting pressurized water or gas into a reservoir to increase oil recovery.

**reprocessing** *Nuclear.* the chemical separation of spent nuclear fuel into constituent parts, typically the contained plutonium, fission products, and uranium.

**reproduction factor** *Nuclear.* the ratio between the number of neutrons inducing fission in a fuel to the number required to maintain the chain reaction.

**reproductive energetics** *Biological Energetics.* the energy expenditure involved with the various aspects of reproduction, including the birth process itself as well as copulation, pregnancy, lactation (in mammals), and various mating and parental behaviors.

**reproductive success** *Ecology.* **1.** the number of surviving offspring produced by an individual organism in its lifetime. **2.** the number of offspring produced by a given population or species over time.

**reproductive toxin** *Health & Safety.* a substance or agent that can have adverse effects on the reproductive system of humans or animals; e.g., lead, DDT.

**reradiation** *Physics.* the emission of electromagnetic radiation that had previously been absorbed by a substance.

**reregulation** *Policy.* a restoration of governmental regulatory controls in an industry (e.g., electrical utilities) that had previously been deregulated, usually because of a perception that free market forces in the deregulated environment have not been efficient and equitable.

**reserve** *Consumption & Efficiency.* an existing accumulation of a resource or material in excess of current usage levels. See PROVED RESERVES and other specific entries.

**reserve additions** *Oil & Gas.* the total quantity of oil and gas added to proved reserves in a year as a result of drilling, changes in economic, geologic, technological, and political factors, and statistical adjustments.

**reserve base** *Consumption & Efficiency.* the portion of an identified resource that meets certain minimum physical and chemical standards according to current extraction and production practices.

**reserve growth** *Oil & Gas.* the increases of estimated recoverable petroleum volume that commonly occur as oil and gas fields are developed and produced, and as extraction technology advances.

**reserve life index** *Consumption & Efficiency.* the hypothetical number of years that it would take to deplete reserves at the current production rate; normally calculated as current reserves divided by current annual production.

**reserve (operating) margin** *Electricity.* the amount of unused available capability of an electric power system (at peak load for a utility system), as a percentage of its total capability.

**reserves** **1.** see RESERVE. **2.** see PROVED RESERVES.

**reservoir** *Environment.* **1.** a large area, natural or artificial, that is used to store water for a municipal water supply or for other purposes such as irrigation or recreation. **2.** a component of the climate system where a greenhouse gas is held. **3.** *Oil and Gas.* a porous, permeable, subsurface sedimentary rock formation in which crude oil or natural gas accumulates. **4.** any pool of crude oil or natural gas. **5.** *Thermodynamics.* any system considered as a repository of heat.

**reservoir elevation** *Hydropower.* the level of the water stored behind a dam.

**reservoir rock** *Oil & Gas.* a subsurface volume of rock that has sufficient porosity and permeability to permit the migration and accumulation of petroleum under adequate trap conditions; an essential element of the petroleum system.

**reservoir souring** *Oil & Gas.* a deleterious increase in hydrogen sulfide concentration in an oil reservoir, caused by the presence of sulfate-reducing bacteria in the reservoir.

**reservoir storage** *Hydropower.* the total volume of water in a reservoir at a given point in time.

**resettlement** see HYDROPOWER RESETTLEMENT.

**residence (time)** *Chemistry.* **1.** the average length of time that a substance spends in a given reservoir at a steady state with respect to processes that add the substance to, or remove it from, the reservoir. **2.** the average time spent by the energy of a substance in a compartment; expressed in units of time as the ratio of the size of the compartment to the flux through it.

**residential** *Consumption & Efficiency.* relating to or involving the consumption of energy by private households, as opposed to industry, agriculture, and so on; e.g., space heating, air conditioning, lighting, refrigeration, cooking, and the like. This excludes institutional living quarters. Thus, **residential energy (consumption), residential sector.**

**residual fuel oil** *Oil & Gas.* a classification for heavier oils that remain after the distillate fuel oils and lighter hydrocarbons are distilled away in refinery operations; included are No. 5, a residual fuel oil of medium viscosity; Navy Special, for use in steam-powered vessels in government service and in shore power plants; No. 6, including Bunker C fuel oil, which is used for commercial and industrial heating, electricity generation, and powering ships.

**residue** *Materials.* **1.** a collective term for byproducts resulting from the processing of biomass that have significant energy potential; e.g., stalks of corn and sugarcane, wheat straw, bark and wood shavings, sawdust, pulping liquors, or food processing wastes. **2.** any materials remaining after a chemical or physical process is complete.

**residue gas** *Oil & Gas.* natural gas from which processing plant liquid products and (in some cases) nonhydrocarbon components have been extracted.

**residuum** another term for RESIDUE.

**resin** *Materials.* any of various solid or semisolid natural or synthetic organic products, usually translucent polymers that do not conduct electricity; used in plastics, textiles, paints, and varnishes.

**resinous fluid** see TWO-FLUID THEORY.

**resistance** *Electricity.* **1.** the ratio of applied electromotive force to the resulting current in a circuit; measured in ohms and following Ohm's law. Resistance opposes the flow of current, generates heat, controls electron flow, and helps supply the correct voltage to a device. **2.** *Ecology.* the degree to which a host can limit the effects of an infection, such as a virus.

**resistance heat(ing)** *HVAC.* a system that produces heat through electric resistance, as when an electric current flows through a material of high resistivity, thus converting the electrical energy into heat that can be transferred to the space to be heated.

**resistive** *Electricity.* **1.** involving or employing resistivity; i.e., resistance to the flow of electric current. **2.** describing a device whose electrical resistance changes in proportion to some other variable; e.g., temperature or flow rate.

**resistive load** *Electricity.* an electrical load without capacitance or inductance, or one in which the inductive portions cancel the capacitive portions at the operating frequency.

**resistive voltage drop** *Electricity.* the voltage developed across a cell by current flow through the resistance of the cell.

**resistivity** *Electricity.* an electrical property of materials, representing the inherent ability of a material to resist current flow.

**resource** *Consumption & Efficiency.* anything that has value for human use, especially a NATURAL RESOURCE such as water, trees, coal, oil, and so on.

**resource curse** *Sustainable Development.* **1.** the negative growth and development outcomes associated with strategies based on mineral and petroleum wealth. **2.** specifically, the inverse relationship between high levels of natural resource dependence and growth rates. See next page.

**resource recovery** see RECOVERY.

**respirable** *Coal.* able to be breathed in; describing minute airborne particles that are capable of being ingested into the lungs, such as those that can result from coal mining, indoor wood combustion, and certain industrial chemical processes. Thus, **respirable dust, respirable particulate.**

**respiration** *Biological Energetics.* the exchange of gases between an organism and the atmosphere. In humans, this is the external and internal processes of breathing. It allows the interchange of oxygen and carbon dioxide within the body cells, in which oxidation of food nutrients provides energy for cell activity and produces the waste materials of carbon dioxide and water.

**respiratory heat** *Biological Energetics.* in plants, the heat evolved during the respiratory process (absorption of oxygen and evolution of carbon dioxide).

**resource curse** Common sense and simple economics suggest that countries blessed with an abundance of natural resources should live long and prosper. Yet over many years, it has been observed that nations rich in oil, gas, or mineral resources have been disadvantaged in the drive for economic progress. This has been labeled "resource curse". In the 1950s and 1960s, concern was based upon deteriorating terms of trade between industrial and developing countries. In the 1970s, it was driven by the impact of the oil price shocks on the oil exporters where windfall revenues seemed to introduce serious distortions to the economies. In the 1980s, the phenomenon of "Dutch Disease"—an overvaluation of the real exchange rate—attracted attention. In the 1990s, it was the impact on government behavior—rent seeking and corruption—that dominated discussions. More recently a revival of interest follows the World Bank's "Extractive Industry Review" and growing concern over corporate social responsibility. Existence of the "curse" is controversial in literature and there is growing evidence that its occurrence is far from inevitable. In many countries natural resource revenues can produce tangible benefit. Where a "curse" has occurred, the main explanation lies in the way large windfall resource revenues affect government behavior. Thus avoidance implies political reform to encourage good governance. Increasingly, analysis addresses how the various players in natural resource projects—international financial institutions, multinational corporations and NGOs—can assist governments at risk to create capacity to manage the problem.

**Paul Stevens**
Dundee University

**respiratory loss** *Biological Energetics.* the energy loss by an organism or system due to the process of respiration.

**response time** *Climate Change.* the time needed for a climate system or its components to reach a new state of balance following a forcing.

**resting energy expenditure (REE)** *Biological Energetics.* the total energy requirements of a person for daily sleep and rest, excluding energy expended in physical activity.

**restoration** *Environment.* a long-term process of reestablishing the original composition, structure, and function of a disturbed environmental site or an ecosystem, as by stabilizing contaminated soil, treating groundwater, removing wastes, planting native vegetation, and so on.

**restructuring** *Electricity.* a reconfiguration of a vertically-integrated electric utility; this usually refers to the separation of the various utility functions into entities that are individually operated and owned. See next page.

**retail** *Economics.* **1.** describing transactions in which the end-use consumer of the good is also the purchaser; e.g., the sale of fuel at the pump to a vehicle owner is a retail purchase. Usually involves sale of items in small quantities and not for subsequent resale. **2.** specifically, having to do with electricity or other utility services sold directly for residential, commercial, or industrial end-use purposes. Thus, **retail competition, customer, market, pricing,** and so on. Compare WHOLESALE.

**retort** *Materials.* **1.** a glass vessel, typically a bulb with a long, downward-pointing neck, used to distill or decompose solid or semisolid substances. **2.** a cylindrical refractory chamber used to heat coal or ore.

**retreat mining (retreating)** *Mining.* a method of coal mining in which the service roadways are created to their predetermined length before any coal is removed from the working face, then mining takes place in a "retreating" direction, back to the original development. The workers mine as much coal as possible from the remaining pillars until the roof falls in; the mined area is then abandoned.

**retrofit** *Consumption & Efficiency.* to modify an existing structure or technology after it has been put in service, as opposed to creating something entirely new; e.g., an older building can be retrofitted with advanced windows to slow the flow of energy into or out of the house. Thus, **retrofitting.**

**retrograde condensation** *Oil & Gas.* the condensation of the vapor phase of a hydrocarbon in contact with its liquid phase, occurring either when the pressure decreases at constant temperature or the temperature increases at constant pressure.

**reuse** *Renewable/Alternative.* the use of material over again instead of disposing of it as

**restructuring** Changes in the regulatory and corporate structure of the electric power industry transforming it from fully regulated monopolies to one in which competition is allowed or required in certain portions of the industry. Traditionally, vertically-integrated monopolies provided generation, transmission, and distribution services to captive customers. Following restructuring, these services were separated, either functionally within one company, or physically with different companies offering different services. Several states that have "restructured" required jurisdictional electric utilities to sell off their generation assets to independent companies. Expectations that competitive forces would lead to more efficient prices than government regulation spurred restructuring. In the U.S., restructuring at the wholesale level began in the mid 1990s with FERC's rule on open access to the transmission system. In response to this rule, and under pressure from large industrial customers seeking lower rates, regulators in several U.S. states issued rules that moved to retail competition in the generation sector. Many municipalities own their own electric systems and were not required to make the same changes to their businesses. There are strongly divergent opinions on the success of restructuring. The term "privatization" refers to the change from government to private ownership as occurred in the electric industry in the U.K. in the early 1990s. The term "deregulation" is sometimes used interchangeably with "restructuring" but implies a greater reduction in regulation than has actually occurred in the U.S. electric industry.

**Jeannie Ramey**
Synapse Energy Economics, Inc.
Cambridge, Massachusetts

waste or recycling it; e.g., using cloth napkins instead of paper or a durable coffee mug instead of a paper cup.

**revegetation** *Ecology.* the restoration of plant life to an area that had previously been devoid of or deficient in vegetation; can be either a natural process or human-assisted.

**Revelle, Roger** 1909–1991, U.S. oceanographer best known for his pioneering work on carbon dioxide balance in the oceans and its effect on climate.

**Revelle factor** see BUFFER FACTOR.

**reverberatory furnace** *Consumption & Efficiency.* a historic type of furnace with a shallow hearth and a roof that deflects the flame and radiates heat toward the hearth or the surface of the charge.

**reverse bias** *Electricity.* a voltage applied to a semiconductor P–N junction to make the P side negative with respect to the N side.

**reverse Brayton cycle** *Refrigeration.* a refrigeration cycle consisting of four processes: a fluid is compressed at constant entropy and then cooled by reversible constant-pressure cooling; the high-pressure fluid expands reversibly in the engine and exhausts at low temperature; the cooled fluid passes through the cold storage chamber and picks up heat at constant pressure; the fluid returns to the suction side of compressor. [Described by U.S. engineer George *Brayton,* 1830–1892.]

**reverse Carnot cycle** *Thermodynamics.* an ideal thermodynamic cycle consisting of four processes: an expansion at constant entropy; a constant-temperature expansion; a compression at constant entropy; and a constant-temperature compression. These processes occur in the reverse order in the standard CARNOT CYCLE.

**reverse current** *Electricity.* the direct current that flows when a reverse bias is applied to a semiconductor device.

**reverse-cycle** *HVAC.* **1.** describing a year-round air-conditioning system in which heat is transferred from the outside of a building to the inside in the winter, by means of a heat pump that extracts heat from the exterior air and transfers it to the interior. In summer the system operates in the conventional manner, shifting heat from the inside to the outside to cool the building. Thus, **reverse-cycle heat (pump); reverse-cycle system. 2.** see REVERSE CARNOT CYCLE.

**reverse osmosis** *Chemistry.* the application of external pressure to oppose the natural process of osmosis, which would normally cause movement from a region of higher concentration (such as pure fresh water) into one of lower concentration (such as a water-salt solution). By the use of a membrane to allow the passage of water molecules but block larger salt molecules, reverse osmosis will cause water to move out of the salt solution, providing a means to remove salt from seawater.

**reverse water gas shift (RGWS)** see WATER GAS SHIFT.

**reversible** *Thermodynamics.* describing an ideal thermodynamic process that can be exactly repeated in the reverse direction by means of an infinitesimal change in the external conditions; the entropy change for the system and its environment is zero. Thus, **reversible change, cycle,** or **process.**

**reversible fuel cell** see REGENERATIVE FUEL CELL.

**reversible turbine** *Conversion.* a hydraulic turbine that can be used either as a pump or as an engine, turbine, water wheel, or other such apparatus to drive an electrical generator; used in pumped-storage plants,

**revisions** *Oil & Gas.* the total quantity of oil and gas added to proved reserves in a year that derive mainly from non-drilling factors, such as the installation of improved recovery techniques or equipment; changes in economic, geologic, technological, and political factors, and correction of prior reporting errors.

**revolution** *Physics.* the motion of a body around a closed path, relative to some point that is internal or external to the moving body. See also ROTATION.

**Reynolds number** *Measurement.* a measure for the importance of viscous friction in a flow, related to the speed of the flow. A lower Reynolds numbers means the flow speed is low and/or the fluid is very viscous, and thus friction is important; a higher number means the speed is high, and friction is unimportant. [Named for British physicist Osborne *Reynolds,* 1842–1912.]

**R-factor** another term for R-VALUE.

**RFC** regenerative fuel cell.

**RFG** reformulated gasoline.

**rhodium** *Chemistry.* a metallic element having the symbol Rh, the atomic number 45, an atomic weight of 102.9, a melting point of 1970°C and a boiling point of 3730°C; a grayish-white, ductile metal that is obtained from platinum ores and gold gravels; used in high temperature alloys with platinum or palladium and in electrical components.

**ribbon** *Materials.* a thin sheet of crystalline or multicrystalline material produced in a continuous process by withdrawal from a molten bath of the parent material. Thus, **ribbon silicon.**

**Ricardian** *Economics.* having to do with or following the theories of David Ricardo (see below).

**Ricardian rent** *Economics.* rent accruing to owners of natural resources due to differing quality, and hence costs, of their holdings. For example, at a prevailing price for oil, the owner of a large, shallow oil field will have lower costs than the owner of a smaller, deeper field, and hence receive a greater Ricardian rent.

**Ricardo, David** 1772–1823, British economist who helped found the classical school of economics; noted especially for describing the principles of DIMINISHING RETURNS and COMPARATIVE ADVANTAGE.

**Ricardo, Harry** 1885–1974, English engineer who established a measurement system that is the basis of today's classification of fuels according to octane rating. He also made important improvements to diesel and airplane engines, most notably a patent for a two-stroke engine design.

**rice coal** see COAL SIZING.

**rich mixture** *Oil & Gas.* an air-fuel mixture that has a relatively high proportion of fuel and a relatively low proportion of air.

**Richter scale** *Earth Science.* a description of earthquake intensity, based on a logarithmic scale ranging from 1 to 9 that expresses the magnitude (**Richter magnitude**) of a given earthquake based on a measurement of the amount of energy dispersed during the event. [Developed by U.S. seismologist Charles F. *Richter,* 1900–1985.]

**Rickover, Hyman** 1900–1986, U.S. naval officer who led the effort to develop the world's first nuclear-powered submarine, *Nautilus,* which was launched in 1954.

**rider** *Coal.* a term for a thinner seam of coal positioned above a thicker one.

**riding carriage** see REITWAGEN.

**right-to-know** *Social Issues.* **1.** the principle that is in the public interest to disclose certain information, especially concerning the actions of the government or other large institutions.

**2.** specifically, the public release of information about the presence of hazardous or toxic chemicals in a community; e.g., from agricultural pesticides.

**Riker, Andrew Lawrence** 1868–1930, U.S. engineer who founded the Riker Electric Vehicle Company (1888), which became one of the country's largest manufacturers of electric cars and trucks.

**Ring of Fire** *Earth Science.* a term for the Circum-Pacific belt, a zone of geothermal activity encircling the Pacific Ocean that is the site of frequent volcanic eruptions and earthquakes.

**Ringelmann chart** *Environment.* a scale traditionally used to make a subjective visual evaluation of the amount of solid matter emitted by a smokestack or other source; the observer compares the color of the smoke with a series of illustrations shaded in various levels of gray. A corresponding number **(Ringelmann number)** is assigned to the shade describing the best match; numbers range from 0 (white) to 5 (black). Named for French engineer Maximilian *Ringelmann,* 1861–1931.

**Rio Declaration** see EARTH SUMMIT.

**riometer** *Measurement.* relative ionospheric opacity meter; a specially designed radio receiver for the continuous monitoring of cosmic noise.

**riparian** *Ecology.* **1.** having to do with or located on the bank of a river or stream. **2.** describing an ecosystem that is transitional between terrestrial and aquatic ecosystems.

**ripper** *Mining.* a device used to break up solid material in a mine; e.g., earth, rock, or a coal seam, usually by means of a series of large teeth that tear away the surface.

**ripple** *Electricity.* the presence of an alternating current component in a direct-current signal; e.g., an undesirable AC component of a pulsating DC current produced by a rectifier or similar power conditioning device. Thus, **ripple current.**

**riser** *Oil & Gas.* **1.** in an offshore drilling operation, the piping that extends from the platform to the hole, through which drilling is conducted. **2.** piping through which gas or liquid may flow upward. **3.** *Electricity.* the cable that connects buried or ground-level electrical cables to overhead power lines.

**risk** *Health & Safety.* the possibility of loss or injury; a finite measure of the severity and likelihood of an occurrence of damage, disease, or any other negative consequence.

**risk analysis (assessment)** *Measurement.* a detailed analysis of the nature and magnitude of unwanted, negative consequences to life, health, property, resources, or the natural environment.

**risk aversion** *Economics.* **1.** the theory or principle that when facing alternate choices with apparently comparable returns, people and organizations will tend to choose the alternative that seems to offer less risk. **2.** *Ecology.* a similar principle applied to animal behavior.

**risk communication** *Health & Safety.* the fact or process of providing the public with information about the risks associated with particular substances, products, activities, or technologies; e.g., prescription drugs, household cleaning products, toxic wastes.

**risk factor** *Health & Safety.* any condition, event, activity, or environmental characteristic that increases the likelihood of a given individual developing a certain disease or experiencing a negative change in his/her health status; e.g., cigarette smoking is identified as the leading risk factor for lung cancer; obesity is regarded as a risk factor for adult-onset diabetes.

**risk loving** see RISK SEEKING.

**risk phrases (R-phrases)** *Health & Safety.* a set of warnings required by the European Union to appear on labels and safety data sheets for hazardous chemicals. These consist of the letter R followed by a number and a phrase describing the type of risk (explosive, flammable, toxic, and so on); e.g., R1: explosive when dry; R11: highly flammable; R25: toxic if swallowed.

**risk proclivity** see RISK SEEKING.

**risk seeking** *Economics.* the converse of the usual principle of RISK AVERSION; the theory that when facing alternate choices with apparently comparable returns, someone will tend to choose the alternative that seems to offer more risk; e.g., investing in speculative rather than blue-chip stocks.

**Risk National Laboratory** a Danish government research institution that is a world leader in wind energy research.

**Rittenhouse, William** 1644–1708, builder of the first American paper mill (1690), near Philadelphia. He also developed an early recycling system in which much of the mill's fiber for hand papermaking was obtained from discarded rags and cotton.

**Rivaz, Francois Isaac de** 1752–1828, Swiss engineer, believed to be the first person to equip a stationary gas engine with electric ignition. In 1813 he fitted an engine of this type to a vehicle and conducted test runs near Lake Geneva.

**R lamp** short for REFLECTOR LAMP.

**road carriage** *History.* a name for early steam-powered passenger vehicles, such as a type developed by engineer Richard Trevithick in the early 1800s.

**road engine** *History.* a name for one of the earliest vehicles powered by an internal-combustion engine, patented by attorney George SELDEN.

**road octane** *Oil & Gas.* the measured antiknock rating of a motor fuel, determined under normal driving conditions.

**road oil** *Oil & Gas.* any heavy petroleum oil, including residual asphaltic oil, used as a dust palliative and surface treatment on roads and highways. It is generally produced in six grades from 0, the most liquid, to 5, the most viscous.

**robbed-out** *Coal.* describing an area of a coal mine from which the supporting pillars have been removed.

**robbing** *Coal.* the process of extracting pillars of coal that had previously been left in place for support

**Robinson, Enders** born 1930, U.S. geophysicist who developed the theory and method of DECONVOLUTION, which would revolutionize the interpretation of seismic data in petroleum exploration.

**robot** *Consumption & Efficiency.* **1.** any industrial device that can be programmed to perform a variety of tasks involving manipulation and movement under automatic control. **2.** specifically, such a device that has somewhat of a humanlike appearance or that operates with humanlike capacities.

**robotics** *Consumption & Efficiency.* the technology of robots; the design, manufacture, and use of mechanical devices that can be programmed to perform various industrial tasks.

**rock bed storage** *Storage.* a type of thermal storage medium used in passive solar heating or cooling, consisting of an underground accumulation of rocks serving as a heat exchanger.

**rock drill** *Mining.* a drill, generally having bits detachable from the drill stem, that is designed to bore relatively short holes through rock for blasting purposes; powered either by electricity, compressed air, or steam. Used to excavate large amounts of earth and rock in tunneling, mining, and quarrying.

**rock dusting** *Mining.* the process of removing a layer of dust, typically composed of powdered limestone, from wall areas in order to contain or minimize explosions, aid in the lighting of the mine, and reduce health hazards

**Rockefeller, John D.** 1839–1937, U.S. industrialist and philanthropist, generally regarded as the founder of the modern oil and gas industry. At its peak, his Standard Oil company controlled about 90% of oil refining in the U.S. This and other business activities made him reputedly the world's richest man,

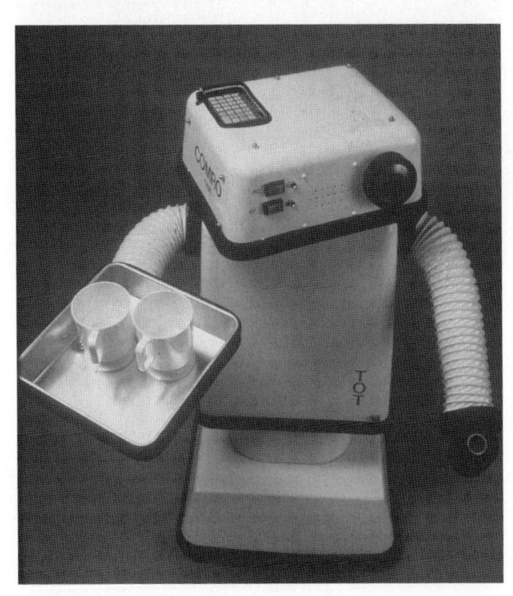


***robot*** *The term "robot" became known in the title of a play by Czech author Karel Capek, in which human-like machines rebel against their living masters.*

and the name "Rockefeller" became a symbol for wealth.

**rocker arm** *Transportation.* a center-pivoted lever in an internal combustion engine that transmits motion from a cam or a push rod to a valve stem.

**rocket** *Transportation.* **1.** a reaction engine that contains within itself everything necessary for the combustion of its fuel, and that therefore does not require an external medium of air for combustion (thus it can operate in outer space). The simplest rocket engine consists of a combustion chamber and a nozzle; a more complex type may have more than one combustion chamber. **2.** a vehicle or projectile propelled by such an engine. Technically, a **rocket engine** is one that employs a liquid propellant, and a **rocket motor** employs a solid propellant.

***rocket*** *The launch of a U.S. Delta II rocket, carrying a GPS (global positioning system) satellite that provides strategic military information.*

**Rocket** *Transportation.* an advanced type of locomotive designed by Robert Stephenson and Company (1829); its basic design principles were utilized from that time until the end of British steam locomotive building in the mid-20th century.

**rocket fuel** *Transportation.* any of various liquid or solid substances used as fuel in rockets, such as liquid hydrogen or gasoline; characterized by the potential for extremely rapid, controlled combustion and subsequent production of large volumes of gas at high pressure and temperature.

**rocketry** *Transportation.* the science and technology of rocket design and operation.

**rocketsonde** *Measurement.* a rocket-borne instrument for measurement and transmission of upper-air meteorological data in the lower 76,000 m of the atmosphere, especially the portion that is inaccessible to radiosonde techniques.

**rockfill dam** *Hydropower.* an embankment formed largely of dumped rock for stability and fitted with an impervious water-face membrane and core-wall.

**rock oil** *History.* an early term for petroleum, based on its presence in rock formations.

**rockwool** *Materials.* a type of insulating material so called because it resembles old grayish wool with dark flecks; spun like fiberglass from the slag from refining metals.

**Rocky Mountain Institute** (est. 1982), a research and advocacy group founded by Amory Lovins in Snowmass, Colorado, aimed at transforming various sectors of the economy to more efficient use of natural resources, especially reduced dependence on fossil fuels.

**Roentgen, Wilhelm** 1845–1923, German physicist famous for his discovery of X-rays (1895). This discovery was the key to a new era of modern physics and it would revolutionize diagnostic medicine.

**roentgen** *Nuclear.* a unit of ionizing-radiation dosage of X-rays or gamma rays, equivalent to the amount that produces one electrostatic unit of electrical charge from secondary ionization in one cubic centimeter of dry air under standard conditions.

**roentgen equivalent man** see REM.

**rolling blackout** *Electricity.* a purposeful series of temporary, controlled power outages carried out by a utility company, with the intent of preventing heavy demand from crashing the electrical grid and causing a full-scale unplanned blackout.

**rolling mill** *Materials.* a sequential arrangement of milling devices used to crush and grind material.

**roof** *Mining.* **1.** the rock, commonly shale and usually carbonaceous, found immediately above a coal seam, ore body, or other tabular deposit. **2.** the overhead surface of a coal working place.

**roof bolt** *Mining.* a long steel bolt driven into the roof of an underground excavations to prevent or limit the extent of roof falls. Thus, **roof bolter.**

**roof fall** *Mining.* a cave-in; a collapse of the roof of a coal mine, especially in a permanent area such as an entry.

**roof pond** *Solar.* **1.** a solar energy collection device consisting of containers of water located on a roof that absorb solar energy during the day so that the heat can be used at night. **2.** a similar system that cools a building by evaporation at night.

**roof rock** another term for SEAL ROCK.

**roof sag** *Mining.* a sinking of the roof of a coal mine, especially in the middle; the equivalent of heave which is a similar dislocation of a mine floor.

**room** *Mining.* **1.** an area abutting an entry or airway in which coal is mined. **2.** any wide working space within a flat mine.

**room-and-pillar mining** *Mining.* a common method of coal mining in which rooms (large open spaces) are cut into the coalbed leaving in place a series of pillars (columns of coal) to help support the mine roof and control the flow of air. As the mining advances, a grid-like pattern of rooms and pillars is formed. Also, **room-and-pillar system.**

**root** *Wind.* a term for the section of a wind turbine blade nearest to the hub; usually the thickest and widest part of the blade.

**rose** see WIND ROSE.

**Rosenfeld, Arthur** U.S. physicist noted for his leadership in the development of energy-efficient technologies, including electronic ballasts for fluorescent lamps, low-emissivity windows, and computer programs for the energy analysis and design of buildings.

**rotary compressor** *Consumption & Efficiency.* a machine having a rotating part that directly compresses fluid in an enclosed housing; the fluid pressure rises as the volume of the closed space decreases.

**rotary drill** *Oil & Gas.* a drilling machine having a rigid, rotating, tubular string of rods connected to a rock-cutting bit; a historic advance (mid 19th century) in drilling deep holes in search of petroleum or gas, as compared to the previous technology of cable tool rigs that relied on pulverizing rock formations.

**rotary engine** *Transportation.* an engine having a thermodynamic cycle mechanism that is powered and designed to move on a circular path; used for steam engines and in some internal-combustion (automotive) engines. See also WANKEL ENGINE.

**rotation** *Physics.* **1.** the motion of a body about an internal axis of symmetry; the axis may be fixed or moving, and actual or theoretical. *Rotation* is the movement of a body in relation to an internal point, and *revolution* is its movement in reference to an external point. Thus the earth *revolves* around the sun, and *rotates* on its own axis. **2.** a description of such rotating motion, expressed as the change in a body's angular orientation over time, with respect to some relevant (usually inertial) frame of reference. **3.** one complete turn of a celestial body about its axis. Thus, **rotational.**

**rotational energy** *Physics.* **1.** the kinetic energy of a body as measured in a nonrotating frame of reference in which its center of mass is stationary. **2.** the minimum work needed to cause an object to rotate at a certain velocity, starting from rest. **3.** the difference between the energy in a molecule with more than one atom, and the energy in a theoretical molecule constructed by stopping the rotation of the atomic nuclei without hindering their vibrational electron movement.

**roundwood** *Biomass.* a term for logs, bolts, or other round sections cut from trees.

**Rover** *History.* the earliest commercially successful version of the modern bicycle, developed by John Kemp Starley of England and marketed by him and fellow bicycle innovator William Sutton through their Rover Company. In full, **Rover Safety Bicycle.**

**Royal Dutch/Shell** (est. 1907), the world's third largest integrated oil company (after BP and Exxon Mobil). Its structure dates back to 1907 from the merger of Royal Dutch, formed in the Netherlands to develop oil fields in Asia, and SHELL.

**royalty** *Economics.* **1.** a payment made for the use of property rights, especially intellectual property. **2.** a payment made in money or kind for a stated share of production from mineral or fuel deposits, paid by the lessee to the lessor; often calculated as a percentage of the gross income from a lease.

**r-per cent rule** another term for the HOTELLING RULE.

**R-phrases** see RISK PHRASES.

**RPM** revolutions per minute.

**RPS** renewable portfolio standard.

**RTG** radioisotope thermoelectric generator.

**RTO** regenerative thermal oxidizer.

**rubber** *Materials.* **1.** a raw gum material characterized by its elasticity, obtained from a certain tree *Hevea brasiliensis,* that is native to tropical America, or from various other tropical trees, such as *F. elastica,* or other trees of the genus Ficus. **2.** any of various synthetic polymers having similar properties to natural rubber.

***rubber*** *The para rubber tree,* Hevea brasiliensis; *the source of the largest amount and highest quality of natural rubber.*

**rubblization** *Nuclear.* a decommissioning technique involving demolition and burial of formerly operating nuclear facilities; so called because above-grade structures are demolished into *rubble* and buried in the structure's foundation below ground. The site surface is then covered for unrestricted use.

**Rubisco** *Biological Energetics.* short for ribulose-1,5-bisphosphate carboxylase, a ubiquitous enzyme that represents the crucial starting point of any food chain, because of its ability to fix inorganic carbon dioxide from the atmosphere.

**rule of law** *Global Issues.* the concept or practice of equal protection under the law for human rights as well as property and other economic rights, and punishment for the violation of such rights.

**Rumford, Count** Benjamin Thompson 1753–1814, British–American scientist noted for his contributions to the modern theory of heat. From his work boring cannon barrels, he noted that the cannon would stay hot as long as the friction of boring continued. This could not be explained by the prevailing CALORIC THEORY of heat as a fluid state of matter, and he concluded that heat is a form of mechanical energy produced by the motion of particles.

**Rumford stove** *History.* an improved stove invented by Count Rumford (1795), an innovative version providing greater heat with less smoke.

**runner** *Conversion.* the rotating part of a turbine that converts the energy of the working fluid (water, steam, or gas) into mechanical energy.

**runoff** *Earth Science.* **1.** the portion of precipitation that flows over the land surface and ultimately reaches streams to complete the water cycle; sources include melting snow and any surface water that moves to streams or rivers through a drainage basin. **2.** wastewater (or other liquids) entering the environment as a result of some industrial activity.

**run-of-mine coal** *Coal.* a heterogeneous material consisting of particles of various sizes, often wet and contaminated with impurities such as rock and/or clay and unsuitable for commercial use without treatment.

**run-of-river** *Hydropower.* describing a hydroelectric project that uses the power of moving water as it occurs, with little or no reservoir capacity for storage; typically employed at a site with a large water flow where major dam construction is not feasible.

**run of the wind** *Wind.* the indicated distance that the wind travels in a certain time, measured by the miles of wind blowing past an anemometer.

**run-or-lose** *Consumption & Efficiency.* a term for energy production that will be lost if not used at the time it is produced; usually refers to solar power and hydroelectric energy that cannot be stored after production.

**rural electric cooperative (REC)** *Electricity.* an independent, locally owned business enterprise in which consumers who receive electrical service are member-owners of the cooperative and share responsibility for its success or failure along with the benefits they receive. Similarly, **rural utilities cooperative.**

**rural electrification** *Social Issues.* the fact or process of utilities providing electrical service to rural communities previously lacking or deficient in this service.

**Rural Electrification Act (REA)** *History.* a measure signed into law by President Franklin D. Roosevelt in 1936, to address the fact that the rural U.S. was still largely without electricity. Within 2 years REA projects provided electricity to 1.5 million farms, and by the mid-1950s virtually all American farms were electrified.

**rural energy** *Social Issues.* energy use by rural households, especially traditional forms such as the burning of wood or crop wastes; usually refers to domestic uses such as heating and cooking, as opposed to the energy used for crop and livestock production.

**ruthenium** *Chemistry.* a rare metallic element of the platinum group having the symbol Ru, the atomic number 44, and an atomic weight of 101.07, with 7 stable isotopes, a melting point of 2310°C, and a boiling point of about 4000°C; it is a grayish-white brittle metal used as a hardener for platinum and palladium, as a catalyst, and in alloys.

**Rutherford, Ernest** 1871–1937, New Zealand born British physicist noted for his studies of radioactivity and his discovery of the atomic nucleus. He coined the terms "alpha", "beta", and "gamma" to classify various forms of rays that were poorly understood before his time. He also devised the "half-life" concept to describe the fact that the intensity of radioactivity declines over time.

**runtime** or **run time** *Storage.* battery life; the total amount of time that a battery will provide power before it must be recharged.

**Ruud heater** *History.* the first device to store and automatically heat water (1889), developed by Norwegian–American engineer Edwin Ruud.

**R-value** *HVAC.* a unit of thermal resistance used for comparing the insulating values of different material, indicating a given material's effectiveness in retarding heat flow; higher R-value numbers indicate greater insulating properties. It is expressed in units of temperature (°F) × hours × square feet per Btu. See below.

**Rydberg constant** *Nuclear.* a fundamental physical constant used in studies of the spectrum of a substance; its value for hydrogen is 109,737.3 $cm^{-1}$. It is regarded as one of the most important physical constants since it has the potential to be determined with high precision and can be regarded as a natural unit of both length and time. [Described by Swedish physicist Johannes *Rydberg* 1854–1919.]

**Rydberg formula** *Nuclear.* a formula used in atomic physics to determine the full spectrum of light emission from hydrogen, later extended to be useful with any element.

**R-value** The term R-value stands for thermal resistance and is a measure of the level of resistance to heat flow a given material or an assembly can offer as a result of suppressing conduction, convection, and radiation. The thermal resistance for a homogeneous material is mainly associated with the conduction of heat and is a function of material thermal conductivity and thickness (the length of heat flow path) and is expressed in F-ft$^2$-h/Btu (K-m$^2$/W). R-value is directly proportional to the material thickness and inversely proportional to its thermal conductivity. Thermal resistance may also be associated with the heat transfer by convection and radiation at a surface. The thermal resistance follows the same analogy as the electrical resistance associated with the conduction of electricity. Therefore, the formulation of heat flows in terms of thermal resistance allows the heat flow through any assembly to be presented as a thermal circuit. Composite components that are characterized by multiple layers with resistances in series and parallel arrangements can be presented by an equivalent thermal circuit. Resistances in series are additive and the overall R-value of an assembly is obtained by summing up the equivalent resistances of all layers comprising that assembly.

**Mohammad Al-Homoud**
King Fahd University of Petroleum and Minerals
Saudi Arabia

**sabkha** *Earth Science.* plural, **sabkhat.** a flat, low area covered by clay, silt, and sand, and often encrusted with salt, found in arid and semiarid environments; e.g., oil-producing regions such as the Arabian peninsula and the Permian Basin of West Texas. [From an Arabic term meaning "salt flat".]

**saccharification** *Chemistry.* the process of converting starch or dextrin into sugar.

**Sachs, Julius Von** 1832–1897, German botanist who was a founder of experimental plant physiology. He discovered that, in the presence of light, chlorophyll catalyzes photosynthetic reactions (1865).

**sacrificial protection** *Materials.* the purposeful corrosion of a less desirable metal so that an adjacent preferred metal can be protected from corrosion. Thus, **sacrificial anode, sacrificial metal.**

**SAE** Society of Automotive Engineers.

**SAE horsepower** *Transportation.* a standard measure of a vehicle's horsepower; the horsepower generated by the engine at the flywheel with all energy-drawing accessories attached, such as an alternator and power steering and air conditioning units. Also called **SAE net horsepower** in contrast with the earlier standard it replaced, **SAE gross horsepower,** which is horsepower measured with no load from a chassis or accessories and with ideal fuel and ignition conditions.

**SAE number** *Transportation.* a system for classifying crankcase, transmission, and differential lubricant according to their viscosities.

**safety injection** *Nuclear.* the rapid insertion of a chemically soluble neutron poison (such as boric acid) into a reactor coolant system to ensure reactor shutdown.

**safety lamp** *Mining.* a coal miner's lamp that is deemed relatively safe for use even in atmospheres that may contain flammable gas.

**safety rod** *Nuclear.* a control rod used in a nuclear reactor to decrease the reactor reactivity in the case of emergencies.

**SAFSTOR** *Nuclear.* safe storage; a method of decommissioning for a nuclear facility in which it is placed and maintained in a condition that permits materials to be safely stored and the facility subsequently decontaminated to levels that permit release for unrestricted use.

**Sagebrush Rebellion** *Policy.* a term for organized resistance in the Western U.S. to government land management policies; applied to various populist movements and protests but generally referring to objections by ranchers and farmers to government restrictions on the private use of Federal land.

**Sahara Blend** *Oil & Gas.* an oil type of Algeria, one of an array or basket of seven crude oil types used by OPEC as reference points for pricing.

**sail** *Transportation.* **1.** a large piece of fabric or similar flexible material that is used to catch or deflect the wind, providing propulsive power for a vessel to move forward. **2.** *Wind.* the extended surface of the arm of a windmill.

**sail wing** *Wind.* a propeller on a wind turbine that is similar to the sail of a boat, with cloth stretched over a metal wire frame to form an airfoil section.

**Saint Elmo's fire** *Earth Science.* a glow accompanying discharges of atmospheric lightning, historically seen at the tops of church steeples or the tall masts of sailing ships, now also seen on the wing tips of aircraft. [From *Saint Elmo,* (Erasmus), patron saint of early Mediterranean sailors.]

**salable ton** *Coal.* the total weight of coal mined and sold, with impurities removed by surface coal processing methods applied at the mine.

**salinization** *Earth Science.* the fact of becoming saline (salty); the process by which salts accumulate in water or soil.

**salinometer** *Measurement.* a device or instrument used to measure the salt content of a solution (e.g., brine), especially one based on electrical conductivity methods.

**salt** *Chemistry.* **1.** NaCl, a crystalline compound, appearing as colorless transparent crystals or a white crystalline powder and occurring as a mineral and as a constituent of seawater; used extensively since ancient times for seasoning and preserving food, and also employed for various industrial purposes. **2.** the compound formed when the hydrogen atom of an acid is replaced by a metal atom or a positive radical, characteristically having an ionic lattice and disassociating completely in a water solution.

**saltbox** *HVAC.* a wood frame house with two stories in front and one behind, and a gabled roof whose rear slope is much longer than its front slope; historically a common design in the New England region of the U.S. because of its inherent energy efficiency.

**salt cake** see SALT SLAG.

**salt cavern** *Earth Science.* a cavern in an underground salt formation that is created in a commercial mining operation through the injection of fresh water and the removal of the salt in solution. Some of these caverns are then used to store hydrocarbons such as crude oil and natural gas, or oil field wastes such as drilling fluids.

**salt dome** *Earth Science.* a dome or anticlinal fold originating from a thick bed of salt up to 5 miles deep below the earth's surface. The domes push their way up through more brittle overlying rocks, are roughly circular, and average up to 1–2 miles in diameter. The tops of these domes can be commercially mined for salt.

**salt gradient pond** *Solar.* a solar pond that consists of three main layers; the top one is near ambient temperature and has low salt content, while the bottom one is very hot and salty and is lined with a dark-colored material. The middle (gradient) zone acts as a transparent insulator, permitting sunlight to be trapped in the bottom layer (from which useful heat is withdrawn). This layer, which increases in brine density with depth, counteracts the tendency of the warmer water below to rise to the surface and lose heat to the air. Also, **salt pond, salt gradient solar pond.**

**salt slag** *Environment.* a waste deposit of salts resulting from an industrial process (e.g., aluminum manufacturing) that constitutes a hazardous waste.

**salt well** *Oil & Gas.* a bored or driven well from which brine is obtained; such wells were an early source of oil in the U.S. oil industry.

**samarium** *Chemistry.* a rare-earth element that is fifth in the lanthanide series in the periodic table, having the symbol Sm, the atomic number 62, an atomic weight of 150.36, a melting point of 1080°C, and a boiling point of 1800°C; a brittle grayish metal used as a dopant for laser materials, in infrared absorbing glass, and as a neutron absorber in certain nuclear reactors.

**samarium oxide** *Chemistry.* $Sm_2O_3$, a cream-colored powder that absorbs moisture and carbon dioxide from the air, insoluble in water and soluble in acids; it is used as a catalyst in ethanol production.

**Samuel, Marcus** 1853–1927, English merchant who founded (1892) Shell Transport and Trading Company, an oil business that would later merge with the Royal Dutch Petroleum Company to form one of the world's largest oil corporations.

**Sandstrom's theorem** *Earth Science.* the argument that surface heating alone is not a sufficient mechanism for supplying energy to the ocean's general circulation, and thus circulation can only be sustained if the heating source is situated at a lower level than the cooling source. [Described by German oceanographer Johan Wilhelm *Sandstrom,* 1874–1947.]

**San Gorgonio Pass** *Wind.* an area near Palm Springs, California that is the location of a large wind energy facility contributing to the electricity needs of Palm Springs and surrounding communities; sited in this location because of the constant winds arising from the VENTURI EFFECT created by the hot desert floor and surrounding mountains.

**sanitary engineering** *Environment.* the sector of civil engineering relating to public health and the environment, such as water supply, sewage, and industrial waste.

**sanitary landfill** *Environment.* a facility used for burying solid wastes, in order to eliminate public health and environmental hazards without contaminating surface or groundwater resources.

**sanitary waste** *Environment.* a classification for solid wastes, such as garbage, that are generated by normal domestic activities and that are not hazardous or radioactive.

**sanitation** *Environment.* the control of all factors in the physical environment that may have harmful effects on public health; the establishment or maintenance of healthy environmental conditions.

**Sarofim, Adel** U.S. engineer responsible for important advances in combustion science that made possible reductions in the release of pollutants from fossil fuel combustion.

***satellite*** *A view of the Syncom IV-5 communications satellite as it passes over the continent of Africa.*

**satellite** *Physics.* **1.** any celestial body, natural or artificial, that revolves in an orbit about a planet. The moon is a natural satellite of the earth. **2.** *Communications.* an artificial body designed to orbit about the earth to relay radio, television, telephone, and other such telecommunication signals to various sites around the world.

**saturated** *Chemistry.* **1.** describing a solution that is not able to dissolve additional solute; i.e., that contains the maximum possible amount of a substance under the given conditions. **2.** describing an organic compound that has atoms attached to each other by single bonds and no free bonds.

**saturated steam** *Geothermal.* steam at a temperature that corresponds to its pressure; the only condition at which true steam can exist.

**saturated vapor** *Thermodynamics.* a vapor that is sufficiently concentrated to exist in equilibrium with the liquid form of the same substance.

**saturation** *Chemistry.* the condition of a fluid holding another substance that cannot hold any additional amount of the substance at the given conditions.

**Saudi Arabian Light** *Oil & Gas.* a crude oil type that formerly was the single reference point for the OPEC pricing structure; now one of an array or basket of seven crude oil types used by OPEC for pricing.

**Saudi Aramco** Saudi Arabian Oil Company (est. 1988); the successor to the Arabian American Oil Company (Aramco), which was the original developer of oil in the country. The Saudi government gradually nationalized this beginning in the early 1970s.

**Sauerbronn, Karl** 1785–1851, German inventor of the Laufmaschine ("running machine"), an early prototype for the modern bicycle.

**Saussure, Horace Bénédict de** 1740–1799, Swiss scientist noted for advances in measuring solar radiation and in the construction of the "hot box", a miniature greenhouse made of wood with glass covers to trap the sun's energy. This became the prototype for solar collectors to cook food and heat homes and hot water.

**savanna** or **savannah** *Ecology.* a regional-scale ecosystem, a plain dominated by grasses interspersed with tall shrubs and some trees in an open formation. It may occur in a variety of tropical and subtropical habitats, but is most common in arid regions or ones with a dry season; e.g., East Africa, where it may have coevolved with the genus *Homo.*

***savanna*** *The East African savanna is noted as the habitat of various large mammals.*

**Savannah** *History.* **1.** a U.S. ship that was the first vessel to make an ocean crossing with steam-assisted power, crossing the Atlantic in 1819. **2.** the world's first nuclear-powered merchant ship, a U.S. vessel launched in 1959.

**Savery engine** *Conversion.* the first practical device to provide mechanical power with steam (1698), designed to lift water for such purposes as keeping mines dry and supplying towns with water. The air in the cylinder was purged by steam once in each working cycle, and was exhausted either together with the water or through a cock valve in the top. [Developed by English engineer Thomas *Savery,* 1650–1715.]

**Savonius turbine** *Wind.* a wind generator employed in various forms but essentially consisting of two offset semicylindrical elements that rotate about a vertical axis; i.e., having roughly an S shape. These can be useful for grinding grain, pumping water, and other such tasks, but generally not for generating electricity because they turn at slow speeds. Thus, **Savonius rotor, Savonius windmill.** [Developed by Finnish engineer S. J. *Savonius.*]

**sawlog** *Biomass.* a log that meets minimum requirements of diameter, length, and lack of defect; i.e., available to be used for commercial purposes.

**sawtimber** *Biomass.* a term for live trees that of suitable size and quality to be harvested for commercial use.

**Saxon wheel** *History.* an advance in the technology of the spinning wheel; it featured a foot pedal, thus leaving both hands free to manipulate the fibers; thought to have been developed in Saxony in the early 1500s.

**SBS** sick building syndrome.

**scale** *Measurement.* **1.** a balance or other such device or instrument used for weighing. **2.** the order of magnitude of some quantity. **3.** a one-to-one correspondence between a measurable physical quantity and a numbering or indexing system to represent this, as in the Celsius temperature scale.

**scale** *Materials.* the deposition of minerals on any solid surface, especially the oxidation product occurring on a metallic surface during heating.

**scale of temperature** see TEMPERATURE SCALE.

**scarcity** *Consumption & Efficiency.* any situation in which demand exceeds supply, especially a situation in which the demand for a resource is significantly greater than the supply.

**scarcity rent** *Economics.* **1.** the added value of a given good or service due to its scarcity. **2.** specifically, the valuation of an extracted, nonrenewable energy resource based on the value that could have been obtained if it had not been extracted until some future date.

**scattering** *Nuclear.* the process by which neutrons collide with atomic nuclei and either gain or lose kinetic energy; neutron scattering from materials allows scientists in diverse fields to explore the structure and dynamics of materials down to atomic scales.

**scatter plug** *Nuclear.* a mechanical device in a fuel assembly, reflecting neutrons back into the core that would otherwise have escaped through the coolant pipe.

**scenario** *Policy.* a plausible and often simplified description of how the future of a given activity or system may develop, based on a coherent and internally consistent set of assumptions about driving forces and key relationships.

**Scheer, Hermann** born 1944, German economist known for his pioneering efforts in the promotion of solar energy technologies in Germany and throughout Europe.

**Scheffler cooker** *Solar.* a type of solar cooker in which a concentrating primary reflector tracks the movement of the sun, focusing sunlight through a hole in the kitchen wall onto a motionless heating surface. [Invented by Austrian physicist Wolfgang *Scheffler.*]

**Schlumberger, Conrad** 1878–1936, French engineer who with his brother **Marcel** invented electrical well logging (1921), a technology that eliminated much of the guesswork associated with determining whether or not a formation held oil or natural gas.

**Scholler process** *Biomass.* a process for converting biomass to ethanol, developed in Germany; it involves a percolation process in which a dilute solution of sulfuric acid is pumped through a bed of wood chips and sawdust.

**Schottky barrier** *Electricity.* a junction between a layer of semiconductor material, such as silicon, and a layer of metal, characterized by hot carriers. Thus, **Schottky barrier diode.** [Named for German physicist Walter Hans *Schottky,* 1886–1976.]

**Schrödinger, Erwin** 1887–1961, Austrian physicist known for his mathematical development of WAVE MECHANICS (1926), and his formulation of the **Schrödinger equation,** which is the most widely used mathematical tool of modern quantum theory.

**Schurr, Samuel H.** 1918–2002, U.S. economist noted for his pioneering analysis of the role of energy in economic systems.

**scientific method** *History.* an organized approach to problem-solving that includes collecting data, formulating a hypothesis and testing it objectively, interpreting results, and stating conclusions that can be independently evaluated and tested by others. The roots of this method are traced to ancient Greece, especially Aristotle, and to the medieval scholar Roger Bacon, but its development is generally associated with 17th-century scientists such as Francis Bacon, Galileo, Descartes, and Newton.

**scientific notation** *Measurement.* a system for indicating very small or very large numbers by writing the number as a decimal number between 1 and 10, multiplied by a power of 10; e.g., 215,000,000 is written as $2.15 \times 10^{8}$; 0.00000215 is written as $2.15 \times 10^{-6}$.

**scintillation** *Nuclear.* a flash of light produced in a transparent material by an ionizing event.

**scintillation counter (detector)** *Nuclear.* an instrument that detects and measures gamma radiation by counting the light flashes (scintillations) induced by the radiation.

**scoop loader** *Coal.* a large loading unit with a scoop-shaped blade, used for removing coal from the working face.

**SCR** selective catalytic reduction.

**scram** *Nuclear.* a term for the sudden shutdown of a nuclear reactor, usually by the rapid insertion of safety rods.

**screw** *History.* **1.** one of the simple machines of antiquity, a device consisting of an inclined plane incised into and spiraling around a tapered cylinder forming a helical groove; used to fasten materials together or to transmit motion and apply pressure. **2.** a small metal fastener of this type, having a tapered body with a spiraling thread and a slotted head.

**scribing** *Photovoltaic.* a process in which a semiconductor wafer is etched with deep groves so that it can be broken into smaller pieces, generally for the purpose of making interconnections.

**scrip** *Coal.* nonlegal tender issued by a coal mining company to workers in place of cash; generally redeemable only at a company-owned store in the historic COMPANY TOWN system.

**scrub** *Biomass.* **1.** a general term for any type of vegetation, such as evergreen shrubs or dwarfed trees, growing in areas with poor soil or low rainfall. **2.** a domestic animal that is considered to be inferior stock.

**scrubber** *Materials.* **1.** any of various devices that use a spray of water or reactant or a dry process to trap pollutants in gaseous emissions. **2.** a mining device in which coarse and sticky material, such as ore or clay, is washed free of adherents or is mildly disintegrated.

**S-curve** *Sustainable Development.* a curve describing population growth, based on a model in which population increases until it reaches the carrying capacity of the ecosystem and then levels off.

**Seaborg, Glenn** 1912–1999, U.S. nuclear chemist noted for his discovery of a number of transuranic elements, including neptunium and plutonium. He also worked on the development of the atomic bomb during World War II.

**sea coal** *Coal.* a form of coal obtained historically, especially in Britain, by gathering it at random on beaches as it washed up on shore from the ocean.

**sealed battery** *Storage.* a term for any type of battery that does not open to the atmosphere under normal conditions; e.g., a typical flashlight battery, as opposed to a typical automobile battery to which additional electrolyte can be added periodically.

**sea level** *Earth Science.* **1.** literally, the ocean surface, considered to be the zero standard for describing heights and depths on earth. **2.** also, **mean sea level,** the average elevation of the ocean surface, taken as the mean level between high and low tides relative to a fixed benchmark on land, calculated over an extended period of time so as to discount the effects of local wind, waves, seasonal effects, and other temporal variations. See also RELATIVE SEA LEVEL.

**sea level pressure** *Earth Science.* the atmospheric pressure at mean sea level, either directly measured or empirically determined from an observed station pressure; considered to be approximately 14.7 pounds per square inch (1 atmosphere) under standard conditions.

**seal rock** *Oil & Gas.* a shale or other impervious rock that acts as a barrier to the passage of petroleum migrating in the subsurface; it overlies the reservoir rock to form a trap or conduit.

**seam** *Coal.* **1.** a vein, stratum, or layer of coal in the earth. **2.** *Mining.* any distinct layer of ore or other valuable material. **3.** a thin rock stratum that separates two distinct layers of different composition.

*seam Kentucky coal miner loads coal from a seam about four feet thick. His expected total load for the day would be 16–17 tons of coal. Photo taken the same year (1946) that singer Merle Travis released the popular coal mining song "Sixteen Tons".*

**seasonal depth** *Photovoltaic.* an adjustment factor used in some system sizing procedures by which the battery is gradually discharged over a period of poor solar insolation; this factor results in a slightly smaller PV array. In full, **seasonal depth of discharge.**

**seasonal energy efficiency ratio (SEER)** *HVAC.* the total cooling output of a system in Btu during its normal annual usage period, divided by the total electric energy input in watt-hours during this period. Used as a measure of the cooling performance of central air conditioners and heat pumps.

**sea surface temperature (SST)** *Earth Science.* the temperature of the layer of seawater (approximately 0.5 m deep) nearest the atmosphere above.

**seaweed** *Ecology.* a popular term for plant life growing in an ocean environment, especially marine algae.

**second** *Measurement.* **1.** the fundamental unit for the measurement of time, equal to one-sixtieth part of a minute; now defined as the duration of 9,192,631,770 cycles of vibration of the radiation emitted at a specific wavelength by a cesium-133 atom. **2.** a unit of angular measurement equal to 1/3600 degree. Sixty seconds equal one minute of arc.

**secondary battery** *Storage.* any battery that can be recharged; i.e., that can receive an additional charge after its primary one. Also, **secondary cell.**

**secondary concentrator** *Solar.* an arrangement of mirrors forming a light funnel, to further concentrate the solar radiation that is incident on a receiver.

**secondary containment** *Nuclear.* an enclosure around a nuclear reactor to provide added protection from the release of radiation in the event of an accident that fails the primary containment system.

**secondary energy** *Conversion.* energy converted from primary natural sources, such as coal, crude oil, or sunlight, into a form that is more easily usable for consumption, such as electricity or refined petroleum products.

**secondary gain** *Solar.* solar radiation absorbed within a building and then re-emitted to the interior of the building, as opposed to radiation transmitted directly through a window to the interior (primary gain).

**secondary production** *Renewable/Alternative.* the production of industrial materials from recovered resources; e.g., metals from scrap.

**secondary radiation** *Nuclear.* radiation resulting from the absorption of other radiation in a material (as opposed to the **primary radiation** produced by the material itself).

**secondary recovery** *Oil & Gas.* the use of methods beyond normal flowing or pumping to recover additional oil from a depleted or nearly depleted reservoir, such as gas reinjection or water flooding.

**secondary refrigerant** *Refrigeration.* a refrigerant substance, usually consisting of a salt solution, used to carry heat from the space to be cooled to the coils that contain the primary refrigerant; utilized in large ice-making equipment, or for a space to be cooled that is remote from the chiller.

**secondhand smoke** see ENVIRONMENTAL TOBACCO SMOKE.

**second-law efficiency** *Thermodynamics.* **1.** the efficiency of an energy-conversion process as described by the second law of thermodynamics (see next). **2.** the design of an energy-conversion device or process so as to obtain maximum efficiency as defined by this law. Also, **second-law optimization.**

**second law of thermodynamics** *Thermodynamics.* a fundamental law of nature that is defined in various ways, such as: (a) heat will not flow spontaneously from a cold object to a hot one; (b) any system that is free of external influences becomes more disordered with time, i.e., it tends toward a state of greater ENTROPY; (c) it is not possible to create a heat engine that extracts heat and converts all this heat to useful work, i.e., no energy-conversion process is 100% efficient. See also THERMODYNAMICS.

**second-round effect** see FIRST-ROUND EFFECT.

**securitization** *Electricity.* the aggregation into one pool of contracts for the purchase of power output from various energy projects, which are then offered as shares for sale in the investment market. Thus, **securitize.**

**sedentism** *Sustainable Development.* a modification of the HUNTING AND GATHERING living pattern, in which the group settles down to live in fixed locations and in more permanent dwellings.

**sediment** *Earth Science.* fragments of organic or inorganic material that are transported by, suspended in, or deposited by wind, water, or ice, and that then are accumulated in unconsolidated layers on the surface of the earth.

**sedimentary** *Earth Science.* **1.** relating to or formed by the deposition of sediment. **2.** describing rock formed by the consolidation of layers of sediment.

**sedimentary basin** *Oil & Gas.* a topographic or structural low area that generally receives thicker deposits of sediments than adjacent areas; the low areas tend to sink more readily, partly because of the weight of the thicker sediments. These are often good prospects for oil exploration; there are about 600 sedimentary basins in the world.

**Seebeck effect** *Electricity.* the generation of an electromotive force by the temperature differences between junctions in a circuit containing two different metals, alloys, or bodies. The thermocouple, an electric temperature-measuring device, is based on this phenomenon. [Discovered by German physicist Thomas *Seebeck,* 1770–1831.]

**seed oils** see OILSEED CROP.

**seep** *Oil & Gas.* **1.** a naturally occurring emergence of liquid petroleum at the surface as a result of a slow, upward migration from its source through minute pores or fissures; typically located at a low elevation where there is an accumulation of water. **2.** of oil, to percolate slowly to the surface in this manner. Thus, **seepage.**

**SEER** seasonal energy efficiency ratio.

**SEIA** socio-economic impact assessment.

**seismic** *Earth Science.* **1.** relating to the study of earthquakes. **2.** relating to or caused by an earthquake or major earth vibration. **3.** relating to an artificially induced earth vibration.

**seismic exploration** *Oil & Gas.* a search for commercially economic subsurface deposits of oil, gas, and minerals by means of artificially induced shock waves in the earth. See next page.

**seismograph** *Earth Science.* an instrument that records vibrations in the ground and determines the location and magnitude of an earthquake.

**seismology** *Earth Science.* the scientific study of earthquakes, including their origin, propagation, energy manifestations, and possible methods of prediction.

**Selden, George** 1846–1922, attorney of Rochester, New York who was the first in the U.S. to apply for a patent for a vehicle

**seismic exploration** The search for commercially economic subsurface deposits of oil, gas, and minerals by the recording, processing, and interpretation of artificially induced shock waves in the earth. Artificial seismic energy is generated on land by shallow borehole explosives such as dynamite, or surficial vibratory mechanisms mounted on specialized trucks; in marine environments, air guns fire highly compressed air bubbles into the water that transmit seismic wave energy into the subsurface rock layers. Seismic waves reflect and refract off subsurface rock formations and travel back to acoustic receivers called geophones (on land) or hydrophones (in water). The travel times (measured in milliseconds) of the returned seismic energy, integrated with existing borehole well information, aid geoscientists in estimating the structure (folding and faulting) and stratigraphy (rock type, depositional environment, and fluid content) of subsurface formations, and facilitate the location of prospective drilling targets. The first known seismic exploration trials were conducted by J. C. Karcher and colleagues, who performed a primitive seismic survey and mapped a shallow limestone bed at Belle Isle, Oklahoma in the summer of 1921. Since then, seismic technologies have evolved into ever more sophisticated techniques through the use of digital computer processing, improved energy sources, advanced acoustic receivers (multi-component), three-dimensional, and four-dimensional (time-lapse) seismic surveys. Seismic exploration is now the most prevalent geophysical technique used in the search for hydrocarbons. Because of potential damage to environmentally sensitive areas such as offshore and arctic regions, seismic exploration activities are strictly regulated by state and federal governments.

**Virginia Pendleton**
Integrity Geophysics
Tulsa, Oklahoma

powered by an internal-combustion engine (1879; granted 1895). He later sued Henry Ford for producing his own automobiles, but the courts ruled the Selden patent too vague to be enforced.

**selective catalytic reduction (SCR)** *Environment.* a process installed downstream of a boiler, involving a catalytic reactor through which the flue gas passes. A reducing agent, typically ammonia, is introduced into the flue gas upstream of the reactor and reacts with the nitric oxide ($NO_x$) in the flue gas to form nitrogen and water, thus reducing harmful $NO_x$ emissions from power-generating equipment.

**selective surface** *Solar.* a solar-absorbing surface that has high absorbance of radiation of one wavelength (e.g., sunlight) but little emission of radiation of another wavelength (e.g., infrared); used as a coating for a variety of solar collection devices. Also, **selective absorber.**

**selenium** *Chemistry.* a nonmetallic element having the symbol Se, the atomic number 34, and the atomic weight 78.96; an essential mineral resembling sulfur, being a constituent of the enzyme glutathione peroxidase and believed to be closely associated with vitamin E and its functions.

**self-combustion** *Transportation.* **1.** a process in which an engine (especially a diesel engine) heats and compresses the fuel mixture to the point where it ignites itself, rather than relying on a separate energy source such as a spark plug. **2.** see SPONTANEOUS COMBUSTION.

**self-discharge** *Storage.* a battery's natural loss of charge over time, as opposed to charge lost through a power demand being placed upon it.

**self-governing** *Wind.* self-regulating; describing a windmill that automatically turns to face changing wind directions and controls its own speed of operation; an important innovation in windmill technology in the mid-19th century.

**self-organization** *Ecology.* **1.** the tendency of an ecosystem to actively shape its evolutionary environment; that is, to use the available energy and matter inputs to achieve higher levels of structure and functioning, given natural selection constraints. **2.** *Economics.* a similar tendency in a human system; e.g., a nation's economy.

**self-rescuer** *Coal.* a small, self-contained breathing apparatus carried by miners underground, used as a filtering device for immediate protection against carbon monoxide and smoke in the event of a mine fire or explosion.

**self-use** another term for PLANT-USE.

**Sellafield** *Nuclear.* the site of Great Britain's first plutonium production reactor, a facility in West Cumbria on the northwest coast of England; also the site of the prototype British Advanced gas-cooled reactor. See also WINDSCALE.

**semi-submersible technology** *Oil & Gas.* an offshore process employing a floating drilling unit with ballasted, watertight pontoons below the sea surface that provide a base for columns to penetrate the water surface and support the operating deck. Thus, **semi-submersible platform, semi-submersible rig.**

**semianthracite** *Coal.* a classification for coal ranking between bituminous coal and anthracite, typically defined as having a fixed-carbon content between 86% and 92%.

**semibituminuous coal** *Coal.* a classification for coal that ranks between bituminous coal and anthracite (or semianthracite); it is harder and more brittle than bituminous coal, has a higher energy content, and burns with relatively little smoke.

**semiconductor** *Electricity.* a solid-state crystalline material having values of electrical resistivity intermediate between metals and insulators. The conductivity of semiconductors can be controlled by adding very small amounts of foreign elements called dopants. Conductivity is facilitated not only by negatively charged electrons, but also by positively charged holes, and it is sensitive to temperature, illumination, and magnetic field.

**sendout** *Storage.* a term for the total amount of gas that is removed from underground storage during a certain time period.

**sensible (cooling) capacity** *HVAC.* the total available refrigerating capacity of an air conditioning unit for removing sensible heat from a space to be cooled.

**sensible heat** *Thermodynamics.* heat energy that causes a rise or fall in the temperature of a substance (gas, liquid, or solid) when added or removed from that substance without a change in state. Compare LATENT HEAT.

**sensible heat storage** *Storage.* a heat storage system that operates on the basis of a change in the temperature (but not the state) of the storage medium (e.g., water).

**sensible (cooling) load** *HVAC.* the load (energy requirement to cool a building space) created by the heat gain of the space due to factors such as solar radiation, air infiltration, the presence of people, lighting, and appliance or equipment use.

**separative work** *Nuclear.* a measure of the effort required in an enrichment facility to separate uranium of a given U-235 or U-238 content into two fractions, one with a higher percentage and one with a lower percentage.

**separator plate** *Renewable/Alternative.* a solid piece of electrically conductive material that is inserted between the cells of a fuel cell stack.

**sequestration** see CARBON SEQUESTRATION.

**SERI** *Solar.* Solar Energy Research Institute; a federally funded institute in the U.S., created in 1974 to conduct research and development of solar energy technologies. Became the National Renewable Energy Laboratory (NREL) in 1991.

**series** *Electricity.* **1.** the arrangement of connecting components in a circuit to form a single path for current. **2.** relating to or operating with such an arrangement. Thus, **series coil, series feed, series modulation,** and so on.

**series circuit** *Electricity.* a circuit in which each component is joined end-to-end successively with the next, so that the same current flows through each component.

**series connection** *Photovoltaic.* **1.** a way of joining photovoltaic cells by connecting positive leads to negative leads; this configuration increases the voltage. **2.** see SERIES CIRCUIT.

**series hybrid** *Transportation.* a type of hybrid electric vehicle that runs on battery power like a pure electric vehicle until the batteries discharge to a set level, when an alternative power unit turns on to recharge the battery.

**series resistance** *Photovoltaic.* resistance to current flow within a solar cell due to factors such as resistance from the bulk of the semiconductor material and the resistance of the contacts and interconnections.

**servomotor** *Conversion.* a low-power electric motor that is used to control mechanical motion.

**setback** *HVAC.* an automatic adjustment of a thermostat to reduce the amount of conditioning provided, thus allowing the temperature to drift naturally to a marginal level; e.g., a lowering (or raising) of the temperature in an office building after the occupants leave in the evening. (The temperature can be readjusted to the desired level before occupants arrive in the morning.) Thus, **setback thermostat.**

**setpoint** *HVAC.* the precise temperature to which a thermostat is adjusted to begin or end the operation of conditioning air.

**settlement price** *Economics.* the official price for oil or gas established by a futures and options exchange (e.g., NYMEX) at the end of the trading day.

**Seven Sisters** *Oil & Gas.* a nickname for the cartel of seven international companies that dominated the world oil market from the period after 1928 until the rise of OPEC in the 1960s; i.e., the U.S. firms Standard Oil of New Jersey (Esso), Socony Vacuum (Mobil), Standard Oil of California (Socal), Texaco, and Gulf, and the European firms Anglo-Iranian Oil (later British Petroleum) and Royal Dutch Shell. [Popularized in the title of a 1968 book by British journalist Anthony Sampson.]

**sewage** *Environment.* waste matter carried off by a pipe, drain, or the like, especially household wastes (e.g., human excreta) as opposed to wastes from industrial processes.

**sewage gas** *Environment.* a combustible gas that is self-generated from the digesting of sewage sludge, having a slow rate of flame propagation.

**shade screen** *Solar.* a screen affixed to the exterior of a window or other glazed opening, designed to reduce the solar radiation reaching the glazing.

**shading** *Solar.* a general term for any protection from the heat gains produced by direct solar radiation.

**shading coefficient** *Solar.* the ratio of solar heat gain through a specific window, with or without shading devices, relative to that occurring under the same conditions through a reference model (a window made of clear, unshaded double-strength glass).

**shadouf** *History.* an ancient device used to raise water from a river or well; a form of lever consisting of a pole balanced on a pivot, with a weight at one end and a bucket at the other. First appearing in Egypt or possibly in India about 1600 BC; versions of this are still in use today. Also spelled **shaduf.**

**shadow pricing** *Economics.* the process of adjusting an absent or distorted market price for a natural resource or environmental good or service, to better reflect the cost to society of obtaining that item.

**shaft** *Conversion.* **1.** a cylindrical piece of metal that rotates or provides an axis of revolution, and upon which rotating machine parts are mounted to transmit power or motion, as in a turbine. **2.** see MINESHAFT.

**shaft horsepower** *Conversion.* a measure of the net mechanical energy per unit time supplied to a turning shaft.

**shaft mine** see MINESHAFT.

**Shah of Iran** see PAHLAVI.

**shale** *Earth Science.* a fine-grained sedimentary rock produced by the consolidation of clay, silt, or mud, and composed roughly of one-third quartz, one-third clay materials, and one-third other miscellaneous materials such as carbonates, iron oxides, feldspars, and organic matter.

**shale gas** *Oil & Gas.* natural gas that is produced from reservoirs predominantly composed of shale with lesser amounts of other fine grained rocks, rather than from more conventional sandstone or limestone reservoirs.

**shale oil** *Oil & Gas.* an unconventional source of oil distilled from oil shale, a kerogen-rich sedimentary shale; it can be refined to yield gasoline, heating oil, and other petroleum products.

**shallow-cycle** *Storage.* describing a type of battery with small plates that cannot withstand many discharges to a low state of charge.

**shallow cycling** *Storage.* the process of repeatedly discharging a secondary battery by only a small proportion of its capacity before recharging again.

**Shannon, Claude** 1916–2001, U.S. mathematician who described how information could be encoded as a series of 1's and 0's (binary format), which set the stage for the development of the modern computer and the digital communication revolution.

**Shannon's index** *Ecology.* a method of measuring the diversity of species in a given community. It is a measure of the average degree of uncertainty in predicting to which species an individual chosen at random from a collection of S species and N individuals would belong. The average uncertainty will increase as the number of species increases.

**Shannon's theory** *Communication.* another term for INFORMATION THEORY, based on the fact that Claude Shannon is generally regarded as having originated this concept in a famous paper written in 1948.

**shape factor** *Solar.* in building design, a measure of the extent to which one surface is visible from another surface; a factor used to analyze or predict the amount of radiation that will leave the surface of one object and reach the surface of another.

**shear** *Physics.* **1.** the movement of parallel surfaces of a solid body, in such a way that they remain parallel to each other; i.e., the surfaces slide against each other. The shape of the body is thus changed, but the volume remains unchanged. **2.** to move in such a manner. **3.** the deformation in shape that results from such a force. Thus, **shear stress.**

**shearer** *Mining.* a large, mechanized cutting machine that is used in longwall coal mining to shear (slice) the material from the face as it moves along a track parallel to the coal face.

**shelf life** *Consumption & Efficiency.* **1.** the length of time, under normal conditions, that an item can be stored without use and still maintain its desired quality (e.g., a battery). **2.** a popular term for the viable or useful period of any activity, trend, product, and so on.

**Shell** a historic oil firm, founded in 1892 as Shell Trading and Transport. It began operation as a small family business in London selling seashells, before growing into an import-export business shipping oil to the Far East. Shell eventually merged with ROYAL DUTCH (1907) to form one of the world's largest oil companies.

**shield(ing)** see RADIATION SHIELD.

**shift reaction** *Hydrogen.* a reaction used in hydrogen manufacture, in which carbon monoxide reacts with steam or water to produce hydrogen and carbon dioxide. Also, **shift conversion.**

**shifting cultivation** *Ecology.* a traditional form of crop production in which the land is allowed to revert to a more natural state once the soil is exhausted or weeding becomes too difficult.

**Shippingport** *Nuclear.* a site in Pennsylvania, the location of the world's first large-scale nuclear power plant.

**shivering thermogenesis** *Biological Energetics.* a process of thermogenesis involving metabolic heat production by muscle contraction.

**shock** *Electricity.* **1.** the effect on the body of an electric current passing through it. **2.** to produce such an effect. **3.** *Physics.* any abrupt motion or force that can initiate a response in a system. **4.** see PRICE SHOCK.

**shock absorber** *Transportation.* a device in a vehicle, such as a spring or hydraulic damper, that is designed to absorb the energy associated with road impact so as to reduce the amount of shock to the frame or support.

**shock wave** or **shockwave** *Physics.* a pressure wave passing through a fluid medium in which the pressure, density, and particle velocity undergo significant changes. The source of the shock wave moves at a speed greater than the speed of sound in the fluid.

**shoe brake** *Transportation.* a brake that operates on friction between an element (shoe) and the surface of a rotating drum.

**shooting** *Oil & Gas.* **1.** the detonation of an explosive device in a wellbore to allow oil or gas to flow to the surface. Thus, **shot well.** **2.** *Coal.* a similar process of detonation to break coal loose in a seam.

**Shoreham** *Nuclear.* a boiling water reactor plant built on the northern shore of Long Island, New York that operated intermittently over a period of two years from 1985–1987 but never reached commercial operation due to intense public opposition.

**short** *Economics.* the market position of a futures contract seller whose sale obligates him to deliver the commodity, unless he liquidates his contract by an offsetting purchase. Compare LONG.

**short circuit** *Electricity.* **1.** a fault that occurs in an electric circuit or apparatus, typically due to imperfect insulation, causing a large current to flow through the system to ground. This has the potential of causing significant damage to the system. **2.** **short-circuit.** to bring about such a current path, usually inadvertently.

**short-circuit current** *Electricity.* current flowing freely through an external circuit that has no load or resistance; the maximum current possible.

**short-rotation forestry** *Biomass.* the practice of growing trees as an energy crop. Also, **short-rotation coppice.**

**short ton** *Measurement.* another term for the traditional ton weight of 2000 pounds; so called because other ton measures are larger.

**shortwall mining** *Mining.* a modification of LONGWALL MINING employing the same basic techniques but involving a working area that is generally smaller. Also, **shortwall system.**

**shortwave** or **short-wave** *Communication.* **1.** a radio wave shorter than that used in AM broadcasting, corresponding to frequencies of over 1600 kilohertz; used for long-distance reception or transmission. **2.** *Physics.* a wave of electromagnetic radiation that is no longer than the length of the radiant energy emitted by the sun in the visible and near-ultraviolet wavelengths. Thus, **shortwave radiation.**

**shot well** see SHOOTING.

**SHP** small hydropower.

**SHS** solar home system.

**Shtokmanovskoye** *Oil & Gas.* one of the world's largest natural gas deposits, located in Russia near the Barents Sea.

**Shuman, Franc** U.S. engineer who operated the world's first solar thermal power station in Meadi, Egypt (1913).

**shunt** *Electricity.* **1.** another term for a parallel connection. **2.** to create such a connection.

**shunt controller** *Storage.* a controller that redirects (shunts) excessive charging current away from a battery.

**shunt load** *Electricity.* an electrical load employed to make use of excess generated power when it is not needed for its primary use; e.g., a residential photovoltaic system that employs power for domestic water heating when it is not needed for purposes such as operating lights or running heating systems. Thus, **shunt loading.**

**shunt regulator** *Photovoltaic.* a type of battery charge regulator in which the charging current is controlled by a switch connected in parallel with the photovoltaic generator; overcharging of the battery is prevented by shorting the PV generator.

**shutdown** *Nuclear.* a decrease in the rate of fission (and heat production) in a reactor, usually by the insertion of control rods into the core.

**shut-in** *Mining.* **1.** to close down operation of a well and mine that is still capable of production; usually refers to a temporary closure. **2.** of or relating to a well or mine having this status.

**shuttle** *Transportation.* **1.** a bus, subway train, airline flight, and so on that regularly travels back and forth along the same (relatively short) route. **2.** see SPACE SHUTTLE.

**shuttle car** *Mining.* a vehicle used for receiving coal from the mining machine and transferring it to an underground loading point, mine railway, or conveyor belt.

**SI** **1.** Système International. **2.** spark ignition (engine).

**sick building syndrome (SBS)** *Health & Safety.* a term for the modern phenomenon that certain specific buildings, and especially mechanically ventilated office buildings and factories, will tend to produce a similar set of health complaints from various occupants, for which a specific cause cannot be identified; e.g., respiratory ailments, nausea, headaches, and irritation of the eyes, nose, and throat.

**sidereal time** *Measurement.* the measure of time as defined by the diurnal motion of the vernal equinox, equal to one complete rotation of the earth relative to the equinox; one **sidereal day** is 23 hours, 56 minutes, and 4.091 seconds.

**sidereal year** *Measurement.* the interval required for the earth to make one absolute revolution around the sun; 365 days, 6 hours, 9 minutes, and 9.5 seconds.

**siemen** *Electricity.* an SI derived unit of measurement for electric conductance, equal to the conductance at which a potential of one volt forces a current of one ampere; the inverse of the ohm. [Named for Werner von *Siemens.*]

**Siemens, Werner von** 1816–1892, German inventor and industrialist who made notable advances in electrical engineering (e.g., the discovery of the DYNAMOELECTRIC principle, an improved electric train) and also co-founded what would become a leading electronics and communications company.

**Sierra Club** a leading U.S.-based environmental advocacy organization, founded in San Francisco in 1892 with John Muir as its first president.

**sievert** *Nuclear.* a standard unit indicating the biological damage caused by radiation, symbol Sv; defined as the quantity of absorbed radiation that induces the same biologic effect in a specified tissue as 1 gray of high-energy X-rays. [Named for Swedish radiologist Rolf Maximilian *Sievert,* 1896–1966.]

**signal** *Electricity.* any transmitted electrical impulse, especially a coded message or text that is conveyed by electronic means.

**Sikorsky, Igor** 1889–1972, Russian-born U.S. aeronautical engineer who built and flew the first multimotored plane (1913) and established the world's endurance record for sustained flight in a helicopter, one of his own design (1941). He also established various aviation companies noted especially for successful helicopter designs.

**silane** *Chemistry.* **1.** $SiH_4$, a colorless gas that has a foul odor and is a dangerous fire hazard; used as a doping agent for solid-state devices. **2.** any of various other compounds of silicon and hydrogen.

**Silent Spring** *Environment.* the title of a 1962 book by U.S. naturalist RACHEL CARSON, regarded as a landmark work in increasing public awareness of the long-term environmental effects of toxic chemicals.

**silica** *Materials.* silicon dioxide ($SiO_2$), occurring widely in nature in various forms; used to make glass, ceramics, and concrete, in pharmaceuticals, and for many other purposes.

**silicate** *Chemistry.* any of a wide variety of compounds containing silicon and oxygen, and one or more metals with or without hydrogen. Most rocks and many minerals are silicates.

**silicate pump** *Earth Science.* an action in water that acts in diatom-dominated communities to increase the transport of silicate from the euphotic zone to deeper water as compared to nitrogen transport; this results in low silicate, high nitrate conditions in the mixed layer.

**silicon** *Chemistry.* a nonmetallic element having the symbol Si, the atomic number 14, an atomic weight of 28.09, a melting point of 1412°C, and a boiling point of about 2480°C; the seventh most abundant element in the universe and the second most abundant in the earth's crust. Large numbers of silicon atoms bond with each other to form a crystal, which makes silicon the most common material used to build semiconductor devices and solar cells. See also AMORPHOUS SILICON; CRYSTALLINE SILICON.

**silicon carbide** *Materials.* SiC, bluish-black crystals with excellent thermal conductivity; a porous ceramic material used in diesel particulate filters, as an abrasive, in light-emitting diodes, and for various other purposes.

**silicon dioxide** see SILICA.

**silicon solar cell** *Photovoltaic.* a type of photovoltaic cell made from a variety of silicon-based materials. See MONOCRYSTALLINE, NANOCRYSTALLINE, POLYCRYSTALLINE.

**silicone** *Materials.* any of a large number of polymers consisting of alternate silicon and oxygen atoms combined with various organic groups; available in various forms and characterized by resistance to water, heat, and the passage of electricity. A widely used industrial material for lubricants, coatings, sealants, insulators, paints, and many other purposes.

**Silliman, Benjamin** 1779–1864, U.S. chemist who originated the process of fractionating petroleum by distillation (1854) and who was one of the first to recognize the potential of petroleum as a valuable energy product.

**silo** *Consumption & Efficiency.* **1.** an aboveground, airtight structure used to store green crops and convert them into silage for livestock feed. **2.** *Nuclear.* an underground missile housing, equipped with facilities for raising the missile to a launch position or for launching directly from the housing.

**silt** *Earth Science.* a very small rock fragment or mineral particle, smaller than a fine grain of sand and larger than coarse clay; usually rich in nutrients needed for agriculture. Typically described as having a diameter from 0.002 to 0.06 millimeter, the smallest soil material that can be seen with the naked eye.

**silver** *Chemistry.* a metallic element having the symbol Ag, the atomic number 47, an atomic weight of 107.87, a melting point of 961°C, and a boiling point of 2210°C; a ductile, white lustrous metal with the highest electrical and thermal conductivity of all metals. It is used in the production of various compounds for photography and catalysis, electrical components, and scientific and medical equipments.

*silver* *Unprocessed silver ore, next to a ten-cent coin showing the relative size. U.S. coins historically contained up to 90% silver but now are made from lesser metals.*

**silver maple** *Biomass.* a type of maple tree, *Acer saccharinum*, characterized by leaves that are light green on top and silvery white below; it has significant potential value as an energy crop.

**silver-oxide battery** *Storage.* a type of alkaline primary battery that uses silver oxide as the positive electrode and zinc as the negative electrode, with an electrolyte of either sodium or potassium hydroxide; used mainly in low-drain applications.

**simple machine** *History.* one of the fundamental devices serving as the basis for all more complex machines; various classifications for this have been made since ancient times; one standard group includes the lever, wheel (and axle), wedge, screw, pulley, and inclined plane.

**simplex** *Electricity.* describing communication in which transmission can go in only one direction at a time.

**Sinclair, Harry** 1876–1956, U.S. oil executive who founded the prominent Sinclair Oil Corporation, later merged into Atlantic Richfield Company, and was a prominent figure in the TEAPOT DOME scandal.

**Sinfelt, John** born 1931, U.S. chemist noted for his invention of a superior catalyst that was an important advance in the effort to produce lead-free, high-octane gasolines cheaply.

**single-crystal silicon** another term for MONOCRYSTALLINE SILICON.

**single-phase** *Electricity.* describing a circuit or device powered by one phase of AC voltage.

**single-stage cycle** *Refrigeration.* a refrigeration cycle in which a compressor pumps a single refrigerant (not a mixture) in a single compression stage from a lower evaporating pressure to a higher condensing pressure.

**sink** *Earth Science.* **1.** an extensive land surface depression, such as the Salton Sink of southeastern California. **2.** *Mining.* a preliminary pit or excavation to be worked and enlarged to the point of becoming a full-sized shaft. **3.** *Physics.* a device that is used to readily absorb an extensive quantity, such as a heat sink that absorbs excessive heat. **4.** *Environment.* a reservoir that uptakes a pollutant from another part of its material cycle.

**sink credit** *Policy.* a framework by which credits can be traded through the emissions trading and joint implementation mechanisms of the Kyoto Protocol; credits could be issued for increases in carbon stocks, or surrendered and canceled for decreases in carbon stocks.

**sintering** *Materials.* the heating of a mass of fine particles (e.g., lead concentrates) below the melting point, causing it to agglomerate to form larger particles.

**SIPH** solar industrial process heat.

**siphon** *Materials.* **1.** a bent tube or pipe used to transfer liquid from a receptacle, in which atmospheric pressure forces the liquid up the shorter leg of the vessel while the weight of the excess liquid in the longer leg causes a continuous downward flow. **2.** to transfer liquid by means of such a device.

**siting** *Policy.* an investigation of an area for the purpose of selecting the location (site) of a large, centralized proposed energy facility; e.g., a dam, nuclear plant, or wind farm.

**SI unit** see SYSTÈME INTERNATIONAL.

**Six-Day War** *History.* a name for the 1967 conflict pitting Egypt, Syria, and Jordan against Israel, so called because of the swift Israeli victory. The war resulted in a Saudi-led oil embargo against the U.S. and Britain, the first Arab oil embargo affecting Western countries.

**sizing** see COAL SIZING.

**skidder** *Biomass.* a machine or device that is used to transport harvested trees or logs from the stump area to the landing or work area.

**skin** *Earth Science.* a layer at the surface of the ocean, having a depth of 1 mm thick or less; used as a context for describing surface temperature **(skin temperature).**

**skipping rope turbine** *Wind.* another term for the Darrieus turbine (vertical axis wind turbine), based on the similarity of the motion of its blades to a child's jump rope.

**skylight** *Solar.* **1.** an opening in a roof that is fitted with translucent or transparent glass or plastic in order to admit sunlight. **2.** visible light that is scattered by molecules in the atmosphere, as opposed to direct radiation.

**sky temperature** *Solar.* the equivalent blackbody radiation temperature of the atmospheric longwave radiation received at a horizontal surface.

**sky view factor** *Solar.* the ratio of the visible portion of the sky at a given location, expressed between 0 and 1; the units are dimensionless.

**slack barrel** *Oil & Gas.* a vessel used to store and ship petroleum paraffin; lighter than the standard oil barrel but similarly shaped, having a capacity of 235–245 lbs. net.

**slag** *Coal.* **1.** the layer covering molten metal during smelting and refining; in refining, it is composed of oxidized impurities; in smelting, it consists of the flux and gangue minerals. **2.** *Coal.* a coarse, granular, incombustible byproduct that remains as a fused layer at the bottom of a furnace after the burning of coal. Thus, **slagging.**

**slamming** *Economics.* a term for the unauthorized switching of a customer's provider of electricity, natural gas, or other energy services.

**slash** *Biomass.* the portions of trees that remain on the forest floor or on the landing after logging operations have taken place; e.g., tree branches, tops of trunks, stumps, branches, and leaves.

**slash-and-burn (farming)** *Ecology.* a traditional form of agriculture in which the existing vegetation in a field is cut away and the field is then cleared by burning; practiced especially in lightly populated rainforest areas. The burning releases nutrients into the soil; typically the field will be cultivated for a few years and then abandoned or left fallow to regenerate.

**Slater, Samuel** 1768–1835, English-born pioneer in the textile industry of the U.S. With Moses Brown, he built a factory in Pawtucket, Rhode Island that was the first successful water-powered textile mill in America.

**slave power** *Biological Energetics.* the use of human slaves as an energy source.

**slip** *Conversion.* the percentage by which the rotor speed of an induction motor falls below the rotation of the magnetic field in which it operates.

**slope mining** *Coal.* an underground coal mining technique in which the mine entry is driven at an angle to reach the coal deposit; employed when the coal deposit is fairly close to the surface but too deep for surface mining. Thus, **slope mine.**

**slow neutron** *Nuclear.* another term for a THERMAL NEUTRON; i.e., a neutron in thermal equilibrium with its surrounding medium; so called in contrast with a fast neutron.

**slow pyrolysis** *Biomass.* the thermal conversion of biomass to fuel, by slow heating to less than 450°C in the absence of oxygen.

**slow wave** *Physics.* any electromagnetic wave whose phase velocity is less than that of the speed of light in a vacuum.

**sludge** *Materials.* **1.** a combination of solids and liquids resulting from sewage-treatment processes, without thickening or physical or chemical pretreatment. **2.** any undesirable solids settled out from a treatment process. **3.** a mixture of cuttings and water formed at the bottom of a borehole after drilling. **4.** a soft, soupy or muddy stream-bottom deposit, especially a black ooze formed on the bottom of a lake.

**sluice** *Hydropower.* **1.** a channel for the passage of water, fitted with a vertically sliding gate **(sluice gate)** for regulating flow control. **2.** the body of water held back or controlled by such a gate.

**slurry** *Materials.* **1.** a thin paste produced by mixing an insoluble substance, such as cement or clay, with enough water or other liquid to allow the mixture to flow viscously. **2.** to prepare such a paste. Thus, **slurrying.** **3.** *Oil & Gas.* a suspension of pulverized solid material in water or oil, pumped or poured into pipelines. **4.** a thin, free-flowing mixture of water and cement pumped into an oil well through the drill pipe, to support the casing and seal the wellbore. **5.** see COAL SLURRY.

**slurry dam** *Coal.* a repository for the coal slurry (mixture of coal and wastes in a liquid medium) from a coal preparation plant.

**small calorie** see CALORIE.

**small hydropower (SHP)** *Hydropower.* a decentralized hydropower facility of limited size; the term can apply to historic sites such as water mills or to contemporary installations. Typically defined as systems with a maximum rated capacity between 25 and 30 megawatts; the actual size designation varies among countries. Smaller units are known as MINIHYDROPOWER or MICROHYDROPOWER facilities.

**Small Island Developing States** a subset of the Alliance of Small Island States that includes only developing country islands. (AOSIS includes some low-lying countries.) These nations are considered to be especially vulnerable to the impacts of climate change.

**smart battery** *Storage.* a term for a battery with internal circuit enabling some type of communication between the battery and the user, e.g., a battery that provides a computer with information about its power status so that the computer can conserve power more efficiently.

**smart growth** *Policy.* a term for alternative forms of development that use land efficiently, mix land uses (such as housing, business, and shopping), and encourage walking or cycling and public transit as well as driving, while protecting surrounding open space.

**smart window** *HVAC.* a modern type of window having special materials whose level of reflectivity can be varied, so that it is possible to control the amount of light (and heat) that passes through the window.

**Smeaton, John** 1724–1792, English engineer who made important early contributions to the understanding of water power and wind power. He also improved Thomas Newcomen's atmospheric steam engine.

**smelting** *Materials.* any of the various metallurgical methods by which ores or concentrates are heat-processed to yield a crude metal, which is then reduced, separated, or refined. Thus, **smelt, smelter.**

**SMES** superconducting magnetic energy storage.

**Smith, Adam** 1723–1790, Scottish political philosopher considered to be one of the most influential economists in history for his book THE WEALTH OF NATIONS. He is best known for his proposal that rational self-interest in a free-market economy leads to improved economic conditions for all.

**Smith, William** 1769–1839, English engineer who helped lay the groundwork for the modern science of geology when he discovered that each layer (stratum) of earth contains a unique assemblage of fossils, and that one formation can be distinguished from another by the characteristic fossils in each.

**smog** *Environment.* **1.** originally, a mixture of smoke and fog, as produced in large urban or industrial centers from the open burning of coal and other fuel. **2.** in subsequent use, a fog contaminated by industrial pollutants, such as carbon monoxide as a by-product from automobile exhaust; especially prevalent in large urban areas surrounded by higher elevations. **3.** in general, any form of urban air pollution, regardless of the source or composition. See next page.

**smoke** *Environment.* a visible mixture of products given off in the burning of an organic substance such as wood, coal, or fuel oil, consisting of a suspension of tiny particles of carbon, hydrocarbons, ash, and so on, as well as vapors such as carbon monoxide, carbon dioxide, and water vapor.

**smokeless coal** *Coal.* a description for coal that burns without smoke, especially semibituminous coal.

**smoke number** *Measurement.* a dimensionless number used to quantify smoke emissions, obtained by comparing the reflectance of a filter paper before and after the passage of a known volume of a smoke-bearing sample.

**smokestack industry** *Social Issues.* **1.** a term for an industrial operation that requires a high level of energy consumption to manufacture its products, such as a paper mill, automobile

**smog** Smog is not a very recent phenomenon: The term, a contraction of smoke and fog, was introduced in 1905 by Dr. H.A. des Vœux, to describe the mixture of soot, sulphuric acid and other pollutants caused by the emissions of many coal furnaces in London, which has affected the population of that city since the times of King Edward Longshanks (around the year 1300). King Edward was the first to try to forbid the use of coal in London. The extreme smog of 1952 in London may have caused the untimely death of 12,000 people and was the direct reason for the first clean air act in England. The combination of stable atmospheric conditions, due to inversion, and the emissions of increasing number of cars has led in Los Angeles since 1943 to a different kind of air pollution; ozone and related compounds formed by the reaction of organic compounds and nitrogen oxides. This new air pollution also was named smog, though the causes were quite different as particles were formed in large numbers which restrict visibility severely, analogous to the classical fog due the emissions of burning coal.

**Sjaak Slanina**
ECN (Energieonderzoek Centrum Nederland)

plant, steel factory, or chemical processing facility. **2.** specifically, such an operation that is a source of air pollution. [From the idea of a traditional factory having large *smokestacks* to carry off the exhaust gases from burning.]

**Smoot, L. Douglas** born 1934, U.S. engineer whose research in fuel combustion led to new insights into the formation and prevention of air pollutants from the burning of coal and other fossil fuels.

**SMR** standard metabolic rate.

**SNAP** *Nuclear.* systems for nuclear auxiliary power; a U.S. program of experimental radioisotope thermonuclear generators (RTGs) and space nuclear reactors flown during the 1960s. A nuclear reactor system was launched by SNAP in 1965, the only nuclear reactor placed in space by the U.S.

**SNG** synthetic natural gas.

**SOC** state of charge.

**social cost** *Social Issues.* **1.** the cost to society as a whole from an event, action, or policy change. **2.** specifically, the sum of external and private costs involved with the use of a given form of energy, especially the use of a traditional form such as coal in comparison with an alternative form such as wind.

**social Darwinism** *Social Issues.* the principle of SURVIVAL OF THE FITTEST as applied to an individual's relative status in a given society, or the relative condition of different societies; a 19th-century belief that cultures develop in a manner supposedly similar to biological evolution and that therefore a society with more advanced technology and greater military and economic power is inherently superior.

**social injustice** *Social Issues.* a lack of justice and equity that violates the rights of an individual or group.

**socialism** *Policy.* governmental policies based on the principle that the state should own and control major industries, using the money acquired by this means to provide benefits to all citizens.

**social justice** *Social Issues.* a condition in which benefits and burdens are fairly distributed throughout society.

**social metabolism** *Social Issues.* an idea adopted by Karl Marx and others in the 19th century, proposing that social systems are analogous to biological systems in that both exist in a permanent state of material and energy exchange with their environment and other systems. Marx largely ignored energy flows, while others employed them as a key concept.

**social multiplier** *Social Issues.* the concept that local political support for a given industry (e.g., an oil company in Texas, agriculture in Iowa) will extend well beyond what might be expected on the basis of the number of people or jobs supported directly by that industry.

**Society of Automotive Engineers** (est. 1905), a non-profit educational and scientific organization dedicated to advancing automotive technology and developing technical information on all forms of self-propelled vehicles.

**socio-economic impact assessment (SEIA)** *Social Issues.* a study assessing, in advance, the social and economic consequences to the human population of the initiation of any industrial activity.

**sociopolitical** *Social Issues.* having to do with both social and political factors; involving various aspects of a society. See below.

**sociotechnical** *Social Issues.* relating to the idea that science is a sociocultural phenomenon and that the technical aspects of an activity should be examined within the context of wider social, economic, and political processes. Thus, **sociotechnical approach, theory,** and so on.

**socket** *Electricity.* a form of electrical jack into which a device with one or more prongs is designed to fit.

**Socony** short for Standard Oil of New York; a historic U.S. oil company established in 1911 after the breakup of John D. Rockefeller's original Standard Oil company; merged with Vacuum Oil Company in 1931 to form Socony-Vacuum; adopted the brand name Mobil in 1954; merged with Exxon (originally Standard Oil of New York) to form ExxonMobil in 1998.

**soda ash** *Materials.* the commercial form of anhydrous sodium carbonate, $Na_2CO_3$, a grayish-white powder or lumps; a widely produced chemical with many industrial uses, as in water treatment, paper and glass manufacture, petroleum refining, and aluminum production.

**Soddy, Frederick** 1877–1956, English chemist noted for his formulation of the concept of isotopes (1913) and for his application of the laws of energy to social and economic theory. He argued that real wealth derives from the use of energy to transform materials into physical goods and services.

**Soddy's displacement law** *Nuclear.* a law stating how much charge a given nuclide will lose during decay, as in alpha decay, when its atomic number will decrease by two.

**sodium** *Chemistry.* a metallic element having the symbol Na, the atomic number 11, an atomic weight of 22.99, a melting point of 97°C, and a boiling point of 883°C; a tetragonal, crystalline, soft, silvery-white solid that does not occur in elemental form in nature due to high reactivity. It has excellent electrical conductivity and high heat-absorbing capacity.

**sodium carbonate** see SODA ASH.

**sodium chloride** *Chemistry.* NaCl, the chemical name for common table salt; see SALT.

**sodium-cooled fast reactor (SFR)** *Nuclear.* an advanced nuclear reactor design in the development stage, featuring a sodium-cooled reactor and a closed fuel cycle. It includes important safety features for the management of high-level wastes.

**sodium hydroxide** *Materials.* NaOH, a white solid that absorbs water and carbon dioxide from the air; corrosive, toxic, and a strong irritant. A widely produced substance used

**sociopolitical collapse** Sociopolitical collapse is the rapid simplification of a society. In historical cases, cities and monuments quickly disappeared, population diminished, and there was pronounced loss of sociopolitical complexity. Famous examples include the collapses of Mycenaean Greece in the 12th century B.C., the Western Roman Empire in the 5th century A.D., and the Classic Maya in the 9th century A.D. The periods that follow are sometimes called Dark Ages. Every society is characterized by a degree of complexity, which describes how extensively it is differentiated by roles, specialization, technology, and information, and what institutions regulate life. More complex societies are most costly to institute and maintain than simpler ones, requiring high levels of energy. Simplified post-collapse societies needed and produced less energy per capita. Before fossil fuels, societies were powered largely by solar energy. In such societies, greater complexity meant that people worked harder. Photosynthesis produces little energy per unit of land (maximum of about 170 kcal/sq m/year). The weakness of ancient complex societies was that peasants, whose taxes funded government and the military, lived on very small margins of net agricultural production. Ancient societies often faced problems, especially war, that required greater complexity and energy production. Taxes compelled peasants to produce more, but excessive taxation could undermine peasants' well-being and ultimately the society itself. Vulnerability to collapse often emerged from the costs ancient societies incurred trying to sustain themselves. Even with fossil fuels, modern societies are also affected by the costs of complexity.

**Joseph A. Tainter**
Rocky Mountain Research Station
USDA Forest Service

in chemical manufacture, soaps and detergents, food processing, and for various other purposes.

**sodium light** *Lighting.* a type of high-intensity discharge light that has the most lumens per watt of any light source; see HIGH-PRESSURE SODIUM.

**sodium-nickel-chloride battery** *Storage.* a battery that uses sodium for the negative electrode and nickel-chloride as the active mass for the positive electrode; it remains in the solid state even at high operating temperatures, and thus must use an intermediate electrolyte that is a molten salt. These batteries represent a potential technology for electric vehicle propulsion.

**sodium nitrate** *Materials.* $NaNO_3$, toxic, colorless, transparent crystals with a slightly bitter taste; a fire risk in contact with organic substances. Used in matches, explosives, and fireworks, in fertilizers, and as a preservative and color fixative for cured meats.

**sodium nitrite** *Materials.* $NaNO_2$, colorless to yellowish crystals or powder; oxidized by air and soluble in water; formerly widely used in cured meat and fish products, but now restricted because of carcinogenic results in laboratory animals.

**sodium peroxide** *Materials.* $Na_2O_2$, a white to yellowish powder that absorbs water and carbon dioxide from the air; an irritant, a strong oxidant, and a dangerous fire and explosive hazard in contact with water and many other substances. Used in bleaches, pharmaceuticals, antiseptics, soaps, and for various industrial purposes.

**sodium-potassium alloy** *Materials.* a liquid metal used as a heat-transfer coolant in industrial processes and some nuclear reactors.

**Sodium Reactor Experiment** *Nuclear.* the first generation of power from a civilian nuclear unit, by a facility at Santa Susana, California. The unit operated from 1957 to 1966.

**sodium-sulfur battery** *Storage.* an advanced battery technology employing molten sulfur and sodium electrodes; ions of sodium flow back and forth between the positive (sulfur) and negative (sodium) electrodes, depending on whether the battery is charging or discharging. Originally developed for use in electric cars and now proposed for various industrial uses.

**soft coal** *Coal.* a term for coal that is not as hard as anthracite coal, especially bituminous coal.

**soft energy** *Renewable/Alternative.* a term for energy sources and processes that are decentralized, renewable, efficient, sustainable, and environmentally safe, especially as contrasted with "hard energy" (traditional industrial energy systems based on coal, oil, or nuclear energy). Thus, **soft energy path, soft technology,** and so on.

**software** *Communication.* a collective term for those aspects of computer technology that are not actual physical objects (hardware), such as operating systems, applications, and utility programs.

**softwood** *Biomass.* one of the botanical groups of trees that in most cases have needle-like or scale-like leaves; the conifers, such as pines, firs, and redwoods. Contrasted with the hardwood or deciduous trees. The *hardwood/softwood* distinction is a general designation and does not literally refer to the hardness of the wood, since some conifers have harder wood than certain hardwoods.

*softwood As a conifer, the redwood tree is technically classified as a softwood, but its wood is durable and is widely used in construction.*

**soil** *Earth Science.* **1.** all loose, unconsolidated, weathered, or otherwise altered surface material lying on the earth above bedrock. **2.** specifically, a natural accumulation of organic matter and inorganic rock material that is capable of supporting the growth of vegetation.

**soil carbon** *Earth Science.* carbon contained in the solid surface layer of the earth; the amount of carbon in the soil is a function of the historic vegetative cover and productivity, which in turn is influenced by climatic variables.

**soil conservation** *Ecology.* any of various methods of land management that seek to protect the soil from erosion and chemical decay, so as to maintain its quality. Similarly, **soil management.**

***soil conservation*** *The technique of strip cropping (alternating different crops) is an effective method of soil conservation because it provides almost total resistance to erosion.*

**soil moisture** *Earth Science.* water stored in or at the continental surface and subject to evaporation.

**soil quality** *Earth Science.* the ability of a given area or type of soil to provide functions of value to humans such as biomass production, water filtration, biodegradation of pollutants, or soil carbon sequestration.

**soil science** *Environment.* the scientific study of the formation, properties, distribution, and classification of soil as a natural resource.

**sol** *Chemistry.* a liquid colloid or mixture in which solid particles, small enough to pass through filter membranes, are dispersed in a liquid phase.

**solar** **1.** of or relating to the sun. **2.** describing those forms of energy derived directly from the sun (e.g., solar thermal heating), as opposed to those ultimately derived from it (e.g., fossil fuels). **3.** describing any renewable form of energy that does not create greenhouse gas emissions or nondegradable toxic wastes.

**solar access** see SOLAR RIGHTS.

**solar activity** *Solar.* a term for variations in the appearance or energy output of the sun; usually associated with the variation of sunspots and other features over the 11-year solar cycle.

**solar air heater** *Solar.* a type of solar thermal system in which air is heated in a collector and either transferred directly to the interior space or to a storage medium.

**solar altitude** *Solar.* the angle between a line from a point on the earth's surface to the center of the solar disc, and a line extending horizontally from that point; the sun's angle above the horizon, as measured in a vertical plane. Also, **solar angle.**

**solar architecture** *Solar.* the conception, design, and construction of buildings and communities so as to make optimal use of incoming solar radiation, as for heating and cooling or power generation.

**solar array** see PHOTOVOLTAIC ARRAY.

**solar-assisted heat pump** see SOLAR HEAT PUMP.

**solar azimuth** *Solar.* the horizontal angle between the sun and due south in the Northern Hemisphere, or between the sun and due north in the Southern Hemisphere.

**solar battery** *Solar.* a battery that is charged through photovoltaic cells.

**solar burst** *Earth Science.* a sudden, transient enhancement of nonthermal radio emission from the solar corona, usually associated with an active region or flare.

**solar cavity receiver** *Solar.* a well-insulated enclosure with a small opening to let in concentrated solar energy, approaching a blackbody absorber in its ability to capture solar energy.

**solar cell** *Solar.* any material that converts sunlight directly into electricity. See below.

**Solar Challenger** *Solar.* a solar-powered airplane that in 1981 flew across the English Channel in 5 hours and 23 minutes; the plane had 16,000 solar cells mounted on the wings that generated 3000 watts of power.

**solar chimney** *Solar.* a power generating facility that converts solar radiation into the kinetic energy of moving air, which is then transformed into electricity by a turbine. A collector is covered by a transparent glazing that heats up the air mass inside it; buoyancy then drives the warmer air into the chimney, where it drives a turbine.

**solar city** *Consumption & Efficiency.* **1.** a term for a city that obtains a significant amount of its total energy needs from solar energy and other renewable resources. Compare FOSSIL CITY. **2. Solar City (program).** an international cooperative program to encourage cities and urban regions to substitute renewable and sustainable forms of energy technology for conventional sources, with the long-term goal of achieving globally acceptable greenhouse gas emission levels and lower reliance on fossil fuels.

**solar cell** A solar cell is any material that converts sunlight directly into electricity. This property of materials is known as the photoelectric effect, first described in 1905 by Albert Einstein in his Nobel prize-winning research. The first solar cell capable of generating substantial amounts of power and the sire to today's booming photovoltaic industry emerged from pioneering semiconductor research done at Bell Laboratories in the 1940s and early 1950s. Out of this research also came the transistor, probably the most important technical discovery of the last century, which has dramatically transformed virtually all aspects of human life. Experimenting with one of these new devices, Gerald Pearson, a Bell scientist, shined light on it and to his amazement recorded a significant amount of electricity produced by the silicon. In just a few short years, silicon solar cells enabled satellites to run almost indefinitely, revolutionizing commercial, military, and scientific space applications essential to modern life. On Earth silicon solar cells have brought abundant clean water, electricity, and telephone services to many in remote regions who had hitherto done without. Roof-top solar cell programs in the developed world promise to revolutionize the way we use electricity with each building becoming its own power plant interacting with the electric utility grid, or acting as a stand-alone system.

**John Perlin**
Historian and Author
Santa Barbara, CA

**solar coating** *Solar.* flat black paint or another highly absorptive substance that is applied to the absorber plate of a solar collector.

**solar collector** see COLLECTOR.

**solar concentrator** see CONCENTRATOR.

**solar constant** *Solar.* the total radiant energy received vertically from the sun, per unit area per unit of time, at a position just outside the earth's atmosphere when the earth is at its average distance from the sun. Its value is approximately 1368 watts per square meter, within an accuracy of about 0.2%. Variations of a few tenths of a per cent are common, usually associated with the passage of sunspots across the solar disk.

**solar cooker** *Solar.* any of various devices that convert radiant energy from the sun into thermal energy for cooking, water purification, and other household uses. A typical version is a concave bowl-shaped dish whose inner surface is made of reflective material; sunlight falling on the inner surface is focused onto a dark cooking pot that is hung or set in front of the cooker.

**solar cooling** *Solar.* the use of the sun's radiant energy to cool a living space, by various means such as a pond of water on a flat roof that cools via evaporation (passive cooling), or the use of solar energy to power a cooling appliance (active cooling).

**solar cycle** *Solar.* the main periodic cycle in the sun's activity that occurs over about an 11-year period; the period is not constant, varying between about 9.5 and 12.5 years. During the cycle, changes occur in the sun's internal magnetic field and in its surface disturbance level.

**solar day** *Measurement.* a day defined as one complete rotation of the earth on its axis in relation to the sun; the 24-hour period between two successive transits of the sun across a given meridian on earth.

**solar declination** *Earth Science.* the apparent angle of the sun either north or south of the earth's equatorial plane. The earth's rotation on its axis causes a daily change in solar declination.

**solar degradation** *Solar.* a deterioration of a material produced by its exposure to solar energy.

**solar desalination** *Solar.* the conversion of brackish water or salt water to useful fresh water by direct utilization of solar energy, such as a solar still.

**solar detoxification** *Solar.* the use of chemical reactions driven by solar radiation to break down contaminant molecules in air or water; e.g., the detoxification of water through thermal chemical reactions driven by the heat generated by concentrating solar collectors.

**solar dish** *Solar.* a system that uses a mirrored dish to collect and concentrate solar radiation onto a receiver, which absorbs the heat and transfers it to fluid within an engine; the heat causes the fluid to expand against a piston or turbine to produce mechanical power, which in turn can be used to run a generator or alternator to produce electricity.

**solar disinfection** *Solar.* a technology employed to improve the quality of drinking water, by the use of solar radiation to destroy pathogenic microorganisms that can cause water-borne diseases.

**solar distillation** *Solar.* a technology employed to purify water by means of solar energy; e.g., a solar collector is used to heat water, which evaporates and then condenses. As the water evaporates, only the water vapor rises, leaving contaminants behind.

**solar drying** *Solar.* the use of solar radiation to accelerate the drying process for various products, especially agricultural crops; goods can be covered and dried through the use of air collectors and solar-powered ventilators.

**solar economy** *Renewable/Alternative.* an economy based on the use of solar power (and other renewable forms of energy) in place of fossil fuels.

**solar electricity** *Solar.* a method of producing electricity from solar energy by using focused sunlight to heat a working fluid, which in turn drives a turbo-generator. Also, **solar electric power.**

**solar elevation** *Solar.* the angle between the center of the sun's disc and the horizon.

**solar energy** *Solar.* **1.** useful energy that is immediately derived from the sun; e.g., a system that collects and uses the heat of the sun to warm a building or to generate electricity. **2.** *Earth Science.* in the larger sense, any energy source that can be ultimately traced to the action of the sun.

**solar energy equivalent** *Solar.* a statement of the amount of energy received from the sun in terms of some conventional energy unit, such as kilowatt-hours, barrels of oil, tons of coal, and so on; used to describe the relative ability of solar energy to fulfill contemporary energy requirements as compared to fossil fuel sources.

**solar engine** *Solar.* a general term for any power source that is driven by solar energy, such as a system of solar panels used to power an electric propulsion system.

**solar facade** *Solar.* a surface or feature of a building that is specifically designed and built to gain solar energy for heat and/or power, as through the use of silicon solar panels.

**solar film** *HVAC.* a window glazing coating, usually tinted gray or bronze, used to reduce building cooling loads, glare, and fabric fading.

**solar flare** *Earth Science.* a region of exceptionally high temperature and brightness that suddenly develops in the solar chromosphere near a sunspot; it is often associated with magnetic field activities.

**solar flux** *Solar.* the flow of solar energy; see IRRADIANCE.

**solar fraction** *Solar.* the percentage of a building's total seasonal energy requirements that can be met by a solar energy device or system.

**solar fuel** *Solar.* **1.** a term for the use of solar energy rather than a conventional fuel as an energy source; e.g., to produce electricity. **2.** the use of a solar energy system to provide high-temperature process heat for the production of storable and transportable fuels; e.g., the production of hydrogen from the thermal dissociation of water.

**solar furnace** *Solar.* a large-scale solar collector that is capable of creating highly concentrated solar energy (typically by means of an array of large curved mirrors) in order to produce extremely high temperatures at a localized site; e.g., the ODEILLO solar furnace.

**solar gain** *Solar.* the amount of energy that a building absorbs due to solar energy striking its exterior and conducting to the interior, or by passing through windows and being absorbed by materials in the building.

**solar greenhouse** *Solar.* a greenhouse that requires minimal use of supplemental energy because it is designed and built to make maximum use of solar energy for both heating and lighting.

**solar heat gain** see SOLAR GAIN.

**solar heat pump** *Solar.* a system that combines the use of a solar energy system with a heat pump, to increase the efficiency of both; e.g., the lesser amount of heat collected by a solar system in winter may be too low in itself for use in direct heating but can be used as a source for a heat pump instead.

**solar home system (SHS)** *Solar.* a stand-alone electrical power supply system for single buildings by means of solar energy, typically used to operate small appliances in rural areas without electrification.

**solar hot box** *Solar.* a wooden box with a clear glass top that allows solar radiation in, and a black inner lining that absorbs this radiation and converts it to heat. A precursor to modern solar thermal collectors, used to heat food and water.

**solar hour angle** *Solar.* the angle at which the earth must turn to bring the meridian of a given point directly under the sun. This is negative in the morning (sun coming from the east) and positive in the afternoon (sun going to the west).

**solar-hydrogen economy** *Hydrogen.* an economy in which direct solar energy would be the primary energy source and hydrogen the secondary energy carrier; power from wind or photovoltaic systems would drive photo-electrolytic hydrogen production.

**solarimeter** *Solar.* an instrument that measures solar radiation.

**solar industrial process heat (SIPH)** *Solar.* the use of solar thermal technologies to produce hot air, water, or steam for industrial purposes, generally at temperatures below 250°C.

**solar insolation** see INSOLATION.

**solar irradiance** see IRRADIANCE.

**solar irradiation** see IRRADIATION.

**solarium** *Solar.* an enclosed room or area designed to make maximum use of sunlight, as by having all or most of the exterior surface composed of clear glass or other materials that do not inhibit the passage of light; e.g., a greenhouse or "Florida room."

**solarization** *Solar.* **1.** the fact of converting or adapting a building or site to the use of solar energy. **2.** a loss of color or transparency in glass due to extensive exposure to sunlight or ultraviolet radiation. **3.** *Ecology.* the use of solar radiation to control soil-borne diseases or pests; the ground is covered with a transparent plastic material to minimize heat and moisture loss and thus provide inhospitable conditions for pest organisms. Thus, **solarize.**

**solar laser** *Solar.* a laser that uses concentrated sunlight for power instead of electricity, typically activated by a solar furnace.

**solar luminosity** *Solar.* the light power output of the sun (or any other star); units are watts or solar luminosities. One solar luminosity (1 Lsun) is the luminosity of the sun, which is about 3.9 x $10^{26}$ W.

**solar maximum** *Solar.* a month during the solar cycle when the 12-month mean of monthly average sunspot numbers reaches a maximum. Similarly, **solar minimum.**

**solar module** see PHOTOVOLTAIC MODULE.

**solar noon** *Solar.* the precise moment of the day that divides the daylight hours for that day exactly in half; the local time of day when the sun crosses the observer's meridian.

**Solar One** *Solar.* a notable early solar power plant operating from 1982 to 1986 in Barstow, California; it was composed of a solar receiver located on top of a tower surrounded by a field of reflectors. Its successor **Solar Two** uses advanced molten-salt technology to overcome energy storage problems that lowered the efficiency of Solar One.

**solar panel** *Solar.* an individual unit in a solar collector (i.e., a device that collects solar energy from incident radiation); typically a sun-oriented box with a transparent cover, containing water tubes or air baffles under a blackened heat-absorbent panel.

**solar parabolic collector** see PARABOLIC COLLECTOR.

**solar parabolic trough** see PARABOLIC COLLECTOR.

**solar pond** *Solar.* a small artificial lake used to collect solar energy, which can then be removed from the pond in the form of useful heat, or an external heat exchanger, or a heat exchanger placed on the bottom of the pond, converts the thermal energy into electricity. See also SALT GRADIENT POND.

**solar power** *Solar.* **1.** see SOLAR ENERGY. **2.** see SOLAR ELECTRICITY.

**solar power satellite system (SPS)** *Solar.* a proposed system to supply power from space for use on the earth. The SPS system would have a huge array of solar cells that would generate electrical power to be beamed to earth in the form of microwave energy sent to a central receiver.

***solar power satellite system*** *Artist's conception of what an SPS system would look like in space.*

**solar power tower** *Solar.* an energy conversion system that uses a large field of independently adjustable mirrors (heliostats) to focus solar rays on a near single point atop a fixed tower (receiver).

**solar preheating** *Solar.* the use of solar energy to partially heat a substance, such as domestic drinking water, prior to heating it to a higher desired temperature with conventional fuel.

**solar propulsion** *Solar.* the concept of using radiant energy from the sun to power spacecraft or satellites, as by means of an array of solar panels placed on the exterior of the craft to generate electricity.

**solar radiation** *Solar.* all the constituents that make up the total electromagnetic radiation emitted by the sun; about 99 percent of solar radiation is contained in a wavelength region from about 300 nanometers (ultraviolet) to 3000 nanometers (near-infrared).

**solar reflectance index (SRI)** *Solar.* a value that incorporates both solar reflectance and solar emittance to represent a material's temperature in the sun; it indicates how hot an actual surface will get relative to standard black and standard white surfaces.

**solar resource** *Solar.* the amount of solar insolation that a given site receives, usually measured in kilowatt hours per square meter per day.

**solar rights** *Solar.* legal issues related to protecting or ensuring access to sunlight to operate a solar energy system, or to use solar energy for heating and cooling.

**solar room** another term for a SOLARIUM.

**solar sail** *Solar.* a principal component of a solar propulsion system, consisting of a large sheet of extremely thin, light fabric coated with aluminum; attached to a spacecraft to permit the use of radiation pressure from the sun to "sail" through space.

**solar satellite (system)** see SOLAR POWER SATELLITE SYSTEM.

**solar simulator** *Solar.* an apparatus that replicates the solar spectrum; used for testing solar energy conversion devices.

**solar spectrum** *Solar.* the part of the electromagnetic spectrum occupied by the wavelengths of solar radiation. The visible region extends from about 390 to 780 nanometers.

**solar still** *Solar.* a device used to desalinate or distill water by means of solar energy; i.e., sunlight heats the water within a receptacle (still) by means of a greenhouse effect.

**solar storm** *Earth Science.* a large-scale solar flare that affects the earth's magnetic field.

**solar thermal collector** see COLLECTOR.

**solar thermal electricity** see SOLAR ELECTRICITY.

**solar thermal energy** *Solar.* the conversion of the radiant energy from the sun into heat, which can then be used for such purposes as space and hot water heating, industrial process heat, or power generation. ☼ See below.

**solar thermal panel** see SOLAR PANEL.

**solar thermal parabolic dish** see PARABOLIC DISH.

**solar thermal power** *Solar.* **1.** see SOLAR THERMAL ENERGY. **2.** see SOLAR ELECTRICITY.

**solar tide** *Earth Science.* the vertical movement of water due to gravitational attraction between the sun and the earth; the solar tide generating force is only about three-sevenths that of the lunar tide, due to the closer distance of the moon to the earth.

**solar time** *Measurement.* **1.** local time as measured by an instrument such as a sundial. **2.** a system of measurement based on the position of the earth in relation to the sun. See also TRUE SOLAR TIME.

**solar transformity** *Solar.* the solar energy required to make a unit of exergy of a product or service; expressed as solar emjoules per joule.

☼ **solar thermal energy** When a dark surface is placed in sunshine, it absorbs solar energy and heats up. A solar thermal collector working on this principle consists of a sun facing surface which transfers part of the energy to a working fluid such as water or air. To reduce heat losses to the atmosphere and to improve its efficiency, one or two sheets of glass are usually placed over the absorber surface and insulation is placed behind the absorber. This simple solar thermal collector is called a flat plate collector, which can achieve temperatures up to about 100°C. To achieve higher temperatures, concentrating collectors, such as, parabolic trough, parabolic dish or a central receiver tower are used. Temperatures as high as 1000°C or even higher can be achieved with concentrating collectors. Solar thermal energy can be used for such applications as, space heating, air conditioning, hot water, industrial process heat, drying, distillation and desalination, and electrical power.

**D. Yogi Goswami**
University of Florida

**solar transmittance** *Solar.* the amount of solar radiation that passes through a given medium, such as a glazing material, usually expressed as a percentage of the total radiation incident on the surface.

**Solar Two** see SOLAR ONE.

**solar wind** *Earth Science.* the outward movement of ionized particles, mostly helium and hydrogen, from the sun through the solar system; produced primarily in the cooler regions of the corona and flowing along open magnetic field lines. Typically, solar wind velocities are 300–1000 km per second.

**solar zenith angle** see ZENITH ANGLE.

**solder** *Materials.* **1.** to join metal objects without melting them, by fusing a metal alloy that has been applied to the joint between them. **2.** any of several alloys used in this process.

**solenoid** *Electricity.* **1.** an electromagnetic coil wound in the shape of a hollow cylinder or spool, typically containing a movable iron core that is pulled into the coil when current flows through the wire, thus allowing it to move other devices such as relays and circuit breakers. **2.** a switch or other device that is activated by such a coil, as in an automobile starting system.

**solid** *Chemistry.* one of the three fundamental states of matter, along with liquids and gases. Of these three, a solid has the greatest tendency to resist forces that would alter its shape; thus its shape and volume tend to remain fixed and are not affected by the space available to it. In comparison with liquids and gases, solids have closely packed molecules; their normal condition is a crystalline structure.

**solid electrolyte** *Electricity.* a term for materials that conduct electricity by ionic diffusion, including crystalline, vitreous, polymeric, or electrolyte-colloidal-particle composites.

**solid electrolyte battery** *Storage.* a battery whose electrolyte is a crystalline salt with primarily ionic conduction capabilities; solid electrolytes do not provide a strong current, but they provide a relatively constant flow of low voltage for extended periods of time.

**solidification** *Chemistry.* the process of becoming solid; the transition of a liquid or gas to the solid phase. Thus, **solidify.**

**solid lubricant** *Materials.* a thin film of solid material interposed between two surfaces to reduce friction and wear under severe operating or environmental conditions; this includes inorganic compounds such as graphite, organic compounds such as soaps and waxes, and metal surface coatings such as oxide films or bonded coatings.

**solid oxide fuel cell (SOFC)** *Renewable/Alternative.* a type of fuel cell in which the electrolyte is a solid, nonporous metal oxide, typically yttria stabilized zirconia (YSZ). Operating temperatures are usually around 1000°C.

**solid polymer fuel cell (SPFC)** *Renewable/Alternative.* a type of fuel cell having an electrolyte that consists of a layer of solid polymer, allowing protons to be transmitted from one face to the other. Operating temperatures are much lower than other fuel cell types (around 90°C), because of the limitations imposed by the thermal properties of the membrane itself.

**solid propellant** *Transportation.* a rocket propellant in solid form that usually contains a combination of fuel and oxidizer.

**solid-state** *Physics.* **1.** of or relating to a device, component, or system (such as a semiconductor, transistor, crystal diode, or ferrite core) that can control electric or magnetic phenomena in solids without heated filaments, moving parts, or vacuum gaps. **2.** relating to or based on solid-state physics.

**solid-state lighting** *Lighting.* a term for lighting applications that use solid-state technology, such as light-emitting diodes, organic light-emitting diodes, or light-emitting polymers. Unlike incandescent or fluorescent lighting, solid-state lighting creates light with virtually no heat, by means of a semiconducting material that converts electricity into light.

**solid-state physics** *Physics.* a branch of physics that is primarily concerned with the study of matter in its solid or condensed phase, especially at the atomic level.

**solidus** *Chemistry.* the points on a phase diagram that indicate the temperature below which a given component will freeze while being cooled, or above which it will melt while being heated.

**solubility** *Chemistry.* the ability or tendency of one substance to dissolve into another substance (at a given temperature and pressure); generally expressed in terms of the amount of solute that will dissolve in a given amount of solvent to produce a saturated solution.

**solubility pump** *Earth Science.* a gas exchange process by which the ocean maintains a vertical gradient in dissolved inorganic carbon (DIC) with the result that DIC is concentrated at lower levels.

**soluble** *Chemistry.* capable of being dissolved in a particular solvent; having the quality of solubility.

**solute** *Chemistry.* a substance that is dissolved in a solvent; the part of a solution that is uniformly dispersed in another substance.

**solution** *Chemistry.* a mixture of two or more substances uniformly dispersed throughout a single phase, so that the mixture is homogeneous at the molecular or ionic level. This may be a solid dissolved in a liquid (e.g., salt in water); a liquid in another liquid (e.g., alcohol in water); a gas in a liquid (e.g., carbon dioxide in water); a gas in another gas (e.g., oxygen in air); a gas in a solid (e.g., hydrogen in platinum); or a solid mixed with another solid (e.g., an alloy of carbon and iron).

**solution gas** *Oil & Gas.* natural gas that is dissolved in crude oil as a result of reservoir pressures.

**solution mining** another term for IN-SITU LEACH MINING.

**solvate** *Chemistry.* a compound formed by the interaction of a solvent and a solute.

**Solvay process** *Materials.* a commercial process used to produce soda ash (sodium carbonate), in which sodium bicarbonate separates as a solid, is calcined, and then is changed to soda ash. [Developed by Belgian chemist Ernest *Solvay,* 1838–1922.]

**solvent** *Chemistry.* the continuous phase of a solution in which a solute is dissolved; water is the most common liquid solvent.

**solvent extraction** *Materials.* a technique used in chemical separations involving two immiscible solvents, in which one or more components are transferred between the solvents.

**solvent naphtha** *Materials.* a mixture obtained by the fractional distillation of coal tar: a yellowish or colorless to white liquid that boils at 160°C; used mainly as a solvent.

**sonar** *Communication.* a system that uses underwater sound waves to determine the location of objects and for navigation and communication. [An acronym for sound navigation and ranging.]

**sonic** *Physics.* **1.** having to do with or caused by the property of sound. **2.** involving speeds at or above the speed of sound.

**sonic barrier** see SOUND BARRIER.

**sonic logging** *Oil & Gas.* a process that determines the size or holding capacity of a well, using a pulse-echo system that measures the distance between a sound-originating instrument and a sound-reflecting surface.

**soot** *Environment.* **1.** a powdery black substance composed chiefly of carbon, formed by the incomplete combustion of wood, coal, oil, or other such material. **2.** specifically, the visible black carbon portion of diesel exhaust. **Diesel soot** is toxic to humans and has also been identified as a major source of global warming.

**sorb** *Chemistry.* to undergo a process of sorption; take up a liquid or gas by absorption or adsorption.

**sorbent** *Materials.* any material, generally a mineral such as a clay or a silicate, that has the capacity to absorb or adsorb another substance,

**sorbent injection** *Environment.* the injection of an absorbent material during a combustion process (such as coal burning) in order to limit the emission of some unwanted substance; e.g., mercury, sulfur dioxide.

**sorghum** *Biomass.* either of two drought-tolerant cereal grasses, *Sorghum vulgare* or *S. bicolor,* having broad, cornlike leaves, a tall pithy stem, and grain borne in a dense terminal cluster; grown mainly for stock feed and syrup but also can be cultivated as a form of biomass energy. Sorghums are noted for their high yields of directly fermentable stalk sugars.

**sorption** *Chemistry.* the interaction of an atom, molecule, or particle with a solid surface, especially the processes of absorption (taking in another substance completely) or adsorption (holding another substance at the surface).

**sound** *Physics.* **1.** a pressure disturbance that propagates through a medium due to the stress or displacement of the medium from its equilibrium state. **2.** the auditory sensation that is induced by such a disturbance, which in humans is perceived by the ears. Sound moves through dry air at a speed of about 331 meters per second (740 miles per hour), assuming standard atmospheric pressure and a temperature of 0°C.

**sound barrier** *Physics.* a sharp, sudden increase in aerodynamic drag on an aircraft approaching the speed of sound; historically perceived as a physical obstacle that might prevent an aircraft from traveling at speeds faster than the speed of sound.

**sound energy** *Physics.* the energy associated with sound waves, expressed quantitatively by the difference between the total energy of a sound field and the energy that would exist in the same region in the absence of sound.

**sour** *Oil & Gas.* referring to gases associated with oil and gas drilling and production that are acidic either alone or when combined with water, such as hydrogen sulfide or carbon dioxide. **Sour gas** and **sour crude** contain significant quantities of such substances.

**source reduction** *Environment.* any of various practices or policies enacted to reduce the amount of a hazardous substance, pollutant, or contaminant at the source; i.e., at the point of generation prior to its entering the environment.

**source rock** *Oil & Gas.* **1.** a sedimentary rock unit whose organic material was transformed over time into liquid or gaseous hydrocarbons under pressure and heat, and that thus is capable of generating and expelling petroleum. **2.** *Earth Science.* the rock from which a sediment is derived.

**souring** see RESERVOIR SOURING.

**Southern Oscillation** *Earth Science.* an oscillation in the surface pressure between the southeastern tropical Pacific and the Australian–Indonesian regions. When the waters of the eastern Pacific are abnormally warm (an EL NIÑO event), sea level pressure drops in the eastern Pacific and rises in the west. This reduction in the pressure gradient is accompanied by a weakening of the low-latitude easterly trades, with global ramifications for weather patterns.

**sovereignty** *Global Issues.* the fact or principle of having legitimate authority over a certain territory and the people and resources therein; the international community recognizes that sovereignty over natural resources is inherent to a nation and is exercised by its representative, the state.

**soybean** *Biomass.* an erect annual dicotyledonous plant of the legume family, *Glycine max*, native to Asia and now widely cultivated in other parts of the world, valuable as a food crop and also as a source of biomass energy.

**space charge** *Electricity.* **1.** an electric charge resident in a cloud of electrons lying between a cathode and plate within an electron tube. **2.** a region of electrical deficiency in the interior of a material that balances the concentration of electrons at the surface. **3.** *Earth Science.* an excess of either negatively or positively charged ions in a layer of the atmosphere.

**space cooler** *HVAC.* a small, self-contained cooling device (typically a compact air conditioner) that is employed in a building to cool one specific room or area of a room, to substitute for (or reduce the operating cost of) central air conditioning.

**space cooling** *HVAC.* **1.** a general term for energy provided to a building or other space for cooling purposes. **2.** the cooling of one specific area of a building, as opposed to a central system that cools the entire structure.

**spacecraft** *Transportation.* any vehicle or other device that travels into or through space, especially one that carries a payload.

**space heat(ing)** *HVAC.* **1.** a general term for energy provided to a building or other space for heating purposes. **2.** the heating of one specific area of a building, as opposed to the central heating of the entire structure.

**space heater** *HVAC.* a small, self-contained heating device (typically an electric heater) that is employed in a building to heat one specific room or area of a room, as opposed to a central heating source such as an oil burner.

**spaceship Earth** *Economics.* a term coined by futurist Buckminster Fuller and later used as the title of a book by economist Kenneth Boulding; the concept that the earth has energy, material, and environmental limits on economic growth, and thus ultimately must establish a "spaceship" economy powered by renewable energy and characterized by the efficient recycling of materials.

**space shuttle** *Transportation.* **1.** also, **Space Shuttle.** any of a series of U.S. space vehicles designed to be launched into space by solid rocket boosters and then return to earth under their own power. **2.** any space vehicle designed to transport people and equipment between earth and a space station.

***Space Shuttle*** *The U.S. Space Shuttle Columbia lands at Edwards Air Force Base in the desert of Southern California.*

**space-time** *Physics.* a four-dimensional space consisting of three dimensions corresponding to the ordinary spatial coordinates of length, width, and height, and a fourth dimension corresponding to time; used in the theory of RELATIVITY, which specifies that the three spatial dimensions cannot be described accurately without consideration of the time of description. Also, **space-time continuum.**

**spallation** *Nuclear.* a nuclear reaction in which a bombarded nucleus breaks up into many particles.

**spark** *Electricity.* a short-lived electric discharge produced by a sudden breakdown of air or another dielectric separating two terminals, accompanied by a momentary flash of light. Thus, **spark discharge.**

**spark gap** *Electricity.* a device consisting of two metal points, tips, or balls separated by a small air gap; a high voltage applied to the electrodes causes a spark (or, with an AC voltage, a train of sparks) to jump across the gap.

**spark-ignition (SI) engine** *Transportation.* an internal combustion engine in which the fuel mixture is ignited by an electrical discharge within the engine. Such an engine runs on an OTTO CYCLE.

**spark plug** or **sparkplug** *Transportation.* a device that is fitted into the cylinder of an internal-combustion engine to provide a pair of electrodes, between which an electrical discharge is passed to ignite an explosive mixture such as gasoline.

**spatial scale** *Measurement.* a term for the dimensional context used to study an environmental phenomenon; e.g., continental scale is above 10 million square kilometers, regional is 100,000 to 10 million sq. km, and local is less than 100,000 sq. km.

**SPD** spectral power distribution.

**special naphtha** *Oil & Gas.* any of various finished products within the naphtha boiling range that are used as paint thinners, cleaners, or solvents.

**special relativity** see RELATIVITY.

**species diversity** see BIODIVERSITY.

**specific conductivity** *Measurement.* the conductivity between two opposite sides of a unit cube of a given material (usually a cube of one centimeter).

**specific energy** *Measurement.* the energy content of a fuel or energy storage device per unit mass.

**specific enthalpy** *Measurement.* the enthalpy of a system divided by its mass.

**specific entropy** *Measurement.* the entropy of a substance per unit mass.

**specific gravity** *Physics.* the ratio of the density of a material at a given temperature and pressure to the density of a standard material. For solids and liquids, the standard is usually pure water at a temperature of 3.98°C and sea level atmospheric pressure; for gases, the standard is often hydrogen, oxygen, or air at a specified temperature and pressure.

**specific heat** *Thermodynamics.* the amount of heat energy required to increase the temperature of a unit mass of a substance, in relation to the amount required to increase the temperature of a unit mass of water by the same increment (e.g., 1 K), assuming constant volume and pressure. Also, **specific heat capacity.**

**specific humidity** *Earth Science.* the weight of water vapor in a gas and water vapor system (e.g., moist air), relative to the total mass of the system.

**specific internal energy** *Thermodynamics.* the internal energy of a system per unit mass of the system, usually expressed as joules per kilogram.

**specific power** *Thermodynamics.* the amount of power per unit mass, usually expressed in watts per kilogram.

**specific speed** *Measurement.* a factor by which the performance of a centrifugal or axial pump or a hydraulic turbine can be computed, given as the speed in revolutions per minute.

**specific volume** *Thermodynamics.* the total volume of a substance divided by the total mass of the substance; the inverse of density.

**spectra** the plural of SPECTRUM.

**spectral** *Physics.* relating to or produced by a spectrum.

**spectral irradiance** *Solar.* the monochromatic irradiance of a surface per unit bandwidth at a particular wavelength, usually expressed in watts per square meter per nanometer bandwidth.

**spectral line** *Physics.* a line in a spectrum due to the emission or absorption of electromagnetic radiation at a discrete wavelength, resulting from discrete changes in the energy of an atom or molecule. Different atoms or molecules can be identified by their sequence of spectral lines.

**spectrally selective** *Materials.* describing materials that filter out much of the heat normally transmitted through clear glass, while allowing a high percentage of light to be transmitted. Thus, **spectrally selective coating, glazing, film,** and so on.

**spectral power distribution (SPD)** *Lighting.* a method of determining the precise color output of a given light source by charting the level of energy present at each wavelength across the visible spectrum. An SPD diagram of sunlight at midday will show that all wavelengths of visible light are present in nearly equal quantities. The closer that an artificial light source matches this distribution, the better it simulates true daylight.

**spectral reflectance** *Solar.* the ratio of energy reflected from a surface in a given spectral waveband to the energy incident in that waveband.

**spectral transmittance** *Solar.* the amount of transmittance per unit interval of wavelength.

**spectrograph** *Measurement.* an instrument that spreads light or other electromagnetic radiation into its component wavelengths (spectrum), and records the results photographically or electronically.

**spectrometer** *Measurement.* a device that produces a spectrum by dispersion and that is calibrated to measure radiant energy with respect to wavelength, intensity, and so on.

**spectroradiometer** *Measurement.* an instrument used to measure and monitor visible and ultraviolet light.

**spectroscopy** *Physics.* the branch of science that is concerned with measurement of the emission and absorption spectra of light and other forms of electromagnetic radiation. Thus, **spectroscope, spectroscopic.**

**spectrum** *Physics.* **1.** the complete range of entities in which some phenomenon or system exists. **2.** the visible portion of all electromagnetic radiation; i.e., the band of colors that is produced by white light passing through a prism, ranging from about 770 to about 380 nanometers and generally perceived as appearing in a sequence of red, orange, yellow, green, blue, indigo, and violet. **3.** the range of wavelengths or frequencies produced when any other form of electromagnetic radiation is dispersed.

**specular** *Solar.* relating to or in the direction in which a mirror reflects light; i.e., the angle of visible incidence is equal to the angle of reflection (in the case of a perfect mirror). Thus, **specular angle, specular surface.**

**specular reflection** *Solar.* reflection from a smooth surface, such as a perfect mirror, in which the light leaves at the same angle at which it arrived. Compare DIFFUSE REFLECTION.

**specular transmission** *Solar.* the process by which incident flux passes through a surface or medium without scattering.

**speculative resources** *Consumption & Efficiency.* energy resources (e.g., coal, uranium) not yet discovered but thought to exist in addition to proved resources, occurring either in known deposits in a favorable geologic setting where no discoveries have been made as yet, or in deposits discoverable with existing exploration techniques.

**speed** *Measurement.* **1.** in general, the time rate at which a physical process takes place. **2.** the time rate of change of position without regard to the direction of motion; i.e., the ratio of the distance traveled to the time elapsed. See also VELOCITY.

**speed of light** see LIGHT.

**speed of sound** see SOUND.

**spent (nuclear) fuel** *Nuclear.* depleted nuclear fuel that has been removed from a nuclear reactor because it has reached the end of its economic lifetime; i.e., it can no longer accomplish power production because it cannot effectively sustain a chain reaction.

**spent fuel pond** *Nuclear.* a temporary storage structure for spent fuel removed from the core of nuclear reactors; these typically are rectangular or L-shaped basins about 40 feet deep, made of thick reinforced concrete walls with stainless steel liners filled with water.

**spent shale** *Oil & Gas.* the mineral waste from a process of retorting oil shale, usually including some char derived from the organic matter.

**spermaceti** *Biomass.* a white, translucent solid that is obtained from the oil in the head of the sperm whale; used in candles.

**sperm whale** *History.* a large, square-snouted, toothed whale, *Physeter catodon*, of the family Physeteridae, characterized by a large closed cavity in its head holding a fluid mixture of spermaceti and oil, and a blubber that yields superior oil. During the 19th century it was extensively hunted as a source of oil **(sperm oil)** for fuel, prior to the advent of electric lighting.

**SPF** spray polyurethane foam.

**SPFC** solid polymer fuel cell.

**sphagnum (moss)** *Biomass.* any moss of the genus *Sphagnum*, in the family Sphagnaceae; it grows in large mats of floating vegetation and is one of the principal plants responsible for the formation of peat. It can maintain a growing part above the water surface while decay occurs below the surface.

**spill** see OIL SPILL.

**spilled water** *Hydropower.* a term for water that passes through a hydropower dam without being used to generate electricity.

**spillway** *Hydropower.* a structure that passes surplus water through, over, or around a dam; e.g., when excess volume is created in the reservoir by heavy rainfall.

**spin** *Physics.* **1.** the rotation of a body about an axis through the body. **2.** the angular momentum possessed intrinsically by elementary particles and nuclei even while at rest. **3.** a quantum number that describes the rotation of an elementary particle on its axis, or of a system of such particles in orbital motion.

**Spindletop** *Oil & Gas.* the name of a small knoll just south of Beaumont, Texas, site of the discovery (1901) of the first giant oil field in the U.S., which transformed the American oil and gas industry. Prominent companies such as Gulf Oil, Texaco, and Hughes Tool Company arose from the effort to exploit Spindletop's oil.

**spinning frame** *History.* a device introduced by English inventor Richard Arkwright in 1769 to produce cotton threads for textile manufacture, considered one of the crucial developments leading to Britain's Industrial Revolution.

*spinning frame The original spinning frame of Richard Arkwright (1769).*

**spinning Jenny** *History.* the first spinning machine to feature more than one wheel; an advance on the traditional spinning wheel developed by English inventor James Hargreaves in 1764; it was capable of spinning eight threads at once and thus provided much greater efficiency.

**spinning reserve** *Electricity.* a term for surplus electrical generating capacity that is operating and able to respond instantly to a sudden increase in the electric load, or to a sudden decrease in power.

**spinning wheel** *History.* an early machine designed to make thread or yarn from raw fiber for the production of cloth, dating from about the 13th century and consisting of a large wheel and a spindle that were connected by a band or belt and powered by a foot pedal.

**spin-spin energy** *Physics.* an energy associated with the interaction between the spin angular momenta of two particles, found to be proportional to the dot product of the angular momentum vectors.

**spiral freezer** *Refrigeration.* a commercial freezer for chilling and freezing food in which a continuous belt carrying food items through the refrigerated enclosure is stacked in an extensive spiraling arrangement, to provide long product residence time within a compact freezer space.

**spiral thermometer** *Measurement.* a thermometer consisting of a pair of metal strips, each of a different material, forming spirals that expand and contract at different rates in response to temperature change.

**spiral-tube exchanger** *Consumption & Efficiency.* a concurrent heat-exchange instrument consisting of a set of spiral coils that are joined by manifolds and arranged concentrically; generally used in air-separation plants for very low temperature exchanges.

**spiral-wound** *Storage.* a design for the electrodes in a lead-acid battery of high surface area, created by winding the electrodes and separator into a spiraling "jelly-roll" configuration.

**spirit** *Materials.* **1.** a solution of a volatile material in alcohol. **2.** any volatile or distilled liquid.

**spirit of coal** *Coal.* a historic term for COAL GAS.

**spirit thermometer** *Measurement.* a thermometer consisting of a closed capillary tube with a liquid-filled bulb at one end; the liquid in the bulb travels up the capillary according to the temperature of the bulb.

**Spitzer, Lyman** 1914–1997, U.S. physicist who helped to develop the foundations of plasma physics in the 1950s and also made major contributions in stellar dynamics and space astronomy.

**splint coal** *Coal.* a hard, dull, blocky, grayish-black, banded bituminous coal that is characterized by a rough, uneven fracture and granular texture and that burns with intense heat.

**split coal** *Coal.* a term for a coal seam that cannot be mined as a single unit because it is separated from the main stream by a parting of other sedimentary rock.

**split-spectrum cell** *Photovoltaic.* a compound photovoltaic device in which sunlight is first divided into spectral regions by optical means. Each region is then directed to a different photovoltaic cell optimized for converting that portion of the spectrum into electricity; this achieves significantly greater overall conversion of incident sunlight.

**split-system** *HVAC.* describing an air conditioner or heat pump system in which the moving fan and cooling-heating coil are separated (split) from the compressor and condenser, with the former being located in an indoor console and the latter in a weatherproof outdoor enclosure. Insulated pipes and electric wiring connect the two separate units.

**spoked wheel** *History.* a major advance in the technology of the wheel, though to have first appeared in Egypt and Palestine about 4000 years ago; being significantly lighter than a solid wheel, it allowed a wheeled cart to carry a heavier load.

**spontaneous** *Physics.* describing a process that occurs because of internal properties, with no external forces required to continue the process, although external forces may be required to initiate it.

**spontaneous combustion** *Chemistry.* **1.** the ignition of a substance brought about by a heat-producing chemical reaction of its own constituents, without exposure to an external source of ignition. **2.** specifically, a process in which coal ignites on its own, because the rate of heat generation by the oxidation reaction exceeds the rate of heat dissipation.

**spontaneous fission** *Nuclear.* fission that occurs spontaneously, that is, not induced by an incident particle.

**spontaneous generation** *History.* the former theory that living matter can come to life spontaneously from a nonliving organic source within a relatively short time period, such as flies or maggots arising from decaying meat; discredited by Francesco Redi in the case of complex organisms and by Louis Pasteur in the case of microorganisms.

*spontaneous generation French scientist Louis Pasteur refuted the theory of spontaneous generation by winning a contest open to anyone who could prove or disprove the theory.*

**spontaneous ignition** see SPONTANEOUS COMBUSTION.

**sport utility vehicle (SUV)** *Transportation.* a contemporary automobile category so called because such vehicles can serve both as large passenger vehicles and as off-road vehicles for recreational purposes. SUVs are currently a source of controversy in the U.S. concerning issues of safety, fuel consumption, emissions, and so on, because they are sold as passenger vehicles but many SUV models are not regulated by the same standards as cars, being classified as light trucks.

**SPOT** *Environment.* System pour L'Observation de la Terre; a program of earth-observing satellites and sensors, used especially for environmental data.

**spot market** *Economics.* a market in which commodities, such as crude oil, are purchased in cash for immediate delivery (i.e., bought "on the spot"). Thus, **spot price, spot purchase.**

**SPR** Strategic Petroleum Reserve.

**sprag** *Mining.* **1.** a short piece of wood placed between the spokes of a wheel or dug into the ground to prevent the movement of a mine vehicle. **2.** a wooden prop positioned on a slant to support a mine roof or prevent coal from flying during blasting.

**sprag road** *Mining.* a mine road so steep that mine cars require sprags between their wheel spokes to slow their descent.

**sprawl** *Social Issues.* a term for land-use patterns characterized by relatively low density of people and economic activity, high use of private vehicles, extensive road networks, and little or no opportunity to travel via mass transit; associated with newly developed areas without formalized growth strategy.

**spray polyurethane foam (SPF)** *Materials.* a polyurethane substance used as an insulating material for rooftops, building spaces, water heaters, and refrigerated transport. Two liquid components are combined and sprayed onto a surface to form SPF; the components react and harden to form a rigid, waterproof insulating layer.

**spray pyrolysis** *Materials.* a deposition process in which heat is used to break molecules into elemental sources that are then spray deposited on a substrate; used for the synthesis of composite nanoparticles that can be used for catalysis, sensors, and electroceramics.

**spreader stoker** *Consumption & Efficiency.* a coal-burning furnace in which a system of mechanical feeders load and evenly distribute a thin layer of fuel across the furnace grate.

**spring** *Consumption & Efficiency.* **1.** a simple machine consisting of an elastic helical coil that when stressed or bent will return to its original form. **2.** a stored energy device or system in a transportation vehicle that absorbs and releases energy to provide a level ride.

**spring tide** *Earth Science.* a tide of greater range that occurs twice a month at the new and full phases of the moon; caused by the earth, sun and moon being almost co-linear, so that the gravitational pull of both the sun and moon reinforce each other. This produces a higher high tide and lower low tide than the average.

**sprocket** *Consumption & Efficiency.* a toothed wheel that engages a cable or power chain.

**SPS** solar power satellite (system).

**spudding** *Oil & Gas.* the act of beginning to drill an oil or gas well.

**Sputnik** *History.* the first spacecraft to achieve and maintain an orbit around the earth, an unmanned artificial satellite launched by the Soviet Union in 1957.

**sputtering** *Photovoltaic.* a process used to apply photovoltaic semiconductor material to a substrate by a physical vapor deposition process; high-energy ions are used to bombard elemental sources of semiconductor material, which eject vapors of atoms that are then deposited in thin layers on a substrate.

**squirrel-cage motor** *Consumption & Efficiency.* an induction motor with a rotor constructed of bars embedded around the periphery of a laminated, iron-core rotor and joined at each end by continuous rings; the most common type of induction-motor rotor (used in many common household devices), due to its inherent low cost and high reliability.

**sr** steradian.

**SRI** solar reflectance index.

**SSC** Superconducting Super Collider.

**SST** sea surface temperature.

**SST** *Transportation.* supersonic transport; a descriptive term for the CONCORDE or another supersonic passenger aircraft.

**St. Louis encephalitis (SLE)** *Health & Safety.* a mosquito-transmitted viral disease that causes inflammation of the brain, named for the U.S. city where the first cases were recognized in 1933. The epidemiology of SLE plays a role in studies of climate change because of the association of this disease with episodes of prolonged drought.

**stable** *Nuclear.* describing atomic particles that have no known mode of decay and are indefinitely long-lived. Thus, **stability.**

**stack** *Renewable/Alternative.* a set of individual fuel cells arranged next to each other and connected in series. Thus, **stacking.**

**stack life** *Renewable/Alternative.* the cumulative period of time that a fuel cell stack may operate before its output deteriorates below a useful minimum value.

**Staebler–Wronski effect** *Photovoltaic.* a term for the tendency of amorphous silicon photovoltaic devices to lose efficiency upon their initial exposure to light. [Described by U.S. engineers David *Staebler* and Christopher *Wronski.*]

**stagflation** *Economics.* an economic downturn characterized by the simultaneous existence of stagnation and persistent inflation.

**stagnation** *Earth Science.* **1.** the condition of a body of water that is not flowing in a stream or is otherwise motionless, being unstirred by currents or waves. **2.** *Physics.* a condition in a flow field about a body at which the fluid has zero velocity relative to the body. **3.** *Economics.* a lack of growth or activity in an economy; can be defined as economic growth of less than 1% in a given year.

**stair pit** another term for BELL PIT.

**stakeholder** *Social Issues.* any person who affects and/or is affected by a decision or policy under consideration; i.e., who has a "stake" in the outcome. The stakeholders of a given energy company might include not only its employees, stockholders, and customers, but also others affected by its activities; e.g., residents of the community in which it operates.

**stall** *Consumption & Efficiency.* an abrupt and unwanted stoppage in the operation or function of a machine, engine, device, and so on; e.g., gross loss of lift in an aircraft or the lack of motion of a wind turbine rotor.

**stand** *Biomass.* a community of trees managed as a productive unit.

**stand-alone** or **standalone** *Electricity.* describing a power source that operates independently of grid-supplied electric power and that is able to deliver a predictable and dependable flow of power; may be a conventional generator or a renewable energy source. Thus, **stand-alone power system (SAPS).**

**standard air** *HVAC.* a reference condition of air used for various comparisons and calculations; typically considered to be dry air weighing 0.075 pounds per cubic foot at sea level pressure and a temperature of 70°F.

**standard atmosphere** *Earth Science.* **1.** a hypothetical, vertical distribution of pressure, temperature, and density that is taken to be representative of the most typical actual conditions of the earth's atmosphere. **2.** see ATMOSPHERIC PRESSURE.

**standard conditions** *Measurement.* **1.** any prescribed set of values considered to be ideal or typical and thus employed for experimental measurements; e.g., a room temperature of 25°C (77°F) used to determine a product's flow rate (such as a fuel). **2.** see STANDARD TEMPERATURE AND PRESSURE.

**standard enthalpy change** *Chemistry.* the change in enthalpy associated with a chemical process or transformation involving substances in their standard states. Thus, **standard enthalpy of formation, combustion,** or **reaction.**

**standard gravity** *Measurement.* the standard value of gravitational acceleration at sea level on earth, accepted as a rate of 9.80665 meters per second per second (or about 32 feet per second per second).

**standard incandescent** *Lighting.* the common type of incandescent bulb that has historically been the most widely used electric light source.

**standard man** see REFERENCE MAN.

**Standard Market Design (SMD)** *Policy.* a U.S. plan for electric utility reform that attempts to simplify the regulation of electricity markets; it emphasizes customer-focused, competitive wholesale power markets, with appropriate flexibility to accommodate regional differences.

**standard metabolic rate (SMR)** *Biological Energetics.* the lowest rate of heat production by a reptile, amphibian, fish, or invertebrate (ectothermic or "cold-blooded" animal) measured at a specified body temperature when costs for activity, temperature regulation, digestion, and other energy expenditures are low or zero.

**Standard Oil** an enterprise founded by John D. Rockefeller in Cleveland, Ohio in 1870 that eventually came to dominate the oil business in the U.S. In 1911 the Standard Oil Company was ruled a monopoly and was broken up into a number of smaller companies, such as Standard Oil of New Jersey (Esso), Standard Oil of New York (Socony), Standard Oil of California (Chevron), Continental Oil (Conoco), and Atlantic Oil (Arco).

**standard pressure** see ATMOSPHERIC PRESSURE.

**standard reaction enthalpy** *Thermodynamics.* the difference between the standard enthalpies of formation of reactants and the products, with each term weighted by the stoichiometric number in the chemical equation.

**standard state** *Physics.* the state of a substance in which it exists in its most pure and stable form at standard pressure and temperature; a set of conditions employed to allow convenient comparison of thermodynamic properties.

**standard temperature and pressure** *Measurement.* a standard set of values for a gas; a temperature of 0°C (273.15 K) and a pressure of 1.00 atmosphere (760 torr).

**standard time** *Measurement.* a system of time measurement based on a stipulated local time within a given band of longitude roughly 15° wide, one of 24 such bands for the globe based on a value of 0 hour for the prime meridian at Greenwich, England. A historic development replacing variations in local civic time, brought about chiefly by the advent of the railroad era, which required standardization of schedules for different locales.

**standby** or **stand-by** *Electricity.* **1.** a power source that supports a utility system and that it is generally running under no load, so that it is available to replace or supplement a facility normally in service. Thus, **standby facility, standby generation. 2.** the state of a battery that is consuming only a minimal amount of power but that can quickly restore a full power supply if called upon.

**standby battery** *Storage.* a battery that has a long shelf life and delivers moderate to high currents when called upon; typically used to supply power to critical systems in the event of a power outage; e.g., emergency lighting in stairwells and hallways, or to substitute for a standard power source in the event of a loss or sag in power.

**standby loss** *Consumption & Efficiency.* the continuous loss of heat from a storage water heater when it is not in use, by conduction through the walls of the tank.

**stand-off mounting** *Photovoltaic.* a technique for placing a photovoltaic array on a sloped roof, by mounting the modules a short distance above the pitched roof and tilting them to the optimum angle.

**Stanley, William** 1858–1916, U.S. engineer who invented the induction coil, a transformer that creates alternating current electricity (1886).

**stannic** *Materials.* describing various compounds of tin, especially those in which the element has a valence of 4.

**stannous** *Materials.* describing various compounds of tin, especially those in which the element has a valence of 2.

**starch** *Materials.* any of a group of polysaccharides that store energy in plants, having the general formula $(C_6H_{10}O_5)_n$, occurring in the form of minute granules in seeds, tubers, and other parts of plants; it forms an important constituent of rice, corn, wheat, beans, potatoes, and other vegetable foods. Starch is a highly amorphous polymer that is readily broken down into glucose by human and animal enzyme systems.

**Starr, Chauncey** born 1912, U.S. physicist who played an important role in the development of nuclear power in the U.S. and also made significant contributions to the understanding of interactions between energy and the environment.

**starved** *Storage.* describing a type of maintenance free lead-acid battery; so called because it contains just enough electrolyte to provide the necessary chemical reaction. Thus, **starved electrolyte (battery).**

**state** *Social Issues.* **1.** the set of political institutions within a determined territory whose concern is with the political and social organization and management of this territory, in the public interest. **2.** *Physics.* the condition or phase of a given system. See STATE OF MATTER.

**state function** *Physics.* a property that depends only on the condition or state of the system, and not on the path used to establish the current conditions. Volume, pressure, and temperature are examples of state functions; heat and work are non-state functions.

**state of matter** *Physics.* the specified physical condition in which a given system or substance exists at a given point in time; the three common states of matter are gases, liquids, and solids. Plasmas are often described as a fourth state of matter.

**state of charge** *Storage.* the available capacity remaining in a battery at any given time, expressed as a percentage of the total rated capacity.

**static** *Physics.* **1.** not experiencing motion; at rest; stationary. **2.** see STATIC ELECTRICITY. **3.** *Communications.* radio or television broadcast interference that is produced by static electricity or atmospheric electricity having a wavelength within that of the broadcast frequencies, usually perceived as a crackling or hissing noise.

**static electricity** *Physics.* energy in the form of a stationary electric charge, such as that stored in capacitors and thunderclouds, or produced by friction or induction.

**static head** *Hydropower.* the vertical distance from the water level of a pump or reservoir to the point of free discharge of the water, as measured when the system is not operating and thus taking into account resistance from gravity but not pressure loss due to friction and turbulence.

**static pressure** *Hydropower.* the pressure produced by an unmoving column of water; determined only by the vertical height of the column, since there is no loss of pressure due to friction or other factors.

**static stability** *Physics.* a condition in which a system is in equilibrium and tends to remain in that state unless an external influence is applied, whereupon the system will react to return to equilibrium.

**stationary battery** another term for STANDBY BATTERY.

**stationary source** *Environment.* a non-mobile facility that releases pollution, such as a factory, power plant, oil refinery, manufacturing operation, and so on. Thus, **stationary-source pollution.**

**statistical mechanics** *Thermodynamics.* the branch of physics that studies the motion of constituent particles in a gas or other system and how this contributes to the whole.

**stator** *Electricity.* the part of a motor, generator, or alternator that does not rotate.

**statvolt** *Electricity.* the unit of voltage in the centimeter-gram-second (CGS) electrostatic system of units; one statvolt is approximately equal to $2.997925 \times 10^2$ volts.

**steady-state** *Physics.* **1.** describing a system in which the conditions do not change with time. **2.** having to do with or based on steady-state theory.

**steady-state theory** *Physics.* the theory that the universe has no beginning and no end but has average properties, constant in space and time; it assumes that new matter is constantly and spontaneously created to maintain these average values in the context of an expanding universe. This theory does not account for certain recent discoveries (e.g., quasars, microwave background radiation), and has generally been supplanted by the BIG BANG theory. Also, **steady-state model.**

**steam** *Chemistry.* **1.** water vapor, especially such vapor having a temperature above the boiling point of water. **2.** water in this vapor form used for the purpose of producing mechanical energy. **3.** in general, the vapor of any liquid.

**steam coal** *Coal.* a coal product having properties that make it suitable for generating steam power; e.g., in historic use for powering steamships and steam locomotives, and in current use as a source of fuel for electric power plants.

**steam distillation** *Oil & Gas.* a petroleum-refining distillation that uses steam to minimize cracking, lowering the boiling point of the oils being distilled by adding the vapor pressure of the steam.

**steam-electrical plant** *Electricity.* an electrical power plant in which the prime mover is a steam turbine.

**steam engine** *Conversion.* **1.** any device that converts the heat energy of steam into mechanical energy. **2.** specifically, a reciprocating engine operating by the force of steam on a piston in a closed cylinder, in which the steam expands from the initial pressure to the exhaust pressure in a single stage. **3.** see STEAM LOCOMOTIVE.

**steam explosion** *Conversion.* a violent boiling or flashing of water into steam, typically occurring when water is superheated.

**steam field** *Geothermal.* a geothermal area with conditions that are favorable for energy use, especially the generation of electricity.

**steam flooding** *Oil & Gas.* an enhanced oil recovery technique in which high-pressure steam is fed through injection wells into a reservoir, in order to reduce viscosity and force greater volumes of oil into the producing wells.

**steam gun** *Consumption & Efficiency.* a machine from which projectiles may be expelled by the elastic force of steam.

**steam jet** *Refrigeration.* a refrigeration system that uses a vapor jet pump instead of a mechanical compressor system, based on the principle of flash evaporation.

**steam locomotive** *Transportation.* a self-propelled railway steam engine and boiler that is mounted on a frame and fitted with wheels driven by the engine.

***steam locomotive*** *A poster circulated in Philadelphia (1839) to discourage the coming of the railroad to the city; perhaps one of the earliest examples of the "NIMBY" mentality.*

**steam point** *Thermodynamics.* the true boiling point of pure water; the temperature at which a mixture of water and steam is in equilibrium at standard atmospheric pressure; i.e., 100°C or 212°F.

**steam reforming** *Hydrogen.* a technique of commercial hydrogen production based on reacting methane with water to produce hydrogen and carbon dioxide; currently the least expensive and most widely employed method of producing hydrogen.

**steam trap** *Consumption & Efficiency.* an automatic device that allows water or air to pass but prevents the passage of steam; used to drain the water of condensation that accumulates in a steam pipe or vessel.

**steam turbine** *Conversion.* any turbine in which high-pressure steam is employed as the working fluid.

**steam wagon** *History.* a steam-powered vehicle developed by French inventor Nicolas-Joseph Cugnot, considered to be the world's first self-propelled mechanical vehicle; built In 1769 with the intent of hauling artillery pieces for the French Army.

**steel** *Materials.* a strong hard metal that is an alloy of iron with small amounts of carbon (less than 2.5%) and lesser amounts of other elements, universally used as a structural material and in manufacturing for its substantial qualities of strength, hardness, and malleability.

**Stefan–Boltzmann law** *Solar.* one of the fundamental laws of radiation, stating that the amount of energy radiated per unit time from a unit surface area of an ideal blackbody is proportional to the fourth power of the absolute temperature of the blackbody. [Named for Austrian physicists Josef *Stefan,* 1835–1893, and Ludwig *Boltzmann,* 1844–1906.]

**Steinmetz, Charles** 1865–1923, German-born engineer who made important contributions to the understanding of electricity, particularly alternating-current systems, thus making possible the expansion of the electric power industry in the U.S.

**Stellarator** *Nuclear.* a device using suitably deformed magnetic fields to confine a high-temperature plasma with a density and confinement time sufficient to sustain a nuclear fusion reaction; an advanced concept for controlled nuclear fusion.

**St. Elmo's fire** see SAINT ELMO'S FIRE.

**step-down** *Electricity.* **1.** to change electricity from a higher to a lower voltage. **2.** such a decrease in voltage. **3.** describing a transformer

*Charles Steinmetz*

that has fewer turns of wire in the secondary winding than in the primary, which causes a decrease of the voltage.

**Stephenson, George** 1781–1848, English engineer who developed the first successful steam locomotive (1815). He made important advances in railway design, including the use of tunnels and embankments, and malleable iron rails instead of cast iron. He also devised a miner's safety lamp at about the same time as did Sir Humphry DAVY.

**step-up** *Electricity.* **1.** to change electricity from a lower to a higher voltage. **2.** such an increase in voltage. **3.** describing a transformer that has more turns of wire in the secondary winding than in the primary, which causes an increase of the voltage.

**steradian** *Measurement.* a unit of solid angular measurement; a solid angle that, having its vertex at the centre of a sphere, cuts off an area of the surface of the sphere equal to that of a square with sides of length equal to the radius of the sphere.

**Stevens, John** 1749–1838, U.S. mechanical engineer who was a pioneer in steamboats, including the world's first sea-going steamboat and steam ferry service. He also was among the first to demonstrate the feasibility of steam locomotion on a rail track (1826).

**Stevin, Simon** 1548–1620, Flemish scientist who introduced decimal fractions to Europe and also made important early contributions to energy mechanics, describing the theory of levers, the inclined plane, and pulleys.

**Stewart, Alice** 1906–2002, British physician who was the first to demonstrate the link between exposure to low-intensity radiation and cancer, through her studies of children who were X-rayed in utero and also of nuclear workers at a site that was part of the Manhattan Project.

**still** *Materials.* **1.** a distilling apparatus used to separate liquids by heating, then cooling to condense the vapor. **2.** in popular use, such an apparatus used in (illegally) distilling alcoholic beverages.

**stillage** *Materials.* residue from a still; grains and liquid effluent remaining after removal of the distilled product; used in animal feed.

**still gas** *Oil & Gas.* a collective term for gases produced in refineries by distillation, cracking, reforming, and other processes; a byproduct in the upgrading of heavy petroleum fractions to more valuable lighter products, consumed internally as refinery fuel and used as petrochemical feedstock.

**Stirling cycle** *Thermodynamics.* an ideal thermodynamic cycle that consists of four processes: a constant-temperature compression; a constant-volume heat transfer to the working fluid; a constant-temperature expansion; and a constant-volume heat transfer out of the system; used as a model for actual devices.

**Stirling (cycle) engine** *Conversion.* **1.** an actual device or system that approximates the ideal Stirling cycle (see above). **2.** specifically, an external-combustion reciprocating engine in which work is performed as air in the cylinder is heated and expands, driving the working piston. Stirling engines historically have not achieved widespread use, but there has been recent interest in this technology for distributed power generation, gas liquefaction, and automotive power. [Developed in 1816 by Scottish inventor Robert *Stirling,* 1790–1878.]

**stochastic** *Measurement.* random; involving random or unpredictable values.

**stock effect** *Economics.* the increase in marginal extraction cost that occurs when a resource stock is depleted, either because it is physically more difficult to extract the resource (e.g., oil well pressure has declined), or because lower quality or higher cost deposits are being extracted.

**Stockholm Conference** *Policy.* the popular name for the United Nations Conference on the Human Environment (UNCHD), held in Stockholm in 1972; considered the first important international conference on the environmental future of the planet.

**stockpiling** *Consumption & Efficiency.* the fact or policy of maintaining an energy supply in reserve to safeguard against future disruptions of supply.

**stock pollutant** *Environment.* a classification for long-lived pollutants that can accumulate over time in the environment because there is little or no capacity to absorb them; e.g., DDT, greenhouse gases.

**stoichiometry** *Chemistry.* **1.** the branch of chemistry that studies chemical processes within the context of the laws of definite proportions and conservation of matter and energy. **2.** the study of the proportional relationships of two or more substances during a chemical reaction. Thus, **stoichiometric.**

**stoker** *Consumption & Efficiency.* **1.** a worker whose job is to tend a furnace and supply it with fuel, especially someone who did this in former times on a coal-powered train or ship. **2.** a continuous mechanical apparatus for feeding coal, coke, or other fuel into a boiler or furnace, handling the byproducts, and controlling the air ventilation to the fuel.

**Stokes, George** 1819–1903, Irish mathematician and physicist whose studies laid the foundation for the modern science of fluid mechanics. He also named and explained the phenomenon of fluorescence (1852) and made important contributions to the wave theory of light, as well as advancing the science of geodesy.

**stope** *Mining.* **1.** an excavation for removing ore in a series of steps, usually from vertical or highly inclined veins. **2.** to excavate ore by driving a series of horizontal workings into a vein, each immediately above or below the previous.

**storage** see ENERGY STORAGE.

**storage battery** *Storage.* any device or system capable of transforming energy from electric to chemical form, and vice versa. The energy can be stored in chemical form and then discharged as electric energy to be employed in an external circuit or apparatus.

**storage cell** *Storage.* a cell that can be recharged with electricity; i.e., an individual component of a storage battery.

**storage reservoir** *Hydropower.* a reservoir that has sufficient space for retaining the water of seasonal runoffs from rainfall and snowmelt.

**storage water heater** *HVAC.* a water heater that releases hot water on demand from the top of the tank; cold water then enters the bottom of the tank to maintain a full tank.

**stove** *HVAC.* an enclosed chamber that can be heated to a high temperature with a fuel-air mixture, for cooking, curing, or other applications.

**stove coal** see COAL SIZING.

**stover** *Biomass.* the dried stalks and leaves of a crop (such as corn) that remain after the grain has been harvested.

**STP** standard temperature (and) pressure.

**straddle plant** *Oil & Gas.* a natural gas processing plant constructed near a transmission pipeline, downstream from the fields where the natural gas in the pipeline has been produced.

**straight-run** *Oil & Gas.* describing petroleum material that has been obtained directly from a distillation unit and has not been cracked or reformed, and which is usually used as a feedstock or as a utility fuel. **Straight-run gasoline** generally must be upgraded to meet current motor fuel specifications.

**strain** *Physics.* **1.** a relative change in the dimensions of a body in response to an applied force, expressed as the ratio of this distortion to some undistorted dimension (not necessarily the same one). **2.** *Materials.* the manifestation of this change in an actual body; deformation of a material under a stress; this may be **elastic strain** (deformation disappears when stress is removed) or **plastic strain** (deformation is permanent).

**strain energy** *Physics.* the energy stored in a body due to an elastic deformation, equal to the work done to produce this deformation.

**stranded** *Economics.* a term for public-interest programs and benefits, such as conservation programs, reliability of supply, fuel diversity programs and low-income ratepayer assistance, that are available under a regulated monopoly system, but which may not be commercially viable in a deregulated, competitive market. Thus, **stranded benefit, stranded cost.**

**stranded gas** *Oil & Gas.* a term for discovered natural gas that is partially or completely isolated from markets, because of distances or insufficient transportation infrastructure, and thus is undeveloped or underdeveloped.

**strange (quark)** *Physics.* the third identified type, or flavor, of QUARK. [So called in contrast with *up* and *down* particles, the two types found in ordinary matter.]

**strata** *Earth Science.* the plural of STRATUM.

**Strategic Petroleum Reserve (SPR)** *Oil & Gas.* an emergency oil store maintained by the U.S. government to compensate for a severe supply interruption. It is located in four underground salt caverns along the Texas and Louisiana Gulf Coast. The SPR was created in 1975 in response to the U.S. energy crisis of 1973–74.

**stratification** *Earth Science.* the fact of a system or physical feature being stratified (divided into layers), such as a process in which sedimentary rocks form, accumulate, or are deposited in layers, or the arrangement of water masses in a body of water into two or more layers having different characteristics.

***stratification*** *Land forms in the Badlands area of South Dakota exhibit the property of stratification.*

**stratigraphic** *Earth Science.* relating to or based on stratigraphy.

**stratigraphic trap** *Oil & Gas.* a sealed oil or gas reservoir that results from changes in the physical properties of a formation, rather than from structural deformation.

**stratigraphic well** *Oil & Gas.* a drilling effort to obtain information that might lead to discovery of an accumulation of hydrocarbons, usually drilled without the intention of being completed for actual production.

**stratigraphy** *Earth Science.* a branch of geology that is concerned with the systemized study, description, and classification of stratified rocks, including their origin, characteristics, distribution, and correlation with one another.

**stratosphere** *Earth Science.* **1.** the region of the upper atmosphere lying above the troposphere and below the mesosphere, characterized by a slight increase in temperature with height and by stable, dry, and cloudless conditions. **2.** in former use, all of the atmosphere above the troposphere.

**stratum** *Earth Science.* a distinct homogeneous layer of rock or unconsolidated sedimentary material that is visibly separate from layers above and below it.

**Strauss, Lewis** 1896–1974, U.S. government official noted for his strong support of America's effort to develop a hydrogen bomb and for his influence on U.S. nuclear policy in general during the Cold War era.

**straw** *Biomass.* the dried stems and leaves of grain crops that remain after threshing; a potential source of fuel.

**streamflow** *Hydropower.* **1.** the act or fact of water flowing in a stream or channel. **2.** the rate at which water passes a given point in a stream, usually expressed in cubic feet per second.

**streamline** *Consumption & Efficiency.* **1.** a line that is parallel to the direction of flow of a fluid at a given point of time. A streamline flow about a body will efficiently follow the contours of the body. **2.** to design or manufacture a vehicle so as to minimize its resistance to motion through a fluid.

**streamlined** *Transportation.* describing a vehicle or vehicle component that is contoured so as to reduce its resistance to motion through a fluid; e.g., the design of certain high-speed trains, such as the premier U.S. trains of the 1930s to 1950s, or current trains of France.

***streamlined*** *The streamlined ICE (InterCity Express), a German-built high-speed train.*

**stream runoff** *Hydropower.* a term for runoff (discharge of water) that occurs naturally in a river, stream, or other such channel of water, without artificial diversions, storage reservoirs, or other human-made structures.

**streetcar** *Transportation.* any public rail car, such as a trolley car, that runs along city streets.

**stress** *Physics.* **1.** an external force that acts on a material and tends to change the dimensions of the material, by compressing it, stretching it, or causing it to shear. **2.** the amount of this effect acting on the surface of a given material, measured in terms of the force exerted per unit area and usually expressed in pascals. **3.** *Biological Energetics.* any external condition that evokes a change in an organism's internal physiological balance (homeostasis), generally leading to an expenditure of energy to restore this balance; e.g., the flight of a prey animal at the sight of a predator.

**stressor** *Biological Energetics.* any physical, chemical, or biological condition that can induce stress.

**stress response** *Biological Energetics.* the physiological, biochemical, reproductive, and other adaptive reaction induced by stress, in a single organism or a whole assemblage.

**strike price** *Economics.* the price at which an underlying futures contract is bought or sold in the event that an options contract is exercised.

**strip mine** *Mining.* an open-cut mine in which the deposits are worked by removing the overburden before extracting the coal or minerals below.

**strip mining** *Mining.* a commonly employed surface system of mining, in which the overlying layer of soil and rock is removed (stripped off) to allow access to the coal or other material of interest. Also, **stripping.** ☼ See below.

**strip (stripping) ratio** *Coal.* the amount of overburden material that must be removed to gain access to a deposit of coal; typically expressed in terms of the thickness, volume, or weight of the overburden in relation to that of the coal, in order to determine the amount

☼ **strip mining** A process of excavating earth, rock, and other material to uncover a tabular, lens-shaped, or layered mineral reserve. The mineral extracted is usually coal or other rocks of sedimentary origin. The mineral reserve is extracted after the overlying material, called overburden is removed. The excavation of the overburden is completed in rectangular blocks in plan view called pits or strips. The pits are parallel and adjacent to each other with each strip of overburden and the mineral beneath extracted sequentially. The mining process using equipment and explosives moves the overburden laterally to the adjacent empty pit where the mineral has been extracted. This lateral movement is called casting or open-casting. The overburden is moved by explosives, draglines, bucketwheel excavators, stripping shovels, dozers, and other equipment. The uncovered mineral is excavated and hauled out of the pit to down-stream processing operations. Filling the adjacent empty pits with the overburden is systemic to the process and therefore insures the genesis of mined-land land reclamation, an advantage of this method of surface mining. Planning strip mining utilizes a cross-section or range diagram of the earth to be removed. Strip mining is also called open-cut mining, open-cast mining, and stripping.

**Andrew P. Schissler**
Pennsylvania State University

of overburden that can be profitably removed to obtain a unit amount of coal.

**stripper** *Oil & Gas.* a term for an oil or gas well that produces at relatively low rates.

**strobe (light)** *Lighting.* **1.** a lamp that is capable of producing an extremely brief and intense flash of light; used in high-speed photography of rapidly moving objects. **2.** a similar bright, flashing light used for visual effects, as for a theatrical performance, concert, and so on.

**stroke** *Transportation.* the linear distance traveled in either direction by a piston or rod in an engine; the standard fuel cycle for a car or truck engine consists of four strokes: intake, compression, power, and exhaust.

**strontium** *Chemistry.* an alkaline-earth metal and element having the symbol Sr, the atomic number 38, an atomic weight of 87.62, a melting point of 770°C, and a boiling point of 1380°C; a silvery metal that rapidly turns yellowish in air, found naturally as a nonradioactive element with 16 known isotopes, 12 of which are radioactive.

**strontium-90** *Chemistry.* a heavy radioactive isotope of strontium with a half-life of 29.1 years, formed as a fission product and present in the fallout from nuclear explosions; it poses a significant hazard in humans and animals because, like calcium, it can be assimilated and deposited in bones.

**structural trap** *Oil & Gas.* a sealed oil or gas reservoir that results from structural deformation, rather than from changes in the physical properties of the formation.

**stückofen** *History.* an important precursor to the blast furnace, in which malleable iron was produced directly from the ore; widely employed in Saxony, Austria, and the Rhineland in the medieval era. [Literally, "wolf oven"; the large mass of iron produced by this oven was known as a *stücke,* or *wolf.*]

**stumpage** *Biomass.* **1.** timber in the form of standing uncut trees; i.e., still on the stump. **2.** the value or rate paid to a property owner for a stand of trees before their harvest.

**Sturgeon, William** 1783–1850, English engineer who built the first practical electromagnet (1825), invented the commutator for electric motors (1832), and made the first moving-coil galvanometer (1836).

**styrene** *Materials.* $C_6H_5CH{=}CH_2$, a colorless, toxic liquid with a strong aromatic aroma. Insoluble in water, soluble in alcohol and ether; it polymerizes rapidly and can become explosive. Used in making polymers and copolymers, polystyrene plastics, and rubbers.

**subatomic** *Physics.* below the atomic level; describing a particle of matter smaller than an atom, such as an electron.

**subbituminous coal** *Coal.* a black to dark brown coal that ranks above lignite and below bituminous coal, used mainly as fuel for steam-electric power generation. It has a higher carbon content and lower moisture content than lignite. The heat content of subbituminous coal generally ranges from 16 to 24 million Btu per ton on a moist, mineral-matter-free basis.

**subcooler** *Refrigeration.* a condenser device that improves the energy efficiency of a chiller by reducing the temperature of the condensed refrigerant liquid.

**subcooling** *Refrigeration.* **1.** the process of creating a decrease in temperature by removing sensible heat from a refrigerant liquid. **2.** see SUPERCOOLING.

**subcritical** *Nuclear.* the condition of a nuclear reactor system when the rate of production of fission neutrons is lower than the rate of production in the previous generation, owing to increased neutron leakage and poisons. Thus, **subcriticality.**

**subcritical mass** *Nuclear.* an amount of fissile material that by its mass or geometry is incapable of sustaining a fission chain reaction.

**subcritical reactor** *Nuclear.* a nuclear reactor in which the number of fissions decreases over time.

**subduction zone** or **boundary** *Earth Science.* a region along which one lithospheric plate descends into the earth's mantle relative to another plate.

**sublimate** *Chemistry.* **1.** to carry out a process of sublimation. **2.** a product obtained from sublimation.

**sublimation** *Chemistry.* **1.** the phase change of a material directly from a solid to a gas without passing through an intermediate liquid phase. **2.** the reverse of this same process.

**submergence** *Earth Science.* **1.** the fact of a site being submerged under water, especially the permanent flooding of land upstream from a dam. **2.** a change in the level of water in relation to the land, so that formerly dry land becomes inundated due to sinking of the land or a rise in water level.

**submersible** *Oil & Gas.* describing oil drilling equipment that can be submerged; see SEMI-SUBMERSIBLE TECHNOLOGY.

**subsidence** *Earth Science.* **1.** a local mass movement in which a portion of the earth's surface gradually settles downward or is displaced vertically, with little or no horizontal movement. **2.** a descent of air in the atmosphere, usually over a wide area. **3.** *Mining.* the sinking of strata, including the surface, as a result of underground excavations such as coal mining or crude oil extraction.

**subsidy** *Economics.* money paid, usually by a government, to either a producer or consumer of a certain product or service that alters its production cost or selling price to the benefit of the recipient; e.g., payments made to farmers to grow or not grow certain crops.

**subsistence farming (agriculture)** *Ecology.* the practice of farming in which the yield is only sufficient to feed the farmer and his family, with little or no surplus that could be sold or traded to others for a profit.

**subsistence fishery (fishing)** *Ecology.* the practice of catching fish (or harvesting other aquatic organisms) for the purpose of immediate household consumption, rather than for sale to others.

**subsonic** *Physics.* **1.** below the speed of sound (about 331 meters per second at standard temperature and pressure). **2.** describing an aircraft or other body moving at less than this speed.

**substrate** *Materials.* the physical support material on which an integrated circuit is constructed or to which it is attached; e.g., a photovoltaic cell.

**subsurface mining** see UNDERGROUND MINING.

**sub-synchronous** *Physics.* describing a device or system that operates at a lower rate relative to something else; e.g., a machine that operates at a speed below that of its power source.

**suburbanization** *Social Issues.* the fact of becoming suburban; a population shift from a large central city to smaller nearby communities. This has significant implications for energy use; e.g., increased dependence on private vehicles for transportation.

**subway** *Transportation.* an underground rail transport system, or a train that is part of such a system. In British English the term *tube* or *underground* is usually used for this, and *subway* refers to an underground passageway, as beneath an urban street.

***subway*** *Variously known as the MTA, MBTA, or simply the T, the Boston subway began service in 1897 and is generally considered North America's oldest subway system.*

**sucker rod** *Oil & Gas.* one of a string of steel rods that connect a downhole oil-well pump to a pumping unit on the surface.

**sucrose** *Materials.* $C_{12}H_{22}O_{11}$, table sugar, a substance in the form of combustible, hard, dry, white crystals; extensively used as a sweetener in foods and beverages.

**suction** *Physics.* **1.** a force that produces a condition of reduced pressure in a given region, so that a fluid in this region will flow to an adjacent region of greater pressure, or a rigid body will be held in place on a surface. **2.** the movement of a fluid, or adherence of a body to a surface, produced by this condition.

**suction head** *Hydropower.* additional energy in a hydropower system, obtained when the vacuum created by a closed outlet system exerts a pulling force on the turbine as the water leaves the system.

**suction pump** *Conversion.* a type of pump in which atmospheric pressure pushes the fluid to be raised into a partial vacuum created under a retreating valved piston on the upstroke, while a valve in the pipe prevents return flow; often used for raising or moving water.

**Suess, Hans** 1909–1993, U.S. chemist who developed an improved method of carbon-14 dating and used it to document that the burning of fossil fuels had a profound influence on the earth's stocks and flows of carbon. (Fossil fuels are so ancient that they contain no C-14.)

**Suess effect** *Climate Change.* a relative change in the ratio of C-14/C or C-13/C for a carbon pool or reservoir; this indicates the addition of fossil fuel $CO_2$ to the atmosphere.

**Suez Crisis** *History.* a military conflict that pitted Egypt against an alliance between France, the United Kingdom, and Israel. The Suez Canal was the principal route for oil shipments from the Middle East to Europe. In 1956, Egypt blockaded the Gulf of Aqaba, closed the Suez canal to Israeli shipping, and announced the nationalization of the canal. This resulted in a sharp increase in the price of oil.

**Suezmax** *Oil & Gas.* Suez maximum; the largest size of fully loaded oil tanker that can navigate the Suez Canal.

**sugar** *Chemistry.* **1.** a family of simple, often sweet, compounds consisting of carbon, hydrogen, and oxygen; obtained from the juice of many plants and particularly from sugarcane and sugar beets. **2.** specifically, SUCROSE, a particular type of this compound.

**sugarcane** or **sugar cane** *Biomass.* a tall grassy plant, *Saccharum officinarum*, characterized by a stout, jointed stalk and cultivated for its sugar content; native to tropical and subtropical regions. Also widely used as a source for biomass fuel production.

**sulfate aerosol** *Climate Change.* particulate matter that consists of compounds of sulfur, formed by the interaction of sulfur dioxide and sulfur trioxide with other compounds in the atmosphere. These aerosols are injected into the atmosphere from the combustion of fossil fuels and the eruption of volcanoes such as MOUNT PINATUBO. Recent studies indicate that sulfate aerosols may lower the earth's temperature by reflecting away solar radiation.

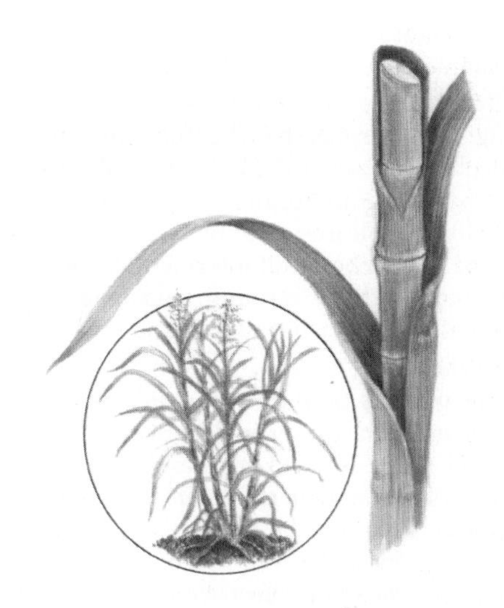

***sugarcane*** *Sugarcane is a food crop and also a widely used source of biomass for fuel.*

**sulfation** *Materials.* **1.** a process in which large crystals of lead sulfate form on a battery plate; a condition that affects unused and discharged batteries, making them difficult to recharge. **2.** any process of formation of lead sulfate.

**sulfur** *Chemistry.* a nonmetallic element having the symbol S, the atomic number 16, and an atomic weight of 32.06; pure sulfur exists in two stable crystalline forms and in at least two liquid forms. The native form of this element is a yellow mineral occurring as thick crystals and as granular to powdery masses, found in volcanic or hot springs deposits, in sedimentary beds, and in salt domes. Sulfur is present at various levels of concentration in many fossil fuels.

**sulfur content** *Oil & Gas.* a categorization of commercial fuels based on the amount of sulfur they contain, with lower sulfur content fuels usually selling at a higher price; e.g., no. 2 distillate fuel is rated as having a 0.05% or lower sulfur content level for on-highway vehicle use, or greater than 0.05% for off-highway use and home heating oil.

**sulfur cycle** *Earth Science.* the cyclic movement of sulfur in different chemical forms; a complex sequence of reactions brought about by bacteria in water and soil, in which sulfur is changed from organic sulfur compounds in plants and animals to elemental sulfur and sulfates, and then eventually returned to organic sulfur.

**sulfur dioxide** *Chemistry.* $SO_2$, a toxic, irritating, colorless gas that is soluble in water and alcohol; freezes at –72.7°C and boils at –10°C; an oxidizing and reducing agent formed naturally by volcanic activity and organic decay. Used as a chemical intermediate, in paper pulping and ore refining, and as a solvent. See also SULFUR DIOXIDE POLLUTION, SULFUR OXIDES.

**sulfur dioxide pollution** *Environment.* a process in which sulfur dioxide ($SO_2$) dissolves in water vapor to form acid, and interacts with other gases and particles in the air to form products that can be harmful to humans and the environment. Over 65% of $SO_2$ released to the air comes from electric utilities, especially those that burn coal. Other sources include petroleum refineries, cement manufacturing plants, or metal processing facilities. $SO_2$ contributes to respiratory illness and aggravates existing heart and lung diseases.

**sulfur hexafluoride** *Chemistry.* $SF_6$, a colorless gas that is slightly soluble in water and soluble in alcohol; it freezes at –50.5°C and sublimes at –63.8°C; used as a dielectric in electronics. One of the six greenhouse gases to be curbed under the Kyoto Protocol, it has the highest known GWP (Global Warming Potential of any gas (23,900 times that of carbon dioxide over a 100-year horizon).

**sulfuric** *Chemistry.* **1.** describing various compounds of sulfur, especially those in which the element has a valence of 6. **2.** relating to or derived from sulfuric acid.

**sulfuric acid** *Chemistry.* $H_2SO_4$, a highly corrosive, dense, oily liquid, colorless to dark brown depending on its purity; its addition to water generates heat and explosive spattering. Used in the manufacture of a wide range of chemicals and materials including fertilizers, paints, detergents, and explosives.

**sulfurous** *Chemistry.* describing various compounds of sulfur, especially those in which the element has a valence of 3.

**sulfur oxides** *Environment.* compounds made up of only sulfur and oxygen. Natural sources include volcanoes and hot springs, while energy use and metal refining are the principal human sources, particularly for sulfur dioxide, a major air pollutant (see above). On a global basis, natural sources contribute about the same amount of sulfur oxides to the atmosphere as human industrial activities.

**sulfur trioxide** *Chemistry.* $SO_3$, a compound occurring as silky, fibrous needles; it decomposes in water, melts at 16.8°C and boils at 44.8°C; a very toxic and a strong irritant that is used in detergents and solar energy collectors. See also SULFUR OXIDES.

**sulphur** another spelling of SULFUR.

**Summerland** *History.* an oil field off the coast of Santa Barbara, California, considered to be the first productive offshore oil facility in the world (1898). Similar offshore wells were reported in the Caspian Sea about five years later.

**sump** *Mining.* **1.** a basin or pit in a mine where water is allowed to collect, as at the bottom of a shaft. **2.** *Environment.* any pit, pool, or other depression in which water collects, especially waste water.

**sun** *Solar.* **1.** also, **Sun.** the star that is the central celestial body in the solar system and the ultimate source of all energy for the earth. **2.** any star that is the source of light and heat for a planetary system.

**Sun Day** *Solar.* a symposium, demonstration, or other event focusing on the use of solar energy as a replacement for fossil fuel sources.

**sundial** *Solar.* a timepiece employed since ancient times to indicate daylight hours by the shadow that the gnomon (a rod or fin) casts on a calibrated dial.

**sunk cost** *Economics.* a cost that has already been incurred, and therefore cannot be avoided by any strategy going forward.

**sunlight** *Solar.* the electromagnetic energy that is visible to the human eye at the earth's surface.

**sun motor** *History.* an early solar energy system developed by Swedish-American engineer John Ericsson (about 1883); it consisted of a displacer type (Stirling) engine powered by a parabolic reflector. Ericsson's system was not adopted commercially at the time, but his parabolic collector became the model for many modern solar systems.

**sun path diagram** *Solar.* a circular projection of the sky vault onto a flat diagram, used to determine solar positions and the shading effects of landscape features for a solar energy system.

**sunroom** another term for a SOLARIUM.

**sunshine duration** *Solar.* the sum of time intervals within a given period, during which the irradiance from direct solar radiation on a plane normal to the sun direction is equal to or greater than 120 watts per square meter.

**sunspace** *Solar.* a type of passive solar heating in which radiant energy from the sun is collected and stored in a space separate from the living space, and is transferred either by natural convection or by fans.

**sunspot** *Solar.* a relatively cool and dark area that can be observed on the sun's photosphere, characterized by a strong magnetic field; sunspots are cyclically variable in number and magnetic polarities. It has been speculated that variations in sunspot activity may affect weather and climate on earth.

**sun-tempered** *Solar.* **1.** describing a building that is specifically designed for high solar energy efficiency; e.g., its longer wall is oriented east to west and the majority of its windows are on the south side. **2.** specifically, a house having south-facing glass making up 7–12% of the total floor area of the house. Thus, **sun-tempering.**

**super cetane** *Biomass.* a biomass-derived product used as a diesel fuel or a diesel additive to improve engine performance; so called because it has a cetane value as high as 100.

**supercapacitor** see ULTRACAPACITOR.

**supercharge** *Transportation.* to force air into an internal-combustion engine at a pressure (significantly) above the atmospheric pressure at the start of the compression stroke; the degree of pressure retained determines the level of supercharging.

**supercharger** *Transportation.* a blower that increases the intake pressure of an engine, so as to make fuel burn more quickly and increase engine power.

**supercomputer** *Communication.* a term for any extremely powerful, large-capacity computer that is capable of processing huge amounts of data in an extremely short time.

**superconducting** *Electricity.* relating to or having the capacity for superconductivity (see below). Also, **superconductive.**

**superconducting magnetic energy storage (SMES)** *Storage.* a technology in which the superconducting characteristics of low-temperature materials produce intense magnetic fields to store energy; proposed as a storage option in photovoltaics to smooth out fluctuations in power generation.

**Superconducting Super Collider (SSC)** *Nuclear.* a massive ring particle accelerator that was planned to be built by the U.S. government near Waxahachie, Texas. Due to concern over the high cost of the project, it was eventually canceled by Congress in 1993 after 22.5 km (14 mi) of its tunnel had already been excavated and $2 billion spent.

**superconductivity** *Electricity.* a phenomenon shown by certain metals, alloys, and other compounds of having negligible resistance to the flow of electric current at temperatures approaching absolute zero. Each material has a critical temperature $T_c$, above which it is a normal conductor, operating as a superconductor only under extreme low-temperature conditions. In addition, certain materials are now known to exhibit superconductivity at temperatures well above absolute zero.

**superconductor** *Electricity.* any material that can exhibit superconductivity (see previous).

**supercooling** *Thermodynamics.* a condition in which a pure substance is cooled below its freezing, condensation, or sublimation point, without experiencing the corresponding change of state that this would normally bring about. Thus, **supercooled.**

**supercritical** *Chemistry.* describing a mobile phase of a substance intermediate between a liquid and a vapor, maintained at a temperature greater than its critical point.

**supercriticality** *Chemistry.* **1.** the fact of being supercritical; i.e., at a temperature above the critical point although no phase change is observed. **2.** *Nuclear.* a condition for increasing the level of operation of a reactor, in which the rate of fission neutron production exceeds all neutron losses, and the overall neutron population increases. Thus, **supercritical reactor.**

**supercritical mass** *Nuclear.* a mass of fissionable material in excess of a critical mass, i.e., an amount of fissionable material greater than that needed to maintain a chain reaction with a constant rate of fission.

**supercritical reactor** *Nuclear.* a nuclear reactor in which the number of fissions increases from one generation to the next generation and the energy released by the chain reaction increases with time.

**supercritical turbine** *Conversion.* a steam turbine in which the feed steam is above the critical point.

**supercritical water-cooled reactor (SCWR)** *Nuclear.* an advanced nuclear reactor design in the development stage that features a high-temperature, high-pressure water-cooled reactor that operates above the thermodynamic critical point of water. It enables a thermal efficiency about one-third higher than current light-water reactors, as well as simplification in the balance of the plant.

**superfluid** *Physics.* a collection of particles that exhibit zero viscosity and zero entropy (roughly speaking, a complete lack of friction); the particles are known to obey the BOSE-EINSTEIN STATISTICS. Thus, **superfluidity.**

**Superfund** see CERCLA.

**superheat** *Thermodynamics.* **1.** to bring a substance to a temperature above that at which a change of state would normally occur, without this change taking place; e.g., to heat a liquid above its normal boiling point without producing vaporization. **2.** the temperature of a substance in this state. **3.** see SUPERHEATED STEAM.

**superheated steam** *Geothermal.* steam heated to a temperature significantly higher than the boiling point corresponding to its pressure, so that it does not recondense to water as readily as ordinary steam; it cannot exist in contact with water or contain water. A natural feature of certain geothermal areas and also created artificially to provide increased efficiency in steam engines.

**superheated vapor** *Thermodynamics.* a vapor that is heated to a temperature exceeding the boiling point of the liquid phase.

**superheating** *Thermodynamics.* **1.** a condition in which a pure substance is heated above its boiling point, melting point, or sublimation point, without experiencing a phase change. **2.** a process in which such a condition is induced as part of a refrigeration cycle.

**superinsulated** or **super-insulated** *Consumption & Efficiency.* describing a type of building that has massive amounts of insulation, airtight construction, controlled ventilation, and other features that maximize the efficiency of its inherent climate control. Thus **superinsulation.**

**superphosphate** *Materials.* a mixture of calcium sulfate and dihydrogen calcium phosphate made by treating bone ash or basic slag with sulfuric acid; used as an agricultural fertilizer. Its development as the first artificial fertilizer (1850s) freed farmers from their historic dependence on animal manure for fertilizer.

**superposition** *Physics.* a principle that may be applied to systems in which individual influences act linearly; the resultant effect on the system is equivalent to the sum of the effects of the individual influences that are acting on the system.

**supersaturation** *Chemistry.* **1.** a condition in which a solution contains more solute than is normally necessary to achieve saturation under the same conditions. **2.** *Hydropower.* such an overabundance of gases in turbulent water (e.g., nitrogen), as at the base of a dam spillway; this can cause a fatal condition in fish similar to the bends in humans. **3.** *Earth Science.* a local atmospheric condition in which the relative humidity is greater than 100%, with more water vapor than is needed to produce saturation of a plane surface of pure water or ice.

**supersonic** *Physics.* **1.** above the speed of sound (about 331 meters per second at standard temperature and pressure). **2.** describing an aircraft or other body moving at greater than this speed. **3.** another term for ULTRASONIC.

**superstrate** *Photovoltaic.* a covering on the sun side of a photovoltaic module, providing protection for the materials from impact and environmental degradation while allowing maximum transmission of the appropriate wavelengths of the solar spectrum.

**super-synchronous** *Physics.* describing a device or system that operates at a higher rate relative to something else; e.g., a machine that operates at a speed above that of its power source.

**supertanker** *Oil & Gas.* a very large oil tanker, typically defined as a vessel designed to transport more than 500,000 deadweight tons of oil. See below.

**supplementarity** *Policy.* in the context of the Kyoto Protocol, the use of flexibility mechanisms such as emissions trading to achieve greenhouse gas reduction goals, while also instituting adequate domestic energy and other policies for the same purpose.

**supply bid** *Electricity.* a bid to an electric power exchange indicating a price at which a seller is prepared to sell energy or ancillary services.

**supertanker** A term originally applied to the class of tankers too large to transit international canals while carrying cargo, and currently defined by two ship classes: Very Large Crude Carriers (VLCCs) between ~200,000 and ~300,000 deadweight tons (dwt) and Ultra Large Crude Carriers (ULCCs) greater than ~300,000 dwt. Supertankers are a remarkable technological response to market conditions that promoted economies of scale without apparent bound in the 1960s, 70s, and 80s. The first modern tanker (tanks integral with hull) was the ~3000 dwt *Glukauf*, built in 1886. Until the 1950s, most crude oil was refined at source and transported to market in products tankers, sized between 12,000 and 30,000 dwt. Larger vessels became economically feasible when oil companies began locating refineries near energy markets, although the Suez Canal restricted tanker size. Energy market shifts and the 1956 closure of the Suez Canal created new routes, removing geographic barriers to construction of the first VLCCs and ULCCs. Fully-loaded supertankers (especially efficient diesel-powered VLCCs built in 1990s) reduced unit shipping costs dramatically, but partial loads could not sustain economies of scale; many were scrapped in the 1980s and 1990s or used for storage. The largest supertanker ever built was the 555,843 dwt *Seawise Giant*, refitted in 2004 as a floating storage and offloading unit named the *Knock Nevis*. Current supertanker sizes are defined by market conditions providing an economic upper bound and geophysical limitations defining the number of routes and ports (or offshore terminals) that these very large vessels can safely serve.

**James J. Corbett**
University of Delaware

**supply elasticity** *Economics.* the response of supply to a change in the price of a commodity.

**supply-side** *Economics.* **1.** having to do with the supply of energy resources to consumers. **2.** relating to or based on the concept that the proper way for a government to stimulate economic growth is by encouraging increases in supply, as through tax cuts and incentives for producers. Compare DEMAND-SIDE.

**suppressed demand** *Economics.* a situation in which the demand for energy services is low due to constraints of income, access, and infrastructure, not because of an absence of consumer interest.

**suppressor** see SURGE SUPPRESSOR.

**surface drift** *Earth Science.* **1.** loose surface material transported from one place and deposited elsewhere by the action of such agents as wind, waves, currents, glaciers, or running water from a glacier. **2.** the flow of waters on the surface of the ocean. **3.** *Mining.* a lateral tunnel (usually inclined) from the surface to a coal seam or ore body to be developed.

**surface energy** *Materials.* the work per unit area required to bring fluid molecules to the interface between two immiscible liquids, or between a liquid and a gas.

**surface (boundary) layer** *Earth Science.* a thin layer of air, extending from near the earth's surface up to the base of the EKMAN CONVERGENCE (less than 300 feet), within which shearing stresses are nearly constant and wind distribution is determined largely by the vertical temperature gradient and the nature and contours of the underlying surface.

**surface mining** *Mining.* a method of mining that involves extraction by removing the layer of earth and/or rock (overburden) at the surface to gain access to a coal bed that is at a relatively shallow depth; the coal is then mined with surface excavation equipment. Thus, **surface mine.**

**surface pipe** *Oil & Gas.* the first casing set into a well, cemented into place as a foundation for subsequent operations and as a means of shutting off shallow water formations from being contaminated by deeper, saline waters.

**surface tension** *Chemistry.* the stretching force required to form a liquid film; it is equal to the surface energy of the liquid per unit length of the film at equilibrium; the force tends to minimize the area of a surface.

**surface warming** *Climate Change.* a process of gradual warming of the earth's surface, caused at least in part by temperature increases resulting from the buildup of greenhouse gases in the atmosphere.

**surge** *Electricity.* a sudden unplanned change in an electrical system's voltage that is capable of damaging electrical equipment, especially an increase in voltage significantly above the designated level; e.g., above 120 volts for U.S. household and office wiring.

**surge capacity** *Electricity.* the maximum power, usually 3–5 times the rated power, that can be provided to an electrical system over a short time without damage to the system.

**surge suppressor** *Electricity.* a component that responds to the rate of change of a current or voltage in order to prevent damage from a sudden fluctuation in power, especially a large increase above a predetermined value; often used to protect computer systems and other electronic equipment. Also, **surge protector.**

**survival of the fittest** *Ecology.* **1.** a popular term for NATURAL SELECTION, the concept that individuals best adapted to their environment will survive to pass on their genes. **2.** *Social Issues.* the application of this principle to human societies; see SOCIAL DARWINISM.

**suspension** *Chemistry.* **1.** a system in which very small particles of solid, semisolid, or liquid material are more or less evenly dispersed in a liquid or gas phase; e.g., fog is a suspension of liquid (water droplets) in a gas (air). **2.** *Transportation.* a system of springs, shock absorbers, or similar devices connecting the axles to the chassis of an automobile, railroad car, or other such vehicle; designed to reduce unwanted motion transmitted from the riding surface.

**sustainability** *Sustainable Development.* the fact of being sustainable; preservation of the overall viability and normal functioning of natural systems.

**sustainable** *Sustainable Development.* **1.** describing activities that make use of the earth's living and physical resources, including humans and their technologies, cultures, and institutions, in a way that does not diminish their ability to support future generations. **2.** specifically, the consumption of energy in a manner that emphasizes renewable sources and the judicious use of non-renewable sources.

**sustainable agriculture** see TRADITIONAL AGRICULTURE.

**sustainable capacity** *Oil & Gas.* the daily amount of oil that an individual oilfield or a group of fields can produce at a rate that can be sustained for more than 90 days.

**sustainable development** a collective term for efforts to develop technological, economic, political, and social systems, so as to provide the goods, services, and amenities that people need or value, at an acceptable cost, while at the same time maintaining the natural environment so that a comparable quality of life will be available to future generations.

**sustainable development indicator** *Sustainable Development.* an indicator that measures or monitors progress toward sustainable development; e.g., $CO_2$ emissions per capita, participation in international environmental treaties, or the equity of income distribution.

**sustainable energy** *Sustainable Development.* energy that is produced and used in ways that will support long-term human development in all its social, economic, and environmental dimensions.

**sustainable tourism** *Sustainable Development.* tourism that suits the needs of current tourists and host regions, while at the same time protecting and enhancing similar opportunities for the future.

**sustaining chain reaction** *Nuclear.* a chain reaction in which an average of exactly one fission is produced by the neutrons released by each previous fission.

**SUV** see SPORT UTILITY VEHICLE.

**Sv** sievert.

**Swan, Joseph** 1828–1914, English inventor noted for his development of the light bulb. He received a British patent for his device in 1878, about a year before Edison's U.S. patent. He later teamed with Edison for the commercial development of electric lamps, under the brand name **Ediswan.**

**Swan lamp** *History.* a pioneering electric lamp developed by Joseph Swan, distinguished by having too small an amount of residual oxygen in the vacuum tube to ignite the filament, thus allowing it to glow almost white-hot without catching fire.

**swap** see COMMODITY SWAP.

**sweet** *Oil & Gas.* having a low level of the acidic gases that are typically associated with oil and gas drilling and production, such as hydrogen sulfide or carbon dioxide. Thus, **sweet gas, sweet crude.**

**sweetening** *Oil & Gas.* the process of removing unwanted gas components from oil and natural gas; e.g., hydrogen sulfide.

**swept area** *Wind.* the area through which the rotor blades spin on a wind energy device, as seen when directly facing the center of the rotor blades.

**swidden (agriculture)** another term for SLASH-AND-BURN.

**switch** *Electricity.* a device that is used to open or close an electric circuit.

**switched-reluctance motor** *Conversion.* a motor in which motion is produced as a result of the variable reluctance in the air gap between the rotor and the stator.

**switchgrass** *Biomass.* a warm-season perennial grass native to North America, *Panicum virgatum,* that has potential as a biomass energy crop.

**symbiosis** *Ecology.* a biological relationship between individuals of two different species in which one member benefits and the other may or may not benefit. Thus, **symbiotic.**

**Symington, William** 1763–1831, English engineer who developed the first steam-powered marine engine (1788). He then developed a successful steam-driven paddle wheel and used it to propel one of the first practical steamboats, the *Charlotte Dundas.*

**synchro-** short for SYNCHRONOUS.

**synchrocyclotron** *Nuclear.* a particle accelerator similar to the cyclotron, in which the frequency of the accelerating voltage is varied in order to track the change in relativistic energy of the particles.

**synchronous** *Physics.* **1.** occurring at the same time or rate; e.g., an artificial satellite with a one-day orbital period that is the same as that of the earth. **2.** *Electricity.* describing a machine operating at the same speed as the existing power supply. Thus, **synchronous motor, synchronous generator, synchronous speed,** and so on. **3.** *Earth Science.* describing geologic features or formations that exist or are deposited at the same time.

**synchrotron** *Nuclear.* a large ring particle accelerator in which the particles move in an evacuated tube at constant radius, accelerated by radio frequency applications with synchronous magnetic field increases to maintain the constant radius.

**syncline** *Earth Science.* a downward fold of stratified rocks in which the sides slope inward toward each other.

**syncrude** *Oil & Gas.* synthetic crude; oil processed from unconventional sources such as oil sands.

**synfuel** short for SYNTHETIC FUEL.

**syngas** *Oil & Gas.* synthesis gas; a mixture of hydrogen and carbon monoxide obtained by the reforming of methane, used as a fuel and as an intermediate in the production of various chemicals.

**synoptic** *Measurement.* **1.** referring to or based on data obtained from a large-scale or overall view. **2.** specifically, referring to weather data obtained simultaneously over a large area in order to provide a simultaneous overall view.

**synthesis** *Chemistry.* **1.** the process of forming chemical compounds from more elementary substances by means of one or more chemical reactions, or by nuclear change. **2.** any unified whole formed by combining constituent elements.

**synthesis gas** see SYNGAS.

**synthetic** *Materials.* **1.** describing a material or item produced by synthesis, especially a compound formed artificially by chemical synthesis. **2.** more generally, describing any product or item that is the result of human technology rather than something that exists in nature in that form.

**synthetic crude** see SYNCRUDE.

**synthetic fuel** *Renewable/Alternative.* a liquid or gaseous fuel derived from a source such as coal, shale oil, tar sands, or biomass, used as a substitute for oil or natural gas. See next page.

**synthetic natural gas (SNG)** *Oil & Gas.* a manufactured product, chemically similar

**synthetic fuel** A generic term applied to any manufactured fuel with the approximate composition and comparable specific energy of a natural fuel. In the broadest definition, a liquid fuel that is not derived from natural occurring crude oil is a synthetic fuel. Modern transportation fuels demand uniform physical properties produced from varying feed stocks with the chemical compositions essentially "synthesized" from petroleum or other fossil fuels. The fraction of petroleum that has boiling points for use in spark-ignition engine is referred to as naphtha. The term "gasoline" implies that many of the components have been synthesized using cracking or reforming techniques, and the term "gasoline" is used rather than synthetic fuel. The term "reformulated gasoline" identifies gasoline containing a greater fraction of reformed petroleum molecules. Fuels from oil sands and heavy oil (naturally de-volatilized petroleum) are referred to using the same terminology as petroleum-based fuels. Liquid fuels produced from coal, peat, natural gas, and oil shale are properly referred to as synthetic fuels. Renewable biomass produced by photosynthesis can be converted to a variety of synthetic fuels. Ethanol, methanol, and biodiesel are appropriately referred to as synthetic fuels independent of origin. Coal gasification and natural gas reforming are sources of "synthesis gas", a mixture of hydrogen and carbon monoxide, the feed for synthetic fuel production. The SASOL complex in South Africa gasifies coal to feed a Fischer-Tropsch process to obtain high quality motor fuels. A range of oxygenated organic chemicals for fuels and commerce are produced with selected Fischer-Tropsch catalysts and adjusted operating conditions.

**Truman Storvick**
University of Missouri

in most respects to natural gas, resulting from the conversion or reforming of petroleum hydrocarbons; may easily be substituted for or interchanged with pipeline-quality gas.

**Système International** *Measurement.* French for the INTERNATIONAL SYSTEM OF UNITS; the official name for this system; abbreviated SI.

**Szargut, Jan** born 1923, Polish engineer noted for his advances in the theory of energy balances of chemical processes, and for important studies of the exergy analysis of thermal, chemical, and metallurgical processes.

**Szilard, Leo** 1898–1964, Hungarian-born physicist who was one of the first to realize that nuclear chain reactions could be used in weapons. He worked with Fermi to develop the first self-sustained nuclear reactor based on uranium fission, and he helped convince Einstein to write to President Roosevelt about the need for an atomic bomb project. After World War II, he actively protested nuclear weaponry.

**Tabor, Harry** Israeli physicist who led his nation's solar industry to international prominence and became one of the world's most respected pioneers in the development of practical applications of solar energy.

**tachometer** *Measurement.* an instrument used to measure the rotational speed of a shaft or machine, as of an engine.

**tachymetabolic** *Biological Energetics.* describing an animal whose metabolism remains at a high level, producing calories and maintaining bodily activity at a high pace. Compare BRADYMETABOLIC.

**tail** see TAILRACE; TAILWATER.

**tail gas** *Oil & Gas.* a term for the lightest hydrocarbon gas released from a refining process.

**tailings** *Mining.* **1.** any of various forms of mining waste, such as refuse from processed ore or the decomposed outcrop of a bed or vein. **2.** see URANIUM TAILINGS.

**tailrace** or **tail race** *Hydropower.* a water outlet that carries water away from a hydropower site (e.g., a water wheel or water turbine), after the water has flowed through or past the energy-conversion device.

**tailwater(s)** *Hydropower.* the water downstream from a dam or other such hydropower site.

**Takayanagi, Kenjiro** 1899–1990, Japanese inventor who was among the leading developers of modern television technology. In 1926 he succeeded in displaying a clear image of a Japanese character on a Braun electric tube; he subsequently demonstrated a cathode-ray tube system and built an all-electronic television set.

**take-back effect** another term for REBOUND EFFECT.

**take-or-pay** *Economics.* a provision in a fuel supply contract that requires the purchaser to pay the supplier for a certain percentage of the fuel under contract, regardless of whether the purchaser actually takes possession of the fuel, uses it, or resells it.

**Talking Telegraph** *History.* the name for a predecessor to the telephone developed by Italian–American inventor Antonio MEUCCI, independently of the device patented by Alexander Graham Bell.

**tallow** *Lighting.* a candle material made from the solid, relatively hard fat of cattle, sheep, or other animals that has been rendered from the surrounding tissue.

**TAME** *Oil & Gas.* tertiary amyl methyl ether; an oxygenate that can be used in reformulated gasoline.

**tandem accelerator** *Nuclear.* a particle accelerator that derives its name from the two stages of acceleration that particles undergo; in one version, heavy negatively charged ions are accelerated through one potential difference before being stripped of more electrons and accelerated again in a system using two or more Van de Graaff generators.

**tanker** *Oil & Gas.* a vessel designed to carry a liquid product, especially one that carries crude oil or other petroleum oil products. Very large oil tankers are by a wide margin the largest vessels ever built.

***tanker*** *Oil tanker in the Gulf of Alaska.*

**tank farm** *Oil & Gas.* an installation used to store crude oil.

**tankless water heater** *Consumption & Efficiency.* a contemporary type of water heater that heats water immediately before it is distributed for end use as required, thus eliminating the standby loss of heat that characterizes a conventional storage tank system. Cold water travels through a pipe into the unit, and either a gas burner or an electric element heats the water only when needed.

**Tansley, Arthur** 1871–1955, British botanist who was a pioneer in the science of plant ecology; he was the first to use the term "ecosystem" in a scientific publication (1935).

**tantalum** *Chemistry.* a chemical element that is a rare metal of the vanadium family, having the symbol Ta, the atomic number 73, an atomic weight of 180.95, a melting point of 299°C, and a boiling point of 5560°C; a malleable gray solid that is resistant to corrosion, used for electric light filaments and as a catalyst.

**Tapline** *Oil & Gas.* Trans-Arabian Pipeline; a pipeline linking oil fields in Saudi Arabia with the Mediterranean Sea at Sidon, Lebanon, a distance of over 1000 miles. It greatly reduces the number of tankers required to transport oil around the Arabian Peninsula through the Suez Canal to the Mediterranean.

**TAPS** Trans-Alaska Pipeline.

**tar** *Materials.* **1.** a dark, viscous, usually pungent substance derived from the destructive distillation of various organic compounds such as wood, coal, or shale; used for various industrial purposes. **2.** a naturally occurring hydrocarbon substance resembling this, such as bitumen.

**tar acid** *Materials.* any of numerous very weakly acidic substances, such as phenol, cresols, and xylenols, that are separated during the industrial fractionation of coal tar.

**tarcrete** *Oil & Gas.* a term for oil deposits on the ground surface that combine with materials such as soot, gravel, and sand to form a hardened, concrete-like layer; this can occur after an oil well fire.

**target organ** *Health & Safety.* the organ that is most likely to be the affected site of a toxic substance that has entered the body.

**tarmac** *Transportation.* tar and macadam; a paved surface consisting of a bituminous substance spread over a layer of rock or stone; used for airport runways.

**tar pit** *Oil & Gas.* a naturally occurring site at which subterranean asphalt has risen to the surface, creating a large pool (or pit) of dark, sticky oil-like matter; e.g., the **La Brea Tar Pits** of downtown Los Angeles, famous for containing the entrapped remains of extinct prehistoric mammals.

**tar sands** see OIL SANDS.

**tar seep** *Oil & Gas.* natural tar deposits near the ground surface that flow slowly through cracks in the earth or from between rocks, frequently forming pits or pools.

**task light(ing)** *Lighting.* focused light; a light source designed specifically to direct light toward a specific work activity performed by a person or machine. Compare AMBIENT LIGHT.

**TATs** thermally activated technologies.

**TCDD** *Health & Safety.* 2,3,7,8-tetrachlorodibenzo-p-dioxin, the most toxic of the DIOXIN family of chemical compounds, a carcinogenic, teratogenic, and mutagenic substance present in defoliants such as **Agent Orange,** which was used in the Vietnam War and has been the subject of investigation (and controversy) concerning its effects on U.S. and Vietnamese personnel who were exposed to it.

**TDEE** total daily energy expenditure.

**Teapot Dome** *History.* the site in Wyoming of a valuable Federal oil reserve; the source of a notorious scandal in the 1920s in which Albert B. Fall, U.S. Secretary of the Interior, was convicted of accepting bribes in the form of "loans" from oil barons Edward Doheny and Harry Sinclair, prior to granting them exclusive rights to the Teapot Dome field and another valuable field at Elk Hills, California.

**technetium** *Chemistry.* an element having the symbol Tc, the atomic number 43, an atomic weight of its isotope with longest half-life (Tc-98) of 97.90; it does not occur naturally on earth but is obtained by deuteron bombardment of molybdenum and in fission products of uranium and plutonium; used in cryochemistry and nuclear medicine.

**technical potential** *Consumption & Efficiency.* the achievable energy savings that result from introducing the most energy-efficient technology at a given time, without taking into account the costs of introduction or the life of the equipment to be replaced.

***Teapot Dome*** *Oil drilling operations near Teapot Dome, Wyoming.*

**Technocracy** *History.* a school of thought originating in the U.S. in the 1930s, arguing that the nation could be rescued from the Great Depression if politicians were replaced by scientists and engineers having the technical expertise to manage the nation's economy and natural resources. Technocracy used growth and decline curves to predict a wide range of societal trends. Thus, **Technocrat.**

**technoeconomic** *Economics.* involving an evaluation of both the technical and economic aspects of something (e.g., resource flows in society, communication networks and services), especially as they relate to one another. Thus, **technoeconomic model.**

**technological optimism** *Consumption & Efficiency.* the idea that technological innovation will be able to overcome or ameliorate natural resource scarcity (especially the depletion of fossil fuels) and environmental degradation.

**technology** *Consumption & Efficiency.* **1.** in general, any use of objects by humans to do work or otherwise alter their environment. **2.** the application of scientific knowledge for practical purposes; the employment of tools, machines, materials, and processes to do work, produce goods, perform services, or carry out other useful activities. Thus, **technological.**

**technology innovation** *Consumption & Efficiency.* a process of invention, innovation, and diffusion by which greater and/or higher quality outputs can be produced using fewer inputs. See below.

**technology transfer** *Global Issues.* the broad set of processes encompassing the movement of knowledge, techniques, capital, and goods among different parties.

**technology innovation** Technology innovation is the process through which new (or improved) technologies are developed and brought into widespread use. In the simplest formulation, innovation can be thought of as being composed of research, development, demonstration, and deployment, although it is abundantly clear that innovation is not a linear process—there are various interconnections and feedback loops between these stages, and often even the stages themselves cannot be trivially disaggregated. Innovation involves the involvement of a range of organizations and personnel (laboratories, firms, financing organizations, etc.), with different institutional arrangements underpinning the development and deployment of different kinds of technologies; contextual factors such as government policies also significantly shape the innovation process. In the energy area, technology innovation has helped expand energy supplies through improved exploration and recovery techniques, increased efficiency of energy conversion and end-use, improved availability and quality of energy services, and reduced environmental impacts of energy extraction, conversion, and use. Most energy innovation is driven by the marketplace, although given the public goods nature of energy services (and reducing their environmental impacts), governments invest significantly in energy research and development programs as well as demonstration and early deployment of selected energy technologies. Still, most investments in energy innovation are targeted towards technologies with clear commercial applications and financial returns, with only marginal investments (at least in relation to the need) towards energy innovation for helping provide modern energy services to the two billion poor people worldwide who do not have access to such services.

**Ambuj Sagar**
Harvard University

**tectonics** *Earth Science.* **1.** a branch of geology that is concerned with the study of the features, deformational movements, and processes of the earth's crust. **2.** see PLATE TECTONICS.

**TEE** total energy expenditure.

**Tehachapi Pass** *Wind.* an area in central California between Barstow and Bakersfield, the traditional overland route from the west Mojave desert to the San Joaquin Valley and the site of a facility that is among the world's largest producers of wind energy.

**TEL** tetraethyl lead.

**telecommunications** *Communication.* **1.** any process that allows a person to communicate over distance by means of an electromagnetic system, with information in the form of sound, text, images, and so on; e.g., a telephone system. **2.** the industry involved with such transmissions.

**telecommuting** *Social Issues.* a contemporary work pattern in which a person works at home, rather than commuting over distance to a place of employment, and communicates with the workplace by means of a personal computer, Internet connection, and so on.

**teleconnection** *Earth Science.* a phenomenon in which weather patterns in one region significantly influence the weather conditions in a distant location, often over an extended time.

**telegraph** *Communication.* a device that transmits information over a distance by means of electricity; its basic form is a key to open and close the circuit and a battery-powered sounder to form the transmission code. The telegraph was the first device to allow immediate communication beyond the range of human sight and hearing, and was the dominant form of such technology in the second half of the 19th century. Modern forms of communication such as the fax machine are adaptations of this system. Thus, **telegraphic.**

**telegraph(ic) code** *Communication.* a system of symbols used for transmitting telegraph messages (such as Morse code); each character is represented by a group of long and short electrical pulses or pulses of opposing polarities, or by time gaps of equal length in which a signal is either present or absent.

**telegraphy** *Communication.* the use of a telegraph system; communication over a distance by means of code signals that are composed of electrical or electromagnetic pulses and that are sent over wires or by radio.

**Teller, Edward** 1908–2003, Hungarian-born physicist known for his contributions to the first demonstration of thermonuclear energy. He became known as "the father of the hydrogen bomb" for his role in the development of this weapon by the U.S. In the 1980s Teller became the leading advocate for the Strategic Defense Initiative, a proposed system for space-based missile defense.

**temper** *Materials.* **1.** to soften and thus toughen hardened steel by reheating it to some temperature below the decomposition temperature. **2.** to carry out a similar process with another material; e.g., glass. **3.** the process of treating a material in this manner. **4.** the degree of hardness and toughness of a metal.

**temperature** *Measurement.* a measure of the average kinetic energy of the molecules in a gas, liquid, or solid; the fundamental quantity measured by a thermometer. Temperature underlies the common notions of "hot" and "cold;" the material or environment with the higher temperature is said to be hotter.

**temperature coefficient** *Measurement.* **1.** the amount that voltage, current, or power output changes due to a change in temperature. **2.** specifically, the ratio of the change of electrical resistance in a wire when the temperature of the wire is increased by 1°C, relative to its resistance at 0°C.

**temperature gradient** *Thermodynamics.* **1.** a difference in temperature per unit distance within the same body of matter, giving rise to a heat transfer from the hotter region to the colder one. **2.** the rate of temperature change over any horizontal or vertical distance.

**temperature inversion** *Earth Science.* **1.** an atmospheric condition in which air temperature increases with altitude rather than the usual decrease; e.g., a situation in which temperature at the top of a mountain is higher than in the valley below. **2.** an anomalous condition in a body of water that causes the temperature of a water layer to increase rather than decrease with depth.

**temperature scale** *Measurement.* any of various standard methods of expressing temperature on the basis of certain reference parameters; e.g., the steam point of water at sea level. See below.

**temperature zoning** see ZONING.

**Tennessee Valley Authority (TVA)** a U.S. federal corporation (est. 1933) that is the nation's largest public power company. President Franklin D. Roosevelt created the TVA during the Depression as part of the New Deal. The TVA embarked on large-scale engineering projects that included dam construction and other water management efforts, to provide services such as electric power production, flood and erosion control, and malaria prevention, over a seven state region.

**temperature scales** Three temperature scales are in common use in science, industry, and everyday life. Two of those scales are SI-metric. The **degree Celsius** (°C) scale was devised by dividing the range of temperature between the freezing and boiling temperatures of pure water at standard atmospheric conditions (sea level pressure) into 100 equal parts, resulting in temperatures of 0°C and 100°C respectively. Temperatures on this scale were at one time known as degrees centigrade, however it is no longer correct to use that terminology. (In 1948 the official name was changed from "centigrade degree" to "Celsius degree" by the 9th General Conference on Weights and Measures (CGPM).) The **kelvin** (K) temperature scale is an extension of the degree Celsius scale down to *absolute zero* (0 K), a hypothetical temperature characterized by a complete absence of heat energy. The kelvin scale is related to the degree Celsius scale through the relationship: degrees Celsius = kelvin – 273.15 K. Temperatures on this scale are called **kelvins**, not degrees kelvin, kelvin is not capitalized, and the symbol (capital K) stands alone with no degree symbol. [In 1967 the new official name and symbol for "Kelvin" were set by the 13th CGPM.] The **degree Fahrenheit (°F)** non-metric temperature scale evolved over time so that the freezing and boiling temperatures of water are whole but not round numbers (32 °F and 212 °F). The zero value (0 °F) is of little real significance other than being a very cold mixture of ice, water, and salt.

**Don Hillger**
Colorado State University

*Tennessee Valley Authority (TVA)* *The Norris Dam on the Clinch River was built in 1933–36 as a major project of the TVA.*

**tensile** *Materials.* **1.** of or relating to tension. **2.** capable of being stretched or drawn out.

**tensile strength** *Materials.* the maximum tensile stress (stretching) that a material can withstand without failure.

**tensile stress** *Materials.* **1.** an external force that acts on a body at both ends, parallel to the length, and that if unopposed will tend to elongate it; i.e., the material will stretch. **2.** a measure of this force per unit area of the material.

**tensiometer** *Measurement.* an instrument used to measure the tension of a solid body, such as a wire or cable, or the surface tension of a liquid.

**tension** *Physics.* **1.** the force exerted at the end points of a stretched object, or the condition of an object being stretched in this manner. **2.** the partial pressure of a gas in a fluid. **3.** *Electricity.* another term for VOLTAGE, as in the term high-*tension* wire.

**tera-** *Measurement.* a prefix meaning "one trillion" ($10^{12}$), as in terahertz; symbol T.

**teratogenesis** *Health & Safety.* an abnormal processes of fetal development leading to birth defects **(terata).** Can be associated with maternal ingestion of certain toxic chemicals.

**teratogenic** *Health & Safety.* causing physical defects in the fetus; preventing normal fetal development.

**teratology** *Health & Safety.* the scientific study of the causes, mechanisms, and manifestations of abnormal fetal development.

**terawatt** *Measurement.* a unit of power defined as one trillion watts; symbol TW.

**terbium** *Chemistry.* an element having the symbol Tb, the atomic number 65, an atomic weight of 158.9, a melting point of 1360°C, and a boiling point of about 2960°C; an extremely reactive rare-earth (lanthanide) element used as a phosphor activator and in solid-state devices.

**TERI** The Energy and Resources Institute (est.1974); an Indian nongovernmental organization noted for its work on the role of energy in sustainable development. It focuses on issues such as rural energy, power sector reform, global climate change, forest conservation, and urban air pollution.

**terminal** *Electricity.* **1.** a point of connection for two or more conductors in an electrical circuit, or a device that is attached to a conductor so as to facilitate connection with another conductor. **2.** *Communication.* a term for an individual computer, especially one that functions as part of a centralized data transmission system. **3.** *Transportation.* a building where passengers embark or disembark, as on a railroad, bus line, or airline.

**terminal velocity** *Physics.* **1.** the constant velocity of a falling body attained when the force of resistance of the medium through which it is falling is equal in magnitude and opposite in direction to the force of gravity. **2.** specifically, the velocity of an object falling through a fluid when there is no net force acting on it, i.e., where the gravity and drag forces are equal.

**ternary fission** *Nuclear.* the breakup of a compound nucleus into three fragments of comparable mass.

**terra cotta lamp** *History.* a historic type of lamp made from an unglazed, lightweight clay material that is typically red in color; believed to have been first used in ancient Greece in the 600 BC.

**terrestrial carbon** *Earth Science.* carbon that is contained in vegetation and soil.

**terrestrial electricity** *Earth Science.* a collective term for all natural electrical phenomena of the earth, including atmospheric electricity.

**terrestrial energy** *Earth Science.* radiant energy emitted by the earth, including its atmosphere.

**terrestrial radiation** *Physics.* **1.** the total infrared energy emitted by the earth and its atmosphere, measured at wavelengths that are determined by their temperature. **2.** *Nuclear.* the radiation that is emitted by naturally occurring radioactive materials in the earth, such as uranium, thorium, and radon.

**terrestrial sequestration** see CARBON SEQUESTRATION.

**territoriality** *Ecology.* the behavior pattern in some animals in which they tend to remain within a certain area, as for the purpose of feeding, mating, and rearing young; typically this area is defended against members of the same species and also sometimes other similar species. Also, **territory maintenance.**

**tertiary recovery** *Oil & Gas.* a third stage of oil recovery beyond primary and secondary recovery; see ENHANCED RECOVERY.

**TES** thermal energy storage.

**Tesla, Nikola** 1856–1943, U.S. engineer, born in Croatia of Serbian descent; he discovered the rotating magnetic field, the basis of most alternating-current machinery. In 1893 Tesla performed experiments with high-frequency electric currents, thought to be the first example of wireless communication. Although Marconi is usually associated with the invention of radio, courts have ruled that Tesla's 1897 patent on radio communication preceded Marconi's patent.

**Tesla coil** *Electricity.* an air-core resonant transformer that can generate extremely high voltages at high frequency; Tesla coils can produce spectacular lightning-like discharges and have often been used in the film industry for special effects.

**tetra-** *Measurement.* a prefix meaning "four."

**tetraethyl lead** *Oil & Gas.* $Pb(C_2H_5)_4$, a combustible, toxic, colorless liquid; insoluble in water and dilute acids. It was widely used as an antiknock agent in gasoline motor fuel from the late 1920s to 1980s, but this use has now significantly declined (and has been prohibited in some areas) because of the pollution it produces; other ingredients (e.g., MTBE) are now added to gasoline to produce the antiknock effect.

**Texaco** *Oil & Gas.* short for Texas Company; a historic pioneer in the U.S. oil industry (founded 1902) and also in oil development in Saudi Arabia through its participation in the Aramco venture (from 1936). Acquired by Chevron in 2001.

**Texas City Disaster** *Health & Safety.* an incident in 1947 at the port of Texas City, Texas; a barge in the harbor (the French vessel *Grandcamp*) was loaded with fertilizer grade ammonium nitrate, which caught fire and exploded, destroying the nearby city and killing an estimated 570 people. This has been described as the most serious industrial accident in U.S. history.

**Texas Railroad Commission** *Oil & Gas.* an organization that was a major influence in the U.S. oil industry beginning in the 1930s, through its ability to regulate drilling, production, and pricing in Texas. The power of TRC waned in the 1970s as local production began to decline.

**TgC** *Measurement.* teragrams of carbon ($10^{12}$ grams of carbon).

**TGV** *Transportation.* Train à Grande Vitesse; a modern high-speed train service in France and other European countries.

*TGV TGV provides high-speed service in France and other countries of Western Europe.*

**theory** for *theory* entries, see the key word; i.e., for *theory of relativity,* see RELATIVITY; for *theory of gravitation,* see GRAVITATION; and so on.

**therm** *Measurement.* a unit of energy equivalent to 100 000 British thermal units, or approximately 1.055 x $10^8$ joules, the industry standard for measurement of natural gas.

**thermal** *Thermodynamics.* **1.** relating to or involving heat or heat transfer. **2.** *Earth Science.* a localized air current that rises aloft when the lower atmosphere is heated enough to produce an instability over a certain area.

**thermal balance** *HVAC.* the level of outdoor temperature at which the heating capacity of a heat pump matches the heating requirements of a building.

**thermal battery** *Storage.* a high-temperature, molten-salt primary battery in which the electrolyte is a solid, non-conducting inorganic salt at ambient temperatures. When power is required, an internal pyrotechnic heat source is ignited to melt the solid electrolyte, thus allowing electricity to be generated electrochemically for periods from a few seconds to an hour. Most often used for military applications such as missiles, torpedoes, and space missions.

**thermal blanket** *Solar.* a floating foam cover that insulates well, but does not allow light to pass through; often used in swimming pools and spas.

**thermal break** *Materials.* a material of low heat conductivity placed between materials of higher conductance to reduce the flow of heat; e.g., certain metal framed windows are designed with thermal breaks to improve their overall thermal performance.

**thermal breeder reactor** *Nuclear.* a reactor using thorium-232 as its basic fuel; it converts this isotope into fissionable uranium-233, which is capable of creating a chain reaction. Thermal breeder technology is simpler than a liquid-metal fast breeder, since ordinary water is employed as a coolant and the fission chain reaction is sustained by thermal neutrons. As yet, no such breeder has been developed that is suitable for high-capacity commercial use.

**thermal bridge** *HVAC.* a component or assembly in a building envelope through which heat is transferred at a significantly higher rate than through the surrounding envelope area.

**thermal building envelope** see THERMAL ENVELOPE.

**thermal capacity** *Storage.* the amount of heat that can be retained by a thermal storage device. Similarly, **thermal discharge.**

**thermal conductance** *Measurement.* the time rate of heat flow through a body (per unit area) from one of its bounding surfaces to the other for a unit temperature difference between the two surfaces, under steady conditions.

**thermal conduction** see CONDUCTION.

**thermal conductivity** *Thermodynamics.* the ability of a system to conduct heat, usually measured in units of thermal conductance.

**thermal convection** see CONVECTION.

**thermal cover** *Ecology.* vegetation of sufficient abundance and height to significantly ameliorate weather effects such as wind, heat, cold, and snow.

**thermal cracking** *Oil & Gas.* a refining process in which gas oils are subject to extreme temperatures under severe pressure, so that the molecules are cracked and broken into new molecules.

**thermal delay** *Measurement.* the time period between the energization of a heat-producing device and the measurable effect of the heat produced.

**thermal depolymerization** *Materials.* a process for the reduction of complex organic materials (usually biomass waste products) into light crude oil.

**thermal desalination** *Conversion.* the removal of salt from seawater by condensing purified vapor to yield a product in the form of distilled water.

**thermal diffusion** **1.** see THERMODIFFUSION. **2.** see THERMAL DIFFUSIVITY.

**thermal diffusivity** *Materials.* the ratio of thermal conductivity to the heat capacity per unit volume for a material; an important criterion for thermal insulators or conductors.

**thermal ecology** *Ecology.* the study of the effects of thermal conditions on living organisms, either in a naturally heated environment or in one affected by human activities.

**thermal efficiency** *Measurement.* **1.** the ratio of the amount of work performed by a heat engine in one cycle to the amount of thermal energy input required to operate the engine over one cycle; a measure of the efficiency of converting a fuel to energy and useful work. **2.** *Biological Energetics.* an expression of the effect of temperature on the rate of plant growth (assuming other conditions are satisfactory).

**thermal emittance** *Physics.* the ratio of the radiant flux emitted by a body to that emitted by a blackbody at the same temperature and under the same conditions. Also, **thermal emissivity.**

**thermal energy** *Thermodynamics.* the kinetic energy associated with the motion of atoms or molecules in a substance; i.e., heat.

**thermal energy storage (TES)** *Storage.* the storage of heat energy by means of sensible or latent heat technologies, in order to provide heating or cooling services at a later time.

**thermal enhancement** *Oil & Gas.* the use of heat (e.g., an injection of steam) to increase the amount of petroleum that can be recovered from a well.

**thermal envelope** *Solar.* a composite structure of building elements separating an interior temperature zone from the exterior.

**thermal equator** *Earth Science.* **1.** an imaginary line around the earth that connects all points having the highest mean annual temperature for their longitudes during a given period; its position varies with the season but is not the same as the literal geographic equator. **2.** the parallel of latitude of 10° north, the region having the highest annual mean temperature of any latitude.

**thermal equilibrium** see EQUILIBRIUM.

**thermal energy** *Consumption & Efficiency.* the exergy associated with a heat interaction; i.e., the maximum amount of work obtainable from a given heat interaction using the environment as a thermal exergy reservoir.

**thermal expansion** *Earth Science.* **1.** an increase in the size of a substance when the temperature of the substance is increased. **2.** *Earth Science.* specifically, an expansion in the volume of the ocean (and thus an elevation of sea level) that results from the warming of ocean waters.

**thermal expansion valve** see THERMOSTATIC EXPANSION VALVE.

**thermal fatigue** *Materials.* fatigue in a material as the result of rapid increases to and/

or decreases from operating temperatures, which can cause distortion or fracture.

**thermal inertia** *Materials.* the ability of a material to store heat and to resist temperature change, dependent on its density and specific heat.

**thermal insulation** *Materials.* a general term for any material or assemblage of materials used to provide resistance to heat flow.

**thermal inversion** *Environment.* a meteorological condition characterized by a temperature increase (rather than a decrease) with altitude. During a thermal inversion, air pollution can increase dramatically as a mass of cold air is held in place below a warmer mass of air. The absence of air circulation prevents the pollution near the earth's surface from escaping.

***thermal inversion*** *Thermal inversion event, New York, late November 1964. The notorious Killer Fog of London (1952) was caused by a thermal inversion at about the same time of year.*

**thermalization** *Nuclear.* the process undergone by high-energy (fast) neutrons as they lose energy by collision.

**thermal lag** *HVAC.* the delay between the absorption of heat by a thermal storage mass in a passive solar heating system and the radiation of that heat into the living space.

**thermal limit** *Electricity.* the maximum amount of power that a transmission line can carry without suffering heat-related damage to line equipment, especially conductors.

**thermal load** *HVAC.* the amount of heat energy added to a system, or that must be removed from a system; measured in watts (joules per second). For example, in power plant cooling the thermal load placed on a receiving water body is the amount of heat discharged into that water body per unit time.

**thermal loss** *HVAC.* the undesirable transfer of heat to the environment from a building or an energy conversion system.

**thermally activated technologies (TATs)** *Consumption & Efficiency.* a diverse group of devices employed to transform heat for a useful purpose, such as energy recovery, heating, cooling, humidity control, or thermal storage. TATs are building blocks for cogeneration systems.

**thermal mass** *HVAC.* a material used to absorb, store, and later release heat, thus retarding the temperature variation within a building space; e.g., concrete, brick, masonry, mortar, rock, water, or any other such materials with high heat capacity.

**thermal neutrality** *Biological Energetics.* a condition of the thermal environment of a homeothermic animal in which its heat production (metabolism) is not increased either by cold stress or heat stress.

**thermal neutron** *Nuclear.* a neutron that is in thermal equilibrium with its surroundings at room temperature.

**thermal parabolic dish** see PARABOLIC DISH.

**thermal (power) plant** *Electricity.* a generating facility that uses heat to produce electrical power.

**thermal pollution** *Environment.* an excessive raising or lowering of water or air temperature above normal seasonal levels, due to an industrial process or other human activity. See also COLD WATER POLLUTION; HOT WATER POLLUTION.

**thermal pond** another term for a ROOF POND.

**thermal radiation** *Thermodynamics.* heat transfer in association with electromagnetic waves.

**thermal reactor** *Nuclear.* a type of nuclear reactor with moderating materials that act to slow neutrons to low velocities to prevent capture of the neutrons by U-238; most power reactors are of this type.

**thermal reflectivity** *Solar.* the fraction of the incident solar radiation on a surface that is reflected from that surface. Also, **thermal reflectance.**

**thermal reforming** *Oil & Gas.* a procedure in petroleum refining in which heat is used to create a molecular readjustment of a low-octane naphtha, in order to produce high-octane motor gasoline without a catalyst.

**thermal resistance** *Measurement.* a quantity expressing the ability of a substance to prevent a heat transfer; equal to the temperature difference across the surfaces of the body divided by the rate of heat transfer.

**thermal runaway** *Transportation.* a term for the uncontrolled burning of a large quantity of soot accumulated in the filter media of a diesel particulate filter.

**thermal shield** *Nuclear.* a layer of high-density material placed between a nuclear reactor vessel and the biological shield, to reduce radiation heating that could damage the biological shield.

**thermal shock** *Consumption & Efficiency.* a cycle of temperature changes resulting in the failure of a metal due to expansion and contraction.

**thermal storage** see THERMAL ENERGY STORAGE.

**thermal storage mass** see THERMAL MASS.

**thermal storage wall** *Solar.* a type of wall used to store heat energy in a passive solar energy system, usually a south-facing glazed wall. Radiant solar energy strikes the glazing and is absorbed into the wall, which conducts the heat into the room over time.

**thermal stratification** *Earth Science.* the (temporary) division of a lake or other body of water into horizontal layers of differing densities, as a result of variations in temperature at different depths.

**thermal stress** *Physics.* stress developed in a restricted material as a result of temperature changes when the material is unable to expand or contract accordingly.

**thermal transmittance** *Measurement.* the time rate of heat flow per unit area and unit temperature difference between a warmer and colder reservoir.

**thermal vapor compression (TVC)** *Consumption & Efficiency.* a process used to remove salt from seawater, by evaporating it through a fine mesh filter that traps the brine particles so that the vapor can then be condensed in a relatively salt-free state.

**thermal wall** see THERMAL STORAGE WALL.

**thermion** *Electricity.* an electron that has been emitted from a heated body, such as the hot cathode of an electron tube. Thus, **thermionic.**

**thermionic converter** or **generator** *Electricity.* a device that converts heat energy into electrical energy through thermionic emission from a heated cathode. Thus, **thermionic conversion** or **generation.**

**thermionic emission** *Electricity.* the emission of electrons from a heated body. Thus, **thermionic current, thermionic diode.**

**thermistor** *Measurement.* a temperature sensor employing the electrical resistance of a semiconductor as its measuring property.

**thermochemical** *Chemistry.* having to do with or based on thermochemistry.

**thermochemical conversion** *Biomass.* the use of heat to produce a chemical reaction; a common means of utilizing biomass energy. Biofuels produced in this manner include **thermochemical alcohol** and **thermochemical ethanol.**

**thermochemical equation** *Chemistry.* an equation representing a chemical reaction that describes both the stoichiometry and the energetics of the reaction.

**thermochemical hydrogen cycle** *Chemistry.* a multiple-step chemical reaction that ultimately results in the overall production of hydrogen (and oxygen) by water decomposition.

**thermochemistry** *Chemistry.* a branch of chemistry studying the heat changes that accompany chemical reactions and changes of state.

**thermochromic** *Materials.* having the ability to change optical properties reversibly and persistently by the action of temperature.

**thermocline** *Earth Science.* a layer in a body of water at which the rate of temperature decrease with depth is at a maximum; e.g., a thin layer of water at the depth range of 300–800 meters in the tropical and subtropical basins, between the deep abyssal waters and the surface mixed layer.

**thermocouple** *Measurement.* a device consisting of two different metallic conductors that are connected at both ends, producing a loop in which heat is converted into electrical current when there is a difference in temperature between their two junctions; used to measure the temperature of a third substance by connecting it to both junctions and measuring the voltage generated between them.

**thermodiffusion** *Chemistry.* a process in which temperature differentiation within a fluid mixture causes one constituent to flow differently than the mixture as a whole.

**thermodynamic cycle** *Thermodynamics.* a series of processes that occur in a cyclic pattern acting on a substance, in which one form of energy is converted to another form. See also CYCLIC PROCESS.

**thermodynamic equilibrium** see EQUILIBRIUM.

**thermodynamic heating** *HVAC.* a term for heating by means of a heat pump; i.e., on the basis of the natural flow of heat energy from a warmer to a cooler body.

**thermodynamics** the branch of science concerned with the study of energy and the relationship of heat transfer and work to other forms of energy; it deals with the behavior of systems in which temperature is a significant factor. See also FIRST, SECOND, THIRD LAW OF THERMODYNAMICS. See below.

**thermodynamic temperature** *Thermodynamics.* a measure expressed in kelvins (K), proportional to the thermal energy of a given body at equilibrium. A temperature of 0 K is called absolute zero and indicates the minimum molecular activity (i.e., thermal energy) of matter that is possible.

**thermoeconomics** *Economics.* the integration of thermodynamics and economics in order to make technical systems efficient, by finding the most economical solution within the limits of what is technically possible.

**thermoelasticity** *Materials.* a condition of elasticity exhibited by a normally rigid material due to an increase in temperature. Thus, **thermoelastic.**

**thermoelectric** *Electricity.* involving or produced by thermoelectricity; i.e., electrical phenomena occurring in conjunction with a flow of heat. Thus, **thermoelectric current, thermoelectric effect, thermoelectric heat(ing),** and so on.

**thermoelectric converter** *Renewable/Alternative.* a device that converts heat energy directly into electrical energy, typically

**thermodynamics** Thermodynamics is the physical science that accounts for the transformations of thermal energy into mechanical energy and its equivalent forms (electricity, self-organization of complex systems), and vice versa. The development of Thermodynamics and the introduction of the concept of entropy, a measure of energy and resource degradation, are rooted into the technological ground of the Industrial Revolution. James Watt's steam engine (1765) paved the way to a massive use of coal to generate heat and then work. The conversion of energy from one form to another was investigated by scientists and technicians in order to deeper understand the nature of heat towards increased efficiency. Thermodynamics was founded between 1850 and 1860 by W. Thomson (Lord Kelvin), R. Clausius and J.C. Maxwell, building on the seminal work of L.S. Carnot "Reflexions sur la puissance motrice du feu et sur les machines propres à developper cette puissance"(1824) and the experiments of R. Mayer and J.P. Joule about the quantitative equivalence between mechanical work and heat. These studies yielded a set of Laws of Thermodynamics, describing the main principles underlying energy transformations: First Law, energy is conserved; Second Law, entropy cannot decrease in isolated systems; Third Law; entropy is zero when absolute temperature is zero. During the 20th century, the laws of thermodynamics were applied to the self-organization of complex, far-from-equilibrium systems (biological, economic, and social). Started as an applied science, Thermodynamics rapidly developed into a more general system of knowledge encompassing almost all branches of life sciences. L. Onsager, I. Prigogine, N. Georgescu-Roegen, A. Lotka, and H. T. Odum, among others, contributed to this research and several statements of Thermodynamics laws were tentatively reformulated or introduced anew.

**Sergio Ulgiati and Carlo Bianciardi**
University of Siena

consisting of a thermoelectric junction in contact with a heat source; the voltage developed across the junction is a function of the temperature of the source. Thus, **thermoelectric conversion.**

**thermoelectric couple** see THERMOCOUPLE.

**thermoelectric generator** 1. see THERMOELECTRIC CONVERTER. **2.** see RADIOISOTOPE THERMOELECTRIC GENERATOR.

**thermoelectricity** *Electricity.* electricity produced by the direct action of heat or the direct conversion of heat into electricity, as in a thermocouple.

**thermoelectron** *Physics.* an electron emitted by a very hot object.

**thermoforming** *Materials.* the process of using air pressure, mechanical energy, or a vacuum to force a heated thermoplastic sheet into the shape of a mold or die; after cooling, the plastic part is removed from the mold and trimmed.

**thermogenesis** *Biological Energetics.* a process of heat production in an organism by means of various physiological processes.

**thermograph** *Measurement.* a self-recording thermometer that measures both air and soil temperature on a continuous display known as a **thermogram.** Thus, **thermography.**

**thermogravimetry** *Measurement.* the measurement of changes in weight of a substance as a function of changes in temperature.

**thermohaline** *Earth Science.* relating to or caused by the joint action of temperature and salinity in a water mass. Thus, **thermohaline system.**

**thermohaline circulation** *Earth Science.* a pattern of global ocean circulation driven by density differences in water.

**thermohaline convection** *Earth Science.* the vertical movement of a layer of water caused by changes in the temperature-salinity relationship between it and the adjoining layers, usually because one layer has become colder or more saline, or both, and thus more dense.

**thermokarst** *Earth Science.* an area in a permafrost region characterized by an irregular land surface, resulting from the melting of ground ice.

**Thermolamp** *History.* an early lamp powered by an illuminating gas (coal gas), patented and exhibited by French chemist Philippe Lebon in 1799.

**thermolysis** *Biological Energetics.* the dissipation of heat from the body.

**thermometer** *Measurement.* any instrument that measures and indicates temperature; the common type consists of a narrow tube filled with a liquid, such as mercury, that rises (expands) as the temperature rises and falls (contracts) as it falls.

**thermometric** *Measurement.* involving or determined by temperature or temperature measurement.

**thermometry** *Physics.* the branch of science concerned with the study and application of temperature measurement.

**thermoneutrality** see THERMAL NEUTRALITY.

**thermonuclear** *Nuclear.* relating to a process in which very high temperatures are used to fuse light nuclei, such as those of the hydrogen isotopes (deuterium and tritium), with the accompanying liberation of energy. Thus, **thermonuclear reaction.**

**thermonuclear device** *Nuclear.* an extremely powerful explosive weapon whose energy release results from thermonuclear fusion reactions. Typically, the high temperatures required are obtained by means of a fission explosion. Also, **thermonuclear bomb.**

**thermophilic** *Biological Energetics.* literally, heat-loving; describing certain microorganisms **(thermophiles)** whose optimal growth occurs at temperatures of 50°C or more. Thermophilic bioprocesses can be employed as an energy source.

**thermophotovoltaic cell (TPV)** *Photovoltaic.* a device concentrating sunlight onto a absorber that heats it to a high temperature; the thermal radiation emitted by the absorber is used as the energy source for a photovoltaic cell that is designed to maximize conversion efficiency at the wavelength of the thermal radiation.

**thermophysics** *Physics.* the study of physical phenomena within the context of heat.

**thermopile** *Electricity.* an array of thermocouples connected in parallel, having greater sensitivity than a single thermocouple; used for converting radiant energy into electrical energy, and for detecting and measuring radiant energy.

**thermoplastic** *Materials.* **1.** a polymer in which the molecules are held together by weak secondary bonding forces that can be softened and melted by heat, then shaped or formed before being allowed to solidify again. **2.** describing textile fibers and resins that have been softened at high temperatures.

**thermoregulation** *Biological Energetics.* the fact of an organism regulating its internal body temperature; this takes place by means of various physiological processes but can also involve behavior; i.e., moving away from a condition of extreme heat or cold. The two most common forms are ECTOTHERMIC and ENDOTHERMIC regulation.

**thermoscope** *History.* the earliest known version of a thermometer, developed by Galileo in 1592, using air in a tube instead of liquid.

**thermoset** *Materials.* describing a group of polymers that soften when initially heated, then harden and condense in bulk and retain a permanent shape; they cannot be softened or reprocessed by reheating.

**thermosiphon** *Solar.* a solar collector system for water heating in which circulation of the collection fluid through the storage loop is provided automatically by a thermosiphoning effect (see next).

**thermosiphoning** *Solar.* a siphoning effect (expansion and upward movement of water or another fluid) due to the natural difference in density between the warmer and cooler portions of the fluid. A process employed in solar collector systems. Thus, **thermosiphon(ing) principle.**

**thermosphere** *Earth Science.* the atmospheric layer in which temperature increases with altitude, constituting essentially all of the atmosphere above the mesosphere; it includes the exosphere and most or all of the ionosphere.

**thermostad** *Earth Science.* a layer of the ocean in which the vertical change of temperature is very slight.

**thermostat** *HVAC.* a device that connects or disconnects a circuit in response to a certain deviation between the calibration of the device and the ambient temperature; used to control the temperature of a room or building by providing warmer or cooler air as needed.

**thermostat setback** see SETBACK.

**thermostatic expansion valve** *Refrigeration.* a refrigerant metering device that maintains a constant evaporator temperature by monitoring the amount of superheat in the refrigerant gas; used in refrigeration and air-conditioning systems.

**thermostatics** *Thermodynamics.* the study of thermodynamic systems that are in thermal equilibrium.

**thermosyphon** another spelling of THERMOSIPHON.

**thermotropism** *Biological Energetics.* the tendency of a plant to grow toward a source of heat.

**thief** *Oil & Gas.* a device used in petroleum sampling, featuring a metal or glass cylinder with a spring valve that can be tripped to obtain (steal) a sample from any depth within a tank.

**thin film** *Materials.* **1.** any of several thin layers (a few microns or less in thickness) of insulating, conducting, or semiconductor material (e.g., copper indium diselenide, amorphous silicon, cadmium telluride, or gallium arsenide) deposited successively on a supporting substrate in precise patterns to collectively form all or part of an integrated circuit. **2.** specifically, such material used to make solar photovoltaic cells. Thus, **thin-film cell, thin-film (integrated) circuit, thin-film module, thin-film semiconductor,** and so on.

**thinning** *Ecology.* the purposeful removal of certain trees from a forest stand, primarily to concentrate growth on fewer stems and thereby increase production of high-value timber, enhance forest health, and so on.

**third law of thermodynamics** *Thermodynamics.* a law of nature stated as: (a) the entropy of a pure crystal is zero at absolute zero (0 K). This reflects the fact that the lower a substance's temperature, the lower its entropy (i.e., the extent of its dispersal of energy); (b) it is impossible to cool a body to absolute zero by any finite process; i.e., a temperature of absolute zero can be closely approached, but not actually reached. See also THERMODYNAMICS.

**third rail** *Transportation.* **1.** a system of electrical traction for an electric rail system in which current is fed from an insulated conductor rail running parallel with the track; used for example in subways. **2.** in figurative use, anything that is extremely dangerous to deal with or encounter (from the idea that a person can be electrocuted by coming in contact with the third rail).

**Thompson, Benjamin** see RUMFORD.

**Thomsen, Julius** 1826–1909, Danish chemist noted for his early work in thermochemistry. He was among the first to adopt the principle of the conservation of energy as the basis of a thermochemical system.

**Thomson, Joseph John** 1856–1940, British scientist who demonstrated (1897) that cathode rays were units of electrical current made up of negatively charged particles of subatomic size, thus identifying the electron.

**Thomson, Robert William** 1822–1873, British engineer who invented the vulcanized rubber pneumatic tire (1845).

**Thomson, William** see KELVIN.

**thorium** *Chemistry.* a radioactive metallic element having the symbol Th, the atomic number 90, an atomic weight (for Th-32) of 232.0; a gray amorphous or crystalline mass that readily burns in air to form thorium dioxide. It forms uranium-233 upon neutron bombardment after several decay steps; used in photoelectric cells and as a nuclear fuel.

**thorium dioxide** *Chemistry.* $ThO_2$, white cubic crystals, insoluble in water and acid; used in nuclear fuel and as a catalyst.

**Three Mile Island** *Nuclear.* a power plant near Middletown, Pennsylvania, the site (1979) of the most serious accident in the history of U.S. commercial nuclear plants. It was caused by a loss of coolant from the reactor core due to a combination of mechanical malfunction and human error. Though it led to no deaths or injuries, this event brought about many changes in areas such as emergency response planning, reactor operator training, and radiation protection.

**three-phase** *Electricity.* describing a circuit operated from three AC voltages that differ from each other in phase by one-third cycle, or 120 electrical degrees; i.e., they are equally out of phase around the 360° cycle. Thus, **three-phase circuit, current, power,** and so on.

**three-way catalytic converter** *Transportation.* a catalytic converter that uses a combination of platinum, rhodium, and palladium to reduce vehicle emissions.

***Three Mile Island*** *Aerial photo of Three Mile Island, taken about two weeks after the March 1979 accident.*

**threshold** *Health & Safety.* **1.** a term for the point at which exposure to a potentially toxic substance becomes sufficient to cause an adverse effect. Thus, **threshold dose. 2.** more generally, the lowest point or value at which a certain action will take place; e.g., the point at which a chemical process begins. Thus, **threshold value.**

**threshold limit value (TLV)** *Health & Safety.* the maximum concentration of a hazardous material to which workers can be safely exposed without an unreasonable risk of disease or injury.

**threshold voltage** *Electricity.* the lowest voltage at which an effect is produced or an indication is observed. Similarly, **threshold current.**

**throttle** *Transportation.* **1.** a choke valve that regulates the flow of fuel to a vehicle's engine to control speed. **2.** to reduce fuel flow by means of such a device.

**throttling** *Thermodynamics.* an irreversible thermodynamic process in which a gas under pressure is allowed to expand by passing into a chamber of lower pressure.

**throughput** *Consumption & Efficiency.* the volume of material that moves through a system in a production process; e.g., the total volume of raw materials processed by an oil refinery in a given period.

**thrust** *Transportation.* **1.** the pushing or pulling force exerted by a power plant, such as an aircraft engine or rocket engine. **2.** *Earth Science.* the upward displacement or overriding movement of one crustal unit over another. **3.** *Mining.* a crushing of coal pillars resulting from excess stress due to the weight of overlying rocks.

**thyratron** *Electricity.* a gas-filled, three-element thermionic electron tube containing a cathode, grid, and anode; a voltage pulse applied to the grid ionizes the gas and causes anode current to flow. Once initiated, the current continues to flow and cannot be changed by grid voltage; the tube can be extinguished only by removing anode voltage or externally opening the anode current path.

**thyristor** *Electricity.* a solid-state semiconductor device similar to a diode, with an extra terminal that is used to turn it on. Once turned on, the thyristor will continue conducting as long as there is a significant current flowing through it; used in such applications as motor speed controls, light dimmers, and pressure-control systems.

**thyroid cancer** *Health & Safety.* malignancy of the thyroid gland that is the most common form of endocrine cancer; radiation exposure is among the cited causes of this disease.

**tidal basin** *Renewable/Alternative.* **1.** a natural body of water that is subject to tidal action, or that is open to another body of water (e.g., a river) subject to such action. **2.** an artificial body of water created as a reservoir to retain incoming tidal water for use as an energy source.

**tidal difference** another term for TIDAL RANGE.

**tidal dissipation** *Earth Science.* **1.** the loss of mechanical tidal energy in the ocean, thought to occur mainly because of friction with the sea bottom or shoreline. **2.** the loss of thermal energy in the ocean due to the action of the tides.

**tidal energy** *Renewable/Alternative.* the use of the kinetic energy of tidal currents as an energy source; e.g., to power turbines that generate electricity. See below.

**tidal mill** *Hydropower.* a historic technology employed from medieval times until the modern industrial age, employing tidal power to grind grain in the manner of a windmill. Typically operated by water from the incoming tide to flow into an upstream pond or reservoir, where it is then dammed behind a sluice gate; when the tide has receded to a sufficiently low level the gate is opened and the dammed water flows outward to power the waterwheel.

**tidal power** *Renewable/Alternative.* **1.** the generation of electricity from the hydraulic energy that is produced by tidal currents of the ocean. Thus, **tidal power plant** or **station.** **2.** see TIDAL ENERGY.

**tidal energy** A term that is used to describe energy which is artificially extracted from tidal motions in the sea. It is recorded historically that tidal mills were constructed in Roman Imperial times to grind cereal in regions where more conventional water wheels were inappropriate, due to low river speeds and relatively flat terrain, such as southern Britain. More recently, there have been proposals to construct artificial barrages across tidal estuaries and utilise the head difference between the sea and the resultant artificial lake to generate electricity. The first commercial construction was in La Rance in Brittany, northern France. This system, which is rated at 240MW, was constructed in the 1960s and incorporates a road link across the estuary. The system has been the subject of a major refurbishment which will extend its operating lifetime for many years. Other tidal barrage systems have been constructed but none are as large as La Rance. Most recent attention has been directed at the development of systems to extract energy from tidal currents. The technology in many cases has a superficial resemblance to wind turbines but designed to extract energy from moving water rather than air. In principle, tidal current systems, often called tidal stream systems, can be constructed in a modular fashion, which means that the resource can be exploited incrementally, unlike tidal barrages where the entire system needs to be constructed before any energy can be extracted.

**Ian G. Bryden**
The Robert Gordon University
Aberdeen

**tidal prism** *Renewable/Alternative.* the total volume of water that flows into a tidal basin with a flood tide and then out again with the ebb; as measured over considerable time, it represents the difference between the basin's mean high-water and low-water volume.

**tidal range** *Earth Science.* the difference in water level between consecutive high and low tides.

**tidal wave** *Earth Science.* **1.** the wave motion of a tide. **2.** an unusually high and destructive ocean wave; a tsunami or storm surge.

**tide** *Earth Science.* the regular cyclical rise and fall of the water level in the oceans and other large bodies of water, caused by the interaction of the rotating earth with various forces, principally the gravitational attraction of the moon and the sun.

**tie line** or **tie-line** *Electricity.* a transmission line connecting two or more power systems.

**tight gas** *Oil & Gas.* a category of unconventional natural gas that is trapped underground in extremely hard rock, or in unusually impermeable sandstone or limestone formation; tight gas requires much greater extraction efforts for acceptable rates of gas flow.

**timbering** *Mining.* the use of wooden beams to provide a roof support in an underground mine.

**time-of-day** *Electricity.* describing an electric rate policy in which the price will vary according to the time of day, to reflect the different costs of providing the service at different times; e.g., rates can be lower for power used during late evening hours than for peak daytime hours. Thus, **time-of-day pricing, time-of-day meter,** and so on.

**time-of-day lockout** *Electricity.* an electric rate feature under which electricity usage is prohibited, or restricted to a reduced level, at fixed times of the day in return for a reduction in the price per kilowatt hour. Also, **time-of-day limit.**

**time-of-use** *Electricity.* similar to TIME-OF-DAY; i.e., a rate system that varies according to the times during which a customer uses electricity, as by on-peak or off-peak or by seasons of the year.

**tin** *Chemistry.* a metallic element having the symbol Sn, the atomic number 50, and an atomic weight of 118.7, occurring as a silvery-white ductile solid or a brittle gray solid; used in alloys and in plating and soldering, and formerly widely used in cans for foods and beverages.

**tipping fee** *Renewable/Alternative.* a fee charged to deliver solid waste to a landfill, waste-to-energy facility, or recycling facility.

**tipple** *Mining.* an elevated structure, located near a mine entrance, that receives coal from mine cars or conveyors and from which the coal is then dumped ("tipped") into larger vehicles for processing and transport.

***tipple*** *At a Harlan County, Kentucky mine, a worker checks the weight of coal moving from the tipple to railway freight cars.*

**tip speed ratio** *Wind.* the relationship between the speed of the wind and the rotation speed of a rotor; a significant factor because if a rotor turns too slowly, it will allow wind to pass too easily through the blades, and if it turns too rapidly it will offer too much resistance to the wind. Thus it is necessary to match the velocity of the rotor to the wind speed for maximum efficiency.

**tissue weighting factor** *Nuclear.* a multiplier of the equivalent dose to an organ or tissue, used for radiation protection purposes to account for the different sensitivities of different organs and tissues to the effects of radiation.

**titanium** *Chemistry.* a metallic element of the first transition series having the symbol Ti, the atomic number 22, an atomic weight of 47.88, a melting point of 1660°C, and a boiling point of 3290°C; a silvery solid or dark gray substance that is highly valued for its favorable ratio of strength to weight.

**titer** *Chemistry.* the reacting strength or concentration of a solution, as determined by titration with a standard.

**titration** *Chemistry.* any of various methods used to determine the concentration of a test substance in a solvent, performed by adding a standard solution to the test solution. The end point (as indicated by a certain reaction, such as a color change) is used to calculate the composition of the sample.

**TLV** threshold limit value.

**TOA** short for TOP-OF-ATMOSPHERE.

**Tobias, Charles** 1920–1996, U.S. engineer whose characterization of nonaqueous electrolytes, and demonstration that reactive metals could be electrodeposited from them, spawned a new field of battery research and led to fundamental advances in lithium batteries.

**TOC** total organic carbon.

**TOE** tonne of oil equivalent.

**tokamak** or **Tokamak** *Nuclear.* a device to contain a controlled thermonuclear fusion reactor, resembling a doughnut-shaped chamber in which plasma is heated and confined by magnetic fields. [Abbreviated from the Russian term for Toroidal Chamber with Magnetic Coil.]

**toluene** *Materials.* $C_6H_5CH_3$, a flammable, toxic, colorless liquid; insoluble in water and soluble in alcohol and ether; it boils at 110°C and freezes at –94.5°C; used in high-octane gasoline, explosives, and organic synthesis.

**Tom Thumb** *History.* the first locomotive in the U.S. to power a passenger train (1830). [From the name of a tiny hero in a popular folk tale.]

**ton** *Measurement.* **1.** a traditional unit of weight used in the U.S. for heavy items, equal to 2000 pounds; equivalent to 907.18 kilograms. **2.** any of various similar large measures of weight, such as the long ton (2240 pounds) or the metric ton (2204.62 pounds).

**ton-hour** *Measurement.* a mass of one ton absorbed or rejected in one hour; a quantity of thermal energy used to describe the capacity of a storage device; e.g., an ice storage system.

**tonnage** *Measurement.* a quantity expressed in tons, especially a measure of the capacity or weight of a ship.

**tonne** *Measurement.* a metric ton (1000 kg), equivalent to about 2205 pounds.

**tonne-kilometer** *Measurement.* the transport of 1 ton of cargo over a distance of 1 km.

**tonne (ton) of oil equivalent** *Measurement.* a means of comparing different forms of energy, based on a standard of the energy produced by the combustion of one metric ton of crude oil.

**ton of cooling** see REFRIGERATION TON.

**ton of refrigeration** see REFRIGERATION TON.

**tool** *Consumption & Efficiency.* **1.** historically, any physical object employed by humans for a specific task or purpose, usually after having been fashioned or altered in some way; e.g., a sharp fragment of stone used to break up or shape softer materials. **2.** in the modern sense, a portable and usually hand-held instrument, either unpowered or powered, that is used to increase the efficiency of a work effort; e.g., a saw, hammer, screwdriver, drill, and so on.

***tool*** *The earliest tools in human history were blades of stone such as these, used as weapons, for shaping or cutting other objects, for scraping hides, and so on.*

**top-down** *Economics.* describing a modeling or analytical approach that arrives at economic conclusions from broad, highly aggregated generalizations about the system, proceeding to regionally or functionally disaggregated details. Thus, **top-down analysis, top-down model(ing)**, and so on. Compare BOTTOM-UP.

**top log** *Biomass.* a term for any log cut from a tree other than the butt log (the one closest to the stump).

**top-of-atmosphere (TOA)** *Solar.* describing the ideal amount of radiation that a location on earth would receive if there were no intervening atmosphere or clouds (i.e., the radiation received in outer space). Thus, **top-of-atmosphere irradiance, top-of-atmosphere radiation.**

**topping** *Oil & Gas.* a distillation process in which crude petroleum is heated and then passed through a tower, where lighter and heavier fractions are separated and drawn off from each other at different levels.

**topping cycle** *Conversion.* a process in a power plant producing electricity and then employing the steam or other thermal energy used for this production for subsequent processes; so called in contrast with a BOTTOMING CYCLE because in this process the electricity is generated at the beginning (top) of the process, whereas in a bottoming cycle it is generated at the end.

**top pressure recovery turbine (TRT)** *Consumption & Efficiency.* an electricity generating system whose energy source is the exhaust pressure and heat from the blast furnace of a steel mill; the electricity generated by this system can then be used to power the electrical equipment of the mill.

**top runner** *Policy.* a means of setting efficiency, emissions, or other standards by identifying representative products (e.g., cars, household appliances) that excel in the characteristic being targeted and using their performance as the basis for setting standards for the entire class of products. [From the *Top Runner* program enacted in Japan in 1999.]

**toroidal** *Electricity.* doughnut-shaped; describing a circuit in which a coil is wound about a circular magnetic core whose permeability is sufficiently high to confine virtually all the magnetic field lines within the core. Thus, **toroidal (magnetic) circuit.**

**torpor** *Biological Energetics.* a state of inactivity in an organism; may be a result of damage to the nervous system, or may be a natural reaction to conserve energy stores.

**torque** *Physics.* **1.** the tendency of a force applied to an object to cause the object to rotate about a given point. **2.** this rotational force used as a basic measure of the propulsive effect of a powered wheel.

**torr** *Measurement.* a unit of pressure, equivalent to the amount of pressure that will support a column of mercury 1 millimeter high at 0°C and standard gravity; equal to 133.32 pascals. [Named for Evangelista *Torricelli.*]

**Torrey Canyon** *Oil & Gas.* a supertanker that ran aground on a reef near the coast of Cornwall in southwestern England in 1967, spilling an estimated 30 million gallons of oil into the sea and thus producing great environmental damage.

**Torricelli, Evangelista** 1608–1647, Italian mathematician who invented the barometer and whose work in geometry aided in the eventual development of integral calculus. He also is the first person reported to have created a sustained vacuum.

**torsion** *Physics.* a twisting or turning motion of a solid body about its axis of symmetry, produced by the application of opposing forces or torques at opposite ends of the body. Thus, **torsional.**

**torsion(al) balance** *Electricity.* an instrument designed to measure weak gravitational, electrostatic, or magnetic forces by determining the amount of torsion that they cause in a wire or filament.

**torus** *Nuclear.* the vacuum vessel used in tokamak fusion research.

**total energy expenditure (TEE)** *Biomass.* the total energy requirements of a person over the course of an entire day, including rest and sleep as well as actual physical activity. Also, **total daily energy expenditure (TDEE).**

**total factor productivity (TFP)** *Economics.* the quantity of output divided by the amount of all inputs used in production. Diverse inputs (labor, capital, energy) typically are aggregated with an indexing procedure, such as one in which the quantity of each input is its share in the total cost of production.

**total fertility rate** *Social Issues.* in a given population, a statement of the total number of children that would be born to a representative woman, if she were to live to the end of her reproductive years and have children at various age levels in accordance with the prevailing age-specific fertility rates.

**total heating value** *Consumption & Efficiency.* the number of British thermal units produced by the combustion at constant pressure of one cubic foot of a fuel, under standard conditions of temperature and pressure.

**total internal reflection** *Photovoltaic.* the trapping of light by refraction and reflection at critical angles inside a semiconductor device, so that it cannot escape the device and must eventually be absorbed by the semiconductor.

**total (solar) irradiance** *Solar.* the solar radiation received at the top of the earth's atmosphere; see TOP-OF-ATMOSPHERE.

**total material requirement (TMR)** *Economics.* the total mass of the physical materials that are mobilized each year to support an economy; this includes not only the direct use of resources for producing goods (e.g., oil and timber harvest), but also "hidden flows" such as mining overburden, processing waste, and soil erosion, as well as the materials embodied in imports.

**total organic carbon (TOC)** *Environment.* the quantity of organic compounds dissolved in water, measured as pure carbon. The TOC of a body of water affects biogeochemical processes, nutrient cycling, biological availability, and chemical transport and interactions. It also has direct implications for drinking water quality and wastewater treatment.

**total suspended particulates (TSP)** *Environment.* the total amount of tiny airborne particles or aerosols present in the atmosphere at a given time. These can be produced by natural processes such as evaporation, forest fires, volcanic eruptions, or pollen dispersal, but TSP measurements usually refer to pollutants from human sources such as fuel combustion, coal burning, municipal waste incineration, and so on. Also, **total suspended particulate matter.**

**total suspended solids (TSS)** *Environment.* all the solid particles in wastewater, effluent, or a natural body of water that will not pass through a filter of a given size; components may include silt, decaying plant and animal matter, industrial wastes, and sewage. A high TSS concentration can have a significant effect on an aquatic system, as by slowing photosynthesis and reducing oxygen levels.

**tower mill** *Wind.* a traditional European mechanical windmill used for grinding grain and other materials as well as pumping water, in which a tower of wood or masonry remains stationary and the cap with the rotor attached turns to face the wind.

**town gas** *Coal.* another term for COAL GAS, because of its historic use for urban lighting and heating.

**toxic** *Health & Safety.* poisonous; harmful to the health of humans and other organisms.

**toxicological study** *Health & Safety.* a controlled laboratory experiment in which a group of animals receive a gradient of doses of compounds with suspected health impacts, and the relationship between dose and disease rate is estimated.

**toxicology** *Health & Safety.* the branch of science that studies the effects of chemicals on living systems, especially the effects of **toxins** (poisons) on health and the environment.

**TPV** thermophotovoltaic cell.

**trace element** *Materials.* a term for an element that appears in very small quantity (typically less than 10 parts per million) in a given medium, such as a mineral or rock.

**trace gas** *Earth Science.* **1.** a minor constituent gas of the atmosphere, such as carbon dioxide, water vapor, methane, or oxides of nitrogen, ozone, and ammonia. Although small in absolute volume, trace gases have significant effects on the earth's weather and climate. **2.** see TRACER GAS.

**tracer gas** *Consumption & Efficiency.* **1.** a gas that is placed within a sealed container or closed system so that leaks can be detected. **2.** a gas used to provide an indication of how the air within a building moves under normal operating conditions.

**tracking array** *Photovoltaic.* a photovoltaic array that follows the path of the sun to maximize the solar radiation incident on the PV surface.

**tracking error** *Solar.* **1.** for a two-axis tracking collector, the angular deviation between the collector-sun line and a line that is normal to the aperture plane. **2.** for a single-axis tracking collector, the angular deviation between two planes that intersect along the axis of rotation.

**traction battery** *Storage.* a battery designed to provide moderate power through many deep discharge cycles, such as those used in golf carts and other small electric vehicles.

**tradable permit** *Economics.* a quantified permit that allows the holder to engage in an action, such as the release of an air pollutant, whose ownership can be transferred among parties.

**traditional** *History.* describing an energy source with extensive use over the course of human history, such as firewood, charcoal, animal dung, human and animal power, or stationary (nonelectric) hydropower. Thus, **traditional energy.**

**traditional agriculture** *Consumption & Efficiency.* a term for the method of agriculture generally practiced in the world prior to (or in the absence of) modern developments such as gasoline-powered farm machinery, chemical fertilizers and pesticides, and artificial hormones and antibiotics, relying instead on human and animal power, organic fertilizers, and biological controls for pests and disease. Compare INDUSTRIAL AGRICULTURE.

**traditional fuel** *Biomass.* a fuel that has been used for an extended period of time throughout human history and that typically does not involve industrial processing, such as wood, charcoal, or animal and vegetable wastes.

**traditional society** *Social Issues.* a descriptive term for a contemporary society that retains many (or all) of the characteristics historically associated with human society in general (e.g., subsistence agriculture, use of human and animal power, burning of biomass for fuel) but replaced in other cultures by industrial processes.

**traffic calming** *Transportation.* a combination of efforts to reduce the negative effects of motor vehicle use and improve conditions for nonmotorized street users; may include such activities as driver education and police enforcement, and also physical measures such as speed bumps.

**traffic engineering** *Transportation.* a branch of engineering concerned with the application of scientific methods and civil engineering principles to the design, construction, and operation of transportation systems.

**tragedy of the commons** *Global Issues.* the principle that if each person pursues his own best interest with respect to the use of a common resource, the resource will eventually become depleted or ruined, unless there is some general societal constraint on individual motivation.

**trailing edge** *Wind.* the edge of a turbine or propeller blade that faces away from the direction of rotation; the opposite of LEADING EDGE.

**trajectory** *Physics.* **1.** the path described by a body as it moves through space. **2.** specifically, the path of a missile, bullet, shell, or other such projectile between its firing point and the point of impact.

**tram** *Transportation.* **1.** a vehicle or cage that is suspended from cables in a tramway system (see next). **2.** a usually open-sided bus used to carry passengers at a low speed for short distances. **3.** a vehicle used to haul loads in a mine.

***tramway*** *Tramway employed to bring skiers to the top of a mountain.*

**tramway** *Transportation.* a suspended cable system along which passengers or freight are transported, especially an aerial system.

**Trans-Alaska Pipeline (TAPS)** *Oil & Gas.* an 800-mile long pipeline extending from

Prudhoe Bay on Alaska's North Slope to the port of Valdez; one of the largest oil pipelines in the world.

**transaxle** *Transportation.* an assembly containing the transmission and differential systems of a motor vehicle.

**Transco** *Electricity.* short for transmission company; an entity engaged solely in providing transmission services, especially on a regional basis. Also spelled **TransCo** or TRANSCO.

**transducer** *Conversion.* **1.** any device that converts an input signal of one form into an output signal of a different form. **2.** specifically, a device that converts electrical energy into acoustic energy, such as a loudspeaker, or that converts acoustic energy into electrical energy, such as a microphone.

**transesterification** *Conversion.* **1.** a transformation of an organic acid ester into another ester of that same acid; e.g., a process that reacts an alcohol with the triglycerides contained in vegetable oils and animal fats to produce biodiesel and glycerin.

**transfer medium** *Solar.* a substance that carries thermal energy from a solar collector to a storage area, or from a storage area to be warmed in a collector; usually either air, water, or an antifreeze solution.

**transformer** *Electricity.* an electrical device used to transfer electric energy from one circuit to another, especially equipment that produces such a transfer with a change in voltage.

**transformity** *Measurement.* a ratio obtained by dividing the total emergy that was used in a process by the energy yielded by this process; a measure of the efficiency of a transformation between energy commodities.

**transient** *Electricity.* **1.** referring to any voltage or current that deviates from the normal steady-state condition. **2.** *Nuclear.* describing a change in a reactor coolant system temperature and/or pressure due to a change in the power output of the reactor.

**transistor** *Communication.* an active semiconductor device, usually made from silicon or germanium and possessing at least three terminals (typically, a base, emitter, and collector); characterized by its ability to amplify current and used in a wide variety of electronic equipment such as amplifiers, oscillators, and switching circuits.

**transit** *Transportation.* **1.** any conveyance or transportation of persons or goods from one place to another. **2.** in popular use, a synonym for MASS TRANSIT (public transportation).

**transition** *Chemistry.* **1.** a phase change between the solid, liquid, or gas states of a substance. **2.** see ENERGY TRANSITION.

**transition layer** *Earth Science.* a dynamic, high-temperature layer of the sun between the chromosphere and the corona.

**transition metal** *Chemistry.* a collective term for the 38 elements in groups III through XII of the periodic table that are both ductile and malleable, and that conduct electricity and heat.

**transmission** *Electricity.* **1.** the movement or transfer of electric energy over an interconnected group of lines and associated equipment. **2.** *Health and Safety.* the transferring of a disease or condition (e.g., a virus) from one person to another. **3.** *Transportation.* the system of gears by which power is transmitted from the engine of an automobile or other motor vehicle to the driving axle or axles.

**transmission line** *Electricity.* a heavy wire carrying large amounts of electricity over long distances from a generating station to a place where the electricity is used.

**transmission reliability margin (TRM)** *Electricity.* the amount of transmission transfer capability necessary to ensure that a transmission network will be secure under a reasonable range of uncertainties in system conditions.

**transmission system** *Electricity.* an interconnected group of electric transmission lines and associated equipment for transferring electric energy in bulk between a point of supply and a point at which it is delivered to consumers or to another electrical system. Similarly, **transmission company.**

**transmissometer** *Measurement.* an electronic instrument system that provides a continuous record of the atmospheric transmission between two fixed points.

**transmittance** *Physics.* the fraction or percentage of a particular frequency or wavelength of electromagnetic radiation that passes through a substance without being absorbed or reflected.

**transmitter** *Communication.* **1.** equipment that produces and modulates radio-frequency energy for the purpose of radio communication. **2.** a device that converts sound waves to electrical waves, such as the microphone of a telephone.

**transmitting** see TRANSMISSION.

**transmitting utility** *Electricity.* a regulated entity that owns and maintains wires used to transmit wholesale power.

**transmutation** *Nuclear.* the bombardment of a nucleus by particles or photons so as to bring about a change in the nucleus that results in a different isotope of the original nucleus, or in different elements; a method to convert radioactive waste into nonradioactive elements, thus reducing radiological hazards and waste disposal problems.

**transnational** *Global Issues.* describing a corporation or other entity that conducts its operations on a global basis; i.e., across national borders.

**transonic** *Measurement.* describing a speed at or near the speed of sound.

**transpiration** *Biological Energetics.* the process by which water in plants passes from the plant to the atmosphere as water vapor, chiefly through stomata (leaf pores).

**transportation** the movement of people and goods over distance by land vehicles and water and air craft.

**transportation sector** *Transportation.* an energy-consuming sector that consists of all vehicles whose primary purpose is transporting people or goods from one physical location to another.

**transshipment** *Oil & Gas.* a method of ocean transport for oil in which ships off-load their cargo to a deepwater terminal, floating storage facility, or smaller tankers, for later transport to the market destination.

**transuranic** *Chemistry.* beyond uranium; describing an element such as plutonium with an atomic number greater than 92, the atomic number of uranium.

**transuranic extraction (TRUEX)** *Nuclear.* a process that uses a solvent to draw the transuranic elements away from the inert materials in nuclear waste.

**transuranic (TRU) waste** *Nuclear.* waste contaminated with alpha-emitting radionuclides with an atomic number greater than 92, half-lives greater than 20 years, and concentrations greater than 100 nanocuries per gram. Such wastes arise mainly from weapons production, and consist of ordinary items such as clothing or tools contaminated with small amounts of radioactive elements, mostly plutonium. Because of the long half-lives of TRU waste, it is not disposed of as either low-level or intermediate-level waste, but it does not have the very high radioactivity of high-level waste.

**trap** *Oil & Gas.* an underground accumulation of petroleum that cannot escape from its reservoir rock; the basic types are STRATIGRAPHIC, STRUCTURAL, and COMBINATION TRAPS. Trap formation is one of the fundamental petroleum system processes.

**trap efficiency** *Environment.* the ability of a screen or filter to trap and retain sediments and nutrients in a flowing stream; e.g., at a dam opening; it is usually expressed as a percentage of the inflowing sediments or nutrients that are retained in the reservoir.

**trapping** *Electricity.* **1.** the confining of light inside a semiconductor material by refracting and reflecting the light at critical angles; trapped light will travel further in the material, increasing the probability of absorption and thus of producing charge carriers. **2.** *Oil & Gas.* the fact of oil being accumulated in a TRAP.

**trash rack** or **trashrack** *Hydropower.* a mesh or set of bars at the inlet of a hydropower system, employed as a guard to trap leaves, branches, and other such debris to prevent obstruction or malfunction of the turbine.

**TRC** Texas Railroad Commission.

**tree-hugger** *Social Issues.* a derogatory term used by opponents of the environmental movement, to characterize environmentalists as being sentimentally devoted to the protection of wildlife and overly concerned about environmental change.

**tree ring** *Climate Change.* a visible indication of the annual growth increment of a tree that indicates, among other factors, the climatic conditions that enhance or limit growth; tree ring widths are studied to investigate solar-terrestrial relationships and climatic cycles and to reconstruct past climates.

**tremor** *Earth Science.* a minor earth movement, usually preceding or following a larger movement (such as an earthquake) or a volcanic eruption.

**Trevithick, Richard** 1771–1833, English engineer who designed and built the *Penydarren,* the world's first steam locomotive to run on iron rails. He also built early steamboats and dredging and threshing machines.

**trichloroethane** another term for METHYL CHLOROFORM.

**trickle charge** *Electricity.* **1.** a small, continuous charge of a storage battery at a rate approximately equal to its losses and suitable to maintain the battery in a fully charged condition. **2.** a similar charge in a circuit to compensate for internal losses and intermittent discharges.

**trickle collector** *Solar.* a type of solar thermal collector in which a heat transfer fluid drips out of a pipe at the top of the collector and then runs down the collector absorber into a tray where it drains to a storage tank.

**trigeneration** *Consumption & Efficiency.* the simultaneous generation of electricity, heating, and cooling.

**triglyceride** *Biological Energetics.* any of a number of naturally occurring lipids formed when three fatty acids replace the three hydrogen atoms in the hydroxyl groups of glycerol, representing the chief constituent of fats and oils, including the fat cells in the human body.

**Trinity site** *Nuclear.* a site in the New Mexico desert at the Alamogordo Bombing Range (now the White Sands Missile Range), the locale of the first testing of a nuclear weapon, on July 16, 1945. A plutonium bomb was tested, the type of weapon later dropped on Nagasaki, Japan.

**triode** *Electricity.* a vacuum tube containing three elements: a cathode, a control grid, and an anode; a device used to amplify weak electric signals. First employed as a detector of radio waves, then as an amplifier for long-distance telephone calls, and eventually as a major component of radio transmitters, sound films, loudspeakers, and many other electronic devices.

**trip** *Consumption & Efficiency.* **1.** an automatic or manual action that shuts down a system. **2.** *Nuclear.* specifically, the insertion of control rods into the fuel core of a nuclear reactor to stop the fissioning process, or the opening of a circuit breaker under abnormal electrical operating conditions. Thus, **tripper, tripping.**

**triple bottom line** *Economics.* a metric of company performance or success based on social and environmental criteria as well as the conventional financial criteria.

**triple point** *Chemistry.* the temperature and pressure at which the gaseous, liquid, and solid phases of a given substance will exist at equilibrium with each other (e.g., water as liquid water, water vapor, and solid ice).

**triptane** *Materials.* $C_7H_{16}$, a flammable, colorless liquid that is insoluble in water and soluble in alcohol; a high-octane aviation fuel.

**tritium** *Hydrogen.* **1.** a radioactive isotope of hydrogen having the symbol T or 3H, with two neutrons and one proton in the nucleus and thus an atomic mass of 3; formed by bombarding lithium with low-energy neutrons. It is the heaviest isotope of hydrogen, and emits a low energy beta particle in its decay to helium-3. **2.** *Earth Science.* this isotope used as a tracer in ocean transport studies, based on its previous production by nuclear weapons testing in the 1950s and 1960s.

**tritium breeding** *Nuclear.* the production of tritium fuel in fusion reactors; source materials include liquid metals, molten salts, and solid breeders.

**trolley** *Transportation.* **1.** a grooved metallic wheel or pulley at the end of a pole that picks up current from an electric wire and carries it to a transit vehicle. **2.** any vehicle that draws power from such a mechanism. Thus, **trolley car.** **3.** any of various small vehicles operating on a track, as in a mine or railroad yard.

**Trombe wall** *Solar.* a wall with high thermal mass, used to store solar energy passively in a solar home; generally it consists of a south-facing glass wall and a blackened concrete wall with an air space between them. The blackened wall is heated by sunlight passing through the glass, and it then releases heat, causing warm air to rise in the space between the glass and the concrete. [Named for French inventor Felix *Trombe,* 1906–1985.]

**trommel** *Mining.* **1.** a revolving cylindrical screen used in the grading of coarsely crushed coal or ore; after material is fed into the screen at one end, the fine material drops through

***trolley*** *Trolley system running in the center of the city of Vienna, Austria. In U.S. usage the term "trolley" is generally considered outdated and is replaced by other terms.*

the holes while the coarse material travels to and is delivered at the other end. Also, **trommel screen. 2.** to separate material into various sizes by means of this process.

**trophic** *Biological Energetics.* having to do with or involving food or feeding.

**trophic dynamics** *Ecology.* a description of the relationships among various species in a given ecosystem, in terms of energy exchange; e.g., the disappearance of a top predator may decrease the population of a certain plant, because a herbivore that is the usual prey of this predator will then increase in population and consume more of the plant.

**trophic level** *Ecology.* a group of organisms of one or more species that occupy the same position on a food chain (food pyramid, food web, etc.); e.g., primary carnivores.

**trophic position** *Ecology.* an indicator of the relative position of a certain group of organisms in the feeding levels of an ecosystem; e.g., the killer whale *(Orca)* can occupy the trophic position of top predator in its habitat.

**trophism** *Biological Energetics.* feeding; the process in which an organism obtains food.

**tropopause** *Earth Science.* the upper limit of the troposphere, a layered band of atmosphere lying just below the stratosphere (about 15–20 km high at the equator, and about 10 km high at the poles), marked by an abrupt decrease of temperature.

**troposphere** *Earth Science.* the lowest layer of the atmosphere, extending to about 15 km above the earth's surface and characterized by a steady drop in temperature with altitude; the part of the atmosphere where weather conditions exist and nearly all cloud formations occur.

**tropospheric ozone** *Environment.* ($O_3$) ozone that is present in the troposphere; a trace gas that plays a controlling role in the oxidation capacity of the atmosphere and also has a significant role in the greenhouse gas effect and urban smog levels.

**tropospheric ozone precursor** see OZONE PRECURSOR.

**troy** *Measurement.* describing a system of weights used for gems and precious metals, based on a pound of 12 ounces.

**TRT** top (pressure) recovery turbine.

**TRU** or **tru** transuranic.

**true airspeed** *Transportation.* the actual airspeed of an aircraft as it moves through the air mass. Because true airspeed will correlate exactly with indicated airspeed only at standard sea level conditions, it is usually calculated by adjusting the indicated speed according to conditions (temperature, density, and pressure).

**true solar time** *Measurement.* a scale of time in which 12:00 noon is the instant at which the sun is at its highest point above the horizon in the sky. Apparent solar time is corrected to true solar time by taking into account the variation in the earth's orbit and rate of rotation.

**TRUEX** transuranic extraction.

**Ts'ai Lun** Chinese inventor who is considered the first producer of paper for writing (about 105–110 AD), reportedly he used the inner bark of a mulberry tree and bamboo fibers placed on a flat piece of cloth.

**Tsiolkovsky, Konstantin** 1857–1935, Russian physicist who laid many of the theoretical foundations for the science of rocketry. In 1903, he described how a reaction thrust motor could allow humans to escape the bounds of earth and suggested the use of reaction vehicles for interplanetary flight. He later presented a design for a multistage rocket and proposed the construction of artificial satellites.

**TSP** total suspended particulates.

**TSS** total suspended solids.

**tsunami** *Earth Science.* an ocean wave of great length and duration, caused by a large-scale movement of the sea floor, such as a volcanic eruption, submarine earthquake, or landslide. Although usually not of great height at sea, its velocity may be as high as 400 knots, so that it travels long distances and in shallow water it can reach heights of 15 meters and cause considerable damage and loss of life. [From a Japanese term for this; literally, "great harbor wave".]

**tube fluorescent** *Lighting.* a term for the common tubular-shaped type of fluorescent lamp.

**tuff** *Earth Science.* a compacted or consolidated deposit of material formed from volcanic ash or dust; **welded tuff** results when ash particles are fused together by heat and pressure.

**Tull, Jethro** 1674–1741, English agriculturist who is one of the founders of scientific agriculture. He advocated using manures, pulverizing the soil, planting with drills, and thorough tilling during the growing period. He also invented a mechanical drill for sowing and a horse-drawn hoe to clear away weeds.

**tumorigenesis** *Health & Safety.* the growth or development of a tumor; i.e., an incidence of cancer.

**tundra** *Ecology.* a vast, nearly level, barren, treeless region located in the Arctic; characterized by very low winter temperatures, short, cool summers, and vegetation consisting of various grasses, rushes, perennial herbs, lichens, and dwarf woody plants.

**tungsten** *Chemistry.* a very hard, highly conductive metallic element having the symbol W, the atomic number 74, an atomic weight of 183.85, a melting point of 3410°C (the highest of all metals), and a boiling point of 5927°C; a white or gray, brittle metallic element that is a member of the chromium family. It is widely used in steel alloys, as filaments for electric light bulbs, in magnets and cemented carbides, and as a heating element in furnaces.

**tungsten carbide** *Materials.* WC, black hexagonal crystals, insoluble in cold water, melting at approximately 2870°C and boiling at 6000°C. It is the strongest structural material known.

**tungsten filament** *Lighting.* a filament used in incandescant lamps and in thermionic vacuum tubes and other tubes requiring an incandescent cathode; those used in larger lamps require an inert gas at a specific pressure.

**tungsten-halogen** *Lighting.* an incandescent lamp employing tungsten as the filament and a halogen gas; see HALOGEN LAMP.

**tungsten lamp** *Lighting.* any incandescent lamp using a tungsten filament; a historic advance (early 1900s) over carbon and other preceding types of filaments.

**tungsten steel** *Materials.* a steel once widely used for cutting tools and magnets but now replaced by high-speed or hot-work steels, which contain tungsten with other alloying elements.

**tunnel diode** *Electricity.* a heavily doped P-N junction diode that is characterized by a negative-resistance region in the forward-bias direction; an increase in applied voltage causes a decrease in forward current, and vice versa; used in ultrahigh-frequency and microwave oscillator and amplifier circuits.

**tunnel freezer** *Refrigeration.* a commercial freezer in which a continuous belt, carrying food items through the refrigerated enclosure, makes a single straight movement through the enclosure.

***tundra*** *Tundra landscape surrounds an Alaska airfield.*

**tunneling** *Physics.* a phenomenon, not explained by classical physics, in which a particle can penetrate and cross a small region where the potential is greater than the particle's available energy. Also, **tunnel effect.**

**Tupolev** *Transportation.* the first supersonic transport plane to take flight (1968), prior to the better-known Concorde. It began service in 1975 and made its last commercial flight in 1978, though later flights of a refurbished version were made in 1996–97.

**turbidity** *Materials.* **1.** any measure of the degree of transparency of a fluid substance; e.g., drinking water. **2.** specifically, a measure of the opacity of the atmosphere. A perfectly clear sky has a turbidity of 0, and a completely opaque sky has a turbidity of 1.

**turbine** *Conversion.* a general term for any machine capable of generating rotary mechanical power by converting the kinetic energy of a stream of fluid (e.g., water, steam, hot gas); turbines operate through the principle of impulse or reaction, or a combination of the two.

**turbine drill** *Oil & Gas.* a type of downhole motor that gives rotation and torque to the bit while transmitting the drill-collar weight to the bit.

**turbocharger** *Transportation.* an exhaust gas-driven device consisting of a shaft with two vaned fan-type wheels at each end; at the hot end, the turbine wheel is driven by the hot pressurized exhaust gases, while at the opposite end, the compressor wheel pressurizes the ambient air supply into the engine intake manifold.

**turbocompressor** *Consumption & Efficiency.* a machine for compressing air or another fluid in order to increase the reactant pressure and concentration.

**turboexpander** *Consumption & Efficiency.* a machine for expanding air or other fluid in order to decrease the fluid pressure and concentration.

**turbofan** *Transportation.* a type of jet engine that derives its thrust primarily by passing air through a large fan system driven by the engine core; the dominant current mode of jet propulsion for commercial aircraft.

**turbojet** *Transportation.* **1.** a jet engine propelled by the simplest form of gas turbine; it utilizes a compressor, combustion chamber, and turbine, with the turbine drawing just

**turbine** Turbines are rotating machines that convert the energy carried by a continuous flow of fluid into shaft power. The word "turbine" (from the Latin "turbo", meaning "vortex") was first proposed in the 1820s by the French scientist Claude Burdin, with reference to hydraulic machines. Primitive hydraulic turbines made of wooden discs carrying straight blades were probably in use in Egypt and Mesopotamia in the 5th century B.C. Some sources report their use in ancient India and China in about the same times. These early machines were used for milling corn and other cereals. In the 1st century A.D., Hero of Alexandria, in his treaty *De Re Pneumatic* (literally, "On fluids"), described what in modern terms would be defined a single-stage, pure reaction steam turbine, but his machine apparently had no practical application, and until the end of the Middle Ages, hydraulic and wind turbines were the sole non-animal source of mechanical power. The modern development of the hydraulic turbine began towards the end of the 18th century where it powered sawmills, textile and manufacturing industries. Industrial development proceeded through a series of successive improvements that greatly increased the power output of turbines. The first steam turbines in commercial service were installed in the northern U.S. by W. Avery in 1831 to power some sawmills. The modern development of steam turbines began in the 1920s when large industrial groups like General Electric, Allis-Chalmers, Westinghouse and Brown-Boveri applied these machines to the generation of electricity. The decisive advance for the gas turbine came from its aeronautic application: the German H. von Ohein (1936–1939) and the Briton F. Whittle (1938–1940) are independently credited with the first successful implementation of the gas-turbine based jet engine. The vast majority of fossil fuel thermo-electrical conversion plants use steam turbines (1–500 MW per unit) and gas turbines (10 kW to 300 MW per unit) as mechanical converters.

**Enrico Sciubba**
University of Rome

enough energy from the gas flow to drive the compressor. **2.** an aircraft employing such an engine; the earliest form of jet aircraft in the late 1930s.

**turbulence** *Physics.* a state of disorder, disarray, or agitation in nature; e.g., an irregular motion of the atmosphere, as manifested by wind gusts and lulls, or a secondary motion of water caused by eddies in a moving flow. Thus, **turbulent flow.**

**turbulent mixing** *Earth Science.* chaotic, non-uniform motions of water or air through the effects of wind, currents, eddies, and so on, serving to transport heat, momentum, and other properties.

**turf** *Biomass.* **1.** a surface layer of earth in which grass is growing, consisting of the grass itself, roots, and the accompanying soil. **2.** the material form of dried peat.

**Turgo turbine** *Conversion.* an impulse turbine similar to the Pelton design but capable of providing greater output in some high-flow hydropower sites; the Turgo runner design allows for more efficient escape of discharged water and thus a larger incoming water jet. As a result, a Turgo turbine can provide equivalent power to a Pelton turbine with a smaller diameter runner.

**Turing, Alan** 1912–1954, British mathematician, one of the founders of modern computer science. In World War II he made a crucial contribution to the British war effort through his success in deciphering Enigma, a previously unbreakable German code. After the war, he designed one of the earliest programmable electronic computers.

**Turing machine** *Communication.* a prediction of the modern computer made by Alan Turing (1936); he described an abstract device that could be made to carry out any form of computing if a tape bearing suitable instructions were inserted into it. The Turing machine is still used in theoretical computer science, especially in complexity theory and the theory of computation.

**Turing test** *Communication.* a means of determining whether a machine can be described as intelligent, presented by Alan Turing; he suggests that computers should be able to "learn" from experience and then modify their operations accordingly. This provided the basis for the field of artificial intelligence.

**turndown** or **turn-down** *Consumption & Efficiency.* the condition of a device operating under less than full load; equipment with good turndown characteristics can be operated efficiently at this level.

**turn-down ratio** *Consumption & Efficiency.* the lowest load at which a device will operate efficiently, as compared to its maximum design load.

**turnkey** *Consumption & Efficiency.* **1.** describing a product or system that can begin successful operation with minimal additional effort once it is installed; i.e., one can simply "turn a key" to begin. **2.** *Oil & Gas.* describing a type of financial arrangement for the drilling of a well in which the contractor assumes full responsibility for the well to some predetermined milestone, such as successful drilling to a certain depth.

**turnover time** *Measurement.* the rate of exit flow of a substance from a given medium, relative to its total mass in this medium, or the time required for a total exit flow of the substance (assuming no further inflow); used to describe material flows in the environment.

**turnpike** *Transportation.* **1.** historically, a roadway on which tolls for passage were collected; so called because the way was blocked by a wooden barrier (*pike*) that would be *turned* to allow passage once the toll was paid. **2.** a former term for a public highway. **3.** a limited-access modern highway, especially one on which tolls are charged.

**TVA** Tennessee Valley Authority.

**TVC** thermal vapor compression.

**TW** terawatt.

**twist** *Wind.* in a wind generator blade, the difference in pitch between the blade root (base) and the tip; typically the pitch is greater at the blade root than the tip.

**two-axis tracking** *Solar.* describing a solar energy system capable of rotating independently about two different axes (e.g., vertical and horizontal), so that the array points directly at the sun at all times. Two-axis tracking arrays capture the maximum possible daily solar energy.

**two-fluid theory** *History.* an 18th-century conception of the phenomenon of electricity (as described by Charles du Fay), according to which electrical charges consist of two separate fluids, one of them vitreous (positive) and the other resinous (negative). Compare ONE-FLUID THEORY.

**two-stage digestion** *Biomass.* a procedure for the anaerobic digestion of organic waste, consisting of an initial thermophilic stage (50–55°C) of shorter duration, and then a longer second stage under mesophilic (35–37°C) conditions.

**two-stroke cycle** *Transportation.* an internal-combustion engine cycle, completed in two strokes, in which the piston compresses the fuel mixture on one side and receives the thrust of the previously compressed gases on the other side during the first stroke, and then draws in a fresh charge on one side while expelling burnt gases on the other side during the second stroke.

**two-stroke engine** *Transportation.* an engine employing the two-stroke cycle (see previous). Developed at around the same time (late 1800s) as the FOUR-STROKE ENGINE that is the standard for passenger vehicles, two-stroke engines are used in lighter engines, as for motorcycles and motorbikes, lawn mowers, chain saws, outboard motors, and so on.

**Tychonic system** *History.* a noted early model for planetary motion devised by the Danish astronomer *Tycho* Brahe, describing the earth as stationary while the moon and the sun revolve around it, and the planets revolve around the sun (a modification of the traditional Ptolemaic system in which all these bodies orbit a stationary earth).

***Tychonic system*** *The Danish astronomer Tycho (Tyge) Brahe, 1546–1601.*

**Tyndall, John** 1820–1893, British physicist who made important studies of energy flows in the earth's atmosphere. He concluded that water vapor is the strongest absorber of radiant heat in the atmosphere, and thus the gas most influencing the earth's surface air temperature. He also speculated that changes in water vapor and carbon dioxide could be related to climate change.

**Type 2 agreement** *Policy.* a policy instrument developed at the World Summit on Sustainable Development (WSSD); an informal agreement involving non-state parties, sometimes among themselves and sometimes with individual governments. Type 2 agreements reflect the increasing role of NGOs and private businesses in international environmental affairs.

**U** uranium; see URANIUM entries.

**UEC** unit energy consumption.

**U-factor** another term for U-VALUE.

**UL** Underwriters Laboratories.

**ULCC** ultralarge crude carrier.

**ulmin** *Coal.* a dark brown to black, amorphous, gel-like constituent of coal, derived from the decay or degradation of vegetable matter.

**ULR** ultralight rail.

**ultimate analysis** *Coal.* the determination by chemical methods of the proportionate amounts of various constituents of coal, by determining the percentages of the elements carbon, hydrogen, oxygen, nitrogen, and sulfur. Other elements that may be present are considered impurities and reported as ash.

**ultimate recovery** *Oil & Gas.* the estimated total volume of oil or gas that a reservoir will yield during its productive lifetime.

**ultracapacitor** *Electricity.* an electrochemical device that stores electricity through capacitive charging and that is capable of providing high power in a compact size and mass.

**ultra-high voltage** *Electricity.* a term for voltages greater than 800 kV.

**ultralarge crude carrier (ULCC)** *Oil & Gas.* a supertanker; an oil tanker of the largest class, having a gross register tonnage in excess of 250,000 tons.

**ultralight rail (ULR)** *Transportation.* a form of light rail transportation employing self-powered individual vehicles of lighter weight and smaller capacity, moving on rails set in existing roadways.

**ultrasonic** *Physics.* having to do with extremely high frequencies of acoustic energy, above the upper range of human hearing (approximately 20,000 hertz), especially sounds that are very far above this range. Thus, **ultrasonics, ultrasound.**

**ultraviolet (UV)** *Physics.* relating to radiation having wavelengths in the range of about 4 to about 400 nanometers, beyond the violet end of the visible spectrum (and thus beyond the range of human vision).

**umbra** *Physics.* **1.** a shadow that obscures all light rays from a specific source, such as the conical section in the shadow of a celestial body. **2.** the dark section of a sunspot.

**UMW** United Mine Workers, a leading U.S. labor union established in 1899 through the merger of two earlier groups.

**UN** see UNITED NATIONS.

**unbundle** *Electricity.* to separate electric utility functions into their basic components and offer different rates for each component; e.g., generation, transmission, and distribution can be unbundled and offered as distinct services. Thus, **unbundling.**

**unbundled rate** *Economics.* a pricing structure in which the various costs for a commodity such as natural gas are broken out and shown separately; employed when gas users buy their own gas and pay the utility only for distribution.

**UNCED** *Policy.* United Nations Conference on Environment and Development; the official name for what is popularly known as the EARTH SUMMIT.

**uncertainty principle** *Measurement.* the premise that it is not possible to measure both energy and time (or position and momentum) with complete accuracy at the same time; thus a description of a mobile particle will contain errors, which can be significant at the atomic scale. Proposed by German physicist Werner HEISENBERG.

**UNCHE** United Nations Conference on the Human Environment; the official name for what is popularly known as the STOCKHOLM CONFERENCE.

**unconditioned space** or **area** *HVAC.* any particular space to which heating or cooling is not supplied, within a structure that is otherwise heated or cooled; e.g., an unheated garage adjoining a home.

**unconformity** *Earth Science.* a break or gap in the geological record, such as an interruption in the normal sequence of deposition of sedimentary rocks, or a break between eroded metamorphic rocks and younger sedimentary strata.

**unconventional** *Renewable/Alternative.* **1.** describing a form of energy that is distinct from the energy sources widely used in the modern industrial world, such as coal and oil; e.g., hydrogen. **2.** describing fossil fuels found in a geologic setting differing from that of conventional deposits of oil or natural gas, and requiring specific technology to develop. Thus, **unconventional energy.**

**unconventional accumulation** *Oil & Gas.* a hydrocarbon deposit of a type that historically has not been produced using traditional development practices, such as a tight gas or oil shale deposit.

**unconventional gas** *Oil & Gas.* a term for natural gas stored in a geologically complex reservoir and thus requiring more advanced exploration and production technology for commercial development; e.g., coalbed methane, shale gas, or gas hydrates.

**unconventional oil** *Oil & Gas.* a term for oil that cannot be economically extracted by traditional methods such as well drilling; e.g., oil obtained from oil sands or oil shale, or from the conversion of natural gas to liquids or from biofuels.

**UNCSD** United Nations Commission on Sustainable Development; a group that oversees the implementation of Agenda 21, the action plan adopted at the Rio Summit which is a blueprint for environmentally sustainable development for the 21st century.

**UNCTAD** United Nations Conference on Trade and Development (est. 1964); the principal agency of the UN General Assembly intended to maximize trade, investment, and development opportunities of developing countries.

**undamped** *Physics.* having a constant or increasing amplitude.

**undercarriage** *Transportation.* **1.** the supporting framework of a vehicle, located beneath the chassis. **2.** an aircraft landing gear.

**underclay** *Coal.* a bed of clay that forms the upper surface of the stratum underlying a coal seam, representing the soil in which trees and other coal-forming vegetation were rooted in a swamp environment.

**undercut** *Mining.* **1.** the process of cutting away a horizontal section or kerf from the bottom of a block of coal so that it will fall more easily. **2.** to cut in the lower part of a coalbed, or beneath one. **3.** *Ecology.* in tree harvesting, a notch that is cut in a tree below the level of the major cut and on the side or direction to which the tree will fall.

**underground** see SUBWAY.

**underground (coal) gasification** *Coal.* the technique of burning coal in place at the mining site, to produce a gaseous mixture that can be used as a fuel gas.

**underground mining** *Mining.* a method of mining that involves extraction by tunneling into the earth to gain access to a coalbed that is relatively far below ground; the coal is then mined with underground mining equipment such as continuous, longwall, and shortwall machines. Underground mines are described according to the type of tunnel used to reach the coal, i.e., drift (level), slope (inclined), or shaft (vertical). Thus, **underground mine.**

**underground storage** *Storage.* **1.** the short-term storage (weeks to months) of thermal energy in a geologic structure. **2.** *Oil & Gas.* the storage of natural gas in underground reservoirs at a different location from the one where it was produced.

**underground thermal energy storage (UTES)** *Geothermal.* the storage of thermal energy underground; e.g., an aquifer storage system that uses a natural underground layer such as sand, sandstone, or chalk as a medium for the temporary storage of heat or cold. Other technologies for underground thermal energy storage are borehole storage, cavern storage, and pit storage.

**underhole** *Mining.* **1.** to mine out or cut away a portion of the bottom of a coal seam so as to leave the top unsupported and ready to be worked. **2.** an excavation made in this manner.

**undershot wheel** *Hydropower.* a vertical water wheel turned on a horizontal shaft by the force of water striking against flat blades attached to its circumference as the water flows beneath the wheel; an older technology that is less efficient than an OVERSHOT wheel but easier to build and suited to a wider variety of sites.

***undershot wheel*** *The historic undershot wheel technology is suited to sites where the stream has a relatively low head (height of flow past the wheel).*

**Underwriters Laboratories (UL)** a private organization that tests and classifies electrical devices (and other equipment) for compliance with any of a large number of safety standards. Founded in Chicago in 1894 and now operating globally.

**undiscovered resources** *Consumption & Efficiency.* energy resources (e.g., coal or oil and gas) postulated from broad geologic information and theory to exist outside of known sites, on the basis of knowledge and theory.

**UNDP** United Nations Development Programme (UNDP) (est. 1965); an agency that provides technical assistance to developing countries via consultants' services, equipment, and fellowships for advanced study abroad.

**UNEP** United Nations Environment Programme (est. 1972), an organization that promotes international cooperation in efforts to promote the sustainable use of the environment.

**unibody** *Transportation.* unit body; a contemporary method of motor vehicle construction in which the body components are fastened onto a steel body shell rather than being bolted to a frame. Compare BODY-ON-FRAME.

**unified field theory** *Physics.* **1.** any theory that combines two or more field theories; e.g., Maxwell's unification of the field theories of electricity and magnetism by developing the theory of electromagnetism. **2.** specifically, the effort by Einstein and others to unify gravitational force and electromagnetic force with a single set of laws and, more generally, to provide a geometrical interpretation for all physical interactions.

**unified principle** *Thermodynamics.* the statement that when an isolated system performs a process after the removal of a series of internal constraints, it will reach a unique state of equilibrium.

**uniflow engine** *Consumption & Efficiency.* a steam-driven engine in which steam enters the cylinder through valves and exits through holes that are uncovered as the piston completes its stroke; it is designed to eliminate initial condensation.

**uniformitarianism** *Earth Science.* the principle that the physical processes that can be observed on earth today have operated in the same manner since the beginnings of the planet; thus ancient geological events can be explained or understood in terms of the phenomena and forces that occur in the present. Contrasted with the earlier theory of CATASTROPHISM.

***uniformitarianism*** *Uniformitarianism indicates that the Grand Canyon was gradually formed over very long periods of time, rather than being carved out by a single unique event.*

**uninterruptible power supply (UPS)** *Storage.* a unit employed to maintain a consistent power supply to an electronic device (e.g., a computer) in the event of an unwanted loss or sag (or a surge) in power; used for critical applications such as network servers, telecommunication systems, or medical, scientific, or military facilities.

**Union of Concerned Scientists** (est. 1969), a noted nonprofit nongovernmental organization that performs research, outreach and advocacy on a range of energy and environmental issues.

**United Nations (UN)** for UN agencies, see the acronym; e.g., United Nations Environment Programme, see UNEP.

**United States** for U.S. agencies, see the name (or acronym); e.g., U.S. Department of Energy, see DEPARTMENT OF ENERGY.

**unit energy consumption (UEC)** *Measurement.* the average annual amount of energy consumed by a certain end-use device or system; e.g., a household appliance or a computer.

**unit train** *Coal.* a term for a railroad train dedicated entirely to hauling coal between a certain coal mine and a specific destination, either a consumer such as an electric power plant or a transfer terminal, typically a long train of 100 cars or more.

**UNIVAC** *History.* Universal Automatic Computer; the earliest electronic computer to store data on magnetic tape and to be mass-produced; first in operation in 1951.

*UNIVAC Cutaway view of the memory unit of a UNIVAC computer, 1950s.*

**universal** *Transportation.* a geared joint placed between two shafts that allows the shafts to turn or swivel at an angle.

**universal coal cutter** *Coal.* a type of coal-cutting machine that is designed to make cuts universally; i.e., at any point between the top and bottom of the coal face.

**universal service obligation** *Electricity.* a commitment by a service provider to provide electricity services to all those who request it within a given area, often accompanied by a commitment to provide the service at a certain price.

**unleaded gasoline** *Oil & Gas.* gasoline that does not contain tetraethyl lead as an anti-knock agent, typically defined as a gasoline that contains not more than 0.05 grams of lead per gallon.

**UNSD** United Nations Statistics Division, an agency that collects energy statistics and updates and maintains the Energy Statistics Database that contains comprehensive energy statistics on more than 200 countries, regions, and areas.

**unstable** *Nuclear.* describing an isotope that is radioactive and gives off radioactivity until stability is reached.

**UPS** uninterruptible power supply.

**upstream** *Consumption & Efficiency.* **1.** referring to the stages of energy production prior to the final stage of processing and/or actual consumption by end users. **2.** referring to an earlier stage in any process with linear progression, such as a production process. Compare DOWNSTREAM.

**uptake** *Environment.* the addition of a substance of concern (such as carbon dioxide) to a sink or reservoir.

**upwelling** *Earth Science.* the movement to the ocean surface of deep, cold, nutrient-laden water, usually caused by a combination of prevailing coastal currents, wind direction, and the Coriolis force acting to pull surface water away from the coast.

**upwind** *Wind.* **1.** in the direction opposite to the way the wind is blowing; against the wind. **2.** describing a wind energy device that operates with the hub and blades facing toward the wind direction; the current majority of wind turbines have this design, which offers the advantage of avoiding the effect of wind shade behind the unit that would be present with a DOWNWIND machine.

**uranium** *Chemistry.* a metallic element having the symbol U, the atomic number 92, an atomic weight of 238.029, a melting point of 1132°C, and a boiling point of 3818°C; a dense silvery solid with three naturally occurring radioactive isotopes. See below.

**uranium-234 (U-234)** *Nuclear.* the least abundant of the three naturally occurring isotopes of uranium (an abundance of 0.006%), which is separated by extraction and used in nuclear research. U-234 decays by alpha emission into thorium-230.

**uranium-235 (U-235)** *Nuclear.* the second most abundant of the three naturally occurring isotopes of uranium (an abundance of about 0.7%). U-235 decays by alpha emission into thorium-231. It is the only isotope existing in nature to any appreciable extent that is fissionable by thermal neutrons. U-235 was the source of energy in the original atomic bomb.

**uranium-238 (U-238)** *Nuclear.* the most abundant and stable of the three naturally occurring isotopes of uranium (more than 99% of natural uranium). U-238 is normally nonfissionable but it absorbs neutrons to produce a radioactive isotope that subsequently decays to the isotope plutonium-239, which is fissionable.

**uranium concentrate** see URANIUM OXIDE CONCENTRATE.

**uranium endowment** *Nuclear.* uranium that is estimated to occur in rock with a grade of at least 0.01% $U_3O_8$. The estimate of the uranium endowment is made before consideration of economic availability and any associated uranium resources.

**uranium enrichment** see ENRICHMENT.

**uranium hexafluoride** *Nuclear.* $UF_6$, a white solid obtained by chemical treatment of $U_3O_8$, forming a vapor at temperatures above 56°C. It is the form of uranium required for the enrichment process.

**uranium ore** *Nuclear.* rock containing uranium minerals in concentrations that can be mined economically, typically 1 to 4 pounds of U-3O8 per ton or 0.05–0.20% U-3O8.

**uranium oxide concentrate** *Nuclear.* the solid product of uranium milling, nominally $U_3O_8$ but actually consisting of mixed

**Uranium** (chemical symbol: U) is a very heavy metal that constitutes a concentrated energy source that is used in nuclear power. Uranium was apparently formed in super novae about 6.6 billion years ago. It occurs in most rocks in concentrations of 2 to 4 parts per million, and in much lower concentrations in seawater. Its radioactive decay provides the main source of heat inside the earth, causing convection and continental drift. Uranium was discovered in 1789 by Martin Klaproth, a German chemist, in the mineral called pitchblende. It was named after the planet Uranus, which had been discovered eight years earlier. Uranium is the heaviest of all the naturally-occurring elements, being 18.7 times as dense as water. The high density of uranium means that it is used in the keels of yachts and as counterweights for aircraft control surfaces, as well as for radiation shielding. 'Natural' uranium as found in the earth's crust is a mixture largely of two isotopes: uranium-238 (U-238), accounting for 99.3% and uranium-235 (U-235) about 0.7%. The isotope U-235 is important because under certain conditions it can readily be split, yielding a lot of energy. When the nucleus of a U-235 atom captures a neutron it splits in two (fissions), releases energy in the form of heat, and emits two or three neutrons. If enough of these expelled neutrons cause the nuclei of other U-235 atoms to split, releasing further neutrons, a fission 'chain reaction' can be achieved. This is the process that occurs in nuclear reactor where the heat is used to make steam to produce electricity. Each fission of a U-235 atom releases about 200 MeV ($3.2 \times 10^{-11}$ joule)—about 50 million times as much energy as burning an atom of carbon. In other terms, a kilogram of natural uranium used in a typical reactor yields around 20,000 times as much energy as a kilogram of coal, and a kilogram of enriched nuclear fuel yields 160,000 times as much. Fission produces hundreds of different kinds of fission products (isotopes of much lighter elements), most of which are radioactive. In addition, a uranium atom may capture a neutron without splitting, leading the formation of a number of radioactive transuranic elements. These byproducts comprise nuclear waste.

**Ian Hore-Lacy**
World Nuclear Association

uranium oxides, hydrides, and impurities; also called *yellowcake* due to its typical color. Uranium is sent from the uranium mill to the refinery in this form.

**uranium reserves** *Nuclear.* estimated quantities of uranium in known mineral deposits of such size, grade, and configuration that the uranium could be recovered at or below a specified production cost, employing current mining and processing technology and under current law and regulations.

**uranium series** *Nuclear.* the 18-member radioactive decay series having uranium-238 as the parent substance and lead-206 as the final stable member. Some relatively long-lived members of this series include uranium-234, thorium-230, and radium-226.

**uranium tailings** *Nuclear.* residue materials left over from the processing of uranium ore in a mill. These residues contain several naturally-occurring radioactive elements, such as uranium, thorium, radium, polonium, and radon.

**urban airshed model** *Environment.* a computational tool that assembles data from various sources to simulate the physical and chemical processes affecting pollution concentrations in a given urban atmosphere; designed to identify key sources of pollutants and ascertain the conditions that negatively affect air quality.

**urban fabric analysis** *Environment.* a method of describing the surface cover characteristics of a given city, by determining the proportions of vegetative, roofed, and paved surfaces cover relative to the total urban surface of the city. This is used to identify areas that might benefit from reflective surfaces and reforestation with the goal of mitigating the heat island effect (i.e., reducing temperature and pollutant levels).

**urban heat island** see HEAT ISLAND.

**urbanization** *Social Issues.* the fact of becoming urban; the process by which the population of a country or region changes from primarily rural to primarily city-dwelling. This has major implications for energy use.

**Ure, Andrew** 1778–1857, British scientist noted as a leading advocate for international free trade and unregulated domestic industry; he argued that protectionism caused antagonisms which often led to war.

***urbanization*** *The skyline of Nairobi, Kenya; urbanization has proceeded at a rapid pace in certain areas of Africa in the past half century.*

**urea** *Chemistry.* $CO(NH_2)_2$, white crystals or powder with a saline taste that occur naturally as a product of protein metabolism; the first organic compound to be synthesized (1824) and still widely used in fertilizers and animal feed and for various industrial purposes.

**Urey, Harold Clayton** 1893–1981, U.S. physicist who led the research group that first isolated deuterium, or heavy hydrogen (1931). Although he subsequently worked on the Manhattan Project, Urey was also an advocate of nuclear arms control.

**used and useful** *Electricity.* a regulatory standard for a utility investment or expenditure that must be met in order to be included in the rate base or recovered as expenses from customers; it specifies that the given expense must be "used and useful" for public service. Thus, **used and useful test.**

**useful energy** *Consumption & Efficiency.* **1.** the actual energy used by a consumer to perform a desired function (heat, lighting, mechanical power, and so on). **2.** in general, any form of energy that serves a valid purpose for humans.

**user cost** *Economics.* **1.** the price paid by the end user of a given good or service. **2.** the value of all future opportunities for use that are lost when a given energy resource is produced and consumed now, rather than at some future time.

**USGS** United States Geological Survey (est. 1879); an agency that assesses and manages the nation's geological studies in four major areas: natural hazards, natural resources, the environment, and information and data management.

**utilitarian** *Social Issues.* describing a theory for the organization of society based on the goal of distributing goods to maximize the total utility of members of the society, with goods being interpreted broadly to include economic benefits, civil rights, freedom, and political power. Thus, **utilitarianism.**

**utility** *Economics.* **1.** an entity that provides a basic service to the public, such as water, energy, transportation, waste disposal, or telecommunications, and that typically is subject to government regulation. **2.** *Social Issues.* the welfare or satisfaction obtained by the public from the use or consumption of goods and services.

**utility-scale turbine** *Wind.* a term for wind turbines of sufficient size to be used to generate power for grid-based distribution and use.

**UV** ultraviolet.

**U-value** *HVAC.* a value that describes the ability of a material to conduct heat, measured as the number of Btu that flows through one square foot of the material in a period of 1 hour, when there is a difference of 1°F in the air temperature inside and outside a building. It is the reciprocal of the R-VALUE.

**V** or **v** volt; voltage.

**vacuum** *Physics.* **1.** in theory, a region that is completely devoid of any form of matter; a totally empty space. **2.** in practical terms, an enclosed region of space in which pressure has been reduced far enough below normal atmospheric pressure that processes occurring within the region are unaffected by the residual matter.

**vacuum evaporation** *Electricity.* the deposition of thin films of semiconductor material by means of the evaporation of elemental sources in a vacuum.

**vacuum gasoil** *Oil & Gas.* a feedstock for fluid catalytic conversion units, used to make gasoline, no. 2 heating oil, and other by-products.

**vacuum heating** *HVAC.* a steam-heating system in which a vacuum pump induces a pressure gradient to induce a return flow of air in the return pipes.

**vacuum tube** *Communication.* an electron tube that has been evacuated so that its electrical characteristics are essentially unaffected by the presence of residual gas or vapor; formerly in widespread use throughout the electronics industry, e.g., in radios and television sets, but now largely replaced by the transistor.

**vacuum zero** *Physics.* the energy of an electron at rest in empty space; used as a reference level.

**Valdez** *Environment.* (Exxon Valdez) a huge oil tanker that ran aground on Alaska's Bligh Reef in 1989 and spilled an estimated 10.8 million gallons of crude oil, fouling the waters of Prince William Sound and eventually affecting more than 1000 miles of beach in south-central Alaska. This spill caused widespread environmental damage, which according to a recent study continues to this day. The cleanup effort cost Exxon $2.5 billion.

**valence** *Physics.* a number indicating the combining power of one atom with others; that is, the number of other atoms with which it can combine.

**valence band** *Electricity.* **1.** the highest energy band in the spectrum of a solid crystal semiconductor or insulator that is occupied by electrons. Electrons cannot conduct in this band. **2.** *Photovoltaic.* an energy band in a semiconductor that is filled with electrons at 0 K.

**valence state** *Physics.* the energy content of an electron in orbit about an atomic nucleus.

**valley fill** *Mining.* the material that is removed from a mine site by a coal company during the process of mountaintop mining, and subsequently deposited in nearby valleys and streams.

**valuation** *Economics.* a procedure to place a monetary value on a good or service, including energy, other natural resources, and sometimes environmental services such as clean air or biodiversity.

**value added** *Economics.* **1.** the difference between the cost of materials purchased by a firm and the price for which it sells goods produced using those materials. **2.** more generally, any change to a good or service at a given stage that increases the amount a customer will be willing to pay for it.

**value-added tax (VAT)** *Policy.* a tax added onto a product during each step of production, from raw material to finished good.

**value-of-service ratemaking** *Economics.* a ratemaking procedure in which electricity is priced on the basis of what the market will bear. It takes into explicit consideration the alternatives available to customers; i.e., if few alternatives are available, customers will be likely to spend more to retain service, and prices can be set higher.

**value theory** *Economics.* a set of ideas that describes how values ascribed to goods and services are formed. Examples include the labor, utility, and energy theories of value.

**valve-regulated** *Storage.* describing a common industrial battery design that operates on the same principle as a sealed battery, in that it requires no replenishment of the water content of the electrolyte solution; so called because a relief valve controls the cell's internal pressure. Thus, **valve-regulated battery, valve-regulated lead acid battery (VLRA).**

**vanadium** *Chemistry.* a rare element having the symbol V, the atomic number 23, an atomic weight of 50.942, a melting point of 1900°C, and a boiling point of about 3000°C; occurring in certain minerals and obtained as a light-gray powder with a silvery luster or as a ductile metal; used in various alloys to increase the shock resistance of steel, as in automobile parts.

**vanadium redox flow battery** *Storage.* an electrochemical system that stores energy in two solutions containing different redox couples with electrochemical potentials sufficiently separated from each other to provide an electromotive force for the oxidation-reduction reactions to charge and discharge the cell; in effect more of a rechargeable fuel cell than a battery.

**Van de Graaff accelerator** *Nuclear.* a particle accelerator in which a charge is transported by an insulating belt to a conductor that builds in voltage as a result of charge collection, similar to the Van de Graaff generators used to demonstrate high voltages. The acceleration tubes for the particles are insulated with compressed gases. [Developed by U.S. physicist Robert J. *Van de Graaff,* 1901–1967.]

**Van de Graaff generator** *Electricity.* an electrostatic generator in which high voltages are produced by a moving belt of insulating material that collects electric charges by induction and discharges them inside a large spherical electrode; used for accelerating electrons, protons, and other nuclear particles.

**Vandergrift, Jacob** 1827–1899, U.S. pioneer of the bulk oil transportation industry; in the 1860s in Pennsylvania he established a profitable business for the shipment of barrels of oil by barge; these boats were the precursors of today's huge oil tankers.

**van der Waals, Johannes** 1837–1923, Dutch physicist known for his research on the gaseous and liquid states of matter.

**van der Waals equation** *Physics.* an equation of state that describes the behavior of real or nonideal gases by using two corrective terms that take into account the size of the molecules and the forces between them.

**van der Waals force** *Physics.* a general term for those forces of attraction between atoms or molecules that are not the result of chemical bond formation or simple ionic attraction; i.e., the relatively brief and weak interactions that neutral, chemically saturated molecules experience.

**vane** *Wind.* a thin, relatively flat object designed to align with an airflow; e.g., to align a wind turbine rotor correctly with the direction of the wind.

**Van Marum, Martinus** 1750–1837, Dutch chemist known for his electrostatic machines and for the discovery of ozone produced by electrical sparks in air. From his experiments, he concluded that the static and the galvanic forms of electricity are equal and have the same origin.

**Van Syckel, Samuel** U.S. oil buyer and shipper who built the first successful major oil pipeline in America (1865), in the Oil Creek, Pennsylvania region.

**van't Hoff's law** *Materials.* a law stating that a dissolved substance exerts the same osmotic pressure as it would if it were an ideal gas that occupied the same volume as the container. [Named for Dutch chemist Jacobus *van't Hoff,* 1852–1911.]

**vapor** *Chemistry.* **1.** a dispersion in air of a substance that is a liquid or a solid in its normal state, such as water. **2.** a gas whose temperature is less than the critical temperature, so that it may be converted to a liquid or solid by compression at constant temperature.

**vapor barrier** see VAPOR SEAL.

**vapor compression cycle** *Refrigeration.* a complete cooling cycle in which a liquid refrigerant is first made to boil into a vapor, thus producing a cooling effect in the medium or space of interest, and then this vapor is recompressed into its liquid state to continue the cycle. Thus, **vapor compression cooling.**

**vapor density** *Earth Science.* another term for absolute humidity; i.e., the mass of water vapor in relation to the unit volume of space that it occupies.

**vapor-dominated** *Geothermal.* describing a high-temperature hydrothermal system in which mixed water and steam exist, and steam is the pressure-controlling fluid phase. Vapor-dominated systems, such as The Geysers in northern California, are relatively rare but have high heat content and are well suited for the production of electrical power, while the more common **liquid-dominated** systems have lower heat content and are suited for direct heat uses.

**vapor engine** *History.* **1.** the name for an early combustion engine developed by U.S. engineer Samuel Morey, regarded as the first successful forerunner to the modern internal combustion engine. **2.** any engine powered by the expanding force of a vaporous fuel.

**vapor lock** *Transportation.* the interruption of a liquid flow by the formation of vapor or gas bubbles in the conduit, especially an obstruction of this type in the flow of gasoline to the engine in a motor vehicle.

**vapor pressure** *Chemistry.* **1.** the pressure at which a liquid and its vapor are in equilibrium at a given temperature. **2.** *Earth Science.* the pressure exerted by water vapor in the atmosphere, independent of any other gases or vapor.

**vapor recovery** *Environment.* the process of capturing chemical vapor, especially for the purpose of reducing pollution; e.g., the gasoline vapor generated by the fueling of vehicles at a service station.

**vapor recovery nozzle** *Transportation.* a special gas pump nozzle that reduces the release of gasoline vapor into the air when vehicles are refueled at a service station.

**vapor seal** *HVAC.* a barrier applied to a surface to prevent moisture from migrating into or out of a given space; a moisture-impervious layer such as plastic film, vinyl, or vapor-retardant paint.

**variable-pitch** *Wind.* describing a type of wind turbine for which the attack angle of the rotor blades can be adjusted, either automatically or manually, in order to respond efficiently to variations in wind speed and direction.

**variable-pitch propeller** *Transportation.* a propeller in which the angle of the propeller blades can be adjusted to suit different flight conditions while maintaining the same engine speed.

**variable-speed** *Wind.* describing a wind turbine in which the rotor speed increases and decreases with changing wind speed, producing electricity with a fluctuating level of power and voltage. Contrasted with a **fixed-speed** turbine, which has a virtually constant rotor speed (typically 1–2% variation).

**variable-torque load** *Conversion.* a load that requires low torque at low speeds and increasing torque as the speed is increased, with very high torque being required at high speeds.

**varistor** *Electricity.* variable resistor; a two-electrode resistor made of semiconductor material and having voltage-dependent nonlinear resistance that drops with an increase in applied voltage; can be used to protect sensitive equipment from power spikes or lightning strikes by shunting the energy to ground.

**VAT** value-added tax.

**VAWT** vertical axis wind turbine.

**vector** *Physics.* a physical quantity that has both a magnitude and a direction, such as velocity, acceleration, or force.

**vector-borne** *Health & Safety.* describing a disease that affects more than one species and that is transmitted by one host (the vector) to other host species. Many vectors are insects or arthropods; e.g., mosquitoes are the vector for malaria. An increase in the incidence of a vector-borne disease can be associated with changes in weather patterns; e.g., a prolonged drought.

**vehicle** *Transportation.* **1.** a self-propelled machine that is designed to transport passengers or goods over land, such as an automobile, truck, bus, or railway car. **2.** more generally, any conveyance that transports people, such as an aircraft, ship, or bicycle.

**vehicle control technology** *Transportation.* a term for various control systems intended to help avoid collisions, prevent or lessen injuries when crashes do occur, and ultimately lead to full vehicle automation; some existing forms include adaptive cruise control, antilock brakes, and electronic malfunction indicators.

**vehicle-mile** *Transportation.* the product of the number of vehicle trips and the average distance traveled, where one vehicle-mile corresponds to one vehicle traveling a distance of one mile, and so on. Similarly, **vehicle-kilometer.**

**vehicle operation stage** *Transportation.* the last stage of a transportation fuel cycle, which begins with refueling of vehicle fuel tanks. Emissions at this stage may come from fuel combustion or conversion, fuel evaporation, and brake and tire wear.

**vehicle-to-grid** *Renewable/Alternative.* describing a system in which power can be provided ("sold") to the electrical power grid by an electric-drive motor vehicle that is connected to the grid when it is not in use for transportation.

**velocipede** *History.* an early type of bicycle propelled by pushing the feet along the ground while straddling the vehicle.

**velocity** *Measurement.* the time rate at which a body changes its position; expressed in units of distance over time, such as miles per hour. Distinguished from *speed* in that velocity is a vector quantity and thus is always described in the context of direction, while speed is a scalar quantity that is described independent of direction.

**vent** *HVAC.* **1.** any opening designed to allow air, water, or pressure to enter or escape from a confined space, as in a building or mechanical system. **2.** to provide such an opening. **3.** *Earth Science.* an opening at the surface of the earth through which volcanic material is ejected, or the conduit through which such material passes. **4.** see DEEP-SEA VENT.

**ventilate** *HVAC.* to carry out or permit a process of ventilation.

**ventilation** *HVAC.* a process of moving or circulating air, so as to supply outside air to an enclosed space and/or remove stale air from the space, for the purpose of cooling, purification, moisture reduction, and so on; this may or may not involve mechanical conditioning.

**ventilation air methane (VAM)** *Coal.* methane released to the atmosphere from the ventilation systems of underground coal mines; because methane is a greenhouse gas, technologies have been developed to degasify the ventilation air through oxidation.

**ventilator** *HVAC.* a device used for ventilation; i.e., to produce an air flow or circulate air currents, such as a fan or blower.

**Venturi effect** *Wind.* **1.** an increased rate of flow of a fluid because of a constriction of the medium through which the fluid is moving; i.e., a "funnel effect", **2.** specifically, such an effect that increases wind velocity; e.g., as hot air currents rise from a valley floor surrounded by mountains. [Named for Italian physicist Giovanni B. *Venturi*, 1746–1822.]

**verified emission reduction** see EMISSION REDUCTION UNIT.

**Verne, Jules** 1828–1905, French author known for his works of science fiction that describe various technologies of the future, such as expanded use of electricity, spaceflights, undersea exploration, and the utilization of hydrogen as a fuel.

**vertical axis** *Conversion.* a turbine axis orientation in which the axis of rotation of the power shaft is perpendicular relative to the ground; an earlier technology than horizontal axis turbines.

**vertical axis wind turbine (VAWT)** *Wind.* a classification for wind turbines in which the axis of rotation is perpendicular to the wind stream and the ground; e.g., the Darrieus turbine.

**vertical integration** *Economics.* the degree to which a firm uses internal transfers rather than market transactions to connect successive stages of production. See next page.

**vertical multijunction (VMJ) cell** see MULTIJUNCTION CELL.

**vertical waterwheel** *Hydropower.* a traditional type of waterwheel consisting of a vertical wheel on a horizontally mounted axle; the undershotwheel, breastwheel, and overshot wheel are the three basic forms of this type.

**very high temperature reactor (VHTR)** *Nuclear.* an advanced nuclear reactor design featuring a graphite-moderated, helium-cooled reactor with a once-through uranium fuel cycle. It supplies heat with core outlet temperatures of 1000°C, which enables applications such as hydrogen production or process heat for a broad range of high-temperature and energy-intensive, nonelectric processes, as in the petrochemical industry.

**very large crude carrier (VLCC)** *Oil & Gas.* a supertanker; an oil tanker having a very large capacity.

**vibration** *Physics.* a process of oscillation (variation from one limit to another) about a position of equilibrium; an alternating or reciprocating motion.

**vertical integration** Vertical integration represents the degree to which a firm uses internal transfers to connect successive stages of production instead of doing it through the market. Since the time of Adam Smith, it has been known that all transactions in the economy are not necessarily most efficiently consummated using the price mechanism. The cost savings attributable to vertical integration (economies of vertical integration, EVI) arise from the technical interdependencies between the stages of production, the avoidance of transaction costs that could occur in a decentralized market (where the primary determinant is "asset specificity"), and the existence of different market imperfections, such as imperfect competition, uncertainty or asymmetric information. If vertical integration or desintegration occur in an industry, welfare may increase or decrease. Public policy then, like antitrust laws or public intervention, by regulating or deregulating certain industries, becomes an important issue. In the case of deregulation, the savings obtained from undertaking different activities together should be kept in mind by the legislator when restructuring an industry. For example, in the electricity industry there are many vertical disintegration experiences which have been carried out in different countries. The avowed aim has been to foster competition in the sector by introducing incentives to reduce costs. The difficulty of this policy lies in the existence of EVI, which stem from powerful technical interdependencies as well as from high market transaction costs. In a decentralised model, the problem of technical interdependencies can be solved with an independent system operator in the transmission stage, one who has authority over individual producers. In any case, the greatest problems for the deregulation process arise in the course of designing and operating the market.

**Javier Ramos Real**
Universidad de La Laguna
Canary Islands, Spain

**Vickrey auction** *Economics.* an auction in which the highest bidder wins but pays only the second-highest bid; the goal is to encourage bidders to bid the largest amount they are willing to pay. Proposed as a method to redesign markets in restructured, deregulated markets, as for electricity and natural gas. [Described by Canadian-born U.S. economist William *Vickrey,* 1914–1996.]

**video** *Communication.* **1.** the visual portion of a television signal, as opposed to the sound portion (audio). **2.** a motion picture or television program presented on magnetic tape **(videotape)** rather than on film.

**view factor** another term for SHAPE FACTOR.

**vinyl** *Materials.* **1.** $CH_2$=CH–, a radical of ethylene that is highly reactive and polymerizes easily; used as the base of many important plastics. **2.** a material made from this substance.

**virgin biomass** *Biomass.* living vegetation that has the potential for use as energy, as opposed to processed or waste materials.

**virgin coalbed methane** see COALBED METHANE.

**virgin fiber** another term for PRIMARY FIBER.

**visbreaking** *Oil & Gas.* a thermal cracking process in which heavy atmospheric or vacuum-still bottoms are cracked at moderate temperatures to increase production of distillate products and reduce viscosity of the distillation residues.

**viscoelasticity** *Materials.* **1.** a condition of a liquid or solid that exhibits viscosity but also memory of past deformation, with the ability to store energy elastically and to dissipate energy due to viscosity of the medium. **2.** the manifestation of this quality in a polymer. Thus, **viscoelastic.**

**viscosity** *Materials.* the internal friction of a fluid; the resistance to flow exhibited by a liquid or gas subjected to deformation.

**viscous** *Materials.* having a high degree of friction between component molecules as they move by each other.

**visible spectrum** see SPECTRUM (def. 2).

**Vision 21** *Sustainable Development.* the concept that future power plants can be operated without the emissions of harmful pollution that current power plants release to the atmosphere. One version of this has been defined by the U.S. Office of Fossil Energy. [So called because it is described as a plan for the 21st century.]

**vis mortua** *History.* dead force; the potential force inherent in a body that is at rest and exerting pressure but doing no active work; a concept employed (as by Gottfried Leibniz) to investigate the fundamental nature of energy. Compare VIS VIVA.

**visual pollution** *Environment.* a subjective term for graffiti, signs, billboards, outdoor advertising, power lines, neon lighting, telephone towers, and other such features of the industrial landscape that are perceived to be offensive, unattractive, or otherwise negatively affecting the visual environment.

**vis viva** *History.* living force; the force of a body in motion as it moves against resistance and does active work; a concept employed (as by Gottfried Leibniz) to investigate the fundamental quantity of motion.

**vitalism** *History.* a (former) theory that life depends on a unique force and cannot be reduced to chemical and physical explanations.

**vitreous fluid** see TWO-FLUID THEORY.

**vitrification** *Nuclear.* **1.** the progressive fusion of a material during the firing process; as it proceeds, glassy bonding increases and the porosity of the fired product decreases. **2.** *Nuclear.* the incorporation of high-level wastes into borosilicate glass, designed to immobilize radionuclides in an insoluble matrix ready for permanent disposal.

**VLCC** very large crude carrier.

**VOC** volatile organic compound.

**void** *Nuclear.* **1.** an empty space that can vary in size from that of one atom to a microscopically visible area within a group of atoms that can be filled with different atoms. **2.** *Nuclear.* a bubble of gas in the coolant of a reactor system, or a larger loss of coolant in the reactor core.

**volatile** *Chemistry.* having the property of volatility; readily converted to a vapor at a relatively low temperature.

**volatile matter** *Materials.* **1.** the products (other than moisture) given off by a material in the form of gas or vapor, as in a process of combustion. **2.** *Coal.* specifically, those products of coal (exclusive of moisture) that are given off in the form of gas and vapor.

**volatile organic compound (VOC)** *Environment.* any of various organic compounds that easily become vapor or gases. They are released from burning fuel, such as gasoline, wood, coal, or natural gas, and from solvents, paints, glues, and other such consumer products. When combined with nitrogen oxides, VOCs react to form ground-level ozone, or smog. Many of these compounds can pose serious health risks to exposed populations.

**volatility** *Chemistry.* **1.** the property of a liquid having a low boiling point and a high vapor pressure at ordinary pressures and temperatures. **2.** a tendency for energy prices to fluctuate widely and unpredictably, especially the price of crude oil.

**volcanism** *Earth Science.* any of the processes in which magma and its associated gases rise up from the earth's interior and are discharged onto the surface and into the atmosphere.

*volcanism Lava and debris from a Hawaiian volcano block a nearby highway.*

**volcano** *Earth Science.* **1.** a vent or fissure in the earth's surface through which magma and its associated materials are expelled. **2.** the large, generally conical structure formed by an accumulation of such expelled material. [From *Vulcan,* the ancient Roman god of fire, especially destructive fire.]

**volcanology** *Earth Science.* the branch of science concerned with volcanoes; the study of volcanic eruptions.

**volt** *Electricity.* a standard unit of potential difference or electromotive force, symbol V;

equivalent to the potential difference between two points requiring one joule of work to move one coulomb of electricity from the point of lower potential to the point of higher potential. Alternatively, one volt is equivalent to the electromotive force that will send a current of one ampere through a resistance of one ohm. [Named for Alessandro *Volta.*]

**Volta, Alessandro** 1745–1827, Italian scientist renowned for his pioneering work in electricity. He investigated the "animal electricity" of Galvani, and found that the current was generated from the contact of dissimilar metals, and that the frog leg was only acting as a detector. Using this information, he constructed the first battery to produce electricity (1800).

**voltage** *Electricity.* the potential difference between two electrodes, as measured in volts.

**voltage depression** *Storage.* an effect in which the peak voltage of a battery drops more quickly than normal as it is used, even though the total power remains almost the same; often caused by repeated overcharging of the battery.

**voltage drop** *Electricity.* the voltage difference between any two specific points in a circuit.

**voltage-keyed** *Storage.* describing a system that incorporates a mechanical identifier on batteries and devices to ensure that only batteries of the correct voltage are connected to the device.

**voltage law** see KIRCHHOFF'S VOLTAGE LAW.

**voltage regulator** *Electricity.* a circuit or device that produces a nearly constant voltage output, even though the voltage input (line) and current output (load) may vary widely. Thus, **voltage regulation.**

**voltage reversal** another term for CELL REVERSAL.

**Voltaic pile** *History.* the first source of a steady electric current, a simple form of electric battery developed by Alessandro Volta in 1799; it consists of alternating zinc and silver disks separated by felt soaked in brine. Also, **Volta's pile.**

**volt-ampere** *Electricity.* a standard unit of apparent power, equal to the product of one volt and one thousandth of an ampere.

**Volta's pile** see VOLTAIC PILE.

**voltmeter** *Measurement.* an instrument that measures voltage; i.e., the potential difference between two points.

**volume** *Measurement.* **1.** a measure of the amount of space that is occupied by a physical object in a three-dimensional context. **2.** *Transportation.* the total flow of traffic past a given point or along a stretch of road, typically expressed in vehicles per hour or per day.

**volumetric** *Measurement.* **1.** having to do with or based on volume. **2.** *Electricity.* describing a rate schedule based on the quantity (volume) of electricity transmitted across the system. Thus, **volumetric rate, volumetric (wires) charge.**

**von** for names with *Von* or *von,* see the last name; e.g., for Julius Von Sachs, see SACHS; for Werner von Siemens, see SIEMENS.

**vortex** *Physics.* a flow field in which fluid particles move in concentric paths.

**vulcanization** *Materials.* **1.** a chemical reaction that causes the cross-linking of rubber with sulfur, generally for the purpose of making the material harder, less soluble, and more durable. **2.** any of various similar cross-linking reactions of polymers. [Named for *Vulcan,* the Roman god of fire, because of the intense heat required for this process.]

**vulcanized rubber** *Materials.* a form of rubber with improved characteristics such as great strength, elasticity, and resistance to solvents and moderate temperature fluctuations. The process involves the formation of cross-linkages between the polymer chains of the rubber's molecules, usually by a process involving combination with sulfur and heating.

**vulnerability** *Climate Change.* the degree to which a population, species, ecosystem, agricultural system, or other biological entity is susceptible to, or unable to cope with, the adverse effects of climate change.

**W** watt.

**wafer** *Electricity.* a thin, flat sheet of semiconductor material (e.g., photovoltaic material) made by cutting it from a single crystal or ingot.

**wagonway** *History.* the earliest known type of railway, a track with wooden or iron rails intended to move wagons of coal easily; built in Britain. Also spelled **waggonway.**

**walking dragline** *Coal.* a large-capacity shoveling machine designed to expose deeply buried coal seams that smaller machines cannot reach; it "walks" along the surface by the alternate movement of large mechanical feet.

**walking plow** *History.* the earliest form of plow, pulled by oxen or horses with a human walking behind to control the animal and maintain the plowing line.

***walking plow*** *The walking plow, now generally replaced by mechanized plows in industrialized countries, is relatively light in weight and usually without wheels.*

**wall energy** *Physics.* a term for the energy per unit area existing between two ferromagnetic domains that have different orientations.

**Wankel engine** *Transportation.* a vehicle engine operating with a three-part rotor that turns in a close-fitting chamber, so that the power stroke is applied sequentially to each of the faces of the rotor. Compared to a reciprocating engine, a Wankel engine is simpler, lighter, contains fewer moving parts, and can burn lower octane fuel. However, it produces incomplete combustion and therefore high emissions and typically consumes more fuel than a piston engine. To date Mazda has been the only carmaker to market this type of engine. [Developed by German engineer Felix *Wankel,* 1902–1988.]

**warm** *Lighting.* a subjective description of the way in which the human eye perceives a certain light source. Colors at the red end of the spectrum are considered to be "warm" and colors at the blue end are considered to be "cool", based on the traditional association of red with heat or fire, and blue with cold or ice. (In terms of COLOR TEMPERATURE, warm sources actually have a lower temperature than cool sources.) Thus, **warm light(ing), warmth,** and so on.

**waste biomass** *Biomass.* a collective term for substances that can be used for biomass energy, including municipal solid waste, sewage, manure, forestry and agricultural residues, and certain types of industrial wastes; contrasted with VIRGIN BIOMASS.

**waste heat** *Thermodynamics.* the portion of the energy input to a mechanical process that is rejected to the environment.

**waste heat recovery** *Renewable/Alternative.* any system or process that actively captures byproduct heat that would otherwise be discharged into the environment, so that it can used for other purposes; e.g., the use of the heat from exhaust gases to heat water for various purposes.

**Waste Isolation Pilot Plant (WIPP)** *Nuclear.* the first underground repository in the U.S. licensed to permanently dispose of transuranic radioactive waste left from the research and production of nuclear weapons; located in the Chihuahuan Desert of southeastern New Mexico.

**waste spark system** *Transportation.* an ignition system that energizes the spark each time the piston is at top dead center, whether it is on the compression stroke or the exhaust stroke; for hydrogen engines, the waste sparks serve as a source of pre-ignition.

**waste-to-energy** *Renewable/Alternative.* describing a process that generates energy, usually power generation, from waste materials, especially by the incineration of municipal solid wastes (MSW). Thus, **waste-to-energy system, waste-to-energy technology.** See below.

**wastewater** *Environment.* used water; a broad term for any water that has been used by a household, farm, business, industry, and so on, and then returned to the environment, typically containing dissolved or suspended matter that can cause pollution if not properly treated.

> **waste-to-energy** The process in which waste is used to generate useful energy—electricity, heat, or both. This is possible (and convenient) when the heat generated by burning the waste is high enough to warrant satisfactory combustion conditions and make available enough energy to overcome losses and auxiliary consumption: in practice, a lower heating value of at least 4 MegaJoules per kg. Waste-to-energy is the offspring of waste incineration, which was originally introduced to sterilize and reduce the volume of waste by combusting it in a furnace. Modern waste-to-energy plants allow the export of energy, with very low environmental impact. The plant comprises four basic sections: waste combustor, recovery boiler, flue gas treatment and steam cycle. The design of the combustor varies widely with the waste characteristics: physical state (solid versus liquid), size distribution, heating value, ash and moisture content, etc. Municipal solid waste typically is burned on a moving grate, where it is kept 20–30 minutes until it is completely combusted. The hot gases generated in the combustor go through the recovery boiler to generate steam, which is used directly as heat carrier or sent to a steam turbine to produce power. Flue gases are treated by adding reactants called sorbents and by filtering the particulate matter. A modern, large plant treating half-million tons of municipal solid waste per year can generate more than 400 million kWh per year, meeting the electricity needs of more than 150,000 families.
>
> **Stefano Consonni**
> Politecnico di Milano, Italy

**water** *Chemistry.* $H_2O$, a colorless, odorless, tasteless liquid having a melting point of 0°C and a boiling point of 100°C at standard atmospheric pressure, and having the allotropic forms of ice (solid) and steam (vapor). The most commonly found substance on earth and present in many other substances, including all organic tissues.

**water clock** *History.* an early timekeeping device based on the uniform flow of water out of or into a graduated vessel.

**water-cooled** see LIQUID-COOLED.

**water-cooled graphite-moderated reactor** *Nuclear.* a pressurized water reactor with individual fuel channels, using ordinary water as its coolant and graphite as its moderator. It differs from most power reactor designs in this respect and in its use for both plutonium and power production. It uses enriched uranium as a fuel. Its design characteristics were shown in the Chernobyl accident to cause instability at low power levels.

**water cycle** see HYDROLOGIC CYCLE.

**waterdrive** *Oil & Gas.* a mechanism to inject water into a reservoir to control the pressure as oil or gas is extracted.

**waterfall** *Hydropower.* **1.** a natural site in a river or stream where the course is interrupted by an abrupt descent in level, causing the water to drop more or less vertically for a significant distance. **2.** any large downward flow of water in a vertical direction.

**water gas** *Coal.* a gas composed primarily of hydrogen and carbon monoxide produced through the interaction of steam with incandescent carbon, usually from anthracite coal or coke; used for lighting (primarily during the 19th century and early 20th century) and as fuel (well into the 20th century).

**water-gas shift (WGS)** *Hydrogen.* the reaction of water and carbon monoxide to produce hydrogen and carbon dioxide; can be carried out as an industrial process to produce hydrogen for use as fuel, as in fuel cell applications. A **reverse water gas shift (RWGS)** converts carbon dioxide into carbon monoxide and water.

*waterfall* *The energy of Niagara Falls has been employed for human use at least since 1759, and the Falls have been a major source of electrical power since the 1880s.*

**water hyacinth** *Biomass.* a floating aquatic plant, *Eichhornia crassipes,* of the family Pontederiaceae, that grows very rapidly on the surface of tropical and subtropical rivers, lakes, and swamps; it has potential as a biomass energy source.

**water jacket** *HVAC.* **1.** a heat-exchanger element in a boiler; water is circulated with a pump through the jacket, so that it picks up heat from the combustion chamber for circulation to heat distribution devices. **2.** a casing filled with circulating water, used in water-cooling an engine or other device.

**waterlogging** *Environment.* the saturation of soil with water, causing the water table to rise high enough to expel normal soil gases and interfere with plant growth or cultivation.

**water mill** or **watermill** *Hydropower.* a milling (grinding) operation that employs water as its energy source, especially the force of falling water to turn a wheel or turbine.

**water pollution** *Environment.* a general term for the discharge or release of harmful or unwanted substances to the ocean, lakes, rivers, and other bodies of water, from such sources as household wastes (e.g., sewage), agriculture (e.g., fertilizers, manure), or runoff from factories, refineries, waste-treatment plants, and the like. This may also include indirect contaminants that enter the water supply from soils or groundwater systems and from the atmosphere via precipitation. Thus, **water pollutant.**

**water power** or **waterpower** *Hydropower.* a general term used to describe both the power generated by the actions of water (streams, rivers, lakes, seas, and so on) and the useful mechanical and electrical energy derived from these actions.

**water-reactive** *Materials.* describing a substance that is considered to be dangerous when it undergoes a chemical reaction with water. This reaction may release a gas that presents a toxic health hazard or that can lead to spontaneous combustion or explosion.

**water turbine** *Hydropower.* a device or machine that generates rotary mechanical power from the energy of a stream of water.

**water vapor** *Earth Science.* atmospheric water that is in its gaseous state, especially when it is in such a state below its boiling point.

**water vapor feedback** *Climate Change.* a positive feedback loop in the atmosphere due to water's strong role as a greenhouse gas; an increase in temperature will increase the water-holding capacity of the atmosphere, which in turn will lead to an increase in the amount of atmospheric water vapor, reinforcing the initial increase in temperature.

**water wheel** or **waterwheel** *Hydropower.* a wheel arranged with floats or buckets so that it can be turned by the force of flowing water; used as a source of energy to drive machinery or raise water.

**Watt, James** 1736–1819, Scottish mechanical engineer known for his highly efficient model of the steam engine. He devised significant improvements to Thomas Newcomen's prior steam engine that resulted in an entirely new type of engine (see WATT ENGINE). Richard Arkwright began using Watt's steam engine in his textile factories in 1783, and by 1800 there were over 500 of these machines in Britain's mines and factories. Watt also coined the term *horsepower,* and he invented the flywheel.

**watt** *Thermodynamics.* a rate of doing work or converting energy, symbol W. It is the metric measuring unit; a rate of doing work of one joule per second is one watt. In electrical power it is equivalent to the current in amperes multiplied by the electrical potential in volts.

**Watt (steam) engine** *Conversion.* an improved type of engine patented in 1769 by James Watt, featuring a separate condensing chamber, an air pump to bring steam into the chamber, and insulation of parts of the engine. It was four times more powerful than the preceding Newcomen engine. The Watt engine is considered one of the driving forces of the Industrial Revolution.

**watt-hour** *Thermodynamics.* a unit of energy equal to the work done by one watt over a time of one hour; symbol Wh.

**wave** *Physics.* **1.** a uniformly advancing disturbance in a medium, in which the moved parts undergo a double oscillation; a collective disturbance that propagates at a definite speed. **2.** any disturbance or sudden change in conditions that moves through a flow. **3.** *Earth Science.* such a disturbance occurring on the surface of the sea or other body of water, taking the form of a moving swell or ridge of water.

**wave energy** *Renewable/Alternative.* energy generated by the ocean's wave currents, especially wind-generated waves; i.e., the kinetic energy of moving water particles and the potential energy of elevated particles.

***wave energy*** *Waves contain large amounts of energy and various efforts have been made in recent years to take advantage of this, though in general the technology is still at the research and development stage.*

**wavelength** or **wave length** *Physics.* the spatial distance between adjacent points of equal phase.

**wave mechanics** *Physics.* a formulation of quantum mechanics developed by physicist Erwin Schrödinger, in which the state of a system is described by a wave function, and physical quantities or their expectation values are obtained by operation on the wave function by the appropriate operators.

**wave theory** *Physics.* the theory that light travels as a wave, originally stated by Huygens (1690), as opposed to the conflicting view that it is a stream of tiny particles, proposed by Newton (1704); modern quantum theory describes light as having both wave and particle characteristics.

**wax** *Materials.* a solid or semi-solid material derived from petroleum distillates or residues by such treatments as chilling, precipitating with a solvent, or de-oiling. It is a light-colored, generally translucent crystalline mass, slightly greasy to the touch, consisting of a mixture of solid hydrocarbons in which paraffin predominates.

**WBCSD** World Business Council for Sustainable Development (est. 1991); a coalition of 170 international companies who share a commitment to sustainable development via economic growth, ecological balance, and social progress. It also includes a regional network of Business Councils for Sustainable Development (BCSDs), mostly located in developing countries.

**WCD** World Commission on Dams.

**WCRP** World Climate Research Programme (est. 1980); a group whose goal is to develop a fundamental understanding of the physical climate system and climate processes, in order to determine to what extent climate can be predicted and the extent of human influence on climate.

**weak force** *Physics.* one of a class of forces between particles that are much weaker than the strong forces that hold the nucleus together; e.g., the neutrino, a weakly interacting particle, will travel through approximately $10^{16}$ cm of iron before scattering, whereas the neutron, a strongly interacting particle, will travel through only about 10 cm before scattering. Also, **weak interaction.**

**weak grid** *Electricity.* a term for a part of an electrical grid far from the main generation units, or for an isolated power system, as on an island.

**Wealth of Nations** *History.* a book published in 1776 by Scottish philosopher Adam Smith; in full *An Inquiry into the Nature and Causes of the Wealth of Nations;* it establishes economics as a distinct field of study and presents the idea that "the invisible hand" of the free market, and not government, is the proper controlling force for a nation's use of resources.

**weapons-grade** *Nuclear.* **1.** describing plutonium containing over 93% of PU-239, the plutonium isotope most suitable for weapons use. **2.** describing uranium containing 90% or more U-235. Also, **weapon-grade.**

**weapons of mass destruction (WMDs)** *Global Issues.* a collective term for various weapons that can cause death and destruction on a large scale, such as nuclear bombs, disease-causing biological agents, and toxic chemicals such as poison gas.

**weather** *Earth Science.* **1.** the short-term state of the atmosphere, as distinguished from the long-term conditions of *climate;* this includes temperature, humidity, precipitation, wind, visibility, and other factors, chiefly considered in terms of their effects on organic life and human activity. **2.** a specific localized atmospheric state at a given time, such as rain or snow.

**weather derivative** *Economics.* a financial vehicle to help utilities manage their weather risk that gives payoffs dependent on the weather; in this context weather usually is measured by a specific deviation in the temperature from a long-term average.

**weathering** *Earth Science.* the natural processes by which the actions of atmospheric and other environmental agents, such as wind, rain, and temperature changes, result in the physical disintegration and chemical decomposition of rocks and earth materials in place, with little or no transport of the loosened or altered material.

**weatherization** *HVAC.* the process of reducing the loss of heating or cooling effects from or into a building; e.g., caulking, weatherstripping, adding insulation, installing storm doors, tinting windows, and so on.

**weatherstripping** *HVAC.* **1.** the process of filling in unwanted cracks or spaces around windows, doors, or other building openings, in order to reduce the passage of air and moisture. **2.** the material employed for this purpose; weatherstripping is marketed in strips or rolls of metal, vinyl, or foam rubber.

**Weber** Ernst Heinrich (1795–1878), Wilhelm Eduard (1804–1891), and Eduard Friedrick Wilhelm (1806–1871), German scientists who made pioneering observations of the energetics and physics of human locomotion. Their work established the mechanism of muscular action on a scientific basis.

**WEC** World Energy Council.

**wedge** *History.* an instrument consisting of two inclined planes that merge in the front to form an edge, used to exert pressure for cutting, propping up, or spreading apart; classified as one of the fundamental simple machines.

**Weibull distribution** *Measurement.* a statistical distribution function with wide applications ranging from material fatigue analysis, reliability, and lifetime modeling to weather forecasting. It often is used to describe the time that the wind spends blowing at a particular speed at a particular location, a key factor in siting wind turbines. [Proposed by Swedish engineer Waloddi *Weibull,* 1887–1979.]

**weight** *Measurement.* **1.** the gravitational force experienced by a body on the earth's surface or in some other gravitational field. See also MASS. **2.** a system for determining the mass of objects; e.g., avoirdupois weight.

**Weinberg, Alvin** born 1915, U.S. physicist known for his leadership in nuclear power and U.S. energy policy, especially in his role as director of Oak Ridge National Laboratory (1955–1973) and the Institute for Energy Analysis (1975–1985).

**weir** *Hydropower.* a low dam designed to permit water to overflow across its entire length, but also to direct water through a specific channel to hydropower turbines when the water level is below this overflow height. This channeling can be done by means of a **weir gate** that is opened to allow the flow of water.

**well** *Mining.* a hole, usually vertical, bored into the earth to gain access to oil, gas, water, or other substances.

**wellbore** *Oil & Gas.* the hole made in the ground for a well.

**wellhead** *Oil & Gas.* **1.** the top level of a well, from which oil pumped out of the well is allowed to flow freely. **2.** the equipment installed at the surface of a well.

**wellhead price** *Oil & Gas.* the price paid for oil or gas at the well site, excluding later charges for processing, transportation, and distribution.

**well log** *Oil & Gas.* a graphic record of the measured physical characteristics of a subsurface rock section as encountered in an oil or gas well. Thus, **well logging.**

**well-to-pump** *Transportation.* describing the aspect of vehicle fuel use including activities from the point at which fuel feedstock is recovered from wells, to the point at which fuels are available for purchase at the pumps of vehicle fueling stations.

**well(s)-to-wheels** *Transportation.* a term for all activities involving motor vehicle fuel beginning with the recovery of the fuel feedstock (the "well" stage) and proceeding through all intermediate steps to the generation and use of energy to move vehicles (the "wheels" stage). See below.

**well-to-wheels analysis** *Transportation.* an analysis involving the entire WELL-TO-WHEELS scope of motor vehicle fuel activities; it focuses in particular on energy use and emissions of energy feedstock production and transportation, on fuel production, transportation, and distribution, and on fuel use in vehicles.

**Welsbach, Carl von** 1858–1929, Austrian chemist and engineer who discovered two rare-earth elements and invented the incandescent gas mantle (WELSBACH MANTLE). He also developed a metal filament light bulb (1898) that was an improvement on the existing carbon filament designs.

**Welsbach mantle** or **burner** *Lighting.* a mantle made of cotton fabric impregnated with rare earth that becomes incandescent when exposed to a gas flame, now used in outdoor and camp lamps. Its development (1885) greatly improved the effectiveness of modern liquid fuel lighting.

**Wender, Irving** born 1915, U.S. chemical engineer noted for his pioneering improvements in the chemistry of coal combustion; a leader in the conversion of coal to liquids and chemicals by indirect liquefaction, and by novel methods of direct liquefaction.

**West Texas Intermediate (WTI)** *Oil & Gas.* a light crude oil of high quality having a low API gravity and low sulfur content; a preferred source for refining into motor gasoline.

**wells-to-wheels** This refers to the analysis of energy and environmental effects of production, storage, and distribution of transportation fuels and their use by motor vehicles. For conventional vehicles powered by petroleum-based gasoline and diesel, such analysis begins with petroleum recovery in oil fields (oil wells) and ends with energy delivered at vehicle wheels. Such analysis is also commonly called fuel-cycle analysis in the transportation field. In consumer products research, similar analyses are often called "life-cycle" or "cradle-to-grave" analyses. For new vehicle propulsion technologies powered with new transportation fuels, the traditional comparison of energy use and emissions related to vehicle operations between them and conventional vehicle technologies does offer a complete comparison, since energy and environmental burdens of some of the new systems may occur during production, storage, and distribution of fuels. This warrants WTW analyses. Earlier WTW analyses were driven mainly by the introduction of battery-powered electric vehicles; and recent WTW analyses stem primarily from interest in hydrogen-powered fuel-cell vehicles. In both cases, while vehicle operations have zero emissions, there are emissions associated with production and distribution of electricity and hydrogen. To allow comparison with traditional analyses covering only vehicle operations, results of WTW analyses are often separated into two groups: wells-to-pumps (WTP) and pumps-to-wheels (PTW). WTP stages start with recovery of energy feedstocks (such as petroleum, natural gas, and coal) and end with fuels available in pumps at vehicle refueling stations. PTW stages cover vehicle operation activities. Recent completed WTW analyses include vehicle propulsion technologies such as international combustion engines, hybrid electric vehicles (powered with international combustion engines), fuel cells, and battery-powered motors. These analyses include gasoline, diesel, ethanol, methanol, compressed natural gas, liquefied petroleum gas, Fischer-Tropsch diesel, hydrogen, and electricity.

**Michael Wang**
Argonne National Laboratory

WTI is a marker crude oil and is the basis for the NYMEX futures price for crude oil. Also, **West Texas Crude.**

**Westinghouse, George** 1846–1914, U.S. inventor and manufacturer, noted for his prolific contributions to railroads and electric power generation and transmission. He formed and directed more than 60 companies to market his and others' inventions during his lifetime, and his Westinghouse Electric Company became one of the largest manufacturing organizations in the U.S.

**Westinghouse (air) brake** *History.* a major advance in safety developed by George Westinghouse (1869); this device for the first time enabled trains to be stopped with accuracy by the locomotive engineer, and it was eventually adopted on the majority of the world's railroads.

**West Nile virus (WNV)** *Health & Safety.* a mosquito-transmitted viral disease that poses significant risk for wildlife, zoo, and domestic animal populations, with occasional spillover to humans. The epidemiology of WNV plays a role in studies of climate change because of a tendency for outbreaks of this disease to be associated with episodes of severe drought.

**wet-bulb temperature** *Measurement.* the temperature to which air will cool when water is evaporated into unsaturated air; wet-bulb temperature and dry-bulb temperature are used to compute relative humidity.

**wet-bulb thermometer** *Measurement.* a thermometer whose bulb is encased by a cloth sleeve saturated with water; used to measure the evaporation of water into the air.

**wet cell** *Storage.* a battery using a liquid electrolyte that is allowed to flow freely within the cell casing. Most automobile and marine batteries are wet cells.

**wet deposition** see DEPOSITION.

**wetland(s)** *Earth Science.* any section of low-lying land that is periodically submerged or whose soil contains a very high level of moisture.

**wet natural gas** *Oil & Gas.* a term for natural gas containing impurities and heavier hydrocarbons, such as propane and butane, which must be removed by distillation to provide a more desirable product.

**wet shelf life** *Storage.* the period of time that a charged battery, when filled with electrolyte, can remain unused before it fails to attain a specified level of performance.

**wet steam** *Geothermal.* a mass of water that is (much) hotter than the usual boiling point, but which remains in liquid form because of high surrounding pressures; a common feature of underground geothermal areas.

**wet steam (power) plant** *Geothermal.* a geothermal power plant in which wet steam (i.e., extremely hot water) is raised to the surface through wells. Thus deprived of the high subsurface pressure that maintained it in the liquid state, it becomes steam to power a turbine directly or to heat another fluid to drive a turbine that generates electricity.

**W-factor** see W-VALUE.

**WGS** water gas shift.

**Wh** watt-hour.

**whale oil** *History.* an oil that is derived from whale blubber, especially from the sperm whale; formerly used as a fuel for lamps.

**wheel** *History.* a circular frame or disk designed to revolve around a central axis; the development of wheeled vehicles in ancient times was a crucial step in the development of human civilization. The earliest known site for the use of the wheel is in present-day Syria, ca. 3200 BC.

**wheelbase** *Transportation.* the distance from the front to the rear wheels of a vehicle, measured between the ground contact centers of each wheel; usually expressed in inches in the U.S.

**wheel horsepower** *Transportation.* horsepower measured at the actual drive wheels of a car, taking into account the load from the chassis and all accessories; this is the most accurate measure of the amount of energy that the car actually generates to move it forward.

**wheeling** *Electricity.* another term for electricity transmission; i.e., the distribution of power from one geographical location to another within an electric power system.

**wheeling-out** *Electricity.* power transmitted out of one transmission area and into another, typically accompanied by a change in ownership of the power.

**White, Leslie Alvin** 1900–1975, U.S. anthropologist known for his writings on the role of energy technology in cultural evolution. He viewed increased energy use as the principal driver not only of material wealth, but also of the norms and values that shape society.

**white coal** *Coal.* a term for an impure form of coal that represents a transitional stage between cannel coal and oil shale.

**whitedamp** or **white damp** *Mining.* a miner's term for the poisonous gas carbon monoxide, a gas that might accumulate in a mine as a product of incomplete burning, as in the case of a blasting operation, or in the aftermath of a mine fire or explosion. [So called because of its colorless property, in contrast with another poisonous mine gas known as BLACKDAMP.]

**white light** *Physics.* **1.** sunlight integrated over the visible portion of the spectrum, so that all colors are blended to appear white to the eye. **2.** any light that is perceived as colorless.

**white spirit** *Oil & Gas.* a term for refined distillate intermediates with a distillation in the naphtha/kerosene range.

**Whitney, Eli** 1765–1825, U.S. engineer famous for his invention of the cotton gin (1794), a machine for separating cotton fibers from the seeds. This invention had immense economic and social effects in the U.S. He went on to establish a firearms factory based on methods comparable to modern mass industrial production, including the use of standardized, interchangeable parts.

**Whittle, Frank** 1907–1996, English aeronautical engineer who invented the turbojet engine (1930), the first application of the gas turbine to jet propulsion. The historic first flight of an aircraft with his turbojet engine took place in 1941.

**whole-body counter** *Health & Safety.* a device used to identify and measure radioactive materials in the body of humans and animals.

**whole-body exposure** *Health & Safety.* relatively uniform exposure to radiation of tissues throughout the body, rather than its being concentrated in a single organ.

**whole-house fan** *HVAC.* a large stationary fan capable of cooling a house by extracting a large volume of warm air when the outside air is cooler; typically placed so as to draw cool air in through open windows and expel warmer air through an attic vent.

**wholesale** *Economics.* **1.** describing transactions in which a merchant purchases goods in large quantities with the intent of offering them for sale in smaller quantities to end-use consumers; e.g., the sale of fuel to a service station by an oil corporation or trading company is a wholesale exchange. **2.** specifically, having to do with large-scale purchase of electricity or other utility services for resale to residential, commercial, or industrial end-users. Thus, **wholesale competition, customer, market, pricing,** and so on. Compare RETAIL.

**WHP** wellhead pressure.

**wicket gate** *Hydropower.* a gate that can be opened and closed to control the flow of water to a turbine.

**Wien's (displacement) law** *Physics.* one of the fundamental laws of radiation, showing that the wavelength for maximum radiative power of a blackbody is inversely proportional to the Kelvin temperature. [Formulated by German physicist Wilhelm *Wien,* 1864–1928.]

**Wigley, Tom** born 1940, Australian physicist and climatologist who made important contributions to climate and carbon-cycle modeling and to climate data analysis.

**Wigner, Eugene Paul** 1902–1995, Hungarian-born U.S. physicist known for his contributions to the theory of the atomic nucleus and elementary particles. He is also considered one of the founders of nuclear engineering through his work at Hanford, Washington during the Manhattan Project.

**Wigner energy** *Nuclear.* in nuclear reactors that use graphite as a moderator, the energy storage created when the graphite is bombarded with neutrons from the core. This causes crystalline dislocations to occur, bringing about swelling of the graphite rods.

**Wild and Scenic Rivers** *Policy.* a U.S. system to preserve certain rivers in their free-flowing condition and protect the water quality of these rivers.

**wildcat** *Oil & Gas.* **1.** an exploration well in an area without known production, especially one involving greater than usual uncertainty or speculation. **2.** to drill such a well. Thus, **wildcatter, wildcatting.** [From a 19th-century American slang term for a risky business venture; said to derive from a certain bank that issued fraudulent notes printed with a symbol for a wildcat.]

**wild well** *Oil & Gas.* an uncontrolled well; a well experiencing a fire or blowout.

**willingness to accept (WTA)** *Economics.* a schedule of prices and quantities indicating the minimum amount that a consumer would accept in exchange for giving up an additional unit, as a function of the number of units the consumer holds.

**willingness to pay (WTP)** *Economics.* a schedule of prices and quantities indicating the maximum amount that a consumer would pay in exchange for obtaining an additional unit, as a function of the number of units the consumer holds.

**willow** *Biomass.* any of numerous deciduous trees and shrubs of the genus *Salix* in the family Salicaceae, having narrow leaves, tassel-like spikes of flowers, and usually flexible twigs; certain types are cultivated as biomass energy feedstocks.

**Wilstätter, Richard Martin** 1872–1942, German chemist known for recognizing that there are two major types of chlorophyll in land plants, which differ from each other in details of their molecular structure and absorb slightly different wavelengths of light.

**wind** *Earth Science.* air that is in motion in relation to the earth's surface, especially in a horizontal direction.

**windage** *Wind.* **1.** the amount of exposure of an object to the force of the wind, or the resistance created by this. **2.** the amount of deflection of a projectile due to the effect of wind, or an allowance in aiming made for this.

**windcharger** *Wind.* **1.** an electricity-generating wind energy device used to charge batteries at a remote site; employed for home electric power in rural areas before the advent of widespread distribution of central station electricity. **2.** a general term for any device that generates electrical power from wind energy.

**wind chill** or **windchill** *Earth Science.* an expression of the combined effects of cold air and wind, derived by subtracting a factor of the wind speed from the air temperature. For example, if the temperature is 10°F and the wind speed is 20 miles per hour, the wind chill is –9, which is regarded as equivalent to a temperature of –9°F. Also, **wind-chill factor.**

**wind-driven circulation** *Earth Science.* a circulation in the upper one-kilometer region of the ocean that is primarily controlled by wind stress; this circulation also depends on buoyancy forcing and mixing.

**wind electric potential** *Wind.* an estimated forecast of the amount of energy that the winds at a given site or in a given area could provide if a wind energy facility were established there.

**wind energy** *Wind.* the energy contained in the movement of air masses; in human energy use traditionally captured by means of the sails of a ship or the vanes of a windmill, and currently by mechanical blades similar to airplane propellers.

**wind farm** or **windfarm** *Wind.* an array or system of multiple wind turbines at a given site, used to capture wind energy for the production of bulk electricity for a grid. [So called because of the sense of "harvesting" wind as if it were a farm crop.] See next page.

**wind farm efficiency** *Wind.* the ratio between the total annual production of a specific wind farm and the production of a corresponding number of isolated wind turbines located in identical but undisturbed wind conditions.

**wind forest** *Wind.* a term for a wind energy facility with an array of turbines located close to each other, giving the (unsightly) appearance of an artificial forest.

**winding** *Electricity.* a conductive path consisting of one or more turns of insulated wire forming a continuous coil; used in transformers, relays, and other electric devices.

**windmill** *Wind.* **1.** a structure or device that converts wind energy to mechanical energy by means of wind-propelled vanes, sails, or slats radiating about a horizontal or vertical shaft; used since early historic times for various purposes, especially to grind grains such as wheat or corn. **2.** a general term for any wind energy device.

**wind park** another term for a WIND FARM.

**wind power** *Wind.* **1.** electricity produced by a wind turbine, measured in watts. Wind power depends on the wind speed raised to the third power (wind speed cubed). **2.** see WIND ENERGY.

**wind power class** *Wind.* a categorization of the quality of the wind resource base at different

**wind farm** A wind farm (often also called a *wind park*) is a cluster of wind turbines that acts and is connected to the power system as a single electricity producing power station. Generally it is expected that a wind farm consists of more than three wind turbines. Modern wind farms may have capacities in the order of hundreds of MW, and are installed offshore as well as on land. Modern wind farms generally are connected to the high voltage transmission system, in contrast to the early application of wind energy for electricity production with wind turbines individually connected to the low-voltage to medium-voltage distribution system. Hence, modern wind farms are considered power plants with responsibilities for control, stability, and power balance. Thus, wind farms are required to contribute to the control of voltage, frequency and reactive power needs in the power system and stay on-line during less critical grid faults, and to help maintain the stability of the power system. While wind farm production cannot exceed the power given by the instantaneous wind resource, capabilities for regulating the power output at time scales consistent with the power system needs, powering up and down, are also included in order to assist with balancing and stabilizing the power system. Most of the other technical issues with wind farms are associated with the close spacing of multiple turbines. The close spacing implies that extraction of energy by wind turbines upwind will reduce the wind speed and increase the turbulence, which may cause reduced efficiency and higher loads on downwind turbines. Another technical issue for large wind farms is the grid connection and the integration into the power system. Large wind farms are very visible, especially at land and in coastal areas and this together with a number of environmental concerns, such as possible disturbance of migrating birds, play an important role in the wind farm planning process and can result in selection of sites with less than optimal wind conditions. However, good wind conditions are essential for the economics viability of any wind project, and methods for accurately predicting wind climates at specific sites worldwide are constantly being improved.

**Erik Lundtang Petersen and Peter Hauge Madsen**
Risø National Laboratory

*windmill* *Traditional type of windmill, a familiar sight of the Dutch landscape and a symbol for the country as a whole.*

sites, based on wind power density at a given height (typically 10 meters). The classes range from 1 to 7, with class 1 having power densities of 0–100 W/m$^2$, and class 7 having power densities of more than 1000 W/m$^2$.

**wind power curve** *Wind.* a graph that depicts the relationship between the power output of a wind turbine, measured as a per cent of the rated output, and the wind speed.

**wind power density** *Wind.* the average wind power over one square meter in a vertical plane perpendicular to the wind direction; expressed in watts per square meter.

**wind power distribution** *Wind.* a description of the percentage of time that the wind blows at various wind speeds over the course of an average year. In lieu of precise data for a site, the WEIBULL or RAYLEIGH distributions are used to estimate wind power.

**wind power plant** *Wind.* a system of wind energy devices employed to generate electricity for public use.

**wind resource** *Wind.* a measure of how much wind energy is potentially available at a given location and height; used to assess the potential power output of a wind turbine.

**wind rose** *Wind.* a map or diagram summarizing the frequency and strength of the wind from different compass directions, as represented by a line drawn in each direction from a common point of origin, with its length an indication of the frequency with which the winds blow from the given direction and its thickness indicating the wind velocity; the number of sectors of the compass may vary from 8 to 32 (16 is typical). Different versions of this design have been used by sailors since ancient times.

**Windscale** *Nuclear.* the name for the air-cooled, graphite-moderated reactor that was the first British weapons grade plutonium-239 production facility. In 1957, a fire at one of the twin Windscale reactors caused a serious nuclear accident, releasing an estimated 20 000 curies of radioactive material. Milk and other produce from the surrounding farming areas had to be destroyed, and news of the accident was suppressed for fear of alarming the public.

**wind shade** *Wind.* a shelter effect that reduces the speed of the wind and thus the amount of energy available to a wind energy device, produced by large objects such as buildings and trees (or by the wind energy device itself).

**windshaft** or **wind shaft** *Wind.* in a windmill, wind turbine, or the like, the iron rod that is attached to and turned by the rotor to provide mechanical power.

**wind shear** or **windshear** *Wind.* the rate at which the windspeed varies at different points in its general direction of motion; this variation can be particularly hazardous to aircraft operations, especially near the ground.

**wind stress** *Earth Science.* a drag or tangential force imposed on the earth's surface by the motion of an adjacent body of air; the dominant driving source for the surface layer of the world's oceans.

**wind turbine** *Wind.* **1.** a wind-driven machine containing curved rotors or blades inside a wheel set vertically on a revolving shaft; wind or air pressure against the blades turns the wheel, and the rotating shaft may then drive a dynamo to produce electrical power. **2.** more generally, any device that converts the energy of the wind to electrical or mechanical power.

**windward** *Wind.* toward the direction in which the wind blows; the opposite of LEEWARD.

**wingwall** *HVAC.* a protruding structural element built onto a building's exterior to improve natural ventilation by accelerating wind speed.

**winning** *Mining.* **1.** the process of excavating, loading, and removing valuable material from the ground, subsequent to development. **2.** a new mine opening. **3.** the part of a coal field that is laid out and ready for working.

**Winogradsky, Sergei** 1856–1953, Russian microbiologist who discovered chemosynthesis, the ability to produce organic compounds using energy contained in inorganic molecules (1887).

**Winston solar collector** another term for a COMPOUND PARABOLIC TROUGH, a type of solar collector invented by U.S. physicist Roland Winston.

**wipmolen** *Wind.* a historic advance in windmill technology (about 1450 in Holland) providing the means for the top portion bearing the sails to be turned to face the wind.

**wire** *Electricity.* **1.** a long, slender strand of drawn conductive metal, either bare or covered with insulation. **2.** a group of such strands. **3.** to connect electrical circuits using a system of such strands. **4.** *Communications.* to send a message by telegraph. **5.** the message so sent.

**wire(s) charge** *Electricity.* a term for charges levied on power suppliers or their customers for the use of transmission or distribution wires.

**wireline logging** *Oil & Gas.* the use of a metallic rope or cable to obtain downhole data in oil and gas wells.

**WMDs** weapons of mass destruction.

**WMO** World Meteorological Organization.

**WNA** World Nuclear Association.

**wood** *Biomass.* the hard substance that is found under the bark of trees. See next page.

**wood alcohol** another term for METHANOL.

**wood-derived fuel** *Biomass.* a fuel that derives its calorific value from wood, but which has been converted to some other type of fuel that does not have a woody physical structure; e.g., black liquor.

**wood fuel** *Biomass.* a collective term for all types of biofuels originating directly or indirectly from wood biomass.

**wood energy** Energy developed by burning woody biomass, i.e., wood, bark, and foliage of trees or their derivates. Trees produce woody biomass through the process of photosynthesis which converts carbon dioxide from the atmosphere and water from the soil to simple sugars and further to cellulose, hemicelluloses and lignin. In the same process energy from solar radiation is converted to chemical energy through photochemical reactions. Carbon, hydrogen, and energy release from dead biomass slowly through natural decomposition or rapidly in burning process. In sustainable forestry, this cycle of carbon is closed, and forests form a renewable reservoir, or sink, for carbon. It follows that substituting wood for fossil fuels helps to mitigate climate change caused by greenhouse gas emissions.The proportion of combustible elements, i.e., carbon and hydrogen, varies slightly between tree components and species. On average, in oven-dry wood the content of carbon is 50% and the content of hydrogen 6%. The effective or lower heating value of dry wood is about 19 MJ/kg and that of fresh wood (moisture content 50%) typically 8–9 MJ/kg. This low energy density relative to fossil fuels limits the practical usefulness of wood. The space required for transporting and storing wood chips is 11–15 times greater than needed for oil and 3–4 times that needed for coal. Therefore, wood is a local fuel which is used usually close to source.The global use of wood approaches 4 billion $m^3$ per annum. About 55% is used directly for fuel in developing countries, particularly in Asia and Africa, where it is the dominant fuel for heating and cooking. The remaining 45% is used as industrial raw material, but some 40% of this ends up as process residues suitable only for energy production. Thus, over 70% of the global wood harvest is either used as a fuel or is potentially available as a relatively clean and renewable source of energy, corresponding to 500 million tons of oil equivalent annually. In addition, low-quality stemwood, branches and stumps that are left unutilized at logging sites form a large potential reserve of wood energy.

**Pentti Hakkila**
VTT Technical Research Centre of Finland

**wood pulp** *Biomass.* pulp that is produced from virgin forest fibers.

**Woods, Granville T.** 1856–1910, U.S. inventor called "the Black Thomas Edison" for his many inventions that provided improved technologies for railroads.

**wood(y) biomass** *Biomass.* biomass from trees, bushes, and shrubs.

**work** *Thermodynamics.* **1.** the (useful) transfer of energy to a body by the application of a force, causing the body to move a certain distance in the same direction as the applied force; in one-dimensional motion, work is the product of the force (F) and displacement (d). Work is not accomplished unless there is a measurable displacement; the application of force alone does not constitute work. **2.** any physical or mental activity that has some practical value, especially an activity that produces revenue or accomplishes some desired goal.

**work function** another term for HELMHOLTZ FREE ENERGY.

**working** *Mining.* **1.** the operation of mining or of exploiting mineral resources in general. **2.** describing ground, rock, or a coal seam under great pressure and emitting creaking noises. 3, see WORKING FACE. **4.** see WORKINGS.

**working face** *Mining.* any place in a coal mine where material is extracted during the mining of a seam, especially a place being actively worked on at a given time.

**working fluid** *Thermodynamics.* the fluid (such as air) that flows through a device to produce work or power. Thus, **working gas.**

**workings** *Mining.* a shaft, quarry, stope, level, or other mining excavation.

**World Commission on Dams** (est. 1998), an independent international process that addresses the controversial issues associated with large dams.

**World Energy Council** (est. 1924), the oldest and most prominent international organization devoted exclusively to energy. It collects and publishes national-level data on all forms of energy resources, production, and consumption, and also holds congresses, workshops, and seminars to facilitate the dissemination of current information related to energy.

**World Meteorological Organization** (est. 1950), a specialized agency of the United

Nations for meteorology (weather and climate), operational hydrology, and related geophysical sciences.

**World Nuclear Association** a global industrial organization that promotes the peaceful use of nuclear power. It is concerned with nuclear power generation and all aspects of the nuclear fuel cycle, including mining, conversion, enrichment, fuel fabrication, plant manufacture, transport, and the safe disposition of spent fuel.

**World Resources Institute** (est. 1982), an independent nonprofit organization that plays a prominent international role in the promotion of the sustainable use of the earth's environment.

**world-systems theory** *Global Issues.* a theory that views the countries of the world as arranged in a hierarchical system and linked through a capitalist economy characterized by patterns of dependence. This hierarchy consists of core states above and peripheral states below, with semiperipheral states that are intermediate between the two.

**World Trade Organization** (est. 1995), an international organization established to replace the institution created by the General Agreement on Tariffs and Trade (GATT). The WTO provides a code of conduct for international commerce and a framework for periodic multilateral negotiations on trade liberalization and expansion.

**wound-rotor induction motor (WRIM)** *Consumption & Efficiency.* an induction motor in which the secondary circuit consists of windings or coils whose terminals are either short circuited or closed through suitable circuits; sometimes used when high starting torque and a low starting speed are required.

**Wp** an abbreviation for peak watt.

**WRDC** World Radiation Data Centre (est. 1964), an organization located in the Main Geophysical Observatory in St. Petersburg, Russia, that serves as a central depository for solar radiation data collected at over 1000 measurement sites throughout the world.

**WRI** World Resources Institute.

**Wright Brothers** **Wilbur** (1867–1912) and **Orville** (1871–1948), U.S. aviation pioneers who achieved the first powered, sustained, and controlled flight of an airplane, on Dec. 17, 1903 near Kitty Hawk, North Carolina. Record-breaking flights in 1908 by Orville in the U.S. and by Wilbur in France brought them worldwide fame.

***Wright Brothers*** *The Wright Brothers' initial flight of 12 seconds; Orville is at the controls while Wilbur stands at right.*

**wrought iron** *Materials.* a commercial iron that contains less than 0.3% carbon and 1.0% or 2.0% slag, giving it ductility and toughness.

**WSSD** World Summit on Sustainable Development; a large-scale international meeting held in Johannesburg, South Africa in 2002 to mark the tenth anniversary of the Rio de Janeiro Earth Summit. The role of energy in sustainable development was a principal theme.

**WTI** West Texas Intermediate.

**WTO** World Trade Organization.

**W-value** *HVAC.* the U-value (measure of the ability to conduct heat) converted into electrical terms for calculations in electric heating. See U-VALUE.

**xenon** *Chemistry.* an element of the noble gas group of the periodic table, having the symbol Xe, the atomic number 54, an atomic weight of 131.3, a melting point of –112°C, and a boiling point of –107°C; a colorless, odorless gas (or liquid) that is chemically unreactive but not inert; used in lamps, luminescent tubes, and lasers.

**X-rays** *Physics.* energy beams of very short wavelengths (0.1 to 1000 angstroms) produced by high-velocity electrons that encroach on various materials, particularly heavy metals. They are commonly created by passing a current of high voltage through a tube with a heated filament, they can penetrate most substances, and they are often used for inner-body examination and treatment. Also spelled variously as **X rays, x-rays,** or **x rays.**

**xylene** *Oil & Gas.* $C_6H_4(CH_3)_2$, a colorless liquid of the aromatic group of hydrocarbons made by the catalytic reforming of certain naphthenic petroleum fractions; used as high-octane motor and aviation gasoline blending agents, solvents, and chemical intermediates.

**Yamani, Ahmed Zaki** born 1930, Saudi Arabian oil minister and former chief spokesman and strategist for OPEC. He helped orchestrate the sharp oil price increases of 1973–74 and 1980–81, when OPEC's control over the market was strong. He eventually became a moderating influence within OPEC, which was the probable cause of his ouster in 1986 after oil prices collapsed.

**yarding** *Biomass.* the initial movement of logs from the original point of felling to an adjacent loading area or landing.

**yaw** *Physics.* **1.** a rotary movement of an object (e.g., an aircraft) around a vertical axis. **2.** to make such a movement; e.g., a wind generator yaws to face winds coming from different directions. Thus, **yaw angle, yaw mechanism.**

**Yeager, Charles (Chuck)** born 1923, U.S. aviator who became the first pilot to fly faster than the speed of sound (October 14, 1947).

***Yeager*** *Legendary pilot Chuck Yeager poses with the aircraft he flew to break the sound barrier.*

**Yeager, Ernest B.** 1924–2002, U.S. chemist known for his pioneering contributions to the understanding of electrochemical reactions and to the development of fuel cell and battery technology.

**yeast** *Biomass.* any of various very small, single-celled fungi of the phylum Ascomycota that reproduce by fission or budding and that are capable of fermenting carbohydrates into alcohol and carbon dioxide; certain yeasts can be utilized to produce an ethanol fuel from biomass.

**yellowcake** *Nuclear.* the popular name for URANIUM OXIDE CONCENTRATE, so called because of its typical yellow color.

**Yergin, Daniel** born 1947, U.S. economist regarded as one of the leading contemporary authorities on the economics and politics of the international oil industry. He received the Pulitzer Prize in 1992 for his work *The Prize: The Epic Quest for Oil, Money and Power.*

**Yerkes, Bill** U.S. engineer noted for his innovations in the design and manufacturing of photovoltaic (PV) cells.

**yield per effort (YPE)** *Measurement.* a measure of the effectiveness of natural resource extraction that compares the quantity discovered or obtained with the effort involved in doing this; e.g., the quantity of crude oil discovered or added to reserves per foot of well-drilled, or the quantity of fish caught per boat-day of effort.

**yocto-** *Measurement.* a prefix indicating $10^{-24}$.

**yotta-** *Measurement.* a prefix indicating $10^{24}$.

**Young, James** 1811–1883, Scottish inventor who became known as "Paraffin" Young because he was the first to extract paraffin from oil-rich shales and coals. He went on to establish a successful industry based on these principles.

**YPE** yield per effort.

**yttrium** *Chemistry.* a metallic element having the symbol Y, the atomic number 39, and an atomic weight of 88.906; melts at 1500°C and boils at 2927°C; a dark gray metal used in nuclear technology, semiconductors, alloys, and microwave filters.

**Y2K Problem** *Global Issues.* a widely publicized potential threat that was predicted to disable the world's communication systems in the year 2000 (Y2K), due to the inability of many computers to accommodate dates beyond 1999. In fact Y2K produced no significant negative effects; analysts have disagreed as to whether this was because of adjustments made beforehand or because the problem had been greatly exaggerated.

**yuca** another name for CASSAVA.

**Yucca Mountain** *Nuclear.* a site in southern Nevada that is proposed to be the location for the United States' long-term repository for high-level nuclear waste.

**Yunus, Muhammad** born 1940, Bangladeshi economist who founded the Grameen Bank, which led to the GRAMEEN SHAKTI power company.

**ZAFC** zinc-air fuel cell.

**Zeeman effect** *Physics.* a phenomenon in which the spectral lines associated with atomic or molecular radiation are split into equally spaced levels (with small separation) when the material is subjected to a static magnetic field. [Named for Dutch physicist Pieter *Zeeman,* 1865–1943.]

**Zeldovich, Yakov** 1914–1987, Russian physicist who first identified the chemical mechanism for producing active nitrogen from molecular nitrogen in the gas phase (1947). His research was fundamental to later work aimed at reducing atmospheric pollution from the combustion of fossil fuels.

**Zeldovich mechanism** *Earth Science.* a description of the production of nitric oxide and nitric dioxide from molecular nitrogen in the gas phase; these reactions are thought to account for much of the active nitrogen formed in hot exhaust gases from combustion sources and following rapid heating of the air during lightning discharges.

**zenith** *Physics.* the point on any given observer's celestial sphere that lies directly above that observer (i.e., that is exactly vertical); the point that is elevated 90° from all points on the observer's astronomical horizon.

**zenith angle** *Solar.* the angle between the center of the sun's disc and the vertical.

**Zeppelin, Ferdinand von** 1838–1917, German military officer who developed the rigid dirigible, a lighter-than-air vehicle (1900). Safety problems that led to accidents, especially the crash of the *Hindenburg* in 1937, brought an end to the use of such craft as passenger vehicles.

**zeppelin** *Transportation.* another term for a dirigible (lighter-than-air passenger vehicle), so called because the first successful craft of this type was developed by Count Ferdinand von *Zeppelin.*

**zepto-** *Measurement.* a prefix indicating $10^{-21}$.

**zero emission(s) vehicle (ZEV)** *Transportation.* a motor vehicle technology that in theory would produce no harmful emissions whatsoever, by using an alternative power source as opposed to petroleum. Similarly, **near-zero (near-ZEV) vehicle.**

**zero governor** *Consumption & Efficiency.* a fuel regulating device in which pressure is reduced to zero at the fuel inlet; when a partial vacuum is created in the fuel line, suction will cause the regulator to open, thus allowing flow to the burner as long as the demand continues. Used for gas heating systems, gas-fired boilers, propane engines, and so on.

**zero-point energy** *Physics.* the nonthermal energy of a system, most evident as its temperature approaches absolute zero.

**zero-point entropy** *Thermodynamics.* the limiting value of the entropy of a system as its temperature approaches absolute zero.

**zeroth law of thermodynamics** *Thermodynamics.* the statement that if two systems are both in thermal equilibrium with a third system, then they are in thermal equilibrium with each other. Though this was formulated after the first, second, and third laws, it is called "zeroth" (zero) to indicate that it logically precedes them, since they are founded on the measurement of temperature and the zeroth law indicates that such measurement is possible. Also spelled **zeroeth.** See also THERMODYNAMICS.

**zetta-** *Measurement.* a prefix indicating $10^{21}$.

**ZEV** zero emission vehicle.

**zinc** *Chemistry.* a metallic element having the symbol Zn, the atomic number 30, an atomic weight of 65.38, a melting point of 419°C, and a boiling point of 907°C; a lustrous, bluish-white transition metal found in ores and used for many industrial purposes such as alloys, galvanized metals, and dry-cell batteries. Zinc is an essential nutrient element in soils and animals.

**zinc-65** *Chemistry.* a radioactive isotope of zinc with a half-life of 244 days, used in alloy-wear studies and in metabolism investigations. It may be found in the radioactive effluent stream from nuclear power stations.

**zinc-air battery** *Storage.* a battery type for which the negative electrode consists of zinc and the positive electrode is a porous body made of carbon with air access. The active mass is thus not contained in the electrode but is taken from the surrounding air as it is needed, and the weight of the battery is reduced accordingly.

**zinc-air fuel cell** *Renewable/Alternative.* a type of fuel cell system in which metal zinc in pellet or particle form is used as fuel instead of hydrogen. Zinc oxide is formed in the electrochemical reaction of zinc with oxygen, which can be recycled; the electrolytes can be liquid or solid fuel.

**zinc-bromide battery** *Storage.* a battery type consisting of a zinc negative electrode and a bromine positive electrode separated by a microporous separator. An aqueous solution of zinc/bromide is circulated through the two compartments of the cell from two separate tanks; the stored amount of energy thus largely depends on the size of the tanks. Developed especially for passenger vehicles.

**zinc-carbon battery** *Storage.* the earliest type of primary cell and battery, still very widely used; it uses carbon and manganese dioxide as the positive electrode and zinc as the negative electrode, with an aqueous solution of ammonium chloride and zinc chloride as the electrolyte. Also known as the Leclanche battery for its inventor Georges Leclanche.

**zinc-chloride battery** *Storage.* an enhanced version of the standard zinc-carbon primary battery that uses zinc chloride as the electrolyte, producing about 50% higher capacity.

**Zinn, Walter** 1907–2000, Canadian physicist who supervised construction of the first experimental nuclear reactor, or atomic pile as it was then called. He also designed the EBR-I, the first nuclear reactor to produce electric power.

**zirconium** *Chemistry.* a metallic element having the symbol Zr, the atomic number 40, an atomic weight of 91.22, a melting point of 1850°C, and a boiling point of 4377°C; used in coatings for nuclear fuel rods and in corrosion-resistance alloys.

**zoning** *HVAC.* **1.** the process of separately heating or cooling individual rooms or spaces (zones) within a building, since a smaller system, and less energy, is required to cool a specific area rather than the entire building. **2.** the physical division of a house into smaller areas in order to facilitate this process.

**Zworykin, Vladimir** 1889–1982, Russian-born U.S. physicist who patented (1928) the use of magnetic fields to guide cathode rays in order to produce images on a fluorescent screen. This is the basis for all conventional television sets and computer monitors. He also developed the first practical television camera and the first electron microscope.

# Notable Quotes on Energy

## Prior to 1800

"It is clear that there is some difference between ends: some ends are *energeia* [energy], while others are products which are additional to the *energeia*."

—Aristotle, Greek philosopher, the first to describe the concept of energy (ca. 325 BC; *energeia* has traditionally been translated as "activity" or "actuality"; some modern texts render it more literally as "in work" or "being at work".)

"Give me a lever and a place to stand, and I can move the whole world."

—Archimedes, Greek mathematician and engineer (attributed, ca. 250 BC; quoted in later sources).

"To the north lies Georgia, near the confines of which there is a fountain of oil which discharges so great a quantity as to furnish loads for many camels. The use made of it is not for the purpose of food, but as an unguent for the cure of cutaneous distempers in men and cattle . . . and it is also good for burning."

—Marco Polo, Venetian merchant and travel author (ca. 1291 AD).

"*E pur (Eppur) si muove.*" ("And yet, it moves" or "Nevertheless, it does move.")

—Galileo Galilei, Italian scientist (attributed, ca. 1633; referring to the motion of the earth around the sun, in contrast with Church teaching of the time which held that the earth was stationary. Galileo is said to have uttered this in secret after he was publicly forced to renounce Copernican theory).

"*Energie* is the operation, efflux or activity of any being: as the light of the Sunne is the energie of the Sunne, and every phantasm of the soul is the energie of the soul."

—Henry More, English philosopher (1642; the first recorded definition of the term *energy* in English).

"All matter attracts all other matter with a force proportional to the product of their masses and inversely proportional to the square of the distance between them."

—Sir Isaac Newton, English scientist (1687; describing his law of gravity in the *Philosophiae naturalis principia mathematica*, regarded by many as the most important book of science ever written).

"I was thinking upon the engine at the time, when the idea came into my mind that as steam was an elastic body it would rush into a vacuum, and if a communication were made between the cylinder and an exhausted vessel it would rush into it, and might be there condensed without cooling the cylinder."

—James Watt, Scottish inventor (1765; depicting the moment that his idea for improving Thomas Newcomen's steam engine was born).

"He snatched lightning from the heavens, then the sceptre from tyrants." (Translation of Latin *Eripuit coelo fulmen, mox sceptra tyrannis.*)

—A. J. R. Turgot, French economist (attributed, ca. 1778; commenting on the achievements of Benjamin Franklin).

"I say it is impossible that so sensible a people [citizens of Paris], under such circumstances, should have lived so long by the smoky, unwholesome, and enormously expensive light of candles, if they had really known that they might have had as much pure light of the sun for nothing."

—Benjamin Franklin, American patriot, author, and scientist (1784; describing the energy benefits of adopting daylight saving time).

"3 Kegs Senica Oil 50 Dllrs"

—An entry in General William Wilson's day book (1797; an inventory of goods from his Pennsylvania general store, one of the earliest records of the price of petroleum).

**1801–1900**

"First, there is the power of the Wind, constantly exerted over the globe . . . . Here is an almost incalculable power at our disposal, yet how trifling the use we make of it!"
—Henry David Thoreau, U.S. author and naturalist (1834).

"Fire is the best of servants, but what a master!"
—Thomas Carlyle, British author (1843).

"Is it a fact — or have I dreamt it — that, by means of electricity, the world of matter has become a great nerve, vibrating thousands of miles in a breathless point of time?"
—Nathaniel Hawthorne, U.S. author (1851).

"The whaleman burns, too, the purest of oil . . . . It is sweet as early grass butter in April. He goes and hunts for his oil, so as to be sure of its freshness and genuineness, even as the traveller on the prairie hunts up his own supper of game."
—Herman Melville, U.S. author (1851; a description in his novel *Moby-Dick* of the use of whale oil for lighting on board the whaling ships of the time).

"Sir, I do not know, but someday you will tax it."
—Michael Faraday, English scientist noted for research in electricity (1854; after being asked by William Gladstone what benefit might come from electricity; as Chancellor of the Exchequer Gladstone increased British tax rates by 50% that same year).

"This lotion is the future wealth of this country, it's the welfare and prosperity for its inhabitants, it's a new source of income for the poor people and a new branch of industry, which shall bear plentiful fruits."
—Ignacy Lukasiewicz, Polish researcher (1854; describing the possibilities of using paraffin for lighting purposes).

"Drill for oil? You mean drill into the ground to try to find oil? You're crazy."
—Reported comment to Col. Edwin L. Drake from an investor whom he asked to support his venture to drill the first commercial oil well in the U.S. (1859).

"Coal is a portable climate. It carries the heat of the tropics to Labrador and the polar circle; and it is the means of transporting itself whithersoever it is wanted . . . coal carries coal, by rail and by boat, to make Canada as warm as Calcutta."
—Ralph Waldo Emerson, U.S. author and philosopher (1860).

"Coal, in truth, stands not beside but entirely above all other commodities. It is the material energy of the country—the universal aid—the factor in everything we do."
—William Stanley Jevons, English economist (1865).

"Coal is everything to us. Without coal, our factories will become idle, our foundries and workshops be still as the grave; the locomotive will rust in the shed, and the rail be buried in the weeds. Our streets will be dark, our houses uninhabitable. Our rivers will forget the paddlewheel, and we shall again be separated by days from France, by months from the United States."
—Editorial, *The London Times* (1866).

"Yes, my friends, I believe that water will one day be employed as fuel, that hydrogen and oxygen which constitute it, used singly or together, will furnish an inexhaustible source of heat and light, of an intensity of which coal is not capable. . . . Water will be the coal of the future."
—Jules Verne, French author and futurist (1874; a prediction by the character Cyrus Harding in the novel *The Mysterious Island*).

"Mr. Watson, come here. I need you."
—Alexander Graham Bell, Scottish inventor (1876; a message to his assistant that was the first demonstration of the electronic transmission of speech).

"Evolution is a change from a less coherent form to a more coherent form, consequent on the dissipation of energy and the integration of matter."

—Herbert Spencer, British philosopher and sociologist (1880).

"At what time does the dissipation of energy begin?"

—William Thomson (Lord Kelvin), British physicist (1885; applying the terminology of his studies of thermodynamics to a question for his wife about their plans for an afternoon walk).

### 1801–1945

"One, two, three, four . . . . Is it snowing where you are, Mr. Thiessen? If it is, would you telegraph back to me?"

—Reginald Aubrey Fessenden, Canadian inventor (1900; believed to be the first voice to be broadcast by radio waves and heard by another person).

"To the electron—may it never be of any use to anybody."

—J. J. Thomson, British physicist; (a favorite toast of his in the early 1900s, commenting on his studies that led to the discovery of the electron).

"I will build a motor car for the great multitude, constructed of the best materials, by the best men to be hired, after the simplest designs that modern engineering can devise, so low in price that no man making a good salary will be unable to own one."

—Henry Ford, U.S. industrialist (1907; referring to the Model T that became the largest-selling car model in the world).

"The progress of science is characterized by the fact that more and more energy is utilized for human purposes, and that the transformation of the raw energies . . . is attended by ever-increased efficiency."

—Wilhelm Ostwald, German physical chemist (1907).

"It was obvious that a new science was in the course of development."

—Marie Curie, Polish chemist (1911; referring to her pathbreaking work in radioactivity with husband Pierre Curie).

"My husband did that on the back of old envelopes."

—Mileva Maric Einstein, Serbian mathematician; first wife of Albert Einstein (1911; commenting on a description of scientific equipment that was said to be able to "probe the deepest secrets of the universe").

"They [coal mine operators] wouldn't keep their dog where they keep you fellows. You know that. They have a good place for their dogs."

—Mary Harris ("Mother Jones"), U.S. labor activist (1912; speaking to striking coal miners in West Virginia).

"There it is. Take it."

—William Mulholland, Irish-born U.S. engineer (1913; the entire text of his speech to a crowd of thousands at the opening of the Los Angeles Aqueduct, which provided the water supply to make the modern city possible).

"The Allies floated to victory on a sea of oil."

—Lord Curzon, British political leader (ca. 1918; summing up the role of energy supply in World War I; also attributed to Winston Churchill).

"Do Middle Eastern conditions justify my putting any money into Mesopotamian oil?"

—Question posed by an American financier to T. E. Lawrence ("Lawrence of Arabia"), to which he replied simply "No." (1920; quoted by author Robert Graves).

"Not only will atomic power be released, but someday we will harness the rise and fall of the tides and imprison the rays of the sun."

—Thomas Alva Edison, U.S. inventor (1921).

"In the struggle for existence, the advantage must go to those organisms whose energy-capturing devices are most efficient in directing available energies into channels favorable to the preservation of the species."

—Alfred James Lotka, U.S. biophysicist (1922).

"The laws of thermodynamics control, in the last resort, the rise and fall of political systems, the freedom or bondage of nations, the movements of commerce and industry, the origins of wealth and poverty, and the general physical welfare of the race."

—Frederick Soddy, English chemist (1922).

"The best way to make a fire with two sticks is to make sure that one of them is a match."

—Will Rogers, U.S. humorist and commentator (ca. 1925).

"I got out of there just in time. He was beginning to ask *me* questions!"

—James Franck, German physicist (1927; describing his role in the Ph.D. oral examination of J. Robert Oppenheimer, later known as the "father of the atomic bomb").

"In fact the number of kinds of elementary particle has shown a rather alarming tendency to increase during recent years."

—Paul Dirac, English physicist (1933).

"The energy produced by the breaking down of the atom is a very poor kind of thing. Anyone who expects a source of power from the transformation of these atoms is talking moonshine."

—Ernest Rutherford, British physicist (born New Zealand) (1933).

"Blazing rockets which in the dark of the night suddenly cast a brief but powerful illumination over an immense unknown region."

—Louis de Broglie, French physicist (1935; describing the impact on the scientific world of Albert Einstein's three landmark papers published in 1905).

"We believe that Japan, having little oil herself, is anxious to seize Borneo, for without oil no empire is secure."

—Gareth Jones, Welsh journalist (1935; quoting an unnamed "leading citizen" from the Netherlands East Indies).

"It is a fearful thought that the labour of crawling as far as the coal face (about a mile in this case but as much as three miles in some mines), which was enough to put my legs out of action for four days, is only the beginning and ending of a miner's day's work, and his real work comes in between. I don't think I shall ever feel quite the same about coal again."

—George Orwell, English novelist and essayist (1936; describing his experiences in coal mines in the north of England).

"It ill behooves one who has supped at labor's table to curse with fine fervor and equal impartiality both labor and its adversaries, when they become locked in deadly embrace."

—John L. Lewis, U.S. union leader (1937; responding to President Roosevelt's comment "a plague on both your houses" in reference to a labor-management conflict in the steel industry).

"Some recent work by E. Fermi and L. Szilard . . . leads me to expect that the element uranium may be turned into a new and important source of energy in the immediate future."

—Albert Einstein, German-born physicist (1939; in a letter to President Roosevelt urging him to provide government support for Enrico Fermi and colleagues in their work on chain reactions).

"For hour after sweating hour, bent double, we worked down there, with the dust of coal settling on us with a light touch that you could feel, as though the coal was putting fingers on you to warn you that he was only feeling you, now, but he would have you down there, underneath him, one day soon when you were looking the other way."

—Richard Llewellyn, Welsh author (1940; describing character Huw Morgan's first day as a coal miner in the novel *How Green was My Valley*).

"I wanted be at my parents' house when the electricity came. It was in 1940. I remember my mother smiling. When the lights came on full, tears started to run down her cheeks. . . . From there I went to my grandmother's house. It was a day of celebration. They had all kinds of parties—mountain people getting light for the first time."

—Clyde T. Ellis, U.S. Congressman from Arkansas (1940; quoted by author Studs Terkel; Ellis served as director of the New Deal program that brought electricity to rural American homes previously without it).

"I have been everlastingly proud of the great contributions TVA has made, which cannot be fully revealed until peace returns to a tortured world."

—George W. Norris, U.S. Senator from Nebraska (1942; referring to the fact that the TVA [Tennessee Valley Authority] supplied much of the electricity for the nation's war effort).

"The taproot of German might."

—Winston Churchill, British political leader (1944; describing the oil fields and refineries near Ploesti, Romania that were Germany's source of military fuel; also quoted as "the taproot of German mechanized power").

"I am become death, the shatterer of worlds."

—J. Robert Oppenheimer, U.S. scientist and director of the Manhattan Project in World War II (1945; quoting the *Bhagavad Gita*, a sacred Hindu text, as he watched the detonation of the first atomic bomb over the New Mexico desert).

## 1946–1980

"The [coal] miner said: 'It's black down there. No light but what you carry on your hat. Occasional lights at the sections. But it is very black.' How can men love darkness rather than the light? How can men choose such an occupation, except that they are forced to it?"

—Dorothy Day, U.S. author and social activist (1946).

"There are no longer problems of the spirit. There is only one question: When will I be blown up?"

—William Faulkner, U.S. novelist (1950; commenting in his Nobel Prize acceptance speech on the challenge of being a creative artist in an age threatened by nuclear war).

"First city in the world to be lit by atomic power. Elevation 5320. Arco. 1951."

—A sign outside the city of Arco, Idaho.

"They've voted to continue the twenty-seven percent tax allowance on oil . . . from now on the whole world is going to be yelling for oil. Texas is booming. The rest of the country is flat."

—Edna Ferber, U.S. author (1952; the character Bick Benedict describes the Texas oil boom in the novel *Giant*).

"For years I thought what was good for our country was good for General Motors, and vice versa."

—Charles Erwin ("Engine Charlie") Wilson, U.S. auto executive, president of General Motors (1953; when asked if he could make decisions as a government official that would be harmful to the interests of GM; now often quoted as "What's good for General Motors is good for the country.")

"It is not too much to expect that our children will enjoy in their homes electrical energy too cheap to meter, will know of great periodic regional famines in the world only as matters of history, will travel effortlessly over the seas and under them and through the air with a minimum of danger and at great speeds, and will experience a lifespan far longer than ours."

—Lewis L. Strauss, first head of the U.S. Atomic Energy Commission (1954).

"Based upon ultimate production of 150 billion barrels the [production] curve must culminate at about 1965 and then must decline at a rate comparable to its earlier rate of growth . . . if we assume [ultimate production of 200 billion barrels] then the date of culmination is retarded only until about 1970."

—M. King Hubbert, U.S. geophysicist (1956; accurately predicting the peak in oil production for the lower 48 U.S. states, a seminal energy and economic event).

"Many times I have been present at gatherings of people who are thought highly educated and who have with considerable gusto been expressing their incredulity at the illiteracy of scientists. I have asked how many of them could describe the Second Law of Thermodynamics. The response was cold: it was also negative. Yet I was asking something which is about the scientific equivalent of: 'Have you read a work of Shakespeare's?'"

—C. P. Snow, British author and scientist (1959).

"It is something like the discipline of surgery—cleanliness, care, seriousness, and practice."

—Willard F. Libby, U.S. chemist (1960; describing the technique of radiocarbon dating, which he pioneered).

"Stars have a life cycle much like animals. They get born, they grow, they go through a definite internal development, and finally they die, to give back the material of which they are made so that new stars may live."

—Hans Bethe, German-born U.S. physicist (1967).

"Oil is seldom found where it is most needed, and seldom most needed where it is found."

—Jan Brouwer, Dutch engineer and senior executive of Royal Dutch/Shell Oil Co. (attributed, ca. 1970).

"This is a sad hoax, for industrial man no longer eats potatoes made from solar energy; now he eats potatoes partly made of oil."

—Howard Odum, U.S. ecologist (1971; referring to the fact that modern agriculture is highly dependent on fossil fuels).

"It will only be when we get a response from nature, in the form of greatly diminished return in the form of surplus energy, that we can expect the present [industrial] revolution to slow down."

—William Frederick Cottrell, U.S. sociologist (1972).

"Let me tell you something that we Israelis have against Moses. He took us 40 years through the desert in order to bring us to the one spot in the Middle East that has no oil!"

—Golda Meir, Prime Minister of Israel (1973).

"I have no doubt that we will be successful in harnessing the sun's energy. . . . If sunbeams were weapons of war, we would have had solar energy centuries ago."

—George Porter, British chemist (1973).

"Oil is much too important a commodity to be left in the hands of the Arabs."

—Henry Kissinger, U.S. Secretary of State (attributed, ca. 1974).

"Let this be our national goal: At the end of this decade, in the year 1980, the United States will not be dependent on any other country for the energy we need to provide our jobs, to heat our homes, and to keep our transportation moving."

—Richard Nixon, U.S. President (1974; after imposing oil price controls in response to the 1973–1974 energy crisis).

"The world can, in effect, get along without natural resources."

—Robert Solow, U.S. economist (1974; reflecting the commonly held view that human ingenuity will be able to substitute for depleted natural resources).

"When we measure energy, the ability to do work with energy, the prime criteria of science through all the ages has been the ability to lift a given weight against gravity and a given height in a given amount of time."

—R. Buckminster Fuller, U.S. inventor and futurist (1975).

"The greatest danger in our bemused drift towards the energy waterfall is that the resulting shock will find us stripped of democratic government by an opportunistic group that comes out on top in the wreckage."

—Earl Ferguson Cook, U.S. geologist (1976).

"Giving society cheap, abundant energy . . . would be the equivalent of giving an idiot child a machine gun."

—Paul Erhlich, U.S. biologist (1978; describing the possible impact of a future abundant source of energy, in light of the environmental damage already brought about by the existing form of cheap energy, fossil fuels).

"Every gallon of oil each one of us saves is a new form of production. It gives us more freedom, more confidence, that much more control over our own lives."

—Jimmy Carter, U.S. President (1979).

"The use of solar energy has not been opened up because the oil industry does not own the sun."

—Ralph Nader, U.S. consumer advocate (1980).

1981 to present

## 1981 to present

"Your grandchildren will likely find it incredible, or even sinful, that you burned up a gallon of gasoline to fetch a pack of cigarettes."

—Paul MacCready, U.S. inventor of innovative flying machines (1981).

"The mastery of fire enabled man not only to keep warm and cook the food, but, above all to smelt and forge metals, and to bake bricks, ceramics, and lime. No wonder that the ancient Greeks attributed to Prometheus (a demigod, not a mortal) the bringing of fire to us."

—Nicholas Georgescu-Roegen, Romanian economist (1982).

"Oil is like a wild animal. Whoever captures it has it."

—Jean Paul Getty, U.S. industrialist, founder of Getty Oil Company (1986; quoted by biographer Robert Lenzner).

"Good evening comrades. Everybody is aware of the terrible setback—the accident at the Chernobyl nuclear power plant. It affects the people of the Soviet Union painfully and has shocked the international community. We are being confronted for the first time with the true power of unbridled nuclear energy."

—Mikhail Gorbachev, Soviet President (1986).

"The Stone Age did not end for lack of stone, and the Oil Age will end long before the world runs out of oil."

—Ahmed Zaki Yamani, Saudi Arabian oil executive; former chief strategist for OPEC (1986).

"We've fetched up on hard ground, north of Goose Island, off Bligh Reef, and evidently leaking some oil and we're gonna be here for a while."

—Joseph Hazelwood, U.S. oil tanker captain (1989; radio message to the Coast Guard immediately after the grounding of the *Exxon Valdez*, which resulted in one of the largest oil spills in U.S. history).

"We nuclear engineers of the first nuclear era have had success, but the generation that follows us must resolve the profound technical and social questions that are convulsing nuclear energy."

—Alvin M. Weinberg, U.S. physicist (1989).

"During all of human existence, people have worried about running out of natural resources: flint, game animals, what have you. Amazingly, all the evidence shows that exactly the opposite has been true. Raw materials—all of them—are becoming more available rather than more scarce."

—Julian Simon, U.S. economist (1994).

"There are no accounting issues, no trading issues, no reserve issues, no previously unknown problem issues. The company is probably in the strongest and best shape that it has ever been in."

—Kenneth Lay, U.S. business executive (2000; speaking as head of the giant energy company Enron, shortly before its collapse).

"Conservation may be a sign of personal virtue, but it is not a sufficient basis for a sound, comprehensive energy policy."
—Richard (Dick) Cheney, U.S. Vice President (2001).

"Affordable energy in ample quantities is the lifeblood of the industrial societies and a prerequisite for the economic development of the others."
—John P. Holdren, U.S. physicist (2001)

"We won't know that we have too many cars—until we have too many cars."
—Eugene Odum, U.S. ecologist (2001).

"Understanding depletion is simple. Think of an Irish pub. The glass starts full and ends empty. There are only so many more drinks to closing time. It's the same with oil. Also, we have to find the bar before we can drink what's in it."
—Colin Campbell, British petroleum geologist (2002).

"Civilization in no immediate danger of running out of energy or even just out of oil. But we are running out of environment—that is, out of the capacity of the environment to absorb energy's impacts without risk of intolerable disruption."
—Vijay Vaitheeswaran, Indian journalist (2002).

"In this house we obey the laws of thermodynamics!"
—Television character Homer Simpson of the animated series *The Simpsons* (2003; commenting on his daughter Lisa's efforts to design a perpetual motion machine).

"The 'soft landing' that many people hope for—a voluntary change to solar energy and green fuels, energy-conserving technologies, and less overall consumption—is a utopian alternative that will come about only if severe, prolonged hardship in industrial nations makes it attractive."
—Joseph Tainter, U.S. anthropologist (2004)

"It's as important to me as the first step [Neil] Armstrong took when he stepped off onto the moon."
—Ted Stevens, U.S. Senator from Alaska (2005; urging his Senate colleagues to vote to allow oil exploration in the Arctic National Wildlife Refuge).

# Illustration Credits

**(Credits continued from p.iv)**

**ebb tide:** HBJ Photo/Cotton Coulson; **Edison:** Courtesy of General Electric Company; **Einstein:** Photo by Lotte Jacobi; **endangered:** U.S. Bureau of Sport Fisheries & Wildlife/Luther Goldman; **energy crisis:** HBJ Photo/Sygma; **ENIAC:** University of Pennsylvania; **environmental injustice:** U.S. Dept. of Agriculture; **erosion:** Morris Books; **fallout shelter:** Federal Emergency Management Agency; **fault:** HBJ Photo; **forest ecology:** American Forest Product Industries; **Ford:** U.S. Environmental Protection Agency; **Franklin:** White House Historical Association; **furnace:** HBJ Photo/C. Aurness; **gas guzzler:** U.S. Environmental Protection Agency; **geyser:** HBJ Photo/Richard Rowan; **glacier:** HBJ Photo/Tom Bean; **Grand Coulee Dam:** U.S. Bureau of Reclamation; **greenfield:** South Dakota Dept. of Highways; **gusher:** Texas Mid-Continent Oil & Gas Association; **heavier-than-air:** U.S. Air Force; **helicopter:** Image After; **Hoover Dam:** Las Vegas Convention Bureau; **hot air balloon:** Image After; **human power:** HBJ Photo; **hunting and gathering:** Morris Books/Nancy Devore; **hydrofoil:** Image After; **hydropower:** Rapho-Guillumette Pictures; **ice shelf:** U.S. Navy; **industrial agriculture:** U.S. Dept. of Agriculture/Jack Dykinga; **irrigation:** U.S. Bureau of Reclamation; **jet engine:** Image After; **Kitty Hawk:** U.S. Air Force; **Krebs:** Courtesy of The Nobel Foundation; **laser:** HBJ Photo/Bruce Frisch; **lava:** HBJ Photo/Randy Hyman; **lighter-than-air:** Image After; **lightning:** The Boston Globe; **Lindbergh:** HBJ Photo; **locks:** U.S. Army; **locomotive:** Association of American Railroads; **Lyell:** HBJ Photo/Metropolitan Museum of Art; **Marconi:** Courtesy of Radio Corporation of America; **mass transit:** Image After; **Maxwell:** National Portrait Gallery; **McCormick:** Courtesy of International Harvester Corporation; **Mercator:** HBJ Photo; **migration:** U.S. Fish and Wildlife Service/Dave Menke; **mill:** U.S. National Archives; **monorail:** HBJ Photo; **nanoscience:** Sandia National Laboratory; **neon:** Morris Books; **Newton:** HBJ Photo; **nuclear:** U.S. Navy; **offshore:** Courtesy of Texas Gas Transmission Corp.; **oil:** Library of Congress; **open-hearth:** Courtesy of U.S. Steel Co.; **open-pit mining:** Courtesy of Hecla Mining Co.; **paddlewheel:** Standard Oil Company of New Jersey; **permafrost:** Morris Books; **photosynthetic:** Illustration by Sharron O'Neill; **phytomass:** Illustration by Sharron O'Neill; **pipeline:** HBJ Photo/Nancy Christensen; **plate tectonics:** U.S. Geological Survey; **platform:** Image After; **public good:** HBJ Photo; **pollution control:** U.S. Environmental Protection Agency; **predator-prey:** Illustration by Barbara Hoopes-Ambler; **Ptolemaic system:** Courtesy of The American Museum of Natural History; **pyrotechnics:** HBJ Photo; **quartz:** HBJ Photo/Manfred Kage; **rainforest:** HBJ Photo/Steve Vidler; **reactor:** HBJ Photo/George Lepp; **reaper:** Courtesy of International Harvester Corporation; **recyclable:** Union Electric Company; **refinery:** Lawrence Migdale/Photo Researchers; **reforestation:** U.S. Forest Service; **remediation:** U.S. Fish and Wildlife Service/R.W. Roach; **renewable:** Image After; **robot:** Morris Books; **rocket:** U.S. Air Force/Carleton Bailie; **rubber:** Illustration by Sharron O'Neill; **satellite:** NASA; **savannah:** HBJ Photo; **seam:** U.S. National Archives; **silver:** HBJ Photo; **softwood:** Illustration by Sharron O'Neill; **soil conservation:** U.S. Dept. of Agriculture; **solar power satellite system:** NASA; **space shuttle:** NASA; **spinning frame:** Science Museum, London; **spontaneous generation:** HBJ Photo; **steam locomotive:** U.S. National Archives; **Steinmetz:** HBJ Photo/Walt Sanders; **stratification:** South Dakota Dept. of Highways; **streamlined:** Image After; **subway:** HBJ Photo/Peter Vandermark; **sugarcane:** Illustration by Sharron O'Neill; **tanker:** HBJ Photo; **Teapot Dome:** U.S. Environmental Protection Agency; **Tennessee Valley Authority:** Tennessee Valley Authority; **TGV:** Image After; **thermal inversion:** HBJ Photo/New York Times; **Three Mile Island:** U.S. Environmental Protection Agency; **tipple:** U.S. National Archives; **tool:** Peabody Museum/Harvard University; **tramway:** HBJ Photo/Richard Rowan; **trolley:** U.S. National Archives; **tundra:** HBJ Photo/Don Loveridge; **Tychonic system:** Morris Books; **undershot wheel:** U.S. Environmental Protection Agency; **uniformitarianism:** Union Pacific Railroad; **UNIVAC:** Courtesy of Remington Rand Corp.; **urbanization:** HBJ Photo/Harold Kinne; **volcanism:** U.S. Geological Survey; **walking plow:** HBJ Photo; **waterfall:** New York State Dept. of Commerce; **wave energy:** Image After; **windmill:** Image After; **Wright Brothers:** Smithsonian Institution; **Yeager:** U.S. Air Force.